AF327411

NITROGEN ECONOMY IN TROPICAL SOILS

Developments in Plant and Soil Sciences

VOLUME 69

The titles published in this series are listed at the end of this volume.

Nitrogen Economy in Tropical Soils

Proceedings of the International Symposium on Nitrogen Economy in Tropical Soils, held in Trinidad, W.I., January 9–14, 1994

Edited by

N. AHMAD
The University of the West Indies
St. Augustine
Trinidad
W.I.

Partly reprinted from *Fertilizer Research*, Volume 42, Nos. 1–3 (1995)

KLUWER ACADEMIC PUBLISHERS
DORDRECHT / BOSTON / LONDON

A C.I.P. Catalogue record for this book is available from
the Library of Congress.

ISBN 0-7923-4094-9

Published by Kluwer Academic Publishers,
P.O. Box 17, 3300 AA Dordrecht, The Netherlands.

Kluwer Academic Publishers incorporates
the publishing programmes of
D. Reidel, Martinus Nijhoff, Dr W. Junk and MTP Press.

Sold and distributed in the U.S.A. and Canada
by Kluwer Academic Publishers,
101 Philip Drive, Norwell, MA 02061, U.S.A.

In all other countries, sold and distributed
by Kluwer Academic Publishers Group,
P.O. Box 322, 3300 AH Dordrecht, The Netherlands.

Printed on acid-free paper

All Rights Reserved
© 1996 Kluwer Academic Publishers
No part of the material protected by this copyright notice
may be reproduced or utilized in any form or by any means,
electronic or mechanical, including photocopying,
recording or by any information storage and retrieval
system, without written permission from the copyright
owner.

Printed in the Netherlands

Contents

* Chapters indicated with an asterisk are reprinted from *Fertilizer Research*, Volume 42, Nos. 1–3 (1995).

Preface

In achieving improved and sustainable agriculture in the Third World, increased productivity from existing farmland is absolutely essential, since additional land is either limited, unavailable, very fragile or must remain in its natural state for strategic purposes and environmental preservation. For this to be achieved, there are several important measures which must be implemented; one of the most important is the use of agro-chemicals to maintain and/or improve soil fertility. Traditional farming systems in this part of the world have generally been exploitive, without consideration for protection of the soil resources. On the other hand, it is important that environmental pollution through excessive and inappropriate fertilizer use must be avoided.

Nitrogen is the most important of all major nutrients and its low inherent status in the soil is probably the major cause of poor crop yields in the developing world. The nitrogen pool can be increased through fertilizer use; however, there are associated problems which are highly dependent on soil properties and the environment. The various soil processes affecting fertilizer nitrogen availability are not well understood, yet knowledge of these is most important in improving nitrogen efficiency. The availability of nitrogen to plants can be influenced by manipulation of the soil and crop biological systems and an understanding of the processes by which it becomes available in this system is critical in the development of sustainable agriculture. These problems have been engaging the Department of Soil Science of the University of the West Indies for some time and with the support of the GTZ (Deutsche Gesellschaft für Technische Zusammenarbeit GmbH) we have been able to utilise advanced fertilizer research techniques such as the use of isotopically labelled nitrogen to provide solutions. This symposium is an integral part of this collaboration with the main aim being to communicate a state of the art in several important scientific areas which affect nitrogen economy in tropical soils and to identify areas in which there are information gaps so that future research can be planned accordingly.

Some of the topics to be addressed include:
- annual fluctuations of native available soil nitrogen
- interactions of important nitrogenous fertilizers with different soils
- the influence of soil type and cropping system on the losses of applied nitrogen by volatilisation, de-nitrification, leaching and run-off
- ways of increasing the persistence of fertilizer nitrogen in the soil system
- the influence of soil micro-flora on applied nitrogen
- the role of the micro-flora in nitrogen availability
- the importance of nitrogen recycling in conservation and efficient use of the nutrient
- the behaviour of applied nitrogen in wet and submerged soils
- the effect of nitrogen fertilizers on soil behaviour
- possible environmental hazards in the Third World from an increased nitrogen fertilizer use
- Socio-economic aspects of nitrogen fertilizer use
- the advantages of using labelled nitrogen in studies on nitrogen economy in soil/plant systems
- Advances in N_2-fixation by *Rhizobium*-legume symbiosis

The organizers are grateful to the many distinguished soil/plant scientists who participated at the Symposium and contributed by preparing and presenting papers. We are much indebted to Kluwer Academic Publishers for agreeing to publish the Proceedings of the Symposium so that the State of the Art in the many aspects of nitrogen economy in soils can be made available to a larger readership.

In organising this important symposium, the Organisers are most grateful to the GTZ (German Agency for Technical Cooperation) for financial support and in addition, the help provided by the following agencies is also hereby graciously acknowledged:

Agricultural Development Bank of Trinidad & Tobago
AMOCO Oil Company of Trinidad & Tobago
Arcadian Trinidad Ltd.
Australian International Development Assistance Bureau
Caroni (1975) Ltd. of Trinidad & Tobago
Commonwealth Foundation
Geest Industries Ltd. of St. Lucia
Hydro-Agri Trinidad Ltd.
Neal & Massey Holdings of Trinidad & Tobago
Organization of American States
Overseas Development Administration of the United Kingdom

N. Ahmad

Chairman, Organizing Committee

Fertilizer Research **42**: 1–11, 1995.
© 1995 *Kluwer Academic Publishers. Printed in the Netherlands.*

1

Forms and nature of organic N in soil

K.R. Kelley[1] & F.J. Stevenson[2]
[1]*Land and Water Sciences, TVA Environmental Research Center, P.O. Box 1010, Muscle Shoals, AL 35660-1010, USA;* [2]*Department of Agronomy, University of Illinois, Urbana, IL 61801, USA*

Key words: nitrogen, ^{15}N humic substances, acid hydrolysis, analytical pyrolysis, NMR

Abstract

Most of the N in surface soils occurs in organic forms. The organic N in soil plays a key role in plant nutrition and soil fertility through its effects on microbial activity and nutrient availability. Typically, about one-third of the fertilizer N applied to temperate-zone soils is immobilized and retained in organic forms at the end of the growing season. A significant portion of this newly immobilized N is no more available to microorganisms and plants than the native humus N. Stabilization processes, probably involving polymerization of amino compounds and polyphenols, result in incorporation of N into humic substances with a concurrent reduction in N availability. This paper presents an account of the forms and nature of organic N in soil, emphasizing possible formation pathways, chemical characterization of humic substances through conventional and solid-state techniques, and the fate and composition of newly immobilized N in soil.

Introduction

It has been well documented that over 90% of the N in most surface soils occurs in organic forms. The forms of organic N in soil can be divided into two broad categories: (i) organic residues, consisting of undecayed plant and animal residues and partial decomposition products, and (ii) soil organic matter or humus. The humus consists of nonhumic, or chemically recognizable substances (e.g. amino acids, carbohydrates, nucleic acids, etc.), and humic substances, consisting of high molecular weight, amorphous, partly aromatic substances formed by secondary reactions. The importance of humus in terms of maintaining or improving soil fertility has been widely recognized and can be attributed to positive effects of humus on physical, chemical, and biological properties of soil.

Organic soil N plays a key role in terms of plant nutrition through direct and indirect effects on microbial activity and nutrient availability. The significance of humic substances to the N fertility of soil arises from the fact that much of the organic N resists attack by microorganisms and is thereby relatively unavailable to plants. Also, as a consequence of mineralization-immobilization turnover by microorganisms, a portion of the fertilizer N applied to soil becomes stabilized by incorporation into humic substances.

Results of field trials using the stable isotope, ^{15}N, have shown that about one-third of the fertilizer N remains behind in organic forms after the first growing season, only a small fraction (<15%) of which becomes available to plants during the second growing season. Stabilization through linkages to or incorporation into the structures of humic substances appears likely. Thus, a knowledge of the chemical nature of N in humic and nonhumic substances will not only provide a better understanding of factors affecting the availability of indigenous soil N but may contribute to the development of improved management practices for the efficient and environmentally acceptable use of fertilizer N.

In this paper, the chemical nature of N in soil humus is discussed, with emphasis being given to incorporation, stabilization, and availability of immobilized fertilizer N in soil. Reviews of soil N availability and of indices of N availability have been compiled [34, 77]; recent reviews of humic substances as related to soil fertility include those of Schnitzer [61, 62], Stevenson [80, 81, 83, 84], and Tate [88].

Formation of organic N complexes in soil

Multiple pathways probably exist for the formation of humic substances and organic N complexes in soil. Early concepts, such as the lignin-protein theory of Waksman [91], theorized that humic materials represent condensation products resulting from reactions involving partially decomposed lignin and microbially synthesized amino compounds. Another popular theory, termed the polyphenol theory, postulates that humic materials are formed through condensation reactions involving amino compounds and polyphenols.

A popular current concept is that humic substances in soil are formed by a multiple stage process that includes: (i) decomposition of all plant polymers, including lignin, into simple monomers, (ii) metabolism of the monomers by microorganisms with an accompanying increase in the soil microbial biomass, (iii) repeated recycling of the biomass C and N with synthesis of new cells, and (iv) concurrent polymerization of reactive monomers into high molecular weight substances. The consensus is that polyphenols, derived from lignin together with those synthesized by microorganisms, are converted to quinones through the catalytic action of polyphenol oxidase enzymes or clay minerals. The quinones then polymerize in the presence or absence of amino compounds (amino acids, NH_3, etc.) to form brown-colored macromolecules. The net effect of this series of reactions is that the N of amino acids is incorporated into complex structures, such as in heterocyclic rings or as a bridge between quinone and aromatic rings. Whereas the relative importance of lignin or microbial synthesis as sources of polyphenols may vary, depending upon soil and climatic conditions, random abiotic polymerization reactions appear to be a central element in the formation of highly complex and heterogenous humic substances in soil.

Principal humification pathways operating during early stages of pedogenesis were evaluated in a recent study of humus formation [43]. The study was conducted using samples collected from an open-cast lignite mine site in which low organic matter (1.9 g C/kg) mine spoil material was used to reclaim mined areas. Reclamation involved liming (1 t/ha), fertilizing (23 kg N, P, K/ha), mulching (300 kg cellulose/ha), and hydroseeding with ryegrass (*Lolium perenne*), fescue (*Festuca rubra*), blue grass (*Poa pratensis*), and white clover (*Trifolium repens*). Soil samples were collected from nearby native pastures and from mined areas 1, 2, 3, 4, 5, and 7 years after application of the reclamation treatments. This provided a series of mine spoils developed from the same parent material and receiving identical management for various periods of time. The C and N contents increased from 1.9 g C/kg and 0.1 g N/kg (raw spoil material) to 37 and 2.2 g/kg, respectively, after 7 years, representing about 49 and 39% of the C and N levels in the nearby undisturbed pasture soils. Conclusions, based on analyses for molecular weight distribution, aromaticity, organometallic complexes, and clay-adsorbed and stabilized humic substances, indicated that humification processes operating in the mine spoil material were similar to those occurring in the native pasture soils. Their findings support current theories of humus formation in that the dominate humification mechanism appeared to involve abiotic polymerization reactions of small organic molecules producing fulvic acids as precursors to humic acids.

Stevenson [78] reviewed several hypothetical structures for humic and fulvic acids and pointed out that humic substances consist of a complex, heterogeneous mixture of organic compounds that cannot be adequately described by a single structural model. Several of the proposed structures contained similar features, however, including N contained in amino acids, peptides, or proteins attached to a central humic core by H-bonding or covalent linkages. Stevenson concluded that significant amounts of the N associated with humic substances cannot be accounted for in known compounds and that this N may occur as: (i) free amino ($-NH_2$) groups, (ii) open chain ($-N-$, $=N-$) groups, (iii) part of a heterocyclic ring, such as an $-NH-$ of indole and pyrrole or the $-N=$ of pyridine, (iv) a bridge constituent linking quinone rings, and (v) amino acids attached to aromatic rings in forms that resist hydrolysis in hot mineral acids.

Characterization of organic N by conventional techniques

The most widely used approach for characterizing organic N in soil or humic substances is by acid hydrolysis. Methods for determining the forms of N in hydrolysates of soil or soil organic matter preparations (i.e. humic acid, fulvic acid, and humin) have been described [79]. In a typical procedure, the sample is heated with 3 N or 6 N HCl for 12 to 24 hours, after which the N is separated into the following fractions:

- Acid insoluble-N: Nitrogen remaining in residue following hydrolysis; usually obtained by difference (total N − hydrolyzable N).
- NH_3-N: Ammonia recovered from the hydrolysate by steam distillation with MgO.
- Amino acid-N: Usually determined by the ninhydrin-NH_3 or ninhydrin-CO_2 methods.
- Amino sugar-N: Steam distillation with phosphate-borate buffer at pH 11.2 and correction for NH_3-N; colorimetric methods are also available.
- Hydrolyzable unknown-N (HUN fraction): Hydrolyzable N not accounted for as NH_3-N, amino acid-N, or amino sugar-N; part of this N occurs as non-α-amino N in arginine, tryptophan, lysine, and proline.

Only from one-third to one-half of the organic N in soils can be accounted for in known compounds, i.e. amino acid-N and amino sugar-N. Soils also contain trace quantities of nucleic acids and other known nitrogenous biochemicals, but specialized techniques are required for their separation and identification.

A large amount of soil N, usually about 25 to 35%, is recovered as acid insoluble-N. At one time, this fraction was thought to be the result of condensation reactions during hydrolysis, but it is now believed that part of this N occurs as structural components of humic substances.

A large proportion of the soil N, about 20 to 25% for surface soils, is recovered as NH_3-N by distillation with MgO. In the older literature, this form of N was referred to as the 'amide-N' of proteins, but it is now known that very little of the NH_3-N is derived from the amino acid amides, asparagine and glutamine. Some of the NH_3-N is derived from indigenous fixed NH_4^+; part comes from partial destruction of amino sugars. It is also known that NH_3 can arise from the breakdown of certain amino acids during hydrolysis. Tryptophan is lost completely. Other amino acids, such as serine and threonine, are partially destroyed. An accounting of all known potential sources of NH_3 in soil hydrolysates shows that about one-half of the NH_3-N, equivalent to 10 to 12% of the total organic N, is still obscure [80].

From 10 to 20% of the soil N is accounted for in the HUN fraction, of which from one-fourth to one-half has been reported to occur as the non-α-amino N of such amino acids as arginine, tryptophan, lysine, histidine, and proline [21]. The possibility that most of the HUN occurs as non-α-amino acid N is remote.

As one might expect, humic and fulvic acids, obtained by classical extraction techniques, contain the same forms of N that are obtained when soils are subjected to acid hydrolysis [2, 7, 14, 37, 56, 89]. However, the N distribution patterns vary somewhat. In comparison with unfractionated soil, lower percentages of the N in humic acids occur as NH_3-N and HUN; higher percentages occur as amino acid-N and as acid insoluble-N.

Part of the amino acid-N of humic acids may exist as peptides or proteins linked to a central core by H-bonding [24]. Peptide-like substances have been determined in humic acids by chromatographic assay of proteinaceous constituents liberated by cold hydrolysis with concentrated HCl [15, 54]. It has also been demonstrated that amino acids are released through the action of proteolytic enzymes [41]. Infrared spectra of some, but not all, humic acids show absorption bands typical of the peptide linkage [51, 82]. Evidence for N in loosely bound forms has come from the observation that the N content of humic acids can be lowered considerably by passage through a cation exchange resin in the H-form [73, 76]. Protein-rich fractions have been obtained from humic acids by extraction with phenol [12, 47, 48].

A major proportion of the N of humic acids, ranging from 30 to 60%, is recovered as acid insoluble-N. This N may occur as a structural component of the molecule, such as a bridge constituent linking quinone groups together. Additional quantities of amino acids can be recovered from humic acids by subjecting the residues from the initial acid hydrolysate to a second hydrolysis with 2.5 N NaOH [22, 54]. Sodium amalgam reduction of the alkali-treated residue has been shown to lead to a further release of amino acids [54]. Acid hydrolysis would be expected to remove amino acids bound by peptide bonds as well as those linked to quinone rings. On the other hand, amino acids bonded directly to phenolic rings may not be released without subsequent alkaline hydrolysis. The effect of alkaline hydrolysis was believed to be due to oxidation of the amino phenol to the quinoneimine form, with subsequent hydrolysis of the imine and release of the amino acid.

An increase in amino acid-N at the expense of the HUN fraction was reported when an acid hydrolysate of humic acids was further heated with 6 N HCl containing 3% H_2O_2 [2]. The increased release of amino acids was attributed to the presence of phenolic-amino acid addition products in the hydrolysate. Hydrolysis of the initial acid-insoluble residue with 3% H_2O_2-6 N HCl led to a further release of amino acids, as well as other N forms. It seems likely, therefore, that part of the acid insoluble-N in humic substances may occur in the form of N-phenyl amino acids resulting from the bonding between amino groups and aromatic rings. For peptides, amino acids directly attached to an aromatic ring may be less available to plants and microorganisms than amino acids of the peripheral peptide chain.

In summary, the N of humic substances exists in several forms, some being labile and readily utilized by microorganisms (e.g. H-bonded peptides and proteins) and some being recalcitrant, or nearly so. Significant amounts of the N associated with humic acids and fulvic acids cannot be accounted for in known compounds.

The assumption is sometimes made that the structural N of humic substances represents the more stable component of soil organic N. On this basis, one would expect major changes to occur in the distribution of the forms of N when soils are subjected to intensive cultivation, namely, large losses of N from amino acids and other more readily available forms at the expense of the acid insoluble-N fraction. In practice, this has not been shown to be the case [80]. Surprisingly, neither long-term cropping nor the addition of organic amendments to the soil greatly affects the relative distribution of the forms of N. All forms, including the acid insoluble-N fraction, appear to be biodegradable [30, 46]. Results of these and other studies show rather convincingly that humic substances are not completely inert but are constantly changing as a consequence of the activities of microorganisms. Under steady-state conditions, mineralization of native humus is compensated for by synthesis of new humus. During humification, the N of amino acids and other amino compounds is incorporated into the structures of humic and fulvic acids and is not readily available to plants.

Characterization of organic soil N by NMR techniques

The development of techniques for characterizing organic matter in intact soil samples represents a significant advance in efforts to further our understanding of the forms and nature of organic N in soil. These techniques, which include ^{13}C and ^{15}N nuclear magnetic resonance (NMR) spectroscopy and analytical pyrolysis (discussed later), allow for structural characterization of organic matter without chemical extraction or fractionation.

The utility of ^{13}C NMR in studies on organic N arises from the fact that one of the main 'chemical shift zones' (i.e. 50 to 60 ppm) in spectra of soil organic matter, and its subfractions, has been assigned to the C of N-alkyl compounds (e.g. amino acids, peptides, and proteins). Results obtained thus far indicate that as little as 2% to as much as 15% of the C in samples analyzed could be accounted for as proteinaceous constituents. It should be noted that the C of methoxyl groups also produces shifts in the 50 to 60 ppm range, but the methoxyl content of humic constituents and well-decomposed organic matter is normally very low. No attempt seems to have been made to interpret the results of ^{13}C NMR studies in terms of organic N composition.

^{13}C NMR spectroscopy has been widely used in studies of humus chemistry. This technique has been applied to liquid and solid-state samples, including humic acids, fulvic acids, and whole soils. The usefulness of ^{13}C NMR for characterizing complex solid-state samples is limited by line broadening and a poor signal-to-noise ratio. Cross polarization and magic angle spinning techniques (CPMAS) have been applied in conjunction with NMR spectroscopy to improve spectral quality and enhance the utility of this approach for analysis of solid-state samples.

CPMAS ^{13}C NMR has been applied in studies covering a wide range of applications and research objectives; specific references can be obtained from comprehensive reviews [25, 44, 45, 92, 93]. Examples of recent applications include: characterization of humic and fulvic acids extracted from soil by different mechanisms (charge repulsion, ligand exchange, or H-bond disruption) [52], relationships between structural characteristics of humic acids and nitrate uptake and growth regulation in plants [53], formation of humic substances in forest soils [11, 38], effects of cultivation on particle size fractions and stability of humic substances [29, 55], effects of vegetation type on structural characteristics of humic acids, fulvic acids, and humin [39, 40], degradation of humic acids by fungi and actinomycetes [17], chemical characterization of fungal melanins [59, 63], modifications and incorporation of straw-derived lignin into humic substances [4], and incorporation of ^{13}C-labeled substrates into humic fractions [6].

Malcolm and MacCarthy [45] recently summarized findings from ^{13}C NMR studies of humic substances and pointed out that the findings give evidence for higher aliphatic contents than previously thought. They also noted that results of ^{13}C NMR studies indicate that substantial differences exist in the reactivity and composition of humic and fulvic acids, that humic acids are more aromatic than fulvic acids, and that distinct differences occur among fulvic acids from diverse environments.

Solid-state ^{13}C NMR analysis is limited by the low C content and low natural abundance of ^{13}C (1.1%) in mineral soils. Baldock *et al.* [6] recently summarized studies in which they added ^{13}C-labeled glucose to a coarse-textured mineral soil to overcome these limitations. After 34 days of incubation, the soil was fractionated on the basis of particle size and density, and the various fractions were analyzed by CPMAS ^{13}C NMR. Over 72% of the residual ^{13}C was found in the clay fraction. Comparisons between ^{13}C NMR spectra of the residual glucose C and spectra of bacteria and fungi, isolated from the same soil, indicated that the glucose was predominately utilized by soil fungi with little or no incorporation into aromatic or phenolic structures.

Of the NMR spectroscopy techniques, CPMAS ^{15}N NMR has the most direct applicability for characterizing the forms and nature of organic N in soil. This technique has been applied in studies involving ^{15}N enrichment and analysis of humic substances in soil [10, 85, 86] and studies of the fate of ^{15}N-labeled glycine [9], ^{15}N-labeled melanoidins [10], and ^{15}N-labeled ammonium sulphate, urea, or wheat straw [8]. Incubation of ^{15}N-labeled glycine in sterile and nonsterile peat led to incorporation of glycine-derived N into amines, secondary amides, and pyrrole-type N forms, possibly due to condensation between sugars and amino compounds through the Mallaird reaction [9]. Further evidence for involvement of the Mallaird reaction was provided by ^{15}N NMR studies in which melanoidins, synthesized from reducing sugars and ^{15}N-labeled amino acids, were found to be quite similar to ^{15}N-labeled humic acids obtained from a mineral soil [10].

Studies involving ^{15}N NMR analysis of ^{15}N-labeled humic substances have been reported recently [85, 86]. Labeling of the humic substances was accomplished by addition of ^{15}N-labeled ammonium sulfate to a low organic matter soil followed by cropping and aerobic incubation for a period of three years. Humic and fulvic acids, extracted by conventional techniques, and humic acids recovered from the humin fraction following treatment with an HF-HCl mixture to dissolve silicate minerals, were isolated and analyzed by CPMAS ^{15}N NMR. Results indicated that most of the N occurred in amide forms (80, 86, and 87% for the humic acid, fulvic acid, and humic acid-from-humin

fractions, respectively); from 9 to 12% and from 4 to 9% of the N occurred in amine and pyrrole-type forms, respectively. In contrast, much lower proportions of the N were accounted for in known forms by conventional acid hydrolysis (amino acids and amino sugars accounted for 16 to 52% of the N). Lower recoveries in known forms by acid hydrolysis were attributed to the occurrence of amide N in nonhydrolyzable or acid-resistant forms and to the release and/or reactions of amide N during hydrolysis. In a follow-up study to characterize the nonhydrolyzable (6 N HCl) residue of humic acids, ^{15}N NMR analysis indicated that most of the 'amide-N' had been liberated as amino acid-N or NH$_3$-N. All of the pyrrole N was accounted for in the HUN fraction, while the N of aromatic amines was accounted for in the HUN or NH$_3$-N fractions [86]. Similarities between the spectroscopic and chemical characteristics of the humic acid and humic acid-from-humin fractions indicated that part of the humin N occurs as humic acids intimately associated with soil colloids.

Characterization of organic soil N by analytical pyrolysis

Analytical pyrolysis is a thermal degradative method that can be applied to solid-state samples in soil organic matter studies. Schulten [66] pioneered a technique combining pyrolysis and soft ionization mass spectrometry in which the sample is pyrolyzed in the mass spectrometer and volatile molecular fragments, or pyrolysis products, are analyzed by field ionization mass spectrometry (Py-FIMS). The technique can be applied to whole soil samples and presents several important advantages over chemical degradative procedures. This technique eliminates the need for extraction, fractionation, and purification, provides for rapid analysis and information on a molecular level for complex organic materials, and is suitable for analysis of mineral soils [70].

Py-FIMS has been applied to the characterization of C compounds in a variety of materials; the reader is referred to recent reviews for additional references [13, 60, 62, 67]. Recent applications include comparisons of whole soils to the humic acid, fulvic acid, and humin fractions [65], characterization of morphologically differentiated horizons in a forest soil [28], humification of primary organic matter inputs (straw and farmyard manures) and dynamics of soil organic matter in agricultural soils [42], characteristics of particle-size fractions as affected by long-term fertilization with farmyard manures [69], effects of soil and crop management on humus properties [23, 68, 70], and influences and limitations due to the composition of the mineral soil matrix [70].

Although this technique has primarily been used to characterize organic C groups, Py-FIMS also offers potential for advancing our knowledge of the structural aspects and forms of organic N in soil. For example, Hempfling *et al.* [28] reported evidence for amines, nitriles, and heterocyclic N fragments released by pyrolysis of a forest soil. They assigned N-containing structural formulas to over 70 of the pyrolysis fragments but did not speculate as to the original structure of the organic N complexes. The structural assignments of the N-containing fragments, ranging in molecular weight from 45 to 410, were based on previous Py-gas chromatography/mass spectrometry (Py-GCMS) studies of humic substances, plant materials, and microbial biomass.

In a recent study involving Py-FIMS, the effects of long-term (> 100 years) fertilization with farmyard manure (FYM) on characteristics and N compounds associated with various particle-size fractions were reported [69]. Results indicated that levels of N compounds increased with decreasing particle size. The fine- and medium-clay fractions and the fine- and medium-silt fractions of the FYM-amended soil were enriched in amino-N and amides. The clay fraction of the intensively managed (unfertilized) soil showed a high relative abundance of pyrroles, pyridines, and indoles, supporting the conclusion that reduced inputs of primary organic matter result in an enrichment of refractory heterocyclic N compounds [42, 68].

The applicability of Py-FIMS for identifying N compounds in humic substances or soil organic matter was recently extended in an investigation involving analysis of 23 amino acid standards [75]. The thermal fragmentation pattern for each amino acid was delineated, and a series of peak signals were selected as markers to identify amino acids in humic substances. The intensities of the amino-marker signals were determined for 27 agricultural soils and found to be correlated with α-amino acid N (r=0.76***) and total N (r=0.65***).

Schulten [67] pointed out that structural investigations by Py-FIMS should include complementary destructive and/or nondestructive techniques. Combinations of Py-FIMS and other analytical approaches have been reported in studies of humic substances and organic N complexes, including ^{13}C NMR [6, 11, 23,

28], Fourier-transformed IR [4, 28, 85], Py-GCMS [75], and classical chemical techniques [68]. Schulten and Schnitzer [71] recently proposed a structural concept for humic acids based on the literature and previous investigations utilizing various combinations of analytical pyrolysis, spectroscopic, chemical, oxidative/reductive degradation, and colloid-chemical and electron microscope techniques. Their model did not include linkages or structures for proteinaceous materials, however.

The application of solid-state techniques appears to offer real potential to advance our knowledge of the structural and functional aspects of organic N complexes in soil. It is not possible to draw inferences as to the original structure of organic N complexes in soil at this time, however, due to our limited knowledge of the forms and nature of organic N in soil.

Fate and chemical nature of newly immobilized N in soil

As noted earlier, a significant fraction of the fertilizer N applied to temperate-zone soils, of the order of 20 to 35%, remains behind in organic forms after the first growing season. No more than 15% of this residual N becomes available to plants during the second growing season, and availability decreases even further for succeeding crops [81, 84]. A key factor affecting the efficiency with which fertilizer N is consumed by crops is the progressive stabilization of N by conversion to more resistant humus forms. The net result is a decrease in the 'labile' pool of potentially available N in the soil. Although a number of explanations can be given for the low availability of residual fertilizer N in soils, there is little doubt that the stability is due in part to incorporation of N into the complex structures of humic substances, for which two mechanisms can be given: (i) the turnover of N through mineralization-immobilization leads to incorporation of N into the melanins of fungi, and (ii) the N of amino acids, peptides, and proteins is stabilized through reactions involving quinones and carbonyl-containing substances.

Several approaches using the stable isotope, ^{15}N, have been utilized to characterize immobilized N in soil, including fractionation based on acid hydrolysis and partitioning into humic and fulvic acids by classical alkali extraction. A noteworthy feature of the acid hydrolysis studies is that much of the N cannot be accounted for in known forms but is recovered in the acid insoluble-N, HUN, and NH_3-N fractions. With time, more and more of the residual N is converted to these unknown forms, and, within a matter of years, the composition of the fertilizer N is indistinguishable from that of the native humus N [3, 74].

Results of the field study of Allen *et al.* [3] are typical in that an average of one-third of the applied ^{15}N was accounted for in the surface soil after the first growing season, the remainder having been taken up by plants or lost through leaching and denitrification. Isotope-ratio analysis revealed that most of the residual N had been incorporated into organic forms. Comparison of the distribution pattern for the fertilizer-derived ^{15}N with that of the native humus N showed that a considerably higher proportion of the fertilizer N occurred in the form of amino acids (59.0 vs. 36.0%) and amino sugars (9.9 vs. 8.0%); lower proportions occurred in the NH_3-N (10.6 vs. 18.1%), acid insoluble-N (10.3 vs. 21.7%), and HUN (10.2 vs. 16.2%) fractions. When the plots were resampled four years later, the fertilizer N remaining in the soil, representing one-sixth of that initially applied, had a composition very similar to that of the native humus N. The mean residence time (MRT), or average life, for the N retained after the first season was about five years. The N retained after this period was postulated to have a MRT of 25 years; thereafter, the MRT would be the same as for the native humus, or an estimated 200 to 800 years. From this, it can be seen that a small fraction of the fertilizer N applied during any given growing season will remain in the soil for a long time, perhaps centuries.

Similar results were reported in which fertilizer N was initially incorporated into such compounds as amino acids and subsequently into more stable forms [74]. As compared to the native soil N, more of the residual fertilizer N occurred in an amino acid-containing fraction, with smaller amounts being accounted for as NH_3-N and acid insoluble-N. Equilibrium with the native soil N had not been achieved in three years.

Residual fertilizer N has also been partitioned into classical humus fractions by alkali extraction [20, 57, 94]. The relative amounts of N recovered in humic and fulvic acids were found to depend on several factors, including form of applied N [94]. Mc Gill *et al.* [48] isolated humic and fulvic acids from ^{15}N-labeled soils and observed a higher degree of labeling in the fulvic acid fraction.

It should be noted that results obtained with ^{15}N-labeled substrates complement those for ^{14}C. From 30 to 37% retention of applied C has been observed at the end of the first growing season, depending on soil and climatic conditions [19, 31, 50, 72]. Experiments with ^{14}C-labeled ryegrass straw showed that approximately 33% of the applied C remained behind in the soil after the first year [31]. This residual C had a MRT of about four years; thereafter, the MRT approached that for the native humus. Much of the N required to maintain a constant C/N ratio will come from applied fertilizer N; part may come from the soil N pool.

Additional information regarding the biological conversion of N to stable humus forms has been provided by studies in which the organic N has been labeled with ^{15}N by short-term incubation of the soil with inorganic ^{15}N and a suitable C substrate. Results of these studies have confirmed that stabilization of N occurs very rapidly and that a significant portion of the immobilized N occurs in forms not readily available to plants or microorganisms. In an early study, soil was incubated for periods of up to 20 weeks with ^{15}NO$_3$ and three C substrates (glucose, straw, and cellulose) [32]. For all C substrates, an initial net immobilization of ^{15}N was followed by a period of net mineralization. At the point of maximum incorporation of N into the biomass, there was a distinct difference in the percentage distribution of the forms of organic N between the newly immobilized ^{15}N and the native humus N, as determined by acid hydrolysis. Higher percentages of the immobilized N occurred as amino acids and in the HUN fraction; a lower percentage was found as acid insoluble-N. With time, the percentages of the organic N as acid insoluble-N and NH_3-N increased while the percentages in the amino acid-N and HUN fractions decreased. Similar trends have been observed by others [1, 36].

Studies involving characterization of newly immobilized ^{15}N in a typical Illinois mollisol have also been reported [26, 27, 35, 36]. The microbial biomass of a Flanagan silt loam (Aquic Argiudoll) was labeled by short-term incubation with glucose and ^{15}N. The labeled soil, collected at the point of maximum conversion of glucose C into microbial biomass, was analyzed by chemical extraction, acid hydrolysis, and alkali fractionation techniques. Utilization of the glucose was accompanied by a rapid immobilization of applied ^{15}N and conversion to organic forms. Several extractants, used as chemical indices of N availability and ranging in severity from hot water to acidified permanganate, were tested to determine the effectiveness by which the newly immobilized ^{15}N could be separated from the native humus N [35]. An inverse relationship was observed between the amount of N extracted and the proportion of the soluble N derived from the labeled biomass, indicating that milder procedures were more selective in extracting the newly immobilized ^{15}N. Distinct differences were also noted between the chemical distribution of the immobilized ^{15}N and the native humus N. Lower proportions of the immobilized N were recovered in the acid insoluble-N and NH_3-N fractions; higher proportions were recovered in the amino acid-N and HUN fractions.

The ^{15}N-labeled soil was also extracted sequentially with 0.15 M $Na_4P_2O_7$ and 0.1 N KOH, following which the N was partitioned into classical humic fractions [26]. Higher percentages of the applied ^{15}N, as compared to the native soil N, were accounted for in the two extracts (22 vs. 16% by 0.15 M $Na_4P_2O_7$ and 27 vs. 22% by 0.1 N KOH, respectively). About 49% of the applied ^{15}N was removed by sequential extraction, as compared to 38% for the native soil N. Although somewhat less than for the native soil N (59%), the percentage accounted for as humin was still surprisingly high (46%). One explanation for this result is that much of the newly immobilized N existed as insoluble cellular components of microorganisms, including fungal melanins that were not extractable under the conditions used. Short-term incubations, as carried out in the study, are known to lead to an initial increase in bacterial numbers, followed by an increase in fungal organisms. A subsequent decline in the fungal population would be expected to be followed by a second increase in bacterial numbers [27]. On a weight basis, the amount of fungal tissue greatly exceeds that for bacteria. Evidence for the presence of humic acid-like substances in alkaline extracts of fungal tissue has been reported [18, 33, 49, 58, 59, 64, 87]. Fungal melanins are not completely solubilized by dilute alkali reagents, and the insoluble residues have properties similar to those of the humin fraction of soil organic matter [58]. The organic matter solubilized by sequential extraction with $Na_4P_2O_4$ and NaOH (extracts combined) was further partitioned into humic and fulvic acids. The percentage of the extracted ^{15}N accounted for in the humic acid fraction (16%) was somewhat less than for the native soil N (24%). These results definitely show that the immobilized ^{15}N was partitioned into all humus fractions and that a significant fraction of the newly immobilized N was indistinguishable from the native humus N.

Results similar to those noted above have been obtained in other studies [5, 16, 48]. It is of interest that distribution patterns of immobilized ^{15}N in the classical humic fractions are very similar to that for the ^{14}C

of labeled C substrates [90], where labeling in all fractions has been observed. Chichester [16] fractionated ^{15}N-labeled soil based on soil particle size and then extracted each particle fraction with 0.5 M $Na_4P_2O_7$ at 100°C. The extracted organic N was further fractionated by acid hydrolysis. Distribution patterns for the fertilizer N and the native soil N were very similar, a result that was attributed to conversion of the fertilizer-derived N into complexes similar in composition and properties to the native humus N. In a study to fractionate immobilized ^{15}N, McGill et al. [48] reported that at the time when the microbial population reached maximum (incubation period of five days), the added ^{15}N extracted by $Na_4P_2O_7$ represented about 20% of the total added ^{15}N and the ratio of the added ^{15}N in fulvic acids relative to that in the humic acids was greater than 2:1. Up to 10% of the total ^{15}N was accounted for in the humic acid fraction, about 50% of which was removed by phenol extraction and assumed to consist of proteinaceous material held to humic acids through H-bonding. He et al. [27] found that aqueous phenol was more selective in removing recently immobilized ^{15}N from soil than reagents typically used for extraction of soil organic matter (i.e. 0.15 M $Na_4P_2O_7$ and 0.1 M NaOH).

Summary and conclusions

Results of ^{15}N studies have clearly demonstrated that from 20 to 40% of the fertilizer N added to agricultural crops of temperate-zone soils is typically incorporated into organic forms during the first growing season. In comparison to the native humus N, more of the newly immobilized N occurs as amino acids and amino sugars and less occurs in unidentified forms. The newly immobilized N is more susceptible to mineralization or extraction, but a significant portion is no more available than the native humus N. Stabilization processes, including polymerization reactions between amino compounds and polyphenols, lead to the incorporation of N into humic structures and a gradual reduction in N availability.

Despite the excellent progress that has been made in recent years, much more needs to be accomplished to advance our knowledge and understanding of the forms and nature of organic N in soil. Numerous studies involving a broad range of conventional and solid-state techniques have been applied to the characterization of organic N complexes, but as much as one-half of the N of humic substances cannot been accounted for in known forms. The nature of this unknown N will remain obscure until more is known about the chemical structures of humic and fulvic acids and the mechanisms whereby N is stabilized by incorporation into humic substances.

References

Ahmad ZY, Yahiro HK and Harada T (1973) Factors affecting immobilization and release of nitrogen in soil and chemical characteristics of the nitrogen newly immobilized. IV. Soil Sci Plant Nutr (Tokyo) 19: 287–298

Aldag RW (1977) Relations between pseudo-amide nitrogen and humic acid nitrogen released under different hydrolysis conditions. In: Soil organic matter studies, pp 293–299. Int Atomic Energy Agency, Vienna, Austria

Allen AL, Stevenson FJ and Kurtz LT (1973) Chemical distribution of fertilizer nitrogen in soil as revealed by nitrogen-15 studies. J Environ Qual 2: 120–124

Amalfitano C, Pignalosa V, Auriemma L and Ramunni A (1992) The contribution of lignin to composition of humic acids from a wheat-straw amended soil during 3 years of incubation in pots. J Soil Sci 43: 495–504

Anderson DW, Paul EA and St Arnaud RJ (1974) Extraction and characterization of humus with reference to clay-associated humus. Can J Soil Sci 54: 317–323

Baldcock JA, Currie CJ and Oades JM (1991) Organic matter as seen by solid state ^{13}C NMR and pyrolysis tandem mass spectrometry. In: Wilson WS (ed) Advances in soil organic matter research: The impact on agriculture and the environment, pp 45–60. Wiltshire, England: Redwood Press Ltd

Batsula AA and Krupskiy NK (1974) The forms of nitrogen in humus substances of some virgin and developed soil of the left bank of the Ukraine. Sov Soil Sci 6: 456–462

Benzing-Purdie L, Cheshire MV, Williams BL, Ratcliffe CI, Ripmeester JA and Goodman BA (1992) Interactions between peat and sodium acetate, ammonium sulphate, urea or wheat straw during incubation studied by ^{13}C and ^{15}N NMR spectroscopy. J Soil Sci 43: 113–125

Benzing-Purdie LM, Cheshire MV, Williams BL, Sparling GP, Ratcliffe CI and Ripmeester JA (1986) Fate of [^{15}N] glycine in peat as determined by ^{13}C and ^{15}N CP-MAS NMR spectroscopy. J Agric Food Chem 34: 170–176

Benzing-Purdie L, Ripmeester JA and Preston CM (1983) Elucidation of the nitrogen forms in melanoidins and humic acid by nitrogen-15 cross polarization-magic angle spinning nuclear magnetic resonance spectroscopy. J Agric Food Chem 31: 913–915

Beyer L, Schulten H-R, Fruend R and Irmler U (1993) Formation and properties of organic matter in a forest soil, as revealed by its biological activity, wet chemical analysis, CPMAS ^{13}C NMR spectroscopy and pyrolysis-field ionization mass spectrometry. Soil Biol Biochem 25: 587–596

Biederbeck VO and Paul EA (1973) Fractionation of soil humate with phenolic solvents and purification of the nitrogen rich portion with polyvinyl-pyrrolidone. Soil Sci 115: 357–366

Bracewell JM, Haider K, Larter SR and Schulten H-R (1989) Thermal degradation relevent to structural studies of humic substances. In: Hayes MHB *et al.* (eds) Humic substances II: In search of structure, pp 181–222. New York: Wiley

Bremner JM (1955) Studies on soil humic acids: I. J Agr Sci 46: 247–256

Cheshire MV, Cranwell PA, Falshaw CP, Floyd AJ and Haworth RD (1967) Humic acid: II. Structure of humic acids. Tetrahedron 23: 1669–1682

Chichester FW (1970) Transformations of fertilizer nitrogen in soils. II. Total and ^{15}N-labelled nitrogen of soil organo-mineral sedimentation fractions. Plant Soil 33: 437–456

Dehorter B, Kontchou CY and Blondeau R (1992) ^{13}C NMR spectroscopic analysis of soil humic acids recovered after incubation with some white rot fungi and actinomycetes. Soil Biol Biochem 24: 667–673

Filip Z, Haider K, Beutelspacher H and Martin JP (1974) Comparisons of IR spectra from melanins of microscopic soil fungi, humic acids and model phenol polymers. Geoderma 11: 37–52

Führ F and Sauerbeck D (1968) Decompostion of wheat straw in the field as influenced by cropping and rotation. In: Isotopes and radiation in soil organic matter studies, pp 241–250. Tech Meeting, FAO/IAEA, Vienna, Austria

Gamzikov GP (1981) Fertilizer nitrogen transformation: An interaction with West-Siberian soils. In: Dutil P and Jacquin F (eds) Colloque: humus-azote, pp 194–197. Station de Science du Sol, INRA, Charlons-Sar-Marne, France

Goh KM and Edmeades DC (1979) Distribution and partial characterisation of acid hydrolyzable organic nitrogen in six New Zealand Soils. Soil Biol Biochem 11: 127–132

Griffith SM, Sowden FJ and Schnitzer M (1976) The alkaline hydrolysis of acid-resistant soil and humic acid residues. Soil Biol Biochem 8: 529–531

Haider K, Gröblinghoff F-F, Beck T, Schulten H-R, Hempfling R and Lüdermann HD (1991) Influence of soil management practices on the organic matter structure and the biochemical turnover of plant residues. In: Wilson WS (ed) Advances in soil organic matter research: The impact on agriculture and the environment, pp 79–91. Wiltshire, England: Redwood Press Ltd.

Haworth RD (1971) The chemical nature of humic acid. Soil Sci 111: 71–79

Hayes MHB (1991) Concepts of the origins, composition, and structures of humic substances. In: Wilson WS (ed) Advances in soil organic matter research: The impact on agriculture and the environment, pp 3–32. Wiltshire, England: Redwood Press Ltd

He X-T, Stevenson FJ, Mulvaney RL and Kelley KR (1988) Incorporation of newly immobilized ^{15}N into stable organic forms in soil. Soil Biol Biochem 20: 75–81

He X-T, Stevenson FJ, Mulvaney RL and Kelley KR (1988) Extraction of newly immobilized ^{15}N from an Illinois Mollisol using aqueous phenol. Soil Biol Biochem. 20: 857–862

Hempfling R, Zech W and Schulten H-R (1988) Chemical composition of the organic matter in forest soils: 2. Moder profile. Soil Sci 146: 262–276

Hopkins DW and Shiel RS (1991) Spectroscopic characterization of organic matter from soil with mor and mull humus forms. In: Wilson WS (ed) Advances in soil organic matter research: The impact on agriculture and the environment, pp 71–78. Wiltshire, England: Redwood Press Ltd

Ivarson KC and Schnitzer M (1979) The biodegradability of the "unknown" soil-nitrogen. Can J Soil Sci 59: 59–67

Jenkinson DS (1965) Studies on the decomposition of plant material in soil I.J. Soil Sci. 16: 104–115

Kai H, Ahmad Z and Harada T (1973) Factors affecting immobilization and release of nitrogen in soil and chemical characteristics of the nitrogen newly immobilized. III. Soil Sci Plant Nutr (Tokyo) 19: 275–286

Kang KS and Felbeck Jr GT (1965) A comparison of the alkaline extract of tissues of *Aspergillus niger* with humic acids from three soils. Soil Sci 99: 175–181

Keeney DR (1982) Nitrogen - Availability indices. In: Page AL *et al.* (eds) Methods of soil analysis, Part 2, Chemical and microbiological properties. Agronomy 9: 711–734. Madison, Wisconsin: Am Soc Agron.

Kelley KR and Stevenson FJ (1985) Characterization and extractability of immobilized ^{15}N from the soil microbial biomass. Soil Biol Biochem 17: 517–523

Kelley KR and Stevenson FJ (1987) Effects of carbon source on immobilization and chemical distribution of fertilizer N in soil. Soil Sci Soc Am J 51: 946–951

Khan SU and Sowden FJ (1972) Distribution of nitrogen in fulvic acid fraction extracted from the black Solonetzic and black Chernozemic soils of Alberta. Can J Soil Sci 52: 116–118

Kögel-Knabner I, Hatcher PG and Zech W (1991) Chemical structural studies of forest soil humic acids: Aromatic carbon fraction. Soil Sci Soc Am J 55: 241–247

Krosshavn, M, Kögel-Knabner I, Southon TE and Steinnes E (1992) The influence of humus fractionation on the chemical composition of soil organic matter studied by solid-state ^{13}C NMR. J Soil Sci 43: 473–483

Krosshavn M, Southon TE and Steinnes E (1992) The influence of vegetational origin and degree of humification of organic soils on their chemical composition, determined by solid-state ^{13}C NMR. J Soil Sci 43: 485–493

Ladd JN and Jackson RB (1982) Biochemistry of ammonification. In: Stevenson FJ (ed) Nitrogen in agricultural soils. Agronony 22: 173–228. Madison, Wisconsin: Am. Soc Agron

Leinweber P and Schulten H-R (1993) Dynamics of soil organic matter studied by pyrolysis-field ionization mass spectrometry. J Anal Appl Pyrolysis 25: 123–136

Leirós, MC, Gil-Sotres F, Ceccanti B, Trasar-Cepeda MC and González-Sangregorio MV (1993) Humification in reclaimed open-cast lignite mine spoils. Soil Biol Biochem. 25: 1391–1397

Malcolm, RL (1989) Application of solid-state ^{13}C NMR spectroscopy to geochemical studies of humic substances. In: Hayes MHB *et al.* (eds) Humic substances II: In search of structure, pp 339–372. New York: Wiley

Malcolm RL and MacCarthy P (1990) The individuality of humic substances in diverse environments. In: Wilson WS (ed) Advances in soil organic matter research: The impact on agriculture and the environment, pp 23–34. Wiltshire, England: Redwood Press Ltd

Malik KA and Haider K (1982) Decomposition of ^{14}C-labelled melanoid fungal residues in a marginally sodic soil. Soil Biol Biochem 14: 457–460

McGill WB and Paul EA (1976) Fractionation of soil and ^{15}N nitrogen to separate the organic and clay interactions of immobilized N. Can J Soil Sci 56: 203–212

McGill WB, Shields JA and Paul EA (1975) Relation between carbon and nitrogen turnover in soil. Organic fractions of microbial origin. Soil Biol Biochem 7: 57–63

Meuzelaar HLC, Haider K, Nagar BR and Martin JP (1977) Comparative studies of pyrolysis-mass spectra of melanins, model phenolic polymers, and humic acids. Geoderma 17: 239–252

Oberländer HE and Roth K (1968) Transformations of ^{14}C-labeled plant material in soils under field conditions. In: Isotopes and radiation in soil organic matter studies, pp 351–361. Tech meeting, FAO/IAEA Vienna, Austria

Otsuki A and Hanya T (1967) Some precursors of humic acid in recent lake sediments suggested by infrared spectra. Geochim Cosmochim Acta 31: 1505–1515

Piccolo A, Campanella L and Petronio BM (1990) Carbon-13 nuclear magnetic resonance spectra of soil humic substances extracted by different mechanisms. Soil Sci Soc Am J 54: 750–756

Piccolo A, Nardi S and Concheri G (1992) Structural characteristics of humic substances as related to nitrate uptake and growth regulation in plant systems. Soil Biol Biochem 24: 373–380

Piper TJ and Posner AM (1968) On the amino acids found in humic acids. Soil Sci 106: 188–192

Preston CM, Newman RH and Rother P (1994) Using ^{13}C CPMAS NMR to assess effects of cultivation on the organic matter of particle size fractions in a grassland soil. Soil Sci 157: 26–35

Rosell RA, Salfeld JC and Sochtig H (1978) Organic components in Argentine soils: 1. Nitrogen distribution in soils and their humic acids. Ochimica 22: 98–105

Rudelov YV and Prynanishnikov DN (1982) Dynamics of immobilized nitrogen. Soviet Soil Sci 14: 40–45

Russell JD, Vaughan D, Jones D and Fraser AR (1983) An IR spectroscopic study of soil humin and its relationship to other soil humic substances and fungal pigments. Geoderma 29: 1–12

Saiz-Jimenez C (1983) The chemical nature of the melanins from *Coprinus* spp. Soil Sci 136: 65–74

Saiz-Jimenez C (1995) The chemical structure of humic substances: Recent advances. In: Piccolo A (ed) Humic substances in terrestrial ecosystems, (in press). Amsterdam, Elsevier Publ

Schnitzer M (1985) Nature of nitrogen in humic substances. In: Aiken GR *et al.* (eds) Humic substances in soil, sediment, and water, pp 303–325. New York: Wiley-Interscience

Schnitzer M (1991) Soil organic matter–the next 75 years. Soil Sci 151: 41–58

Schnitzer M and Chan YK (1986) Structural characteristics of a fungal melanin and a soil humic acid. Soil Sci Soc Am J 50: 67–71

Schnitzer M, Ortiz de Serra MI and Ivarson K (1973) The chemistry of fungal humic acid-like polymers and of soil humic acids. Soil Sci Soc Am Proc 37: 229–236

Schnitzer M and Schulten H-R (1992) The analysis of soil organic matter by pyrolysis-field ionization mass spectrometry. Soil Sci Soc Am J 56: 1811–1817

Schulten H-R (1977) Pyrolysis-field ionization and field-desorption mass spectrometry of biomacromolecules, microorganisms, and tissue material. In: Jones CER and Cramers CA (eds) Analytical Pyrolysis, pp 17–28. Amsterdam: Elsevier Publ

Schulten H-R (1987) Pyrolysis and soft ionization mass spectrometry of aquatic/terrestrial humic substances and soils. J Anal Appl Pyrolysis 12: 149–186

Schulten H-R and Hempfling R (1992) Influence of agricultural soil management on humus composition and dynamics: Classical and modern analytical techniques. Plant and Soil 142: 259–271

Schulten H-R and Leinweber P (1991) Influence of long-term fertilization with farmyard manure on soil organic matter: Characteristics of particle-size fractions. Biol Fertil Soils 12: 81–88

Schulten H-R and Leinweber P (1993) Pyrolysis-field ionization mass spectrometry of agricultural soils and humic substances: Effect of cropping systems and influence of the mineral matrix. Plant and Soil 151: 77–90

Schulten H-R and Schnitzer M (1993) A state of the art structural concept for humic substances. Naturwissenschaften 80: 29–30

Shields JA and Paul EA (1973) Decomposition of ^{14}C-labelled plant material under field conditions. Can J Soil Sci 53: 297–306

Simonart P, Batistic L and Mayaudon J (1967) Isolation of protein from humic acid extracted from soil. Plant Soil 27: 153–161

Smith SJ, Chichester FW and Kissel DE (1978) Residual forms of fertilizer nitrogen in field soils. Soil Sci 125: 165–169

Sorge C, Schnitzer M and Schulten H-R (1993) In-source pyrolysis-field ionization mass spectrometry and Curie-point pyrolysis-gas chromatography/mass spectrometry of amino acids in humic substances and soils. Biol Fertl Soils 16: 100–110

Sowden FJ and Schnitzer M (1967) Nitrogen distribution in illuvial organic matter. Can J Soil Sci 47: 111–116

Stanford G (1982) Assessment of soil nitrogen availability. In: Stevenson FJ (ed) Nitrogen in agricultural soils. Agronomy 22: 651–688. Madison, Wisconsin: Am. Soc Agron

Stevenson FJ (1982) Humus Chemistry. New York: Wiley-Interscience

Stevenson FJ (1982) Nitrogen - Organic forms. In: Page Al *et al.* (eds) Methods of soil analysis, Part 2, Chemical and microbiological properties. Agronomy 9: 625–641. Madison, Wisconsin: Am Soc Agron

Stevenson FJ (1982) Organic forms of soil nitrogen. In: Stevenson FJ (ed) Nitrogen in agricultural soils. Agronomy 22: 67–122. Madison, Wisconsin: Am Soc Agron

Stevenson FJ (1985) Cycles of soil: C, N, P, S, micronutrients. New York: Wiley-Interscience

Stevenson FJ and Goh KM (1971) Infrared spectra of humic acids and related substances. Geochim Cosmochim Acta 35: 471–483

Stevenson FJ and He X-T (1990) Nitrogen in humic substances as related to soil fertility. In: MacCarthy P *et al.* (eds) Humic substances in soil and crop sciences; Select readings, pp 91–109. Madison, Wisconsin: Am Soc Agron

Stevenson FJ and Kelley KR (1985) Nitrogen transformations in soil: A perspective. In: Malik KA *et al.* (eds) Nitrogen and the environment, pp 7–26. Faisalabad, Pakistan: Nuclear Institute for Agriculture and Biology

Su-Neng Z and Qi-Xiao W (1992) Nitrogen forms in humic substances. Pedosphere 4: 307–315

Su-Neng Z, Qi-Xiao W, Li-Juan D and Shun-Ling W (1992) The nitrogen form of nonhydrolyzable residue of humic acid. Chinese Science Bulletin 37: 508–511

Tan KH, Sihanonth P and Todd RL (1978) Formation of humic acid like compounds by the entomycorrhizal fungus, *Pisolithus tinctorius*. Soil Sci Soc Am J 42: 906–908

Tate III RL (1987) Soil organic matter. New York: Wiley-Interscience

Tsutsuki K and Kuwatsuka S (1978) Chemical studies of soil humic acids: III. Soil Sci Plant Nutr 24: 29–38

Wagner GH (1968) Significance of microbial tissue to soil organic matter. In: Isotopes and radiation in soil organic matter studies, pp 197-205. Tech. meeting, FAO/IAEA, Vienna, Austria

Waksman SA (1932) Humus. Baltimore, Maryland: Williams and Wilkins

Wilson MA (1989) Solid-state nuclear magnetic resonance spectroscopy of humic substances: Basic concepts and techniques. In: Hayes MHB *et al.* (eds) Humic substances II: In search of structure, pp 309–338. New York: Wiley

Wilson MA (1990) Application of nuclear magnetic resonance spectroscopy to organic matter in whole soils. In: MacCarthy P *et al.* (eds) Humic substances in soil and crop sciences; Selected readings, pp 221–260. Madison, Wisconsin: Am Soc Agron

Wojcik-Wojtikowiak D (1978) Nitrogen transformations in soil during humification of straw labeled with ^{15}N. Plant Soil 49: 49–55

Fertilizer Research **42**: 13–26, 1995.
© 1995 *Kluwer Academic Publishers. Printed in the Netherlands.*

Seasonal fluctuations of native available N and soil management implications

M.T.F. Wong[1] & S. Nortcliff[2]
[1]*CSIRO Division of Soils, Private Bag PO, Wembley, Western Australia 6014;* [2]*Department of Soil Science, The University of Reading, Whiteknights, Reading, RG6 2DW UK*

Key words: CERES models, fluctuations, lignin, microbial motility, mineralization, nitrate, nitrate retention, polyphenol, synchrony, water potentials

Abstract

The concentration of native available N in tropical soils fluctuates considerably in response to seasonal changes in soil water potential. Such fluctuation reflects the net effect of inputs of N from mineralization, fertilizers and the atmosphere, and removal by plant uptake, immobilization, leaching and gaseous losses. The greatest concentrations normally occur during the transition between the dry and wet seasons. In East-Africa, up to 184 kg mineral N ha^{-1} has been measured in the 0–40 cm soil layer and in Trinidad, 143 kg ha^{-1} was found in the 0–10 cm layer. Release and accumulation of mineral N occur as a result of the influence of soil water potential on microbial activity. This is due to changes in microbial motility, solute diffusion, microbial survival and the release of protected organic matter. A quantitative understanding of these processes should increase the efficiency of use of this valuable N resource by crops. Current methods of forecasting mineralization under field conditions include measurement of the soil mineralization potential, the release of N from seasonal inputs of litter and model predictions. Litter quality is important. Its composition, in particular its nitrogen, lignin and soluble polyphenol content has a major impact on its N mineralization rate.

Crop uptake, gaseous and leaching losses decrease the concentration of soil mineral N during the wet season. These losses are important under moist tropical conditions. For example, at Port Harcourt and Ibadan in Nigeria, leaching losses were large in spite of NO_3^- adsorption which decreased the depth of NO_3^- leaching relative to through-flow. To minimise these losses, it is essential to synchronise plant nutrient demand with supply by mineralisation. This is particularly important at the start of the tropical rainy season when high rates of mineralisation often in excess of the relatively low levels of crop demand, are observed. Fertilizer recommendation, the time table for cropping and the farming system used therefore need to take into account the seasonal availability of N. The CERES model simulates crop growth and development and the N-cycle. As development and validation continue, such models should provide a strong basis for better soil, crop and fertilizer management practices. A better understanding of the processes should provide a strong basis for futher development of such models.

Introduction

Nitrogen is often the most important nutrient limiting the yield of crops in the tropics [55]. The largest proportion is found in the soil organic matter and availability to plants is dependent on its mineralization to NH_4-N. Under favourable conditions, NH_4-N is nitrified to NO_3-N. Both ammonification and nitrification are biological processes and have basic requirements of adequate soil water content and temperature. In North-West Europe, mineralization occurs predominantly in spring and autumn when the soil is warm and moist. The rate generally decreases in mid and late summer due to lower soil water potentials . It essentially stops during the winter months as the soil temperature falls below the threshold required for microbial activity [50]. Fluctuations of soil mineral N concentrations reflect the net effect of inputs from mineralization, fertilizers and the atmosphere, and removal

14

by plant uptake, immobilization, leaching and gaseous losses.

Seasonal temperature variation is less in tropical environments and minimum temperature thresholds for soil biological activities are nearly always achieved except at high altitudes. Soils of these regions normally have limited water stroage capacities and the timetable for cropping is mostly set by the seasonal rainfall which subjects the soil to large fluctuations in water potential. Suitable periods for crop growth can be determined using a notional water balance showing only rainfall and potential evapotranspiration. The main features of such water balances have been described for contrasting regions of the tropics [13]. Close to the equator, rain occurs throughout the year and deficit with respect to evapotranspiration is met from water stored in the soil. In the zone of bimodal rainfall, there are two relatively short dry seasons. In the seasonally arid tropics, there is a single dry season. The amount and temporal distribution of rainfall strongly influence soil microbial activity and the fluctuation of mineral N in tropical soils. An understanding of the processes is vital in attempts to increase the efficiency of use of this important plant nutrient by matching the supply of mineral N with crop demand.

Fluctuations of soil mineral N

The greatest concentrations of native mineral-N occur during the transition between the dry and wet seasons. Hardy [24] measured the fluctuations of NO_3-N at the University of West Indies Farm in Trinidad. The soil at the site is sandy with 1.6–2.7% organic C. The rainy season extends from June to December with a small decrease in rainfall in September or October. The mean rainfall is 1740 mm a^{-1}. The dry season starts at the end of January in most years. The soil water content during the dry season is generally greater than the soil water content at wilting point, implying that the soil water potential is greater than -1.5 MPa. NO_3-N gradually accumulates in the top-soil layers during the dry season, leading to a sizeable reserve of available N at the start of the rainy season. In the driest year, 143 kg NO_3-N ha^{-1} was found in the 0–10cm layer.

In another detailed field experiment on forest and savannah sites in Ghana, soil (0–10 cm) was sampled at fortnightly intervals to determine the fluctuations of NO_3-N concentration [20]. The forest and savannah sites had bimodal rainfall distributions with a short dry season from July to August and a longer one from

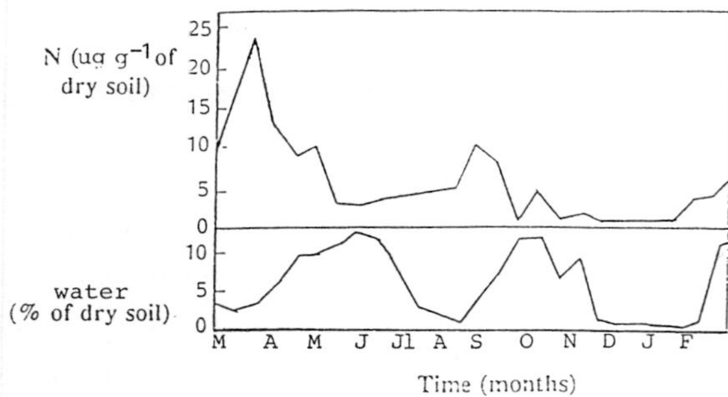

Fig. 1. Fluctuations of soil nitrate-N and water in the 0–10 cm layer of a cultivated savannah site, Ghana [20].

December to March. Intense drying decreased the total soil water content to less than 2% during the dry seasons. The pattern of NO_3-N fluctuations for the cultivated plot in the savannah region is shown in Fig. 1.

The NO_3-N concentration was low throughout the main rains from April to July. It increased during the short dry season. The concentration decreased with the start of the second rains in October and then increased gradually during the main dry season. In contrast to the wetter site in Trinidad, a more rapid increase in mineral N concentration was measured at the start of the rainy season in March. At its peak, the amount of NO_3-N measured in the 0–10 cm layer was 30 kg ha^{-1} in the cultivated savannah soil. This pattern of NO_3-N fluctuation was found to be typical of cultivated and bare fallow plots at all the sites measured. At both the Trinidad and Ghana sites, the mineral-N concentration under grass remained low throughout the year. This is typical of grassland where uptake by grass and inputs of organic materials of wide C:N ratios result in rapid immobilization of the mineralized N [51]. The immobilized N is released rapidly once the grass is removed. Similar fluctuations of mineral-N typified by a gradual accumulation during the dry season and a rapid build up at the start of the wet season have been measured at several sites in East Africa [56]. Between 13 to 184 kg mineral N ha^{-1} were found in the 0–40 cm layer during the first rains. Recent measurements continue to demonstrate the importance of the flush of mineral-N during the transition between the dry and wet seasons [35]. Soil drying and rewetting increase the rate of organic matter decomposition and release of mineral-N (Fig. 2). The effect is also obtained with the decomposition of plant materials [1]. The amounts mineralized are smaller on sites that have been cultivated for long periods. At a savanna site which had

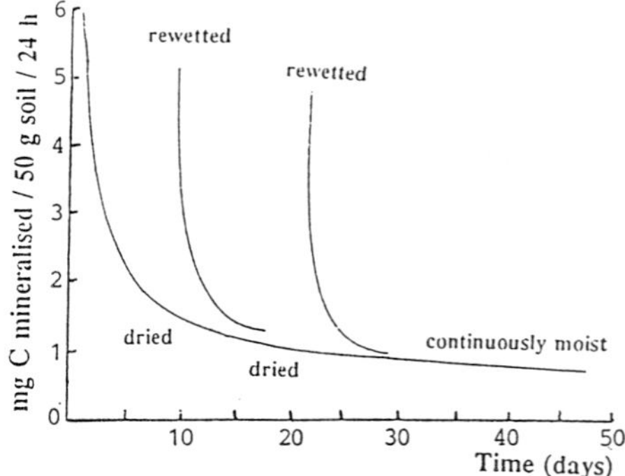

Fig. 2. Decomposition of organic matter following successive air-drying and rewetting of a soil [5].

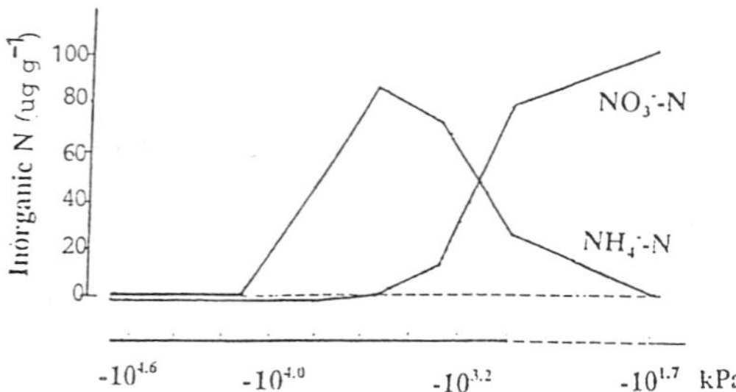

Fig. 3. Influence of soil water potential on the build-up of inorganic nitrogen after a 28-day incubation [11].

been cultivated for about 20 years in northern Nigeria, mineral-N ($NH_4^+ + NO_3^-$) was low at the time of planting. The largest flush of mineral-N was measured in plots that received farmyard manure [72]. The pattern of mineral-N fluctuation indicates that soil water plays a determining role in the mineralization of organic matter.

Effect of soil water on organic matter decomposition

Soil drying and microbial activity

Microbial motility and solute diffusion are limited by the size of the soil pores that are filled with water and the thickness of water films on soil surfaces. Both are determined by the soil matric potential (Ψ_m). The equivalent radius (r) of the water filled pores can be calculated from the capillary rise equation: $r = -147/\Psi_m$, where r is in μm and the matric potential is in kPa. Soil type is also important since the particle size and packing arrangement will determine the pore size distribution and number of pores of the ideal size that are filled with water [46].

Bacteria and protozoa are sensitive to low matric potential since they can only move in water filled pores and water films that are thicker than their own size. As the matric potential is decreased, the continuity of water filled pores is decreased, paths become more tortuous and organisms can be trapped in pores occluded by narrow necks [25]. Bacterial movement becomes negligible as soil dries to between -20 and -100 kPa matric potential. Activities decrease sharply as the matric potential decreases to between -50 and -300 kPa and are negligible at -1.5 MPa [23]. Protozoa are predatory and water-filled pores >3 μm (matric potentials > -200 kPa) are likely to be required for their activities [32]. Fungi and actinomycetes are less sensitive since they able to take up nutrients from water films that are thinner than their own dimensions. Continuous water paths are not required since hyphae can cross an open air space. This enables the organisms to function at water potentials even as low as -10 MPa due to the advantage of their morphology [59].

The severity of the drying effect is influenced by type of food. With solid food such as pieces of roots, the relative humidity of the air around the food particle is high until the soil is quite dry. This will ensure enough water films around the particle for activity. With soluble food, diffusion to microbial surfaces will be slower as the radius of pores are smaller and the paths of the solute become more tortuous. The resistance to diffusion is approximately inversely proportional to the cube of the soil water content [46]. Nitrification is more likely to be inhibited during the dry season due to the inability of bacteria to move to enriched microsites and of substrate to diffuse at an adequate rate to meet the needs of the bacterial cells.

Ammonification occurs in relatively dry soils (Fig. 3) because solid food can be reached by actinomycetes and fungi which are able to remain active at lower water potentials [71, 11]. A gradual accumulation of NH_4-N is often measured during the dry season when plant uptake and leaching losses are expected to be small. In Kenya, ammonification in the Kikuyu red loam produced about 55 kg NH_4-N ha^{-1} in the 0–60 cm layer during the dry season. Nitrification occurred when the matric potential increased to more than -1.5 MPa [53].

The ability to predict the rate of mineralization as the soil's water potential is decreased is clearly valuable. The rate of gross mineralization in an Oxisol from Kenya was 0.08 μg N g^{-1} d^{-1} at -5.91 MPa and 2.12

16

μg N g^{-1} d^{-1} when the soil was incubated at -63 kPa [48]. Decreased microbial motility and solute diffusion and the need of microbial cells to achieve water potential equilibrium with the environment result in decreased rates of organic matter decomposition at lower water potentials. The optimum water potential for organic matter decomposition is between -20 to -50 kPa. The relative rate of decomposition of soil organic matter and litter in soils from different sites can be described using a log-linear relationship between the water potential (Ψ in kPa) of the drying soil and decomposition rate [59]:

$$K/K_m = 116 \log \Psi + 0.87, r^2 = 0.77,$$

where K is the actual rate of decomposition and K$_m$, is the rate at optimum water potential for the experimental conditions. Since total soil N is intimately associated with soil C, a similar relationship was obtained for the relative rate of N release:

$$K/K_m = 32 \log \Psi + 0.744, r^2 = 0.90,$$

This type of simple regression equation is able to fit more recent data [44]. A decrease in water potential from -10 to -20 kPa resulted in a 10% decrease in microbial activity in the relatively moist soil. This shows the sensitivity of the process to changes in the soil water potential which should be controlled together with temperature in laboratory incubation experiments on organic matter decomposition. Plant residues give a similar log-linear relationship between soil water potential and decomposition rate [63]. At 20 °C, 25% of ^{14}C labelled wheat was lost at -1.5 MPa in 30 days and even at -5.0 MPa, 10% was lost. These simple equations use water potential which is a thermodynamic scale independent of soil type. Linear relationships are obtained with the soil water content [70]. Such equations are used to modify the rate of organic matter decomposition and nutrient release in models [28]. There is a marked hysteresis in soils undergoing repeated drying and wetting cycles and different equations may be required when the dry soil is wetted. Orchard and Cook [44] obtained a log-linear relationship between the difference in water potential of the dry and moist soil sample and the rate of organic matter decomposition. An increase of 5 MPa gave a 40 fold increase in microbial respiration. In a loamy sand and a sandy loam sample of an Ultisol, the half life of the pool undergoing decomposition following rewetting varied from 0.041 to 0.064 h^{-1}, implying that about 97% of N contained in the rapidly mineralizable N pool was released within 4 days of rewetting [9].

Microbial survival at low water potential

Microbes (algae, actinomycetes, bacteria, fungi, protozoa, viruses and yeasts) are the main soil organisms responsible for organic matter decomposition and mineralization. Other organisms, for example termites are also involved [64] but the mass of the larger organisms is smaller than the microbial biomass. Bacteria and actinomycetes are most numerous and active but fungi and protozoa are also important. The microbial biomass contains between 1% and 5% of the soil organic C and N [12]. In a soil containing 2% C and 0.2% N and with a bulk density of 1.3 g cm^{-3}, this is equivalent to 520-2600 kg C ha^{-1} and 52-260 kg N ha^{-1} in a 20 cm layer. The turnover of the microbial biomass is rapid with the dead or damaged cells forming a source of readily available substrate for the surviving microorganisms. The effect of microbial damage and death by heat or antiseptic on the release of mineral-N has been known since the beginning of this century [54]. A proportion of the soil microbial biomass is also killed by drying and the death rate depends upon the drying intensity [57]. In an experiment reported by Marumoto *et al.* [36], drying of a Parabrown Earth and a Chernozem released on average 37% of the microbial biomass-N following a 28-day incubation at 22 °C. Mineralization of the biomass-N accounted for 55% of the flush of N mineralised after remoistening the air-dried soil. The remainder came from non-biomass N. Drying tends to kill the more active biomass first and between 1/3 to 1/4 was killed by slow drying of a grassland soil [6]. The death of the organisms that are more sensitive to dehydration results in a shift in the population composition. Measurements of bacterial and actinomycetes populations in Kenya showed that the number of actinomycetes increased from 30% of the population to more than 90% as the drying season proceeded [38]. Actinomycetes have wider C:N ratios than bacteria. When bacteria are used as a substrate, release of mineral N should occur due to the lower N contents of the actinomycetes.

Decomposition of dead or damaged microbial cells can provide a significant proportion of mineral-N at the beginning of the rainy season. Moist soil will also allow protozoa to be active and to release mineral-N by predation of bacteria [33]. The calculations are given in Table 1 for the mineralization of 20% and 60% of the biomass-N. The amounts mineralized from 20% of the microbial biomass represent 0.2 to 1.0% of the total N and those from 60% of the biomass represent 0.6 to 3.0% of the total N.

Release of protected organic matter

The amount of organic matter rendered decomposable by heat or irradiation is slightly more than that obtained by chloroform fumigation [27]. The small additional amount may be due to the removal of the waxy coating protecting the organic matter, increased surface area of the organic matter or increased solubility in water due to irradiation or heating. Desorption of organic matter from clay surfaces and the rupturing of microaggregates to expose physically protected soil organic matter are important processes during soil drying and rewetting cycles [57, 12, 70]. Tillage disrupts the soil physical conditions and results in an increased rate of organic matter decomposition and nutrient release [4]. The rate of mineralization is higher with samples containing higher organic matter contents and is increased as the intensity of drying is increased [5, 43].

Effect of temperature

The effect of soil water is magnified by higher temperatures since the rate of decomposition increases as the soil temperature is increased. For a number of calcareous soils in South Australia, the pattern of decomposition was similar to that measured at Rothamsted, UK, but the rate of decomposition was twice as fast and was attributed to an 8–9 °C difference in temperature [34]. The pattern of decomposition on an Alfisol in Ibadan, Nigeria was also similar to that at Rothamsted but the rate was four times as fast due to higher temperatures [29]. The rate seems to double for approximately every 10 °C rise in temperature in accordance to the van 't Hoff's law if other factors affecting decomposition are held constant (Fig. 4). Similarly, the average rates at 12 sites in Costa Rica were 2–3 times greater than those of 4 sites in Germany [19]. It appears that for measurements made at relatively long time intervals (Fig. 4), the effect of seasonal soil water fluctuations are not clearly visible since the effect of drying may be offset by the rapid increase in decomposition obtained upon soil wetting. Seasonal fluctuations are important in nutrient dynamics and the behaviour of the soil organic matter and biomass under fluctuating water potentials

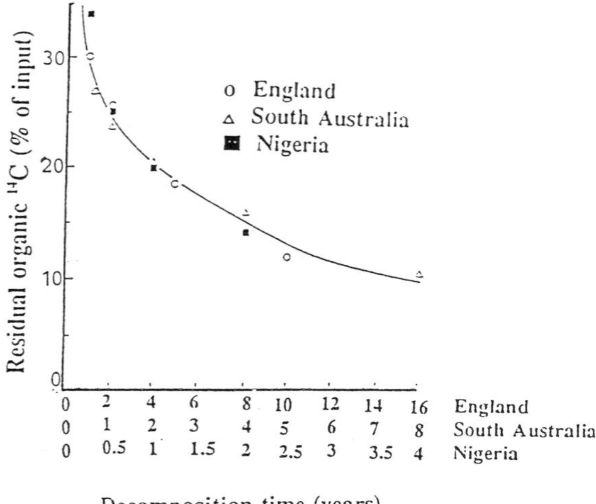

Fig. 4. Organic ^{14}C residues from plant materials decomposing in soils in three climatic regions [34].

need to be taken into account because of their influence on the temporal availability of nitrogen to plants.

Quality of organic nitrogen

The size of the soil water and temperature effects depends on the amount of readily mineralizable N. The death of the microbial biomass and the release of protected non-biomass organic matter increase the amount of readily mineralizable N. The mineralization capacity can be determined by incubation to determine the soil mineralization potential N_0 and rate constant K for mineralization [62]. The rate constant can be adjusted for the soil temperature (T) and water content (θ). This method is widely used as a simple and successful means of predicting mineralization in soils. The method assumes that the soil organic N consist of a single homogenous pool mineralizing according to first order kinetics. The long term mineralization capacity can be expressed as ($dN/dt = N_0 KT\theta$). The use of soil water content is more specific to soil type than the soil water potential which improves the general validity of the equation. As expected, values of N_0 vary widely between soils and their history of use whereas the rate constant varies little among a wide range of soils. The model can be extended to accommodate a fast and a slow decomposing organic matter pool which are either both decomposing according to first order kinetics or with one pool decomposing according to zero order

Table 1. Calculated amounts of N mineralized (kg ha^{-1}) at 0–20 cm from microbial biomass-N at four levels of total N of which 1 to 5% are found in biomass and at two rates of mineralization of biomass-N

% Total N in biomass	20% of biomass N mineralized Total N g N kg^{-1}				60% of biomass N mineralized Total N g N kg^{-1}			
	0.5	1.0	2.0	3.0	0.5	1.0	2.0	3.0
1	3	5	10	16	8	16	31	47
2	5	10	21	31	16	31	62	94
3	8	16	31	47	23	47	94	140
4	10	21	42	63	31	62	125	188
5	13	26	52	78	39	78	156	234

kinetics [58, 9]. The simple one pool model may overestimate mineralization in the field possibly due to the use of artificially dried soil samples during incubation or immobilization in the presence of plants. It may also underestimate mineralization in soils undergoing repeated drying and wetting cycles due to the rapid increase in the pool of mineralizable N. Testing of the method standardised for ease of comparison between sites is required for tropical soil conditions to assess its effectiveness in predicting N-mineralization in the field.

The major input of organic matter into soils is through plant materials. The input fluctuates in arable lands due to seasonal application of crop residues and tree litter and root death. Input of plant residues on arable land in the seasonally arid tropics is estimated to be about 3.5 t ha^{-1} a^{-1} and in the humid regions, the estimate is about 5.0 t ha^{-1} a^{-1} [22]. Assuming a N content of 1.5%, this is equivalent to inputs of 53 and 75 kg N ha^{-1} a^{-1}. In low input systems, leguminous and non-leguminous litter provide a valuable source of N, the availability of which depends on the chemical composition of the decomposing material. The N released supplements that which is mineralized from soil organic matter. The initial phase of decomposition is rapid due to the decomposition of water soluble organic matter such as sugars and starch, cellulose, hemicellulose and amino acids followed by a slower phase due to the decomposition of lignin and other more resistant materials [65, 41]. This pattern of decomposition of plant materials is often described as L-shaped. In Tenney and Waksman's [65] experiment, the rate of decomposition was increased as the N content was increased up to 1.7% N. Net mineralization occurred when the N content of the plant materials was

more than 1.7%. In the same year, Jensen [31] related decomposition to the C:N ratio of plant materials. It is now accepted that C:N ratios <20–25 are generally required for net mineralization of the organic material. If the ratio exceeds 25, mineralization is delayed until decomposition brings the ratio closer to 25. Mineralization generally then proceeds according to 1st-order kinetics. The decomposition rate is sometimes better related to the N content than to the C:N ratio and a critical level of 1.7% N has again been suggested for net mineralisation [16].

In some plant materials, both N content and the C:N ratio do not forecast mineralization rates accurately. This is especially true for plant materials containing high lignin and water/alcohol-soluble polyphenol contents. For example, failure of incubated *Desmodium intortum* to release N inspite of its high N content (5.0% N) was attributed to its high content of soluble polyphenols [66]. The polyphenols are able to form complexes with proteins and increase their resistance to microbial decomposition. Palm and Sanchez [45] found that mineralization rates of leaves of tropical legumes and rice straw were better correlated with their content of soluble polyphenols than with their N or lignin contents; they obtained a log-linear relationship between the polyphenols to N ratio and net mineralization rate. The highest mineralisation rates were obtained with three leguminous materials with polyphenols to N ratio less than 0.5. The remaining leguminous materials with ratios between 0.5 to 1.5 did not mineralize inspite of their high N content (>2.0%). The use of (lignin + polyphenols) to N ratios improved the prediction of mineralization rates [15]. The simple equation would be useful to synchronise the release of mineral-N with crop uptake if it is found to be generally applicable. The use of standardised methods will help in this context [15].

Residues of the more recently applied organic materials can also cause predictions of mineral-N by the Stanford and Smith [62] method to be underestimated. The residues can be separated from the soil on the basis of specific gravity. Those with specific gravity <2 consist of a mixture of humified and non humified materials derived from less decomposed plant and animal residues [10]. The light fraction (specific gravity < 1.6) contains materials with high carbon content and wide C:N ratio and consists mostly of structural litter that has not undergone significant humification. The fraction with specific gravity between 1.6 and 2.0 consists of cell debris and humifed materials which are extractable with NaOH [58]. These fractions are more labile than the more humifed materials (SG > 2) adsorbed on mineral surfaces [14]. Large seasonal fluctuation indicates that these fractions are an important reservoir of labile C and N [60]. Measurements of these fractions should improve our prediction of mineralization in the field.

Models of N-mineralisation

Determination of N mineralization based on measurements of N_0, K and the rate of litter application is widely used. Another approach is to use models of N mineralization. Such models should assist in managing the soil mineralized nitrogen effectively. Models vary in complexity; for a years-to-centuries timespan simple single homogenous compartment models have been used:

$$\mathrm{d}N/\mathrm{d}T = A - kN$$

where N is the quantity of N in the compartment, A is the annual input of fresh organic-N and k is the fraction decomposed each year.

Upon integration we obtain:

$$N = N_E + (N_0 - N_E)\mathrm{e}^{-kt}$$

where N_E is the equilibrium content.

The single homogenous compartment model described adequately the long term mineralisation of organic N at several sites in west Africa. The rate constant K ranges from 0.02 to 0.6 a^{-1}, depending on the depth of the soil sampled, soil type, organic matter content and history of use [21, 28].

The one compartment model can fit long term data well but in fact the soil organic matter is not homogenous. Fresh plant materials decay at a fast rate in the initial weeks, followed by a much slower decomposition [65, 41]. The decay can be described by a 2 compartment process where each comparanent decomposes at its own rate by a first order process as for the single compartment model. The model is good for the initial years but it underestimates the equilibrium content. The model should therefore contain at least one more compartment which is decomposing more slowly. More realistic models of organic matter decomposition and N mineralisation use several compartments, each decomposing by 1st order kinetics as before. The rate constant can be modified to take account of temperature, soil water content, tillage, crop cover and clay content [28]. Incoming organic C added monthly is partitioned into decomposable (DPM) and resistant plant material (RPM). Both decompose to the same products viz: CO_2, microbial biomass and humus. It is assumed that biomass and humus are always formed in the same proportion (0.15). At each month each compartment (Y) decays as follows:

$$Y = Y_0\mathrm{e}^{-kt}, t = \text{months}.$$

The model uses a stepwise transfer matrix to convert the quantity of organic matter in the compartrnent at the beginning of the month to that at the end. We shall not attempt a thorough discussion of models. The model described by Van Veen *et al.* [69] stresses the importance of frequent drying and wetting on organic matter decomposition and the turnover of the soil microbial biomass-N. The model of Parton *et al.* [47] and its later developments are also widely used in the tropics. A deeper understanding of processes should provide a strong basis for further development of these models to improve the management of soil organic matter and N in the tropics.

To model fluctuations in mineral-N concentration, a comprehensive model is required to take account of the major processes affecting the behaviour of N in a soil/crop system. A simple model is that developed by Bradbury *et al.* [8] for a cereal/soil system to take account of fertilizer and atmospheric inputs of N, mineralization, crop uptake, immobilization and gaseous and leaching losses.

Crop uptake and losses

The mineralized N is an important resource and the chief management objective is to maximise its efficiency of use. Losses can be severe because of lack of syn-

20

chrony especially in freshly cultivated soils where N mineralization is fast compared with crop requirement at the start of the rainy season. In a detailed field experiment at Port Harcourt in South East Nigeria, twelve undisturbed monolith lysimeters were used to trace the path of ^{15}N labelled urea and native mineralized N in soil, crops and drainage water [67]. Four lysimeters were uncropped in the first year and the rest were cropped with maize in the first and upland rice in the second cropping season. The urea was applied at 138 kg N ha^{-1} to maize and 92 kg N ha^{-1} to rice. Drainage water was collected quantitatively from each lysimeter over the two-year period and analyzed separately. The urea was rapidly hydrolysed, and only traces of NH_4-N were found in the drainage water. Maize and rice gave average grain yield of 2500 and 320 kg ha^{-1} respectively. The maize yields were considered normal for the region but the rice yields were low. Only 35% of the ^{15}N-labelled urea applied to the first season maize was harvested in the plant biomass over the two-year period (Table 2). The first maize harvest accounted for the highest proportion of the recovery and the subsequent crops contained little ^{15}N. A similar pattern of ^{15}N-labelled fertilizer uptake was obtained with rice but the recovery over the two-year period was only 19%. The true recovery of the fertilizer was probably greater because of the effect of mineralization/immobilization turnover [26, 30]. At the end of the experiment, 27 percent of the labelled N applied to maize and 29 percent of that applied to rice was recovered in the soil.

As the soil used in the lysimeters was a coarsely textured Ultisol and average annual rainfall was 2420 mm, large losses were expected in the drainage water. The recovery of unlabelled N in the drainage water and crops implied that 280 kg N ha^{-1} was mineralized from the soil organic matter during the two-year period. This represents the mineralization of 1.9% of the total soil N content. In the cropped lysimeters, 58% of the mineralized N was recovered in crops and 42% was lost by leaching in the first year. A similar rate of mineralization and leaching loss of mineralized N was reported under a field-grown crop at Ibadan [39]. The lysimeters that were cropped also lost between 22 and 29% of the ^{15}N-labelled nitrogen in the drainage water. The loss of NO_3-N was accompanied by an equivalent loss of exchangeable Ca, Mg and K (Table 3). In the first year, 27% of the exchangeable Ca content of the cropped lysimeters was leached. The corresponding values for Mg and K were 29% and 10% respectively [77]. As expected, the losses were more severe when the lysimeters were uncropped. The drastic depletion of the soil

reserve of base cations will result in a rapid decrease in the base saturation of the soil and the release of Al into the soil solution.

Direct measurements of denitrification have been reported for the tropical rain forests of Costa Rica [52]. The highest gaseous losses of N occurred in primary forest (13 kg ha^{-1} yr^{-1}) and early successional sites (19 kg ha^{-1} yr^{-1}) and least in mid-successional forest (4 kg ha^{-1} yr^{-1}). In a lysimeter experiment in Trinidad, about 20% of the ^{15}N-labelled fertilizer (400 kg N ha^{-1}) was lost in the 0-60 cm layer during part of the wet season from the beginning of June to the beginning of October. Leaching losses were small and the loss was attributed to denitrification in the moist soil [42]. At Port Harcourt, the total recoveries of labelled N in crop, soil and drainage water was between 70 and 87%, implying dentrification losses.

The water content of the lysimeters at field capacity was 160 dm^3. It was expected that the NO_3-N would be quickly leached below 1.35 m, the depth of the lysimeters, with a net rainfall of 160 dm^3. The actual volume required was 400 dm^3. The ratio 400/160 = 2.5 pore volumes meant that 2.5 times as much rain was needed to displace the NO_3-N than was expected [73]. Laboratory tests showed that the soil contained positive charges and adsorbed NO_3^-. NO_3-N adsorption has since been measured in a wide range of tropical acid soils [74].

For a given value of drainage W_D, the depth of NO_3-N penetration in a homogenous soil containing no cracks and channels and initially at field capacity is $W_D/0$ where θ is the soil volumetric water content at field capacity. Positive charges retard the movement of NO_3-N by a factor largely determined by the adsorption coefficient (b) of NO_3^- on the soil surface [75]:

$$1/R_F = 1 + bp/\theta$$

where R_F is the NO_3^- retention factor and p is the soil bulk density. The depth of NO_3-N penetration with adsorption in then $W_D/\theta R_F$. The calculated amounts of water required to displace NO_3-N through a range of tropical acid soil were 2 to 27 times the amounts required without NO_3-N adsorption. In soils where NO_3-N adsorption occurs, increases in pH and competition by Cl^-, SO_4^{2-}, $H_2PO_4^-$ and organic anions are expected to increase the rate of nitrate leaching.

The capacity to adsorb NO_3-N ranged from 0.01 to 1.74 cmol$_c$ kg^{-1} in a range of tropical acid soils. The smaller charge was measured in a sandy Ultisol from North Thailand and the greatest charge in an Andept from Colombia [75]. Significant amounts

Table 2. Percentage recovery of ^{15}N from labelled urea in harvested plant biomass, drainage water and soil [67]

Treatment	Recovery over two years			
	Biomass	Drainage	Soil	Total
^{15}N applied to maize in 1st year	35	29	27	82
^{15}N applied to rice in 1st year	19	22	29	70
^{15}N applied in 1st year, uncropped, cropped in 2nd year	2	65	20	87

Table 3. Amounts of cations leached in the first year in kg ha^{-1} and as a percentage of the initial amounts in soil after liming and fertilizer application [77]

	Initial amounts (kg ha^{-1})	Loss under crop		Uncropped lysimeters	
		(kg ha^{-1})	(%)	(kg ha^{-1})	(%)
Ca	1160	312	27	399	34
Mg	108	31	29	40	37
K	328	33	10	29	9

of NO$_3$-N can be retained by the positive charges. A one metre layer of soil weighing 13000 t ha^{-1}, with NO$_3$-N accounting for half of its positive charge, will hold 91 kg NO$_3$-N for each 0.1 cmol$_c$ kg^{-1}. We measured 450 kg NO$_3$-N ha^{-1} in a 20–80 cm layer of the Colombian Andept (Table 4). The storage will minimise leaching losses provided that the NO$_3$-N can be retrieved by deeper rooting plants.

Soil management implications

Synchronisation of supply of N and crop requirement

Management of the soil mineral-N should aim at synchronising the crop demand for N with N supply by mineralization and fertilizer application [2]. Planting early may decrease the leaching losses of the mineralized N by increasing the demand for N in the first weeks of the growing season [5]. The risk of crop failure due to unreliable rain and low soil water reserves during the early stages of crop establishment should be minimised by allowing a minimum soil water reserve or net rainfall to accumulate before sowing. Minimum tillage should decrease the rapid release of N in excess of crop demand especially in freshly cultivated soil.

Hardy [24] suggested sowing a temporary grass ley before the end of the dry season and ploughing the ley before cropping to exploit the high efficiency of N conservation under grass. This is technically feasible if drought resistant varieties and labour are available since the immobilized N is rapidly released on cultivation. Agroforestry may be a more practical alternative since it is possibly less labour intensive and the use of leguminous trees may improve the nitrogen content of the system. The permanent root network of the agroforestry trees is expected to take up NO$_3$-N and leaching losses may be lessened by deeper active roots acting as a safety net. Prunings of the trees are used to release N at a more controlled rate than that obtained during the transition between the dry and wet seasons. Development of such cropping system will require the evaluation of methods such as that of Fox *et al.* [15] for predicting the rate of release of mineral-N from tree prunings.

The rooting depth required to recover nutrients from the subsoil with a specified efficiency can be calculated using a simple leaching model [68]. The application of such a model requires the growth and

Table 4. Amounts of exchangeable nitrate on the anion exchange capacity (AEC) measured in two soils fom the humid tropics

Soil	Depth (cm)	AEC (cmol$_c$ kg^{-1})	Exchangeable N* (kg ha^{-1})	% AEC (as nitrate)
Munchong (Oxisol)	0–20	0.20	49	70
	20–35	0.83	176	84
	35–50	1.09	136	50
Popayan (Andept)	20–40	1.56	102	47
	40–60	1.74	181	74
	60–80	1.66	166	72

*Exchangeable nitrate = (KCl extractable - soil solution nitrate).

normal functioning of the tree roots in the subsoil. Acidity may, however, be a barrier to root growth. Trees that are resistant to subsoil acidity may be particularly effective. The treatment of subsoil acidity with organic matter or $CaSO_4$ is also possible [76]. As positive charge increases in the subsoil, NO_3-N adsorption must be taken into account when calculating the depth of leaching of the anion since its effect on the rate of leaching is large [75, 7].

McGill and Myers [37] examined the synchrony of cropping systems by comparing the time course of crop demand for N with that of supply of available N (Fig. 5). Under humid tropical conditions in Indonesia, mineralization occurs from December to May when the soil is warm and moist. The rate declines in the dry season. The simulated N demand for three crops shows that the pattern of mineralization coincides well with N demand for the first two crops. The third crop is more likely to be N deficient than the first two crops. The local practice is to grow a legume at this time. More important losses of NO_3-N are expected to occur between the first and second crop. In the semi-arid tropics at Hyderabad, India, mineralization will be rapid when the dry soil is moistened during the short wet season. Mineralization coincides with the requirement of the first crop but the second crop is more likely to benefit from fertilizers. Alternatively, a legume may be grown as the second crop.

Comprehensive models such as CERES-Rice and CERES-Maize simulate crop growth and development on a daily basis using genetic, edaphic and climatic information. The N-component of the model simulates mineralization/immobilization of N from the soil organic matter and crop residues. It traces the path of N in crop and losses to the atmosphere and in the drainage

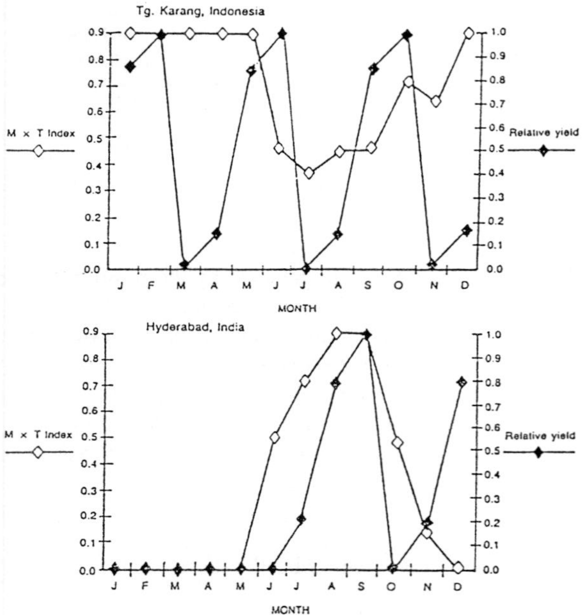

Fig. 5. Mineralization index for moisture × temperature compared to relative crops growth rates (yield) for two tropical sites; Tanjong Karang in the humid tropics and Hyderabad in the semi-arid tropics [37].

water [18, 7]. The N submodel of CERES-Maize simulates the effect of legume green manure incorporation. For some legumes, the release of N is overestimated presumably due to the lignin and soluble polyphenols content of the materials [15]. An important aspect of development by Bowen *et al.* [7] with regard to tropical acid soils is the inclusion of NO_3-N retention in the model. As development and validation continue, such models should provide a sound basis for better fertilizer and crop management practices.

Improving the release of N

Even when losses are small, the mineralisation of soil-N is not always sufficient to support a crop. This problem is more severe after several years of cultivation, contributing to the pressure on the farmer to abandon the land for a fresh plot. The size of the flush of mineralized N depends upon the C content of the soil [5]. This controls the size of the biomass that the soil can support. After several years of cultivation the level of soil C may be insuffcient to release an adequate amount of N to meet crop requirement. The conversion of a dry deciduous forest on an Ultisol in India to cropped land halved the organic C and N and biomass C and N content of the soil in only a few years [61]. The decline can be decreased with applications of crop residues which improves the soil total C, N and biomass C [3]. The microbial biomass-N can then be released when the soil conditions result in the death and mineralization of soil microbes [49]. The importance of organic matter on the amount of N released at the start of rain is illustrated in Fig. 6 for a field site at Ibadan, Nigeria [39].

The Jenkinson [28] model predicts the changes in the soil organic matter levels following clearance of natural vegetation, soil management and plant residue applications. Thus 11.0 t C ha^{-1} yr^{-1} are required to maintain the equilibrium level of organic matter at 43 t ha^{-1} in a humid tropical forest. Converting the forest to arable land would decreases the annual input of organic C. Inputs of 2.0 t C ha^{-1} typical of arable lands in humid tropical forest regions are predicted to decrease the soil organic matter content to 25 t ha^{-1} in 15 years. Similar calculations can be made for drier savanna regions which generally have lower soil organic matter content [22]. The model is valuable in determining the level of organic matter application required to optimise the productivity of the farming system.

Fertilizer management

With finite limits to arable land, the problem of food shortage in many parts of the tropics and especially in Africa can be overcome by yield increases through more intensive agriculture. Fertilizers will play a key role in this strategy. Trials in West Africa demonstrated the good response that can be obtained from the application of fertilizers, and economic analysis of the experimental results showed that the increased yield also translated into increased profits for the farmer [78]. The fertllizers are however expensive and losses can be

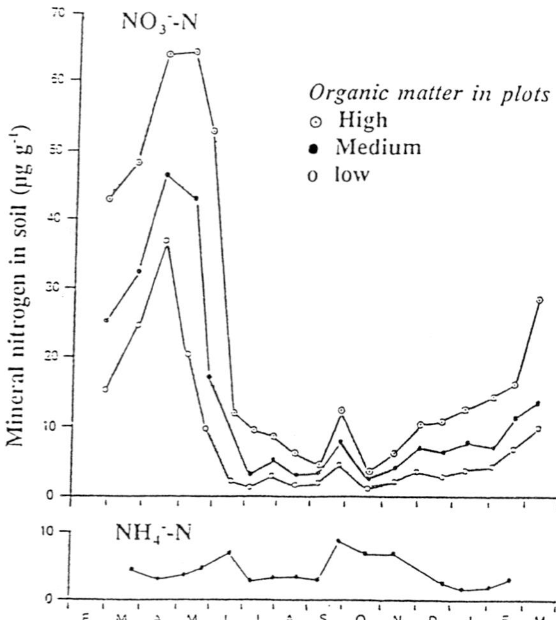

Fig. 6. Nitrate nitrogen in soil (0–10 cm) from high, medium and low organic matter plots [39].

environmentally damaging. The efficiency of use must be maximised by applying the right amount at the right time to take into account the mineralization of both soil and plant residue-N and the crop requirement for the targeted yield. The amounts mineralized are significant especially in freshly tilled soils and soils with high organic matter contents or when litter rich in N is used. The same consideration applies for other mineralized nutrients such as S and P [17]. Adjustment of the fertilizer recommendation is done by measurement of the soil mineral-N content and N mineralization potential or by model prediction of N mineralization prior to fertilizer recommendation. Myers [40] used a sigmoid growth curve to determine the crop demand for N and calculated the N mineralization rate from the soil mineralization potential taking into account the soil water potential and temperature. The fertilizer requirement was then calculated as the difference between the mineralization rate and the crop requirement. A comparison of the time course of N mineralization and crop demand allowed an assessment of the extent of synchrony in the system. Comprehensive models such as CERES-Maize and CERES-Rice take inputs from mineralization as well as losses to the atmosphere and by leaching and immobilization into account [7,18]. The rapid growth of fertilizer use in Africa and other parts of the tropics will increase the demand for mole accurate fertilizer recommendation. Accurate models simulating crop growth and development and N dynamics will be required to predict crop responses to fertilizers and the efficiency of fertilizer use under tropical soil conditions. Further research and a better understanding of processes should provide a strong basis for the continued development of such models.

Acknowledgements

We are grateful to the Overseas Development Administration (ODA) for sponsoring our participation at the symposium and to Dr. N. Ahmad for inviting us to the symposium.

We thank Professor D. Greenland for his friendly advice and assistance in literature search with SOIL-CD. The work of Wong presented in this paper was funded by the ODA.

References

1. Amato A, Jackson RB, Butler JHA and Ladd JN (1984) Decomposition of plant material in Australian soils. II Residual organic ^{14}C and ^{15}N from legume plant parts decomposing under field and laboratory conditions. Aust J Soil Res 22: 331–341.

2. Anderson JM and Ingram JSL (1989) Tropical soil biology and fertility: a handbook of methods. CAB International, Oxon, UK

3. Ayanaba A, Tuckwell SB and Jenkinson DS (1976) The effect of clearing and cropping on the soil organic reserves and biomass of tropical forest soils. Soil Biol Biochem 8: 519–525

4. Balesdent J, Meriotti A and Boisgontier D (1990) Effect of tillage on soil organic carbon mineralization estimated from ^{13}C abundance in maize fields. J Soil Sci 41: 587–596

5. Birch HF (1958) The effect of soil drying on humus decomposition and nitrogen availability. Plant and Soil 10: 9–31

6. Bottner B (1985) Response of microbial biomass to alternate moist and dry conditions in a soil incubated with ^{14}C- and ^{15}N-labelled plant material. Soil Biol Biochem 17: 329–337

7. Bowen WT, Jones JW, Carsky RJ and Quintana JO (1993) Evaluation of the nitrogen submodel of CERES-Maize following legume green manure incorporation. Agron J 85: 153–159

8. Bradbury NJ, Whitmore AP, Hart PBS and Jenkinson DS (1993) Modelling the fate of nitrogen in crop and soil in the years following application of ^{15}N-labelled fertilizer to winter wheat. J Agric Sci (Camb.) 121: 363–379

9. Cabrera ML (1993) Modelling the flush of nitrogen mineralization caused by drying and rewetting soils. Soil Sci Soc Am J 57: 63–66

10. Christensen BT (1992) Physical fractionation of soil and organic matter in primary particle size and density separates. Adv Soil Sci 20: 1–90

11. Dommergues Y, Garcia JL and Ganly F (1980) Microbiological considerations of the nitrogen cycle in West African ecosystems. In: Rosswall T (ed) Nitrogen Cycling in West African Ecosystems, pp 55-72. IITA, Ibadan, Nigeria.

12. Duxbury JM, Smith MS and Doran JW with Jordan C, Szott L and Vance E (1989) Soil organic matter as a source and sink of plant nutrients. In: Coleman *et al.* (eds.) Dynamics of Soil Organic Matter in Tropical Ecosystems, pp 33–67

13. Elston J and Dennett MD (1981) Climates of the tropics in relation to crop productivity. In: Greenland DJ (ed) Characterisation of Soils, pp 262–275. Clarendon Press, Oxford

14. Ford GW and Greenland DJ (1968) The dynamics of partly humified organic matter in some arable soils. Trans 9th Int Congr Soil Sci 2: 403–410

15. Fox RH, Myers RJK and Vallis I (1990) The nitrogen mineralization rate of legume residues in soil as influenced by their polyphenol, lignin, and nitrogen contents. Plant and Soil 129: 251–259

16. Frankenberger WT and Abdelmagid HM (1985) Kinetic parameters of nitrogen mineralization rates of leguminous crops incorporated into soil. Plant and Soil 87: 257–271

17. Ghani A, McLaren RG and Swift RS (1990) Seasonal fluctuations of sulphate and soil microbial biomass-S in the surface of a Wakanui soil. NZ J Agric Res 33: 467–472

18. Godwin DC, Singh U, Buresh RJ and DeDatta SK (1990) Modelling of nitrogen dynamics in relation to rice growth and yield. In:Trans 14th Int Congr Soil Sci Kyoto, Japan. Vol IV, pp 320–325

19. Gonzalez MA and Sauerbeck DR (1982) Decomposition of ^{14}C-labelled plant residues in different soils and climates of Costa-Rica. Proc Reg Coll on Soil Organic Matter Studies, Piracicaba, Brazil. pp 141–146

20. Greenland DJ (1958) Nitrate fluctuations in tropical soils. J Agric Sci 50: 82–92

21. Greenland DJ (1980) The nitrogen cycle in West Africa-Agronomic considerations. In: Rosswall T (ed) Nitrogen Cycling in West African Ecosystems, pp 73–81. SCOPE/UNEP, Sweden

22. Greenland DJ, Wild A and Adams D (1992) Organic matter dynamics in soils of the tropics-From myth to complex reality. In: Lal R *et al.* (eds) Myths and Science of Soils of the Tropics. SSSA special publication No.29, pp 17-33. SSSA, ASA, Madison, WI, USA

23. Griffin DM (1981) Water potential as a selective factor in the microbial ecology of soils. In: Kral DM and Cousin MK (eds) Water Potential Relations in Soil Microbiology. SSSA Publication No. 9, pp 141–151. SSSA, Madison, WI, USA

24. Hardy F (1946) Seasonal fluctuations of soil moisttire and nitrate in a humid tropical climate (Trinidad, B.W.I.). Trop Agric (Trinidad) 23: 40–49

25. Harris PJ (1988) Ecology of the soil population. In: Wild A (ed) Russell's Soil Conditions and Plant Growth, pp 472–499. Longman, UK

26. Jansson SL and Persson J (1982) Mineralization and immobilization of soil nitrogen. In: Stevenson FJ (ed) Nitrogen in Agricultural Soils, pp 229–252. ASA, Madison, Wisconsin, USA

27. Jenkinson DS (1966) Studies on the decomposition of plant material in soil II. Partial sterilization of soil and the soil biomass. J Soil Sci 17: 280–302

28. Jenkinson DS (1990) The turnover of organic carbon and nitrogen in soil. Phil Trans Royal Soc, London, B 329: 361–368.

29. Jenkinson DS and Ayanaba A (1977) Decomposition of carbon-14 labelled plant material under tropical conditions. Soil Sci Soc Am J 41: 912–915

30. Jenkinson DS, Fox RH and Rayner JH (1985) Interaction between fertilizer nitrogen and soil nitrogen..the so called "priming" effect. J Soil Sci 36: 425–444

31. Jensen HL (1929) On the influence of the carbon : nitrogen ratios of organic material on the mineralization of nitrogen. J Agric Sci 19: 71–82

32. Kuikman PJ, Jansen AG and van Veen JA (1991) 15N-nitrogen mineralization from bacteria by protozoan grazing at different soil moisture regimes. Soil Biol Biochem 23: 193–200

33. Kuikman PJ, van Vuuren MMI and van Veen JA (1989) Effect of soil moisture regime on predation by protozoa of bacterial biomass and the release of bacterial nitrogen. Agric Ecosys Environ 27: 271–279

34. Ladd JN, Amato M, and Oades JM (1985) Decomposition of plant material in Australian soils. III. Residual organic and microbial biomass C and N from isotope-labelled legume material and soil organic matter decomposing under field conditions. Aust J Soil Res 23: 603–611

35. Marrs RH, Thompson J, Scott D and Proctor J (1991) Nitrogen mineralization and nitrifcation in terra firme forest and savanna soils on Ilha de Maraca, Roraima, Brazil. J Trop Ecol 7: 123–127

36. Marumoto T, Anderson JPE and Domsch KH (1982) Mineralization of nutrients from soil microbial biomass. Soil Biol Biochem 14: 469–476

37. McGill WB and Myers RJK (1987) Controls on dynamics of soil and fertilizer N. In: Follett R, Cole CV and Stewart JWB (eds) Soil Fertility and Organic Matter as Critical Components of Production Systems. Soil Sci Soc Am Spec Publ No. 19, pp 71–98. ASA, Madison

38. Meiklejohn J (1957) The number of bacteria and actinomycetes in Kenyan soils. J Soil Sci 8: 240–247

39. Mueller-Harvey I, Juo ASR and Wild A (1989) Mineralization of nutrients after forest clearance and their uptake during cropping. In: Proctor J (ed.) Mineral Nutrients in Tropical Forest and Savanna Ecosystems, pp 315–324. Blackwell Scientific Publishers, Oxford, UK

40. Myers RJK (1988) Nitrogen management of upland crops: From cereals to food legumes to sugarcane. In: Wilson JR (ed) Advances in Nitrogen Cycling in Agricultural Ecosystems, pp 257–273. CAB International, Oxon, UK

41. Newman AS and Norman AG (1943) The activity of subsurface soil population. Soil Sci 55: 377–391

42. Nkrumah M, Griffith SM and Ahmad N (1989) Lysimeter and field studies on ^{15}N in a tropical soil. II. Transformation of $(NH_2)_2CO$-^{15}N in a tropical loam in lysimeter and field plots. Plant and Soil 114: 13–18

43. Norman DH (1972) Mixed cropping in Northern Nigeria III. Mixtures of cereals. Exp Agric 15: 41–48.

44. Orchard VA and Cook FJ (1983) Relationship between soil respiration and soil moisture. Soil Biol Biochem 15: 447–453.

45. Palm CA and Sanchez PA (1991) Nitrogen release from the leaves of some tropical legumes as affected by their lignin and polyphenolic contents. Soil Biol Biochem 23: 83–88,

46. Papendick RI and Campbell GS (1981) Theory and measurement of water potential. In: Kral DM and Cousin MK (eds) Water Potential Relations in Soil Microbiology. SSSA Publication No. 9, pp 1-22. SSSA, Madison, WI, USA

47. Parton WJ, Anderson DW, Cole CV and Stewart JBW (1983) Simulation of soil organic matter formations and mineralization in semi arid agro ecosystems. In: Lawrence *et al.* (eds) Nutrient Cycling in Agricultural Ecosystems. The Univ Georgia Spec Publ No. 23, pp 533–550

48. Pilbeam CJ, Mahapatra BS and Wood M (1993) Soil matric potential effects on gross rates of nitrogen mineralization in an Orthic Ferralsol from Kenya. Soil Biol Biochem 25: 1409–1413

49. Power JF and Doran JW (1988) Role of crop residue management in nitrogen cycling and use. In: Hargrove WL (ed) Cropping Strategies for Efficient Use of Water and Nitrogen. ASA Special Publication N0. 51, pp 101–113. ASA, CSSA, SSSA. Madison, WI, USA

50. Powlson DS (1993) Understanding the soil nitrogen cycle. Soil Use Manage 9: 86–94

51. Robertson FA, Myers RJK and Saffigna PG (1993) Carbon and nitrogen mineralization in cultivated and grassland soils in subtropical Queensland. Aust J Soil Res 31: 611–619.

52. Robertson GP and Tiedje JM (1988) Deforestation alters denitrification in a lowland tropical forest. Nature 336: 756–759

53. Robinson JBD (1957) The critical relationship between soil moisture content in the region of wilting point and the mineralization of natural soil nitrogen. J Agric Sci 49: 100–105

54. Russell EJ and Hutchinson HB (1909) The effect of partial sterilisation of soil on the production of plant food. J Agric Sci 3: 111–144

55. Sanchez P (1976) Properties and management of soils in the tropics. John Wiley and Sons, New York

56. Semb G and Robinson JBD (1969) The natural nitrogen flush in different arable soils and climates in East Africa. East Afr Agric For J 34: 350–370.

57. Seneviratne R and Wild A (1985) Effect of mild drying on the mineralization of soil nitrogen. Plant and Soil 84: 175–179

58. Skjemstad JO, Vallis I and Myers RJK (1988) Decomposition of soil organic nitrogen. In: Wilson JR (ed) Advances in Nitrogen Cycling in Agricultural Ecosystems, pp 134–144. CAB International, Oxon, UK

59. Sommers LE, Gilmour CM, Wildung RE and Beck SM (1981) The effect of water potential on decomposition processes in soils. In: Kral DM and Cousin MK (eds) Water Potential Relations in Soil Microbiology. SSSA Publication No. 9, 97–117. SSSA, Madison, WI, USA

60. Spycher G, Sollins P and Rose S (1983) Carbon and nitrogen in the light fraction of a forest soil: vertical distribution and seasonal patterns. Soil Sci 135: 79–87

61. Srivastava SC and Singh JS (1991) Microbial C, N and P in dry tropical forest soils:effects of alternate land uses and nutrient flux. Soil Biol Biochem 23: 117–124

62. Stanford G and Smith SJ (1972) Nitrogen mineralization potential of soils. Soil Sci Soc Am J 36: 465–472

63. Stott DE, Elliott LF, Papendick RI and Campbell GS (1986) Low temperature or low water potential effects on the microbial decomposition of wheat residue. Soil Biol Biochem 18: 577–582

64. Swift MJ and Boddy L (1984) Animal-microbial interactions in wood decomposition. In: Anderson JM, Rayner ADM and Wilson DWH (eds) Invertebrate-microbial Interactions, pp 89–131. Cambridge University Press, Cambridge

65. Tenney FG and SA Waksman (1929) Composition of natural organic materials and their decomposition in the soil: IV. The nature and rapidity of decomposition of the various organic complexes in different plant materials, under aerobic conditions. Soil Sci 28: 55–84

66. Vallis I and Jones RJ (1973) Net mineralization of nitrogen in leaves and leaf litter of *Desmodium intortum* and *Phaseolus atropurpureus* mixed with soil. Soil Biol Biochem 5: 391–398

67. Van der Kruijs ACBM, Wong MTF, Juo ASR and Wild A (1988) Recovery of ^{15}N-labelled fertilizer in crops, drainage water and soil using monolith lysimeters in south-east Nigeria. J Soil Sci 39: 483–492

68. Van Noordwijk M (1989) Rooting depth in cropping systems in humid tropics in relation to nutrient use efficiency. In: van der Heide J (ed) Nutrient Management for Food Crop Production in Tropical Farming Systems, pp 129–144. Institute of Soil Fertility, Haren and Universitad Brawijaya, Malang

69. Van Veen JA, Ladd JN and Frissel MJ (1984) Modelling C and N turnover through the microbial biomass in soil. Plant and Soil 76: 257–274

70. West AW, Sparling GP, Feltham CW and Reynolds J (1992) Microbial activity and survival in soils dried at different rates. Aust J Soil Res 30: 209–222

71. Wetselaar R (1968) Soil organic nitrogen mineralization as affected by low soil water potentials. Plant and Soil 29: 9–17

72. Wild A (1972) Mineralization of soil nitrogen at a savanna site in Nigeria. Exp Agric 8: 91–97

73. Wong MTF, Wild A and Juo ASR (1987) Retarded leaching of nitrate measured in monolith lysimeters in south east Nigeria. J Soil Sci 38: 511–518

74. Wong MTF, Hughes R and Rowell DL (1990a) The retention of nitrate in acid soils from the tropics. Soil Use Manage 6: 72–74

75. Wong MTF, Hughes R and Rowell DL (1990b) Retarded leaching of nitrate in acid soils from the tropics: measurement of the effective anion exchange capacity. J Soil Sci 41: 655–663

76. Wong MTF and Swift RS (1994) Amelioration of aluminium phytotoxicity with organic matter. In: Date RA, Grundon NJ, Rayment GE and Probert ME (eds) Proceedings of the 3rd International Symposium on Plant-Soil Interactions at Low pH: Principles and Management. Brisbane, Australia

77. Wong MTF, Van Der Kruijs ACBM and Juo ASR (1992) Leaching loss of calcium, magnesium, potassium and nitrate derived from soil, lime and fertilizers as influenced by urea applied to undisturbed lysimeters in south-east Nigeria. Fert Res 31: 281–289.

78. Wong MTF, Wild A and Mokunye AU (1991) Overcoming soil nutrient constraints to crop production in West Africa: Importance of fertilizers and priorities in soil fertility research. Fert Res 29: 45–54

Fertilizer Research **42**: 27–32, 1995.
© 1995 *Kluwer Academic Publishers. Printed in the Netherlands.*

Effect of cultivation on the nitrogen fertility of selected agro-ecosystems in South Africa

C.C. du Preez[1] & M.E. du Toit[2]
[1]*Department of Soil Science, University of the Orange Free State, P.O. Box 339, Bloemfontein 9300, Republic of South Africa and* [2]*Institute for Soil, Climate and Water, Private Bag X79, Pretoria 0001, Republic of South Africa*

Key words: crop production, mineralizable nitrogen, organic matter, tillage operation, total nitrogen

Abstract

Surveys of soils on commercial farms practising rainfed agriculture in South Africa, found large losses of N fertility. Information regarding the pattern of this loss is not available for different agro-ecosystems. Thus, the aim was to quantify the effect of cultivation on the N fertility of soils in five agro-ecosystems with virgin soils serving as reference, find the relation between the period of cultivation and the decrease in N fertility for each agro-ecosystem and determine whether this decline differed between the agro-ecosystems. Five to six sites with different cultivation periods were sampled within each agro-ecosystem. At each site a cultivated soil and its virgin counterpart were sampled to 200 mm depth and total and mineralizable N were determined as indices of N fertility. Cultivation, irrespective of the period, caused a significant decrease in the N fertility of all five agro-ecosystems and the rate of loss was rapid during the first few years of cultivation. Thereafter, the rate decreased until an equilibrium was approached. In agro-ecosystems from the warmer, drier regions, an equilibrium was approached much sooner than in agro-ecosystems from the cooler, wetter regions, but the percentage of N fertility loss was larger in the latter regions. The percentage loss of mineralizable N was larger than that of total N. This depletion of N fertility in cultivated soils may result in unsustainable crop production.

Introduction

It is generally assumed that the N fertility of soil in natural agro-ecosystems reaches a maximum equilibrium level under specific environmental conditions where organic matter inputs equal losses (Tate, 1992). This level of equilibrium is influenced in order of importance by: climate > vegetation > topography = parent material > time, with all five soil forming factors being partially interactive (Stevenson, 1986).

In agro-ecosystems the intervention of man changes this equilibrium through effects of cultivation, which accelerates the oxidation of soil organic matter at or near the surface (Campbell & Souster, 1982). This results in loss of the potentially mineralizable N fraction of organic matter that serves as a reservoir of N available to crops. It is widely accepted that after a certain period of cultivation the organic matter approaches a new equilibrium at a lower level,

provided constant management practices are employed (Smith & Elliot, 1990). This decomposition rate to the new, lower equilibrium depends on climatic conditions, soil properties and cultivation periods (Rasmussen & Collins, 1991).

Surveys of soils on commercial farms practising rainfed agriculture in South Africa found large losses of organic matter and a consequent reduction in the N fertility near the surface (Du Toit *et al.*, 1994; Prinsloo *et al.*, 1990), which is in agreement with similar surveys elsewhere in the world (Dalal & Mayer, 1986; Haas *et al.*, 1957). Information regarding the pattern of this N loss which includes total and mineralizable N, is not available for agro-ecosystems in the central parts of South Africa where dryland crop production is common.

Thus, the aim of this investigation was to quantify the effects of cultivation on N loss in five different agro-ecosystems with virgin soils serving as reference, to

28

Table 1. Climatic data and soil characteristics of the selected agro-ecosystems

	Agro-ecosystem				
	1	2	3	4	5
Climatic data[1]					
Mean annual					
Temperature (°C)	17.9	16.6	14.9	16.0	13.8
Precipitation (mm)	455	563	670	516	624
Evaporation (mm)	2275	2346	2310	1564	1733
Aridity index (P/E)	0.20	0.24	0.29	0.33	0.36
Soil characteristics[2]					
Type	Quartzipsamment	Plinthustalf	Haplustert	Plinthustalf	Plinthustalf
Textural class	Fine sand	Fine loamy sand	Clay	Fine sandy loam	Fine sandy loam
Particle size distribution (g kg^{-1})					
Coarse sand (2–0.5 mm)	83	33	29	19	20
Medium sand (0.5–0.25 mm)	106	66	73	28	55
Fine sand (0.25–0.1 mm)	408	320	98	197	316
Very fine sand (0.1–0.05 mm)	272	377	114	410	276
Coarse silt (0.05–0.02 mm)	30	65	78	105	72
Fine silt (0.02–0.002 mm)	31	42	131	90	132
Clay (<0.002 mm)	69	95	474	145	138
pH (H$_2$O)	6.0	5.8	6.7	6.1	5.2
CEC (mmol$_c$ kg^{-1})	40.2	27.8	387.2	77.1	88.2
Organic C (g kg^{-1})	3.83	6.25	16.68	10.77	19.20
Total N (mg kg^{-1})	392	618	1174	1085	1391
C:N	9.77	10.11	14.21	9.93	13.81

[1]Obtain from the climate statistics of South Africa (Weather Bureau, 1986).
[2]Classification according to soil taxonomy (Soil Survey Staff, 1987) and analyses with standard methods (The Non-affiliated Soil Analysis Work Committee, 1990). Only mean values of virgin soils are given.

find the relation between the period of cultivation and the decrease in N for these agro-ecosystems and determine whether the decline in N status, as a result of cultivation, differed between the selected agro-ecosystems.

Material and methods

Five distinctive agro-ecosystems used for commercial crop production in the central parts of South Africa (Fig. 1) were selected to obtain a variation in climatic conditions and soil properties (Table 1). Such an agro-ecosystem was defined as a region where the three environmental factors which affected yield, namely climate, slope and soil, were for practical purposes, homogeneous. The variation of these factors is not sufficient to influence significantly the crops that can be produced, the yield potential of the crops and the production techniques.

The main crops produced on agro-ecosystems 1, 3 and 4 were maize, sorghum and wheat, respectively. On agro-ecosystems 2 and 5 either maize or wheat were the main crops. Maize and sorghum are grown in summer when most of the rain falls. In winter, wheat is grown mainly on stored water in the soil. These crops are fertilized by the farmers with N in accordance with potential grain yield. The grain yield of maize, sorghum and wheat on the five agro-ecosystems varied between 1.5–4.0, 2.0–4.5 and 1.0–3.0 ton ha^{-1}, respectively (Scheepers *et al.*, 1984). In most cases only one crop is grown annually with a fallow period of five months. Sometimes the fallow period is extended to 10 months for water storage, which causes also a crop rotation. At all five agro-ecosystems, the primary tillage operation comprises either mouldboard or chisel ploughing after harvesting when the

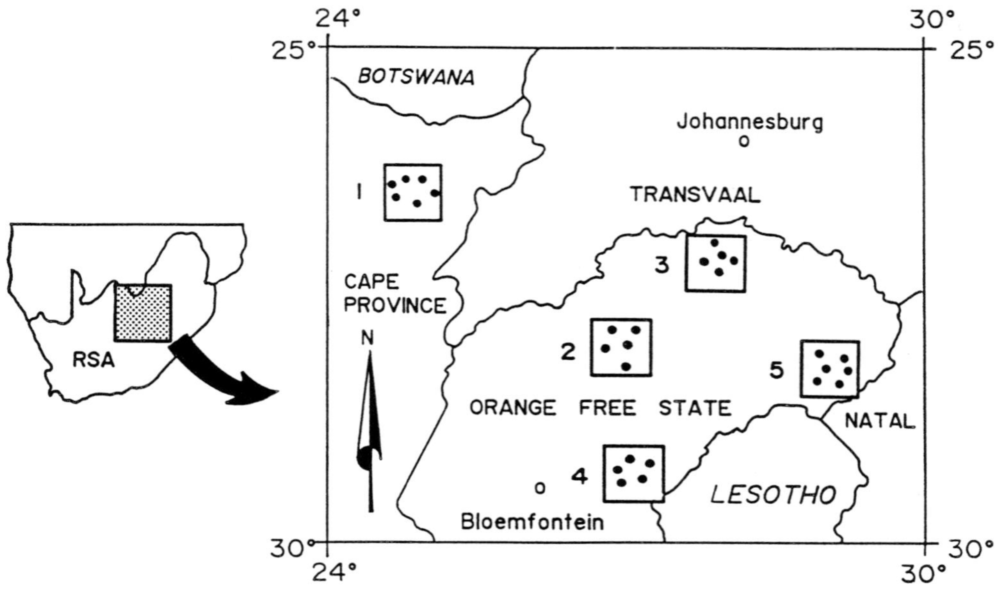

Fig. 1. Distribution of the five agro-ecosystems in the area of investigation.

soil is sufficiently moist. The depth of tillage varies between agro-ecosystems, depending on the soil texture. These tillage operations are normally deeper in the more sandy agro-ecosystems due to severe soil compaction than in the more clayey agro-ecosystems, but in all cases, the depth of tillage is at least to 200 mm. Shallow secondary tillage operations, which are sometimes necessary, are much more variable.

Five to six sites with different cultivation periods were sampled within each agro-ecosystem. At each site, six plots, each 10×10 m, of both a cultivated soil and its virgin counterpart were sampled. Twenty subsamples were collected at each plot to a depth of 200 mm and thoroughly mixed to obtain a composite sample. These composite samples, were dried in a ventilated room heated to 35 °C, crushed, passed through a 2 mm sieve and stored in glass bottles for analysis.

Total N was determined on these stored samples by a modification of the Kjeldahl method (Bremner & Mulvaney, 1982). Subsamples of the six cultivated plots as well as of the six virgin plots of each site were thoroughly mixed to obtain two composite samples for the determination of N mineralization by an aerobic incubation method (Prinsloo *et al.*, 1990).

Mineralizable N was determined by mixing thoroughly 10 g soil (clay < 400 g kg^{-1}) with 20 g filtersand and 8 g soil (clay > 400 g kg^{-1}) with 24 g filtersand. The filtersand was previously treated with

hydrochloric acid and distilled water, and then sieved to between 0.25 and 0.60 mm. The mixtures of soil and pre-treated filtersand were packed in triplicate, into incubation tubes of 15 mm internal diameter and 245 mm length, with one end drawn to a fine jet. The mixtures were retained in the tubes by cotton wool plugs on both ends. Before incubation, residual inorganic N was leached with 75 cm^3 0.01 M CaCl$_2$ and excess liquid was sucked off through the narrow end of the tube to a moisture tension of -33 kPa. Then the wide end of the tube was sealed with Parafilm to prevent water loss, while the narrow end was left open to ensure aeration during horizontal incubation in darkness at 30 °C in a humid incubator. The mineralized N(NH$_4^+$, NO$_2^-$ and NO$_3^-$) was leached as before following 2, 5, 9, 14 and 22 weeks of incubation. An automatic flow system was used to determine the NH$_4^+$-N (Technicon, 1977) and the NO$_2^-$-plus NO$_3^-$-N (Technicon, 1978) in the leachates. The N mineralized over the 22 week period was calculated from the sum of NH$_4^+$-, NO$_2^-$- and NO$_3^-$-N measured in the eluate at each incubation interval.

The following equation (Jenkinson, 1981) best described the decline in the two N fertility indices, viz. total and mineralizable N due to period of cultivation at each of the five agro-ecosystems:

$$X_t = X_e + (X_0 - X_e)\mathrm{e}^{-Xrt} \qquad (1)$$

where X_t is the N fertility after a certain cultivation period, X_e the expected equilibrium of N fertility after a long cultivation period, X_0 the mean N fertility of the virgin soils and X_r the decline rate of N fertility and t the cultivation period. Either total or mineralizable N can be used as an index for N fertility in this equation.

Results and discussion

The five agro-ecosystems differed significantly from one another regarding climatic conditions and soil properties (Table 1). Agro-ecosystem 1 has the lowest aridity index indicating warmer, drier conditions and agro-ecosystem 5 has the highest aridity index indicating cooler, wetter conditions. Climatic conditions at the other three agro-ecosystems are intermediate as illustrated by their aridity index values. The botanical composition, basal ground cover and biomass production of the natural vegetation which is typical of grassland, change in accordance with the climate. The topography varies from a flat landscape in the west to a rolling landscape in the east. There is also large variation in soil texture between the five agro-ecosystems, viz. from fine sand in agro-ecosystem 1 to clay in agro-ecosystem 3.

Thus, as expected, the N fertility of the five agro-ecosystems also varied largely before cultivation started as indicated by the total and mineralizable N of the virgin soils, respectively (Fig. 2). The N fertility of the virgin soils at agro-ecosystem 1 was the lowest and that at agro-ecosystem 5 the highest with the other three agro-ecosystems somewhere in between.

In the area of investigation it was found (Du Toit & Du Preez, 1993) that the total N content of the virgin soils increased linearly with increasing fine silt-plus-clay content (r = 0.91; n = 41), and exponentially, with increasing aridity index values (r = 0.89; n = 41). The total N content was found to be lower in the warmer, drier regions than in the wetter, cooler regions. In a region with a given index of aridity, the total N content increased as the fine silt-plus-clay content increased.

Cultivation, irrespective of the period, caused a significant decrease in the N fertility of soils from all five agro-ecosystems as demonstrated by the total and mineralizable N, respectively (Fig. 2). This results in a need for higher N fertilizer applications which may in turn accelerate depletion (Jansson & Persson, 1982). As expected, in a specific agro-ecosystem, the pattern

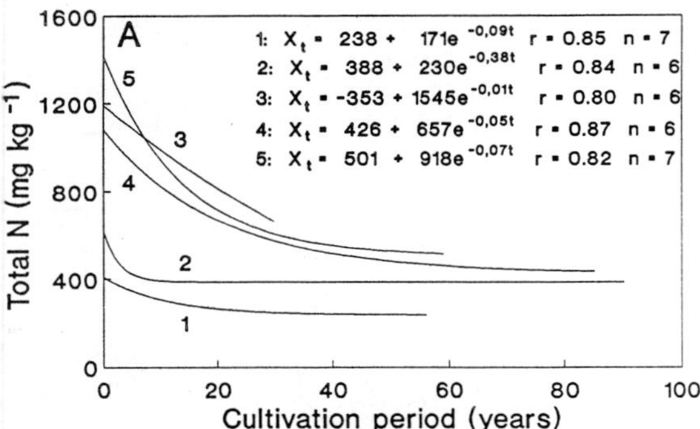

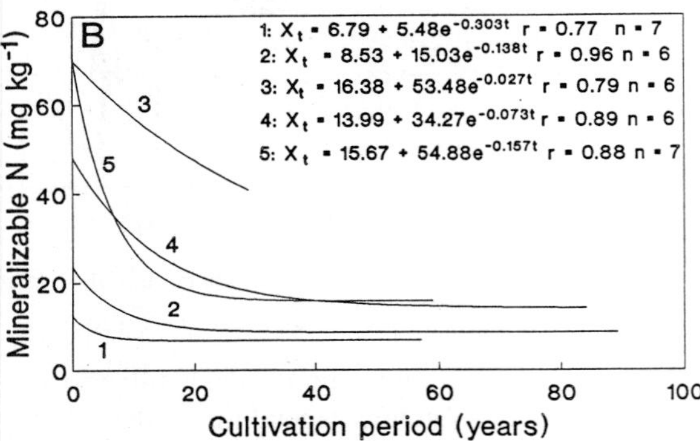

Fig. 2. Effect of cultivation period on the amount of total N (**A**) and mineralizable N (**B**) in soils of the five agro-ecosystems.

of loss for mineralizable N was very similar to that of total N, since the latter serves as a reservoir for the former.

The rate of N fertility loss in agro-ecosystems 1, 2, 4 and 5 was rapid during the first few years of cultivation (Fig. 2). Thereafter, the rate decreased until an equilibrium was approached. In agro-ecosystem 1 and 2 from the warmer, drier regions, the equilibrium was approached before 20 years of cultivation and in agro-ecosystems 4 and 5 from the cooler, wetter regions, the equilibrium was approached after 40 years of cultivation. After this very little or no loss of N fertility occurred. No conclusion can be made as to when an equilibrium will be approached in agro-ecosystem 3 since no site with a cultivation period of longer than 30 years could be found.

The high NO_3^--N values reported occasionally for subsoils (Du Preez & Burger, 1986) and for groundwaters (Henning & Stoffberg, 1990) in the area of investigation may have resulted directly or indirectly from this pattern of N loss. During the first few years of cultivation, N is mineralized in excess of crop demand and leaching of NO_3^- may follow, especially during

Table 2. Percentage loss of total and mineralizable nitrogen from agro-ecosystems 1, 2, 4 and 5 after a long cultivation period when a new equilibrium is reached

Agro-ecosystem	Aridity index	Fine silt-plus-clay (g kg^{-1})	Total N (mg kg^{-1})		Loss (%)	Mineralizable N (mg kg^{-1})		Loss (%)
			Virgin[1]	Cultivated[2]		Virgin[1]	Cultivated[2]	
1	0.20	100	392	238	39	11.8	6.8	42
2	0.24	137	618	388	37	21.7	8.5	61
4	0.33	235	1085	426	61	48.5	14.0	71
5	0.36	270	1391	501	64	70.3	15.7	78

[1] Values measured in laboratory
[2] Values estimated with equation 1.

fallow which in all cases lasts at least five months. In order to maintain an optimum yield after several years of cultivation, there is a need for higher N fertilizer application, which may be subject to NO_3^- leaching if not used by the crop.

The total N values measured in the virgin soils of agro-ecosystems 1, 2, 4 and 5 have a much wider range than the quilibrium total N values estimated with Equation 1 for the cultivated soils (Table 2). This observation, which is also valid for the mineralizable N fraction, implies that the percentage loss of N increased with increasing N in the virgin soils. Thus, it seems that the intervention of man through the effects of long-term cultivation, irrespective of the practices, has an equalizing effect on the N status.

The estimated percentage loss of mineralizable N on account of cultivation tends to be higher than the estimated percentage loss of total N, especially in agro-ecosystems 2, 4 and 5 (Table 2). This is in agreement with the findings of other researchers (Campbell & Souster, 1982; Prinsloo *et al.*, 1990) that mineralizable N is more vulnerable than total N to such losses.

Some of the measures proposed to maintain or restore N fertility have been tested since 1979 in a wheat cultivation experiment situated in agro-ecosystem 5 (Wiltshire & Du Preez, 1993). After 12 years, a comparison of the least and most conservational treatments in the trial, viz. mouldboard ploughing versus no-tillage, indicates a difference in N fertility of only about one tenth. About 40% of the N accumulated under grassland had been lost from the cultivated area at the start of the trial. Thus, once N fertility is depleted, restoration through conservation practices is a slow process in agro-ecosystem 5, probably due to a shortage of crop residues.In the other agro-ecosystems, which are warmer and drier, restoration of N may be even slower.

Unfortunately, the derived constants for Equation 1 as shown in Figure 2 for each agro-ecosystem are not validated yet due to a lack of independent data sets. However, a few checks (data not shown) indicated that the constants for an agro-ecosystem are fairly representative. None the less, proper validation is necessary if these constants will be used for other purposes such as modelling.

In conclusion, intensive cropping practices over many years severely depleted the N fertility of all five agro-ecosystems, despite N fertilization. This depletion has significant implications for sustainable crop production in the central parts of South Africa. Cultivation practices of conservational nature should be employed to restore the soil N status of these areas. Such practices where grain crops are rotated with suitable annual legumes would probably enhance restoration as shown (Agenbag & Maree, 1989) for the wheat growing region near the south-west coast of South Africa with a Mediterranean climate.

Acknowledgements

The authors are grateful to the editor of the South African Journal of Plant and Soil for permission to use selective information from two of their papers in order to compile this paper for the International Symposium on Nitrogen Economy in Tropical Soils at Trinidad. The financial contributions by both the Institute for Soil, Climate and Water and the University of the Orange Free State are acknowledged. The authors also thank those farmers who provide the soils for sampling and persons who were willing to help with the sampling of the soils.

References

Agenbag GA & Maree PCJ (1989) The effect of tillage on soil carbon, nitrogen and soil strength of simulated surface crusts in two cropping systems for wheat (*Triticum aestivum*). Soil Till Res 14: 53–65

Bremner JM & Mulvaney CS (1982) Nitrogen - total. In: Page AL (eds) Methods of Soil Analysis, Part 2. Chemical and microbiological properties, pp 595–624. American Society of Agronomy, Madison, Wisc, USA

Campbell CA & Souster W (1982) Loss of organic matter and potential mineralizable nitrogen from Saskatchewan soils due to cropping.Can J Soil Sci 62: 651–656

Dalal RC & Mayer RJ (1986) Long-term trends in fertility of soils under continuous cultivation and cereal cropping in Southern Queensland. II. Total organic carbon and its rate of loss from the soil profile. Aust J Soil Res 24: 281–292

Du Preez CC & Burger R du T (1986) Inorganic nitrogen in maize soils of the Orange Free State and Transvaal at planting time, 1984. S Afr J Plant Soil 3: 148–150 (In Afrikaans)

Du Toit ME & Du Preez CC (1993) Relationship between organic matter content of certain virgin orthic topsoils, soil properties and climatic data in South Africa. S Afr J Plant Soil 10: 168–173 (In Afrikaans)

Du Toit ME, Du Preez CC, Hensley M & Bennie ATP (1994) Effect of cultivation on the organic matter content of selected dryland soils in South Africa. S Afr J Plant Soil 11: 71–79 (In Afrikaans)

Haas HJ, Evans CE & Miles EF (1957) Nitrogen and carbon changes in Great Plains soils as influenced by cropping and soil treatments. USDA Tech Bull no 1164, Washington

Henning JAG & Stoffberg DJ (1990) Effect of tillage depth on yield and root development of maize in a Bleeksand soil with a water table. Appl Plant Sci 4: 17–20 (In Afrikaans)

Jansson SL & Persson J (1982) Mineralization and immobilization of soil nitrogen. In: Stevenson FJ (ed) Nitrogen in Agricultural Soils, pp 229–252. American Society of Agronomy, Madison, Wisc, USA

Jenkinson DS (1981) The fate of plant and animal residues in soils. In: Greenland DJ & Hayes MBH (eds) The Chemistry of Soil Processes, pp 505–561. John Wiley & Sons, New York, USA

Prinsloo MA, Wiltshire GH & Du Preez CC (1990) Loss of nitrogen fertility and its restoration in some Orange Free State soils. S Afr J Plant Soil 7: 55–61

Rasmussen PE & Collins HP (1991) Long-term impacts of tillage, fertilizer and crop residue on soil organic matter in temperate semi-arid regions. Adv Agron 45: 93–134

Scheepers JJ, Smit JA & Ludick BP (1984) An evaluation of the agricultural potential of the Highveld Region in terms of dryland cropping and livestock production. Dept of Agric Tech Commun no 185, Pretoria

Smith JL & Elliot LF (1990) Tillage and residue management on soil organic matter dynamics in semiarid regions. Adv Soil Sci 13: 69–88

Soil Survey Staff (1987) Keys to Soil Taxonomy. SMSS Tech Monograph no 6, Ithaca, New York, USA

Stevenson FJ (1986) Cycles of soils: carbon, nitrogen, phosphorus, sulfur, micronutrients. John Wiley & Sons, New York, USA

Tate RL (1992) Soil Organic Matter: Biological and Ecological Effects. Krieger Publishing Company, Malabor, Florida, USA

Technicon (1977) Individual/simultaneous determination of nitrogen and/or phosphorus in BD acid digests. Industrial method no 329–74 W/B. Technicon Industrial Systems. Tarrytown, New York, USA

Technicon (1978) Nitrate and nitrite in water and wastewater. Industrial method no 100–70 W/B. Technicon Industrial Systems. Tarrytown, New York, USA

The Non-affiliated Soil Analysis Work Committee (1990) Handbook of standard soil testing methods for advisory purposes. Soil Science Society of South Africa, Pretoria, South Africa

Weather Bureau (1986) Climate of South Africa: climate statistics up to 1984, WB 40. Government Printer, Pretoria, South Africa

Wiltshire GH & Du Preez CC (1993) Long-term effects of conservation practices on the nitrogen fertility of a soil cropped annually to wheat. S Afr J Plant Soil 10: 70–76

Fertilizer Research **42**: 33–41, 1995.
© 1995 *Kluwer Academic Publishers. Printed in the Netherlands.*

Assessment of biological nitrogen fixation

S.K.A. Danso
FAO/IAEA Division, Wagramerstrasse 5, P.O. Box 100, A-1400 Vienna, Austria

Key words: acetylene, ^{15}N, nodules, reference plant, ureide

Abstract

The four commonly used methods for measuring biological nitrogen fixation (BNF) in plants are: the total nitrogen difference (TND) method, acetylene reduction assay (ARA) technique, xylem-solute (or ureide production) method and the use of ^{15}N labelled compounds.

The TND method relies on a control non-N_2-fixing plant to estimate the amount of N absorbed by the fixing plant from soil. It is one of the simplest and least expensive methods, but works best under low soil N conditions. The ARA technique measures the rate of acetylene conversion to ethylene by the N_2-fixing enzyme, nitrogenase. The ethylene produced can then be converted into N_2 fixed, using a conversion ratio, originally recommended as 3. Although the method is inexpensive and highly sensitive, its major disadvantages are, the short-term nature of the assays, the doubtful validity of always using a conversion ratio of 3 and the auto-inhibition of acetylene conversion to ethylene. The ARA technique is therefore not a method of choice for measuring BNF.

The xylem-solute technique can be used to measure BNF for those species that produce significant quantities of ureide as product of BNF. Although simple and relatively inexpensive, it is an instantaneous assay and also needs to be calibrated against a known method. The most serious limitation is, that only a small proportion of N_2-fixing plants examined are ureide exporters, and the method is therefore not widely applicable.

The ^{15}N methods, classified into the isotope dilution and A-value methods, appear to be the most accurate, but also the most expensive. They involve labelling soil with ^{15}N fertilizer and using a non-N_2-fixing reference plant to measure the ^{15}N/^{14}N ratio in the soil. The ^{15}N isotope dilution approach is both operationally and mathematically simpler than the A-value approach. To limit potential errors in the selection of reference crops, it is recommended to use ^{15}N labelled compounds or soil labelling methods that result in the slow release of ^{15}N or the slow decline of ^{15}N/^{14}N ratio in the soil. Additionally, the use of several reference plants rather than a single one can improve the accuracy of the results.

Introduction

Nitrogen is a major nutrient element required by crop plants, and its scarcity in soil significantly affects crop yields. On the other hand excessive amounts in soil could result in undesirable environmental effects. Thus, both the gains and losses of N in agricultural soils are important processes which need to be measured.

Depletion of soil N in agricultural soils occurs primarily through plant N uptake. This is often replenished through fertilizer N additions or at least in part through natural processes, the most important of which is biological nitrogen fixation (BNF). It has been estimated that on a global scale, BNF may contribute some 90 million tons ha^{-1} yr^{-1} in agricultural systems of which the *Rhizobium*-legume symbiosis is estimated to contribute about 40 million tons [25]. Accurate measures of BNF are important as a prerequisite in determining how environmental factors can be managed for higher contributions from BNF, and to get a good estimate of N balance in various soil-plant systems.

Several methods have been devised for measuring BNF. However, some of these are only qualitative and not very useful for quantification purposes [8]. The most suitable methods are those that can distinguish between the amounts or proportions of plant N derived from atmospheric N_2 fixation, distinct from the N contribution from soil, and where relevant, from fertilizer applied N. Details of the various methods for measuring BNF have been sufficiently given in several reviews [7-9, 14, 16, 32, 50, 54]. The present paper will therefore not attempt to duplicate these details. Instead, it is intended to critically assess the strengths and weaknesses of those that have been most often used, and to discuss situations under which the methods could most usefully be applied, as well as, efforts to find solutions to some of the problems identified.

Total nitrogen difference (TND) method

This is one of the oldest and simplest methods, and has provided many valuable estimates of N_2 fixed, upon which several management practices have been based. The TND method measures BNF as the difference between the total N contents of plants that fix N_2 and those that do not derive N from fixation. In essence, the method is based on the assumption that both the N_2 fixing and non-fixing control plants absorb equal amounts of soil N for growth [46]. However, this assumption may not hold under all situations, as it requires that the different plants, in addition to being similar in root morphology and in several physiological attributes, should also absorb their N from similar depths and horizons [11]. The fulfilment of the underlying assumption is therefore the greatest limitation of the TND method [8].

Despite the stated limitations, the TND method has on several occasions provided BNF estimates that are not significantly different from those obtained using more sophisticated and expensive techniques [22, 43]. These similarities therefore justify a close examination of the strengths of the TND method, and when it is most likely to provide satisfactory results. As suggested by Danso [9], and supported by the studies of Paterson and LaRue [40] and Rennie [45], the TND method will often give reliable estimates of N_2 fixed in plants grown in soils or systems in which the initial N content is low. This is because BNF is high under these conditions, in contrast to soil N uptake, and therefore potential errors introduced by the inability of a control plant to correctly assess the amount of soil N absorbed by the fixing plant is of limited effect (compared to where soil N uptake is high and BNF is low). Several such situations exist, e.g., in areas consisting of sand dunes, systems undergoing primary colonization, and in many semi-arid areas. It is therefore not surprising that the values of N_2 fixed assessed by Dommergues [15] for *Casuarina* stands growing on the sand dunes in Cape Verde are still widely quoted, while the

BNF studies by Gauthier *et al.* [21] in a sandy soil in Senegal generally showed no significant differences between TND estimates of N_2 fixed and those obtained using the more expensive and more sophisticated ^{15}N methodologies.

The acetylene reduction assay (ARA) technique

The ARA technique is based on the fact that nitrogenase, the enzyme involved in N_2 fixation is also able to catalyse the reduction of acetylene to ethylene [17]. The ARA technique as used today to measure BNF is based on procedures outlined by Hardy et al. [26]. Simply, the method involves incubating the sample to be assayed, e.g., detached nodules, decapitated plant with roots and nodules attached, whole plants, etc., in a gas-tight chamber containing 0.03 to 0.1% (v/v) acetylene for periods of time ranging from a few minutes to several hours. At the end of the incubation period, a gas sample from the incubation vessel is injected into a gas chromatograph fitted with a Porapak N or P column and assayed for ethylene production [24].

The amount of ethylene produced could itself be used as a measure of nitrogenase or relative N_2-fixing activity. Otherwise, the quantity of ethylene formed can be converted into total amount of nitrogen fixed by multiplying ethylene produced by a conversion factor of 3 [24]. The rationale is, that stoichiometrically, 3 pairs of electrons are used up during the conversion of N_2 to NH_4 compared to the single pair of electrons used in the conversion of acetylene to ethylene, i.e.,

$$N_2 + 8H^+ + 12ATP + 6e^- \longrightarrow \qquad (1)$$
$$2NH_4^+ + 12ADP + 12Pi$$

$$C_2H_2 + 2H^+ + XATP + 2e^- \longrightarrow \qquad (2)$$
$$C_2H_4 + ADP + XPi$$

The ARA technique therefore indirectly measures BNF by estimating enzyme activity based on electron flux through nitrogenase. The use of the ARA technique has been facilitated by the fact that no end products other than ethylene have been identified in this reduction reaction [24]. Also, the ethylene produced is stable, and can be stored, thus making it possible to analyse the gas samples at a later period after sampling. These advantages, together with simplicity, low cost and high sensitivity made the ARA technique a method of choice, particularly in the 1970s, and together with the oil crisis, generated great interest in BNF studies.

Although the method is still in use, it is no longer that popular, to the extent that many reviewers and editors of journals often tend to question or reject papers that base their interpretations on ARA measurements [53]. What then could have contributed to this paradox? Detailed accounts of the problems associated with the ARA technique have been provided in several critical reviews [8, 16, 338, 58], and only the major ones will be outlined briefly in this paper.

1. Probably the most apparent problem is that the ARA technique, as frequently used involves short-term assays, in contrast to the process it is intended to measure, BNF that proceeds over long durations in crops. ARA measurements therefore have to be extrapolated to cover several periods over which no measurements were made. The incorrect implication here is, that N_2 fixation rates are constant over long periods. In fact rates of N_2 fixation are known to exhibit very wide diurnal [4, 44] and seasonal variations [61, 62]. To minimise these errors, frequent sampling is required, making the method more tedious than originally intended.

2. Very few ARA measurements have been performed directly on whole plants growing in the field. However, it has been noted that any disturbance in the N_2-fixing system induces an increased resistance to the flow of oxygen into nodules and this adversely affects the rate of acetylene conversion into ethylene [36, 58]; the greater the disturbance, the greater the effect. For different samples, Vessey [53] has thus ranked the inhibition of acetylene conversion as follows: whole nodulated root with shoot intact < detached, whole nodulated roots < partially nodulated roots < nodulated root segments < detached nodules < nodule slices; even the shaking of nodules to remove adhering soil is reported to have adverse effects [58].

3. Most often the sample used for ARA assays consists of uprooted plants. The problem here is, that the amount of ethylene produced will be governed by the proportion of the active nodules that remained intact on the plant in the uprooting process. Utmost care is therefore needed in the uprooting process itself. However, even in normal friable soil or in pot studies, it is difficult to recover 100% of intact nodules, especially where a substantial number of them are located on the lateral roots. Full recovery of nodules becomes an almost impossible task where plants are uprooted from dry and diffi-

cult to work soils. Also, as plants grow older, the most active nodules tend to be the newly formed, small nodules located on the distal ends of roots, often the small lateral roots that are at the greatest risk of detaching with the slightest disturbance. This may in part contribute to some of the reported declines in ARA estimates of N_2 fixation, much earlier, e.g. [55] than has been detected when the ^{15}N labelling method was used [32].

4. Serious doubts with the ARA technique as a measure of BNF have been raised; this has been attributed to the validity in using the conversion factor of 3 to derive N_2 fixed. This ratio was actually suggested only on theoretical grounds (see equations 1 and 2). In actual practice, some of the conversion ratios that have been reported for different N_2-fixing systems range from as low as 1.5 to as high as 25 for some non-plant systems [26]. Therefore, there is great uncertainty as to the correct conversion ratio to be used, unless a prior calibration (e.g., by using $^{15}N_2$ gas as a standard) has been done for each system. Even then, for the same *Rhizobium*-plant symbiosis, the conversion ratio does not always remain the same; it has been shown to vary under different environmental conditions [58].

5. The original suggestion to incubate samples in closed containers containing 10% acetylene results in an acetylene-induced inhibition of ethylene production [38]. To reduce such errors, it is recommended to use an open flow-through gas system in which air containing 10% acetylene is passed continuously through the assay vessels and the effluent gas analysed for ethylene [58]. The open flow-through system contrasts with the normal assay procedure in that it measures the rate of ethylene production rather than its accumulated concentration [58]. This again makes the ARA technique more tedious and methodologically more demanding than originally perceived.

Given the above-stated difficulties, several researchers have raised questions as to the best possible use of ARA data. Vessey [53], among many others have suggested the possibility of using ARA measurements for the relative ranking of BNF in different treatments. In doing so, the assumption would have to be, that whatever imperfections there are in the ARA technique, would affect the different treatments equally. This assumption has been refuted by Witty and Minchin [58] and Minchin *et al.* [37]. For example, Minchin *et al.* [37]

called attention to the fact that the acetylene-induced inhibition of ethylene production is inversely related to the stress imposed on the treatment. Thus, it is likely that an incorrect ranking can be made by comparing ARA measurements for different treatments (i.e. under different stresses), one of which is normally an unstressed control.

Analysis of N solutes in xylem exudate and plant parts.

This method is based on the determination of the composition of nitrogen compounds in plant tissues or the N flowing through the xylem sap to the shoot. The solute method was developed as a method for differentiating between fixed N and soil-derived nitrate N in plants [34].

In most agricultural soils, nitrate is the predominant form of N for plant growth [51] and the form of nitrogen most readily assimilated by growing plants. The nitrogen fixed in the nodule is exported via the xylem stream to the shoot either as amides, predominantly asparagine and glutamine or as the ureides, allantoin and allantoic acid [39]. However, substantial ureide production is restricted to some tropical legume species [42] and none of the temperate legumes examined so far has been found to produce ureides. Because ureide-exporting species have low nitrate reductase activity in their roots [1], much of the absorbed nitrate is transported to the shoot unchanged. Therefore the composition of the N compounds in the xylem exudate of nodulated ureide-producing plants change progressively from one dominated by ureides to one dominated by nitrate and amino compounds as the plant's dependence on N_2 fixation decreases in response to increasing contributions from soil N uptake [27, 34]. These changes are so specific that it has been possible to use changes in xylem exudate and in shoot extracts as an indication of the N_2-fixing ability of a legume, and even some quantitative measurements have been made [27, 28, 34]. For measuring the proportion of the plant's N derived from fixation, a calibration relating xylem solute composition in the presence of different levels of nitrate N with measured values of N_2 fixation using e.g., ^{15}N labelling is needed [27].

The sampling and analytical techniques have been described in detail by Peoples et al. [42] and will not be repeated here. The sample collection may be done by root bleeding or tissue extraction, but the original method of decapitating plants and collecting the xylem exudate has proven to be difficult under some circumstances, e.g., when the soil is dry [27, 52]. Pre-watering is therefore sometimes necessary but not always successful. An alternative method developed by Herridge [29] originally using soybean has proved very successful. It involves applying a mild vacuum to the lower end of the detached shoot of the plant and cutting successively from the top end of the shoot to allow the xylem contents to be drawn through. The exudate is trapped in a 5-ml vacutainer connected in series to the source of the vacuum and the shoot.

Compared to the ^{15}N labelling methods, the xylem exudate method is simpler, less expensive, and the method has great potential in many studies, such as in the screening of large germplasm collections for relative N_2 fixation ability. It was initially hoped that it would become a potential tool for measuring N_2 fixation in trees. However, the method is severely hindered by the fact that only a small proportion of known N_2-fixing plants are ureide exporters. For example, of 35 nitrogen fixing trees examined, only two showed a high abundance of ureides in the xylem sap [30], and Sanginga et al. [49] could not find any significant correlation between ureide production and BNF in *Leucaena*. Besides, the xylem-exudate method is an instantaneous assay and unlike the ^{15}N soil-enrichment methods, requires the interpolation of several independent estimates of solute composition at different growth stages before a time-integrated estimate of BNF in plants over a growing season may be made.

The ^{15}N labelling methodologies

The ^{15}N methods may be classified into :
  – Use of $^{15}N_2$-labelled gas
  – The isotope dilution method
  – The A-value method

The common principle behind these three methods is that the N_2-fixing plants or systems are grown in soil or an atmosphere containing $^{15}N/^{14}N$ ratio measurably different from the almost constant $^{15}N/^{14}N$ ratio of 0.3663% present in the atmosphere. The incorporation of N_2 from fixation will thus result in a different $^{15}N/^{14}N$ ratio in plant tissue than that of the substrate on which the plant is growing. In the case of fixing plants incubated under $^{15}N_2$ labelled gas, the N in the plant tissues of a N_2-fixing plant will have a significantly increased $^{15}N/^{14}N$ ratio, in contrast to where available soil N rather than the N_2 in air is labelled. These methodologies have all been extensively reviewed [7, 8, 9, 14, 86, 57]. Details on how to use these methods will not therefore be given in the present paper. Rather, an attempt will be made to give a critical assessment of their limitations and strengths.

The use of $^{15}N_2$ gas is the only direct method for detecting BNF, but has limited practical or field application in terms of quantifying BNF. The limited field applicability stems from the fact that plants can be enclosed for only short durations in sealed containers, and therefore under conditions often completely different from those in the field. Replacing all the air in the field by $^{15}N_2$ labelled gas is an impossible task. The method is therefore, most useful in exploratory studies of plants in ecosystems where not much is known about their N_2 fixing capabilities.

Both the isotope dilution [20, 33] and A-value methods [18] involve growing plants on soil containing higher $^{15}N/^{14}N$ ratio than that of the N_2 fixed from the atmosphere with both methods, the $^{15}N/^{14}N$ ratio of the N_2 that is fixed results in plant tissues accumulating N of a lower $^{15}N/^{14}N$ ratio than that assimilated from the labelled soil. The most important and difficult requirement for these methods is, how to accurately determine the integrated $^{15}N/^{14}N$ ratio of soil -derived N. The difficulty stems from the fact that the $^{15}N/^{14}N$ in a labelled soil does not remain constant with time [19]. In addition, the amounts of this N of changing $^{15}N/^{14}N$ ratio absorbed by a plant can vary drastically with plant age, or time. The estimation of the integrated $^{15}N/^{14}N$ ratio absorbed by a fixing plant from soil cannot therefore be adequately established simply by using instantaneous chemical extraction methods [14]. What has therefore been adopted as an alternative approach is to use the $^{15}N/^{14}N$ ratio in selected non-N_2-fixing plants that accumulated their N from only soil to represent the integrated $^{15}N/^{14}N$ ratio of the soil N absorbed by the fixing plant. The higher the N_2 fixation that occurred, the greater the dilution that will occur in the $^{15}N/^{14}N$ ratio of the tissues of the fixing plant, compared to the N in the reference plant [9].

The A-value method involves applying higher rates of ^{15}N enriched fertilizer to the reference plants (to allow good growth, particularly in N-poor soils) than to the N_2-fixing plant (so as not to severely depress N_2 fixation). The assumption here is, that the available, native (unlabelled) soil N (A_s) is not altered through the addition of different rates of a ^{15}N enriched fertilizer, and thus the A_s assessed by the reference plant is equal to A_s assessed by the fixing crop. Since the A

value measured for the fixing crop also includes that for N_2 fixed ($A_s + A_a$), then the A value for fixed N_2, A_a is equal to $A_s + A_a$ (fixing crop) $- A_s$ (reference crop). The proportion of N derived from atmospheric N_2 fixation (% Ndfa) can then be calculated using the following equation [14]:

$$\% \, Ndfa = \frac{\% \, Ndff}{R} \times Aa$$

The A-value method has more underlying assumptions and is more complicated conceptually and mathematically. Therefore, it is not surprising that the A-value method has been criticised more often than the isotope dilution method (14). Consequently, the A-value method has been less often used to measure BNF, compared to the isotope dilution method. The discussion of the ^{15}N soil labelling approach for measuring N_2 fixation will therefore focus largely on the isotope dilution method. However, there are situations where the A-value method gives more reliable estimates than the isotope dilution method [48]. Details of the A-value method for measuring N_2 fixation, including the equations used have been provided elsewhere [9, 12, 18].

The isotope dilution method can be classified into two: (a) one that relies on the inherent higher $^{15}N/^{14}N$ ratios in some soils than that of atmospheric N_2 [50] or (b) where a ^{15}N enriched inorganic or organic N source has been deliberately added to soil to artificially increase any discrepancies between the $^{15}N/^{14}N$ ratio of soil N and that of atmospheric N_2 [20, 33]. In both cases, what is measured is the extent to which the $^{15}N/^{14}N$ ratio of plant tissue N is lowered (relative to soil N) as a result of the assimilation of unlabelled N_2 [14]. A major difference between the two approaches, is that for the ^{15}N natural abundance approach (in contrast to where ^{15}N labelled fertilizer is added), the $^{15}N/^{14}N$ ratios of soil N and atmospheric N_2 are not vastly different. This therefore necessitates far greater precautions and more sophisticated and expensive instrumentation than is necessary when ^{15}N labelled materials are directly applied to soil [14]. These requirements of the ^{15}N natural abundance method tend to limit its use to the more advanced laboratories, mostly in developed countries. For example, emission spectrometers are more rugged and considerably cheaper than mass spectrometers but cannot detect these low levels of differences in $^{15}N/^{14}N$ ratios or $\delta^{15}N$ natural abundance. Besides, as will be discussed later, the application of label to soil is of greater utility in soil N balance studies compared to the ^{15}N natural abundance method which is used mostly to only measure N_2 fixation. However, a major advantage of the natural abundance method is, that it may be less prone to serious errors from reference plants than where a ^{15}N label is applied to soil, due to the more gradual decline in the natural $^{15}N/^{14}N$ ratios of soils [41].

The accuracy of BNF measurements using the isotope dilution method depends very much on how accurate the soil $^{15}N/^{14}N$ ratio assessed by the reference plant reflects that of soil-derived N in the fixing plant. Consequently, this is the greatest source of error with these ^{15}N methods, especially for the many studies that may have been conducted without prior selection of a suitable reference plant and where the criteria suggested by Fried *et al.* [19] have not been satisfied. In any case prior selection may not work always, as a particularly reference plant may not be satisfactory under all environments [6, 10]. Even some non-nodulating isolines have been found to nodulate in some soils (Danso, unpublished data), and it should therefore not be taken for granted that non-nodulating isolines can be used as non-fixing reference plants without examining their roots for nodules.

There is a need to examine the sources of the problem with obtaining reliable reference plants and to devise remedial measures. The major cause underlying reference plants giving erroneous measures of $^{15}N/^{14}N$ ratio absorbed by the fixing plant has been identified as the rapid decline of the $^{15}N/^{14}N$ ratio with time that occurs in soils into which ^{15}N labelled fertilizers have been added [19, 59]. Further experiments [60] revealed that by using practices that result in fairly stable $^{15}N/^{14}N$ ratios in soil with time, mis-matches between the $^{15}N/^{14}N$ ratio of the N in the reference plant and the true $^{15}N/^{14}N$ ratio of soil sampled by the fixing plant are drastically reduced. Although the ideal situation, a completely stable $^{15}N/^{14}N$ ratio over time is not practically possible in most situations, the aim should be to adopt ^{15}N labelling materials and methods that result in more gradual declines in the $^{15}N/^{14}N$ ratio in soil [14]. These include the use of organic matter-labelled ^{15}N, slow-release inorganic N formulations [19, 60], or the application of the required amount of ^{15}N fertilizer in small splits over time rather than as a single dose [11, 14]. Further improvement in the accuracy of the ^{15}N methods has resulted from using the mean of the $^{15}N/^{14}N$ estimated by several reference plants [3] and possibly in conjunction with practices that result in a slowly declining $^{15}N/^{14}N$ ratio of the soil [5]. However, it is essential to note that reference crop induced errors in BNF measurement are of little importance at high levels of N_2 fixation, in contrast to when N_2 fixation is low [12, 23].

A disadvantage of the ^{15}N methods which has received considerable attention is the high cost in material and equipment. Danso *et al.* [14] have made suggestions on where cuts in the cost of using ^{15}N labelled fertilizers to measure BNF can be made. They pointed out that the use of small isotope sub-plots instead of the traditional yield plots and using fertilizer enriched with 1 to 10 atom % ^{15}N excess rather than highly enriched fertilizers may lower costs without significantly decreasing the accuracy of the determination. Furthermore, emission spectrometers, several times less costly than mass spectrometers are now available, and should in most cases (unless the ^{15}N natural abundance method is used) be capable of providing reliable measures of BNF.

There are several advantages of the ^{15}N methods. These include:

1. Giving truly integrated estimates of BNF for whole growing seasons or up to desired periods.

2. Direct applicability in the field.

3. Ability to distinguish the nitrogen derived from fixation from that derived from soil or fertilizer. This is important for assessing soil N balances or for comparing varieties, species or treatments that accumulate e.g., similar amounts of N but differ in N_2 fixation, or where differences in the genetic potentials for total N accumulation bear no relationship with relative N_2-fixing abilities.

4. Where necessary, treatments can be ranked for BNF ability simply on the basis of differences in their $^{15}N/^{14}N$ ratios. The lower the $^{15}N/^{14}N$ ratio, the higher the N_2 fixing ability.

5. The ^{15}N methods are often more reliable and provide more accurate results.

6. In addition to providing estimates of BNF, the use of ^{15}N labelling methods provides many additional useful agronomic information, such as the fertilizer use efficiency of the legume or reference crop (particularly if it happens to be an important crop). Other information or benefits that can be obtained at no extra cost include the use of such labelled plant material to also label soil N for further BNF studies or to assess ease of decomposition of different organic matter sources and rates of N mineralization. Useful data on N transfer between plants have also been reported from experiments intended to measure BNF. Thus, the often cited high cost of the ^{15}N methodology should always

be viewed in terms of the secondary benefits that can be obtained.

Conclusion

None of the methods for measuring BNF can be singled out as being near perfect. They each have their strengths and weaknesses that need to be considered in deciding when to use what method. The TND method is relatively very simple and inexpensive, and is most reliable in soils with low available N content. Many estimates of BNF have unfortunately been made using more sophisticated and expensive methods in situations where the simpler and cheaper TND method could have been used. Another advantage of the TND method is, that for most experiments total N is determined as one of the routinely examined parameters, in which case TND determinations may not involve extra work other than the growth and total N determination for the control. Experiments are needed to examine whether the reliability of TND estimates may not be improved by using the mean total N from several potential control crops rather than from a single one.

The ARA technique certainly has lost the favour it enjoyed in the early 1970s despite its simplicity and low cost. Although there appears general agreement that the ARA technique may not be valid for quantifying N_2 fixation, ARA could still be a useful tool in some studies. For example, it could be used to provide highly useful indications of potential N_2-fixing species, especially in novel or unexplored environments. It has other qualitative uses, such as for establishing the time of onset of fixation or for screening *Rhizobium* for ineffective strains. One potential area where advantage may be taken of the rapidity, sensitivity and low cost of the ARA technique is, for screening large collections of plants, e.g., in breeding programmes for increasing BNF in crops; the ARA technique should at least be capable of identifying and thus eliminating obvious non- (or very poor) N_2-fixers prior to using more time-consuming and expensive methods on the remaining plants. The ARA technique has also been used as a standard method for screening potential non-N_2-fixing crops to be used as reference plants to estimate BNF by the ^{15}N soil labelling methods.

The xylem-exudate method, like the ARA technique is simple and inexpensive and very useful for comparing the N_2-fixing abilities of different plant genotypes. However, it suffers from being an instantaneous assay, and would need the extrapolation of several independent measurements to obtain integrated values of N_2 fixed. Besides, the method is only limited to N_2-fixing plants that produce ureide.

The ^{15}N methods have great potential, and in general are capable of providing more accurate data on BNF than most other methods. Their major limitation, that of the accuracy of the reference crop in assessing the integrated uptake of $^{15}N/^{14}N$ ratio of soil N, particularly when N_2 fixation is low can be overcome or reduced to tolerable levels. This may be done by decreasing the rapid rate at which the $^{15}N/^{14}N$ ratio of soil N declines, e.g. by incorporating ^{15}N labelled fertilizers that release their ^{15}N slowly in soil, using ^{15}N labelling procedures that result in an operationally steady level of ^{15}N with time, and better still in conjunction with the use of several rather than only one reference crop. The ^{15}N methods give an integrated measure of BNF, can be applied directly in the field, are able to distinguish N_2 fixed from soil-derived N, and are good for many studies on N turnover in soils, in addition to their being used to measure BNF.

References

1. Atkins CA, Pate JS, Griffiths GJ and White ST (1980) Economy of carbon and nitrogen in nodulated and non-nodulated NO_3-grown cowpea [*Vigna unguiculata* (L.) Walp.]. Plant Physiol. 66: 978–903

2. Attewell J and Bliss F A (1985) Host plant characteristics of common bean line selected using indirect measures of N_2 fixation. In: Evans HJ, Bottomley PJ and Newton WE (eds) Nitrogen Fixation Progress, pp 3–9 Dordrecht, the Netherlands: Martinus Nijhoff Publishers.

3. Awonaike KO, Danso SKA and Zapata F (1993) The use of a double isotope (^{15}N and ^{34}S) labelling technique to assess the suitability of various reference crops for estimating nitrogen fixation in *Gliricidia sepium* and *Leucaena leucocephala*. Plant and Soil 155/156: 325–328

4. Ayanaba A and Lawson TL (1977) Diurnal changes in acetylene reduction in field-grown cowpeas and soybeans. Soil Biol Biochem 9: 125-129

5. Boddey RM, Urquiaga S and Neves MR (1990) Quantification of the contribution of N_2 fixation to field grown grain legumes - a strategy for the practical application of the ^{15}N isotope dilution technique. Soil Biol Biochem 22: 649–655

6. Chaiwanakupt P, Siripaibool C and Snitwongse P (1991) Evaluation of the appropriate non-N_2-fixing crops to quantify nitrogen fixation by soybean using the ^{15}N isotope dilution method. In: Proceedings Symposium on Stable Isotopes in Plant Nutrition, Soil Fertility and Environmental Studies, pp 89–99. STI/PUB/845, IAEA, Vienna

7. Chalk PM (1985) Estimation of N_2 fixation by isotope dilution: An appraisal of techniques involving ^{15}N enrichment and their application. Soil Biol Biochem 17: 389-410

8. Danso SKA (1985) Methods of estimating biological nitrogen fixation. In: Ssali H and Keya SO (eds) Biological Nitrogen Fixation in Africa, pp 213–244. MIRCEN, Nairobi

9. Danso SKA (1988) The use of ^{15}N enriched fertilizers for estimating nitrogen fixation in grain and pasture legumes. In: Beck DP and Materon LA (eds) Nitrogen Fixation by Legumes in Mediterranean Agriculture, pp 345–357 Dordrecht, the Netherlands: Martinus Nijhoff Publishers

10. Danso SKA (1991) Natural and artificial methods of ^{15}N labelling of soil to estimate biological nitrogen fixation: Review of symposium papers. In: Proceedings Symposium on Stable Isotopes in Plant Nutrition, Soil Fertility and Environmental Studies pp 147–152. STI/PUB/845, IAEA, Vienna

11. Danso SKA, Bowen GD and Sanginga N (1992) Biological nitrogen fixation in trees in agro-ecosystem. Plant and Soil 141: 177–196

12. Danso SKA, Hardarson G and Zapata F (1986) Assessment of dinitrogen fixation potentials of forage legumes with ^{15}N technique. In: Haque I, Jutzi S, and Neate PJH (eds) Proceedings of a Workshop on Potentials of Forage Legumes in Farming Systems of Sub-Saharan Africa, pp 26–57. ILCA, Addis Ababa, Ethiopia

13. Danso SKA, Hardarson G and Zapata F (1988) Dinitrogen fixation estimates in alfalfa-ryegrass swards using different nitrogen-15 labelling methods. Crop Sci 28: 106–110

14. Danso SKA, Hardarson G and Zapata F (1993) Misconceptions and practical problems in the use of ^{15}N soil enrichment techniques for estimating N$_2$ fixation. Plant and Soil 152: 25–52

15. Dommergues YR (1963) Evaluation du taux de fixation de l'azote dans un sol dunaire reboise filae (*Casuarina equisetifolia*). Agrochimica 105: 179-187

16. Denison RF, Weisz PR and Sinclair TR (1983) Analysis of acetylene reduction rates of soybean nodules at low acetylene concentrations. Plant Physiol 73: 648-651

17. Dilworth M (1966) Acetylene reduction by nitrogen-fixing preparations from *Clostridium pasteurianum*. Biochem Biophys Acta 127: 285–294

18. Fried M and Broeshart H (1975) An independent measurement of the amount of nitrogen fixed by a legume crop. Plant and Soil 43: 707–711

19. Fried M, Danso SKA and Zapata F (1983) The methodology of measurement of N$_2$ fixation by non-legumes as inferred from field experiments with legumes. Can J Microbiol 29: 1053–1062

20. Fried M and Middelboe V (1977) Measurement of amount of nitrogen fixed by legume crop. Plant and Soil 43: 713–715

21. Gauthier D, Diem HG, Dommergues YR and Ganry F (1985) Assessment of N$_2$ fixation by *Casuarina equisetifolia* inoculated with *Frankia* ORSO21001 using ^{15}N methods. Soil Biol Biochem 17: 375–380

22. Ham GE (1978) Use of ^{15}N in evaluating N$_2$ fixation of field-grown soybeans. In Isotopes in Biological Dinitrogen Fixation, Proceedings Advisory Group Meeting, pp 151–161. STI/PUB/478, IAEA, Vienna

23. Hardarson G, Zapata F and Danso SKA (1988) Dinitrogen fixation measurements in alfalfa ryegrass swards using nitrogen-15 and influence of the reference crop. Crop Sci 28: 101–105

24. Hardy RWF, Burns RC and Holsten RD (1973) Applications of the acetylene-ethylene assay for measurement of nitrogen fixation. Soil Biol Biochem 5: 47–48

25. Hardy RWF and Havelka UD (1975) Nitrogen fixation research - A key to world food? Science 188: 633–643

26. Hardy RWF, Holsten RD, Jackson EK and Burns RC (1968) The acetylene-ethylene assay for N$_2$ fixation: Laboratory and field evaluation. Plant Physiol 43: 1185–1207

27. Herridge DF (1982a) Relative abundance of ureides and nitrate in plant tissues of soybean as a quantitative assay of nitrogen fixation. Plant Physiol 70: 1–6

28. Herridge DF (1982b) Use of the ureide technique to describe the nitrogen economy of field-grown soybeans. Plant Physiol 70: 7–11

29. Herridge DF (1984) Effects of nitrate and plant development on the abundance of nitrogenous solutes in root-bleeding and vacuum-extracted exudates of soybean. Crop Sci 25: 173–179

30. Kessel C van, Nakao P, Roskoski JP and Kevin K (1988) Ureide production by N$_2$-fixing leguminous trees. Soil Biol Biochem 20: 891–897

31. Knowles R (1981) The measurement of nitrogen fixation. In: Gibson AH and Newton WE (eds) Current Perspectives in Nitrogen Fixation. pp 26–57. Amsterdam: Elsevier Press

32. Kumarasinghe KS, Danso SKA and Zapata F (1992) Field evaluation of N$_2$ fixation and partitioning in climbing bean (*Phaseolus vulgaris* L.) using ^{15}N. Biol Fertil Soils 13: 142–146

33. McAuliffe C, Chamblee DS, Uribe-Arango H and Woodhouse Jr WW (1958) Influence of inorganic nitrogen on nitrogen fixation by legumes as revealed by ^{15}N. Agron J 50: 334–337

34. McClure PR and Israel DW (1979) Transport of nitrogen in the xylem of soybean plants. Plant Physiol 64: 411–416

35. McClure PR, Israel DW and Volk RJ (1980) Evaluation of the relative ureide content of xylem sap as an indicator of N$_2$ fixation in soybeans. Plant Physiol 66: 720–725

36. Minchin FR, Sheehy JE and Witty JF (1986) Further errors in the acetylene reduction assay: effects of plant disturbance. J Exp Bot 37: 1581–1591

37. Minchin FR, Witty JF and Mytton LR (1994) Reply to "Measurement of nitrogenase activity in legume root nodules: In defense of the acetylene reduction assay" by J.K. Vessey. Plant and Soil 158: 163–167

38. Minchin FR, Sheehy JE and Muller M (1983) A major error in the acetylene reduction assay: decreases in nodular nitrogenase activity under assay conditions. J Exp Bot 34: 641–649

39. Pate JS and Atkins CA (1983) Nitrogen uptake, transport and utilization. In: WJ Broughton (ed) Nitrogen Fixation Vol 3. Oxford: Clarendon Press

40. Patterson TG and LaRue TA (1983) Nitrogen fixation by soybeans: Seasonal and cultivar effects and comparison of estimates. Crop Sci 23: 488–492

41. Peoples MB, Bergersen FJ, Turner GL, Sampet C, Rerkasen B, Bhromsiri A, Nurhayati DP, Faizah AW, Sudin MN, Norhayati M and Herridge DF (1991) Use of natural enrichment of ^{15}N in plant available soil N for the measurement of symbiotic N$_2$ fixation. In: Proceedings Symposium on Stable Isotopes in Plant Nutrition, Soil Fertility and Environmental Studies, pp 117–129 STI/PUB/845, IAEA, Vienna

42. Peoples MB, Faizah AW, Rerkasem B and Herridge DF (1989) Methods for Evaluating Nitrogen Fixation by Nodulated Legumes in the Field, Monograph No. 11. ACIAR, Canberra

43. Phillips DA, Jones MB, Center DM and Vaughn CE (1983) Estimating symbiotic nitrogen fixation by *Trifolium subterraneum* (cultivar Woogenellup) during regrowth. Agron J 75: 736–741

44. Rainbird RM, Atkins CA and Pate JS (1983) Effect of temperature on nitrogenase functioning in cowpea nodules. Plant Physiol 73: 392–394

45. Rennie RJ (1984) Comparison of N balance and ^{15}N isotope dilution to quantify N_2 fixation in field-grown legumes. Agron J 76: 785–790

46. Rennie RJ and Rennie DA (1983) Techniques for quantifying N_2 fixation in association with non-legumes under field and greenhouse conditions. Can J Microbiol 29: 1022-1035

47. Rennie RJ, Rennie DA and Fried M (1978) Concepts of ^{15}N usage in dinitrogen fixation studies. In: Isotopes in Biological Dinitrogen Fixation, pp 107–130. IAEA, Vienna

48. Sanginga N, Danso SKA, Zapata F and Bowen GD (1990) Influence of reference trees on N_2-fixation estimates in *Leucaena leucocephala* and *Acacia albida* using ^{15}N labelling techniques. Biol Fertil Soils 9: 37-42

49. Sanginga N, Mulongoy K and Ayanaba A (1988) Nodulation and growth of *Leucaena leucocephala* (Lam.) de Wit as affected by inoculation and N fertilizer. Plant and Soil 112: 129–135

50. Shearer G and Kohl H (1986) N_2-fixation in field settings: Estimations based on natural ^{15}N abundance. Aust J Plant Physiol 13: 699–757

51. Stevenson FJ (1986) Nitrogen in Agricultural Soils. ASA Monograph No 22, 940 pp Madison, Wisconsin: American Society Agronomy

52. Streeter JG (1979) Allantoin and allantoic acid in tissues and stem exudate from field-grown soybean plant. Plant Physiol 63: 478–480

53. Vessey JK (1994) Measurement of nitrogenase activity in legume root nodules. In: Defense of the Acetylene Reduction Assay. Plant and Soil 158: 151–162

54. Weaver RW (1986) Measurement of biological dinitrogen fixation in the field. In: Hauck RD and Weaver RW (eds) Field Measurement of Dinitrogen Fixation and Denitrification, pp 1–10. SSSA Special Publication No 18

55. Westermann DT, Kleinkopf GE, Porter LK and Leggett GE (1981) Nitrogen sources for bean production. Agron J 73: 660–664

56. Witty JF (1983) Estimation of N_2-fixation in the field using ^{15}N-labelled fertilizer: Some problems and solutions. Soil Biol Biochem 15: 631–640

57. Witty JF and Day JM (1978) Use of ^{15}N$_2$ in evaluation asymbiotic N_2 fixation. In: Isotopes in Biological Dinitrogen Fixation. Proceedings Advisory Group Meeting, pp 135–150. STI/PUB/478, IAEA, Vienna

58. Witty JF and Minchin FR (1988) Measurement of nitrogen fixation by the acetylene reduction assay: myths and mysteries. In: Beck DP and Materon LA (eds) Nitrogen Fixation by Legumes in Mediterranean Agriculture, pp 331-344. Dordrecht: Martinus Nijhoff Publishers

59. Witty JF, Minchin FR, Sheehy JS and Minquez MI (1984) Acetylene-induced changes in the oxygen diffusion resistance and nitrogenase activity of legume root nodules. Ann Bot 53: 13–20

60. Witty JF and Ritz K (1984) Slow-release nitrogen-15 fertilizer formulations to measure nitrogen fixation by isotope dilution. Soil Biol Biochem 16: 657-661

61. Zapata F, Danso SKA, Hardarson G and Fried M (1987a) Nitrogen fixation and translocation in field-grown fababean. Agron J 79: 505-509

62. Zapata F, Danso SKA, Hardarson G and Fried M (1987b) Time course of nitrogen fixation in field-grown soybean using nitrogen-15 methodology. Agron J 79: 172–176

Fertilizer Research **42**: 43–59, 1995.

© 1995 *Kluwer Academic Publishers. Printed in the Netherlands.*

High productivity analysis of ^{15}N and ^{13}C in soil/plant research

A. Barrie, S.T. Brookes, S.J. Prosser & S. Debney
Europa Scientific Ltd., Electra Way, Crewe, CW1 1ZA UK

Key words: nitrogen-15 analysis, carbon-13 analysis, isotope ratio mass spectrometry, continuous-flow, ANCA-MS, GC-IRMS

Abstract

On-line sample preparation and analysis enables faster testing of hypotheses in biological research, particularly in field experiments where many samples must be processed to integrate spatial variability. Soil scientists were first to recognise the need for a fast, easy-to-use ^{15}N analyser to replace the isotope ratio mass spectrometer (IRMS) and Kjeldahl-Rittenberg sample preparation. Development has since led to a variety of 'continuous-flow' methods. Today's instruments accept solid, liquid or gaseous samples in their naturally occurring form (leaves, soil, algae, soil gas) and process 100–125 samples per day. Automated sample preparation and analysis takes 3–10 minutes per sample.

By 1984, three groups had interfaced an automated nitrogen analyser to an IRMS. Continuous-flow combustion converted N in plant tissue or soil to a pulse of N_2 gas, which was taken to the mass spectrometer by a helium carrier. Throughput increased from 20 to 100 analyses per day and only 5 μg N was required compared with 50 μg N for Kjeldahl-Rittenberg preparation and IRMS analysis. 1984 was also the year in which a gas chromatograph (GC) was first used on-line with an IRMS for isotopic analysis of individual compounds separated from a mixture (GC-IRMS).

In 1987, a software-controlled Automated Nitrogen Carbon Analyser-Mass Spectrometer (ANCA-MS) was developed for solid and liquid samples. Both ^{15}N and ^{13}C were measured to (SD, n = 5) 0.0002 atom%, enough for tracer studies. Less hardware made it portable, enabling use at remote sites. Software control enabled submicrogram quantities of N to be measured by moving the O_2 pulse, with its N_2 blank, out of phase with the sample. Software also allows an 'accelerated mode' for ^{15}N, doubling throughput for a small loss in precision.

Several workers have developed micro-diffusion methods for ANCA-MS determination of NH_4^+ and NO_3^- in KCl extracts. In 1990, a 'dual isotope' mode was added for sequential analysis of ^{13}C and ^{15}N from the same sample. In 1992 a new, 120° deflection, mass analyzer was designed to improve accuracy and precision at the high pressures used in CF-IRMS. Its precision of < 0.1‰ ^{13}C for > 100 μg C and < 0.3‰ ^{15}N for > 50 μg N extends ANCA-MS to 'natural abundance' applications.

Gaseous samples were first analyzed by CF-IRMS by manually injecting into the carrier flow of an ANCA-MS instrument. The initial interest was for ^{13}C breath tests – radiation-free non-invasive tests of metabolic function. Demand for these led to the development in 1989 of an automated GC-IRMS method to separate and measure ^{13}C in CO_2 from human breath. The instrument was further developed for isotopes in other gases: $^{13}CO_2$, $^{15}N_2$, $^{15}N_2O$, and $^{13}CH_4$ above soil; ^{18}O from water equilibrated with CO_2 and ^{15}NO from NO_3^- in KCl plant extracts. The 'dual isotope' mode allows ^{15}N to be measured in both N_2 (mass 28) and N_2O (mass 44) separated from the same sample. Recently GC-IRMS has been used to determine ^{15}N in N_2O produced from NO_2^- or NO_3^- in KCl soil extracts.

We predict that these simpler, automated instruments and the growth of analytical services will encourage more widespread use of stable isotope techniques by agricultural researchers.

Introduction

Being chemically identical, isotopes make ideal tracers of natural processes and stable isotopes are safe both for the user and the environment with no legal restrictions on disposal. However, researchers used to the simplicity of measuring radioisotopes with a liquid scintillation counter are generally surprised, if not disappointed, by the lack of routine analysis procedures when they begin working with stable isotopes.

However, new methods are being introduced in many laboratories which are increasing productivity five to ten-fold and enabling researchers to spend less time collecting results and more time interpreting them. In 1990, this subject was comprehensively reviewed [1] and so the focus of this paper will be on developments since then.

Traditional instruments can only analyze samples after conversion to gases such as N_2 and CO_2. The isotope ratio mass spectrometer (IRMS) was developed by geochemists [2] in the 1950s to detect part per thousand differences in the natural abundance of stable isotopes. Commercial instruments followed, designed to meet their need for highly precise data. Although a modern IRMS has automated gas handling and data processing, skilled, manual conversion of samples to gases is the limiting step and analysis of as few as 10–20 samples a day is not uncommon.

In soil science, factors such as biological and spatial variability over a site and sampling errors limit the use we can make of high precision stable isotope instrumentation. An alternative approach is to average out these variations by analyzing many more samples with adequate precision. In 1985, consultation with authorities in the field revealed a consensus view that a precision (SD) of 0.0005–0.001 atom% ^{15}N (1–3‰ $\delta^{15}N$) was good enough for tracer studies and better than 0.2‰ $\delta^{13}C$ would be desirable for natural abundance measurements.

New methodologies have developed around Continuous Flow-Isotope Ratio Mass Spectrometry (CF-IRMS) [3, 4]. This automates sample preparation and removes the need for extensive operator training.

In plant and soil research, the most useful form of CF-IRMS is Automated Nitrogen Carbon Analysis-Mass Spectrometry (ANCA-MS) which became commercially available in 1987. ANCA-MS accepts solid or liquid samples in a fairly 'raw' state, such as dried and ground leaves or soil, solutions and oils. Automated sample preparation and analysis takes 3–8 minutes per sample giving a throughput of 100–125 samples per day so increasing laboratory productivity five-to ten-fold over the traditional IRMS.

ANCA-MS applications with ^{15}N tracers include fertilizer-use efficiency, biological nitrogen fixation (isotope dilution and gas uptake methods), translocation of N in crops, mineralisation and immobilisation of N in soils, nitrification and leaching of nitrate. Whilst these areas benefited from the ability to analyse many more samples at lower precision, recent developments in instrumentation have improved the precision to $\pm$ 0.1‰ $\delta^{13}C$, $\pm$ 0.3‰ $\delta^{15}N$ at natural abundance without sacrificing sample throughput. This means that the technique is beginning to be used to study biological processes that induce isotope fractionation, such as photosynthesis, biological N fixation (natural abundance method) and the N and C dynamics of soil organic matter.

Changes in research directions, driven by concern over global warming and the need for sustainable agriculture, have raised new analytical requirements such as different gas species and/or much smaller samples. Instrumental developments to tackle these problems have begun with a new, 120° deflection mass spectrometer design [5]. This has greater sensitivity because it can let more sample into the analyser without sacrificing stability and accuracy.

Atmospheric N_2O and CH_4 account for over 20% of global warning. They are respectively 200 and 30 times more radiatively active than CO_2 with atmospheric concentrations rising at 0.3% yr^{-1} for N_2O and 0.9% yr^{-1} for CH_4. There is a need to quantify their fluxes and to understand the mechanisms of their production and removal.

A specialised instrument [6] couples a GC to the IRMS for on-line gas purification. Known as GC-IRMS, this measures ^{15}N and ^{13}C in the gases N_2, N_2O, CH_4 and CO_3. The development of a thermal desorption trapping interface [7] for the GC-IRMS allows atmospheric concentrations of N_2O and CH_4 to be analysed, with sufficient precision for tracers.

GC-IRMS has also been used for other applications: measurement of ^{18}O from water [8] equilibrated with CO_2; ^{15}N in NO gas [9] from NO_3^- in KCl plant extracts; and recently ^{15}N has been analysed [10] in N_2O produced from NO_2^- or NO_3^- in KCl soil extracts.

Continuous flow isotope ratio mass spectrometry

As the development and validation of these techniques from 1983 to 1990 has been reviewed previously [1] we will only outline this period before concentrating on later developments.

In the 1980s traditional dual inlet-IRMS instruments became more automated [11] but sample preparation remained the major obstacle to complete automation of the analytical procedure. The breakthrough in increasing analytical productivity came in 1983 when Preston and Owens [12] proved the feasibility of automated 'continuous-flow' sample preparation for ^{15}N in solids and liquids by Dumas combustion in a He carrier.

This was followed by system automation and validation for ^{15}N in an inter-laboratory comparison [13] and extension to agricultural materials [14]. Validation for ^{13}C then followed [15]. This early work formed the basis for today's ANCA-MS instruments. Preliminary investigations have since been carried out on S isotopes using modified ANCA-MS systems [16, 17].

In 1984, an alternative continuous flow method (GC-IRMS) used gas chromatography to separate compounds from a mixture, then combust these to CO_2 for analysis of the individual compounds [18, 19]. A different version of this method was applied to $^{13}CO_2$ in gas mixtures in 1991 [8]. This was later extended to ^{15}N and ^{13}C in trace gases, N_2O and CH_4 from soils [7]. Off-line HPLC separation has been used with ANCA-MS [4]. The feasibility of using HPLC for on-line mixture separation was recently demonstrated [20].

These two approaches to CF-IRMS are compared in Fig. 1.

Sample preparation options

In 1983, the main requirement was for ^{15}N measurements. The two preparation methods [21, 22] in use were the Kjeldahl-Rittenberg procedure and Dumas combustion.

Dumas seemed best for interfacing to a stable isotope analyser. Automated elemental analysers were already using the continuous flow principle which creates a pulse of a sample-derived gas in an inert gas stream flowing to a detector. These instruments [23] combusted samples in an He carrier and took the combustion products to a thermal conductivity detector (TCD). One model (Automated Nitrogen Analyser, model 1400, Carlo Erba, Milan, Italy), designed for total N analysis and in use since the late 1970s, looked promising as an automated sample converter.

The alternative Rittenberg procedure, for conversion of Kjeldahl digests to N_2 gas, was automated in the early 1980s and connected to a stable isotope analyser [24–27]. However, it is not continuous flow and needs a multi-valve inlet system with a liquid nitrogen trap to transfer purified N_2, to the mass spectrometer.

Choice of analyser

Stable isotopes can be analysed by either Optical Emission Spectrometry (OES) or mass spectrometry. Although OES would have solved the immediate problem for ^{15}N, it would have been unsuitable for other isotopes, particularly ^{13}C, unlike the mass spectrometer. Although OES is simpler, faster and analyses smaller samples than conventional IRMS, its precision is too low for many applications. Poor precision means more isotope must be added in tracer experiments and natural abundance measurements are not possible. For field experiments, this would have meant unreasonably high costs for ^{15}N tracer. OES has been interfaced [28] to automated Dumas combustion for ^{15}N but the best reported precision is only 0.006 atom% ^{15}N.

With IRMS, the major source of cost and complexity is the dual-inlet. However, this can be removed when a continuous flow automated sample converter is connected on-line. On balance, mass spectrometry was chosen because it meets the required precision of < 0.0005 atom% ^{15}N at natural abundance and may be used for other isotopes.

At this point it is worth raising the commonly asked question "why do we need another mass spectrometer – what's wrong with the one we have?" The answer is summarised in Fig. 2.

Essentially, an isotope mass spectrometer is a 'y-axis' instrument, needing exceptional stability in the signals measured but little resolution, only enough to separate simple gases in the mass range 25 to 70. It gets this stability by deliberately using wide slits to degrade the resolution and give the characteristic 'flat-topped' peaks. It also simultaneously detects three ion beams using a triple collector and makes frequent comparison with a reference material. Only a magnetic sector instrument can operate with 'flat-topped' peaks, so a quadrupole mass spectrometer is ruled out.

A magnetic sector organic mass spectrometer is an 'x-axis' instrument, optimised for the best possible resolution to separate closely spaced peaks. This needs a narrow slit which prevents flat-topped peaks. This and

Continuous - Flow Techniques

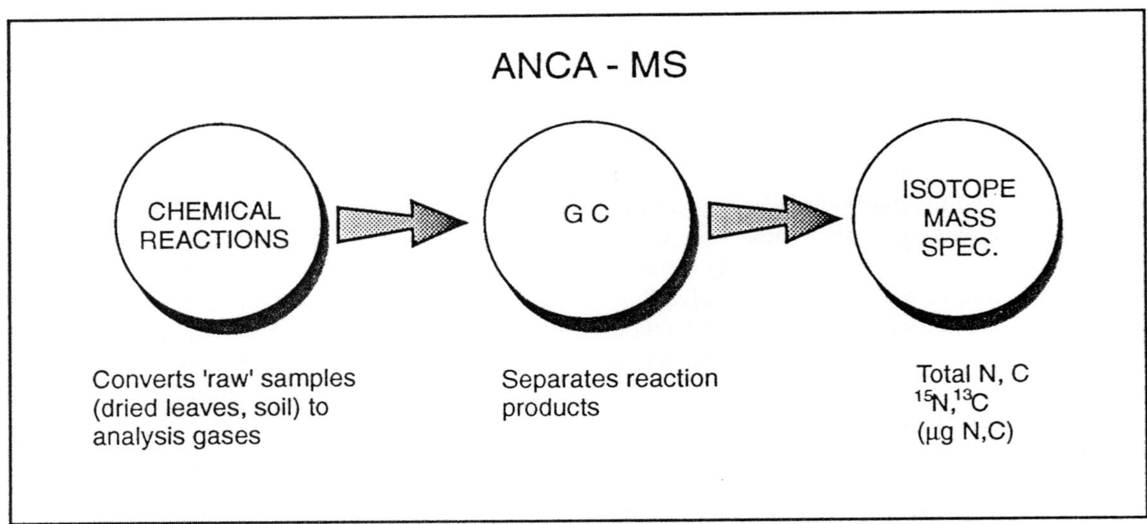

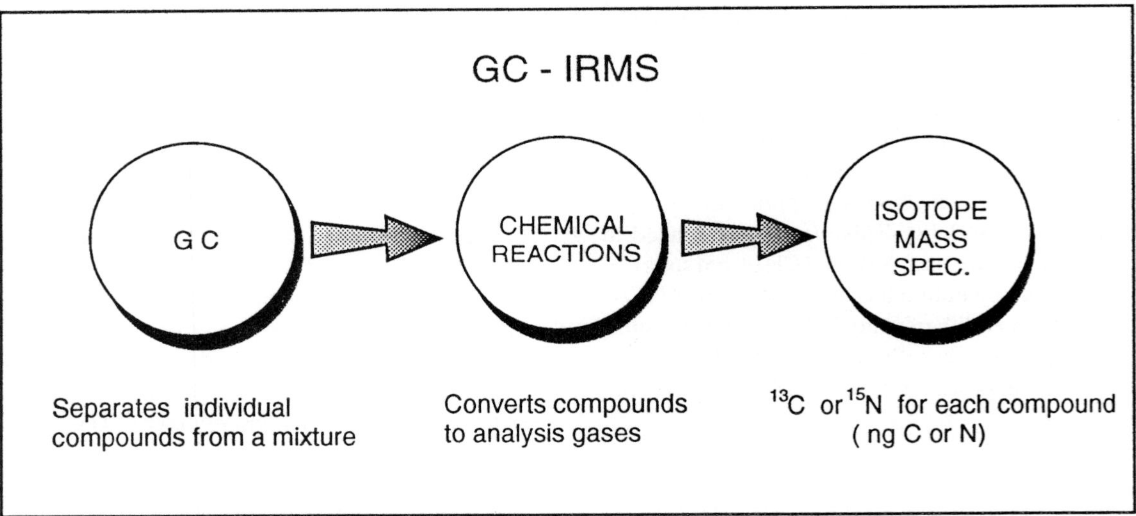

Fig. 1. The two branches of CF-IRMS. Automated Nitrogen Carbon Analysis-Mass Spectrometry (ANCA-MS) is widely used in soil science. Biological material is combusted and the reaction products are separated by GC to give pulses of pure N_2 and CO_2 for analysis of ^{15}N and ^{13}C, total N and total C in the bulk material. Gas Chromatography-Isotope Ratio Mass Spectrometry (GC-IRMS) measures ^{13}C or ^{15}N in individual compounds separated from a mixture by GC then converted to pulses of pure analysis gases.

the use of a single collector restrict the achievable precision to 5‰ $\delta^{13}C$ in the best reported measurements [29, 30] compared with 0.5‰ for the equivalent experiment [19] using IRMS.

Automated NC analyser-mass spectrometry

The pioneers of the experiments reported in 1983/84 identified opportunities for improvement and during 1986 collaborated to produce a dedicated instrument [3] which integrated sample preparation with mass spectrometric analysis [Automated $^{15}N^{13}C$ Analyser (ANCA), Europa Scientific, UK]

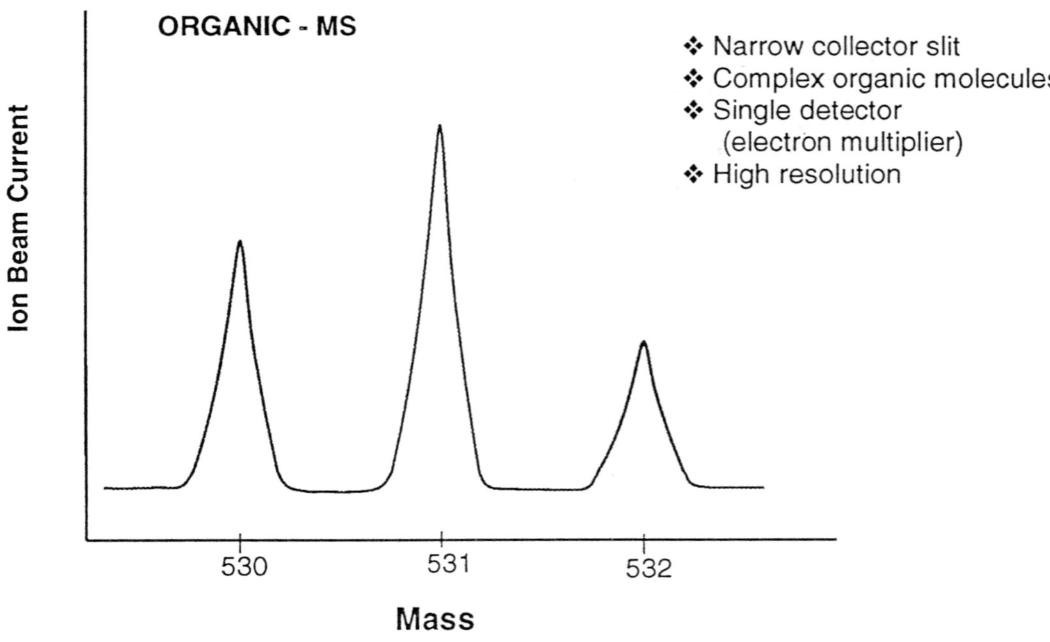

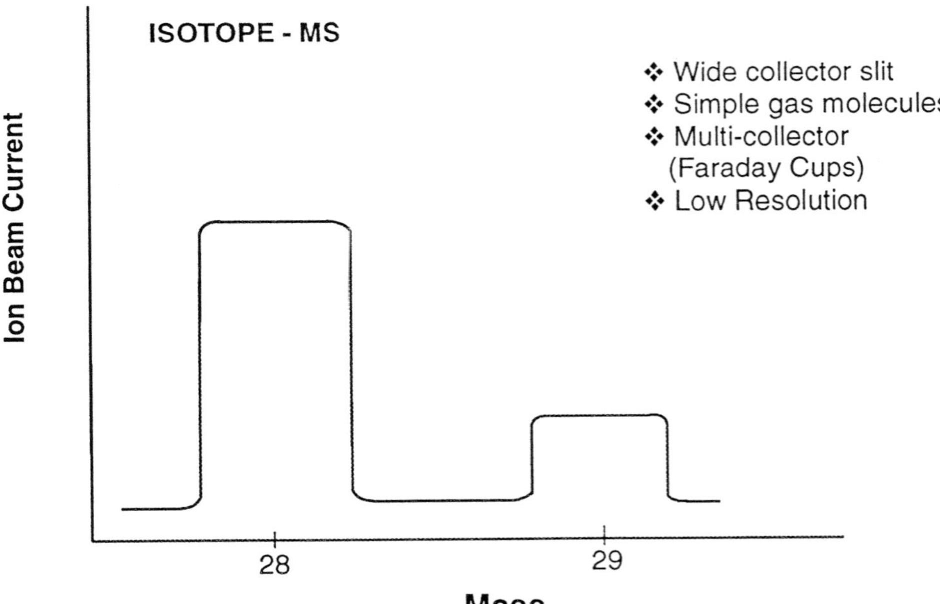

Fig. 2. Comparison of mass spectra and features of an isotope mass spectrometer and an organic mass spectrometer. Beam stability is more important than resolution for isotopic measurements.

The aim was a routine instrument for those biologists who saw the value of stable isotopes in their research but had no wish to become mass spectrometry experts. The dual inlet was omitted as was the TCD since the mass spectrometer measures total N and total C by summing the areas of the three beams. The high vacuum system was automated and software control introduced to simplify operation, improve reliability and give flexibility for future applications. This resulted in a compact, bench-top system [3] which has been used on board ships [31].

'State-of-the-art' ANCA-MS systems [ANCA-SL, Europa Scientific] are functionally very similar. The main differences are more digital control and measurement in the combustion preparation module (pressure, flow, temperature) and the use of a higher sensitivity, differentially pumped mass spectrometer. The Windows™ software requires microprocessors in both the preparation module and mass spectrometer to handle local functions and talk to the central PC through an RS232 network. Higher sensitivity has improved precision at natural abundance to 0.1‰ δ ^{13}C for > 100 μg C and 0.3‰ δ ^{15}N for > 50 μg N compared with 0.2 and 0.8 respectively for the original ANCA.

Sample preparation

Operation of the ANCA-SL instrument is shown schematically in Fig. 3. Solid or liquid samples are sealed into tin capsules and loaded into a 66-place autosampler (6 mm high × 4 mm diameter; 8 × 5 mm and 12 × 5 mm capsules are available). A capsule falls into the vertical combustion tube (quartz, 21 mm diameter) as a pulse of O_2 enters. An oxidation catalyst of Cr_2O_3 granules at 1020 °C is followed by chopped CuO wire to oxidize hydrocarbons while Ag wool traps S and halogens. Flash combustion of the tin raises the local temperature to ~1700 °C for complete oxidation. Combustion products (CO_2, N_2, NO_X and H_2O) are swept over Cu wires at 600 °C to reduce oxides of N to N_2 and remove excess O_2. Water of combustion is removed by a $Mg(ClO_4)_2$ trap. If only ^{15}N and total N are being analysed, a Carbosorb (Merck, Poole, UK) trap is automatically selected to remove CO_2. A GC column (0.5 m, 6 mm diameter proprietary packing) at 100 °C separates N_2 from trace impurities. This temperature also gives enough separation between N_2 and CO_2 to allow measurement of ^{15}N and ^{13}C from the same sample. In this case the CO_2 trap is by-passed. If only ^{13}C analysis is required, elution of CO_2 can be speeded up by increasing the GC temperature to 125 °C.

In contrast to the original elemental analysers, in a dedicated instrument high purity O_2 only flows during injection and the size of the O_2 pulse can be matched to the type and size of sample using software. An excess of 25–50% above that estimated for combustion of the sample and tin capsule is added. In this way O_2 is not wasted, the N_2 blank is minimised, and the

Cu packing lasts longer. This allows combustion and reduction packings to be changed together, say, after 1000 analyses or 9 days if running at full capacity. These measures reduce the running costs for O_2 and Cu to one-third of that for an elemental analyser based ANCA-MS.

Analysis

Only a small portion ($\approx$ 1%) of the He/N_2 effluent is admitted to the mass spectrometer to stop the pressure rising too high (max. 10^{-4} mbar). To minimise fractionation, a 0.15 mm diameter capillary is used with a T-piece and crimp. This is set to confine pressure in the analyser to the range 2×10^{-6} to 10^{-5} mbar, depending on sample size. The main effluent passes to atmosphere through a coil of 1 mm i.d. tubing to minimise back diffusion of air.

The mass spectrometer is preset for different groups of three masses, usually 28/29/30 or 44/45/46. It measures ^{15}N, total N, ^{13}C and total C in the pulses of N_2 and CO_2 in the He carrier. The TracerMass used in the original ANCA had an 11 cm radius, 90° sector, normal geometry analyzer. A single turbomolecular pump was used to make the vacuum system safe and easy to use. Frequency of filament changing was reduced by using a cold cathode gauge for vacuum measurement and a ThO_2-coated Ir filament in the ion source.

The TracerMass has been replaced in the ANCA-SL by a new analyser, the '20–20'. This uses 120° deflection, 20 cm dispersion, extended geometry. Its novel ion optics came out of a design study [5] to find the best geometry for continuous flow use and to improve sensitivity for new applications.

At the high He pressures used in CF-IRMS, ions are more likely to collide with He atoms in the analyser, be deflected into the 'wrong' collector and give an error (called 'abundance sensitivity') in the isotopic measurement. A short path length reduces these collisions and the 120° optics has a shorter path length for a given dispersion than any other design. Separate pumping of the source and analyser also reduces collisions.

The 20–20 analyser focuses in two directions to give 100% transmission and this, combined with a new ion source, leads to higher sensitivity, good linearity and better precision than earlier designs. The 120° gometry is also more stable and less sensitive to magnet positioning than the conventional Cross [36] extended geometry design. In addition, the 120° sector has a much smaller 'footprint', allowing it to fit in the same

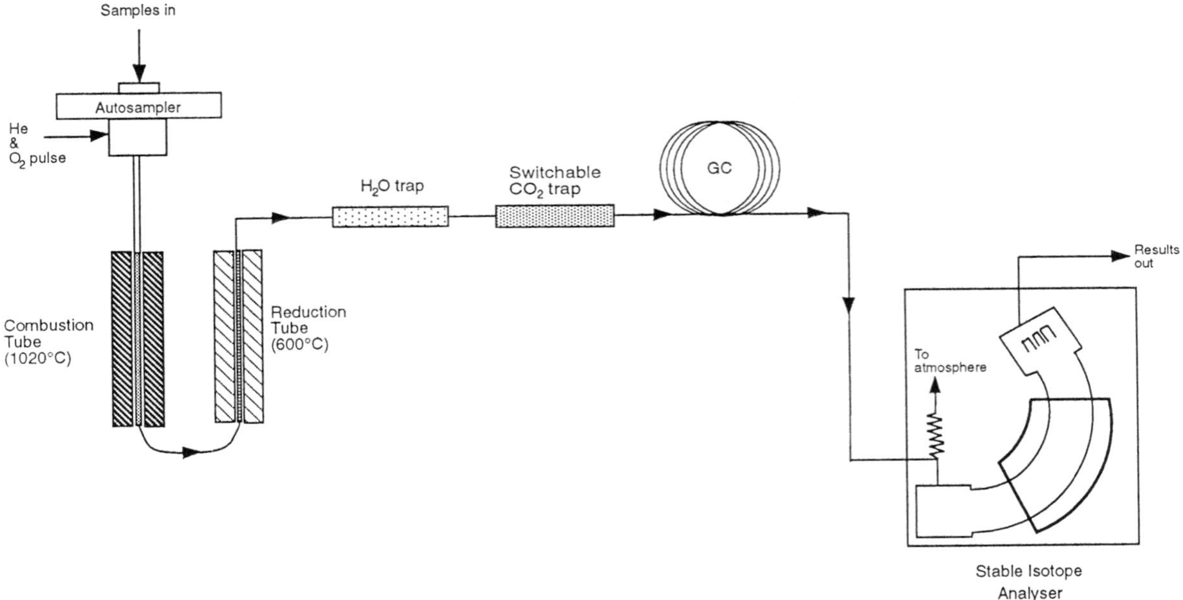

Fig. 3. Schematic of a modern ANCA-MS system. Natural materials are combusted, using a pulse of pure O_2 in a helium carrier, to N_2, NO_X CO_2 and H_2O. Hot Cu reduces NO_X to N_2 and removes excess O_2. A packed-column GC separates N_2 and CO_2 for analysis of ^{15}N and total N; ^{13}C and total C.

space as a normal 90° analyser of the same radius, continuing the bench-top ANCA tradition.

Operating procedures

Weighed samples are analysed in batches that include reference materials of known isotopic and elemental composition, placed within the autosampler carousel. These calibrate the measured isotope ratios, elemental composition and mass spectrometer drift by measuring the same reference every n^{th} sample. This method calibrates both physical and chemical stages of the analysis process.

Blanks for ^{15}N analysis are measured by running without a capsule, prior to a batch run, to measure the small amount of N_2 impurity in the O_2 pulse. These are then subtracted from sample/reference measurements during batch analysis. Software controls the analytical cycle and data collection. Analytical cycle time for each sample is typically 4 min for ^{15}N and total N. The details of the integration and calibration procedure have been described previously [1, 3].

The various operating modes that can be used with a dedicated ANCA-MS are listed in Table 1. Carbon is analysed in a similar way to N but takes longer, about 7 min. For ^{13}C the blank comes from the cap-

sule rather than the O_2 but this is negligible except for the smallest samples. For ^{15}N analysis of samples with low O_2 demand (< 10 μg N), the blank can be effectively removed using the 'small sample' mode [33, 4]. Because the sample is combusted without added O_2, there is no blank. Oxygen is injected later in the cycle to replenish the catalyst and timed to place the blank in the gap between sample peaks. Using this mode, ANCA-MS can analyse samples 70 times smaller than OES with better precision [1]. The 'accelerated' mode [3, 34], which almost doubles sample throughput, is popular with users who can accept a small loss in precision. The first cycle is increased to 5 min but subsequent samples are dropped early giving analyses 2.5 min apart.

Results obtained with the above modes illustrating the achievable precision and accuracy have been presented previously [1]. It should be noted that elemental analyser systems can only use the standard ^{15}N mode [Table 1]. If the GC column is changed they can be used in standard ^{13}C mode also.

Recent hardware and software changes have allowed measurement of total ^{15}N, and total ^{13}C from a single sample combustion [35]. This 'dual isotope' mode is both a time-saver and useful where sample is

Table 1. ANCA-MS operating modes

Mode	Gas analysed	CO_2 trap	GC temp (°C)	Cycle time (s)	Sample range (μg element)
Standard [15]N	N_2	in	100	240	10–100
Standard [13]C	CO_2	out	125	420	10–2000
Accelerated [15]N	N_2	in	100	150	10–250
Small sample [15]N	N_2	in	100	240	0.2–10
Dual isotope	$N_2 + CO_2$	out	75	480	10–2000

Table 2. [15]N and ^{13}C analysis of wheat flour in a single cylce

Replicate	$\delta\ ^{15}N$ (‰)	Total N (%)	$\delta\ ^{13}C$ (‰)	Total C (‰)
1	2.93	1.64	−25.37	39.88
2	2.27	1.67	−25.32	40.80
3	2.41	1.68	−25.31	41.03
4	2.52	1.69	−25.31	40.05
5	2.61	1.67	−25.25	41.25
mean	2.55	1.67	−25.31	40.60
SD	0.22	0.02	0.04	0.54
%RSD (CV)	0.086	0.012	0.0016	0.013

Flour was measured against itself as a reference. $\delta^{15}N$ values are relative to air and $\delta^{13}C$ values are relative to PDB.

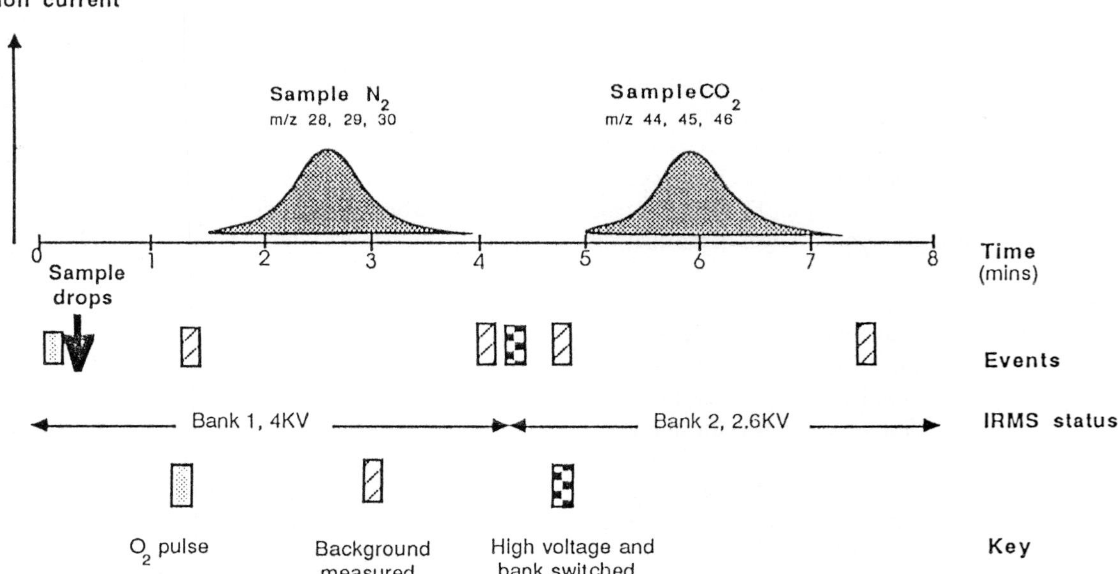

Fig. 4. ANCA-MS in 'dual isotope' mode to measure ^{15}N and ^{13}C from the same sample. Two sets of ion source parameters are stored to optimise the mass spectrometer for N_2 and CO_2. For samples with high C/N ratios, a high electron current is used for the small N_2 peak and then switched to a low value before the large CO_2 peak arrives. This way, both isotopes are measured with high sensitivity. A similar method is used in GC-IRMS with denitrification samples. Here the opposite applies, a large N_2 peak is followed by a very small N_2O peak.

limited. Results for wheat flour are shown in Table 2. The type of trace produced is shown in Fig. 4.

Solid samples

Oven dried (80 °C) and finely ground ($\leq$ 250 μm) soil and plant material is required to ensure homogeneity and hence a representative measurement. Kelley [36] has recently discussed the variability resulting from poor grinding and the problems of cleaning some types of grinder to avoid cross contamination. He has developed an improved version of the batch grinding roller mill originally devised by Hauck [37] which overcomes these difficulties and processes 55 samples in 12 hours.

Loading tin capsules requires two pairs of curved forceps, a small spatula and a clean surface. A mirror tile is ideal as any material which 'misses' the capsule shows up clearly. The capsule is held on the tile with one pair of forceps while it is filled. Then the second pair of forceps is used to fold over the open end twice to stop the contents falling out. Finally both pairs of forceps are used to squeeze and turn the capsule until it makes a ball.

Samples containing N in the range 10 to 1000 μg N can be analysed routinely although a high C/N ratio will reduce the upper limit. Over 20 μg is recommended for good precision. Typical amounts required are 4 mg grain, 12 mg straw, 30 mg soil. Care should be taken when grinding fibrous material such as straw and roots. It is good practice to load samples in increasing order of enrichment to minimise memory. With samples such as soils which contain non-combustible material, the combustion tube will need to be 'de-ashed' every 60 to 300 analyses. This is done by lifting off the autosampler and sucking out the ash through a 6 mm diameter stainless steel tube connected to a trap to protect the vacuum pump from debris.

Liquids

Most liquids are homogeneous and so give good reproducibility. Any that separate into two phases or contan precipitates may give variable results. Solutions are a convenient way of introducing standards. They can be accurately pipetted as standards for total N or C as well as isotopic abundance.

Constant sample size gives better reproducibility because of the 'pressure effect', a dependence of isotope ratio on sample size which affects all mass spectrometers [5]. Source design and tuning can reduce this effect but, for the most precise work, control of sample size helps [38].

Solutions are either pipetted onto an inert adsorbent, such as Chromosorb, or freeze-dried directly into a capsule. Freeze-drying avoids the ash contribution of the adsorbent and allows up to 100 μl of solution to be used. Viscous liquids such as oils are easily dispensed at constant volume using an automatic, positive displacement pipette. Typically 2 μl is used.

Conventional steam distillation methods for concentrating inorganic N in soil extracts for ^{15}N analysis are time-consuming with a high risk of cross contamination between samples. For ^{15}NH$_4^+$ and ^{15}NO$_3^-$ analyses of low concentration solutions such as KCl extracts, Brooks et al. [39] have developed a simple diffusion method for ANCA-MS which concentrates up to 200 μg N from up to 60 ml of solution. Disposable 140 ml containers are used to avoid cross contamination and after 6 days at room temperature all the NH$_3$-N has been collected as NH$_3$ vapour on 7 mm discs of filter paper acidified with KHSO$_4$ (Fig. 5). The discs are then loaded into tin capsules. Blanks are prepared using the same technique as for the samples, but instead of the extract, KCl is poured through the filter paper.

To analyse NO$_3$-N, NH$_4^+$ is first removed, then Devarda's alloy is added to reduce NO$_3^-$ to NH$_4^+$ and the above procedure repeated.

For samples below 100 μg N, allowance must be made for the blank, which can be 2–5 μg N. As the blank is at natural abundance it will dilute an enriched sample, lowering the measured isotope ratio. Kelley et al. [34] found that a correction could be applied for samples as small as 20 μg N. For those not into 'disposables', Sachir et al. [40, 41] have developed a procedure with reusable food preserving jars and shown 100% recovery of N. Mulvaney [42] has given a useful 'recipe' for the procedure and details of precautions to be taken with micro-diffusions.

Around 100 samples per day can be processed, including the time to prepare soil extracts, remove aliquots for analysis and start the diffusion procedure.

An off-line High Performance Liquid Chromatography (HPLC) method for ANCA-MS analysis of ^{15}N and ^{13}C labelled metabolites has been developed [4] for biomedical studies of protein and fat metabolism. Because only small quantities are collected, the blank must be minimised. The small sample mode is recommended for ^{15}N, as are aluminium capsules, cleaned at 450 °C for 3 h, to reduce the C blank. These methods should be applicable to studies of plant metabolism.

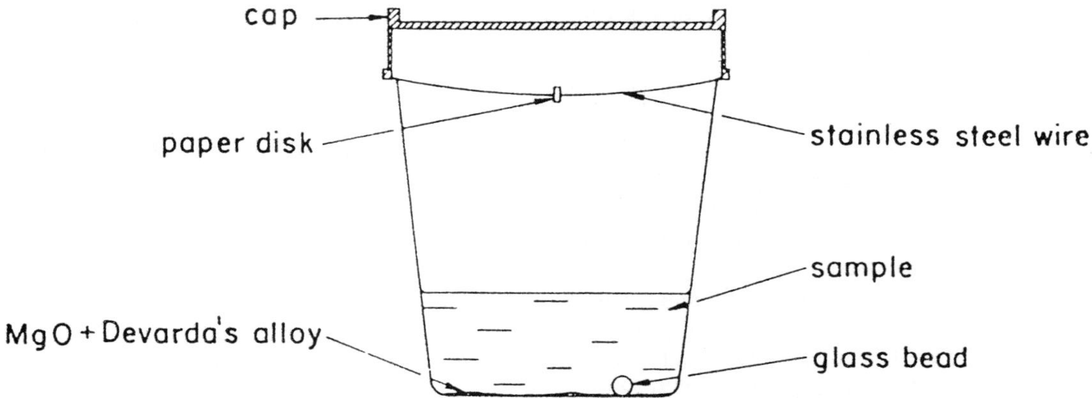

Fig. 5. Diffusion apparatus used to prepare 2 *M* KCl soil or plant extracts for ANCA-MS. Ammonia-N is trapped on the paper disc which is then wrapped in a tin capsule for ^{15}N and total N analysis.

Gas chromatography-IRMS

The first form of GC-IRMS to be developed [18] is similar to GCMS but offers 10–100 times better precision for isotopic measurement. It uses a high resolution capillary GC in much the same way as GCMS to separate compounds dissolved in an organic solvent. However, taking the case of ^{13}C/^{12}C, instead of analysing them as intact molecules by peak jumping between masses M for ^{12}C and M + 1 for ^{13}C, they are combusted to CO_2 for simultaneous analysis using a triple collector.

It is also possible to measure ^{15}N in a similar way to ANCA-MS by using a reduction stage after combustion. This is more difficult, however, because the abundance of N_2 in air is some 2000 times greater than the atmospheric composition of CO_2 making it sensitive to even small leaks. Sensitivity is lower since organic molecules contain far fewer N than C atoms. Also the gas analysed, N_2, has two atoms of N compared with one C in CO_2 leading to dilution of ^{15}N and the ionisation efficiency for N_2 is only $\sim$70% of that for CO_2.

To our knowledge, this approach has not been applied to plant and soil sciences although potentially it could be used to study biochemical pathways in plants, turnover of soil organic matter and degradation of organic compounds in soils.

In 1990 a version of GC-IRMS was developed for gas mixtures which was later to prove more useful in soil science. This was first used [8] in medical diagnostic tests to separate and purify CO_2 from human breath. In these tests, the patient is given a ^{13}C labelled compound which can break down to $^{13}CO_2$, indicating a healthy or diseased state depending on the specific test.

The important gaseous samples in agricultural research are the trace gases implicated in global warming, N_2O, CH_4 and CO_2. Soil is the primary source for both N_2O and CH_4 and soils are also important sinks for CH_4. Since agricultural management can significantly alter the source/sink relationship between these gases [43], it is important to study many different agricultural systems to make global estimates. It is estimated [44] that 20–45% of global CH_4 emissions come from flooded rice fields where addition of fertilizer N can increase the CH_4 flux [45] Even small fluxes can be accurately determined from ^{15}N and ^{13}C measurements. However, isotopic analysis of atmospheric concentrations of N_2O, CH_4 and CO_2 by conventional dual inlet IRMS requires litres of sample, complex glass manifolds and hours of preparation per sample by skilled technicians [46–48].

These difficulties in sample preparation have been overcome by the system shown schematically in Fig. 6, which has been evolving over the past four years. It can analyse N_2O, CH_4 and CO_2 at atmospheric concentrations [7] and also at the higher concentrations [6] encountered in some plant and soil applications.

Operation

The GC-IRMS shown schematically in Fig. 6 [ANCA-G, Europa Scientific, UK] allows ^{15}N or ^{13}C to be measured in N_2 and one of N_2O, CO_2 or CH_4 from a single 13 ml gas sample in less than 10 minutes. Methane is analysed after combustion to CO_2. The 'dual isotope' mode is used to analyse N_2 at m/z 28,

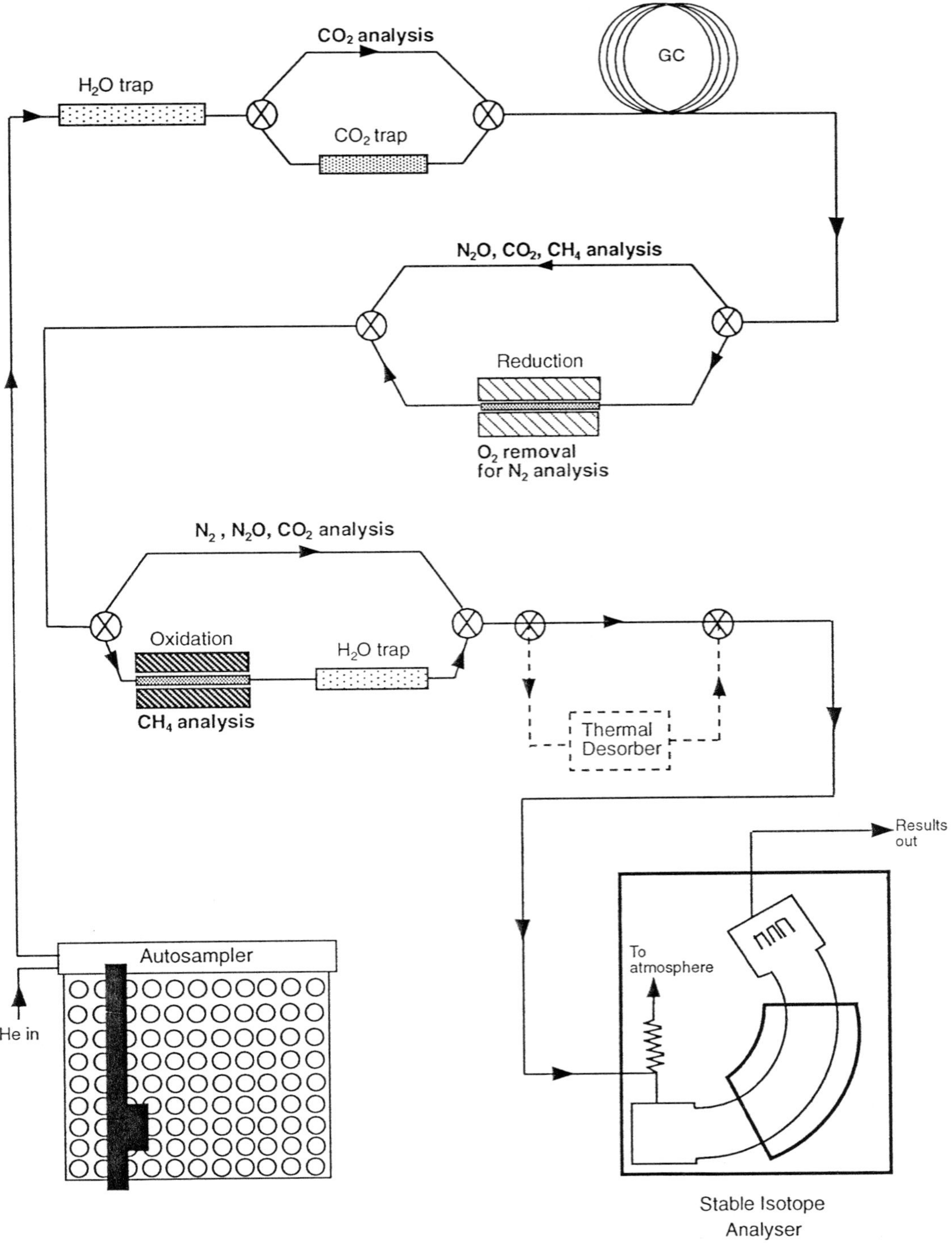

Fig. 6. Schematic of a GC-IRMS system for analysing trace gases. Chemical traps, oxidation and reduction furnaces and the thermal desorber (if fitted) are switched in and out of circuit by software to suit the gas being purified. N_2, N_2O and CO_2 in a sample are analysed after purification, CH_4 is converted to CO_2 before analysis. The thermal desorber is used to boost the signal from gases close to atmospheric concentration or to increase precision at higher concentrations.

54

29, 30 followed by either N_2O or CO_2 at m/z 44, 45, 46. To measure more gases means filling more sample vials.

The addition of a switchable thermal desorption module allows these gases to be measured at concentrations ranging from ambient to several percent by volume. At higher concentrations, thermal desorption may not be necessary, thus reducing the analysis time to 6 min. Gas samples are collected in preevacuated septum-sealed vials and loaded onto a 220-position autosampler. At the start of the analytical cycle the sample is flushed into the He carrier (80 ml/min) through a double hole concentric needle.

Analysis of N_2 + N_2O
The gas is first dried by a $Mg(ClO_4)_2$ trap and then passed through a software-selected CO_2 trap (Carbosorb). Dual column GC separation gives a choice of packing volume depending on the sample size and gas concentration and separates N_2 from N_2O. The N_2 elutes first and passes to the IRMS through a reduction furnace (Cu granules, 600 °C) which removes any co-eluting O_2. This prevents the reaction between O_2 and N_2 in the ion source to form NO and the associated problem of an artificially high m/z 30 signal [6]. The reduction furnace is then by-passed for the slower eluting N_2O.

If ambient levels of N_2O are being analysed, the carrier flow is diverted to the thermal desorption module where a selective zeolite absorbs the N_2O. The zeolite is then rapidly heated to 'focus' the desorbed N_2O as a sharp peak in a slow carrier flow (0.5 ml/min) which passes to the IRMS for isotopic analysis.

Analysis of N_2 + CO_2
In this mode, the CO_2 trap is bypassed but otherwise operation is as for N_2 + N_2O.

Analysis of N_2 + CH_4
As CH_4 is converted to CO_2 for analysis it is important to first remove CO_2 from other sources. The flow from the sample vial is first passed through the CO_2 trap. Then N_2 is separated from O_2 by the reduction furnace as above. The CH_4 passes on to an oxidation furnace (Pt/CuO granules, 950 °C). A second trap takes out water of combustion and the CO_2 passes through the thermal desorption module to the IRMS. A CO_2 blank from the furnace packings is measured and used for software correction.

Performance

Precision of measurement (SD, n = 5) varies significantly with gas, concentration and whether or not thermal desorption is used. Fig. 7 shows a compilation of data from mainly unpublished work by Brooks, Atkins and Stevens and their coworkers using the Tracer Mass analyser. Very recently, an improved version of this instrument has been developed using the 20–20 analyser and this is expected to give improved precision.

Thermal desorption can either be used to extend measurements to very low concentrations or to improve precision 2–3 fold at higher concentrations where GC separation would normally be sufficient. It is the CF-IRMS analogue of the cold finger in conventional dual inlet-IRMS.

Precision for N_2O is quite satisfactory for tracer experiments. With thermal desorption, it is 0.0025 atom% ^{15}N (~7%) at atmospheric concentration (0.31 ppm). Without thermal desorption, the same precision is reached at 50 ppm. At 100 ppm, precision with thermal desorption improves to 0.0002 atom% ^{15}N (0.5‰), almost good enough to detect natural abundance differences.

As the atmospheric concentration of CO_2 is several orders of magnitude greater than that of N_2O, analysis is easily performed without thermal desorption. Here the benefit of thermal desorption is an improvement in precision from 0.6‰. $\delta\,^{13}C$ to 0.23‰. $\delta\,^{13}C$.

Typical CH_4 precision ranges from 50‰ $\delta\,^{13}C$ at atmospheric concentration to 0.2‰ $\delta\,^{13}C$ at 500 ppm. Although there are applications at elevated concentrations, there is clearly a need for further development at atmospheric concentration. By 10 ppm, precision is satisfactory (5‰.) for tracer experiments.

Applications

Gas samples from soil covers, growth chambers or reaction vessels are collected by syringe and injected into 13 ml pre-evacuated autosampler vials.

Denitrification
GC separation without thermal desorption is suitable for denitrification studies at moderate N_2O concentrations (5–1000 ppm). Although reasonable precision is only reached above 20 ppm N_2O, laboratory incubations of soil cores typically yield this concentration. The 'isotope pairing method' [49], developed to mea-

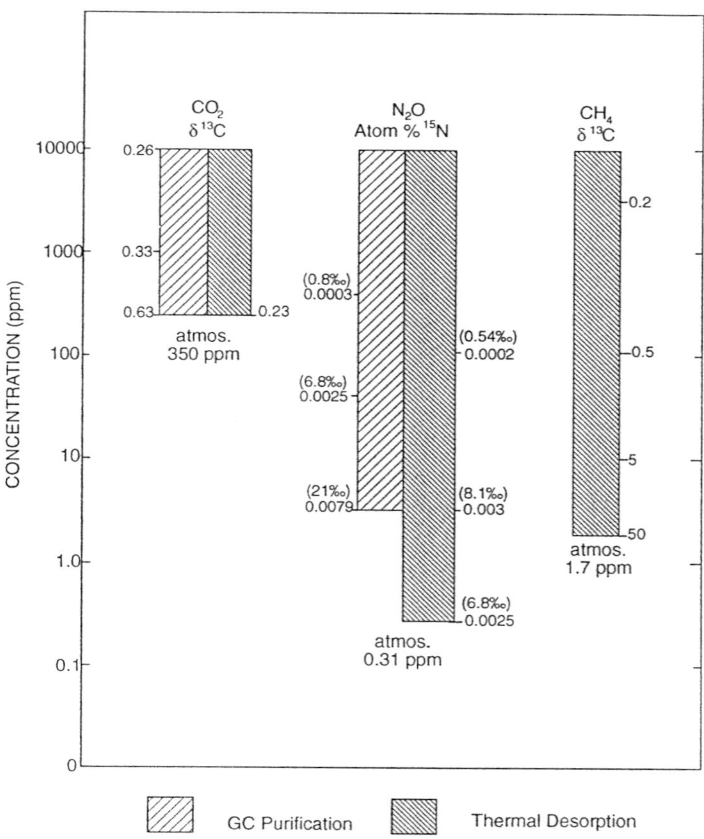

Fig. 7. Comparison of concentration range and precision obtainable for trace gases using GC-IRMS with GC separation alone and combined with thermal desorption.

sure N_2O fluxes from both denitrification and coupled nitrification-denitrification in sediments, can also be used with GC separation alone. Applications are, however, limited for field experiments. Some field studies, such as denitrification from cow slurry [50], do yield sufficient N_2O. Acetylene blocking techniques may also bring the N_2O concentration, within the range of this instrumentation.

More preferable, however, is the ability to measure sub-ppm concentrations of N_2O directly. The thermal desorption method allows isotopic analysis of trace N_2O with precision suitable for tracer work. Even when the concentration of N_2O does not increase discernibly during a study, increases in its ^{15}N abundance are easily detected allowing very low denitrification rates to be calculated [51]. The enrichment of N_2 is simultaneously measured, giving information on another product of denitrification. Replicate analyses of air samples are shown in Table 3. External precision (SD, n = 5) for N_2 of 0.03‰ $\delta^{15}N$ was as good as any

Table 3. ^{15}N analysis of N_2O and N_2 by GC-IRMS with thermal desorption (at.% ^{15}N)[a]

Replicate	N_2O	N_2
1	0.36064	0.36628
2	0.36442	0.36631
3	0.36218	0.36628
4	0.36767	0.36629
5	0.36676	0.36630
6	0.36412	0.36631
7	0.36544	0.36629
Mean[a]	0.3645	0.36629
SD	0.0025	0.00001
%RSD (CV)	0.68	0.003

[a]Units: 0.00037 atom% ^{15}N = 1‰ difference.

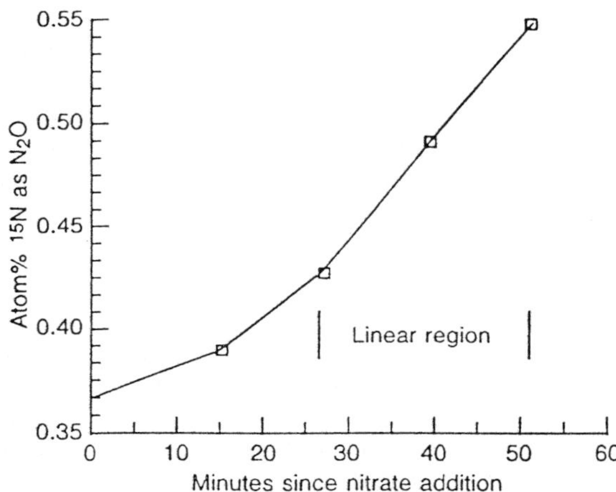

Fig. 8. Increase in ^{15}N abundance of N$_2$O in a 8.3 l cover over soil treated with 1.0 kg·ha^{-1} NO$_3^-$-N at 50 at.% ^{15}N. Using the linear part of the plot the N$_2$O flux was calculated to be 9.6 mg N$_2$O-N ha^{-1}d^{-1} but GC-IRMS can measure rates ten times lower.

dual inlet IRMS, while that for was N$_2$O sufficient for tracer studies.

Figure 8 shows the results of an N$_2$O flux experiment by Brooks et al. [51]. On a summer's day in Berkeley, California, very dry soil inside a 25 cm PVC ring was first moistened with 1 l of water. One hour later, 1.0 kg ha^{-1} of NO$_3$-N at 50% ^{15}N enrichment was added, and the 8.3 l cover replaced. 20 ml gas samples were periodically removed by syringe for ^{15}N$_2$ and ^{15}N$_2$O analysis by GC-IRMS with thermal desorption. This was a very small addition considering that typical fertilizer application rates for crops are 100–200 kg ha^{-1}.

Although there was no measurable increase in N$_2$O concentration, the ^{15}N abundance of N$_2$O increased linearly after the first 25 min [Fig. 8]. The linear part of the plot was used to calculate the N$_2$O production rate. In this case it was 400 μg ha^{-1} h^{-1} but the technique can measure rates as low as 40 μg ha^{-1} h^{-1}, a thousand-fold improvement over conventional methods [52].

The ability to sequentially measure N$_2$ and N$_2$O automatically from a single 13 ml air sample gives quick feedback between experimental design and analysis of results. Allowing one hour to load the autosampler and type in sample names, and assuming one reference for every ten samples, the 10 minute cycle for sample purification and analysis means 125 samples can be processed each day.

Growth Chambers

Although the GC-IRMS instrument shown in Fig. 6 is too new to have many published examples of its use, we can see areas where it would be useful. One is in the ^{15}N$_2$ incorporation method, generally considered to be the 'gold standard' for measuring biological nitrogen fixation [53]. The whole plant, or perhaps only parts such as the root system, is exposed to a ^{15}N$_2$-enriched atmosphere and incubated for a week or so. At the end of the experiment, gas samples are analysed as well as parts of the plant to determine ^{15}N uptake. GC-IRMS would greatly simplify the measurement of gas samples, with ANCA-MS being used for the plant material.

A similar application would be ^{13}CO$_2$ uptake in studies of carbon isotope discrimination, carbon turnover and partitioning [49].

^{15}N-nitrate in NO from plant/soil extracts

A batch procedure has been used to selectively reduce NO$_3^-$ in plant and soil extracts to NO gas for ^{15}N analysis by a GC-IRMS instrument [9]. The extract is placed in an autosampler vial with 100 μl Hg, the septum cap re-fitted and the vial flushed with He. Then 1 ml of degassed 10 M H$_2$SO$_4$ is added and the batch of vials placed in an ultrasonic bath for 30 min. The vials are then loaded on the autosampler and the NO analysed for ^{15}N. Since the GC is by-passed in this application, the cycle time is only 2 min/sample.

The advantages of this method are that it avoids distillation and diffusion and is very fast, allowing the analyst to prepare and analyse 50 samples in under 4 h.

^{15}N-nitrite and ^{15}N-nitrate in N$_2$O from plant/soil extracts

GC-IRMS has also enabled separate determination of both ^{15}N-NO$_2^-$ and ^{15}N-NO$_3^-$ in plant and soil extracts using batch reactions which produce N$_2$O as an analysis gas [10].

For ^{15}N in NO$_2^-$, 50 ml m of 2 M KCl extract is first placed in a 100 ml septum-capped bottle. Then 1 ml of 1 M HCl is added along with 0.5 ml 0.04 M NH$_2$OH. The cap is replaced and the bottle shaken for 16 h at 120 rpm. Finally 13 m of the headspace is transferred to an autosampler vial and the N$_2$O analysed for ^{15}N.

To analyse ^{15}N in NO$_3^-$ the NO$_2^-$ is first removed as N$_2$ by hand shaking for 5 s with 2.5 ml 0.2 M sulphamic acid. Five ml acetate/acetic acid buffer are added to raise the pH to 4.7 followed by shaking for 16 h at 120 rpm with Cu/Cd foil to reduce NO_3^- to N_2O for ^{15}N analysis.

Like the 'NO method', this also avoids distillation and diffusion but is better in that it distinguishes between NO_2^- and NO_3^- and avoids the use of Hg. Over 100 samples can be prepared and analysed in 24 h. The ability to determine NO_2^- directly, rather than by difference is important for fertilized soils where NO_2^- is low.

The method can measure ^{15}N with a precision (CV) of < 1% for 0.25–5.0 μmol NO_2^- (3.5–7.0 μg N) at ≥ 1 at.% excess and also for 5 μmol at 0.1 at.% excess. The authors suggest this approach will be most useful for studies on NO_2^- dynamics where tracers are applied at 10 at.% ^{15}N. In NO_3^- analyses, they found a CV of $\leq$ 1% for 5–50 μmol (70–700 μg N) at ≤ 1 at.% excess. At 0.1 at.% excess, 25 μmol (350 μg N) were needed to meet this precision. This method is linear from 0 to 20 atom% excess.

Conclusion

Since its discovery 10 years ago, CF-IRMS has changed the way stable isotopes are analysed. Samples are accepted in their natural state, requiring minimal sample preparation, and software control has simplified training of technicians and ease of use.

With care, ANCA-MS can now be as precise as dual-inlet IRMS, when sample preparation errors are included. Sample size can be reduced by between ten and several hundred times that necessary for dual-inlet IRMS. With the 'small sample' mode, ANCA-MS can be more precise than OES for samples 70 times smaller.

Micro-diffusion sample preparation with ANCA-MS avoids the sample cross-contamination problems associated with the conventional Kjeldahl-Rittenberg approach. The five to ten-fold increase in productivity has enabled both larger scale experiments and more rapid testing of hypotheses. Laboratories using ANCA-MS can publish more research with lower running costs and fewer technicians.

Some laboratories sell their excess capacity and there are also specialist analytical service companies. Using an analytical service is an inexpensive way to start in stable isotopes.

The fluxes of the trace gases N_2O, CO_2 and CH_4 in field experiments can now be measured using isotope tracers with a new form of GC–IRMS for gas mixtures. This allows 125 small (13 ml) gas samples to be automatically purified and analysed daily. This routine operation with small samples will greatly increase the productivity of researchers investigating denitrification mechanisms. Samples may be aliquots of air drawn from soil covers or gaseous products (NO, N_2O) of reactions with NO_3^- or NO_2^-. In the future we may see the capillary GC form of GC-IRMS used to study biochemical pathways such as synthesis of amino acids from ^{15}N labelled NO_3^-.

Most of the processes of the N-cycle can now be quantified using ^{15}N tracers with CF-IRMS analysis of raw soil or plant material; NH_4^+, NO_3^- and NO_2^- from extracts; N_2O and N_2 gases. One gap in our tool kit is the separation of NO for studies of denitrification. Although CF-IRMS has successfully analysed NO gas from reactions, further development is needed to separate NO from the other gases with which it is produced, N_2O and N_2. This is important because in some situations, notably dry soils, NO concentration can be several times greater than that for N_2O.

Until now, development has focussed on multi-purpose instruments for laboratories with a wide range of research interests. This means ^{13}C as well as ^{15}N analysis and gaseous samples as well as solids and liquids can be processed. However, these are relatively expensive and 'overkill' for those researchers in tropical countries who only want to study N-fertilizer efficiency. The micro-processor technology used in the latest instruments could be adapted to produce a '^{15}N only instrument'. This would process plant and soil samples for ^{15}N and total N for not much more than the price of an optical emission spectrometer, but be easier to use and more precise, saving on the cost of tracer ^{15}N.

In summary, the availability of these easy-to-use, high productivity instruments together with the growth of analytical service facilities should encourage more widespread use of stable isotope methods in agricultural research.

[1]Precision figures are usually quoted at natural abundance. When using tracers, larger numbers apply for enriched samples and smaller numbers for depleted samples. A convenient way to account for this is to report results with a percent relative standard deviation, %RSD, also known as coefficient of variation, CV.

References

1. Barrie A (1991) New methodologies in stable isotope analysis. In: Stable Isotopes in Plant Nutrition, Soil Fertility and Environmental Studies, 3–25. IAEA, Vienna
2. Hoefs J 1973 Stable Isotope Geochemistry. Berlin: Springer Verlag
3. Barrie A, Davies J E, Park A J and Workman CT (1989) continuous-flow stable isotope analysis for biologists. Spectroscopy 4: 42–52
4. Preston T and McMillan D (1988) Rapid sample throughput for biomedical stable isotope tracer studies, Biomed Environ Mass Spectrom 16: 229–235
5. Prosser SJ (1993) A novel magnetic sector mass spectrometer for isotope ratio determination of light gases. Int J Mass Spectrom Ion Processes 125: 241–266
6. Stevens RJ, Laughlin RJ Atkins GJ and Prosser SJ (1993) Automated determination of nitrogen-15 labeled dinitrogen and nitrous oxide by mass spectrometry. Soil Sci Soc Am J 57: 981–988
7. Atkins GJ Barrie A, Prosser SJ, Brooks PD and Herman DJ (1991) Rapid isotopic methods to study the fluxes of radiatively active trace gases. Applications of isotopes and radiation in conservation of the environment, 3–25. IAEA, Vienna
8. Prosser SJ, Brookes ST, Linton A and Preston T (1991) Rapid automated analysis of ^{13}C and ^{18}O of CO_2 in gas samples by continous-flow, isotope ratio mass spectrometry, Biol Mass Spectrom 20: 724–730
9. Prosser SJ, Gojon A and Barrie A (1993) Fast automated analysis of nitrogen-15 nitrate from plant and soil extracts. Soil Sci Soc Am J 57: 410–414
10. Stevens RJ and Laughlin RJ (1994) Determining nitrogen-15 in nitrite or nitrate by producing nitrous oxide. Soil Sci Soc Am J 58: accepted for publication
11. Hachey DL, Wong WW, Boutton TW and Klein PD (1987) Isotope ratio measurements in nutrition and biomedical research. Mass Spectrometry Reviews 6: 289–328
12. Preston T and Owens NJP (1983) Interfacing an automatic elemental analyser with an isotope ratio mass spectrometer: the potential for fully automated total nitrogen and nitrogen-15 analysis, Analyst 108: 971–977
13. Barrie A and Workman CT (1984) An automated analytical system for nutritional investigations using N-15 tracers. Spectros Int J (Canada) 3: 439–447
14. Marshall R and Whiteway J (1985) Automation of an interface between a nitrogen analyser and an isotope ratio mass spectrometer. Analyst 110: 867–871
15. Preston T and Owens NJP (1985) Preliminary C13 measurements using a gas chromatograph interfaced to an isotope ratio mass spectrometer. Biomed Mass Spectrom 12: 510–513
16. Pichlmayer F and Blochberger K (1988) Isotopenhäufigkeitsanalyse von Kohlenstoff, Stickstoff und Schefel mittels Gerätekopplung Elementaranalysator-Massenspektrometer. Fresenius Z Anal Chem 331: 196–201
17. Haystead A (1991) Measuring natural abundance variations in sulphur isotopes by Dumas elemental analysis-mass spectrometry. Stable Isotopes in Plant Nutrition, Soil Fertility and Environmental Studies, 3–25. IAE, Vienna
18. Barrie A, Bricout J and Koziet J (1984) Gas chromatography-stable isotope ratio analysis at natural abundance levels. Biomed Mass spectrom 11: 583–588
19. Barrie A, Bricout J and Koziet J (1984) Gas chromatography-stable isotope ratio analysis, a technique for ^{13}C tracer studies at low isotopic enrichments. Spectros Int J (Canada) 3: 259–268
20. Caimi RJ and Brenna TB (1993) High precision liquid chromatography - combustion isotope ratio mass spectrometry. Anal Chem 65: 3497–3500
21. Huck RD (1982) In: Methods of Sol Analysis, Part 2. Chemical and Microbiological Properties Agronomy Monograph No. 9. American Society of Agronomy 735–779
22. Fiedler R and Proksch G (1975) The determination of nitrogen-15 by emission and mass spectrometry in biochemical analysis. Anal Chim Acta 78: 1–62
23. Pella E and Colombo B (1973) Study of carbon, hydrogen and nitrogen determination by combustion gas chromatography. Mikrochim Acta (Wien): 697–719
24. McInteer BB and Montoya JG (1981) Automation of a mass spectrometer for nitrogen isotope analysis. In: (1981) Recent Frigerio A (ed) Developments in Mass Spectrometry in Biochemistry, Medicine and Environmental Research 7, pp 343–349 Amsterdam: Elsevier
25. McInteer BB, Montoya JG and Starks EE (1984) An automated mass spectrometer grows up, Spectros. Int J (Canada) 3: 226–234
26. Mulvaney RL, Fohringer CL, Bojan VJ, Michlik MM and Herzog LF (1990) A commercial system for automated nitrogen isotope ratio analysis by the Rittenberg technique. Rev Sci Instru 61: 897–903
27. Mulvaney RL and Liu P (1991) Refinement and evaluation of an automated mass spectrometer for nitrogen isotope ratio analysis by the Rittenberg technique, J Automatic Chem 13: 273–280
28. Therion JJ, Human HGC, Classe C, Mackie RI and Kistner A (1986) Automatic nitrogen-15 analyser for use in biological research. Analyst 111: 1017–1021
29. Matthews DE and Hayes JM (1978) Isotope-ratio-monitoring gas chromatography-mass spectrometry. Anal Chem 50: 1465–1473
30. Otsuki A, Ino Y and Fujii T (1983) Simultaneous measurements and determinations of stable carbon and nitrogen isotope ratios, and organic carbon and nitrogen contents in biological samples by coupling of a small quadrupole mass spectrometer and modified carbon-nitrogen elemental analyzer. Intl J Mass Spectrom Ion Phys 48: 343–346
31. Owens NJP (1988) Rapid and total automation of ship board N-15 analysis: examples from the North Sea. J Exp Mar Biol Ecol 122: 163–171
32. Cross WG (1951) Two-directional focussing of charged particles with a sector-shaped, uniform magnetic field. Rev Sci Instrum 22: 717–722
33. Owens NJP and Rees AP (1989) Determination of nitrogen-15 at sub-microgram levels of nitrogen using automated continuous-flow isotope ratio mass spectrometry. Analyst 114: 1655–1657
34. Kelley KR, Ditsch DC and Alley MM (1991) Diffusion and automated nitrogen-15 analysis of low-mass ammonium samples. Soil Sci Soc Am J 55: 1016–1020
35. Brookes ST, Davies JE, Scrimgeour CM, Smith K and Watt PW (1992) Single cycle analysis of ^{15}N and ^{13}C enrichment in proteins. In: Takai K (ed) Frontiers and New Horizons in Amino Acid Research, pp 411–415. Amsterdam: Elsevier
36. Kelley KR (1994) Conveyer belt apparatus for fine grinding of soil and plant materials. Soil Sci Soc Am J 58: 144–146

37. Hauck RD (1986) Personal communication. ASA Annual Meeting, New Orleans
38. Handley LL, Daft MJ, Wilson J, Scrimgeour CM, Ingleby K and Sattar MA (1993) Effects of the ecto- and VA-mycorrhizal fungi *Hydnagium carneum* and *Glomus clarum* on the δ ^{15}N and δ ^{13}C values of *Eucalyptus globulus* and *Ricinus communis*. Plant, Cell and Environment 16: 375–382
39. Brooks PD, Stark JM, McInteer BB and Preston T (1989) Diffusion method to prepare soil extracts for automated nitrogen-15 analysis. Soil Sci Soc Am J 53: 1707–1711
40. Sachir NS, Mulvaney RL and Azam F (1993) Determination of nitrogen by microdiffusion in Mason jars: I. Inorganic nitrogen in soil extracts. Commun Soil Sci Plant Anal 24: 1745–1762
41. Sachir NS, Mungwari FP, Mulvaney RL and Azam F (1993) Determination of nitrogen by microdiffusion in Mason jars: II. Inorganic nitrogen in soil extracts. Commun. Soil Sci. Plant Anal 24: 2747–2763
42. Mulvaney RL (1993) Mass Spectrometry. In: (eds) Nitrogen Isotope Techniques, Knowles R and Blackburn TH pp 11–57. Academic Press San Diego:
43. Mosier A, Schimel D, Valentine D, Bronson K and Parton W (1991) Methane and nitrous oxide fluxes in native, fertilized and cultivation grasslands. nature 350: 330–332
44. Bouwman AF (1990) The role of soils and land use in the Greenhouse effect. In: Bouwman AF (ed) Soils and the Greenhouse Effect. Chichester, UK: John Wiley & Sons
45. Cicerone RJ and Shetter JD (1981) Sources of atmospheric methane: measurements in flooded rice paddies and a discussion. j Geophys Res 86: 7203–7209

Fertilizer Research **42**: 61–75, 1995.
© 1995 *Kluwer Academic Publishers. Printed in the Netherlands.*

Adapting the potentially mineralizable N concept for the prediction of fertilizer N requirements

C.A. Campbell, Y.W. Jame, O.O. Akinremi & M.L. Cabrera[1]
*Agriculture and Agri-Food Canada, Research Station, P.O. Box 1030, Swift Current Saskatchewan, S9H 3X2,
Canada; [1]Department of crop and Soil Sciences, College of Agricultural and Environmental Sciences, University
of Georgia, Athens, Georgia, 30602-7272, USA*

Key words: potentially mineralizable N, CERES model, LEACHM model, fertilizer N requirements, crop rotations

Abstract

Quantification of N dynamics in the ecosystem has taken on major significance in today's society, for economic
and environmental reasons. A major fraction of the available N in soils is derived from the mineralization of
organic matter. For decades, scientists have attempted to quantify the rate at which soils mineralize N, but the
complexity of the N cycle has made this a major task. Further, agronomists have long sought soil test methods that
are practical, yet will provide accurate means of predicting the amounts and rates of release of N from soils. Such
tests would allow us to make more precise fertilization decisions. This paper discusses the potentially mineralizable
N concept, first promoted by Stanford and colleagues [61, 62, 64], and suggests how it may be incorporated into
deterministic models, such as CERES and LEACHM, so as to provide more accurate estimates of N mineralization
under field conditions. We also suggest how the potentially mineralizable N concept may be coupled to quick,
routine laboratory methods of determining available soil N, such as the hot 2 M KCl extracted NH_4-N method
recently developed by Gianello and Bremner [35], and used together with deterministic N models, such as CERES,
for predicting probable fertilizer N requirements.

Introduction

Nitrogen (N) is one of the most important plant nutrients. Crops obtain N primarily from the decomposition
of soil organic matter but, also from added fertilizers,
manures, and through N fxation by legumes, and from
dry and wet deposition. Throughout the world, soil
degradation associated with improper management has
resulted in the diminution of the N-supplying capacity
of many soils; this, plus the desire to increase production, has led to an increase in N fertilizer requirements
and use [25]. In turn, the increasing use of fertilizers
has resulted in considerable concern by society regarding the possible negative influence that N can have on
the environment and health. The scientific community
has responded by seeking crop management methods
that producers may adopt to increase the efficient use
of N.

Because soils are major sources of plant available
N, and because excess N may impair the environment,
it is imperative that we be able to quantify N mineralization. Due to the very dynamic nature of N, this has
been a very challenging problem for scientists throughout the decades. This paper discusses: (i) how the concept of potentially mineralizable N may be adapted
into deterministic models that have been designed to
quantify all aspects of the N cycle, and (ii) how this
concept may be adapted to provide more accurate soil
tests for N fertilizer requirements.

Balance method of estimating N fertilizer requirements

The amount of N fertilizer to be applied depends on the
attainable dry matter yield (Y_{dm}), the N concentration
of the crop (N_y), the amount of residual mineral N

present in the soil at seeding (N_{rm}), and the amount of N that the soil will mineralize from organic matter during the growing season (N_s) [58]. The crop does not recover all of the fertilizer N (N_f) nor soil N that is available to it (i.e., efficiency (E) of use of the various sources of N is less than 100%). Therefore, E will influence the quantity of N that must be applied in order to achieve a certain yield. If plant N at harvest is given by $Y_{dm} \times N_y$,

$$Y_{dm}N_y = E_{rm}N_{rm} + E_sN_s + E_fN_f \qquad (1)$$

Rearranging gives:

$$N_f = [Y_{dm}N_y - E_{rm}N_{rm} - E_sN_s]/E \qquad (2)$$

We can set a target yield and measure N_{rm} at seeding. As well, we can make reasonable estimates of the efficiencies even though these will vary depending on weather, fertilizer and crop management factors [53, 58]. The greatest difficulty in solving equation 2 is quantifying N_s.

Long before fertilizer use became commonplace, scientists have sought quick, routine tests for quantifying the N-supplying capacity of soils so as to determine the influence of crop rotations, legumes, and manure additions on fertilizer N requirements [10, 59]. This need is even more pressing today because, as producers adopt better management principles, they are likely to gradually increase the N-supplying capacity of their soils over time [18, 26, 27, 45, 51]. As well, where poor management is practised, the fertility of soils will be degraded [18]. It is unlikely that empirical methods such as the NO_3-test that is commonly used on the Canadian prairies and in some regions of the USA to determine fertilizer N requirements will adequately reflect the positive or negative changes in soil fertility. At present we have no simple soil tests that will reflect these changes in soil fertility but we will make some suggestions later in this paper. It is worth noting that, most of the routine soil tests used to assess available N have proven inadequate because they do not measure the potential of the soil to mineralize N over the growing season. Furthermore, such indices do not allow us to quantify N_s as a function of weather conditions. This is a prerequisite of any system if it is to be used in a predictive manner.

Estimation of N mineralization from potentially mineralizable N

Although there are several thousand kilograms N per hectare in most fertile agricultural soils, only 1 to 2% of this becomes available to crops each year [10]. The N that is mineralized comes from a heterogeneous pool of components varying in stability [10]. The organic components in soil include fresh crop and animal resiidues, microbial biomass, microbial metabolites and cell wall constituents adsorbed to colloids, physically protected material in small pores, and the very stable humus. Although fresh residues and microbial biomass are important as mineralizable substrates, specific compounds in the mineralizable pool have not been directly identified.

Stanford and Smith [64] proposed an incubation-leaching technique for use in quantifying the active N pool of soil organic matter. Stanford and co-workers [61, 62 64] advanced the concept of potentially mineralizable N, denoted as N_0, and a related mineralization rate constant (k), for use in characterizing soil-available N. Since then this concept has also been used to assess soil N availability in cropland, forests, and waste-disposal sites [7–9, 16, 29, 34, 55].

Stanford and Smith [64] suggested that N_0 and k could be estimated by statistical techniques if the soil was incubated at optimum moisture and temperature conditions for mineralization and the N mineralized (N_s and time of incubation (t) measured. They assumed that organic N mineralization under optimum conditions followed first-order kinetics, i.e.,

$$N_s = N_0[1 - e^{-kt}] \qquad (3)$$

Stanford [58] suggested that the potentially mineralizable N concept could be used to estimate N_f by replacing N_s in equation 2 by $N_0 (1 - e^{-kt})$. We could then modify k as a function of temperature and soil water [15] and calculate the probable N_s by using historical, long-term records of temperature and precipitation. Other soil characteristics that are required to allow estimation of N_s are field capacity and wilting point, and if these are not available they can be estimated from soil texture [65].

Use of this type of model requires that values of N_0 be known for each soil and it is desirable that these values do not change markedly from year to year. Few experiments have been conducted in which N_0 has been measured at regular intervals. In one such study conducted in Australia [30], N_0 decreased rapidly from 200 to 105 mg kg^{-1} soil during the first 20 yr after breaking and cultivating a Waco soil; however, during the following 50 yr of cultivation N_0 was essentially constant.

Equation 3 only holds when the mineralization process is conducted under optimum conditions for mineralization [64] for a long enough period ($\geq$ 24 wk) [14, 33]. If the incubation period is shorter than about 20 wk, the possibility of obtaining an inverse relationship between N_0 and k is great [32, 50]. Some scientists, in France [39] and also in Canada [32] did not obtain a first order relationship; rather, they obtained a zero order relationship for net N mineralization. However, Houot et al. [39] did not cite their temperature of incubation, while Ellert [32] incubated soil at 28 °C (might be too low) and worked with a forest soil where immobilization may have contributed to the observed departure from the first order relationship most frequently reported for cultivated soils. Low temperatures tend to favour immobilization over mineralization, especially early in the incubation, leading to an S-shaped type of curve.

The rate constant (k) in equation 3 must be adjusted for environmental conditions experienced *in situ* in order to predict the actual amount of N mineralized. For North American soils, optimum temperature for mineralization is assumed to be about 35 °C and optimum soil water content is field capacity (-0.03 to -0.01 MPa). The k is temperature-dependent [16, 19, 62]. Stanford et al. [62] found Q_{10} to vary between 1.8 and 1.9 for 39 US soils and suggested a Q_{10} of 2 could be used to represent most soils. The most precise results will be obtained if Q_{10} determined for a particular soil is used [15]; however, Q_{10} of a soil zone, or a Q_{10} of 2, will provide acceptable results if specific values for the soil are not available.

Prediction of N_s in the field also has to take into account the effect of soil water content. Stanford and Epstein [61] suggested that if the values of cumulative N_s were expressed on a relative basis with respect to maximum N_s, then the effect of water content could be described as:

$$Relative\ N_s = Soil\ water\ content/$$
$$Optimum\ Water\ content \qquad (4)$$

Similar results were obtained for nine Chilean soils [28]. However, the above expression was not suitable for 5 Australian and 32 Canadian soils [46]. Myers et al. [46] proposed a model similar to that in equation 4, except that they divided the available water in the soil by the possible available water for the soil, where the lower limit of available water was assumed to equal the water held by the soil at -4.0 MPa and the upper limit was taken as water held at -0.01 to -0.03 MPa. Myers et al. [46] obtained a linear relationship between N_s and this water quotient in the soils they tested, except for one Australian and nine Canadian soils, in which N_s showed a curvilinear response to the water quotient. Thus, in certain cases, the simple approach proposed by Stanford and Epstein [61] may not be adequate to represent the effect of soil water content.

There is some debate concerning the method used to correct N_s for soil water content. In some studies [15, 38] the water content factor was used to correct the rate constant k, but others [47] suggest that the water factor should be used to correct N_0. There is insufficient experimental data available to guide us in rationalizing this problem. Intuitively, however, it seems reasonable to expect such intensity factors as temperature, water and aeration to influence the rate of reaction rather than N_0. The potentially mineralizable N, which is a capacity factor, is a discrete quantity for each particular soil; its value is determined by the conditions (climate, edaphic, management) under which the soil has developed so that the moisture conditions in an incubation or in the field in a particular year merely determines what proportion of this factor will be mineralized in unit time. In any event, adjusting k for moisture resulted in only slightly larger predictions than when the correction was applied to N_0 [15, 38].

The correction by temperature and soil water content has been performed using hourly [15, 16], daily [7], weekly [44, 57], monthly [48, 52, 60], and yearly [66] time steps. Most workers have used Euler' s integration method, whereas Campbell et al. [15, 16] used the integrated form of the model to estimate N mineralized within each time step.

Stanford [59] summarized the first studies that attempted predictions of N mineralized from soil organic matter under greenhouse [63] and under field conditions [52, 60]. We shall report on some of the subsequent studies. Campbell et al. [16] used measured N_0 and k values for the top 2.5 cm of a loam soil to predict N mineralized in plastic bags in the laboratory and buried in the field, taking into account soil moisture and temperature conditions. The model predicted well the amounts of N mineralized in the laboratory but tended to under-predict the amounts of N mineralized in the field at the two lowest water contents. These underpredictions were attributed to the accidental entry of water into some bags, which could have caused wetting and drying cycles and induced flushes of N mineralization [16].

In another study [15], a more rigorous field evaluation of Stanford and Smith's approach was conducted with moisture effect calculated according to Myers et al. [46]. The N_0 and k values for the upper two 7.5-cm layers of a loam soil were used together with estimated temperature and soil water content, to predict N mineralized in lysimeters (0.15 m diameter, 1.2 m long) subjected to summer fallow, dryland spring wheat (*Triticum aestivum* L.) and irrigated spring wheat treatments. Predicted N mineralized closely agreed with measured values during the first 45 to 60 days, but tended to under predict observed values afterwards. Under predictions for the entire growing season amounted to 16, 26, and 31% of irrigated wheat, summer fallow, and dryland wheat treatments, respectively. The authors explained the under predictions as due to the model not accounting for the flushes in N mineralization that typically occur as a result of wetting/drying cycles on the Canadian Prairies.

In a Japanese study, N mineralized from soil organic matter during the corn (*Zea mays* L.) growing season in four soils was measured [55]. N_0 and k values were estimated and the dependence of k on temperature evaluated so as to predict N mineralized with adjustment for field temperature. In general, predicted amounts of N mineralized agreed closely (-1% to 8% of measured values) with observed amounts when the measured values ranged from 52 to 85 kg N ha^{-1}. Predicted amounts tended to be larger (1% to 51%) than measured amounts when mineralized N ranged from 103 to 226 kg N ha^{-1}. The authors attributed part of the over predictions to the existence of dry conditions which decreased N mineralization in the field. Because the model used did not make adjustments for soil water content, it would be expected to over predict N mineralized under dry field conditions.

Cabrera and Kissel [7] evaluated Stanford and Smith's approach in three Kansas soils during two sorghum (*Sorghum bicolor* [L.] Moench) growing seasons. Using the upper three 15-cm layers (0-45 cm total depth), they obtained over predictions ranging from 114 to 343% of the actual amounts of N mineralized (32 to 51 kg N ha^{-1}). When only the upper 15 cm layer was considered, the over predictions ranged from 9 to 181% [6]. It was hypothesized that the over predictions could have been due to an improper function of adjustment of the amounts of N mineralized based on soil water content, and to the crushing of the clay soil (to pass 2-mm sieve) before incubation [7]. In a

follow-up study, it was found that, even after subtracting the initial N flush, samples that had been dried and crushed to pass through a 2-mm sieve showed higher N mineralization than undisturbed soil samples [9]. It was concluded that field-moist samples, as undisturbed as possible, should be used for determination of mineralization parameters.

This concept has been used, modified, criticized and discussed in great detail [5, 7-10, 15, 19, 31, 33, 43, 46-47, 50]. However, the foregoing research results indicate that there is a good chance we may be able to predict N mineralized under field conditions by following Stanford and Smith's approach. Much research remains to be conducted so as to improve the accuracy of predictions by this technique [14]. This type of model has been criticized for various shortcomings discussed elsewhere [14], but most of these may either not be quantitatively signifcant [15], or can be overcome if proper sample handling, incubation and calculation procedures are employed [14]. For example, soil samples should be air-dried and stored near freezing until analyzed. The incubation temperature is best at 35 °C, and the length of incubation should be at least 24 wk [32]. When the model proposed by Campbell et al. [15] is used for estimating N_s, two factors that can lead to low estimates of N_s is the failure to account for the N accruing from the most recent crop residues, and N lost by denitrification. However, these can be estimated by various commonly used simulation models such as LEACHM, CERES, EPIC and others.

Estimating N mineralization in long-term rotation studies

The use of the potentially mineralizable N concept to estimate N_s under laboratory, greenhouse and field conditions has been generally encouraging; however, most field studies have been one-season experiments. If the potentially mineralizable N concept is to be useful in a practical sense, it must perform credibly for systems that are degrading or aggrading as a consequence of crop management over time.

To test the usefulness of this concept under natural field conditions over several years requires the use of dynamic models that will allow accounting for the main components of the N cycle. Numerous deterministic models, appropriate for this task, have been developed (e.g., EPIC, NTRM, NLEAP, CERES, LEACHM). At Swift Current, Saskatchewan, we have tested LEACHM and CERES for simulating N min-

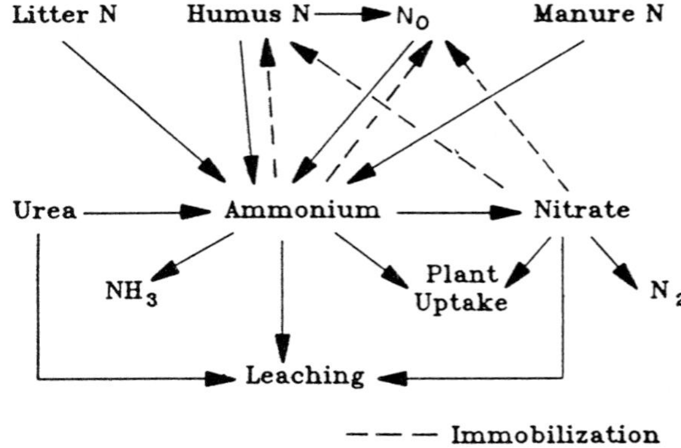

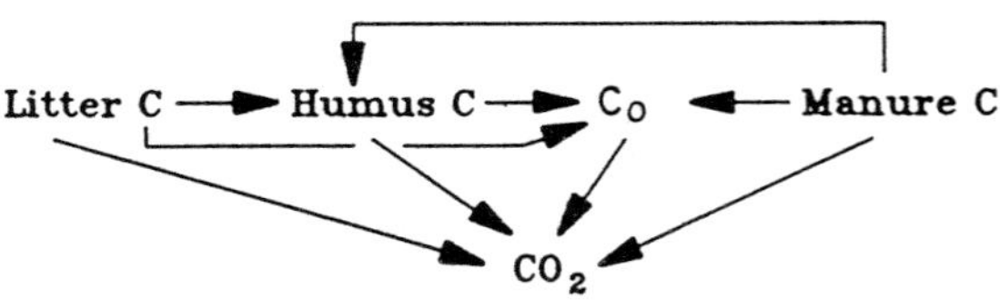

Fig. 1. Nitrogen submodel of LEACHMN [40] after modification [1], showing potentially mineralizable nitrogen (N_0) and potentially mineralizable carbon (C_0). (The CERES N submodel [37] was also modified as shown here).

eralization in two crop rotations from our long-term crop rotation study [23, 26]. We used the NO_3 leaching version of LEACHM (i.e., LEACHMN) to model N dynamics in the well-fertilized continuous spring wheat (Cont W) and fallow-spring wheat (F-W) rotations over the period 1967 to 1991 [1]. In the case of CERES, we only simulated a wet and a dry year.

The mechanisms describing mineralization and immobilization processes in CERES and LEACHMN are very similar [37, 40]. Both models simulate the decay of two types of organic matter: fresh organic matter (FOM) which includes crop residues or green manure, and a humic pool (HUM). However, in CERES, the FOM is further divided into three pools: carbohydrate, cellulose and lignin. Each pool has a different decay rate. In LEACHMN, FOM is treated as a single pool with one decay rate. Input data required for the models are the amount of straw added, its C/N ratio, its depth of incorporation into soil and the amount of root residue from the previous crop. These data are used to initialize FOM and the N contained within the fresh organic matter (FON), for each soil layer.

The mineralization routine also requires the amount of soil organic C in each layer as an input. This is used with a simplifying assumption of bulk soil C/N ratio of 10 to estimate the N contained in the humus pool. In testing CERES and LAECHMN models, our preliminary results showed that a single humus pool was inadequate. Therefore, both models were modifed by splitting the humus pool into two components: an active component (potential mineralizable-N), and a slow-N release fraction (humus-N). The characteristics of N_0 and k in the active pool were measured for this soil in a previous study (15, 16). The amount of humus-N was the total N contained in the soil less N_0. In the modified models (Fig. 1), the immobilized N and C were sent into the active N_0 pool, instead of into the FOM pool as it was done in the original models. The simulation results obtained from this modified model were much more plausible than that obtained from the original unmodified models.

Simulation with LEACHMN model

The water movement sub-routine of LEACHMN model was calibrated using data gathered in 1976 in an experiment conducted to measure Cl movement in lysimeters in the field (Campbell, unpublished data) and was validated using data from another lysimeter experiment [13]. The results obtained for water, Cl and NO_3 distribution in the soil profile throughout the growing season, under summer fallow and cropped systems, indicated a realistic representation of the system had been achieved using the modifed model [1].

We used this modified, calibrated model to simulate water and N in the rotation study. Total soil N (0.185%), assumed to be the starting value [4], was used to initialize the model for the year 1967. This was split into two fractions, an active fraction (potentially mineralizable-N) and a stable fraction (humus-N) based on the ratio of N_0/total N measured on the same soil in 1982 [16]. In that study, the ratio of N_0 to total N was 0.08, hence the initial value of total N in 1967 of 0.185 (or 1850 mg kg^{-1}) was split into 148 mg kg^{-1} as N_0 and 1702 mg kg^{-1} as humus in the 0-0.15 m layer. A similar calculation was made for the 0.15-0.30 m depth. For C, this was initialized using the value of organic C measured in 1967 and assuming the C/N ratio of the potentially mineralizable fraction was 10. The potentially mineralizable carbon (C_0) was subtracted from the total C to derive the humus C at the beginning of the experiment. It was assumed that very little N mineralization occurred below the 0.30 m depth [11].

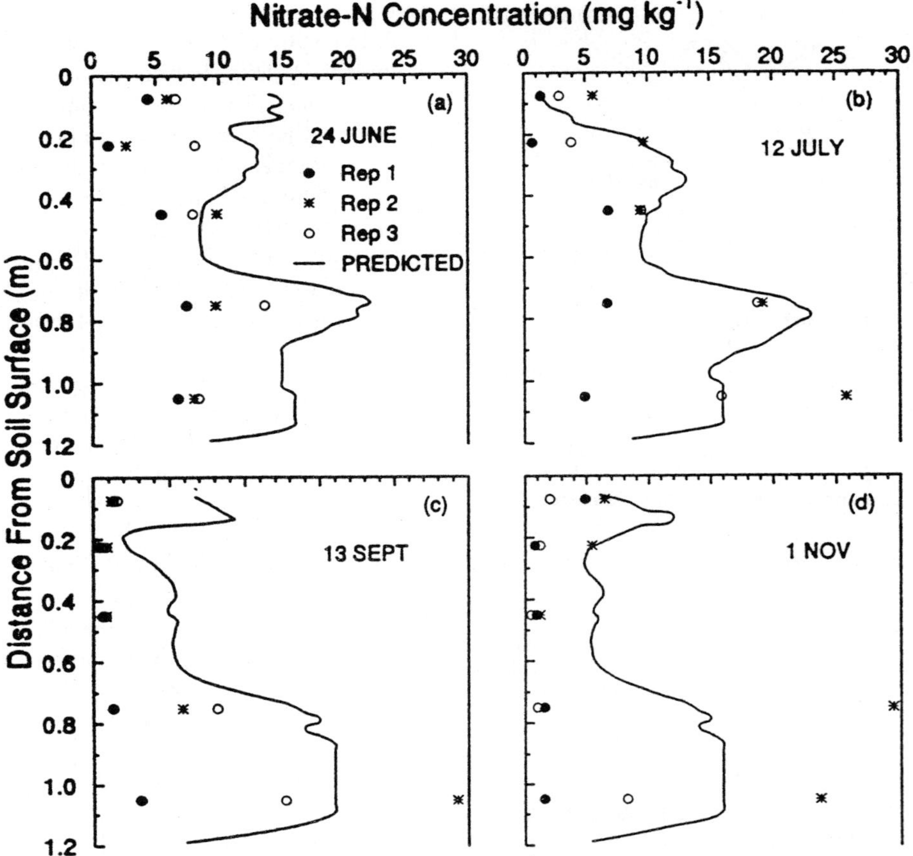

Fig. 2. Simulated soil nitrate distribution under Cont W in 1982 at Swift Current, Saskatchewan (wet year). Rainfall from 6 May to 24 June = 11 cm, 24 June to 12 July = 8.4 cm, 12 July to 13 September = 9.4 cm, and 13 September to 1 November = 6.7 cm [1].

The rate constant (k) associated with N_0 for the 0-0.15 m layer was taken as 0.013 day^{-1} [16]. The k value for the 0.15-0.30 m layer was assumed to be one-tenth of that for the surface 0.15 m [1]. The rate constant for the mineralization of humus-N (k_h) was not measured; it was obtained through calibration of the model (k_h = 0.00035 day^{-1}). It was assumed that the rate constant did not vary between growing seasons nor during a growing season.

The amount of straw-N available at the start of the experiment was estimated from straw yield of an adjacent field in 1966 [21]. Akinremi et al. [1] assumed that 50% of the 1966 straw remained at the start of the experiment in 1967 and that the straw was uniformly distributed within the plow layer (0-0.15 m) with about one-tenth located in the 0.15-0.30 m layer. The measured straw dry weights and straw N concentrations were used in the model calculations. The straw plus roots was assumed plowed into the soil and made avail-

able for decomposition a day after harvest as specified for this model. Root dry matter was estimated from the grain and straw dry matter by assuming that the root constitutes 16% of total plant weight at harvest [13]. Straw and root C concentrations were assumed to be 45% and root N concentration assumed to be 1% [12].

When the modified model was used to simulate water and N disposition in the soil in the two designated crop rotations at Swift Current, the model performed well for water (data not shown) and reasonably well for NO_3 (Fig. 2). But, as is readily seen in Fig. 2, even in this apparently uniform site, spatial variability of NO_3-N in the field can be quite large, especially in the 0.6-1.2 m depth. Therefore, striving for extreme precision in these types of analyses does not seem warranted.

We used the modified LEACHMN to estimate changes in N_0 during the growing season. Few studies have been conducted to follow changes in N_0 during the course of the growing season. In Sweden, the N mineralized in a 13-wk incubation for soils taken in April, June, August and October, decreased during the growing season but increased again after harvest [5]. The latter increase was attributed to the input of crop residues after harvest. Our simulation results for 1982 in the continuous wheat system (Fig. 3, top) followed the trend observed by Bonde and Rosswall [5]. The year 1982 was wet and this favoured N-mineralization; accordingly, the model predicted that N_0 declined gradually during the growing season until harvest (13 September), after which N_0 increased in response to the fresh residues. But, as shown by the results for the same rotation in 1978 (Fig. 3, bottom), the pattern of change in N_0 during a growing season will depend upon the amount of crop residues occurring in the previous year. Because N_0 is reduced by mineralization and increased through immobilization, the process that predominates during the growing season will dictate the trend of N_0. In years when crop residues are low, the build-up of N_0 may not compensate for its decline. In such situations N_0 will decline throughout the growing season, then increase after harvest when inputs of crop residues occur. These hypotheses remain to be tested in field studies.

The simulations predicted that, over the 24-yr period, N_0 will decline by 40% under F-W, but by only 10% under continuous wheat (Fig. 4). The rapid decline in N_0 in the first 4 yr of the experiment was due to low yields during the first 3 yr. While N_0 recovered in Cont W after the fourth year, N_0 in the F-W rotation continued to decline for about 10 yr. The higher annual production of crop residues under Cont W and the rapid decomposition of residues during the fallow phase of the two-year rotation account for this difference. These results are in accord with reports in the literature [12].

The model predicted a decline in the humus-N of about 2% under continuous wheat and 4% under F-W during the 24-yr period, with the trend being similar to that predicted for N_0 (data not shown). Although the humus fraction declined very little during the period simulated, the model suggests that the contribution of this fraction to the mineralized N was very important. In several years, mineral N produced from humus-N constituted about 50% of net mineralized N because of the magnitude of the humus fraction. Studies of N mineralization-immobilization using ^{15}N, showed that the relative contribution of various N fractions to the mineralized-N pool were: biomass 24%, metabolites 4%, active-N 32%, and stabilized-N, 40% [49].

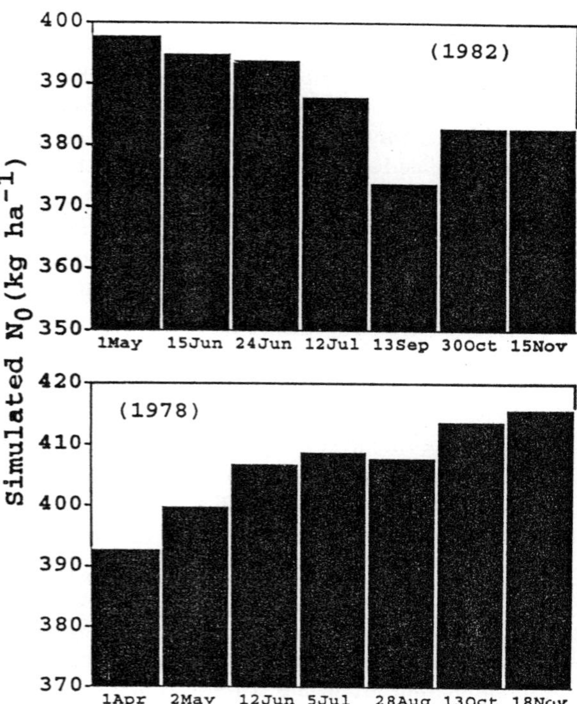

Fig. 3. Changes in simulated potential mineralizable nitrogen (N_0) in the 0–0.3 m depth, under Cont W at Swift Current, Saskatchewan, during the growing season of two dissimilar years [1].

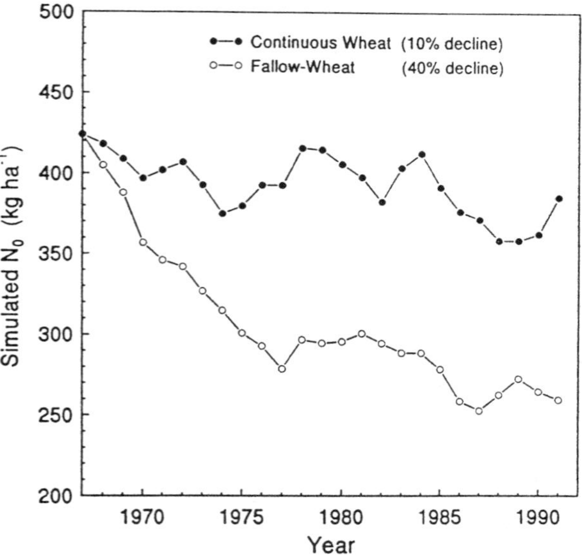

Fig. 4. Trends in simulated potentially mineralizable N (N_0) during 24 yr under F-W and Cont W rotations at Swift Current, Saskatchewan values are for 0–0.3 m depth) [1].

The first 3 pools can be assumed to be approximately equivalent to N_0 in our model. The relative contribution of humus-N to the mineral-N fraction as predicted by this modified LEACHMN model, was therefore in reasonable agreement with that reported [49].

The simulated amount of N_s for the period between spring thaw and freeze-up each year was estimated for Cont W, the fallow phase of F-W [(F)-W], and the wheat phase of F-W, [F-(W)] (Table 1). The results for F-(W) and Cont W were reasonable compared to estimates of N_s made for this soil by balance sheet methods [15, 20, 22, 24, 26]. However, values for (F)-W were lower than for F-(W). We had expected the (F)-W, with its longer period of moist soil conditions, to accumulate the most mineral N in this period. The F-(W) system should accumulate more N_s than Cont W in years when the previous fall and winter were dry such that stored water in surface soil of fallow exceeded that in surface soil of Cont W; otherwise, both of these systems should produce similar N_s. Our simulation results support the latter hypothesis. We have not yet identified the reason for the apparent inadequacy in simulating N_s in the (F)-W system. However, it could be because all the residues from the previous years' crop are being soil-incorporated at harvest making it immediately available for decomposition. (In Saskatchewan this rarely happens unless there is fall tillage.) This could result in overestimation of N immobilization in the short term. A more realistic accounting of residue decomposition, such as that used in CERES [67], might improve N simulation in LEACHMN.

Simulations with CERES model

The LEACHMN model simulates most of the major soil N transformations as well as N uptake by the crop; however, it is defcient because crop growth and yield responses are not incorporated in the model. It uses an empirical function to simulate plant N uptake. To simulate N dynamics adequately in a range of diverse cropping environments, a model capable of describing the major soil N transformations, as well as the plant component, is required. The CERES-wheat model satisfies the latter constraint. CERES also describes most of the major water and N cycles but does so less rigorously than LEACHMN [54].

Since 1987, Jame *et al.* [42] have been testing CERES at Swift Current and have calibrated and modified it for use on the Canadian Prairies. We used this modified version of CERES to estimate the net N mineralization during a wet year (1982) and a dry year (1973) for the same two rotations tested with LEACHMN.

The crop was seeded on 15 May in 1973 and on 5 June in 1982; however, we simulated periods from Spring soil sampling (3 May, 1973 and 6 May, 1982) to the date of crop physiological maturity (9 August, 1973 and 3 September, 1982). The soil water contents in the profile (0–1.2 m), measured on the date of sampling, were used as the initial values in the simulation. These water contents were much higher in 1973 (250–270 mm) than in 1982 (131–212 mm). Total precipitation received during the simulation periods were 62 mm in 1973 and 310 mm in 1982.

In the simulation with CERES, we assumed that 70% of the straw produced in the previous crop year were present at the start of the simulation run for (F)-W and Cont (W) (cropped the previous year) and 15% for F-(W) (cropped 2 yr previously). We also assumed that the straw was uniformly incorporated within the 0.1 m layer during seedbed preparation. Root dry matter was estimated as 16% of the straw weight [13]. The C concentration of straw and root was assumed to be 45%, and N concentration of the straw assumed to be 0.6% [13].

CERES requires that the soil organic C in each soil layer be specified. This is used to calculate the initial humic pool for each layer and, together with a simplifying assumption of bulk soil C/N ratio of 10, to estimate the N associated with the humic pool (NHUM). In CERES it is assumed that 20% of the gross amount of N mineralized each day is transferred from FON pool into the humic pool. As organic matter decomposes some N is required by the decay process and this amount is incorporated into microbial biomass which is included in the FON pool. The balance between the N immobilized and the N mineralized from FON and NHUM determines whether net mineralization or immobilization occurs.

When we used CERES without modifying the N sub-model, the largest amount of net N mineralized during the growing season from the three cropping systems was only 21 kg N ha^{-1} [(F)-W in 1982] and the smallest amount was 5 kg N ha^{-1} for Cont W in 1973 (Figs 5 and 6). These values are substantially lower than values generally reported for this soil [15, 20, 24, 26]. These low estimates of N_s were mainly associated with the very small k value for humus (k_h = 0.000083 day^{-1}). This value was adopted from the PAPRAN model [56]. In the SOIL-SOILN model, a

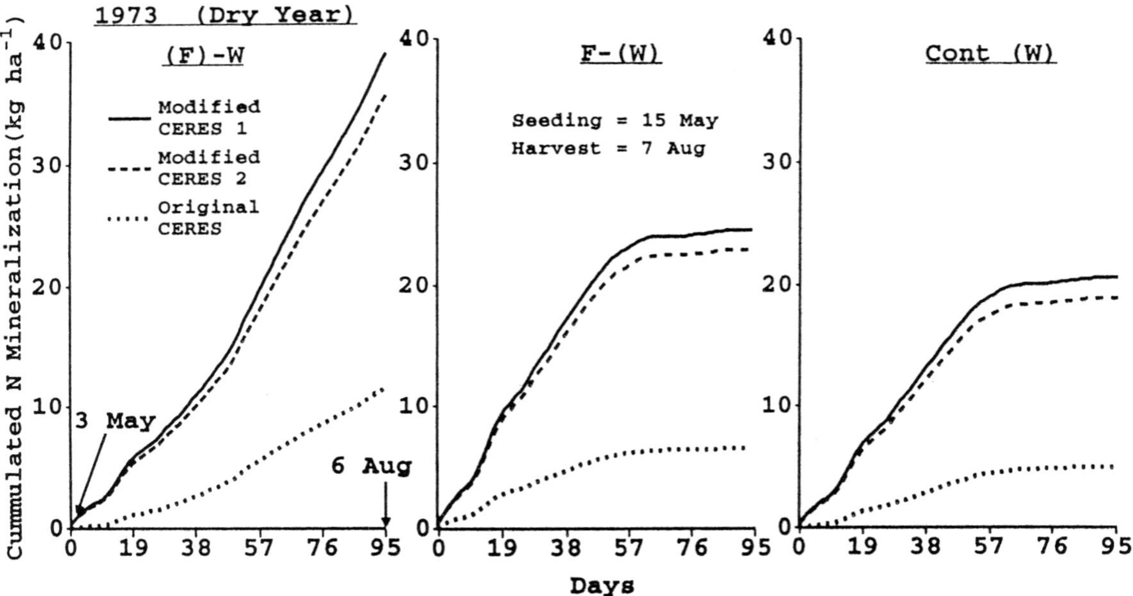

Fig. 5. Simulated N mineralized in fallow-spring wheat (F-W) and continuous wheat (Cont W), during the growing season of a dry year (1973), at Swift Current, Saskatchewan. [The initial water content in the 0–1.2 m soil depth was 248 mm in (F)-W, 262 mm in F-(W) and 270 mm in Cont W. (Modifed CERES 1 refers to case where N and humus pool used, and modified CERES 2 uses only N_0.) Precipitation received between 4 May and 9 August was 62 mm].

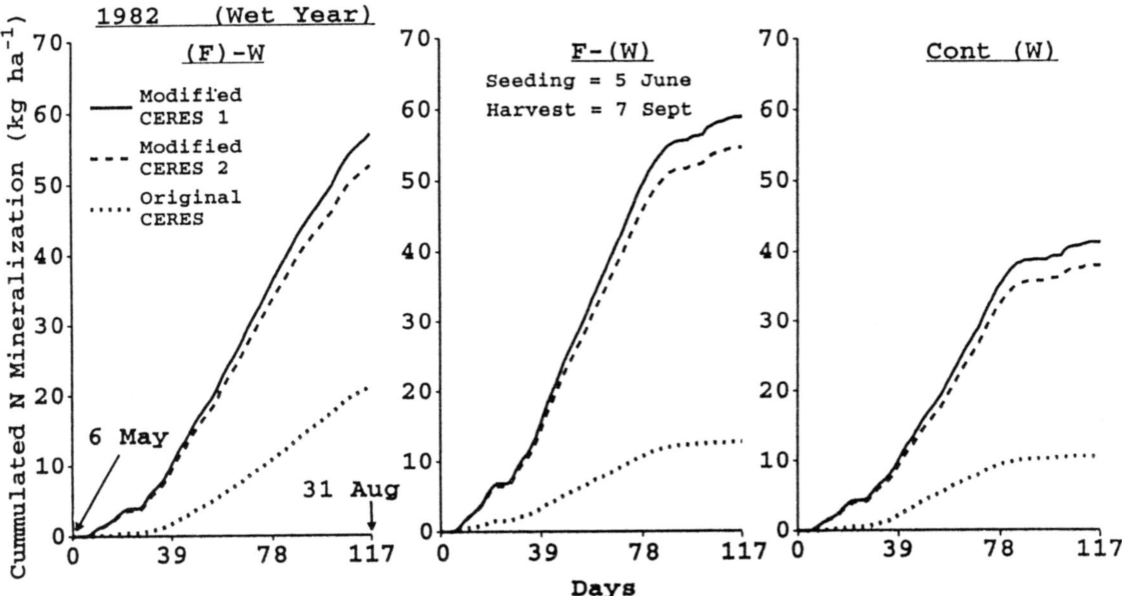

Fig. 6. Simulated N mineralized in fallow-spring wheat (F-W) and continuous wheat (Cont W), during the growing season of a wet year (1982), at Swift Current, Saskatchewan [The initial water content in the 0–1.2 m soil depth was 131 mm in (F)-W, 212 mm in F-(W) and 163 mm in Cont (W). Precipitation received between 7 May and 2 September was 309 mm. (Modified CERES 1 refers to case where N_0 and humus pools are used, and modifed CERES 2 uses only N_0].

Table 1. Net *N* mineralized (N_s) from Spring thaw to wheat harvest for fallow-wheat (both phases) and continuous wheat as estimated by simulation with the LEACHMN model [1].

Year	(F)-W[a] (kg ha^{-1})	F-(W)[a] (kg ha^{-1})	Cont W (kg ha^{-1})
1967	47	27	27
1968	26	40	18
1969	41	42	34
1970	48	55	40
1971	28	37	22
1972	33	29	23
1973	22	36	36
1974	34	41	47
1975	23	40	27
1976	26	43	23
1977	25	46	25
1978	18	36	5
1979	27	30	24
1980	26	37	30
1981	24	33	32
1982	43	48	49
1983	15	36	9
1984	21	21	5
1985	42	28	37
1986	41	37	43
1987	33	33	27
1988	20	27	30
1989	37	29	37
1990	40	32	24
1991	22	45	15
Mean	31	36	28

[a] Each phase of the fallow-wheat rotation was present each year. The values pertain to the phase in parenthesis. (F)-W means N mineralized during fallow phase; F-(W) means N mineralized during wheat phase.

greater value for the rate constant k_h (0.0003 day^{-1}) was used [3].

We modified the N sub-model of CERES as we did with LEACHMN (Fig. 1). The rate constant associated with N_0 for the surface layer (0–0.1m) was 0.013 day^{-1} as discussed previously, and k_h was 0.000083 day^{-1} as used in the original CERES model. However, the rate constant for the 0.1 to 0.2 m layer was assumed to be 0.3 of that in the 0.0 to 0.1 m layer and for the 0.2 to 0.4 m layer it was assumed to be 0.1 of the rate constant of the 0.0 to 0.1 m layer. We assumed that no N mineralization occurred below 0.40 m [11].

When this modified CERES model was used to predict N_s during the growing season, it predicted 39, 25, and 21 kg N ha^{-1} produced by (F)-W, F-(W) and Cont(W), respectively, in the dry 1973 (CERES 1, Fig. 5); LEACHMN had predicted the corresponding values of 22, 36, and 36 kg N ha^{-1} (Table 1). In the wet year (1982), the corresponding values for the modified CERES simulation were 58, 59, and 42 kg N ha^{-1} (CERES 1, Fig. 6), while LEACHMN had predicted 43, 48, and 49 kg N ha^{-1}, respectively (Table 1). The results obtained with CERES are more plausible, both in terms of the relative size of N_s for (F)-W compared to the other two systems tested, and in terms of N_s in the wet year compared to the dry year.

We re-examined our premise for revising the LEACHM and CERES N submodels (Fig. 1) and reasoned that because the measured N_0 was derived from all available pools of N (except litter, most of which is usually removed and discarded during soil preparation for incubation), then we should really only be replacing the "humus" in Fig. 1 with N_0. That is, in Fig. 1 we are in essence double-counting the humus fraction. If we wished to use the result from incubation in the manner shown in Fig. 1, then we should have partitioned the active pool of N into a more labile and a more stable N fraction, as has been done by others [8, 31].

We therefore conducted another simulation with the CERES model modified to replace humus with N_0. This simulation resulted in N_s over the growing season being about 5 kg N ha^{-1} less than in CERES 1 (CERES 2, Figs 5 and 6). This indicates that even if the revised model in Fig. 1 is invalid, the error is small because k_h in CERES is so small (0.000083 day^{-1}).

When the rate constants for biochemical constituents of crop residues used in the original CERES (Table 2) were used to estimate the amount of N mineralized, we obtained values of 8.6 kg ha^{-1} for (F)-W, 5.3 for F-(W) and 4.9 for Cont(W) in the dry 1973 growing season while for the moist 1982 the corresponding values were 18.2 kg ha^{-1} for (F)-W, 3.7 for F-(W) and 9.0 for Cont(W). Recently, Vigil *et al.* [67], working with grain sorghum, concluded that the decomposition rate constants of the residue constituents are an order of magnitude smaller than those originally used in CERES (Table 2). When the constants of Vigil *et al.* [67] were used in CERES, the values obtained for N mineralized from (F)-W, F-(W) and Cont (W) in 1973 were 6.6, 2.3, and 2.1 kg ha^{-1} respectively and corresponding values in 1982 were 13.6, 2.6, and 4.8 kg ha^{-1} respectively. Whether the lower values of Vigil et al. [67] were due to the difference in crop (Sorghum vs. wheat) is uncertain.

The validity of the N_s values obtained in the simulations is not easily verified because most balance sheet estimations are very crude, using various gross assumptions. However, the values obtained by LEACHMN and CERES were generally well within the range of N_s estimated for this soil in previous studies [15, 20, 24, 26]. The advantage of using these dynamic models, as compared to the simple model used by Campbell et al. [15], is that we are able to include N mineralized from recent crop residues, and also to account for N lost by leaching, gaseous means and N immobilized. These models require further fine-tuning, but the results we have obtained by incorporating the N_0 concept into the N submodel appear to have result-

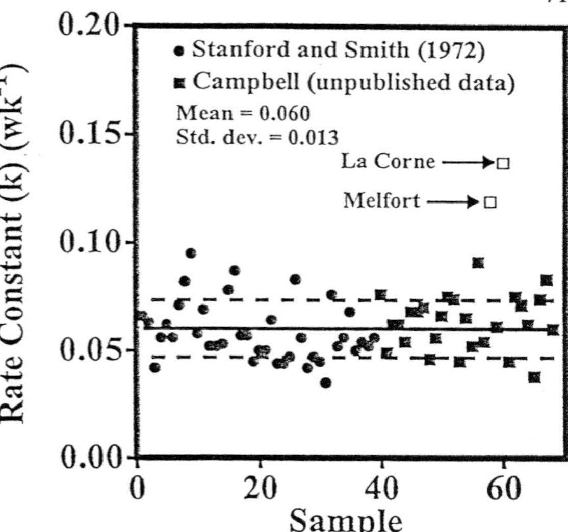

Fig. 7. Relationship between rate constant (k) as measured in a recent Saskatchewan study (C.A. Campbell, unpublished data) compared to k reported for an array of US soils [64].

ed in more credible outputs than when a balance sheet approach is used, especially in the case of CERES.

Estimating potentially mineralizable N by chemical methods for soil testing purposes

In the foregoing discussion we demonstrated that the concept of potentially mineralizable N can be used in simulation models to quantify the amount of N that a soil will mineralize during a growing season under field conditions. Of the factors required to make this estimate, the most difficult to obtain is N_0 because it differs for each soil [16, 64], and because its determination requires long-term incubation. The latter would make this technique impractical for use in soil testing laboratories. Estimating the k value is less difficult because it appears to be less variable among soils, averaging about 0.06 for cultivated land in North America (Fig. 7). In general, a Q_{10} of 2.0 is a reasonable approximation for most soils, even though if the Q_{10} for the specific soil is available, the results will be more accurate [15].

If the potentially mineralizable N approach is to be used by soil testing laboratories, a quick, effective method for estimating N_0 must be developed. Two promising methods have been proposed [35, 36]. One involves determination of the NH_4-N produced when the soil is digested with 2 M KCl at 100 °C for 4 h [35].

Table 2. Crop residue decomposition rate constants used in CERES compared to other values used by others.

Source	Carbohydrates	Cellulose	Lignin
		(day^{-1})	
Seligman and Van Keulen [56]	0.8(0.2)^a	0.05	0.0095
Vigil *et al.* [67]	0.05	0.0043	0.00095
Beek and Frissel [2]	0.075	0.005	0.00095

^{a} CERES model uses Seligman and Van Keulen's values, except for carbohydrates where 0.2 is used.

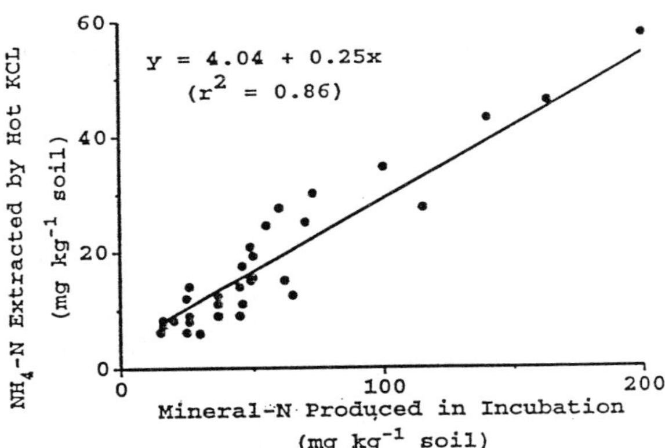

Fig. 8. Relationship between NH$_4$-N extracted with hot, 2 *M* KCl and mineral N produced by aerobic incubation of soil at 35 °C for 14 days in 33 Brazilian soils [35].

The second involves measuring NH$_4$-N produced by steam distillation of soil with pH 11.2 phosphateborate buffer solution for 8 min [36]. Both methods were tested using 33 Brazilian soils [35] and 30 diverse Iowa soils [36] with encouraging results. Recent unpublished research with 29 diverse Canadian prairie soils supported these conclusions (C.A. Campbell, unpublished data). Thus, excellent agreement was obtained [35] between the hot, KCl-extracted N and the amount of N mineralized during the first 14 days of aerobic incubation at 35 °C for the 33 Brazilian soils (Fig. 8). A close relationship was also found between the chemically extracted N and N mineralized under anaerobic incubation [36]. The latter was confirmed by others using some infertile upland soils in Thailand [41].

Although these results are promising, the relationship was not made relative to N_0 which is what is required. If a close relationship between N_0 and the NH$_4$-N extracted with hot 2*M* KCl or phosphate-borate buffer can be demonstrated, then it should be possible to determine N_0 for major soils of a region, develop the relationship between the chemically extractable N and N_0 in these soils, and store this information in the computer files of soil test laboratories for discrete soil regions. We could then determine the chemically extracted N on the soil of a farmer once every 3 to 5 years and estimate N_0 from the known chemically extractable-N vs. N_0 relationship. All the characteristics for each farmer's fields could be determined once, and stored in computer files for use in estimating fertilizer N requirements each year based on the soil test mineral N, the yield goal, and the degree of risk the farmer is willing to take (i.e., in terms of precipitation and temperature). From the grain yields of the previous year it would be possible to estimate the amount of N that would be derived from the most recent crop residues. Some account may need to be made for the timing of release of N from the latter source, especially in colder climates where very limited decomposition occurs between harvest and seeding. Crops such as Spring cereals which take up most of their N early in their growth (by heading stage) may only use a portion of the N from the previous year's residues for yield; by time the remaining N becomes available it may more likely go to increase grain protein instead of yields.

We used data from two previous studies [16, 64] to determine the relationship we might expect to find between N_0 and N mineralized during the 0 to 2-wk period of aerobic incubation. The results (Fig. 9) were very encouraging, with about 85% of the variability in N_0 being accounted for by N mineralized in 0-2 wk. Thus, it seems probable that a close relationship may also exist between N_0 and the hot KCl and/or phosphate-borate extractable NH$_4$-N; however, this remains to be determined experimentally.

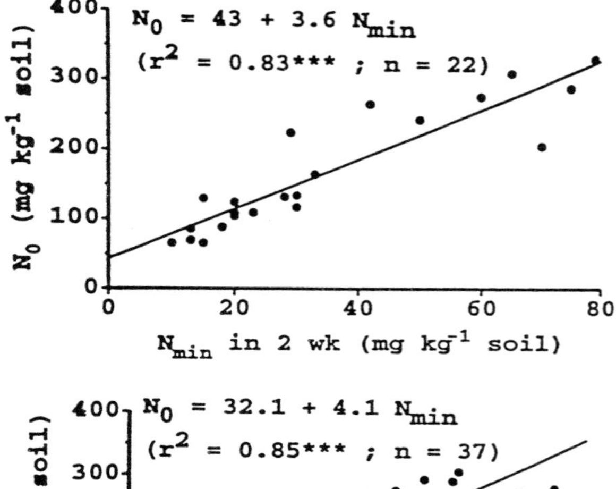
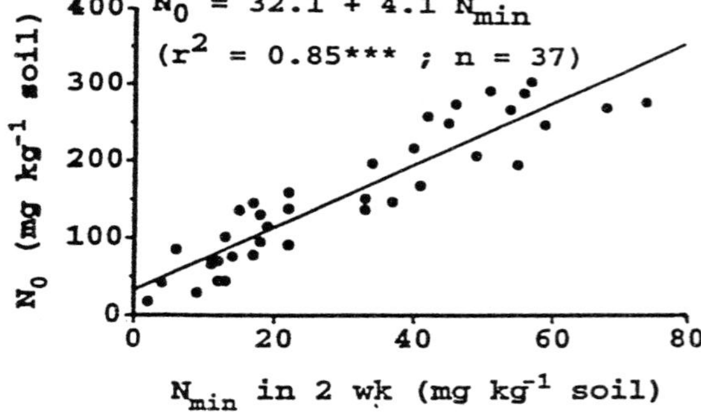

Fig. 9. Relationship between N_0 and N mineralized during the first 2 weeks of aerobic incubation at 35 °C: (top) adapted from Campbell et al. [16]; (bottom) adapted from Stanford and Smith [64].

Conclusions

The need to develop accurate methods for quantifying the rate at which soils will mineralize N cannot be overemphasized. A solution to this problem is required if farmers are to make more efficient use of N fertilizer, thereby maximizing net returns and reducing environmental pollution. Scientists have similar goals to producers but they have the added obligation of trying to improve understanding of the mechanisms that control N behaviour in the soil-plant-air system so as to facilitate our predictions of N behaviour in the environment.

This paper has demonstrated that significant progress is being made towards achievement of these goals. It demonstrates that the potentially mineralizable N concept can be used, together with deterministic models such as LEACHMN and CERES, to estimate N mineralization in field experiments. The paper further suggests ways in which it might be possible to couple the potentially mineralizable N concept with routine laboratory methods of measuring available N, such as the hot KCl or phosphate-borate extractable

N procedures proposed by Gianello and Bremner, so as to provide an improved soil test for N. If this latter approach proves fruitful, it offers the advantage over current soil tests in that it would allow modification of estimates of N mineralization to be made based on probable temperature and soil water for an area. It would also prove advantageous in humid areas where such commonly used tests as NO_3-N in soil are not feasible because of leaching.

Much research remains to be done, however, both to improve the N sub-models of the deterministic models, as well as to establish whether the routine chemical extraction procedures are effective on a universal basis, and if they can be quantitatively related to potentially mineralizable N.

References

1. Akinremi OO, Campbell CA, Jame YW, Zentner RP and Chang C (1993) Simulating nitrogen dynamics and nitrate leaching using LEACHM model. Pub No 379M0083. Research Branch, Agriculture Canada, Research Station, Swift Current, SK
2. Beek J and Frissel MJ (1973) Simulation of nitrogen behaviour in soils. PUDOC, Wageningen, The Netherlands
3. Bergstrom L and Jarvis NJ (1991) Prediction of nitrate leaching losses from arable land under different fertilization intensities using the SOIL-SOILN models. Soil Use and Management 7: 79–85
4. Biederbeck VO, Campbell CA and Zentner RP (1984) Effect of crop rotation and fertilization on some biological properties of a loam in southwestern Saskatchewan. Can J Soil Sci 64: 355–367
5. Bonde TA and Rosswall T (1987) Seasonal variation of potentially mineralizable nitrogen in four cropping systems. Soil Sci Soc Am J 51: 1508–1514
6. Cabrera ML (1986) Studies on the prediction of nitrogen mineralized from soil organic matter under field conditions. Ph.D. Dissertation, Kansas State University, Manhattan, Kansas
7. Cabrera ML and Kissel DE (1988a) Evaluation of a method to predict nitrogen mineralized from soil organic matter under field conditions. Soil Sci Soc Am J 52: 1027–1031
8. Cabrera ML and Kissel DE (1988b) Length of incubation time affects the parameter values of the double exponential model of nitrogen mineralization. Soil Sci Soc Am J 52: 1186–1187
9. Cabrera ML and Kissel DE (1988c) Potentially mineralizable nitrogen in disturbed and undisturbed soil samples. Soil Sci Soc Am J 52: 1010–1015
10. Campbell CA (1978) Soil organic carbon, nitrogen and fertility. In: Schnitzer M and Khan SU (eds) Soil Organic Matter. Developments in Soil Science Vol 8, pp 173–272. Amsterdam: Elsevier
11. Campbell CA and Biederbeck VO (1982) Changes in mineral N and numbers of bacteria and actinomycetes during two years under wheat fallow in southwestern Saskatchewan. Can J Soil Sci 62: 125–137
12. Campbell CA, Biederbeck VO, Zentner RP and Lafond GP (1991a) Effect of crop rotations and cultural practices on soil

organic matter, microbial biomass and respiration in a thin Black Chernozem. Can J Soil Sci 71: 363–376

13. Campbell CA, Cameron DR, Nicholaichuk W and Davidson HR (1977) Effect of fertilizer N and soil moisture on growth, N content and moisture use by spring wheat. Can J Soil Sci 57: 289–310

14. Campbell CA, Ellert BH and Jame YW (1993a) Nitrogen mineralization in soils. In: Carter MR (ed) Can Soc Soil Sci. Soil Sampling and Analytical Methods, pp 341–349. Boca Raton, Fl: Lewis Pub

15. Campbell CA, Jame YW and DeJong R (1988) Predicting net nitrogen mineralization over a growing season: model verification, Can J Soil Sci 68: 537–552

16. Campbell CA, Jame YW and Winkleman GE (1984) Mineralization rate constants and their use for estimating nitrogen mineralization in some Canadian prairie soils. Can J Soil Sci 64: 333–343

17. Campbell CA, Lafond GP, Leyshon AJ, Zentner RP and Janzen HH (1991b) Effect of cropping practices on the initial potential rate of N mineralization in a thin Black Chernozem. Can J Soil Sci 71: 43–53

18. Campbell CA, Lafond GP and Zentner RP (1993b) Spring wheat yield trends as influenced by fertilizer and legumes. J Prod Agric 6: 564–568

19. Campbell CA, Myers RJK and Weier KL (1981) Potentially mineralizable nitrogen, decomposition rates and their relationship to temperature for five Queensland soils. Aust J Soil Res 19: 323–332

20. Campbell CA and Paul EA (1978) Effect of fertilizer N and soil moisture on mineralization, N recovery, A-values under spring wheat grown in small lysimeters. Can J Soil Sci 58: 39–51

21. Campbell CA, Pelton WL and Nielsen KF (1969) Influence of solar radiation and soil moisture on growth and yield of Chinook wheat. Can J Plant Sci 49: 685–699

22. Campbell CA, Read DWL, Biederbeck VO and Winkleman GE (1983a) First 12 years of a long-term crop rotation study in southwestern Saskatchewan - nitrate-N distribution in soil and N uptake by the plant. Can J Soil Sci 63: 563–578

23. Campbell CA, Read DWL, Zentner RP, Leyshon AJ and Ferguson WS (1983b) First 12 years of a long-term crop rotation study in southwestern Saskatchewan - yields and quality of grain. Can J Plant Sci 63: 91–108

24. Campbell CA, Stewart DW, Nicholaichuk W and Biederbeck VO (1974) Effects of growing season soil temperature, moisture, and NH_4-N on soil nitrogen. Can J Soil Sci 54: 403–412

25. Campbell CA, Zentner RP, Dormaar JF and Voroney RP (1986) Land quality, trends and wheat production in western Canada. In: Slinkard AE and Fowler DB (eds) Wheat Production in Canada - a Review. Proc of the Canadian Wheat Production Symposium, Saskatoon, SK, pp 318–353

26. Campbell CA, Zentner RP, Selles F, Biederbeck VO and Leyshon AJ (1992) Comparative effects of grain lentil-wheat and monoculture wheat on crop production, N economy and N fertility in a Brown Chernozem. Can J Plant Sci 72: 1091–1107

27. Campbell CA, Zentner RP, Selles F, McConkey BG and Dyck FB (1993c) Nitrogen management for spring wheat grown annually on zero-tillage: yields and N use effciency. Agron J 85: 107–114

28. Cavalli JC and Rodriguez J (1975) Effect of moisture on nitrogen mineralization of nine soils of Santiago Province. Ciencia Inv Agr 2: 101–111

29. Chae YM and Tabatabai MA (1986) Mineralization of nitrogen in soils amended with organic wastes. J Environ Qual 15: 193–198

30. Dalal RC and Mayer RJ (1987) Long-term trends in fertility of soils under continuous cultivation and cereal cropping in southern Queensland. VII. Dynamics of nitrogen mineralization potentials and microbial biomass. Aust J Soil Res 25: 461–472

31. Deans JR, Molina AE and Clapp CE (1986) Models for predicting potentially mineralizable nitrogen and decomposition rate constants. Soil Sci Soc Am J 50: 323–326

32. Ellert BJ (1990) Kinetics of nitrogen and sulfur cycling in Gray Luvisol soils. Ph.D. Thesis, Dept Soil Sci Univ of Saskatchewan, Saskatoon, SK

33. Ellert BH and Bettany JR (1988) Comparison of kinetic models for describing net sulfur and nitrogen mineralization. Soil Sci Soc Am J 52: 1692–1702

34. Fyles JW and McGill WB (1987) Nitrogen mineralization in forest soil profiles from central Alberta. Can J For Res 17: 242–249

35. Gianello C and Bremner JM (1986a) A simple chemical method of assessing potentially available organic nitrogen in soil. Commun Soil Sci Plant Anal 17: 195–214

36. Gianello C and Bremner JM (1986b) Comparison of chemical methods of assessing potentially mineralizable organic nitrogen in soil. Commun Soil Sci Plant Anal 17: 215–236

37. Godwin DC and Jones CA (1991) Nitrogen dynamics in soil-plant systems In: Hanks J and Ritchie JT (eds) Modeling Plant and Soil Systems. Agronomy 31, pp 287–321 Madison, Wisconsin

38. Griffin GF and Laine AF (1983) Nitrogen mineralization in soils previously amended with organic wastes. Agron J 75: 124–129

39. Houot S, Molina JAE, Chaussod R and Clapp CE (1989) Simulation by NCSOIL of net mineralization in soils from the Deherain and 36 Parcelles Fields at Grignon. Soil Sci Soc Am J 53: 451–455

40. Hutson JL and Wagenet RJ (1991) Simulating nitrogen dynamics in soils using a deterministic model. Soil Use and Management 7: 74–78

41. Ila'ava VP and Waring S (1992) Comparison of two chemical methods of assessing potentially available organic nitrogen in soils. Commun Soil Sci Plant Anal 23: 157–164

42. Jame YW, Liick J and Thick WA (1993) A User's Guide to the Modified CERES-Wheat Model for Use on the Canadian Prairies. Research Branch, Agriculture Canada. Swift Current, SK

43. Juma NG, Paul EA and Mary B (1984) Kinetic analysis of net nitrogen mineralization in soil. Soil Sci Soc Am J 48: 753–757

44. Marion GM, Kummerow J and Miller PC (1981) Predicting nitrogen mineralization in chaparral soils. Soil Sci Soc Am J 45: 956–961

45. Mason MG and Rowland IC (1992) Effect of amount and quality of previous crop residues on the nitrogen fertilizer response of a wheat crop. Aust J Exp Agric 32: 363–370

46. Myers RJK, Campbell CA and Weier KL (1982) Quantitative relationship between net nitrogen mineralization and moisture content of soils. Can J Soil Sci 62: 111-124

47. Olness A (1984) Re: nitrogen mineralization potentials, N_0 and correlations with maize response. Agron J 76: 171–172

48. Oyanedel C and J Rodriguez S (1977) Estimation of N mineralization in soils. Ciencia Inv Agr 4: 33–44

49. Paul EA and Juma NG (1981) Mineralization and immobilization of soil nitrogen by microorganisms. In: Clark FE and Ross-

wall T (eds) Terrestrial Nitrogen Cycles. Processes, Ecosystem Strategies and Management Impacts. Ecol Bull (Stockholm) 33: 179–194

50. Paustian K and Bonde TA (1987) Interpreting incubation data on nitrogen mineralization from soil organic matter. In: Cooley JH (ed) Soil Organic Matter Dynamics and Soil Productivity. Proceedings from an INTECOL Workshop. INTECOL Bull. 15, International Association for Ecology, Athens, GA

51. Power JF, Doran JW and Wilhelm WW (1986) Uptake of nitrogen from soil, fertiliser and crop residues by no-till corn and soybean. Soil Sci Soc Am J 50: 137–142

52. Prado O and Rodríguez J (1978) Estimation of nitrogen fertilizer requirements of wheat. Ciencia Inv Agrr 5: 29–40

53. Rennie DA, Campbell CA and Roberts TL (1993) Impact of macronutrients on crop responses and environmental sustainability on the Canadian prairies - a review. Ottawa, ON: Can Soc Soil Sci

54. Ritchie JT (1985) A user-orientated model of the soil water balance in wheat. In: Day W and Atkin RK (eds) Wheat Growth and Modelling, New York: Plenum Press

55. Saito M and Ishii K (1987) Estimation of soil nitrogen mineralization in corn-grown fields based on mineralization parameters. Soil Sci Plant Nutr 33: 555–566

56. Seligman NC and van Keulen H (1981) PAPRAN. A simulation model of annual pasture production limited by rainfall and nitrogen. In: Frissel MJ and van Veen JA (eds) Simulation of Nitrogen Behaviour of Soil-Plant Systems, pp 192–221. PUDOC, Wageningen, the Netherlands

57. Smith SJ, Young LB and Miller GE (1977) Evaluation of soil nitrogen mineralization potentials under modifed field conditions. Soil Sci Soc Am J 42: 74–76

58. Stanford G (1973) Rationale for optimum nitrogen fertilization in corn production. J Environ Qual 2: 159–166

59. Stanford G (1982) Assessment of soil nitrogen availability. In: Stevenson FJ (ed) Nitrogen in Agricultural Soils. Agron 22, pp 651–688. ASA, CSSA, SSSA, Madison, WI

60. Stanford G, Carter JN, Westermann DT and Meisinger JJ (1977) Residual nitrate and mineralizable soil nitrogen in relation to nitrogen uptake by irrigated sugarbeets. Agron J 69: 303–308

61. Stanford G and Epstein E (1974) Nitrogen mineralization-water relations in soils. Soil Sci Soc Am Proc 38: 103–107

62. Stanford G, Frere MH and Schwaninger DH (1973a) Temperature coefficient of soil nitrogen mineralization. Soil Sci 115: 321–323

63. Stanford G, Legg JO and Smith SJ (1973b) Soil nitrogen availability evaluations based on N mineralization potentials of soils and uptake of labelled and unlabelled nitrogen by plants. Plant and Soil 39: 113–124

64. Stanford G and Smith SJ (1972) Nitrogen mineralization potentials of soils. Soil Sci Soc Am Proc 36: 465–472

65. Tietje O and Tapkenhinrichs M (1993) Evaluation of pedotransfer functions. Soil Sci Soc Am J 57: 1088–1095

66. Verstraete W and Voets JP (1976) Nitrogen mineralization tests and potentials in relation to soil management. Pedologie 26: 15–26

67. Vigil ME, Kissel DE and Smith SJ (1991) Field crop recovery and modeling of nitrogen mineralized from labeled sorghum residues. Soil Sci Soc Am J 55: 1031–1037

Fertilizer Research **42**: 77–87, 1995.
© 1995 *Kluwer Academic Publishers. Printed in the Netherlands.*

Field application of the ^{15}N isotope dilution technique for the reliable quantification of plant-associated biological nitrogen fixation

Robert M. Boddey, Octávio C. de Oliveira, Bruno J.R. Alves & Segundo Urquiaga
EMBRAPA - Centro Nacional de Pesquisa de Agrobiologia, Km 47, Seropédica, Itaguaí, 23851-970, Rio de Janeiro, Brazil

Key words: biological nitrogen fixation, control crops, isotope dilution technique, legumes, ^{15}N slow-release fertilizer

Abstract

To apply the isotope dilution (ID) technique, it is necessary to grow the "N$_2$-fixing" crop in a soil where the mineral N is labelled with ^{15}N. Normally the "N$_2$-fixing" crop and a suitable non-N$_2$-fixing control crop are grown in the same labelled soil and the ^{15}N enrichment of the control crop is assumed to be equal to the ^{15}N enrichment of the nitrogen (N) derived from the soil in the "N$_2$-fixing" crop. In this case the proportion of unlabelled N being derived from the air via biological N$_2$ fixation (BNF) in the "N$_2$-fixing"crop will be proportional to the dilution of the enrichment of the N derived from the labelled soil.

To label the soil, the technique most often used is to add a single addition of ^{15}N-labelled N fertilizer shortly before, at, or shortly after, the planting of the crops. Data in the literature clearly show that this technique results in a rapid fall in the ^{15}N enrichment of soil mineral N with time. Under these conditions, if the control and the "N$_2$-fixing" crops have different patterns of N uptake from the soil they will inevitably obtain different ^{15}N enrichments in the soil-derived N. In this case the isotope dilution technique cannot be applied, or if it is, there will be an error introduced into, the estimate of the contribution of N derived from BNF.

Several experiments are described which explore different strategies of application of the ID technique to attempt to attenuate the errors involved. The results suggest that it is wise to use slow-release forms of labelled N, or in some cases, multiple additions, to diminish temporal changes in the ^{15}N enrichment of soil mineral N. The use of several control crops produces a range of different estimates of the BNF contributions to the "N$_2$-fixing" crops, and the extent of this range gives a measure of the accuracy of the estimates. Likewise the use of more than one ^{15}N enrichment technique in the same experiment will also give a range of estimates which can be treated similarly. The potential of other techniques, such as sequential harvesting of both control and test crops, are also discussed.

Introduction

Of the techniques available to quantify the contribution of biological N$_2$ fixation (BNF) to nodulated legumes and other "N$_2$-fixing" plants, only techniques which utilize the stable isotope ^{15}N can provide direct estimates of the quantity of biologically-fixed N incorporated into plant tissue. The use of ^{15}N-labelled N$_2$ gas is usually only feasible for short-term experiments under controlled conditions, but the ^{15}N isotope dilution (ID) technique can be applied to field studies over the whole plant growth cycle.

To apply the ID technique it is necessary to grow the "N$_2$-fixing" legume in a soil where the mineral N is labelled with ^{15}N. If the ^{15}N enrichment of the N being absorbed from the soil by this plant is known, then the amount of unlabelled N being derived from the air via BNF will be proportional to the dilution of the enrichment of the N derived from the labelled soil.

Normally the "N$_2$-fixing" crop and a suitable non-N$_2$-fixing control crop are grown in the same labelled soil and the ^{15}N enrichment of the control crop is regarded as the ^{15}N enrichment of the N in the legume derived from the soil. In fact, if the ^{15}N enrichment of the labelled soil N can be determined by direct analysis of the ^{15}N enrichment of the soil mineral N [12,19,38], then the use of control crops can be dispensed with.

In the case where a non-N_2-fixing control crop is utilized, the basic assumption made in applying the technique is that the ^{15}N enrichment of the N derived from the soil by the "N_2- fixing" crop is equal to the ^{15}N enrichment of the N in the control crop. Alternatively this can be expressed as: The ratio (R) of labelled fertilizer N to unlabelled soil N accumulated by the plants is the same for the "N_2-fixing" crop and the control crop.

Labelling the soil N with ^{15}N

Addition of soluble labelled N fertilizer

The technique most often used is to add a single addition of soluble ^{15}N-labelled N fertilizer (eg. ammonium sulphate or urea) to the surface of the soil shortly before, at, or shortly after, the planting of the crops. This form of ^{15}N addition results in a rapid fall in the ^{15}N enrichment of soil mineral N, as the ^{15}N-labelled N is added to the soil mineral N pool which is continuously being replenished by unlabelled N from the mineralization of soil organic matter [17,38,49]. This is illustrated by data from a recent field experiment where ^{15}N-labelled ammonium sulphate (10 kg N ha^{-1}) was added 1 day after planting of *Phaseolus vulgaris* (Fig. 1 - Boddey *et al.*, unpubl. data). Under these conditions if the control and the legume crops have different patterns of uptake of N from the soil they will inevitably obtain different ^{15}N enrichments in the soil-derived N. As is evident from the basic assumption of the technique this mean that the isotope dilution technique cannot be applied, or if it is, there will be an error introduced into the estimate of the contribution of N derived from BNF [3,11,14,35,45,50,51,52].

The obvious solution to this problem would appear to be to select a non-N_2-fixing control crop which has the same soil-N uptake pattern as the "N_2-fixing" crop. While the uptake of soil N by a non-N_2-fixing crop can be studied by sequential harvests, this is not possible in the case of the "N_2-fixing" crop as there is no way to distinguish between unlabelled N derived from soil and that derived from BNF.

Ledgard *et al.* [28] developed a technique to compare the ratio (R) of fertilizer N to soil N in "N_2-fixing" and control crops which depended on the use of increasing additions of labelled N and the measurement of the natural ^{15}N abundance in both crops in a treatment where no N fertilizer was added. The technique requires great care to be taken to avoid contamination

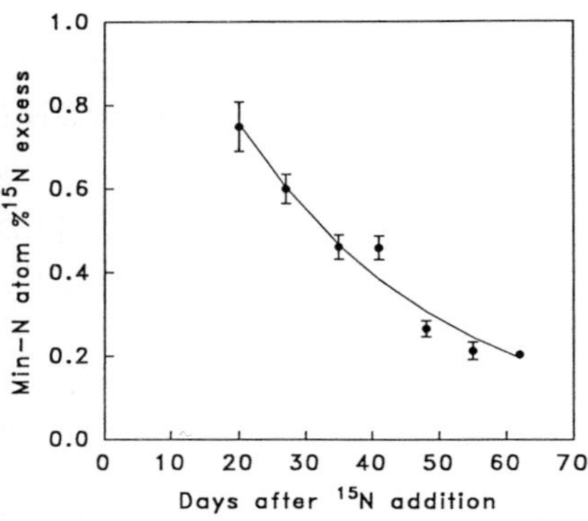

Fig. 1. Decline in ^{15}N enrichment of soil mineral N (KCl extract) in soil amended with 10 kg N ha^{-1} of ^{15}N-labelled ammonium sulphate (10.1 atom % ^{15}N). Vertical bars indicate standard errors of means (20 replicates). Boddey *et al.*, (unpubl. data)

of the unfertilized plots with enriched N and the use of a sensitive (double inlet) mass spectrometer. While the technique can theoretically be used to select appropriate control crops for any particular legume crop (any control crop which attained a ^{15}N enrichment similar to the enrichment of the N derived from the soil by the legume), the extra work involved and its low sensitivity in soils with low natural ^{15}N abundance [3], has not encouraged its application by other workers.

A further technique was explored by Wagner and Zapata [46] based on the interesting idea that if the ratio of added labelled to native unlabelled sulphur taken up from the soil by the two crops is equal, then the ratio of labelled to unlabelled N should also be equal. As both S and N are controlled in the soil by similar microbiological processes the idea was logical. However, in two experiments designed to test this technique, Hamilton *et al.* [20,21] found that there was no fixed relationship between the ratios of labelled to unlabelled S and labelled and unlabelled N among the different control crops and legume crops and thus the ratio of labelled to unlabelled S could not be used as evidence for equal (or unequal) ratios of labelled to unlabelled N derived from soil sources by the different crops.

Rennie and Thomas [36] suggested that it was possible to test if the ^{15}N enrichment of the soil N taken up by the "N_2-fixing" and control was equal, by calculating the soil 'A-value' for each of the crops from the % Ndff and % Ndfs (%N derived from fertilizer and soil, respectively). They quite correctly stated that if the 'A_s-value' was equal for the "N_2- fixing" and control crops then the two crops did remove N from the soil with the same ^{15}N enrichment. However, to calculate the %Ndfs in the "N$_2$-fixing" crop it is necessary to apply the ^{15}N ID technique and assume that the basic assumption of equality of (R) for both crops holds [7]. As the technique is based on circular logic, it follows that the ratio R will always be the equal for all crops tested and hence the "test" is no test at all.

It is apparent that testing the basic assumption of the ID technique is extremely difficult and the question arises of the magnitude of possible errors involved in relying on this assumption without testing it. While most investigators appear to be conscious of the risks involved, most studies, even recent ones, generally use a single addition of ^{15}N labelled fertilizer and a single harvest of the "N$_2$-fixing" and control crops (the single ^{15}N addition/single harvest technique) without testing whether soil N uptake patterns of the two crops were the same.

Magnitude of errors involved in the use of the "single ^{15}N addition single harvest technique"

As it is difficult to discover the ^{15}N enrichment of soil-derived N in a "N$_2$-fixing" crop, how can it be determined whether there are significant differences between this enrichment and that derived from the soil by the control crop? The answer is by inference: it is logical to assume that if there are large differences in ^{15}N enrichment between different control crops, then it is possible that there will be large differences between the ^{15}N enrichment of the control crop and that of the soil-derived N in the legume.

Only a small proportion of the ID studies to quantify BNF contributions to legumes have used more than one control crop. However, in almost all of them where the soil was amended with a single dose of soluble labelled fertilizer, significant differences between the ^{15}N enrichments of different control crops were recorded [15,19,21,46,50]. This is illustrated by data from a recent field experiment conducted at our Centre, where 4 non-N$_2$-fixing control crops (non-nod soybean, rice, okra and sorghum) were used in an experiment to quantify the BNF contribution to soybean inoculated with two different *Bradyrhizobium japonicum* strains (Boddey *et al.*, unpubl.).

Figure 2 shows the contrasting uptake patterns of labelled and unlabelled N by the non-nod soybean and the sorghum which resulted at the final harvest (73 days after planting - DAP) of ^{15}N enrichments of 0.1491 and 0.1012 atom % ^{15}N excess in the two controls, respectively. The estimates of the BNF contributions

Table 1. Estimates the contribution of biologically fixed nitrogen to nodulating soybean inoculated with either 29W or CB 1809 strains of *Bradyrhizobium japonicum* using 4 different non-N$_2$-fixing control crops

Controls	%dfa[b] of nodulating soybean	
	29W	CB 1809
Non nod soybean	47.2	58.3
Sorghum	15.7	32.5
Rice	50.2	59.7
Okra	13.7	31.5
LSD (Student)	30.6	25.0

[a] Soil labelled at planting with a single addition of ^{15}N-labelled ammonium sulphate. Data are means of 4 replicates. Boddey *et al.* (unpubl. data).
[b] % N derived from BNF.

(% N derived from BNF - %Ndfa) to the nodulated soybean ranged from 13.7 to 50.2%, and 31.5 to 59.7%, for the soybean inoculated with the strains 29W and CB 1809 of *B. japonicum*, respectively, depending on which control was used (Table 1).

Occasionally differences in N uptake patterns between legume and control crops are so large that application of the ID calculations yields clearly erroneous negative estimates of contributions of BNF [29,33,37,51].

As has been pointed out by several authors when the %Ndfa in the "N$_2$-fixing" crop is high, the estimate of %Ndfa is relatively insensitive to mismatching of soil N uptake patterns of the different crops [7,22,23]. This is well illustrated. by a simulation of the influence of a 5, 10 or 20% variation in ^{15}N enrichments of different control crops on the estimates of Ndfa of the "N$_2$-fixing" crop (Fig. 3).

It follows that if it is thought that %Ndfa is low, or could be low, then the application of the ID technique using a single dose of ^{15}N labelled fertilizer at planting with a single harvest at crop maturity, will not yield a reliable estimate of the BNF contribution to the "N$_2$-fixing" crop.

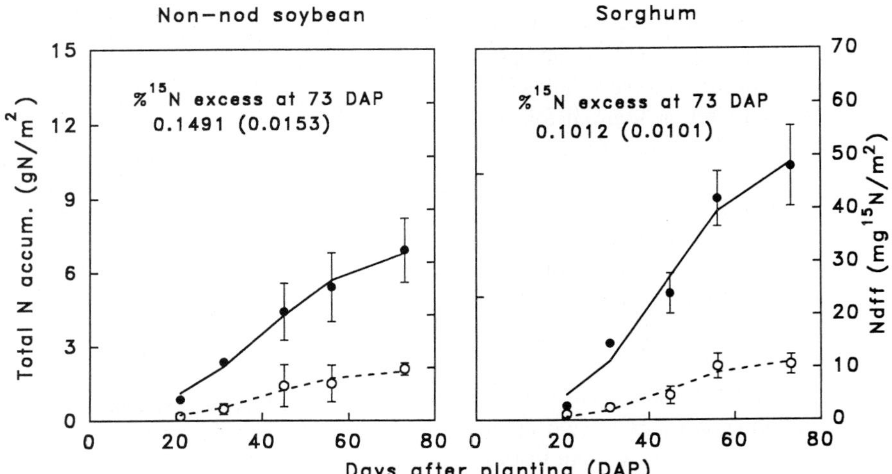

Fig. 2. Accumulation of labelled (- - - 0 - - -) and total N (——•——) by 2 non-N$_2$-fixing control plants (non-nod soybean and sorghum) in soil amended with 2.0 kg N ha^{-1} of ^{15}N-labelled ammonium sulphate (21.3 atom % ^{15}N) at planting. Vertical bars indicate standard errors of means (4 replicates). Boddey *et al.,* (unpubl. data).

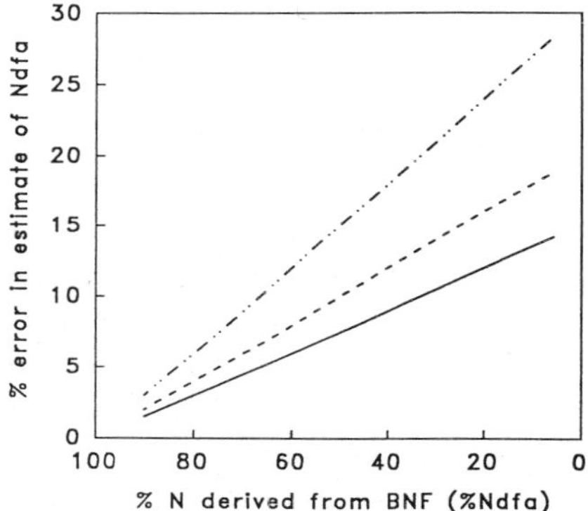

Fig. 3. Simulation of the influence of 5(————), 10 (- - -) or 20% (----------) variation in the ^{15}N enrichment of different control crops on the estimates of N derived from BNF, at decreasing proportions of N derived from BNF (%Ndfa). For the simulation it was assumed that there was a 10% variability in the ^{15}N enrichment of the "N$_2$-fixing" crop.

Strategies to apply the ID technique to provide reliable estimates of BNF contributions

Labelling the soil to diminish temporal variations of ^{15}N enrichment of soil mineral N

From the above discussion it is apparent that when the soil is labelled with a single dose of ^{15}N-enriched fertilizer at planting, different crops can obtain very different ^{15}N enrichments in the N they accumulate from the soil. An obvious solution to this problem is to attempt to label the soil in such a manner that temporal variations in ^{15}N enrichment of soil mineral N are diminished. Possible techniques were discussed by Chalk [11] and Boddey [3] although just about all techniques have their disadvantages:

a. Pelleting ^{15}N-labelled ammonium sulphate with gypsum was used by Witty [49] and later by others [7,18], but these latter authors showed that it was not very effective in slowing down ^{15}N release.

b. Immobilizing the mineral N from these fertilizers by adding sugars, cellulose, sawdust or straw has also been tried [9,18,30,33,39,40]. The favourable effect of immobilizing the labelled fertilizer was shown in an experiment where 3 control crops were utilized to quantify the BNF contribution to field grown soybean in an experiment performed in the field near Brasília [4]. The data show that in the treatment where sugar was mixed with the ^{15}N-labelled ammonium sulphate fertilizer in a C:N ratio of 10:1, rates of decline of ^{15}N enrichment of the plants with time were considerably reduced (Fig. 4). In the treatment where the labelled fertilizer was immobilized with sugar, the differences between the ^{15}N enrichments of the different control crops were considerably less than when no sugar was added (Table 2).

However, even in cases where such techniques were used to immobilize the added labelled N, significant differences in ^{15}N enrichment between different control crops have been reported [7,46,49].

Table 2. [15]N enrichment and total N accumulation of nodulated soybean and 3 non-N_2-fixing control crops grown in soil amended with ^{15}N-labelled ammonium sulphate at planting, with or without the addition of sugar at a C:N ratio of 10:1[a]

Crop	Total N accumulation (g m^{-2})		[15]N enrichment (Atom % [15]N excess)	
	-Sugar	+Sugar	-Sugar	+Sugar
Nodulated soybean	17.71	20.36	0.3616	0.3238
Non-nod soybean	15.81	15.96	0.4314	0.3770
Sorghum	6.63	6.18	0.4272	0.3253
Sunflower	16.70	20.93	0.6217	0.4778
LSD (Student)	7.36	11.60	0.1362	0.1481

[a]After Boddey and Urquiaga [4].

The other disadvantage is that adding carbon sources is likely to reduce the availability of soil N to the plants thus affecting (probably positively) the amount of N derived from BNF by the legume. This does not seem to have been the case in the study described above [4] where total N yield of the nodulated soybean and control crops was not significantly affected by the addition of sugar (Table 2).

c. Additions of ^{15}N labelled plant material or compost. This has also been used quite frequently, especially by our group in Rio [5,6,7,24,41]. Again initially ^{15}N enrichment of soil N declines quite quickly, although less quickly than if soluble labelled fertilizer is added. This treatment also immobilises soil N (see b. above) but in our (tropical) conditions after about 18 months soil ^{15}N enrichment becomes virtually stable.

d. Multiple small additions of labelled soluble N fertilizer. This technique has been favoured by many authors [5,6,16,29,42,43], but has been criticized by Rennie [34] on the basis that if test and control crops assimilate different quantities of mineral N between harvests then the residual mineral N in the soil at harvest will be different for the two crops. A further addition of labelled N to this pool will then cause a different initial enrichment in the mineral N for the two crops at the start of the second growth period. However, if the quantity of labelled N added is small compared with plant uptake then it is unlikely that significant amounts of residual mineral N are available at the time of harvest, which as Rennie admits, invalidates this criticism. The evidence suggests that this technique is more effective in reducing temporal changes in ^{15}N enrichment of soil mineral N than the single addition technique, and has the advantage that the ^{15}N enrichment of the soil mineral N is still high towards the end of the crop growth cycle when BNF inputs are usually largest. If small amounts are used this does not significantly alter soil mineral N availability, unlike techniques b. and c. above, so that it is a good technique to use if the objective of the study is to estimate BNF in an actual farming system. The disadvantage comes in the labour involved in the additions and making sure that no significant amounts of ^{15}N salts are absorbed by the leaves of the developing or mature plants. This latter problem can be attenuated by applying the ^{15}N material dried onto sand particles which can be brushed off the leaves and then subsequently washed into the soil by irrigation [2,10].

Use of several control crops

Recently we suggested that a good strategy to apply the ID technique is to use several controls in conjunction with a technique to reduce temporal changes in ^{15}N enrichment of soil mineral N [7]. None of the above techniques will produce a completely stable and uniform ^{15}N enrichment of soil mineral N unless the fertilizer is mixed throughout the whole volume of the soil and then left to equlibriate for many months [41]. As for normal field experiments this is not practical, the idea behind this strategy is that if there are still considerable changes in soil mineral N enrichment with time then each different control crop can give an independent estimate of the BNF contribution to the legume. If the range of these estimates is small, then it is probable that the estimate is fairly accurate, and the range of the different estimates gives an idea of the accuracy. The data in our paper illustrates the use of this approach.

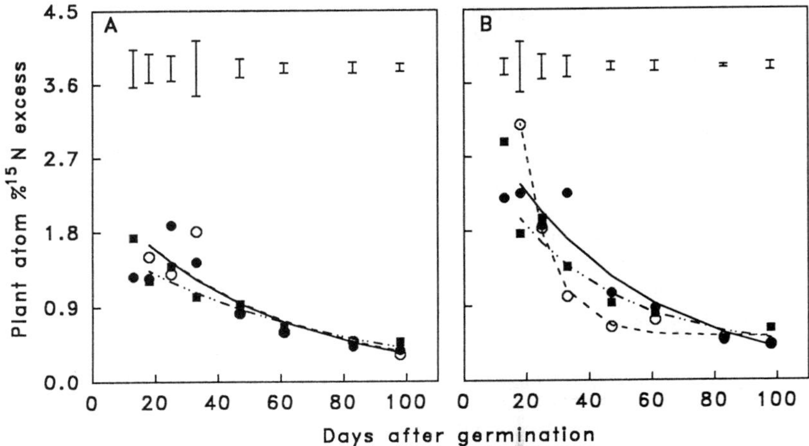

Fig. 4. Decline in ^{15}N enrichment of 3 non-N$_2$-fixing control plants grown in soil amended with ^{15}N-labelled ammonium sulphate with (A) or without (B) the addition of sugar in a C:N ratio of 10:1. ——•—— non-nod soybean, - - 0 - - - sorghum, ---■---- sunflower. Vertical bars indicate standard errors of means (4 replicates). After Boddey and Urquiaga [4].

Naturally more work and analyses are involved but a much more reliable estimate of the BNF contribution to the legume is obtained.

Use of two different methods of applying [15]*N*

This technique was suggested by Hamilton *et al.* [20] and its use demonstrated in a recent paper of Viera-Vargas *et al.* [44]. It is recommended that contrasting techniques of application of ^{15}N to the soil are utilized. By contrasting it is meant that they cause different types of changes in ^{15}N enrichment in the soil mineral N with time. Probably this means that either a single addition of soluble labelled fertilizer, or a slow- release form (options b. or c. above) should be used together with the multiple additions technique. The former techniques cause a decrease in ^{15}N enrichment of soil mineral N with time, and the latter can cause an overall increase of enrichment of this N with time. It is preferable also to use several control crops with this technique, although perhaps just 2 may be sufficient instead of 3 or so for the multiple controls technique described above (section: *Use of several control crops*). If the legume and one of the control crops have identical patterns of soil N uptake in the two ^{15}N labelling treatments then the estimates of N derived from BNF by the legume will also be identical. It follows that the control crop which gives the closest agreement in BNF estimate in the two contrasting ^{15}N labelling treatments is that which is giving the estimate closest to the true contribution.

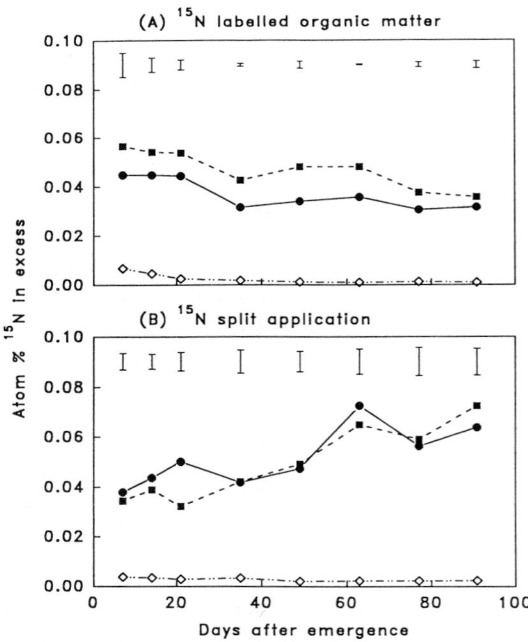

Fig. 5. [15]N enrichment of the forage legume *Centrosema* (hybrid Itaguaí ----◊----) and the non-N$_2$-fixing control crops, *Panicum maximum* (cv IR 442 ——•——) and *Sorghum bicolor* (cv BR 301 - - -■- - -), grown in a concrete tank filled with Itaguaí series soil (Typic Hapludulf) amended with either (A) ^{15}N-labelled organic matter or, (B) with unlabelled organic matter and split applications of ^{15}N-labelled ammonium sulphate. Vertical bars indicate standard errors of the means (4 replicates). After Viera Vargas *et al.* [44].

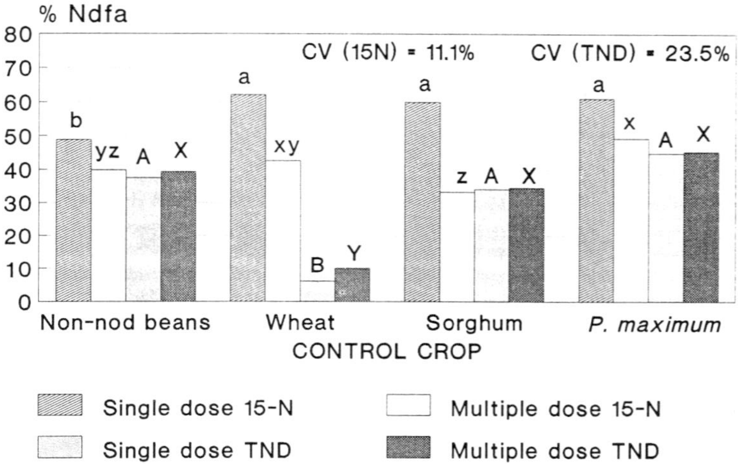

Fig. 6. Estimates of the proportion of plant nitrogen derived from N_2 fixation (% Ndfa) by nodulated common beans (*Phaseolus vulgaris* cv. Negro Argel) using 4 different non-N_2-fixing control crops and either the ^{15}N isotope dilution (ID) or total N difference (TND) techniques. ID estimate of % Ndfa = 100. (1 - (Atom % ^{15}N excess in beans/Atom % ^{15}N excess of control)). TND estimate of % Ndfa = 100. (Total N of beans - Total N of control)/(Total N of beans). CV = Coefficient of variation. Different capital or lower case letters above bars indicate significant differences between means (4 replicates) at $p = 0.05$ (Tukey). After Viera Vargas *et al.* [44].

Typical results showing the contrasting changes in plant ^{15}N enrichment are shown in Figure 5, and when it was applied to quantify the BNF contribution to nodulated *Phaseolus vulgaris* grown in pots, it was found that the non-nodulating mutant of *P. vulgaris* gave similar estimates of N derived from BNF regardless of the labelling technique used (Fig. 6). This technique probably requires even more work, more experimental units and more ^{15}N analyses than the multiple control methods (section: *Use of several control crops*), but it may be worthwhile in some circumstances.

Time course measurements of crop ^{15}N enrichment

This technique can be applied with just one control crop and with a single addition of soluble labelled fertilizer at planting, but better quality results will almost certainly be obtained if a slow-release form of N fertilizer is used and perhaps even more than one control. In this approach several (probably at least 5) complete harvests of each crop (legume + controls) are taken during the season. The harvests should be spaced so that more of them are taken during the time of maximum N_2 fixation by the legume, if this can be predicted. At each complete harvest the material is analysed for total N and ^{15}N enrichment and from these data the curve of the recovery of labelled fertilizer (or excess ^{15}N) can be plotted for each crop. If the curve of the recovery of labelled fertilizer is very similar for the control crop

(or one of the control crops) then the estimate of BNF in the legume derived from this control crop should be closest to the true BNF contribution. The equations of growth curve of Hamilton *et al.* [19] can be used to fit the accumulation of labelled fertilizer by the different crops:

$$T\ ^{15}N - c.\ \exp\ (Nu \cdot t)$$

where T ^{15}N is the ^{15}N accumulation by the crop, Nu is the ^{15}N uptake constant and t is the time in days. If the constant Nu and the proportionality constant c are equal the curves can be considered to be of the same shape. Even visually this is usually evident from the graphs. This kind of approach was used by Boller and Nosberger [8], and also by Pareek *et al.* [32], Watanabe *et al.* [48] and Watanabe [47]. Even if the curves are not very similar a good idea of the size of the error in the estimate for the BNF contribution to the legume can be obtained from the data. Again more pots, or plots are required and hence more ^{15}N-labelled material is necessary, as well as a considerable number of extra analyses for N and ^{15}N enrichment.

A modelling technique which theoretically requires no control plant

If the ^{15}N enrichment of the soil mineral N is completely stable with time, space and depth in the pots/plots then theoretically any truly non-N_2-fixing control plant, or even none at all, can be used to quantify the BNF contribution to a legume. The approach was first suggested by Kohl and Shearer [26] and applied (with a control crop) by our group to quantify BNF contributions to 11 *Panicum maximum* ecotypes [31].

If no control is used then the ^{15}N enrichment of the soil mineral N is estimated from KCl (or similar) extracts of the soil. This was first tried by Chalk *et al.* [12].

By estimating the decline in ^{15}N enrichment of soil mineral N in a soil amended with a single dose of labelled fertilizer using KCl extracts, and evaluating the total N and ^{15}N accumulation in the legume crop, and fitting curves to these parameters, it is theoretically possible to estimate the amount of N and its ^{15}N enrichment taken up by the legume for each day during crop growth, and from this estimate the overall ^{15}N enrichment accumulated by the legume during plant growth. In the two studies published recently [19,38] the authors claim that this approach was successful for both a pot study on soybean and a field study on lupin, respectively. In the first study on soybean one worrying aspect was that the ^{15}N enrichment of the soil mineral N declined much faster under the soybean plants than under two control crops (non-nod soybean and ryegrass). This would suggest that there is an interaction of the crop type with the rate of decline of ^{15}N enrichment in the soil, which if true, would invalidate the whole concept of the ID technique. This was not discussed by the authors, and it may be that it was due to excretion of fixed (unlabelled) N into the soil by the soybean and was an artifact of the very small pots used. In the field study [38] there was a poor fit (R^2 = 0.60) of the curve of labelled fertilizer accumulation under the lupin which may have caused some error in the BNF estimate. In both studies control crops were incorporated in the experimental design for verification purposes and BNF estimates derived from these controls and the modelling method were in reasonably good agreement. In neither case did the authors predict the total N accumulation of the control crops using their technique and compare it to the actual N accumulation observed. This seems like a missed opportunity and would have provided the technique with a very good verification procedure.

The technique requires a considerable number of soil samples to be taken from each plot throughout the growth cycle (the above authors sampled at 6 occasions during crop growth) and each must be extracted with KCl and the ^{15}N enrichment of the mineral N from these samples measured by mass or emission spectrometry. In the soils used by these workers mineral N levels were apparently reasonably high throughout plant growth making such measurments of ^{15}N enrichment feasible. However, it is our experience that such analyses on soils of low mineral N content often encountered in tropical soils are extremely difficult, at least for analyses with a mass spectrometer where more than 500 μg N sample^{-1} are usually required. In this case large quantites of soil have to be extracted, the extracts distilled and the large volumes of distillate dried, which means using large quantities of reagents all which may have traces of mineral N in them and cause major errors in the estimates of soil mineral ^{15}N enrichment. These problems have been discussed by Chen *et al.* [13] and should be less acute if an emission spectrometer is used for the ^{15}N analyses or a modern, continuous-flow ANCA-IRMS instrument [1] where less than 50 μg of N are required for analysis.

To summarise, the technique is fairly laborious and difficult to apply in tropical soils of very low mineral N content. It still requires more verification, but it could be a very valuable technique for studies where soil mineral N levels are high and it is not desirable to lower them during the study. As mentioned above (section: *Labelling the soil to diminish temporal variations...*) some of the techniques aimed to slow down temporal changes of ^{15}N enrichment of soil mineral N can immobilize soil N and thus cause changes in the N_2-fixing activity of the legume. If it is desirable that a single dose of labelled fertilizer is used, for example in a study on the effect of fertilizer N on BNF of the legume, then this modelling technique could be much more satisfactory than any other.

Seed nitrogen

If the quantity of N in the seed is a significant proportion of the total N accumulated by the legume, or the control crop, this must be taken into account. This will apply particularly to studies of BNF throughout plant ontogeny where at early harvests seed N constitutes a large proportion of plant total N. Normally this seed N is unlabelled and most of it is incorporated into plant tissue. If the control crop has far less seed N than the legume crop (or vice versa) as may be the case with a grass or cereal being used as a control for soybean (or even worse *Canavalia* spp.) then considerable overestimates of BNF can be recorded. This is apparent in the work of Kucey [27] and has been fully discussed by Jensen *et al.* [25], Hamilton *et al.* [19] and Smith *et al.* [38].

Conclusion

The main conclusion is that there is no quick and simple method to reliably quantify the contribution of BNF to legume or other "N_2-fixing" crops in the field. As mentioned above (section: *Addition of soluble labelled N fertilizer*) the technique where just one control crop is used with a single addition of soluble ^{15}N labelled fertilizer at, or near planting, with a single harvest at crop maturity, is that most frequently used. This strategy requires few plots, and hence only a small area to be labelled, and only a few samples have to be analysed for ^{15}N enrichment, but as argued above (section: *Magnitude of errors involved...*) the results are of limited value, sometimes useless, especially if BNF contributions are small. All the other techniques described above (sections: *Strategies to apply...* and *A modelling technique...*) require more work, the use of more plots, more ^{15}N enriched material, and all generate more samples to be analysed for ^{15}N. However, these techniques can yield estimates of BNF contributions to plants growing under field conditions for which the limits of accuracy are known with some confidence.

Among the techniques discussed in the sections: *Strategies to apply...* and *A modelling technique...* that requiring least work and expense is the use of a slow-release form of ^{15}N with several non-N_2-fixing control crops, as described by Boddey *et al.* [7]. The major disadvantage of this technique is that such slow-release forms of ^{15}N can often immobilize soil mineral N, sometimes to the point where overall soil N availability is decreased such that the BNF contribution to the legume is increased. If this can prejudice the objectives of the experiment then it is recommended that small frequent additions of labelled fertilizer should be used (section: *Labelling the soil to diminish...* option d.) or the modelling technique described in the section: *A modelling technique...* [19,38].

References

1. Barrie A (1991) New methodologies in stable isotope analysis. In: Stable Isotopes in Plant Nutrition Soil Fertility and Environmental Studies, pp 3–25. Int. Atomic Energy Agency, Vienna

2. Bergersen FJ and Turner GL (1983) An evaluation of ^{15}N methods for estimating nitrogen fixation in a subterranean clover-perennial ryegrass sward. Aust J Agric Res 34: 391–401

3. Boddey RM (1987) Methods for the quantification of nitrogen fixation associated with gramineae. CRC Crit Rev Plant Sci 6: 209–266

4. Boddey RM and Urquiaga S (1990) Quantification of the contribution of BFN to field grown plants: The use of the N isotope dilution technique - problems and some solutions. In: Gueye M, Mulonguy K and Dommergues Y (eds) Maximiser la Fixation Biologique de L'azote pour la Production Agricole et Forestière en Afrique, Vol 2(2), pp 298–316, Collection Actes de l'Inst Sénégalais Recherches Agricoles, Dakar, Senegal

5. Boddey RM, Chalk PM, Victoria RL, Matsui E and Döbereiner J (1983) The use of the ^{15}N dilution technique to estimate the contribution of associated biological nitrogen fixation to the nitrogen nutrition of *Paspalum notatum* cv. batatais. Can J Microbiol 29: 1036–1045

6. Boddey RM, Chalk PM, Victoria RL and Matsui E (1984) Nitrogen fixation by nodulated soybean under tropical field conditions estimated by the ^{15}N isotope dilution technique. Soil Biol Biochem 16: 583–588

7. Boddey RM, Urquiaga S, Neves MCP, Suhet AR and Peres JR (1990) Quantification of the contribution of N_2 fixation to field-grown grain legumes - A strategy for the practical application of the ^{15}N isotope dilution technique. Soil Biol Biochem 22: 649–655

8. Boller BC and Nosberger J (1988) Influence of dissimilarities in temporal and spatial N-uptake patterns on ^{15}N-based estimates of fixation and transfer of N in ryegrass-clover mixtures. Plant and Soil 112: 167–175

9. Broadbent FE, Nakashima T and Chang CY (1982) Estimation of nitrogen fixation by isotope dilution in field and greenhouse experiments. Agron J 74: 625–628

10. Cadisch G, Sylvester-Bradley R and Nosberger J (1989) ^{15}N-based estimates of N_2 fixation of eight tropical forage legumes at two levels of P:K supply. Field Crops Res 22: 181–194

11. Chalk PM (1985) Estimation of N_2 fixation by isotope dilution: An appraisal of techniques involving ^{15}N enrichment and their application. Soil Biol Biochem 17: 389–410

12. Chalk PM, Douglas LA and Buchanan SA (1983) Use of ^{15}N enrichment of soil mineralizable N as a reference for isotope measurements of biologically fixed nitrogen. Can J Microbiol 29: 1046–1052

13. Chen D, Chalk PM and Freney JR (1991) External-source contamination during extraction-distillation in isotope-ratio analysis of soil inorganic nitrogen. Anal Chim Acta 245: 49–55

14. Danso SKA (1988) The use of ^{15}N enriched fertilizers for estimating nitrogen fixation in grain and pasture legumes. In: Beck DP and Materon LA(eds) Nitrogen Fixation by Legumes in Mediterranean Agriculture, pp 345–348. Martinus Nijhoff, Dordrecht, The Netherlands

15. Duque FF, Neves MCP, Franco AA, Victoria RL and Boddey RM (1985) The response of field grown *Phaseolus vulgaris* to Rhizobium inoculation and the quantification of N_2 fixation using ^{15}N. Plant and Soil 83: 333–343

16. Edmeades DC and Goh KM (1978) Symbiotic nitrogen fixation in a sequence of pastures of increasing age measured by a ^{15}N dilution technique. N Z J Agric Res 21: 623–628

17. Fried M, Danso SK and Zapata F (1983) The methodology of measurement of N_2 fixation by non-legumes as inferred from field experiments with legumes. Can J Microbiol 29: 1053–1062

18. Giller KE and Witty JF (1987) Immobilized ^{15}N fertilizer sources improve accuracy of field estimates of N_2-fixation by isotope dilution. Soil Biol Biochem 19: 459–463

19. Hamilton SD, Smith CJ, Chalk PM and Hopmans P (1992) A model based on measurement of soil and plant ^{15}N enrichment to estimate N_2 fixation by soybean (*Glycine max* L. Merril) grown in pots. Soil Biol Biochem 24: 71–78

20. Hamilton SD, Chalk PM, Smith CJ and Hopmans P (1992) Evaluation of reference plants by labelling with ^{35}S in the estimation of N_2 fixation by ^{15}N isotope dilution. Plant and Soil 145: 177–185

21. Hamilton SD, Hopmans P, Chalk PM and Smith CJ (1993) Field estimation of N_2 fixation by *Acacia* spp. using ^{15}N isotope dilution and labelling with ^{35}S. For Ecol Manage 56: 297–313

22. Hardarson G, Danso SKA and Zapata F (1988) Dinitrogen fixation measurements in alfalfa-ryegrass swards using nitrogen-15 and the influence of the reference crop. Crop Sci 28: 101–105

23. Hardarson G, Danso SKA, Zapata F and Reichardt K (1991) Measurements of nitrogen fixation in fababean at different N fertilizer rates using the ^{15}N isotope dilution and 'A value' methods. Plant and Soil 131: 161–168

24. Henzell EF, Martin AE, Ross PJ and Haydock KP (1968) Isotopic studies on the uptake of nitrogen by pasture plants. IV. Uptake of nitrogen from labelled plant material by Rhodes grass and siratro. Aust J Agric Res 19: 65–77

25. Jensen ES, Andersen AJ and Thomsen JD (1985) The influence of seed-borne N in ^{15}N dilution studies with legumes. Acta Agric Scand 35: 438–443

26. Kohl DH and Shearer G (1981) The use of soils lightly enriched in ^{15}N to screen for N_2-fixing activity. Plant and Soil 60: 487–489

27. Kucey RMN (1989) Contribution of N_2 fixation to field bean and pea uptake over the growing season under field conditions in southern Alberta. Can J Soil Sci 69: 695–699

28. Ledgard SF, Morton R, Freney JR, Bergersen FJ and Simpson JR (1985) Assessment of the relative uptake of added and indigenous soil nitrogen by nodulated legumes and reference plants in the ^{15}N dilution measurement of N_2 fixation. I. derivation of method. Soil Biol Biochem 17: 317–321

29. Ledgard SF, Simpson JR, Freney JR and Bergersen FJ (1985) Field evaluation of ^{15}N techniques for estimating nitrogen fixation in legume-grass associations. Aust J Agric Res 36: 247–258

30. Legg JO and Sloger C (1975) A tracer method for determining symbiotic nitrogen fixation in field studies. In: Klein ER and Klein PD (eds) Proc. 2nd Int. Conf. Stable Isotopes, pp 661-666. Oak Brook, USA

31. Miranda CHB and Boddey RM (1987) Estimation of biological nitrogen fixation associated with 11 ecotypes of *Panicum maximum* grown in nitrogen-15-labeled soil. Agron J 79: 558–563

32. Pareek RP, Ladha JK and Watanabe I (1990) Estimating N_2 fixation by *Sesbania rostrata* and *S. cannabina* (syn. *S. aculeta*) in lowland rice soil by the ^{15}N dilution method. Biol Fertil Soils 10: 77–88

33. Patterson TG and LaRue TA (1983) Nitrogen fixation by soybeans: seasonal and cultivar effects, and comparison of estimates. Crop Sci 23: 488–492

34. Rennie RJ (1985) Theoretical errors in sequential N-15 labelling of soil to estimate N_2 fixation by isotope dilution. Crop Sci 25: 891

35. Rennie RJ (1986) Advantages and disadvantages of nitrogen-15 isotope dilution to quantify dinitrogen fixation in field-grown legumes - a critique. In: Field Measurement of Dinitrogen Fixation and Denitrification, Soil Sci Soc Am Spec Publ 18, pp 43–58. Am Soc Agron, Madison, Wisconsin, USA

36. Rennie RJ and Thomas JB (1987) ^{15}N-determined effect of inoculation with N_2-fixing bacteria on nitrogen assimilation in Western Canadian wheats. Plant and Soil 112: 183–193

37. Sanginga N, Danso SKA, Zapata F and Bowmen GD (1990) Influence of reference trees on N_2-fixation estimates in *Leucaena leucocephala* and *Acacia albina* using ^{15}N-labelling techniques. Biol Fertil Soils 9: 341–346

38. Smith CJ, Chalk PM, Hamilton SD and Hopmans P (1992) Estimating N_2 fixation by field-grown lupins (*Lupinus angustifolius* L.) using soil and plant ^{15}N enrichment. Biol Fertil Soils 13: 235–241

39. Talbott HJ, Kenworthy WJ and Legg JC (1982) Field comparison of the nitrogen-15 and difference methods of measuring nitrogen fixation. Agron J 74: 799–804

40. Talbott HJ, Kenworthy WJ, Legg JO and Douglass LW (1985) Soil nitrogen accumulation in nodulated and non-nodulated soybeans: a verification of the difference method by a ^{15}N technique. Field Crops Res 11: 55–67

41. Urquiaga S, Cruz KHS and Boddey RM (1992) Contribution of nitrogen fixation to sugar cane: nitrogen-15 and nitrogen balance estimates. Soil Sci Soc Am J 22: 105–114

42. Vallis I, Haydock KP, Ross PJ and Henzell EF (1967) Isotopic studies on the uptake of nitrogen by pastures. III. The uptake of small additions of ^{15}N-labelled fertilizer by Rhodes grass and Townsville lucerne. Aust J Agric Res 18: 865–877

43. Vallis I, Henzell EF and Evans TR (1977) Uptake of soil nitrogen by legumes in mixed swards. Aust J Agric Res 28: 413–426

44. Viera-Vargas MS, Oliveira OC de, Souto CM, Cadisch G, Urquiaga S and Boddey RM (1994) Use of different ^{15}N techniques quantify the contribution of biological nitrogen fixation to legumes. Soil Biol Biochem (In press)

45. Vose P and Victoria RL (1986) Re-examination of the limitations of nitrogen-15 isotope dilution technique for the field measurement of dinitrogen fixation. In: Field Measurement of Dinitrogen Fixation and Denitrification, Soil Sci Soc Am Spec Publ 18, pp 23–41. Am Soc Agron, Madison, Wisconsin, USA

46. Wagner GH and Zapata F (1982) Field evaluation of reference crops in the study of nitrogen fixation by legumes using isotope tecniques. Agron J 74: 607–612

47. Watanabe I (1991) Errors related to the ^{15}N dilution method for estimating nitrogen fixation. In: Stable Isotopes in Plant Nutrition Soil Fertility and Environmental Studies, pp 83–88. Int Atomic Energy Agency, Vienna

48. Watanabe I, Chui GY and Yoshida T (1990) Estimation of N_2 fixation in soybean and cowpea by using residual ^{15}N. Soil Sci Plant Nutr 36: 375–381

49. Witty JF (1983) Estimating N_2 fixation in the field using ^{15}N-labelled fertilizer: some problems and solutions. Soil Biol Biochem 15: 631–639

50. Witty JF and Giller KE (1991) Evaluation of errors in the measurement of biological nitrogen fixation using ^{15}N fertilizer. In: Stable Isotopes in Plant Nutrition, Soil Fertility and Environmental Studies, pp 59–72. Int Atomic Energy Agency, Vienna

51. Witty JF and Ritz K (1984) Slow release ^{15}N fertilizer formulations to measure N_2-fixation by isotope dilution. Soil Biol Biochem 16: 657–661

52. Witty JF, Rennie RJ and Atkins CA (1988) ^{15}N addition meth-
     ods for assessing N_2 fixation under field conditions. In: Sum-
     merfield RJ (ed) World Crops: Cool Season Food Legumes, pp
     715–730. Kluwer, Dordrecht, The Netherlands

Fertilizer Research **42**: 89–97, 1995.
© 1995 *Kluwer Academic Publishers. Printed in the Netherlands.*

Rational nitrogen fertilization in intensive cropping systems

Charles W. Rice[1], John L. Havlin[1] & James S. Schepers[2]
[1]*Department of Agronomy, Kansas State University, Manhattan, KS 66506-5501, USA and* [2]*USDA-ARS, University of Nebraska-Lincoln, Lincoln, NE 68583-0915, USA*

Key words: N mineralization, N use efficiency

Abstract

The objective of a rational N fertilization program is to account for the sources and fate of N while estimating crop N needs. Efficiency of N use will vary with cropping systems and N sources. Management technologies that affect N use efficiency include the amount of N applied, timing and placement of N fertilizer, and use of inhibitors. One of the main problems in making a fertilizer N recommendation is to account for the contribution of N mineralization to plant available N. Most laboratory procedures do not account for the environmental factors that affect N mineralization and only estimate the size of the mineralizable N pool. However, changes in soil moisture and temperature can dramatically affect the amount and rate of release of mineralized N. Field and modeling techniques are two possible techniques to estimate N mineralization. Field techniques can be divided into soil and plant approaches. Soil incubations in the field provide a quantitative approach while soil nitrate tests during the growing season provide a qualitative approach to estimating N mineralization. The plant is the ultimate integrator of N mineralization. Plant N uptake by an unfertilized crop can provide a quantitative approach with certain precautions. This approach may be costly, labor intensive, and site specific. Crop N uptake during the growing season can be estimated by measuring the tissue N content or using a chlorophyll meter. The chlorophyll meter measures the greenness of the plant and has been shown to be positively correlated to plant N status. Modeling may provide another option by including the factors that affect the rate of N mineralization from a known pool. The two most important variables include soil moisture and temperature. Realistic yield expectations and accounting for existing and projected amounts of available N can improve the accuracy of N recommendations.

Introduction

Production of food and its protein content are key components to sustaining the world population. In many cropping systems, N frequently limits crop yield and protein levels and additional N inputs are required to optimize productivity and profitability. Nitrogen availability to crops is regulated by a biologically dynamic soil N cycle; thus, it is subject to the environmental factors that regulate the activity of microorganisms. In general, a portion of the N applied in excess of crop requirements ultimately will be converted to NO_3^-, which may be leached to groundwater. Thus, it is important to develop rational practices for N fertilizer to ensure sufficient but judicious use of N for economic and environmental reasons. Previous N recommendations were based on inexpensive fertilizer, a lack of appreciation for spatial variability, and minimal environmental concerns. In some environmentally sensitive areas, future N recommendations need to be as site specific as possible.

Accurate recommendations for N fertilizer require knowledge of plant N requirements, external N sources, and N losses from the soil system. These components are represented by the equation:

$$N_f = N_c - (N_{sources} - N_{losses}) \qquad (1)$$

where:

$$N_f = \text{N fertilizer}$$
$$N_c = \text{N needed by the crop}$$
$$N_{sources} = \text{N sources}$$
$$N_{losses} = \text{N losses}$$

This apparently simple equation is very complex. The amount of N needed by the crop (N_c) is a function of dry matter yield and N concentration of the crop. The estimate of crop N requirement influences N recommendations more than any other factor. Overestimating the yield goal will result in a N recommendation that exceeds the actual crop N requirement; thus, establishing a realistic yield goal is important for reducing the environmental impact of N use (Jackson *et al.*, 1987). Power and Broadbent (1989) suggested using a yield goal that is 5 to 10% greater than average yield of the previous 5 to 7 years. It is unrealistic to expect precise estimates because of yearly variations in growing season weather. If the yield includes the entire plant, then the N concentration of that yield will determine the crop's N needs. In the case where only a portion of the crop is harvested (i.e. grain), an estimate of the N requirement for the entire crop should be used.

Exogenous N inputs include fertilizer, manure, and biologically fixed N. The contribution of biologically fixed N becomes more important in tropical systems, where free-living and associative N_2 fixation increases. Associative N_2 fixation with tropical grasses, including sugar cane, has been reported to be as much as 100 kg N ha^{-1} (Neyra and Dobereiner, 1977; Dobereiner, 1978). In rice production, the N contribution of *Azolla* has been reported to be equivalent to 30 kg urea-N ha^{-1} (Watanabe, 1987). Inputs from irrigation and precipitation also must be considered. Nitrogen inputs from precipitation normally are small, but localized sources can make a significant contribution. Nitrogen inputs from irrigation also are normally small, but when NO_3^- contaminates the groundwater, significant amounts of N can be added through irrigation. Schepers and Mosier (1991) reported that > 100 kg N ha^{-1} can be added during one growing season in the Platte River valley of Nebraska in the US, where the NO_3^--N concentration can exceed 30 mg N L^{-1}.

Internal inputs include residual soil inorganic N and N mineralized from crop residues, soil organic matter, and manure that decompose during the growing season. In perennial crops, root N that will be translocated to the shoots also should be considered.

Nitrogen losses include leaching of NO_3^-, gaseous N losses from soil through denitrification, volatilization of N from fertilizers and vegetation, and erosion. The quantity of N lost through these pathways greatly depends on the local conditions of soil, crop, and climate

Our simple equation now takes the form:

$$N_f = N_c - [(N_m + N_{in} + N_r + N_{fix}) - (N_l + N_d + N_v + N_c) - N_{im}] \quad (2)$$

where,

$$
\begin{aligned}
N_m &= \text{N mineralization from soil organic matter} \\
N_{in} &= \text{residual inorganic soil N} \\
N_r &= \text{N from residue} \\
N_{fix} &= \text{biologically fixed N} \\
N_l &= \text{N loss from leaching} \\
N_d &= \text{N loss from denitrification} \\
N_v &= \text{N loss from volatilization} \\
N_c &= \text{N loss from erosion} \\
N_{im} &= \text{temporary N loss from immobilization}
\end{aligned}
$$

The ultimate goal of a N management program is to accurately supply the crop requirement, while minimizing N losses through leaching, denitrification, volatilization, and erosion. Minimizing N losses and maximizing crop recovery of applied N will enhance N use efficiency and reduce environmental risk inherent with the use of external N inputs.

Efficiency of N use

Implicitly included in a N fertilizer recommendation is the efficiency of N use by the crop. Each of the N sources identified in Equation 2 will have an associated efficiency that is a function of losses. Management systems that dramatically change N use efficiency require that the N recommendation model be recalibrated. Nitrogen use efficiency is a function of climate; soil properties; crop and soil management; and management of the N applied (time, form, and placement). Although we often have very little control of climate, changes in precipitation amounts and timing will alter N transformations and subsequent availability. Myers (1988) reported N fertilizer efficiencies ranging from 12 to 74% for arable crops in the tropics. Poor recovery could be due to leaching and denitrification in wet climates and poor N uptake in dry climates. Crop and soil management practices are important considerations in management of N inputs, e.g. the effect tillage systems have on fertilizer transformations and microbial activity. For example, Phillips *et al.* (1980) reported that no-tillage corn production required more N fertilizer compared to plowed systems. This conclusion is based

Crop Growth - Maize

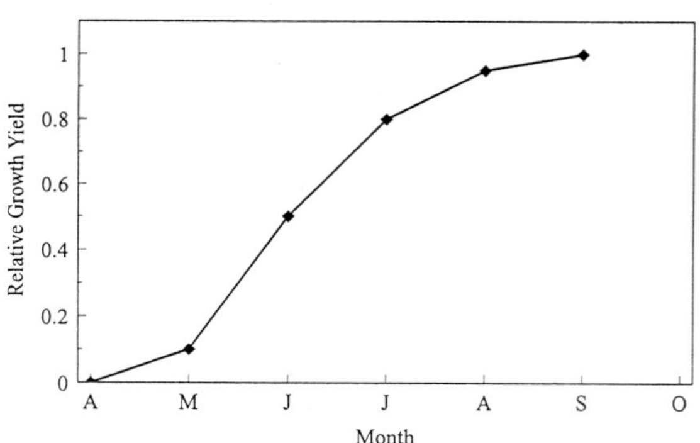

Fig. 1. Typical growth of maize in the central US.

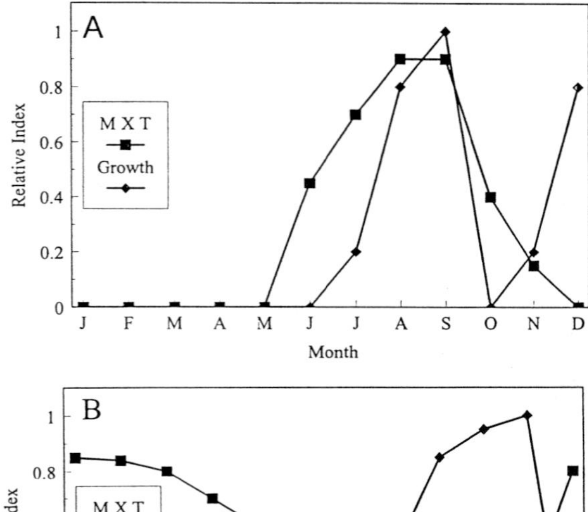

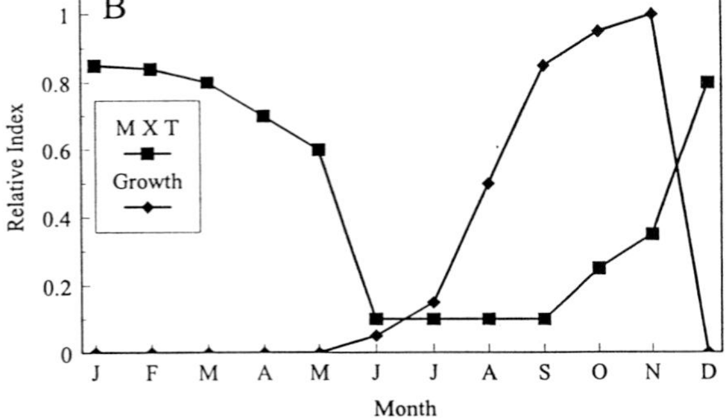

Fig. 2. Moisture-temperature index (■) for N mineralization and relative growth (♦) for (**A**) a semi-arid, tropical site with *kharif* and rubi at Hyderabad, India and (**B**) a subtropical site with wheat at Darling Downs, Australia (McGill and Myers, 1987).

on the premise that fertilizer N is used less efficiently under a no-tillage system. In reality, no-tillage systems generally exhibit higher denitrification (Rice and Smith, 1982) and greater potential for immobilization (Rice and Smith, 1984) and leaching (Thomas *et al.*, 1973). Crop recovery of applied N sometimes can be enhanced by adopting new technologies that improve the synchrony between crop needs and N supply.

Nitrogen management technologies have been developed to reduce N contact with soil microorganisms, N leaching, and N volatilization. Reducing N contact with soil microorganisms can reduce denitrification and immobilization. Nitrification inhibitors also can be used with N sources to temporarily reduce microbial activity responsible for the oxidation of NH_4^+. Nitrogen management technologies incompass application timing and placement of fertilizers and manures, use of inhibitors and N amount.

Timing

Maximum crop recovery and use of applied N require that N available be synchronous with plant growth and N demand (Fig. 1). Nitrogen synchrony often is not practical or possible. Supplying N after the crop reaches full canopy without damage to the crop can be difficult unless it is supplied by fertigation, a high-clearance vehicle, or aerial application. The timing of N availability must consider N mineralization as well as external N inputs. McGill and Myers (1987) discussed this relationship using a moisture-temperature index as an indicator of microbial activity and N mineralization. For example, the semi-arid tropical climate

of Hyberabad, India, creates conditions conducive for synchrony between N mineralization and kharif and rabi crops (Fig. 2A). An example of a nonsynchronous relationship between N mineralization and crop N demand occurs in sub-tropical Queensland Australia with spring wheat (Fig. 2B). To improve the recovery of fertilizer N, some producers will split their N fertilizer application or delay the application into the growing season. This practice can successfully increase N fertilizer use efficiency and potentially reduce N fertilizer inputs. Studies have shown a 10 to 30% decrease in the amount of N fertilizer needed by the crop from this practice (Wells, 1984).

Placement

Adoption of N placement technologies can reduce N losses and increase plant N uptake. Placement of N becomes more important as reduced tillage systems are adopted. In the United States of America, common placement methods for maize include either sur-

Table 1. Fertilizer N recovered by corn as affected by tillage (1985–1986 average)

N Placement	Time Applied	Tillage System (%)	
		Plowed	No-till
Broadcast	At planting	46	36
	Delayed	56	44
Injected	At planting	59	52
	Delayed	56	56

Pierce and Rice (pers. commun.).

Table 2. Grain yield response of maize to urease inhibitor

Treatment	Irrigated	Dryland
	(Mg ha^{-1})	
Urea	7.02a	3.16a
Urea + NBPT	8.28b	4.30b
NH_4NO_3	8.28b	3.85b

Lamond *et al.* (1994).

face or subsurface banding of the N fertilizer (Lamond *et al.*, 1991). Placement of fertilizer N is even more critical with the use of urea in rice production. Volatilization losses are extremely high (up to 96%) unless deep placement of urea is used (Watanabe *et al.*, 1981; Zhao-liang, 1981). Subsurface placement of manure N can reduce N volatilization losses of surface-applied manure. Often, timing and placement interact positively (Table 1). These data show greater fertilizer N recovery by maize when the N application was delayed to the sixth leaf stage or subsurface banded at planting. In this example, no benefit resulted from subsurface banding with a delayed application. Other studies have shown the benefit of banded fertilizer on recovery of applied N (Murphy *et al.*, 1978; Mengel, 1982).

Inhibitors

Specific inhibitors of biological N processes have been developed and can be used to increase N use efficiency and decrease leaching losses. The most common and effective products inhibit nitrification and urea hydrolysis. Nitrification inhibitors temporarily inhibit the nitrification process when crop N uptake is low and denitrification and leaching potentials are high. Several compounds have been reported to inhibit nitrification, but only three have been developed commercially: nitrapyrin, N serve, [2-chloro-6-(trichloromethyl)pyridine]; etridiazol, Dwell, [5-thoxy-3-(trichloromethyl)-1,2,4-thiadiazaole]; and dicyandiamide, DCD, (Peterson and Frye, 1989). The extent of nitrification inhibition depends upon the amount and time of N application, soil temperature, and precipitation patterns (Frye, 1981). Generally, conditions conducive for leaching and denitrification will enhance the effectiveness of nitrification inhibitors. A more complete discussion

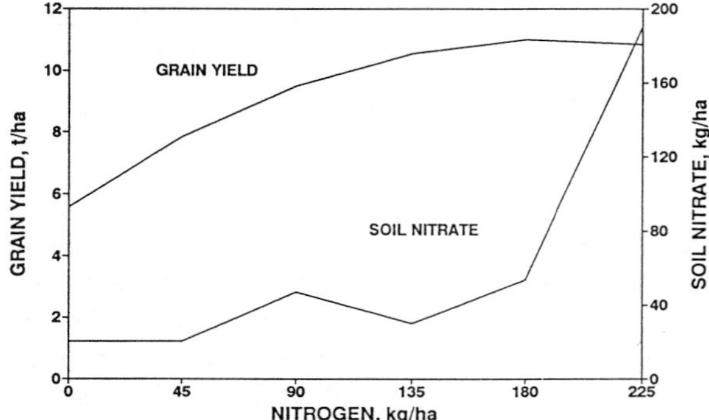

Fig. 3. Nitrogen response curve for maize. Soil nitrate concentrations in the surface 3.2 m. N rate for economic optimum yield was about 180 kg N ha^{-1}.

of nitrification inhibitors is presented by Sahrawat (1994).

Urease inhibitors reduce urease activity, resulting in delayed hydrolysis of urea or urea-based fertilizers that will increase transport of surface-applied urea into the soil. Movement of urea into the soil can significantly reduce volatilization losses of applied urea. This can translate to greater grain yields (Table 2) (Lamond *et al.*, 1994). A more complete discussion of urease inhibitors is presented by Bremner (1994).

Amounts of N applied

Identifying the precise amount of external N applied to a crop is extremely important in maximizing crop recovery of N and minimizing environmental risk associated with N use. A typical N response curve is illustrated in Figure 3. The largest incremental response to applied N occurs at the lower amounts of N application. More importantly, the final 10% yield response to applied N uses approximately 35% of the N fertilizer required for maximum profit (Schlegel and Havlin, 1994). These data also show that increasing the amount of N applied beyond the economic optimum amount dramatically increases profile N content and NO_3^- leaching potential.

One of the main problems in developing an accurate N recommendation is to quantify the contribution of N mineralization, because it is an important source of plant N. Accurately assessing N mineralization is difficult and has been the subject of research for almost 100 years. Universal procedures to quantify N mineralization still have not been developed. Early research focused on developing a laboratory test to determine the N mineralization potential of soils. Laboratory procedures can be divided into chemical and biological indices. These are reviewed and discussed elsewhere (Keeney, 1982; Meisinger, 1984; Rice and Havlin, 1994; Stanford, 1982). Generally, the laboratory procedures have not been adopted widely, because they do not predict N mineralization under field conditions. These laboratory procedures generally have been unsuccessful, because they fail to consider the dynamics of the factors that regulate the rate of N mineralization. These factors are substrate quality; moisture; substrate accessibility (clay content, tillage; wetting and drying cycles, freeze-thaw cycles); temperature; and pH. A complete discussion of these factors is given by Rice and Havlin (1994).

The nature and complexity of these regulatory factors complicate assessment of N mineralization in the laboratory. Field techniques and modeling are two other possible choices. Field techniques include soil sampling and incubations over time and plant sampling. Field techniques integrate many of the regulatory factors noted above, as well as crop and soil management factors that affect N mineralization. A disadvantage of the field techniques is that they are often site specific. Field techniques can be divided into quantitative or qualitative approaches and have been discussed in greater detail by Schepers and Meisinger (1994).

Assessment of nitrogen mineralization

Soil incubations

A quantitative approach to field N mineralization is the use of fallow soil. This approach avoids the disturbance that occurs with laboratory techniques. The lack of disturbance can emulate conditions that occur in the field with differing soil management systems, such as reduced tillage and residue management. Other physical conditions also are preserved. The soil in conservation tillage systems, often is more dense, cooler, and wetter than plowed soil. To eliminate large leaching and denitrification losses, the fallow technique can include a cover (Rice *et al.*, 1987). Rice *et al.* (1987) reported differences between tillage systems and soil drainage classes under the covered fallow soils that were not shown by laboratory techniques. One potential disadvantage of covering the soil is that it would not undergo the same wetting and drying cycles that occur under normal field conditions. A more detailed discussion of this technique is presented by Schepers and Meisinger (1994).

Soil nitrate tests

Soil nitrate tests provide a relative index of *in situ* N mineralization. Specific soil NO_3^- tests have many variations that are dependent on climatic conditions, crop, and soil type. In the humid temperate regions of the US, the pre-sidedress nitrate test (PSNT) for maize is being evaluated. This test measures the amount of NO_3^- in the top 30 cm of the soil profile when the corn is at the 6-leaf stage or approximately 30 cm tall (Bock *et al.*, 1992). The PSNT was proposed first by Magdoff *et al.* (1984) to identify situations where fertilizer would not be needed to attain optimum maize yields. The PSNT integrates residual NO_3^- from the previous season, N losses, and spring N mineralization until the time sampled. The intent of this test is to allow N mineralization to proceed as long as possible before the accumulated NO_3^- is measured and yet allow enough time to apply sidedress N before the crop grows too tall to prevent access to the field. The advantage is that, in the eastern two thirds of the US, the PSNT can separate sites that have sufficient N (> 20–25 mg NO_3^--N kg^{-1}) from those that require additional fertilizer N inputs for maize production. The PSNT does not quantitatively measure N mineralization. For the PSNT to be useful, a standard sampling time and rapid analysis are important so that sidedress fertilizer N can be applied in a timely manner. Presently, the PSNT has been evaluated only for maize and needs to be evaluated for other crops. Neteson *et al.* (1989) reported that soil profile inorganic N currently is being used in the Netherlands to adjust N fertilizer recommendations for several crops. Including soil profile inorganic N was either equally effective or decreased N fertilizer requirement when compared to using a fixed N rate.

Plant N status

The plant is the ultimate integrator of environmental variables controlling the N transformations within the soil N cycle and, ultimately, the N available to it. Two approaches will be discussed: a qualitative approach involving chlorophyll meters and a quantitative approach using N uptake by an unfertilized crop. Crop N uptake provides an estimate of N availability during the growing season. The N status of the crop can be quantified by determining the tissue N content or using a chlorophyll meter. The chlorophyll meter measures the chlorophyll content or greenness of the plant that is correlated positively with crop N status. The chlorophyll meter is ideal for real time analysis of the crop N status and indicates N stress. The principle is discussed further by Schepers *et al.* (1992a,b) who had success comparing chlorophyll meter readings of an area requiring N to an area where N is nonlimiting because of application of adequate N fertilizer. Some of the disadvantages discussed by Schepers and Meisinger (1994) are that the data generated are site and plant (variety) specific. This makes calibration of the meters difficult unless a sufficiency index is calculated.

Plant N uptake

Nitrogen uptake by an unfertilized crop represents the best method for quantifying net N mineralization. The advantages of this approach are that it integrates field temperature, moisture, and aeration conditions that greatly influence N mineralization potential and can be adapted easily to include many crop and soil management practices. Nitrogen uptake in unfertilized plots also includes the normal rooting depth of the crop being studied and integrates spatial and temporal influences on N mineralization. Accurate measures of crop N uptake must include total biomass N and not just grain yield. One problem with estimating total N mineralization is accounting for root N content. A constant shoot:root N ratio cannot always be assumed and must be validated. Other disadvantages of this technique are costs, labor requirements, and site specificity. Information derived from one site may not be transferable to another site, unless some other nonsite-specific parameters can be correlated to N mineralization or can be used to predict it.

Models

Modeling techniques are other options that could directly incorporate the factors affecting microbial activity and N mineralization, as well as the other factors affecting N availability and losses. Current attempts to model N mineralization for field situations include the rate factor (k) and the size of the mineralizable pool (N_0). Most studies suggest use of first order kinetics (Broadbent, 1986; Juma *et al.*, 1984; Stanford and Smith, 1972). A few studies have suggested a two-pool model to predict N mineralization (Molina *et al.*, 1980; Deans *et al.*, 1986). For use under field conditions, the rate then is adjusted for moisture and temperature (Campbell *et al.*, 1981; Myers *et al.*, 1982). However, Campbell *et al.* (1988) reported that their model underestimated N mineralization during the growing season. They attributed the underestimate to flushes of N mineralization during rewetting of dry soils. For temperature, Honeycutt *et al.* (1988) suggested using accumulated heat units to predict N mineralization; however, the heat unit concept is unable to predict N mineralization under dry soil conditions (Doel *et al.*, 1990). Such models need to include climatic parameters and soil properties. Neeteson *et al.* (1989) have successfully used a N fertilizer recommendation model for potatoes in the Netherlands.

Fertilizer N recommendation

Fertilizer N recommendations need to be calibrated across broad areas, even though site-specific information is used to develop the appropriate algorithms. To calibrate N recommendations, knowledge of the crop response to N fertilizer is required. Critical variables for determining the N rate need to be identified and measured. Important site variables include texture, pH, and organic matter content. Management variables include tillage systems, irrigation, and rotations. Once the crop response and the important variables are identified, then the recommendation model should be tested over a range of environmental conditions and N components.

For making fertilizer N recommendations, at least two strategies emerge, depending on other characteristics of the production system. The main distinction between these strategies is the ability of the producer to apply N fertilizer easily at any time during the growing season versus situations that essentially require all the fertilizer to be applied while the plants are small

Table 3. An example data set from Nebraska to determine N fertilizer recommendations for corn with two irrigation systems

Data	Conventional	Center-Pivot
Residual soil N		
0–20 cm (mg kg^{-1} nitrate-N)	13.5	12.3
20–90 cm (mg kg^{-1} nitrate-N)	7.2	4.5
0–90 cm (wt. average, mg kg^{-1} nitrate-N)	8.8	6.2
Organic matter		
0–20 cm (%)	1.77	1.9
Irrigation water (mg l^{-1} nitrate-N)	30	30
Expected yield (mg ha^{-1})	12.6	12.6
Estimated water application (cm)	25	25
Estimated crop N need (kg ha^{-1})	309	309
21.4 (expected yield) + 39		
N credits		
Residual soil N (kg ha^{-1})	79	56
9 (mg nitrate-N kg^{-1})		
Organic matter (kg ha^{-1})	56	60
2.5 (expected yield)(% organic matter)		
Irrigation water (kg ha^{-1})	75	75
0.1(cm depth)(mg l^{-1} nitrate-N)		
Others		
Soybeans (50 kg ha^{-1})	0	0
Alfalfa (135–170 kg ha^{-1})	0	0
Manures (variable	0	0
Fertilizer N required (kg ha^{-1})	99	118

to avoid mechanical damage. Situations that are limited to preplant or sidedress N application involve a greater degree of anticipation on the part of the farmer, because they provide little opportunity to compensate for atypical climatic conditions. In contrast, situations where fertigation or other means of N application (i.e. high clearance vehicle, foliar application, etc.) are possible throughout the growing season allow producers to move closer to spoon-feeding their crops. As it turns out, the greater flexibility a producer has in terms of N application times and forms, the more decisions that must be made about fertilizer costs, labor, and convenience. The two scenarios for N recommendations that follow are for irrigated corn in Nebraska (Table 3). One field is under conventional furrow irrigation and the other adjacent field has been under center-pivot irrigation for several years. Because of the methods of irrigation and past fertilizer N management practices,

the levels of residual soil N are different for the two fields.

Conventional field

Consider applying approximately 20–30 kg N ha^{-1} as a starter fertilizer at planting and the remainder as preplant or sidedress application. Local regulations require preplant applications in excess of 90 kg N ha^{-1} use a nitrification inhibitor. The credit given for the amount of N supplied in irrigation water is conservative, because the amount of water applied by furrow irrigation can be quite variable within a field, ranging from 60–90 cm. However, because a considerable portion of the irrigation water is applied to corn after silking, a portion of the N in the irrigation water will not be utilized by the crop. The value of 25 cm water application is based on the evapotranspiration minus long-term average precipitation during the growing season.

Center-pivot field

Consider applying approximately 20–30 kg N ha^{-1} as a starter fertilizer at planting and 50–75% of the remainder as preplant or sidedress application. Use chlorophyll meters or other means to monitor crop N status during the growing season. If an N deficiency appears to be developing, apply 20–30 kg N ha^{-1} as urea ammonium nitrate in the irrigation water as needed. No N should be applied in the water beyond 3 weeks after silking.

One future N management scenario under development involves the use of equipment to apply variable rates of N. This approach will depend on a field location system, typically a global positioning system (GPS), to identify the position of the fertilizer applicator in the field. The variable-rate map of fertilizer N recommendations will be generated based on a variety of factors, such as residual soil N, organic matter content, and yield goal, for each part of the field. This is an emerging technology that offers the potential to reduce N applications in fertile areas of the field and maintain productivity in less fertile areas.

Conclusions

Our ability to make fertilizer recommendations is becoming more sophisticated, as we learn more about the soil/plant/water system. In the past, yield was the only measure of crop N need; however, this has contributed to some environmental problems. We now know that accounting for existing and projected amounts of available N in the root zone can improve the accuracy of N recommendations. Biologically fixed N, from associative and free-living systems, needs to be quantified, especially in tropical systems. In most cases, maximum yield is not desirable, because it results in economic and environmental risks. Realistic yield expectations should be stressed when making N fertilizer recommendations. We should realize that we cannot achieve or expect 100% efficiency from N fertilizers. Some losses through denitrification and leaching must be expected, and, in some years, unusual weather events will create situations that can significantly reduce or increase N use efficiency.

Acknowledgments

This paper was partly supported by a specific cooperative agreement from the USDA-ARS No. 58-5440-2-105. Contribution No. 94-577-J from the Kansas Agric. Exp. Stn., Manhattan, KS.

References

Bock BR, Kelley KR and Meisinger J (1992) Predicting N fertilizer needs for corn in humid regions: Summary and future directions. *In*: Bock BR and Kelley KR (eds) Predicting N fertilizer needs for corn in humid regions. Bull. Y-226, pp 115–127. National Fertilizer and Environmental Research Center, Tennessee Valley Authority, Muscle Shoals, AL

Bremner J (1994) Problems in the use of urea as a nitrogen fertilizer. *In*: Ahmad N (ed) International Symposium on Nitrogen Economy in Tropical Soils. 9–14 Jan. 1994. Port of Spain, Trinidad. Univ. West Indies, Trinidad, West Indies (*In press*)

Broadbent FE (1986) Empirical modeling of soil nitrogen mineralization. Soil Sci 141: 208–213

Campbell CA, Jame YW and De Jong R (1988) Predicting net nitrogen mineralization over a growing season: Model verification. Can J Soil Sci 68: 537–552

Campbell CA, Myers RJK and Weier KL (1981) Potentially mineralizable nitrogen, decomposition rates and their relationship to temperature for five Queensland soils. Aust J Soil Res 19: 323–32

Deans JR, Molina JAE and Clapp CE (1986) Models for predicting potentially mineralizable nitrogen and decomposition rate constants. Soil Sci Soc Am J 50: 323–326

Dobereiner J (1978) Potential for nitrogen fixation in tropical legumes and grasses. *In*: Dobereiner J *et al.* (eds) Limitations and Potentials for Biological Nitrogen Fixation in the Tropics, pp 13–34 Plenum Publishing Corp., New York

Doel DS, Honeycutt CW and Halteman WA (1990) Soil water effects on the use of heat units to predict crop residue carbon and nitrogen mineralization. Biol Fertil Soils 10: 102–106

Frye WW (1981) Nitrification inhibitors. Soil Sci News and Views 2 (10). Dept. of Agronomy, Univ. of Kentucky, Lexington

Honeycutt CW, Zibilske LM and Clapham WM (1988) Heat units for describing carbon mineralization and predicting net nitrogen mineralization. Soil Sci Soc Am J 52: 1346–1350

Jackson G, Keeney D, Curwen D and Webendorfer B (1987) Agricultural management practices to minimize goundwater contamination. Environ Resources Center, Univ WI, Dept. Nat Resources, Madison, WI. 114 p.

Juma NG, Paul EA and Mary B 1984 Kinetic analysis of net nitrogen mineralization in soil. Soil Sci Soc Am J 48: 753–757

Keeney DR (1982) Nitrogen availability indices. *In*: Page AL *et al.* (eds) Methods of Soil Analysis, part 2, 2nd Ed. Agronomy 9: 711–733

Lamond R, Whitney DA, Hickman JS and Bonczkowski LC (1991) Nitrogen rate and placement for grain sorghum production in no-tillage systems. J Prod Agric 4: 532–535

Lamond RE, Whitney DA, Maddux LD, Gordon B, Key O and Koch S (1994) Nitrogen management for no-till corn and grass sorghum production. *In*: Lamond RE (ed) Kansas Fertilizer Research 1993. Rep. Pro. 697, pp 93–97 Kansas Agric Exp Stn. Manhattan, KS

Magdoff FR, Ross D and Amadon J (1984) A soil test for nitrogen availability to corn. Soil Sci Soc Am J 48: 1301–1304

McGill WB and Myers RJK (1987) Controls on dynamics of soil and fertilizer nitrogen. *In*: Follett RF *et al.* (eds) Soil Fertility and Organic Matter as Critical Components of Production Systems. Soil Sci Soc Am Spec Pub 19, pp 73–99. Soil Sci Soc Am, Madison, WI

Meisinger JJ (1984) Evaluating plant-available nitrogen in soil-crop systems. *In*: Hauck RD (ed) Nitrogen in Crop Production, pp 391–416. Am Soc Agron Madison, WI

Mengel DB, Nelson DW and Huber DM (1982) Placement of nitrogen fertilizers for no-till and conventional till corn. Agron J 74: 515–518

Molina JAE Clapp CE and Larson WE (1980) Potentially mineralizable nitrogen in soil: The simple exponential model does not apply for the first 12 weeks of incubation. Soil Sci Soc. Am. J 44: 442–443

Murphy LS, Leikam DF, Lamond RE and Gallagher PJ (1978) Applying N and P at the same time into the same soil shows promise for winter wheat. Better Crops 62: 16–23

Myers RJK (1988) Nitrogen management of upland crops. From cereals to food legumes to sugar cane. *In*: Wilson JR (ed) Advances in Nitrogen Cycling in Agricultural Ecosystems. CAB International, Walling Ford, UK

Myers RJK, Campbell CA and Weier KL (1982) Quantitative relationship between net nitrogen mineralization and moisture content of soils. Can J Soil Sci 62: 111–124

Neeteson JJ, Dilz K and Wijnen G (1989) N-fertilizer recommendations for arable crops. *In*: Germon JC (ed) Management Systems to Reduce Impact of Nitrates, pp 253–264 Elsevier, New York NY

Neyra CA and Dobereiner J (1977) Nitrogen fixation in grasses. Adv Agron 29: 1–38

Peterson GA and Frye WW (1989) Fertilizer nitrogen management. *In*: Follett RF (ed) Nitrogen Management and Ground Water Protection, pp 183–219 Elsevier, New York

Phillips RE, Blevins RC, Thomas GW, Frye WW and Phillips SH (1980) No tillage agriculture. Science 208: 1108–1113

Power JF and Broadbent FE (1989) Proper accounting for N in cropping systems. *In*: Follett RF (ed) Nitrogen Management and Ground Water Protection, pp 160–181. Elsevier, New York

Rice CW, Grove JH and Smith MS (1987) Estimating soil net nitrogen mineralization as affected by tillage and soil drainage due to topographic position. Can J Soil Sci 67: 513–520

Rice CW and Havlin JH (1994) Integrating mineralizable N indices in fertilizer N recommendations. *In*: Havlin *et al.* (eds) New Directions in Soil Testing for Nitrogen, Phosphorus, and Potassium. Soil Sci Soc Am Spec Publ 40, pp 1–13. Soil Sci Soc Am, Madison, WI (*In press*)

Rice CW and Smith MS (1982) Denitrification in no-till and plowed soils. Soil Sci Soc Am J 46: 1168–1173

Rice CW and Smith MS (1984) Short-term immobilization of fertilizer nitrogen at the surface of no-till and plowed soil. Soil Sci Soc Am J 48: 295–297

Sahrawat KL (1994) Nitrification inhibitors and persistence of fertilizer nitrogen in the soil system. *In*: Ahmad N (ed) International Symposium on Nitrogen Economy in Tropical Soils. 9–14 Jan. 1994. Port of Spain, Trinidad. Univ. West Indies, Trinidad, West Indies (*In press*)

Schepers JS, Blackmer TM and Francis DD (1992a) Predicting N fertilizer needs for corn in humid regions: Using chlorophyll meters. *In*: Bock BR and Kelley KR (eds) Predicting N Fertilizer Needs for Corn in Humid Regions. Bull. Y-226, pp 103–114. National Fertilizer and Environmental Research Center, Muscles Schoals, AL

Schepers JS, Francis DD, Vigil M and Below FE (1992b) Comparison of corn leaf nitrogen and chlorophyll meter readings. Commun Soil Sci Plant Anal 23: 2173–2187

Schepers JS and Meisinger JJ (1994) Field indicators of nitrogen mineralization. *In*: Havlin *et al.* (eds) New Directions in Soil Testing for Nitrogen, Phosphorus, and Potassium. Soil Sci Soc Am Spec Publ 40, pp 31–47. Soil Sci Soc Am, Madison, WI (*In press*)

Schepers JS and Mosier AR (1991) Accounting for nitrogen in nonequilibrium soil-crop systems. *In*: RF Follet *et al.* (eds) Managing Nitrogen for groundwater Quality and Farm Profitability, pp 125–138. Soil Sci Soc Am, Madison, WI

Schlegel AJ and Havlin JL (1994) Corn response to long term nitrogen and phosphorus fertilization. J Prod Agric (*In press*)

Stanford G (1982) Assessment of soil nitrogen availability. *In*: Stevenson FJ (ed) Nitrogen in Agricultural Soils, pp 651–688. Soil Sci Soc Madison, WI

Stanford G and Smith SJ (1972) Nitrogen mineralization potentials in soils. Soil Sci Soc Am Proc 36: 465–472

Thomas GW, Blevins RL, Phillips RE and McMahon ME (1973) Effect of killed sod mulch on nitrate movement and corn yield. Agron J 65: 736–739

Watanabe I (1987) Summary report of the *Azolla* program of the international network on soil fertility and fertilizer evaluation for rice. *In*: *Azolla* Utilization. Int Rice Res Inst, Los Banos, Philippines

Watanabe I, Craswell ET and App AA (1981) Nitrogen cycling in wetland rice fields in south-east and east Asia. *In*: Wetselaar R *et al.* (eds) Nitrogen Cycling in South-East Asian W Monsoonal Ecosystems, pp 4–17. Aust Acad Sci, Canberra, Australia

Wells KL (1984) Nitrogen management in the no-till system. *In*: Hauck R (ed) Nitrogen in crop production, pp 535–550. Am Soc Agron, Madison, WI

Zhao-liang Z (1981) Nitrogen cycling and the fate of fertilizer nitrogen in rice fields of the Suchow district, Jiangsu Province, China. *In*: Wetselaar R *et al.* (eds) Nitrogen Cycling in South-East Asian Wet Monsoonal Ecosystems, pp 73–80. Aust Acad Sci, Canberra Australia

Fertilizer Research **42**: 99–107, 1995.
© 1995 *Kluwer Academic Publishers. Printed in the Netherlands.*

Nitrogen assimilation by legumes - processes and ecological limitations

S. Schubert
Institute of Plant Nutrition, University of Hohenheim, Fruwirthstr. 20, 70599 Stuttgart, Germany

Key words: dinitrogen fixation, drought, legumes, membrane transport, nitrogen metabolism, soil acidity

Abstract

Legumes may feed on three different sources of nitrogen: nitrate, ammonium, and, due to symbiotic N_2 fixation, atmospheric dinitrogen. In all three cases ammonium is finally assimilated by the glutamine synthetase (GS) / glutamate synthase (GOGAT) system. NH_4^+ produced by nitrogenase in symbiosomes of legume nodules is released into the host cell cytosol where it is incorporated into amino acids and amides. The release of NH_4^+ into the cytosol appears to occur purely by diffusion. Therefore, the activity of the GS / GOGAT enzymes is decicive to avoid product inhibition of nitrogenase by NH_4^+. No information is available on the mechanism of xylem loading with amides or ureides, a process that may play a key role in avoiding accumulation of amino acids in infected nodule cells. The same applies to phloem unloading of sucrose. Both transport processes, however, may determine the efficiency of N_2 fixation by legumes.

There is no convincing evidence that N_2 fixation by legumes is generally limited by energy supply to nodules. On the other hand, N_2 fixation is often restricted by environmental constraints. Environmental stresses may limit N_2 fixation of legumes at four different levels: *Rhizobium (Bradyrhizobium)* multiplication in soil, rhizobial infection of roots, nodulation, and N_2 fixation. There is increasing evidence that, sufficient infection by effective rhizobial strains provided, N demand of the host plant determines the potential of N_2 fixation. Various environmental stresses and supply of mineral N reduce nodulation and nitrogenase activity without affecting total N concentration of the plant tissue. Stress-induced reduction of plant growth, however, results in an accumulation of free amino acids, amides, or ureides in shoots, roots, and nodules which may be responsible for the regulation of nodulation and nitrogenase activity via a feedback system. This implies that enhancement of N_2 fixation by legumes can be realized in two different ways: either by improvement of stress resistance and dry matter accumulation or by uncoupling of the feedback control.

Introduction

Legumes may feed on three different sources of nitrogen: NO_3^-, NH_4^+, and, due to symbiotic N_2 fixation, atmospheric N_2. In terms of metabolic processes, NH_4^+-N nutrition is the simplest case. Although NH_4^+ may be readily oxidized to NO_3^- in soil, there is evidence that, particularly in tropical soils, available NH_4^+ concentrations may be comparable with NO_3^- concentrations [81]. Nitrification may be inhibited by high temperature and low pH [8]. Consequently, NH_4^+ nutrition may play a major role in tropical soils whereas in temperate soils N is predominantly taken up as NO_3^- [5, 2, 43].

The incorporation of N from the three sources into plant constituents involves metabolic differences that largely affect plant growth. This is especially the case when N_2 is symbiotically fixed by legumes. The purpose of this paper is to describe the major metabolic and transport processes that play a role in N assimilation by legumes. In a second part, the effect of soil acidity, drought, and mineral N on N_2 fixation will be treated. Emphasis will be laid on processes in those legumes that are sufficiently infected (and nodulated) by effective rhizobial bacteria. The effect of soil acidity, drought, and mineral N on rhizobial population, root infection, and nodulation is recognized but will not be discussed here.

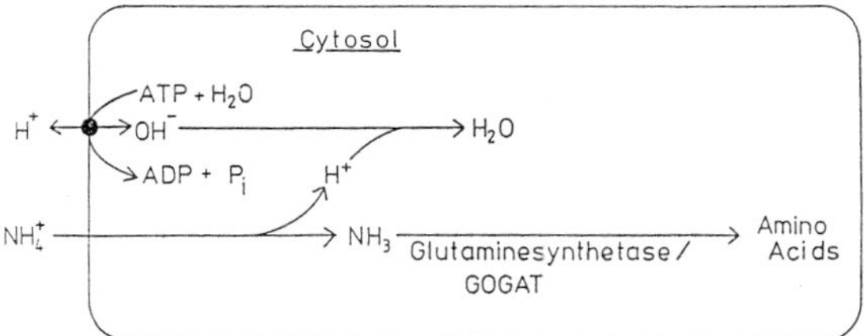

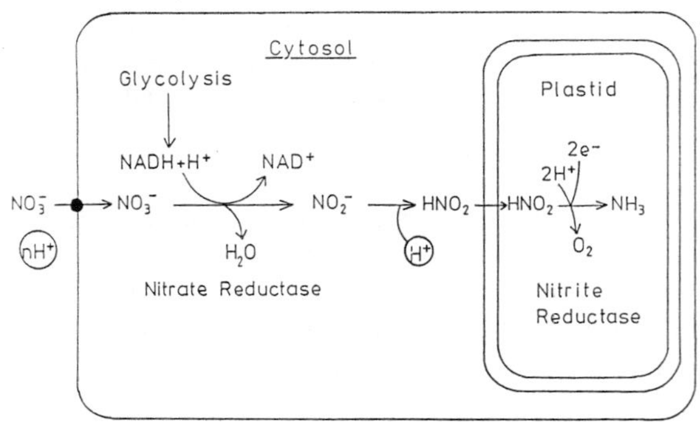

Fig. 1. Model of ammonium uptake and assimilation by root cells. Depolarization of the membrane potential and dissociation of ammonium to ammonia and proton stimulates the plasmalemma ATPase that is responsible for rhizosphere acidification.

Fig. 2. Model of nitrate uptake and reduction by root cells. Nitrate/proton co-transport and nitrate reduction are proton-consuming processes that are responsible for rhizosphere alkalinization.

Nitrogen assimilation

Uptake and assimilation of ammonium and nitrate

A strong driving force for the uptake of NH_4^+ by root cells is established by a plasmalemma-bound H^+ ATPase (Fig. 1) which generates both an electrical and a pH gradient [31, 45, 67, 77]. As a cation NH_4^+ is attracted by the negative charge and enters the cell passively via a specific transport system [9, 36]. Both the partial depolarization of the plasmamembrane and the liberation of protons upon incorporation of NH_3 into glutamate stimulates net proton release by ATPase [22, 45]. Therefore, legumes (and other plants) that depend on NH_4^+ nutrition will strongly acidify their rhizosphere [40, 61].

In the cytosol, NH_4^+ is incorporated into glutamate by glutamine synthetase (GS) forming glutamine. A transferase reaction catalyzed by glutamine oxoglutarate aminotransferase (GOGAT, glutamate synthase) synthesizes two molecules of glutamate by transferring the amino group from glutamine to oxoglutarate [64]. This is the major pathway of NH_4^+ incorporation into amino acids for all three sources of nitrogen. Assimilation of NH_4^+ into oxoglutarate by glutamate dehydrogenase is of minor importance due to the high k_m of this enzyme [12]. According to other results [76], however, the activity of glutamate dehydrogenase may increase under water stress conditions.

In contrast to NH_4^+, NO_3^- has to be transported actively across the plasmalemma of root cells, i.e. against an electro-chemical gradient (Fig. 2). Transport is assumed to be a proton/nitrate symport [78, 80]. Symport of protons with NO_3^- consumes protons in the rhizosphere and therefore increases the pH [49]. This is in contrast to uptake of NH_4^+ or N_2 fixation where generally acidification is observed [6, 29, 35, 46, 59].

Nitrate reduction, catalyzed by nitrate reductase occurs in the cytosol and generates NO_2^-. Because NO_2^- is reduced to NH_4^+ in plastids, NO_2^- has to overcome two membranes of the organelle envelope. This is realized by protonation of NO_2^- to HNO_2 which, as a weak acid, may readily diffuse through membranes in its undissociated form [30]. The consumption of protons in the cytosol adds up to alkalinization and therefore inhibits proton release by ATPase. This is a second reason why there is an increase in rhizosphere pH with NO_3^- nutrition [2, 49].

Dinitrogen assimilation

The third source of N is atmospheric N_2 which can only be made available to legumes by a symbiosis with rhizobia. Within the family of Rhizobiaceae four different genera are distinguished today [19] with *Rhizobizium* and *Bradyrhizobium* being the most important ones from an agricultural point of view. Rhizobial infection by effective strains leads to the formation of nodules which are the specialized organs for N_2 fixation. They derive from a secondary meristem in the root cortex tissue [7, 84].

Nodules consist of three major tissues: peripheral cortex, vascular bundles, and central infection zone [33, 74, 75]. Surrounded by the cortex and infected cells, vascular bundles are connected with root phloem and xylem strands. Energy and C skeletons are delivered as sucrose into the infection zone via the phloem. There is uncertainty about the route of sucrose transport into infected cells. Noninfected cells may play a role in the translocation of C and metabolites to infected cells as judged from the high frequency of plasmodesmatal connections between infected and non-infected cells [74]. Low concentration of sucrose and absence of fructose in the nodule apoplast favor the concept of symplastic unloading of sucrose from phloem into infected cells [74]. For root tissue there is evidence that phloem unloading may occur via a symplastic pathway [24, 50]. Alternatively, phloem unloading of sucrose into the apoplast and hydrolysis by invertase may represent a possibility of selective uptake of hexoses into infected cells (Fig. 3). However, experimental evidence for the nature of sugar import from phloem into infected nodule cells is still missing.

Hexoses are degraded in the glycolytic pathway and organic acids are imported into the bacteroids. Like mitochondria and chloroplasts bacteroids are surrounded by a second membrane which originates from the host cell plasmalemma. Thus bacteroids, surrounded by the peribacteroid membrane, form an organelle similar to chloroplasts and mitochondria. This organelle is called 'symbiosome' [84]. The symbiosome is responsible for N_2 fixation by catalytic activity of the nitrogenase. The product of N_2 fixation, NH_4^+, is released into the cytosol by passive diffusion across the symbiosome membranes [36, 73].

In the host cell cytosol NH_4^+ is assimilated by the GS/GOGAT enzymes [25, 44, 76]. Amides (or in many tropical legumes ureides) are exported to the shoot via the xylem [64]. Again nothing is known about the transport processes at the plasmalemma of infected cells which are responsible for amide or ureide export into the xylem. Active transport mechanisms may contribute to low cytosolic amino acid concentrations in infected cells which are required for efficient NH_4^+ assimilation. This in turn generates the diffusion gradient for NH_4^+ between symbiosome and cytosol. A steep gradient enhances the release of NH_4^+ into the cytosol and avoids product inhibition of nitrogenase by NH_4^+ [76].

In many agro-ecosystems of the tropics or subtropics N_2 fixation by legumes makes an important contribution to the N economy of soils [56]. Therefore, breeding programs have been initiated that intend to increase the potential for N_2 fixation [e.g. 10, 14]. Identifying the rate-limiting steps in the supply with sucrose and/or the export of reduced N may represent a key to further success in these efforts. Transport processes at the plasmalemma of infected cells may be an important factor in the efficiency of N_2 fixation in legumes. However, there are deep gaps in our knowledge of membrane processes of carbon import into and nitrogen export out of infected cells.

Soil acidity

Several ecological constraints may limit the N_2 fixation potential of legumes [11, 28, 79]. The identification of the process that is most sensitive to the stress situation promises the greatest success in breeding programs or in an improvement of agronomical practices [1]. In humid areas legume growth may be adversely affected by soil acidity [3]. Soil acidity is a complex of soil factors that may all limit growth to varying degree depending on the conditions.

High proton activity (low pH) may directly affect plant growth. [34, 39]. However, low pH also promotes the solubilization of toxic ions e.g. Al and Mn. In addition, in soils of low pH deficiencies of Ca, Mg, P, Mo, and S are likely to occur [20]. For an N_2-fixing legume the situation becomes even more complicated because all these constraints may not only become apparent at the host plant level but also at the level of rhizobial multiplication, infection, and nodulation [25, 51]. Improvement of N_2 fixation first requires an understanding of which of the four different levels is most sensitive to soil acidity. The best parameter to decide this question is N concentration of the plant tissue [52]. If N_2 fixation is directly impaired by low pH, N deficiency will occur. On the other hand, if growth of the host plant is more sensitive, total N concentration will either increase or will remain unchanged.

Field bean (*Vicia faba*) is extremely sensitive to low pH. In a series of experiments it has been demonstrated that this sensitivity is independent of Al or Mn toxicity and nutrient deficiency, although these factors may come into play when pH values of less than 5 are considered [63, 65, 85]. However, plant growth may be significantly reduced at moderately low pH (6.2) [65]. This is of particular importance in soils of poor buffer capacity because N_2 fixing beans strongly decrease their rhizosphere pH [65].

Low rhizosphere pH prevents cell elongation but not cell division [70, 85]. Recent evidence suggests that poor growth of field bean at low pH is caused by a lack of net proton release by ATPase activity ([85], Fig. 4). An insufficient net proton release by root cells apparently results in a decrease of cytoplasmic pH [23, 85]. This in turn may not only have serious consequences for root cell metabolism due to changed enzyme activities [17] but may specifically change gene activation [22].

Despite strong inhibition of nitrogenase activity, N concentrations in the plant tissue were optimal, indicating that inhibition of N_2 fixation was not limiting

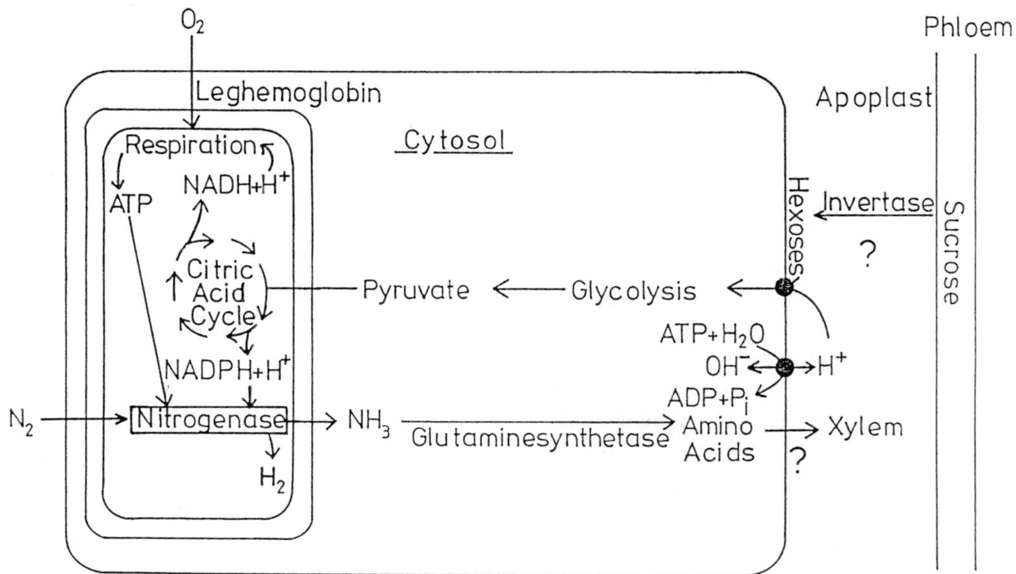

Fig. 3. Model of metabolic and transport pathways in infected cells of legume nodules.

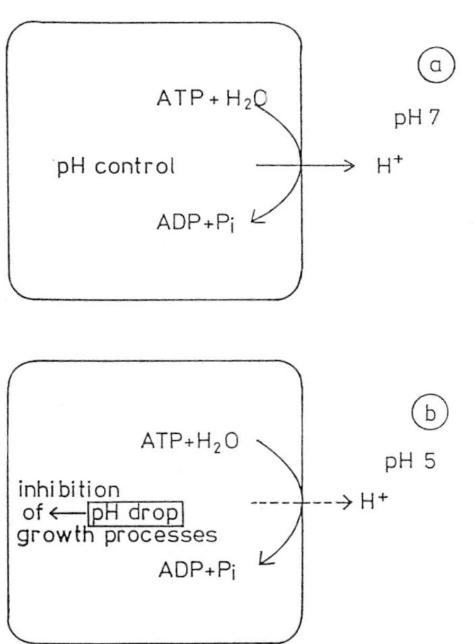

Fig. 4. Model of inhibition of root cell elongation by low medium pH. a: sufficient net proton release by ATPase activity, b: inhibited net proton release.

plant growth at low pH [63]. This example not only shows that growth of field beans is more pH-sensitive than the establishment and functioning of the symbiosis, it also infers that there must be a feedback control of N_2 fixation by plant growth.

Drought stress

Drought stress severely inhibits nitrogenase activity [26], N_2 fixation [54], and nodulation [54]. Various hypotheses have been proposed to explain the effect of water stress on N_2 fixation. During the seventies it was suggested that water stress inhibits photosynthesis by stomatal closure and thereby decreases the supply with sucrose to nodules. This would result in an energy shortage of N_2 fixation. Although at severe stress primary inhibition of photosynthesis cannot be ruled out [18], there is convincing evidence that drought stress does not generally limit the supply of nodules with carbohydrates [13, 32, 57, 58].

In the eighties an alternative hypothesis was advanced that explains reduced nitrogenase activity in terms of an increased resistance of the variable O_2 diffusion barrier in the nodule inner cortex [32, 33, 48, 68]. According to this concept various stresses alter the diffusion of O_2 in the inner cortex tissue thus inhibiting carbon metabolism and nitrogenase activity. Alternatively, it was proposed that the increased gas resistance may also lead to an accumulation of H_2 which may inhibit nitrogenase activity [32].

There is a need for efficient control of the O_2 supply of infected cells because of the extreme O_2 sensitivity of the nitrogenase enzyme [25, 71]. However, the question arises whether O_2 supply by variation of the gas diffusion barrier is the immediate point of metabolic control of nitrogenase activity. The fact that O_2 concentration in the infection zone limits nitrogenase activity and that increasing O_2 concentrations stimulate nitrogenase activity [32] does not necessarily imply that nitrogenase activity is under primary control of the variable O_2 diffusion barrier. Furthermore, increasing O_2 concentrations did not completely abolish the inhibition of nitrogenase by water stress [26].

Some authors have reported on an accumulation of reduced N in the form of NH_4^+, amino acids, and ureides in shoots, roots, and even nodules under stress conditions [57, 83]. An example is shown in Figure 5. Drought stress (-0.5 MPa), induced by PEG in nutrient solution, reduced the leaf water potential of alfalfa from -0.5 MPa to -1.2 MPa. This significantly inhibited shoot growth as indicated by a significant difference in shoot dry weight between control and stress treatments after 5 days of stress (not shown). Total amino acid concentrations in shoots, roots, and nodules were significantly increased by the stress (Fig. 5) whereas total N concentrations remained unchanged (not shown) in agreement with other findings [54].

The accumulation of amino acids indicates that growth reduction was not due to N deficiency but probably to a drop of turgor [4] or cell wall extensibility [82]. The accumulation of amino acids suggests that O_2 supply did not primarily limit nitrogenase activity. In contrast, this suggests that nitrogenase activity was limited by product inhibition in the form of NH_4^+ which may accumulate when its incorporation into amino acids and/or amides is inhibited. Because growth of young vegetative parts (meristems) requires the supply of amino acids for growth processes, the inhibition of cell extension and especially cell division will result in an accumulation of amino acids [44]. The reduced demand for amino acids will then in turn result in an accumulation in roots and nodules. The decrease of glutamine synthetase and increase in glutamate dehydrogenase activity during water stress applied to soybean [76] indicates that under water stress conditions ammonium may indeed accumulate in infected cells.

The proposed model of the regulation of nitrogenase activity during reduced nitrogen demand under water stress conditions suggests that nitrogenase activity is under control of the NH_4^+ concentration of the host cell cytosol. According to this concept the increase in resistance of the variable O_2 diffusion barrier is a regulation mechanism superimposed on the control of O_2 supply to the infected zone. The latter may respond to the lower O_2 demand of infected cells when nitroge-

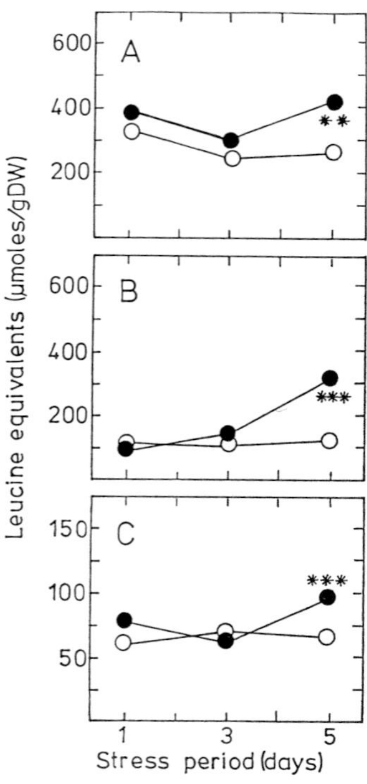

Fig. 5. Effect of water stress (-0.5 MPa) induced by PEG on the concentrations of amino acids in nodules (**A**), roots (**B**), and shoots (**C**) of alfalfa in nutrient solution. o control, ● stress treatment. Significant differences at **$p = 1\%$ and 0.1% level, respectively. (Schubert, Serraj, Plies-Balzer, and Mengel, unpublished results)

nase activity is reduced. Thus the superimposed regulation mechanism of O_2 diffusion avoids inhibitory oxygen concentrations when O_2 utilization is reduced under stress conditions. To test this model, comparative data of O_2 concentrations and substrate concentrations in infected cells under stress conditions is needed.

Two alternative hypotheses have recently been proposed to explain the accumulation of nitrogenous compounds in the nodule infection zone. According to Streeter [75] the morphological structure of legume nodules makes it appear unlikely that reduced N is pulled into the xylem by the transpiration stream. The vascular bundles come to a dead end in the nodule. Therefore water leaves the nodule via the xylem, whereas water import occurs via phloem (and cortex). Since water import and export are in an equilibrium, it is possible that disturbance of this equilibrium results in a build-up of reduced N.

Available water within the nodule would then be a key factor for the potential of N_2 fixation. In this context, it is important to understand the mechanism that drives water through the nodule. Because water movement in plants generally follows a potential gradient established by solute transport across membranes, the identification of the responsible active transport mechanism is of crucial importance in the understanding of N_2 fixation limitation under water stress conditions. In principle, two major possibilities of active transport are conceivable: active phloem unloading of sugars into infected cells and active xylem loading of reduced N (amides or ureides). In any case, the plasmalemma of cells in the infection zone is predestined for either of the two alternatives.

A second hypothesis advanced by Parsons *et al.* [53] proceeds on the assumption that reduced N that is not consumed in the shoot tissue will be recycled through phloem. If plant growth is restricted by a stress condition the accumulation of amino acids will decrease the C/N ratio in the phloem sap. Thus amino acids in phloem sap that arrive at the nodule may regulate N_2 fixation and possibly also nodule growth. In amide exporters glutamate, serine and proline may be of particular importance in this respect. Interestingly, proline accumulated severalfold in field bean exposed to water stress [57].

In ureide exporters glutamine and asparagine may regulate nodule activity in a feedback control. Possibly these metabolites are also involved in changes of the gas diffusion barrier and thereby regulate the oxygen availability in the infection zone [53]. An important implication of this hypothesis appears to be a lack of selectivity of phloem unloading. Symplastic unloading of sugars from phloem should prevent cells of the infection zone from discriminating between sugars and amino acids. On the other hand, apoplastic unloading would allow the exclusion of amino acids at the plasmalemma of infected cells.

An exact understanding of the inhibition of nitrogenase activity requires a profound understanding of the membrane transport processes that are responsible for phloem unloading of sucrose and xylem loading with amides and ureides. In this context two aspects are of particular importance. First, which are the active transport steps at the plasmalemma of infected cells? Second, does selectivity of phloem unloading and/or xylem loading explain the feedback control of the regulation of nitrogenase activity?

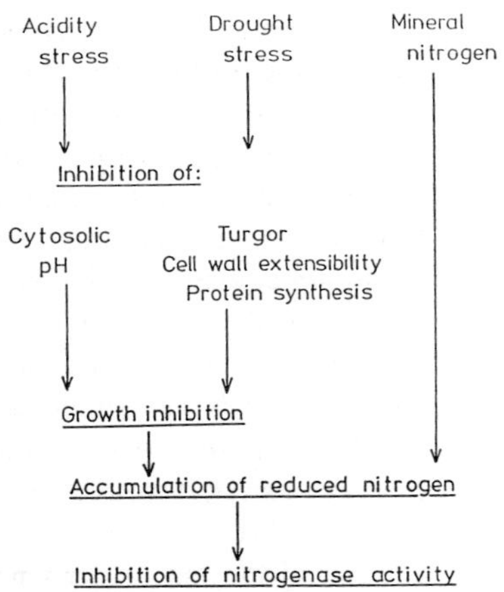

Fig. 6. Model of inhibition of N_2 fixation by various environmental constraints.

Mineral nitrogen

There are large differences in N_2 fixation efficiency among the various legume – rhizobium symbioses [15, 37, 56]. Generally, forage legumes harvested in the vegetative state are more efficient than grain legumes. Whereas the former meet 80 to 90% of their N demand by N_2 fixation, the latter show large differences in efficiency [28]. This has been explained in terms of source-sink relationships. In contrast to field bean (*Vicia faba*) and pigeon pea (*Cajanus cajan*), soybean, pea, and white lupin show poor N_2 fixation capabilities after flowering. The latter three species develop strong sink activity in pods thereby drawing away N from photosynthesizing leaves [28, 38, 41, 47, 69]. This ends up with a lack of sucrose supply to nodules (or alternatively: a low C/N ratio in phloem sap [53]) which then decreases nodule activity. Therefore, mineral N application to legumes with poor N_2 fixation capacity may increase yield performance [41, 47], particularly if applied via the leaves [16]. Legumes with high N_2 fixation capacity (e.g. *Vicia faba*) will take up available mineral N from soil and correspondingly reduce N_2 fixation [41, 60, 62].

On the other hand, there are many reports on an inhibition of nitrogenase activity after mineral N fertilization [25, 66, 72]. Although the reduced nodule activity is related to the increase of the variable O_2 diffusion resistance [48] the mechanism is still under debate [32]. One possible explanation for the reduction of nitrogenase activity with NO_3^--N supply may be the feedback mechanism described above [5]. After uptake, NO_3^- may be reduced in roots or shoots and may then meet the N demand of plant growth. Amino acids from N_2 fixation not consumed for plant growth will then be recycled to nodules via phloem where they suppress nitrogenase activity [5].

Higher N_2 fixation capabilities of legumes in the presence of NO_3^- (e.g. supernodulating mutants [27]) may be explained in terms of uncoupling of N_2 fixation from the feedback control. Mechanistically this may be realized by inclusion of NO_3^- or reduced N compounds in vacuoles or other cell compartments where these metabolites are hindered from participation in the recycling and feedback control. This may also play a role in genotypes with higher leaf weight where a higher percentage of N may be transiently stored in proteins which can be mobilized during pod filling [55].

Conclusions

There is increasing evidence that, in effectively nodulated legumes, N_2 fixation is under tight metabolic control of the host plant. Soil acidity, water stress, and other environmental constraints (e.g. low light intensity or potassium deficiency [21]) may decrease N_2 fixation regulated by a feedback control system which is triggered by disturbed plant growth and accumulated reduced N (Fig. 6).

Two possibilities may be envisaged to increase N_2 fixation by plant breeding. Because N_2 fixation appears to be controlled by feedback inhibition, inefficient fixation in some legume species may be caused by inefficient membrane transport processes for phloem unloading of sugars and/or xylem loading of ureides or amides. In efficiently fixing legumes, it is possible to further increase N_2 fixation by uncoupling it from the feedback control. From an ecological point of view this approach appears to be questionable because it increases N concentrations in the tissue. After harvest N losses due to fast mineralization of the N-rich material provokes environmental hazards. For this reason it appears to be more appropriate to improve the stress tolerance of legumes. This not only improves yield and yield stability but, at the same time, increases the amount of N fixed.

References

1. Adams F (1981) Nutritional imbalances and constraints to plant growth on acid soils. J Plant Nutr 4: 81–87

2. Allan S and Raven JA (1987) Intracellular pH regulation in *Ricinus communis* grown with ammonium or nitrate as N source: The role of long distance transport. J Exp Bot 38: 580–596

3. Andrew CS (1976) Effect of calcium, pH and nitrogen on the growth and chemical composition of some tropical and temperate pasture legumes. Aust J Agric Res 27: 61–623

4. Arneke WW (1980) Der Einfluss des Kaliums auf die Komponenten des Wasserpotentials und auf die Wachstumsrate von *Phaseolus vulgaris* L. Ph. D. thesis, Justus Liebig University Giessen, Germany

5. Barraclough PB (1986) The growth and activity of winter roots in the field: nutrient uptakes of high-yielding crops. J Agric Sci Camb 106: 45–52

6. Bashan Y and Levanony H (1989) Effect of root environment on proton efflux in wheat roots. Plant Soil 119: 191–197

7. Bauer WD (1981) Infection of legumes by rhizobia. Ann Rev Plant Physiol 32: 407–449

8. Beck T (1983) Die N-Mineralisierung von Böden im Laborbrutversuch. Z Pflanzenernähr Bodenkd 146: 243–252

9. Bertl A, Felle H and Bentrup FW (1984) Amine transport in *Riccia fluitans*. Cytoplasmic and vacuolar pH recorded by a pH-sensitive microelectrode. Plant Physiol 76: 75–78

10. Bliss FA (1993) Breeding common bean for improved biological nitrogen fixation. Plant Soil 152: 71–79

11. Bliss FA (1993) Utilizing the potential for increased nitrogen fixation in common bean. Plant Soil 152: 157–160

12. Caba JM, Lluch C and Ligero F (1993) Genotypic differences in nitrogen assimilation in *Vicia faba*: Effect of nitrate. Plant Soil 151: 167–174

13. Chen CL and Sung JM (1983) The effect of water stress on enzymes of carbon metabolism in cytosol of soybean nodules. J Agric Ass China 121: 28–34

14. Chloupek O, Babinec J and Malá M (1992) Breeding of lucerne for higher symbiotic nitrogen fixation and testing in hydroponics with limited mineral nitrogen. Bodenkultur 43: 109-113

15. Danso SKA, Hardarson G and Zapata F (1993) Misconceptions and practical problems in the use of ^{15}N soil enrichment techniques for estimating N_2 fixation. Plant Soil 152: 25–52

16. Da Silva PM, Tsai SM and Bonetti R (1993) Response to inoculation and N fertilization for increased yield and biological nitrogen fixation of common bean (*Phaseolus vulgaris* L.). Plaint Soil 152: 123–130

17. Davies DD (1973) Control of and by pH. Symp Soc Exp Biol 27: 513– 529

18. Durand JL, Sheehy JE and Minchin FR (1987) Nitrogenase activity, photosynthesis and nodule water potential in soyabean plants experiencing water deprivation. J Exp Bot 38: 311–321

19. Elkan GH (1992) Taxonomy of the rhizobia. Can J Microbiol 38: 446–450

20. Fageria NK, Maligar VC and Wright RJ (1989) Growth and nutrient concentrations of alfalfa and common bean as influenced by soil acidity. Plant Soil 119: 331–333

21. Feigenbaum S and Mengel K (1979) The effect of reduced light intensity and suboptimal potassium supply on N_2 fixation and N turnover in *Rhizobium* infected lucerne. Physiol Plant 45: 245–249

22. Felle H (1989) pH as a second messenger in plants. In: Second Messengers in Plant Growth and Development, pp 145–166. Alan Liss Inc. New York

23. Gerendás J, Ratcliffe RG and Sattelmacher B (1990) ^{31}P nuclear magnetic resonance evidence for differences in intracellular pH in the roots of maize seedlings grown with nitrate or ammonium. J Plant Physiol 13: 125–128

24. Giaquinta RT, Lin W, Sadler NL and Franceschi VR (1983) Pathway of phloem unloading of sucrose in corn roots. Plant Physiol 72: 362–367

25. Giller KE and KJ Wilson (1991) Nitrogen Fixation in Tropical Cropping Systems. CAB International, Oxon, UK

26. Guerin V, Trinchant JC and Rigaud J (1990) Nitrogen fixation (C_2H_2 reduction) by broad bean (*Vicia faba* L.) nodules and bacteroids under water-restricted conditions. Plant Physiol 92: 595–601

27. Hansen AP, Yoneyama T, Kouchi H and Martin P (1993) Respiration and nitrogen fixation of hydroponically cultured *Phaseolus vulgaris* L. cv. OAC Rico and a supernodulating mutant. I. Growth, mineral composition and effect of sink removal. Planta 189: 538–545

28. Hardarson G (1993) Methods for enhancing symbiotic nitrogen fixation. Plant Soil 152: 1–17

29. Haynes RJ (1983) Soil acidiflcation induced by leguminous crops. Grass For Sci 38: 1–11

30. Heber U and Heldt HW (1981) The chloroplast envelope: Structure, function, and role in leaf metabolism. Ann Rev Plant Physiol 32: 139–168

31. Hodges TK (1973) Ion absorption by plant roots. Agronomy 25: 163–207

32. Hunt S and Layzell DB (1993) Gas exchange of legume nodules and the regulation of nitrogenase activity. Ann Rev Plant Physiol Plant Mol Biol 44: 483–511

33. Iannetta PPM, De Lorenzo C, James EK, Fernandez-Pascual M, Sprent JI, Lucas MM, Witty JF, De Felipe MR and Minchin FR (1993) Oxygen diffusion in lupin nodules. Visualization of diffusion barrier operation. J Exp Bot 44: 1461–1467

34. Islam AKMS, Edwards DG and Asher CJ (1980) pH optima for crop growth. Results of a flowing solution culture experiment with six species. Plant Soil 54: 339–357

35. Israel DW and WA Jackson (1978) The influence of nitrogen nutrition on ion uptake and translocation by leguminous plants. In: Andrew CS and Kamprath EJ (eds) Mineral Nutrition of Legumes in Tropical and Subtropical Soils, pp 113–128. CSIRO, Australia

36. Kleiner D (1981) The transport of NH_3 and NH_4^+ across biological membranes. Biochim Biophys Acta 639: 41–52

37. LaRue TA and Patterson TG (1981) How much nitrogen do legumes fix? Adv Agron 4: 15–38

38. Lawrie AC and Wheeler CT (1975) Nitrogen fixation in the root nodules of *Vicia faba* L. in relation to the assimilation of carbon. New Phytol 74: 429–436

39. Mahler RL and McDole RE (1987) Effect of soil pH on crop yield in northern Idaho. Agron J 79: 751–755

40. Marschner H and Römheld V (1983) *In vivo* measurement of root-induced pH changes at the soil-root interface: effect of plant species and nitrogen source. Z Pflanzenphysiol 111: 241–251

41. Martin P (1990) Einfluß von Mineralstoffen auf das symbiontische N_2-Bindungssystem bei Leguminosen. Kali-Briefe (Büntehof) 20: 93–110

42. Mengel K (1977) Factors of nutrient availability and their relevance for crop production. In: Proc. of the International Seminar on Soil Environment and Fertility Management in Intensive Agriculture, pp 788–796. Tokyo, Japan

43. Mengel K (1991) Available nitrogen in soils and its determination by the "N_{min}-method" and by electroultrafiltration (EUF). Fert Res 28: 251–262

44. Mengel K (1994) Impact of the nutrient status of the host plant on symbiotic dinitrogen fixation. Z Pflanzenernähr Bodenkd 157: 233-241

45. Mengel K and Schubert S (1985) Active extrusion of protons into deionized water by roots of intact maize plants. Plant Physiol 79: 344–348

46. Mengel K and Steffens D (1982) Beziehung zwischen Kationen/Anionen-Aufnahme von Rotklee und Protonenabscheidung der Wurzeln. Z. Pflanzenernähr Bodenkd 145: 229–236

47. Merbach W and Schilling G (1980) Wirksamkeit der symbiontischen N_2- Fixierung der Körnerleguminosen in Abhängigkeit von Rhizobienimpfung, Substrat, N-Düngung und ^{14}C-Saccharoselieferung. Zentralbl Bakteriol II Abt 135: 99–118

48. Minchin FR, Sheehy JE and Minguez MI (1984) Acetylene-induced changes in the oxygen diffusion resistance and nitrogenase activity of legume root nodules. Ann Bot 5: 13–20

49. Moorby H, Nye PH and White RE (1985) The influence of nitrate nutrition on H^+ efflux by young rape plants. Plant Soil 84: 403–415

50. Mühling KH, Schubert S and Mengel K (1993) Role of plasmalemma H^+ ATPase in sugar retention by roots of intact maize and field bean plants. Z Pflanzenernähr Bodenkd 156: 155–161

51. Munns DN (1986) Acid soil tolerance in legumes and rhizobia. Adv Plant Nutr 2: 63–91

52. Munns DN, Hohenberg JS, Righetti TL and Lauter DJ (1981) Soil acidity tolerance of symbiotic and nitrogen-fertilized soybeans. Agron 73: 407–410

53. Parsons R Stanforth A, Raven JA and Sprent JJ (1993) Nodule growth and activity may be regulated by a feedback mechanism involving phloem nitrogen. Plant Cell Environ 16: 125–136

54. Peña-Cabriales JJ and Castellanos JZ (1993) Effects of water stress on N_2 fixation and grain yield of *Phaseolus vulguris* L. Plant Soil 152: 151–155

55. Peña-Cabriales JJ, Grageda-Cabrera OA and Kola V (1993) Time course of N_2 fixation in common bean (*Phaseolus vulgaris* L.). Plant Soil 152: 115–121

56. Peoples MB and Herridge DF (1990) Nitrogen fixation by legumes in tropical and subtropical agriculture. Adv Agron 44: 155–223

57. Plies-Balzer E (1992) Ertagsleistung, N_2-Fixierung und osmotische Anpassung an Wasserstress bei *Vicia faba*. Ph. D. thesis Justus Liebig University Giessen, Germany

58. Plies-Balzer E, Kong T, Schubert S and Mengel K (1994) Effect of water stress on plant growth, nitrogenase activity and nitrogen economy of four different cultivars of *Vicia faba* L. Eur J Agron (*In press*)

59. Raven JA and Smith FA (1976) Nitrogen assimilation and transport in vascular land plants in relation to intracellular pH regulation. New Phytol 76: 415–431

60. Santos PJA, Edwards DG Asher CJ and Dart PJ (1993) Responses of *Bradyrhizobium*-inoculated mungbean (*Vigna radiata*) to applied nitrogen. In: Barrow NJ (ed) Plant Nutrition - From Genetic Engineering to Field Practice, pp 443–446. Kluwer Academic Publishers, Dordrecht, The Netherlands

61. Schaller G and Fischer WR (1985) pH-Änderungen in der Rhizosphäre von Mais- und Erdnußwurzeln. Z Pflanzenernähr Bodenkd 148: 306–320

62. Scherer HW and Danzeisen L (1980) Der Einfluß gesteigerter Stickstoffgaben auf die Entwicklung der Wurzelknöllchen, auf die symbiontische Stickstoffassimilation sowie auf das Wachstum und den Ertrag von Ackerbohnen (*Vicia faba* L.). Z. Pflanzenernähr Bodenkd 143: 464–470

63. Schubert E, Mengel K and Schubert S (1990) Soil pH and calcium effect on nitrogen fixation and growth of broad bean. Agron J 82: 969–972

64. Schubert KR, (1986) Products of biological nitrogen fixation in higher plants: Synthesis, transport, and metabolism. Ann Rev Plant Physiol 37: 539–574

65. Schubert S, Schubert E and Mengel K (1990) Effect of low pH of the root medium on proton release, growth, and nutrient uptake of field beans (*Vicia faba*). Plant Soil 124: 239–244 (1990)

66. Serraj R, Drevon JJ, Obaton M and Vidal A (1992) Variation in nitrate tolerance of nitrogen fixation in soybean (*Glycine max*) - *Bradyrhizobium* symbiosis. J Plant Physiol 140: 366–371

67. Serrano R (1989) Structure and function of plasma membrane ATPase. Ann Rev Plant Physiol Plant Mol Biol 40: 61–94

68. Sheehy JE, Minchin FR and Witty JF (1983) Biological control of the resistance to oxygen flux in nodules. Ann. Bot 52: 565–571

69. Shibles R, Anderson IC and Gibson AH (1975) Soybean In: Evans LT (ed) Crop Physiology, pp 151–189. Cambridge University Press, Cambridge, UK

70. Smith DL and Krikorian AD (1992) Low external pH prevents cell elongation but not multiplication of embryonic carrot cells. Physiol Plant 84: 495–501

71. Sprent JI and Sprent P (1990) Nitrogen Fixing Organisms. Pure and Applied Aspects. Chapman and Hall, London, UK

72. Streeter JG (1985) Nitrate inhibition of legume nodule growth and activity. I Long term studies with a continuous supply of nitrate. Plant Physiol 77: 321–324

73. Streeter JG (1989) Estimation of ammonium concentration in the cytosol of soybean nodules. Plant Physiol 90: 779–782

74. Streeter JG (1992) Analysis of apoplastic solutes in the cortex of soybean nodules. Physiol Plant 84: 584–592

75. Streeter JG (1993) Translocation - A key factor limiting the efficiency of nitrogen fixation in legume nodules. Physiol Plant 87: 616–623

76. Sung JM and Chen CL (1983) The effect of water stress on enzymes of ammonia assimilation in cytosol of soybean nodules. J Agric Ass China 121: 20–27

77. Sussman MR and JF Harper (1989) Molecular biology of the plasma membrane of higher plants. Plant Cell 1: 953–960

78. Thibaud JB and Grignon C (1981) Mechanism of nitrate uptake :in corn roots. Plant Sci Lett 22: 279–289

79. Tsai SM, Bonetti R, Agba SM and Rossetto R (1993) Minimizing the effect of mineral nitrogen on biological nitrogen fixation in common bean by increasing nutrient levels. Plant Soil 152: 131–138

80. Ullrich CI and Novacky AJ (1990) Extra- and intracellular pH and membrane potential changes induced by K^+, Cl^-, $H_2PO_4^-$, and NO_3^- uptake and fusicoccin in root hairs of *Limnobium stoloniferum*. Plant Physiol 94: 1561–1567

81. Uloro Y and Mengel K (1994) Response of ensete (*Ensete ventricosum* W.) to mineral fertilizer applications under representative growth conditions in the vicinity of Hossaina, southwest Ethiopia. Fert Res 37: 107–113

82. Van Volkenburgh E and Boyer JS (1985) Inhibitory effects of water deficit on maize leaf elongation. Plant Physiol 77: 190–194

83. Venkatesvarlu B and Rao AV (1987) Quantitative effects of field water deficits on N_2 (C_2H_2) fixation in selected legumes grown in the Indian desert. Biol Fert Soils 5: 18–22

84. Werner D, Ahlborn B, Bassarab S, Kape F, Kinnback A, Mellor RB, Mörschel E, Müller P, Parnike M, Schmidt P and Schultes A (1992) Nodule development and nitrogen fixation in the *Rhizobium/Bradyrhizobium* system. In: Mengel K and Pilbeam DJ (eds) Nitrogen Metabolism of Plants, pp 17-30. Proc Phytochem Soc Europe 33. Oxford Science Publications, UK

85. Yan F, Schubert S and Mengel K (1992) Effect of low root medium pH on net proton release, root respiration and root growth of corn (*Zea mays* L.) and broad bean (*Vicia faba* L.). Plant Physiol 99: 415–421

Fertilizer Research **42**: 109–115, 1995.
© 1995 *Kluwer Academic Publishers. Printed in the Netherlands.*

Fate of $CO(NH_2)_2$-^{15}N applied to taro (*Colocasia esculenta* var. *esculenta*) in an acid Vertisol of Trinidad

N. Ramnanan, N. Ahmad & S.M. Griffith
Department of Soil Science, University of the West Indies, St. Augustine, Trinidad and Tobago, West Indies

Key words: Colocasia esculenta var. *esculenta*, crop uptake, ^{15}N, soil management, tropical soils, urea, Vertisols

Abstract

Vertisols are an important natural resource, particularly in the developing world, but a greater understanding of their chemical characteristics with special reference to the fate of applied N is needed for their greater sustainable use. The Bejucal clay (*Chromic Dystraquerts*, very fine, mixed, acid) is an important agricultural soil in North Trinidad and a series of studies were undertaken on it to determine the optimum rate and the best time of application and to quantify the efficiency of applied N using $CO(NH_2)_2$-^{15}N and taro, commonly called dasheen locally, (*Colocasia esculenta* var. *esculenta*), as a test crop.

The optimum rate was found to be 280 kg N ha^{-1} for corm production from a field experiment, laid out in a randomised incomplete block design. This corresponded to an apparent recovery at 10 per cent when $CO(NH_2)_2$ was applied at 14 days after planting (DAP). Generally, a decrease in N uptake with increasing levels and later application times was observed. Another field experiment which employed ^{15}N applied at 14, 56, and 112 DAP showed that ^{15}N applied at 112 DAP corresponded to the highest ^{15}N level in the plant tissue.

A green house experiment conducted to determine the efficiency of the use of the applied N showed the highest recovery of approximately 49 per cent at a rate equilavent to 280 kg N ha^{-1} with efficiencies of 43 and 40 per cent at the 140 and 420 kg N ha^{-1} of applied N levels, respectively. A single application at 14 DAP resulted in an efficiency of 41 per cent compared to 44 and 47 per cent for the 2 split and 3 split applications, respectively. Split applications increased N uptake but without a corresponding increase in yield.

These studies showed that N applied as $CO(NH_2)_2$ at 280 kg N ha^{-1} at 14 DAP is optimum for upland taro production under conditions of the Bejucal clay. Under normal field conditions surface washing may be the most important N loss pathway but when excessive fertilizers are used then the gaseous pathways may also contribute to N losses.

Introduction

Large areas of heavy clay soils exist throughout the world and remain underutilised as a result of the lack of appropriate management practices, especially with reference to N. Since Vertisols potentially represent a significant land resource base for the developing world to increase agricultural production, there is an urgent need for appropriate strategies to manage these soils.

The main problems of N management on the Bejucal clay (*Chromic Dystraquerts*, very fine, mixed, acid) series are: low native N content, restricted drainage due to flat relief, fine soil texture and heavy rainfall for approximately six months of the year followed by a severe dry spell, which makes land preparation at any time difficult, resulting in poor soil conditions for root growth. In Trinidad as in other areas where these soils occur, there are limited soil fertility recommendations except for plantation crops such as sugar cane (*Saccharum officinarum*) and commercial crops such as rice (*Oryza sativa*).

A research project was undertaken to determine the efficiency of N applied to this heavy clay and methods which can be employed to improve this. The taro crop (*Colocasia esculenta* var. *esculenta*), referred to locally as dasheen, was chosen as the test crop in these studies because of its ability to withstand flooded and waterlogged or saturated soil conditions (Sivan, 1980), which may occur locally, for at least four months of the year.

Before the efficiency of the applied N could be measured, the optimum rate of N application had to be determined since this apparently varies from 40 kg N ha^{-1} (Igobokume & Ogbannaya, 1980) to 1120 kg N ha^{-1} (De la Pena & Plucknett, 1967), depending on the farming systems and the particular regions of the world. A series of experiments were therefore undertaken with the aim firstly to determine the optimum rate and best suited application time for the crop, secondly, to determine the time of most active N uptake and thirdly, to quantify the per cent uptake and the per cent residual soil N.

Materials and methods

Optimum rate and best suited application time

This experiment sought to determine the optimum application rate of $CO(NH_2)_2$-N and the most appropriate time of application in the cropping cycle of taro, which would affect N uptake and yield.

The experiment was laid out in a randomised incomplete block design. The treatments were 4 levels of N with 2 levels of P and K, respectively, replicated 3 times and confounded in blocks of 24 units. N was not applied to the control treatment of each block. Three application times at 14, 56, and 112 days after planting, (DAP) respectively, were used as subplots of N and applied to each plot. These times corresponded to the initial vegetative growth phase, rapid vegetative growth and development phase, and commencement of rapid bulking of corms, respectively. Applied rates of N were 0, 280, 560, and 1120 kg N ha^{-1}. Basal applications of triple superphosphate (TSP) and muriate of potash (KCl) were applied at 0 and 560 kg ha^{-1} P_2O_5 and K_2O, respectively. The fertilizers were applied in a circular band around the plant, approximately 15 cm from its base, then covered over using a hoe, the process referred to as moulding.

Plants were established at a density of approximately 37500 plants per hectare (pph) and at 46, 75, 103, 125, 140, and 180 DAP; various plant data were collected at a 3 weekly interval during active growth and development and a longer time period during crop senescence. Measurements were made according to Chapman (1964) for the estimation of leaf area. The plant height and leaf number were also recorded for the same, randomly chosen, three plants per plot on which all observations were conducted. At seven months after planting all the plants were harvested, cleaned and separated into corms and cormels and the fresh weights per plot were recorded.

Determination of the most suited N application time

This field experiment was conducted to determine the most active time of N uptake by the taro crop. A rate of 300 kg N ha^{-1} was applied in three equal splits at 14, 56, or 112 DAP with plots receiving ^{15}N at one of these particular times and ^{14}N at the other two times. The other treatments were surface application and surface application around the plant followed by moulding operation which covered the fertilizer. The treatments were laid out in randomised blocks replicated four times. Procedures for soil and plant sampling were carried out according to Barrie (1991). Total N and ^{15}N abundance in the plant and soil samples were determined using a Carlo Erba ANA 1500 automatic N analyser interfaced with a Europa scientific stable isotope analyser. The combination of the instruments used the principle of contineous flow-istope ratio mass spectrometry (CF-IRMS) as described by Barrie (1991).

Quantifying crop uptake and residual soil N

This experiment was conducted to determine the percentage of applied N that was taken up by the crop, that which remained in the soil and indirectly estimate the losses due to gaseous pathways. This experiment was conducted in the green house due to the high cost of the ^{15}N enriched $CO(NH_2)_2$ and the high expected losses due to surface washing that normally occur in the field. A randomised block design with 3 levels of N and 3 split application regimes replicated four times was used. Plant and soil samples were treated as above.

Results

Optimum rate and best suited application time

Leaf area, leaf number and plant height showed similar trends and only leaf area responses will be discussed here.

The leaf area response of taro to varying levels of applied N (Fig. 1) revealed that all plots treated with N had a significantly ($p<0.01$) greater leaf area than

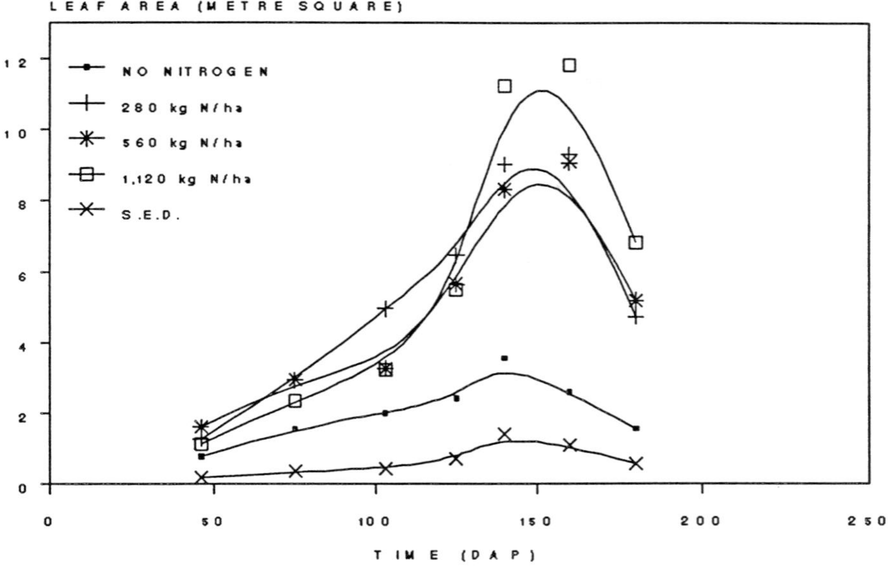

Fig. 1. Leaf area response of Taro crop to varying levels of applied N

Table 1. Yield response of Taro to N rates and times of application

N Levels	Application Time	Corm weight (kg/plot)	Corm weight (tn ha^{-1})	Cormel weight (kg/plot)	Cormel weight (tn ha^{-1})	Total yield (kg/plot)	Total yield (tn ha^{-1})
N_0	T_1	5.20	12.00	0.77	1.78	5.96	13.80
(No N)	T_2	5.40	12.50	0.80	1.86	6.26	14.49
	T_3	5.27	12.19	0.76	1.75	6.02	13.94
N_1	T_1	14.19	32.84	3.58	8.28	17.75	41.02
280 kg	T_2	10.49	24.27	2.98	6.90	13.48	31.19
N ha^{-1}	T_3	6.50	15.05	1.34	3.10	7.85	18.18
N_2	T_1	11.59	26.84	2.30	6.84	15.93	36.87
560 kg	T_2	8.52	19.73	2.47	5.72	11.03	25.53
N ha^{-1}	T_3	6.19	14.34	1.76	4.08	8.03	18.60
N_3	T_1	10.13	23.45	2.41	5.59	12.33	28.55
1120 kg	T_2	7.62	17.65	2.18	5.05	9.81	22.70
N ha^{-1}	T_3	6.29	14.56	1.77	4.09	8.06	18.65
SED			0.688		0.285		0.744
F-value (N)			61.72[a]		21.38[a]		133.97[a]
F-value (T)			74.33[a]		26.02[a]		109.18[a]
F-value (T×N)			10.63[a]		5.10[a]		17.32[a]

[a] significant at the 0.1% level.

tn: tonnes.

the control plots to which N was not applied. Leaf area developed at the 280 kg N ha^{-1} level of applied N, increased steadily from 1.28 m^2 at 46 DAP to 9.30 m^2 at 160 DAP, then rapidly declined. At the 560 and 1120 kg N ha^{-1} levels, a similar trend was generally followed, but at the 140 to 180 DAP period, the 1120 kg N ha^{-1} level was significantly ($p<0.01$) greater than the other levels of applied N, with a correspondingly high leaf production at the 560 kg N ha^{-1} level over the 280 kg N ha^{-1} level. This means that 280 kg N ha^{-1} is the optimum level as this level of applied N produced a normal shoot production curve for this crop (Fig. 1). The results also suggest that levels above the optimum rate result in excessive shoot production especially at a time when corm bulking is expected to be most active. Consequently, inter leaf competition may result in a possible reduction of photosynthates to the corms with an expected reduction in yield.

The yield response to applied levels and times of N are presented in Table 1. The data reveal that all levels of N significantly out-yielded the control plots. The 280 kg N ha^{-1} produced the greatest yield of both corms and cormels and was followed by the 560 and 1120 kg N ha^{-1}, respectively. For all levels of applied N, the earlier application (14 DAP) resulted in the greatest yield. This was suprising as it was felt that N applied at the most active time of corm bulking may have resulted in the greatest corm yield.

Apparent recoveries of N ranged from less than one per cent to about 10 percent. The general trend observed with apparent recovery was a decrease in recovery as applied N increased from 280 to 1120 kg N ha^{-1} and at a particular N level, it further decreased with later N application Times.

No significant response on corm yield or uptake of N was observed with applications of P and K. As these soils are adequately supplied with P and K and the taro crop possesses and extensive root system which develops early in crop growth.

Determination of the most suited N application time

The above data show that ^{15}N applied at 56 DAP resulted in a 65% higher recovery than the application at 14 DAP. The application at 112 DAP resulted in almost 95% higher recovery than that applied 14 DAP and approximately 20% higher than the application at 56 DAP. Under the prevailing growth conditions and the normal growth pattern of taro, the 112 DAP period corresponds to a time of rapid growth with high dry

matter production by the shoot system and the commencement of active bulking of corms and supposedly high N demand and uptake. The data from this experiment seems to contradict the results from the previous experiment; however, it should be noted that in the previous experiment apparent recoveries were studied which were influenced by yield.

Covering of the applied fertilizers resulted in a 26 per cent higher recovery which was highly significant ($p<0.01$); placing of the fertilizer in this manner might

Table 2. Apparent N recovery with varying levels and times of application

N level (kg N ha^{-1})	Time of N application (DAP)	Apparent recovery (Per cent)
280	14	10.31
	56	7.56
	112	2.95
560	14	5.20
	56	2.68
	112	1.08
1120	14	3.73
	56	1.29
	112	0.83

Table 3. ^{15}N recovery by the taro crop in response to time of N application

Time of application	Per cent ^{15}N recovery
14 DAP	16.6
56 DAP	27.4
112 DAP	32.1
SED	1.70
F value (2,23)	43.95***

*** Significant at the 0.1% level.

Table 4. ^{15}N recovery by the taro crop in response to various methods of application

Method of application	Per cent ^{15}N recovery
Surface	22.6
Moulding (covering)	28.4
SED	1.39
F value (2,23)	19.00***

*** Significant at the 0.1% level.

Table 5. Recovery of applied N by taro crop as influenced by split application regimes

Split regimes	Per cent ^{15}N crop recovery
Straight 14 DAP	41.1
1/3 14 DAP, 2/3 56 DAP	44.1
1/3 14 DAP, 1/3 56 DAP, 1/3 112 DAP	46.8
SED	2.13
F value (2,23)	3.66*

*Significant at 5% level.

Table 6. Recovery of applied $(NH_2)_2CO$-^{15}N by the taro crop as affected by N levels

N levels (ppm)	Per cent ^{15}N crop recovery
70 (N$_1$)	43.4
140 (N$_2$)	48.6
210 (N$_3$)	40.1
SED	1.75
F value (2,35)	11.74***

***Significant at 0.1% level.

Table 7. Mean corm dry matter response to increasing levels of applied $CO(NH_2)_2$

N levels (ppm)	Corm dry matter (g/pot)
70 (N$_1$)	97.66
140 (N$_2$)	128.18
210 (N$_3$)	110.76
SED	9.713
F value (2,33)	4.97*

*Significant at the 5% level.

Table 8. Mean corm dry matter response to split regimes of applied N

Split regimes	Corm dry matter (g/pot)
Straight 14 DAP	125.41
1/3 14 DAP, 2/3 56 DAP	104.74
1/3 14 DAP, 1/3 56 DAP, 1/3 112 DAP	106.44
SED	4.684
F value (2,33)	12.00***

***Significant at the 0.1% level.

have resulted in less surface washing in the covered plots compared to the surface applied plots.

Quantifying crop uptake and residual soil N

Table 5 shows an increase in efficiency with splitting of the applied N. The split application resulted in the highest recovery of approximately 47 per cent by the taro crop followed by 44 per cent with two split applications. This suggests that split applications may effectively increase fertilizer N uptake by the crop when compared to a single application at 14 DAP.

Crop uptake of ^{15}N increased from the N$_1$ to the N$_2$ level (considered optimum) then decreased. This is in keeping with the trend observed in the first experiment.

As seen in Table 7, the N$_2$ level of applied N significantly ($p<0.05$) out yielded the other two N levels with the N$_3$ level being significantly higher than the N$_1$.

According to the data in Table 8, a single N application yielded significantly ($p<0.001$) higher corm dry matter than the split regimes despite the higher recoveries with the split regimes.

The data in Table 9 show a significant ($p<0.01$) decrease in the per cent ^{15}N recovered in the soil after cropping with an increase in the levels of applied N. The data when compared to that in Table 6 suggests that at the lower than optimum level a higher residual soil N occurs compared to the optimal and super optimal levels which approximately equals that of crop uptake.

Residual soil ^{15}N in response to split regimes

The overall mean values for the residual ^{15}N for the various split application regimes were 46, 49, and 44

Table 9. ^{15}N recovery in the soil with varying levels of applied N

N levels (ppm)	Residual soil ^{15}N (Per cent)
70 (N$_1$)	51.78
140 (N$_2$)	47.84
210 (N$_3$)	39.37
SED	3.525
F value (2,35)	6.47**

**Significant at 1% level.

Table 10. Interaction of N levels and split regimes on the N lost

	N_1	N_2	N_3	Split means
		(per cent)		
Single	7.62	5.24	31.93	14.93
Two splits	6.02	5.16	18.61	9.93
Three splits	4.80	7.99	13.00	8.60
N level means	6.14	6.13	21.18	

Standard errors of mean:	Overall N: 1.663
(32 d.f.)	Overall Split: 1.985
N × Split:	3.438

per cent respectively for the single, two split and three split application regimes respectively. However, these values were not significantly different ($p>0.05$).

Covering of the fertilizer by moulding resulted in a 26 per cent higher uptake which was highly significant ($p<0.01$) but there were no significant differences in yield.

Table 10 shows a significant ($p<0.01$) interaction between N levels and split regimes which is significant ($p<0.01$) only at the N_3 level of applied N. At the N_3 level where the optimum level of N was exceeded, a single application resulted in the greatest percentage of unaccounted N (approximately 32 percent). A significant ($p<0.01$) reduction in losses occurred when 1/3 of the N_3 was applied at 14 DAP followed by 2/3 at 56 DAP. A further significant reduction in losses occurred with three split applications of applied N at 14, 56, and 112 DAP, respectively.

Discussion and conclusion

Data from the three experiments suggest that N application rates lower than, or exceeding the optimum rate of 280 kg N ha^{-1} resulted in lower yields. Lower N rates resulted in sub-optimal shoot growth which probably resulted in a lower net assimilation of photosynthates produced, than at the 280 kg N ha^{-1}; also, at above optimal N levels, the lush shoot growth resulted in excessive leaf production, which mutally shade each other, resulting in a higher percentage of photosynthetic consumption by the shoot system.

The best time of application for the highest yields seem to be soon after planting, that is, 14 DAP. However, a single application 14 DAP resulted in the lowest ^{15}N uptake, whereas the split N applications signif-

icantly increased uptake. This suggest the existence of a critical N content in the shoot or a critical ratio of shoot to corm N, and when this is not optimum, the synchrony between the shoot growth and corm growth is disturbed, resulting in a decrease in corm yield; this conclusion is further supported by Ramnanan (1993).

In acid Vertisols such as the Bejucal clay, the most important loss mechanism for applied N is surface runoff (Ahmad, 1983). An almost 100 per cent accountability of applied N obtained in the enclosed system in the green house for levels less than 280 kg N ha^{-1} suggests this may be the case. However, even in the enclosed system, a rate of 210 ppm which is equivalent to 420 kg N ha^{-1} resulted in a significant loss of N, as much as 30 per cent, which suggests that the gaseous pathway may become important at higher application rates.

Under field conditions the primary factors controlling denitrification of applied N are O_2, NO_3^- and C concentrations. In N-fertilized soils, NO_3^- is not likely to be limiting to denitrification (Myrold & Tiedje, 1985).

Chesney (1967) demonstrated that coarse textured soils of Trinidad had a greater denitrifying potential than fine textured soils and that within a textural group, denitrification increased with increasing pH. The Bejucal clay being very fine and acid therefore does not provide ideal conditions for denitrification to take place.

Soil water affects NO_3 supply by controlling the activities of mineralising and nitrifying organisms. In a water saturated soil, as occurs in the Bejucal clay for much of the growing season, NO_3 levels are exteremly low due to inhibition of nitrification by anaerobiosis. For conditions in Trinidad it was postulated (Hardy, 1946) that NO_3 is likely to accumulate in the rooting zone when the monthly rainfall is less than 6 cm and lost by leaching when it exceeds 10 cm has been accepted as a working basis. This therefore means that under field condition NO_3 concentrations may rarely be present, at suitable levels for any significant period of time, to make denitrification important from a soil management point of view.

The other gaseous pathway, volatilisation, may not be significant under field conditions for this soil. In a study of volatilisation under Trinidad's conditions (Fullerton, 1990) found that soils such as the acid Vertisols, with good buffering capacities had low levels of volatilisation. Although his study of this soil entailed only 100 kg N ha^{-1} compared to 420 kg N ha^{-1} where gaseous losses probably occurred volatilisation is not expected to be a significant pathway of loss for this soil.

References

Ahmad N (1983) Vertisols. In: Wilding LP, Smeck NE & Hall GF (eds). Pedogenesis and Soil Taxonomy II. The Soil Orders. Developments in Soil Science IIB, pp 91–123. Elsevier, Amsterdam, The Netherlands.

Barrie A (1991) New methodologies in stable isotope analysis. In: Stable istopes in Plant Nutrition, Soil Fertility and Environmental Studies. Proceeding of a Symposium, Vienna, October 1–5, 1990, pp 3–26. IAEA, Vienna, Austria.

Chapman T (1964) A note on the measurement of leaf area of the Tannia (*Xanthosoma sagittifolium*). Trop Agric (Trin). 41: 351–352.

Chesney HAD (1967) Denitrification in some Trinidad Soils. M Sc Thesis, Department of Soil Science, UWI, St. Augustine, West Indies.

De la Pena RS & Plucknett DL (1967) The response of taro (*Colocasia esculenta* L. Schott) to NPK fertilization in Hawaii. In: Symposium of the International Society of Tropical Root Crops, 1st, St. Augustine, 1967. Proceeding, pp 70–85. University of the West Indies, St. Augustine, West Indies.

Fullerton TP (1990) Volatilization of ammonia from urea in contrasting Trinidad Soils. Ph D Thesis. Department of Soil Science, UWI, St. Augustine, West Indies.

Hardy F (1946) Seasonal fluctuations of soil moisture and nitrate in a humid tropical climate (Trinidad, B.W.I.). Trop Agric (Trin) 23: 40–49.

Igobokume MC & Ogbannaya JC (1980) Yield and Nitiogen uptake by cocoyam as affected by nitrogen application and spacing. In: Terry ER *et al.* (ed), Tropical Root Crops: Research Strategies for the 1980s, pp 255–258. IDRC, 1981.

Myrold DD & Tiedje JM (1985) Diffusional constraints on denitrification in soil. Soil Sci Soc Am J 49: 651–657.

Ramnanan N (1993) Use of Urea on dasheen (*Colocasia esculenta* var. *esculenta*) in an acid Vertisols of Trinidad. M. Phil. thesis, Soil Science Department, U.W.I. Trinidad.

Sivan S (1980) Review of Taro research and production in Fiji. In: Chandra S (ed), Edible Aroids, pp 52–63. Oxford Univ. Press, Oxford, UK.

Fertilizer Research **42**: 117–121, 1995.
© 1995 *Kluwer Academic Publishers. Printed in the Netherlands.*

Brown leaf spot disease and fertilizer interaction in irrigated rice growing on different soil types

R.H. Phelps & C.R. Shand
Caroni Research Station, Caroni (1975) Limited, Waterloo Road, Caprapichaima, Trinidad & Tobago, West Indies.

Key words: Cochliobolus miyabeanus, rice brown leaf spot, phosphorus, disease-fertilizer interaction

Abstract

Studies have been undertaken in a 1000 ha area of irrigated rice (*Oryzica sativa*) at Caroni (1975) Limited to determine the effect of the fertilizer programme on the incidence of important diseases. Over a period of three years higher levels of brown leaf spot (*Cochliobolus miyabeanus*) on rice varieties Oryzica 1 and Oryzica 5 on three different soil types were associated with increasing levels of leaf P, from a low of 0.149% of dry matter (DM) to a high of 0.396% DM. On the Washington silty clay loam series (Inceptisol) brown leaf spot incidence was lowest when leaf P was between 0.135% and 0.149% of DM. However, disease incidence was higher when leaf P levels fell to 0.133% of DM or rose above 0.149%, under conditions where N was more than adequate. The moderate levels of the disease experienced over the period had no effect on yield, as grain infection was minimal. The results support the conclusion that the incidence of brown leaf spot on irrigated rice at Caroni is influenced by sub-optimal levels paticularly of P. Careful monitoring and managememt of P nutrition is seen as an important part in the overall strategy for controlling the disease.

Introduction

Irrigated rice is grown by Caroni (1975) Limited in Trinidad in an area south of the Caroni River which comprises four soil series, *viz.*, Bejucal clay (Vertisol), Cunupia clay (Inceptisol), Frederick (Mollisol) clay and the Washington silty clay loam (Inceptisol). These soils are flooded during the wet season. They have a moderately acid reaction of pH 5.0 to 5.5, the acidity increasing with soil depth. They are moderately high in all major nutrients. Whenever these soils are drained, rice and vegetable crops have been grown successfully. The profiles are not known to contain sodium, except for the subsoil of Frederick clay which is potentially acid sulphate in nature, mainly through underground seepage of saline water from the Gulf of Paria.

Investigations conducted at the Caroni Rice Project have identified the most important diseases as hoja blanca, brown leaf spot caused by *Cochliobolus miyabeanus* (Ito and Kuribayashi) Drechsler ex Dastur and sheath blight caused by *Rhizoctonia solani*

Kuhn [13]. Brown leaf spot is widely distributed and is known to cause severe losses in many parts of the world [12]. Losses are incurred through effects on seedling survival as well as reductions in yield and grain quality caused by inhibited root and shoot elongation, in addition to the more obvious leaf and kernel infection [12].

Investigations in several countries have shown a relationship between severity of brown leaf spot and nutritional status. From extensive investigations in Japan, it was concluded that the disease is associated only with abnormal or poor soil and that it serves as an index of such conditions. On the other hand, plants were less susceptible when P levels were low [12]. Disease was more severe when N was deficient after the middle of the growth period. Seedling blight and leaf spot symptoms have been shown to increase under conditions of deficiency and excess of NH_4^+-N [3].

In Japan, leaf spot symptoms on plants deficient in K were reported to be similar to those due to *C. miyabeanus*, although the fungus was not found in the lesions. A deficiency of K enhanced susceptibility to brown leaf spot, apparently by inducing a reduction in the oxidation-reduction (Eh) potential of the cell sap [12]. Deficiencies in Mn, Si and Mg had similar effects on the Eh of cell sap and, by implication, similar influences on susceptibility to disease. In India, micronutrients were also shown to influence the incidence brown leaf spot. Low levels of disease occurred in response to foliar applications of $FeSO_4$ and $CuSO_4$, although grain yields were not affected [11]. On the Histosols of Florida, Si content is low. Applications of $CaSiO_2$ slag led to a 32% reduction in the incidence of brown spot [4]. The effect on the disease was attributed only to the added Si and not to the Ca. It is generally accepted that brown leaf spot disease is associated with rice growing under conditions of some form of nutrient or other stress.

Studies have been in progress at the Caroni Rice Project with the objective of managing brown leaf spot by optimizing the fertilizer programme. They were instituted because of apparent adverse effects of the disease on both yield and quality of the grain. These investigations have revealed higher levels of disease to be associated with higher concentrations of leaf P on the Bejucal clay and Washington loam series [14,15]. This report describes the studies on fertilizer/brown leaf spot interactions on three soil series on which irrigated rice is grown by Caroni (1975) Limited.

Materials and methods

Experiments were conducted between 1990–1992 in selected fields on three soil series at the Caroni Rice Project, *viz.*, Bejucal and Cunupia clays and Washington silty clay loam. Fertilizer was applied to individual pl ots (10×5 m) with four replications in a randomized complete block design. In the experiment on Bejucal clay in 1990 varying amounts of N (in the form of urea) were applied at 0, 4 and 9 weeks, as indicated in Table 1. In 1991 N was applied to Washington loam at 3, 6 and 9 weeks. In both trials 58 kg ha^{-1} of P_2O_5 was added at seeding. In the experiments on Washington loam and Cunupia clay in 1992, each plot, except the zero fertilizer control, received a total of 60 kg ha^{-1} of N, but in varying proportions applied at 3, 6 and 9 weeks (Table 2). Three levels of P - 25, 50 and 75 kg ha^{-1} were applied at seeding. In all experiments each treatment, except T8 in 1992, received a standard dressing of 25 kg ha^{-1} of muriate of potash at 6 and 9 weeks. The severity of brown leaf spot was determined by assigning infected plants to the following grades devised by Aluko [1]:

Description

1. Very few, if any, scattered lesions (0–20 per leaf) on leaves or few to many undeveloped specks with no necrotic spots.
2. Few scattered but typical necrotic lesions (20–40 per leaf), mainly on upper leaves.
3. Few to moderate numbers of necrotic lesions (40–60 per leaf) usually scattered over a high proportion of the leaves.
4. Moderate numbers of necrotic lesions (60–80 leaf) occurring well separated on leaves.
5. Abundant necrotic lesions (above 80 per leaf) on all leaves but with negligible degree of coalescence.
6. Very abundant necrotic lesions (above 100 per leaf) on all leaves and usually coalescing to leave portions of affected leaves dead or rendered moribund.

Assessments were made at the milk stage in 20 plants selected at random from each plot and the assigned grades were used to derive a disease index. Data were subjected to ANOVA and means were compared using Duncan's Test ($p = 0.05$).

Leaf analysis

Leaf samples from laminae were taken at random in each plot at approximately six weeks after seeding and were analysed for percentage N, P and K in the dry matter (DM). N and P were determined spectrophotometrically using a Coleman Junior IIA Linear Absorbance Spectrophotometer. K was determined by flame photometry with a Corning Eel Flame Photometer. The levels of adequacy of the three nutrients were assessed as follows:

Deficient	Critical	Adequate
1. % N < 2.9	1. % N = 2.9	1. % N >2.9
2. % P < 0.12	2. % P 0.13 - 0.21	2. % P > 0.21
3. % K < 0.84	3. % K 0.84 - 1.96	3. % K > 1.96

Yields

The rice from individual plots (5×10 m), which were surrounded by guard rows of 1 m wide, was harvested manually. Data were subjected to ANOVA and means

Table 1. Brown leaf spot disease index, leaf N, P and K content and rice yield from rice variety Oryzica 1 on Bejucal clay (1990) and Washington loam (1991) following applications of varying amounts of urea at different stages of growth*

Soil series	Nitrogen rate (kg ha^{-1})		Disease index	% DM			Yield (kg ha^{-1})
				N	P	K	
Bejucal	T1.	60, 60, 60	2.6 bc	3.76	0.226	2.50	1, 820 a
clay	T2.	0, 90, 90	2.4 ab	3.93	0.221	1.95	1, 660 a
	T3.	40, 40, 40	2.7 c	4.28	0.246	1.94	1, 470 a
	T4.	0, 60, 60	2.2 a	3.59	0.208	1.98	1, 870 a
	T5.	20, 20, 20	2.6 bc	4.09	0.231	2.00	1, 790 a
	T6.	0, 30, 30	2.2 a	3.85	0.208	2.02	2, 250 a
	T7.	0	2.8 c	3.71	0.224	1.98	2, 170 a
							CV:12.8%
Washing-	T1.	20, 80, 80	0.6 a	4.34	0.135	2.02	2, 770 a
ton loam	T2.	20, 160, 0	1.2 b	4.68	0.169	2.14	2, 810 a
	T3.	20, 50, 50	0.6 a	4.36	0.145	2.13	2, 710 a
	T4.	20, 100, 0	0.8 a	4.78	0.149	2.20	2, 700 a
	T5.	20, 20, 20	1.1 b	4.38	0.153	2.15	3, 055 a
	T6.	20, 40, 0	0.5 a	4.58	0.146	2.23	3, 055 a
	T7.	20, 0, 0	1.4 b	4.09	0.133	2.10	2, 975 a
							CV:83.4%

*Means are of four replications. Those followed by the same letter are not significantly different according to Duncan's Test ($p = 0.05$).

Results and discussion

The incidence of brown leaf spot did not exceed moderate levels in any of the experiments and, consequently, the disease was not considered to be a factor in yield of grain. Aluko [1], in Nigeria, has shown that adverse effects on grain nurnbers and weight generally do not occur unless the disease index is over 3 at the reproductive stage of growth. These observations by Aluko [1] confirmed earlier work in India which demonstrated that the pathogen was most destructive on grain yield and quality when infection occurred at the flowering and milk stages [5, 6]. In the studies reported here disease indices varied between 0.5 and 2.8 (Tables 1 and 2), on the scale used by Aluko [1]. The absence of an effect on yield due to disease could explain the lack of a relationship between yield and severity of disease. Although the highest yield on the Bejucal series was obtained from treatment 6, with the lowest level of disease, the second highest yield came from the control

were compared using Duncan's Multiple Range Test ($p = 0.05$).

plot which registered the highest disease index (Table 1). Similar responses were recorded on Washington loam where heaviest yields were given by treatments 5 and 6 which had significantly different disease indices (Table 1).

The high levels of leaf N in all but the trial on Cunupia clay made it impossible to relate disease severity to N status. The low disease incidence in plots on Cunupia clay showing both deficient amounts of leaf N and high levels of leaf P (Table 2), which might be expected to induce enhanced disease, could have been due to the higher resistance to brown leaf spot of variety Oryzica 5 as well as to unfavourable environmental conditions. The incidence of brown leaf spot has been shown to be influenced by variety [12, 16] and by environmental conditions [5, 7]. Leaf K was either adequate or marginal in the three soil series and could not have contributed to the incidence or increased severity of brown leaf spot.

Our studies revealed that, on the Bejucal clay and Washington loam, higher brown spot indices were associated with higher levels of leaf P (Tables 1 and 2). This observation held true except for treatment 7 on Washington loam where a higher disease index was

Table 2. Brown leaf spot disease index, leaf N, P and K content and rice yield from rice variety Oryzica 5 on Washington loam and Cunupia clay, following applications of varying amounts of TSP at seeding and of urea at 3, 6 and 9 weeks after seeding.*

| Soil series | | Fertilizer rate (kg ha^{-1}) at weeks | | | | Disease index | % DM | | | Yield (kg ha^{-1}) |
| | | P | N | | | | N | P | K | |
		0	3	6	9					
Washington	T1.	25	20	20	20	1.1 b	3.55	.251	1.11	2,425 a
loam	T2.	25	0	30	30	1.2 b	3.68	.245	1.59	1,920 a
	T3.	50	20	20	20	0.5 a	3.20	.226	1.75	2,140 a
	T4.	50	0	30	30	1.2 b	3.39	.248	2.04	2,060 a
	T5.	75	20	20	20	0.4 a	3.00	.193	1.70	1,965 a
	T6.	75	0	30	30	0.6 a	3.15	.219	1.75	2,485 a
	T7.	0	0	30	30	1.2 b	3.05	.266	1.59	2,675 a
	T8.	0	0	0	0	1.1 b	3.64	.258	1.13	2,015 a
										CV:80.8%
Cunupia	T1.	25	20	20	20	1.3 a	2.63	.258	1.29	1,860 a
clay	T2.	25	0	30	30	1.2 a	2.68	.296	1.28	1,745 a
	T3.	50	20	20	20	1.3 a	2.88	.291	1.43	1,905 a
	T4.	50	0	30	30	1.1 a	3.11	.286	1.54	1,920 a
	T5.	75	20	20	20	1.2 a	2.80	.281	1.35	1,805 a
	T6.	75	0	30	30	1.2 a	2.98	.298	1.44	1,695 a
	T7.	0	30	30	30	1.3 a	2.85	.316	1.45	1,880 a
	T8.	0	0	0	0	1.3 a	3.11	.313	1.47	1,830 a
										CV:11.6%

*Means are of 4 replications. Those followed by the same letter are not significantly different according to Duncan's Test. (P = 0.05).

recorded in plants having the lowest leaf P of 0.133% (Table 1). However, the incidence of brown leaf spot was lowest where leaf P was between 0.135% and 0.149% of DM. The disease was more severe on plants with leaf P levels above 0.149% (Table 1). The data indicate that, on the Washington soil series, when leaf P content falls to 0.133% or rises above 0.149% of DM the effect of brown leaf spot disease on the host is enhanced. The results suggest, therefore, that on some soil types the disease is more severe when leaf P levels either fall below or rise above some critical value. This conclusion is supported by research conducted in India which demonstrated that, in pot experiments, additions of 50 ppm of P to natural soils reduced the levels of brown leaf spot, whereas applications of 100 and 200 ppm of P led to higher disease incidence [9]. There was evidence that the effect of P on disease expression was related to its interaction particularly with Ca and Mn. At the Caroni Rice Project trace elements were not included in the fertilizer programme and their lev-

els of deficiency or adequacy were never addressed. It is unclear, therefore, what would be the effect of P, as related to its interaction especially with Ca, Mn and Si, on disease expression.

It is widely accepted that brown leaf spot is of importance only when rice is cultivated on soils that have some severe chemical deficiencies and/or where there are poor water management practices. Such conditions cannot be said to prevail at the Caroni Rice Project, yet the disease occasionally causes significant leaf and panicle spotting. Even where infection is severe enough to cause significant yield loss, the disease can be responsible for substantial reduction in grain quality. Reduced yields in some fields have been caused by poor land levelling (which resulted in bare soil areas) and low plant populations due to variations in seeding rates that were related to varietal changes. Our results suggest that the incidence of the disease on any particular variety is influenced by subtle, suboptimal levels of key nutrients, particularly P.

Environmental conditions favourable to infection at critical stages of growth are known to affect disease severity. Infection of the grain has been shown to cause more severe yield loss than leaf infection [10]. Similar differential effects due to foliar and grain infections have been reported [8]. Foliar infection did not show a simple relation to grain yield losses in a mangrove swamp ecosystem in Sierra Leone. Yield losses were attributed to grain infections induced by favourable long dew periods and moderate diurnal and nocturnal temperatures. In these studies N at $40 \, kg \, ha^{-1}$ reduced the disease in one year but not in the two subsequent years [8]. Environmental conditions favouring infection led to significant yield losses despite adequate N applications.

The studies in Sierra Leone support our conclusions that brown leaf spot is not solely a disease of poor soil nutrient status. Rather, it is capable of causing losses even when the crop is adequately fertilized, if weather conditions favour infection of the leaf and/or the grain. In Trinidad and Tobago, management of the disease, therefore, would seem to require optimizing fertilizer application, especially of P, for the specific soil type, in addition to the use of appropriate measures of disease resistace (if available), manipulating varieties and planting dates and chemical protection at critical stages of growth. Further studies will seek to develop these refinements.

References

1. Aluko MO (1970) The measurement of brown leaf spot on rice. PANS 16: 86–81
2. Aluko MO (1975) Crop losses caused by the brown leaf spot disease of rice in Nigeria. Plant Dis Rep 59: 609–613
3. Chattopadhyay SB and Dickson JG (1960) Relation of nitrogen to disease development in rice seedlings infected with *Helminthosporium oryzae*. Phytopathology 50: 434–438
4. Datnoff LE, Raid RN, Synder GH and Jones DB (1991) Effect of calcium silicate on blast and brown spot intensities and yields of rice. Plant Dis 75: 729–732
5. Fazli SFI, and Schroeder HW (1966) Kernel infection of Bluebonnet 50 rice by *Helminthosporium oryzae*. Phytopathology 56: 507–509
6. Fazli SFI and Schroeder HW (1966) Effect of kernel infection of rice by *Helminthosphorium oryzae* on yield and quality. Phytopathology 56: 1003–1105
7. Fomba SN and Singh N (1990) Crop losses caused by rice brown spot disease in mangrove swamps of northwestern Sierra Leone. Trog Pest Manag 36: 387–393
8. Fomba SN and Singh N (1991) Observations on the incidence of rice brown spot, leaf scald and leaf smut diseases in a tidal mangrove swamp at Rokupr, Northwestern Sierra Leone. Trop Pest Manag 37: 349–355
9. Kaur P, Kaur S and Padmanabhan SY (1982) Relationship of phosphorus concentration in the host and brown spot disease of rice. Indian Phytopathol 35: 393–398
10. Kulkarni S, Ramakrishnan K and Hedge RK (1980) Ecology, epidemiology and supervised control of rice brown leaf spot. Int Rice Res Inst Newsl 5: 13–14
11. Muthusamy S, Mariappan V, Narasimhan V, Muthusamy M and Eswaramoorthy S (1988) Effect of micronutrient on rice brown spot (BS) incidence. Int Rice Res Inst Newsl 13(6): 32–33
12. Ou SH (1987) Rice diseases. 2nd edn CAB International, UK 380 p
13. Phelps RH (1990) Monitoring of disease incidence levels in commercial fields of rice at Caroni Rice Project in 1989. Caroni Res Sta Tech. Rep 18: 1–3
14. Shand CR and Phelps RH (1991) Brown leaf spot disease x fertilizer interaction on rice. Caroni Res Sta Tech Rep 24: 18–22
15. Shand CR and Phelps RH (1992) Brown leaf spot disease and fertilizer interaction on irrigated rice in Caroni Rice Project. Caroni Res Sta Tech Rep 27: 63–67
16. Singh RN (1988) Status of brown spot (BS) and narrow brown leaf spot (NBLS) in eastern Uttar Pradesh, India Int Rice Res Inst Newsl 13: 35–36

Fertilizer Research **42**: 123–128, 1995.
© 1995 *Kluwer Academic Publishers. Printed in the Netherlands.*

Use of ^{15}N for fertilizer N recovery and N mineralization studies in semi-arid Kenya

C.J. Pilbeam & G.P. Warren
Department of Soil Science, The University of Reading, Whiteknights, Reading, RG6 6DW UK

Key words: fertilizer, mineralization, nitrogen-15, soil, water

Abstract

Maize and beans were grown on a ferralsol at Kiboko, Kenya, with up to 120 kg N ha^{-1}. Within the 10 kg N ha^{-1} plots, ^{15}N labelled fertilizer was applied in microplots. There was no significant response in yield to fertilizer N and labelled N recovery was low, being 7.5% or less in one season and 17.7% or less in the second season. Samples of Kiboko soil at four different water contents were incubated and the rate of gross N mineralization over 7 days was calculated, utilizing ^{15}N labelling of the mineral N. Gross N mineralization increased greatly with soil moisture and a fitted relationship between gross N mineralization rate and soil water content was obtained. Using measurements of soil water content at the field site, daily values of the soil N supply by gross mineralization were calculated. On average, modelled gross soil N mineralized could supply much (> 69%) of the N removed from the plots. It is suggested that the lack of response to fertilizer N may be explained by the coincidence of a high rate of N mineralization, and increased crop demand, caused by the onset of rain.

Introduction

Nitrogen deficiency is a major fertility constraint in tropical soils. However, there are many cases of poor and unreliable responses to fertilizer N in soils of semi-arid Kenya. For example, maize at Katumani failed to respond to 70 kg N ha^{-1} N in long rain seasons which followed drought in the preceding short rains (Nadar & Faught, 1984). In another nitrogen and phosphorus factorial experiment replicated at six sites, there was no significant response to N at two sites, including Kiboko (Okalebo, 1987). Further examples are to be found in other semi-arid regions, such as at Matopos, Zimbabwe, where sorghum failed to respond to up to 90 kg N ha^{-1} in three seasons out of six (H Ssali, pers. commun.). Similarly, insignificant responses to N fertilizer were also found in Antigua and Barbados, drier regions of the Caribbean (Ahmad, 1978). No explanation is given in any of these cases for the absence of a response to N fertilizer. Work at Sadore, Niger (Christianson and Vlek, 1991) has shown that the response of millet grain yields to fertilizer N is affected by the amount and distribution of rainfall. In dry years crop growth was limited more by water supply than by N, and so in these years the crop N demand was low, and therefore the yield response to increasing levels of N fertilizer was small or in some cases absent.

Birch (1958, 1959, 1960) offered an alternative to this 'crop N-demand' explanation for the absence of a yield response to N fertilizer. He has shown that the amount of nitrate nitrified by a soil is greater if it experiences repeated cycles of wetting and drying, rather than being wetted up from a dry state, and maintained at a constant high moisture content (Birch, 1958). Furthermore, he has also shown that the amount of nitrate nitrified upon wetting a dry soil is dependent upon the length and intensity of drying in the preceding dry period (Birch, 1959; 1960). Consequently, in semi-arid areas nitrate production will be greater in seasons of low and intermittent rainfall than in seasons of high and more continuous rainfall. For this reason, responses to nitrogen may be more apparent in wet seasons than dry seasons (Birch, 1958).

The use of labelled N fertilizer has led to a much improved understanding of the N cycle in soils of temperate climates. Labelling makes it possible to assess immobilization of mineral N and gross mineralization of organic N, whereas ordinary incubation techniques can only measure net mineralization, the difference between gross mineralization and immobi-

Table 1. Physical and chemical characteristics of two soil horizons from the field site at Kiboko, Kenya (unpubl. data, Kenya Soil Survey)

	0–19 cm	19–35 cm
% Sand	74	70
% Silt	5	5
% Clay	21	25
pH (KCl)	6.0	5.8
% C	0.6	0.17
% N	0.09	0.07

lization. However, there has been little such investigation of soils of semi-arid Africa. In an attempt to redress this, a pulse dilution technique for soil to which labelled N has been added (Barraclough, 1991; Barraclough et al., 1985; Geens et al., 1991) was applied to a soil incubation experiment with soil sampled from Kiboko (Pilbeam et al., 1993), a location typical of semi-arid Kenya. The rate of gross N mineralization in Kiboko soil was strongly dependent on soil water content (Pilbeam et al., 1993). In contrast to common temperate conditions, the start of the growing season in the semi-arid tropics is often, although not always, marked by a large and rapid increase in soil moisture. Consequently, the N availability through mineralization may be expected to vary greatly during the growing season.

In this paper, measurements of gross N mineralized in Kiboko soil are related to assessments of N fertility in the field, to seek an explanation for unreliable responses to fertilizer N.

Materials and methods

Field experiment

The Kiboko site experiences an annual rainfall of about 650 mm, bimodally distributed, and the soil is an Acriorthic Ferralsol. Details of chemical and physical characteristics are given in Table 1. Sole stands of maize (Zea mays L. var. Makueni Composite) and common bean (Phaseolus vulgaris L. var. B9) were grown on land previously cropped for over 15 years but without recent fertilization. Maize was grown at a plant density of 4.4 plants m^{-2}, while the density of beans was 12.5 plants m^{-2}. Fertilizer N was applied as calcium

ammonium nitrate; 5 different rates up to 120 kg ha^{-1} were used, namely 10, 20, 40, 80 and 120 kg N ha^{-1}. For both maize and beans, except at the lowest (10 kg N ha^{-1}) level, N was applied as a split dressing, half to the seed-bed at sowing, and the other half 26 days after sowing. In the short rains of 1990, only the lowest (10 kg N ha^{-1}) and highest (80 and 120 kg N ha^{-1}, for maize and bean respectively) rates were applied. Within the 10 kg N ha^{-1} plots, the labelled fertilizer experiment was conducted in microplots of 1 m^2. ^{15}N was applied as (^{15}NH$_4$)$_2$SO$_4$ (at 10 kg N ha^{-1}) with 5.3 atom% ^{15}N abundance at 12 days after sowing. The harvest was made 79 days after sowing for beans and 98 days for maize. Samples of all plant fractions (grain, haulm and stover) were dried, ground and their total N concentration and ^{15}N abundance measured by mass spectrometry.

Soil incubation experiment

Soil was collected from the surface horizon (0–20 cm) in a random selection of the plots at Kiboko, prior to cultivation in January 1991, and mixed thoroughly. This experiment has been described fully elsewhere (Pilbeam et al., 1993). Briefly, soil samples (80 g) were incubated for 24 days with treatments comprising 4 water contents (10, 15, 20 and 25% by weight). Then a solution of (^{15}NH$_4$)$_2$SO$_4$ (7.5 mg N per sample), with 10.366 atom% ^{15}N abundance, was added to all samples and the incubation continued for another 2 days in half the samples, and 9 days in the remainder. NH$_4^+$ and NO$_3^-$ were extracted with 1 M KCl and the amounts of N and ^{15}N abundances measured by mass spectrometry. To obtain estimates of the gross mineralization of N, these data were fitted to the following equation, derived from a model of soil N turnover (Barraclough, 1991; Barraclough et al., 1985):

$$A_t^* = A_0^* / (1 + \theta t / A_0)^{m/\theta} \qquad (1)$$

where A_0 is the initial size of the NH$_4^+$ pool (i.e. after 2 days' incubation with ^{15}N), A_0^* the ^{15}N abundance of that pool, A_t^* the ^{15}N abundance of the final NH$_4^+$ pool (after 9 days' incubation), t the incubation time (7 days), θ the rate of change of the NH$_4^+$ pool size (mg N d^{-1}) and m the rate of gross N mineralization. Net N mineralization rate was calculated from the initial and final sizes of the NH$_4^+$ and NO$_3^-$ pools.

Table 2. Grain yields (kg ha^{-1}) of maize and bean crops receiving different levels of N fertilizer grown at Kiboko, Kenya in the long rains 1990 and short rains 1990. Standard error of the mean in parenthesis

| | | Fertilizer rate (kg N ha^{-1}) | | | | | |
		10	20	40	80	120	Mean
Long rains 1990	Maize	2305	3059	2532	2237		2500 (447.3)
	Beans	1022		1100	960	995	1020 (138.3)
Short rains 1990	Maize	1679			1588		1634 (225.6)
	Beans	1044				1041	1043 (96.8)

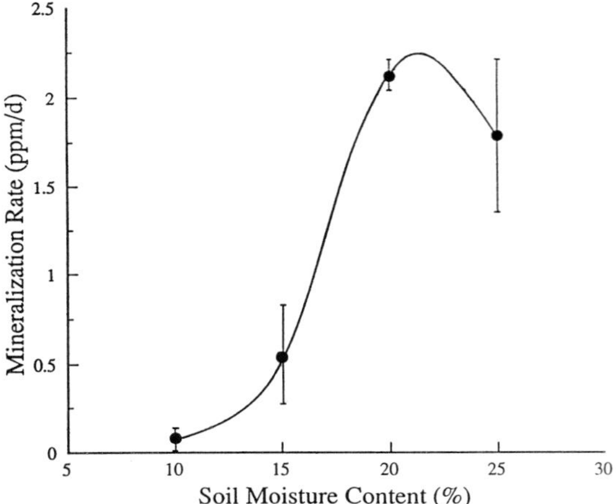

Fig. 1. Relationship between the rate of gross N mineralization and soil water content for soil from Kiboko, Kenya.

Results

Field experiment

Yields of maize were around 2500 kg ha^{-1} and 1600 kg ha^{-1} in the long and short rains 1990 respectively (Table 2), while yields of beans were approximately 1000 kg ha^{-1} in both seasons (Table 2). There was no response to applied fertilizer N. Fertilizer N recovery in the whole plant was low in the long rains (from March) 1990, being only 6.1% for maize and 7.5% for beans (Table 3). Recovery was higher in the short rains (from October) 1990, at 15.8% for maize and 17.7% for beans. Generally, fertilizer N was ineffective and the crop derived most of its N (> 96%) from the soil only (Table 3).

Soil incubation experiment

Gross N mineralization depended strongly on soil water content and peaked at 2.1 mg N kg^{-1} day^{-1} at 20% water content (Table 4). This relationship was fitted to a curve (Fig. 1) that could be used to estimate gross N mineralized in the field. Net N mineralization declined with the first two increases of water content. Since gross N mineralization increased with water content, the decline in net mineralization is thought to be due to increased immobilization and/or possible gaseous losses of nitrate. It is apparent that net N mineralized could not be used to relate soil N mineralization in laboratory conditions to the N supply from soil in the field.

Modelling of N mineralization in the field

Daily rainfall at Kiboko was measured and the soil water content measured by neutron probe at approximately weekly intervals during the rainy seasons, commencing on 3 May 1990, about the end of the long rains 1990. Daily soil water content values for each 10 cm layer beginning at the soil surface and extending to 30 cm were assumed to equal the value of the previous measurement until halfway between two readings, after which the value of the subsequent measurement was used. Using the relationship between gross N mineralization and soil water content (Fig. 1), daily values of gross N mineralization were estimated for each 10 cm layer to a depth of 30 cm. These daily values were summed to give an estimate of the amount of nitrogen mineralized in a 30 cm profile over 2 rainy seasons (short rains 1990 and long rains 1991).

Table 3. Total crop N, source of crop N and recovery of labelled fertilizer in the grain and straw of maize and bean crops at Kiboko, Kenya in the long rains 1990 and short rains 1990. ^{15}N (5.3 atom%) was applied as $(^{15}NH_4)SO_4$ at 10 kg N ha^{-1} 12 days after sowing

		Mean atom % ^{15}N in dry matter	% Fertilizer recovery		Amount of N (kg ha^{-1}) from (kg N ha^{-1})		Total N
			Grain	Straw	Fertilizer	Soil	
Long rains 1990	Maize	0.458	5.30	0.79	0.60	31.96	32.56
	Beans	0.473	6.15	1.33	0.75	34.05	34.80
Short rains 1990	Maize	0.490	9.86	5.90	1.58	61.83	63.41
	Beans	0.519	14.09	3.58	1.77	53.87	55.64

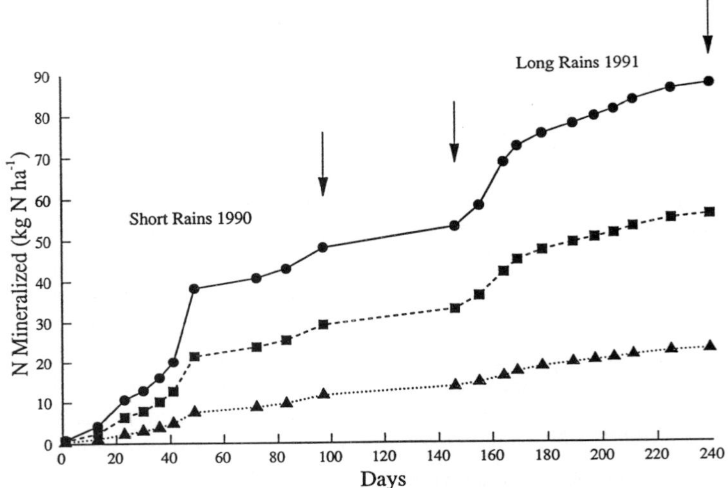

Fig. 2. Modelled cumulative gross N mineralized in the 0–30 cm (●), 10–30 cm (■) and 20–30 cm (▲) soil layer in the field at Kiboko for the short rains 1990 and the long rains 1991.

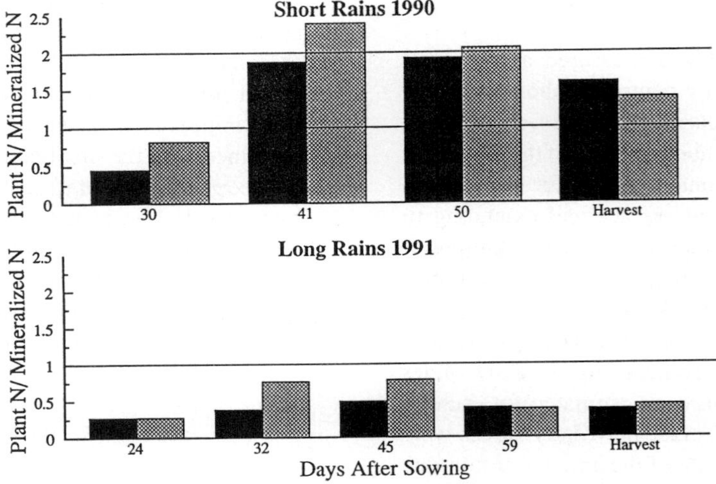

Fig. 3. Plant N content at different stages of growth as a proportion of the modelled cumulative N mineralized during the growth of maize (▨) and bean (■) crops grown at Kiboko, Kenya in the short rains 1990 and long rains 1991.

Table 4. Gross and net N mineralized during 7 days' incubation of Kiboko soil, with single value standard errors in brackets

Soil water content (%)	Gross mineralization rate (mg N kg^{-1} d^{-1})	Net mineralization rate (mg N kg^{-1} d^{-1})
10	0.08 (0.007)	0.90 (0.161)
15	0.54 (0.276)	0.46 (0.161)
20	2.12 (0.070)	-0.33 (0.16l)
25	1.79 (0.409)	2.20 (0.161)

Modelling of N mineralization in the field

Gross N mineralized in the field increased greatly on wetting of the soil in each cropping season. This can be seen in Figure 2, where at some time after the first significant rains of the season, there is a sharp increase in the gradient of the plot of cumulative gross N mineralized against date in both the 0–10 and 10–20 cm layers. Gross N mineralized in the 10–20 cm layer was similar to that in the 0–10 cm layer, and 50% more than in the 20–30 cm layer. Gross N mineralized in the 20–30 cm layer was relatively constant because water content in this layer did not fluctuate widely.

The cumulative total of gross N mineralized in the top 30 cm of soil over 2 seasons was about 90 kg ha^{-1} (Fig. 2), mineralization being greater in the wetter short rains (48.2 vs. 34.2 kg N ha^{-1} in the short rains 1990 and long rains 1991 respectively). For the yields obtained in the short rains 1990, the N offtake by grain was approximately 40 to 50 kg N ha^{-1} (Table 5). Thus it appears that on average, mineralization of soil N can supply much, or in some cases all, of the N removed in the grain from these plots.

In the short rains 1990 N mineralization could always provide more than 50% of the total plant N during the growth of the bean crop (Fig. 3) and often more than 45% of the total plant N during the growth of the maize crop (Fig. 3). N mineralization in the long rains 1991 could provide all the N for the maize and bean crops during their growth (Fig. 3).

For the short rains in 1990, a comparison of estimated N supplies and crop uptake was possible. Gross soil N mineralized was calculated for the appropriate length of season for each crop. At the low fertilization rate (10 kg N ha^{-1}), gross N mineralized could account for all of the N offtake in grains and 77% of the total N uptake (Table 5). Additional sources of N would be (i) initial soil mineral N produced by dry

season mineralization, (ii) wet and dry deposition of N, (iii) decomposition of crop residues (excluded from the sieved soil samples) and (iv) non-symbiotic N fixation.

The higher rates of N fertilization gave a significant increase in N uptake by beans although grain yields were not increased, suggesting that other factors were limiting bean yield. For maize, stover N uptake, but not grain N, was increased by N fertilizer (Table 5).

Discussion and conclusion

The site had been cropped for a prolonged period compared to local farming practice, so it was expected that the soil N reserves would be depleted, and a response to applied N seemed likely. However, there was no response and little labelled fertilizer N was taken up by the crops. It was estimated that much of the N required by the plant could be supplied from mineralization of soil organic N, based on gross N mineralization rates at a range of soil water contents. It is recognized that this calculation of N supply in the field is subject to several approximations: (i) the soil incubation experiment was not performed for each horizon separately; (ii) soil water measurements in the field were not made daily; (iii) the mineralization – water content relationship is based on four water contents only. Nevertheless, the method described provides an independent means of assessing the rate of N supply from a major part of the soil, based on an understanding of the soil N turnover. In conjunction with detailed measurements of other sources of N such as atmospheric deposition, and initial mineral N in the profile, a more complete budget for the field N cycle could be obtained.

This approach also reveals in principle the possibility that a high supply of N from the soil occurs at the time when it is required by a crop, when soil moisture is high (Fig. 2). We offer the explanation for the low response to fertilizer N that a high rate of N mineralization occurred at the times of high crop demand through vegetative growth: in the semi-arid climate, both take place when the soil is wetted after a dry period. Thus N supply coincides with N demand and fertilizer N does not provide a useable addition to the N being made available by rapid mineralization. Estimates of N supply based on net N mineralization during the incubation could not indicate this explanation because the N mineralized in incubation was partly reimmobilized instead of being diverted to plant uptake.

Table 5. Comparison of crop N uptakes and soil N supplies for the short rains 1990, with standard errors of differences in brackets

	Source N supplied (kg N ha^{-1})		N in crop (kg N ha^{-1})			
	Fertilizer	Gross soil N mineralized	Grain	Stover	Pod wall	Total
Maize	10	48.2	41.5	21.9		63.4
	80	48.2	37.1	33.2		70.3
			(10.4)	(7.7)		
Bean	10	43.0	42.8	9.2	3.6	55.6
	120	43.0	55.4	17.6	9.4	82.4
			(3.6)	(2.7)	(0.9)	

Furthermore, this high rate of N mineralization during early rapid plant growth also has implications for the nodulation of legumes. Nodulation is repressed by high mineral N concentrations in the soil (Munns, 1977). This may explain why the beans in this experiment failed to nodulate (Pilbeam *et al.*, 1995).

The surplus fertilizer N may be lost through leaching and denitrification, although with the moderate rainfall in these seasons, this is thought to be unlikely. Volatilization is unlikely at this site because of the low soil pH. Consequently, immobilization in soil organic matter is the most likely fate, and then the immobilized fertilizer N could still benefit crops in the longer term by maintaining the soil organic N pool.

Acknowledgements

Financial support of the European Community (for CJP) and the Natural Resources Institute, United Kingdom (for GPW) is gratefully acknowledged.

References

Ahmad N (1978) Soil management for maize production in the Caribbean. *In*: Symposium on Maize and Peanut, November 13–18, 1978. Agricultural Experiment Station, Paramaribo, Suriname

Barraclough D (1991) The use of mean pool abundances to interpret ^{15}N tracer experiments. I. Theory. Plant and Soil 131: 89–96

Barraclough D, Geens E L, Davies G P & Maggs J M (1985) Fate of fertilizer nitrogen. III. The use of single and double labelled ammonium nitrate to study nitrogen uptake by ryegrass. J. Soil Sci. 36: 593–603

Birch H F (1958) The effect of soil drying on humus decomposition and nitrogen availability. Plant and Soil 10: 9–31

Birch H F (1959) Further observations on humus decomposition and nitrification. Plant and Soil 11: 262–286

Birch H F (1960) Nitrification in soils after different periods of dryness. Plant and Soil 12: 81–96

Christianson C B and Vlek P L G (1991) Alleviating soil fertility constraints to food production in West Africa: Efficiency of nitrogen fertilizers applied to food crops. *In*: Mokwunye A U (ed.) Alleviating Soil Fertility Constraints to Increased Crop Production in West Africa: 45–57. Kluwer, Dordrecht, The Netherlands

Geens E L, Davies G P, Maggs J M & Barraclough D (1991) The use of mean pool abundances to interpret ^{15}N tracer experiments. II. Application. Plant and Soil 131: 97–105

Munns D N (1977) Mineral nutrition and the legume symbiosis. *In*: Hardy R W F & Gibson A H (eds.) A Treatise on Dinitrogen Fixation Section IV: 353–391. John Wiley, Chichester, UK

Nadar H M & Faught W A (1984) Maize yield response to different levels of nitrogen and phosphorus fertilizer application: a seven season study. East Afr. Agric. For. J. 44: 147–156

Okalebo J R (1987) A study of the effect of phosphorus fertilizers on maize and sorghum production in some East African soils. Ph.D Dissertation, University of Nairobi, Kenya

Pilbeam C J, Mahapatra B S & Wood M (1993) Soil matric potential effects on gross rates of nitrogen mineralization in an orthic ferralsol from Kenya. Soil Biol. Biochem. 25: 1409–1413

Pilbeam C J, Wood M & Mugane P G (1995) Nitrogen use in maize-grain legume cropping systems in semi-arid Kenya. Biol. Fertil. Soils 20: 57–62

Fertilizer Research **42**: 129–138, 1995.
© 1995 *Kluwer Academic Publishers. Printed in the Netherlands.*

Significance of timing and method of N fertilizer application for the N-use efficiency in flooded tropical rice

H.F. Schnier*
*The International Rice Research Institute P.O. Box 933, Manila, Philippines (*present address: P.O. Box 41607, Nairobi, Kenya)*

Key words: ammonia volatilization, denitrification, deep placement, ^{15}N-recovery, band placement, urea supergranules

Abstract

N-use efficiency in flooded tropical rice is usually low. Fertilizer N losses result mainly from losses of volatile NH_3 after broadcast application of urea into floodwater between transplanting and early tillering which is a common practice of farmers. Losses appear predominantly during the first week after urea application. With broadcast and incorporation of N into soil at transplanting losses may be reduced but are still substantial. Deep placement of urea supergranules (USG) has not been adopted by farmers because it is very laborious. A new application technique, namely injection of dissolved urea into the upper soil layer, was developed by which fertilizer N losses were effectively minimized while at the same time allowing flexible timing of application independent of crop stage and water management. It provides N-use efficiency equal to that achieved by USG point placement but is less labor-intensive.

Introduction

More than 90% of the rice (*Oryza sativa* L.) in the world is produced in Asia (IRRI, 1986). Within Asia, rice production has increased considerably over the last decade as a result of adoption of higher yielding cultivars, increased irrigation and an increased use of mineral fertilizers. An estimated 24% of the increase in Asian rice production from 1965 to 1980 has been attributed to use of fertilizers, mainly N (Barker *et al.*, 1985). Nitrogen represents approximately 70% of the total fertilizer nutrient ($N + P_2O_5 + K_2O$) consumption in Asia (FAO, 1987), and approximately 60% of the N fertilizer consumed in Asia is used on rice (Stangel and De Datta, 1985). Urea is the major N fertilizer for rice and it accounts for more than 75% of the total N fertilizer consumption in Asia. Potential environmental degradation associated with the use of fertilizer N to the biosphere and hydrosphere, and the expense of N in production make the efficient use of N for crop production imperative.

Efficiency of N fertilizer is generally low, with normally less than 40% of the applied N taken up by the flooded rice (Craswell and Godwin, 1984; Craswell and Vlek, 1979; De Datta, 1986b; Vlek and Byrnes, 1986). This situation challenges soil scientists and those with expertise in plant nutrition and crop production to develop strategies to increase N fertilizer use efficiency. The greatest economic benefits from increased fertilizer use efficiency in rice can be expected in the irrigated areas of Asia because of its dominance in global rice production and because of the production of over 70% of the world's rice (De Datta and Buresh, 1989).

To increase crop recovery of urea-N which will also lead to the improvement of agronomic and physiological efficiency of fertilizer use, two major issues are important. These are (i) timing and (ii) method of N application. This paper addresses recent research findings on these issues.

Table 1. Summary of ammonia volatilization losses from urea by direct measurements and total N loss by ^{15}N balance technique for urea applied to flooded rice at various locations in the Philippines

Timing and method of urea application	Reference	Location	Urea applied (kg N ha^{-1})	Ammonia volatilization (% of N applied)	Total N-loss (% of N applied)
Basal incorporation	Fillery *et al.*, 1986b	Munoz	80	15	18
	Fillery *et al.*, 1986b	Los Banos	60	13	26
Broadcast at 10–21 DT	Fillery *et al.*, 1986b	Munoz	80	47	45
	Fillery and De Datta, 1986	Munoz	58	36	41
	Fillery *et al.*, 1986b	Los Banos	60	27	60
	Fillery and Byrnes, 1984	Los Banos	60	53	55
	IRRI, 1987	Calauan	80	54	64
Broadcast at 5–7 DBPI	Fillery *et al.*, 1986b	Munoz	40	11	13

DT = days after transplanting; DBPI = days before panicle initiation.

Table 2. Recovery of ^{15}N labeled fertilizer by rice as affected by time and method of N fertilizer application (58 kg N ha^{-1}) (Schnier *et al.*, 1988)

Treatment	Recovery of ^{15}N labeled fertilizer				
	Straw	Root	Grain	Soil	Unaccounted for
	(% of applied N)				
First application					
Broadcast and incorporated, TP	11.0	1.8	18.8	20.4	48.0
Broadcast, 14 DT	14.5	1.9	24.8	23.5	35.3
Band-placed, 14 DT	23.2	3.5	41.4	22.4	9.5
USG, TP	25.8	2.7	44.6	20.4	6.5
LSD 0.05	3.7	–	8.3	4.6	7.0
Second application					
Broadcast, 5–7 DBPI	25.8	2.6	43.3	18.1	10.2
Band-placed, 5–7 DBPI	25.0	2.1	54.7	16.3	1.9
LSD 0.05	5.3	–	7.1	4.9	11.0

TP = transplanting; DT = days after transplanting; DBPI = days before panicle initiation; USG = urea supergranules.

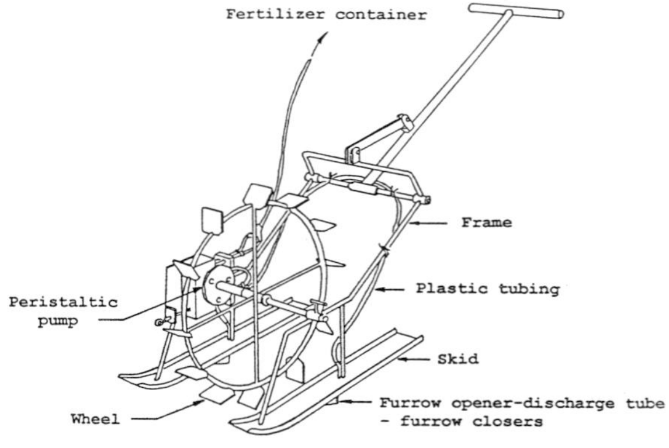

Fig. 1. Liquid urea injector with peristaltic pump.

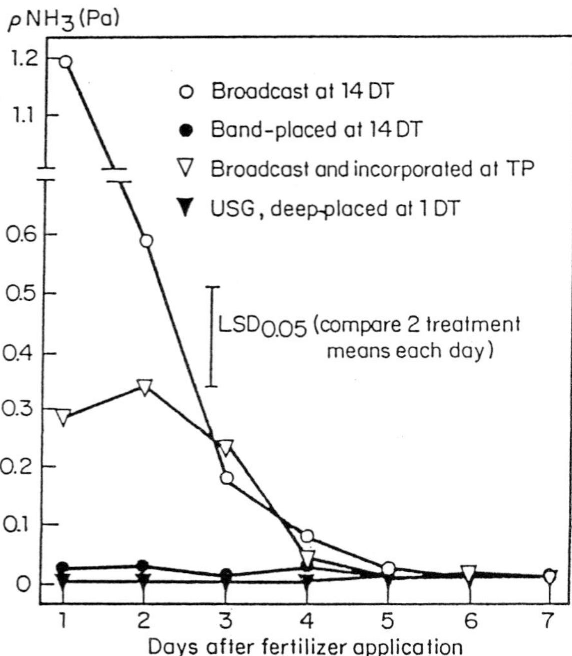

N-losses and efficiency of N fertilizer

The poor utilization of N fertilizer by rice is due to high N losses from the soil/plant system. Ammonia volatilization, denitrification, leaching, and runoff are mechanisms of N loss (Craswell and Vlek, 1979; De Datta and Patrick, 1986; Savant and De Datta, 1982), among which the first two are thought to be the more important. Research has shown that NH_3 volatilization is a major cause of N loss from urea applied to flooded rice (Filler *et al.*, 1984; Mikkelsen *et al.*, 1978). The magnitude of NH_3 loss depends on wind speed (air turbulences at the water surface), temperature, rainfall, ammoniacal-N concentration, pH and cation exchange capacity (Freney *et al.*, 1983). In rice fields, rapid hydrolysis of urea and subsequent high floodwater ammoniacal-N concentrations following application, high temperatures and elevated floodwater pH resulting from photosynthetic activity of the aquatic flora, produce an environment favorable for NH_3 loss (Fillery *et al.*, 1986a; Mikkelsen *et al.*, 1978; Vlek and Craswell, 1981). NH_3 fluxes began within 2–4 hours after urea application (Fillery *et al.*, 1984; Fillery and De Datta, 1986; Simpson *et al.*, 1984) after which NH_3 flux followed a diurnal pattern following the dynamics of pH and temperature of the floodwater. NH_3, fluxes ceased 7–10 days after N-application, when the floodwater was essentially depleted of ammoniacal-N.

Direct measurements by micrometeorological methods have shown that losses due to NH_3 volatilization ranged from 11 to 54% of the applied urea-N (Table 1). Ammonia loss tended to be greater from urea broadcast into the floodwater at 10–21 days after transplanting (DT) than from urea either broadcast and incorporated before transplanting or broadcast at 5–

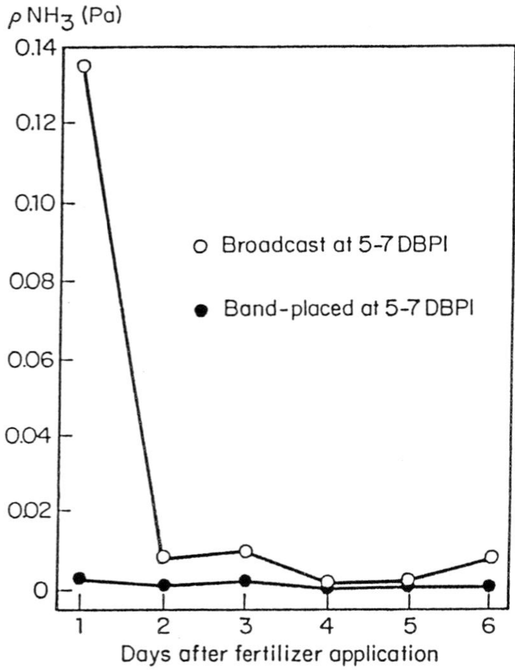

Fig. 2. Partial pressure of NH_3 (pNH_3) in floodwater as affected by method of application. (Schnier *et al.*, 1988).

Table 3. Grain yield and agronomic efficiency of IR64 rice as affected by time and method of N fertilizer application (58 kg N ha^{-1}) (Schnier *et al.,* 1988)

Treatment	Grain yield (t ha^{-1})	Agronomic efficiency (kg grain kg^{-1} N applied)
No fertilizer	5.0	–
2/3 broadcast and incorporated, TP + 1/3 broadcast 5–7 DBPI	6.4	24
2/3 broadcast, 14 DT + 1/3 broadcast 5–7 DBPI	6.4	24
2/3 band-placed, 14 DT + 1/3 broadcast 5–7 DBPI	7.2	38
2/3 band-placed, 14DT + 1/3 band-placed 5–7 DBPI	7.5	43
USG point placed, TP	7.2	38
LSD 0.05	0.6	–

TP = transplanting; DBPI = days before panicle initiation; DT = days after transplanting; USG = urea supergranules.

7 days before panicle initiation (DBPI). Buresh and De Datta (1990) found that 76% of the total N loss during the vegetation period occurred within the first 6 days after N application. Similar data were obtained by others (Buresh and Austin, 1988; Reddy *et al.*, 1990; Schnier *et al.*, 1990). These results highlight the significance of NH_3 volatilization as a loss mechanism.

Besides NH_3 volatilization, fertilizer N may also be lost by denitrification. As compared to NH_3 volatilization, however, little information is available about the magnitude of denitrification losses of fertilizer N because suitable methods for direct field measurements are lacking. Indirect estimates of denitrification range from 0–33% of applied N (Fillery *et al.*, 1986b; Fillery and De Datta, 1986; IRRI, 1987). Recent direct measurements of denitrification by Buresh and De Datta (1990) showed losses of only 0.1–2.2% of applied N in contrast to total N losses of 10–56% determined by ^{15}N balance in the same experiments. These measurements, however, did not account for N_2 and N_2O gases entrapped in the soil (Letey *et al.*, 1980; Reddy *et al.*, 1990) or lost through the rice plant (Buresh *et al.*, 1989; Mosier *et al.*, 1990), nor was the method ever proved. According to Reddy *et al.* (1976) the loss by denitrification is limited by the rate of nitrifi-

cation. Lindau *et al.* (1990) did not measure any N_2 and N_2O emissions within the first 7 days following urea application. These findings suggest that high N losses during the first week after N application are caused mainly by ammonia volatilization (Buresh and De Datta, 1990).

The magnitude of N losses and thus the nitrogen use efficiency is significantly influenced by timing and method of N fertilizer application (Table 1).

Method of N application

Deep placement of urea

Research on deep placement of N fertilizers started in the 1940's (Mitsui, 1956) and many studies at the International Rice Research Institute (IRRI) and other institutions (De Datta, 1987) have demonstrated that fertilizer use efficiency in flooded rice production could be significantly increased by placing fertilizer N in the reduced soil layer at depths ranging from 5–15 cm (Obcemea *et al.*, 1984), except on soils with high percolation rates (Kaytal *et al.*, 1985; Savant and Stangel, 1985). Various types of N fertilizers, and various types

Table 4. Grain yield and agronomic efficiency for direct seeded rice crops as affected by time and method of urea application (87 kg N ha^{-1}) (Schnier *et al.*, 1990b)

Treatment	Grain yield (t ha^{-1}	Agronomic efficiency (kg grain kg^{-1} N applied)
Dibbled rice in hills		
No fertilizer	3.3	–
2/3 broadcast and incorporated, S + 1/3 broadcast 5–7 DBPI	7.7	51
1/3 broadcast and incorporated, S + 2/3 broadcast 5–7 DBPI	7.2	45
1/3 band-placed, 20 DS + 2/3 band-placed 5–7 DBPI	8.5	60
USG point placement, 20 DS	7.5	48
Row seeded rice		
No fertilizer N	3.6	–
1/3 broadcast and incorporated, S + 2/3 broadcast 5–7 DBPI	7.3	46
1/3 band-placed, 20 DS + 2/3 band-placed 5–7 DBPI	8.1	55
LSD 0.05	0.4	

S = seeding; DBPI = days before panicle initiation; DS = days after seeding; USG = urea supergranules.

of deep placement (e.g. mudballs, urea supergranules, briquets) of N fertilizer were tested to increase N fertilizer efficiency (De Datta, 1975, 1977; Tanaka *et al.*, 1959). In spite of more than 40 years of research, however, farmers have not adopted deep placement at the farm level since fertilizer materials like urea briquets or urea supergranules (USG) are not commercially available, would be more expensive to purchase, and most importantly laborious to deep-place (Kaiser, 1984). The labor requirements for USG point placement are high, amounting to 80–110 man-hours per ha (R. Diamond (IFDC), pers. communication). In attempts to mechanize deep placement, applicators have been developed for prilled urea (PU) and USG (Khan *et al.*, 1984; Prins *et al.*, 1984; Savant, 1988). However, the performance of these machines in the field has been inconsistent, mainly due to technical problems e.g. caking of PU.

Timing of N application

Since the common farmers' practice of broadcasting urea into the floodwater between transplanting and early tillering leads to high volatilization losses, it was recommended to broadcast and incorporate urea into drained soil before transplanting (De Datta *et al.*, 1987a). With this practice, ammonia volatilization may be reduced but still remains quite high because of difficulty of incorporation of the very soluble urea (De Datta *et al.*, 1987a, 1988, 1989; Padilla *et al.*, 1990) (see also Table 2). It is important that no standing water is present so the urea can be thoroughly incorporated

134

Table 5. Recovery of ^{15}N labeled urea by rice as affected by time and method of N fertilizer application (87 kg N ha^{-1}) (Schnier *et al.*, 1990)

Treatment	Recovery of ^{15}N labeled fertilizer				
	Straw	Root	Grain	Soil	Unaccounted for
			(% of applied N)		
First application					
Transplanted					
Broadcast and incorporated, TP	11	2	16	26	45
Band-placed, 14 DT	19	3	43	25	10
USG, TP	21	2	52	21	4
Direct-seeded					
Broadcast and incorporated, S	10	2	21	29	38
Band-placed, 20 DS	17	2	48	28	5
USG, 20 DS	22	3	53	20	2
LSD 0.05	4	–	8	4	8
Second application					
Transplanted					
Broadcast, 5–7 DBPI	14	3	47	19	17
Band-placed, 5–7 DBPI	22	2	56	17	3
Direct seeded					
Broadcast, 5–7 DBPI	18	2	48	24	8
Band-placed, 5–7 DBPI	21	2	56	19	2
LSD 0.05	3	–	6	5	4

TP = transplanting; DT = days after transplanting; DBPI = days before panicle initiation; S = seeding, DS = days after seeding, USG = urea supergranules.

into the soil. Padilla *et al.* (1990) demonstrated that the effectiveness of fertilizer incorporation is influenced by the method and implements used. Conventional water buffalo- and single axle tractor-drawn comb harrows were equally, but only partially, effective in reducing floodwater N by 42–56% of the values without incorporation. The movement of transplanters through the field did not effectively incorporate urea into the soil.

Despite the recommendation of basal application (broadcast and incorporation), farmers commonly apply the first N dose into the floodwater after crop establishment. Many farmers in the Philippines and elsewhere in Asia are reluctant to apply N fertilizer as a basal dose prior to transplanting or direct seeding (Fujisaka, 1993). Reasons given by farmers for not adopting the practice are that such an application was unnecessary and an extra cost, that it leads to more weed growth, and often did not increase yield. Since most farmers don't have rotary tillers incorporation of urea is not very thorough and therefore gives only little benefit. It also requires capital early in the season when transplanting and other costs have to be met (Fujisaka, 1993).

Besides the socio-economic considerations mentioned above, the question arises whether a basal N application is necessary for the nutrition of the rice crop. Timing of fertilizer application generally is determined by crop demand at specific growth stages and the corresponding availability of soil N.

Transplanted rice suffers from a physiological transplanting shock (Dingkuhn *et al.*, 1990) for 10–14 days following transplanting during which the root systems have little sink capacity for N (Schnier *et al.*, 1987). Thus, fertilizer N applied at transplanting cannot readily be taken up by the plants and is subject to ammonia loss, immobilization, denitrification and fixation.

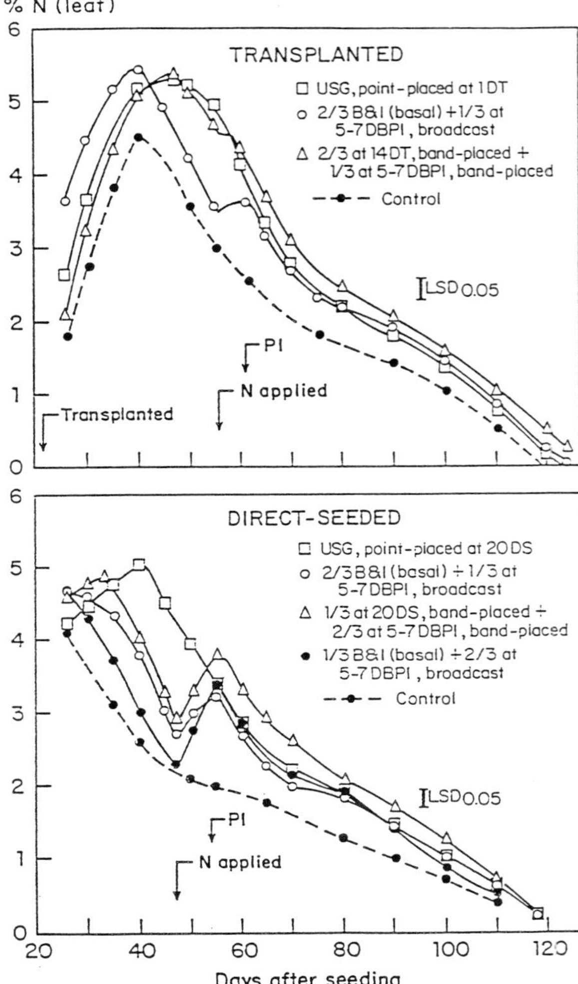

Fig. 3. Leaf N content at various growth stages of transplanted and direct seeded IR64 rice as affected by fertilizer application time and method (87 kg N ha^{-1}). (Schnier *et al.*, 1990). USG = urea supergranules, B & I = basal broadcast and incorporated, DT = days after transplanting, DS = days after seeding, DBPI = days before panicle initiation.

Furthermore, the N supplying capacity of soils following mineralization of N from organic matter after soil puddling (during land preparation) is usually sufficient to meet crop demands during the early stage. Keerthisinghe *et al.* (1985) observed an increase of exchangeable NH_4^+-N in the beginning of the season even without N application. Exchangeable NH_4^+-N amounts in a number of Philippine rice soils were between 18 and 55 kg N ha^{-1} at transplanting (Schön, 1982; Schön *et al.*, 1985). Thus, there seems to be lit-

tle need for early-season N applications on soils with relatively high N fertility. A basal N application may be required only on soils with a very low N status to promote early crop growth.

With regard to the timing of N application in relation to crop demand, farmers' practice of delayed application, therefore, is reasonable. The method of broadcasting fertilizer N to the floodwater which is usually the only option, however, has the disadvantage causing high losses by ammonia volatilization, since urease activities and algal populations are established which promote losses, but the plant canopy and sink capacity are not well enough developed to curtail losses.

Therefore, a fertilizer management practice that combines the farmers' usual timing of N application and water management with the advantages of deep placement is needed. Schnier *et al.* (1988) developed a band-placement method by which urea solution (up to 70% PU solution) is band-placed 5–6 cm deep in the soil with a two-row manually pushed injector (Bautista and Schnier, 1989; Fig. 1). This method allows flexible N applications independent of crop growth stage and water management with row transplanted rice. Labor requirements vary between 12 and 16 man-hours per ha, depending on soil conditions. The urea solution can easily be prepared in the field by mixing PU with water from the irrigation ditch and carried in a knapsack sprayer. The injector has been tested in various soil types and consistently given good results.

The NH_3 partial pressure in the floodwater was significantly reduced when liquid urea was band-placed at 14 DT as compared to broadcast application at 14 DT (farmers' practice) and broadcast and incorporated before transplanting (Fig. 2). Similar results were found for a second N application made at 5–7 days before panicle initiation (Fig. 2b). The generally lower pNH_3 and its fast decline is attributed to a higher absorbance of fertilizer N by the well-developed root system in the surface soil at this growth stage and the rapid water movement into the soil.

About 20% of the fertilizer N applied at the first application was recovered in the soil at harvest. regardless of treatment (Table 2). Fertilizer recovery in the plant, however, was almost double in the band-placed and USG treatments than in the broadcast and incorporated treatments (Table 2). Consequently, the unaccounted fertilizer N for the band placement and USG treatments was less than 10% of the applied N. Similarly, fertilizer recovery in the plant with band placement of the second application was significantly greater than with broadcast application although the effect was less

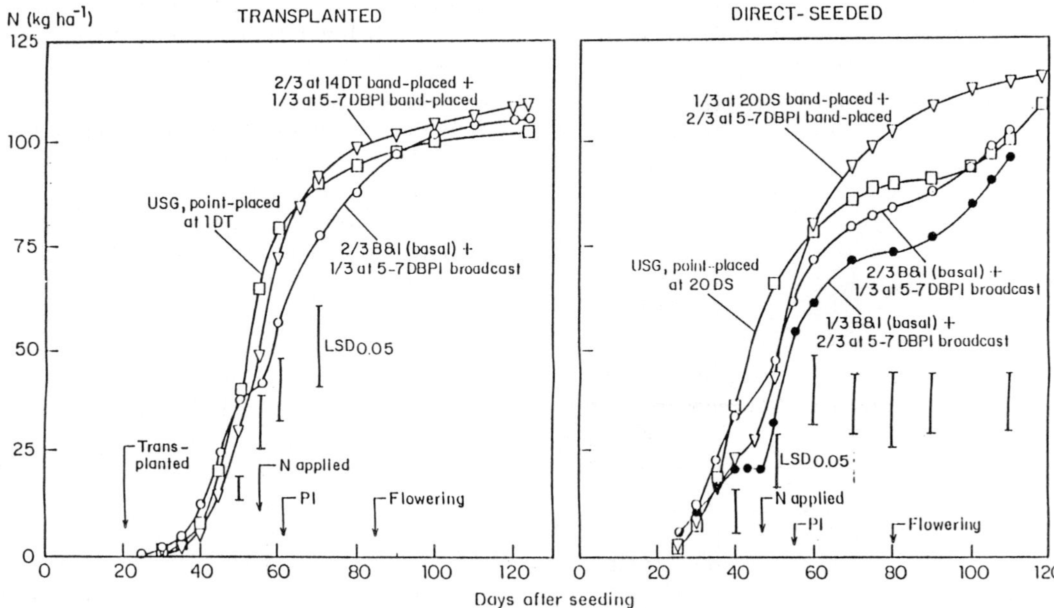

Fig. 4. Plant N uptake at various growth stages of transplanted and direct seeded IR64 rice as affected by fertilizer N application time and method (87 kg N ha^{-1}). (Schnier *et al.*, 1990). For abbreviations, refer to Fig. 3.

pronounced than in the first N application (Table 2). The increased fertilizer N recovery by the plants was well reflected in higher grain yield and agronomic efficiency (Table 3).

The efficiency of N fertilizer management practices can also be affected by the rice establishment method. Transplanting has been the more popular crop establishment method in Asia's irrigated and rainfed lowland rice-growing areas (De Datta, 1981), but it is increasingly replaced by direct seeding as labor becomes more costly (De Datta, 1986a).

Direct seeded rice exhibits a different growth pattern than transplanted rice (Dingkuhn *et al.*, 1990). Because of the absence of a transplanting shock, direct seeded rice exhibits a more vigorous vegetative growth phase leading to excessive tiller formation and biomass production that often leads to N deficiency at later growth stages (Schnier *et al.*, 1990). Leaf N concentrations are usually lower in direct seeded than in transplanted rice (Fig. 3). Tissue N concentration just before panicle initiation affects yield potential through the number of spikelets formed while at later growth stages N concentration in the plant affects the conversion of biomass at grain filling (Dingkuhn *et al.*, 1991).

Thus, a 1/3–2/3 split band placement of N led to a more balanced N nutrition (Fig. 4) and higher grain yield than the 2/3–1/3 split used for transplanted rice (Table 4). Band placement was superior to broadcast application. As N losses were minimized by band placement (Table 5), the application of only 1/3 of the dose was adequate in supplying the crop with N during the vegetative stage without causing excessive growth, while the remaining 2/3 of the dose could be applied to raise the N concentration in the plant tissue at the panicle initiation stage.

The optimum split of band placed fertilizer N in transplanted rice may vary from location to location. Studies by Schnier *et al.* (1993) on various soils in Bangladesh and Indonesia showed that in some cases a 2/3–1/3 gave higher yields than a 1/3–2/3 split while in other cases the inverted split was superior or both splits were equal. The optimum split ratio is likely to be dependent on the growth duration of the cultivar, N rate, N response, and N supply of the soil. These aspects require more site specific investigations. Band placement of the full amount in a single application resulted in yields similar to those with USG point placement but were generally lower than split applications indicating that a single application led to an unbalanced N nutrition of the crop. In these experiments, yields with the best band application were between 18 and 55% greater than those obtained with the farmers' practice of broadcasting at 15–21 DT.

The costs of the injector is approximately 50 US\$. With an average additional yield increase of $0.5\,t\,ha^{-1}$, the investment of the injector would be recovered in the first cropping season (cost of labor for application not included).

Conclusions

N loss by NH_3 volatilization is the main cause for low N-use efficiency of flooded tropical rice and is strongly affected by timing and method of fertilizer application. Both in transplanted and direct seeded rice, band placement significantly increased N recovery by the rice crop and grain yield. For high grain yields, the total N uptake is not only important but also the N status in the plant at specific growth stages such as panicle initiation. This was especially evident in direct seeded rice where band placement as a 1/3–2/3 split led to a more balanced N nutrition and gave the greatest grain yield. Band placement of a urea solution proved to be an effective application method to minimize NH_3 volatilization losses while at the same time, allowing flexible timing of application independent of crop stage and water management. Band placement provides N use efficiency equal to that achieved by USG point placement while being less labor-intensive (24–32 hours for split band application vs, 80–110 hours for point placement). Adoption of band placement by farmers will depend on promotion, demonstration and further testing by extension groups. It has the advantage over USG in that a new fertilizer product does not need to be introduced.

References

Barker R, Herdt RW and Rose B (1985) The rice economy in Asia. Resources for the Future, Washington, DC. 324 p

Bautista EU and Schnier HF (1989) Injecting fertilizer solution in wetland paddies. Agric Mechanization in Asia, Africa and Latin America (AMA) 20: 23–27

Buresh RJ and Austin ER (1988) Direct measurement of dinitrogen and nitrous oxide flux in flooded rice fields. Soil Sci Soc Am J 52: 681–688

Buresh RJ and De Datta SK (1990) Denitrification losses from puddled rice soils in the tropics. Biol Fertil Soils 9: 1–13

Buresh RJ, De Datta SK, Samson MI, Phongpan S, Snitwongse P, Fagi AM, Tejasarwana R and Wetselaar R (1989) Denitrification loss from urea applied to transplanted rice in Thailand, Indonesia and the Philippines. Agron Abstr, Am Soc Agron, Madison, WI

Graswell ET and Vlek PLG (1979) Fate of fertilizer nitrogen applied to wetland rice. In: Nitrogen and rice, pp 175–192. International Rice Research Institute, Los Banos, Laguna, Philippines

Craswell ET and Godwin DC 1984 The efficiency of nitrogen fertilizers applied to cereals in different climates. In: Tinker PB and Lauchli A (eds) Advances in plant nutrition, Vol 1, pp 1–55. Praeger, NY

De Datta SK (1975) Increasing fertilizer efficiency in rice. Unpublished paper presented at the Plenary Session on Relevant Rersearch for Increasing Rice Production, IRRI, 21–24 April 1975

De Datta SK (1977) Fertilizer management for efficient use in wetland rice soils. Unpublished paper presented at the Symposium, Soil and Rice, IRRI, 20–23 September 1977

De Datta SK (1981) Principles and Practices of Rice Production. John Wiley, New York

De Datta SK (1986a) Technology development and the spread of direct seeded flooded rice in Southeast Asia. Exp Agric 22: 417–426

De Datta SK (1986b) Improving nitrogen fertilizer efficiency in lowland rice in tropical Asia. Fert Res 9: 171–186

De Datta SK (1987) Nitrogen transformation processes in relation to improved cultural practices for lowland rice. Plant Soil 100: 47–69

De Datta SK and Patrick WH Jr (eds) (1986) Nitrogen Economy of Flooded Rice Soils. Martinus Nijhoff, Dordrecht, The Netherlands

De Datta SK and Buresh RJ (1989) Integrated nitrogen management in irrigated rice. Adv Soil Sciences, Vol. 10. Springer, NY

De Datta SK, Obcemea WN, Chen Ry, Calabio JC and Evangelista RC (1987a) Effect of water depth on nitrogen use efficiency and nitrogen-15 balance in lowland rice. Agron J 79: 210–216

De Datta SK, Fillery IRP, Obcemea WN and Evangelista RC (1987b) Floodwater properties, nitrogen utilization, and nitrogen-15 balance in a calcareous lowland rice soil. Soil Sci Soc Am J 51: 1355–1362

De Datta SK, Buresh RJ, Samson MI and Kai-Rong W (1988) Nitrogen use efficiency and nitrogen-15 balances in broadcast seeded flooded and transplanted rice. Soil Sci Soc Am J 52: 849–855

De Datta SK, Trevitt ACF, Freney JR, Obcemea WN, Real JG and Simpson JR (1989) Measuring nitrogen losses from lowland rice using bulk aerodynamic and nitrogen-15 balance methods. Soil Sci Soc Am J 53: 1275–1281

Dingkuhn M, Schnier HF, De Datta SK, Wijangco E and Dörffling K (1990) Diurnal and developmental changes in canopy gas exchange in relation to growth in transplanted and direct-seeded flooded rice. Aust J Plant Physiol 17: 119–134

Dingkuhn M, Schnier HF, De Datta SK, Dörffling K and Javellana C (1991) Relationships between ripening-phase productivity and crop duration, canopy photosynthesis and senescence in transplanted and direct-seeded lowland rice. Field Crops Res 26: 327–345

FAO (1987) 1986 FAO fertilizer yearbook, Vol 36. Food and Agriculture Organization of the United Nations, Rome, Italy

Fillery IRP and Byrnes BH (1984) Concurrent measurement of ammonia loss and denitrification in flooded rice fields. Agron Abstr, American Society of Agronomy, Madison, WI. 204 p

Fillery IRP and De Datta SK (1986) Ammonia volatilization from nitrogen sources applied to rice fields: I,. Methodology, ammonia fluxes, and nitrogen-15 loss. Soil Sci Soc Am J 50: 80–86

Fillery IRP, Simpson JR and De Datta SK (1984) Influence of field environment and fertilizer management on ammonia loss from flooded rice. Soil Sci Soc Am J 48: 914–920

Fillery IRP, Roger PA and De Datta SK (1986a) Ammonia volatilization from nitrogen sources applied to rice fields: II. Floodwater properties and submerged photosynthetic biomass. Soil Sci Soc Am J 50: 86–91

Fillery IRP, Simpson JR and De Datta SK (1986b) Contribution of ammonia volatilization to total nitrogen loss after applications of urea to wetland rice fields. Fert Res 8: 193–202

Freney JR, Simpson JR and Denmead OT (1983) Volatilization of ammonia. In: Freney JR and Simpson JR (eds) Gaseous loss of nitrogen from plant-soil systems, pp 1–32. Martinus Nijhoff/Dr W Junk Publishers, The Hague, The Netherlands

Fujisaka S (1993) Were farmers wrong in rejecting a recommendation? The case of nitrogen at transplanting for irrigated rice. Agric Syst 43: 271–286

IRRI (1986) World rice statistics 1985. International Rice Research Institute, Los Banos, Laguna, Philippines

IRRI (1987) Annual report for 1986. International Rice Research Institute, Los Banos, Laguna, Philipppines

Kaiser K (1984) Possibilities to increase the technical and economic efficiencies of nitrogen fertilizers in irrigated rice in the Philippines. PhD dissertation, University of Hohenheim, Stuttgart, Germany

Kaytal JC, Bijay Singh, Vlek PLG and Craswell ET. (1985) Fate and efficiency of nitrogen fertilizers applied to wetland rice II. Punjab, India. Fert Res 6: 279–290

Keerthisinghe G, Mengel K and De Datta SK (1985) Importance of exchangeable and non-exchangeable soil NH_4 in nutrition of lowland rice. Soil Sci 140: 194–201

Khan AU, Kiamco LC, Tiangco VM, Camacho IR, Diestro MS and Bautista EU (1984) Applicators for improved fertilizer use efficiencies in wertland paddies. Philipp J Crop Sci 9: 206–216

Letey J, Jury WA, Hadas A and Valoras N (1980) Gas diffusion as a factor in laboratory incubation studies on denitrification. J Environ Qual 9: 223–227

Lindau CW, DeLaune RD, Patrick WH Jr and Bollich PK (1990) Fertilizer effects on dinitrogen, nitrous oxide, and methane emissions from lowland rice. Soil Sci Soc Am J 54: 1789–1794

Mikkelsen DS, De Datta SK and Obcemea WN (1978) Ammonia volatilization losses from flooded rice soils. Soil. Sci Soc Am J 42: 725–730

Mitsui S (1956) Inorganic Nutrition, Fertilization and Soil Amelioration for Lowland Rice., 2nd ed. Yokendo Press, Tokyo

Mosier AR, Mohanty SK, Bhadrachalam A and Chakravorti SP (1990) Evolution of N2 and N2O from the soil to the atmosphere through rice plants. Biol Fertil Soils 9: 61–67

Obcemea WN, De Datta SK and Broadbent FE (1984) Movement and distribution of fertilizer nitrogen as affected by depth of placement in wetland rice. Fert Res 5: 125–148

Padilla JL, Buresh RJ, De Datta SK and Bautista EU (1990) Incorporation of urea in puddled rice soils as affected by tillage implements. Fert Res 26: 169–178

Prins WH, van Brakel GD and van der Sar T (1984) Pneumatic injector for deep placement of urea in wetland rice soils. Paper presented at 2nd International Conference on the Development of the Agricultural Machinery Industry in Developing Countries, Amsterdam, The Netherlands

Reddy KR, Patrick WH Jr and Phillips RE (1976) Ammonium diffusion as a factor in nitrogen loss from flooded soils. Soil Sci Soc Am J 40: 528–533

Reddy KR, D'Angelo E, Lindau C and Patrick WH Jr (1990) Urea losses in flooded soils with established oxidized and reduced soil layers. Biol Fertil Soils 9: 283–287

Savant NK (1988) How to use the IFDC dispenser method of hand. deep placement of urea supergranules in transplanted rice. International Fertilizer Development Center, Muscle Shoals, AL

Savant NK and De Datta SK (1982) Nitrogen transformations in wetland rice soils. Adv Agron 35: 241–302

Savant NK and Stangel PJ (1985) Suggestions for efficient use of urea supergranules for wetland rice. In: Stangel PJ et al. (eds) Proceedings of the workshop on urea deep placement technology. IFDC Spec Publ. SP-6 pp 133-142. International Fertillzer Development Center, Muscle Shoals, AL, USA

Schnier HF, De Datta SK and Mengel K (1987) Dynamics of ^{15}N labelled ammonium sulfate in various inorganic and organic soil fractions of wetland rice soils. Biol Fertil Soils 4: 171–177

Schnier HF, De Datta SK, Mengel K, Marqueses EP and Faronilo JE (1988) Nitrogen use efficiency, floodwater properties, and nitrogen-15 balance in transplanted lowland rice as affected by liquid urea band placement. Fert Res 16: 241–255

Schnier HF, Dingkuhn M, De Datta SK, Marqueses EP and Faronilo JE (1990a) Nitrogen-15 balance in transplanted and directseeded flooded rice as affected by different methods of urea application. Biol Fertil Soils 10: 89–96

Schnier HF, Dingkuhn M, De Datta SK, Mengel K, Wijangco E and Javellana C (1990b) Nitrogen economy and canopy carbon dioxide assimilation of tropical lowland rice. Agron J 82: 451–459

Schnier HF, De Datta SK, Fagi AM, Eaqub M, Faruque Ahmed, Tejasarwana R and Mazid A, (1993) Yield response of wetland rice to band placement of urea solution in various soils in the tropics. Fert Res 36: 221–227

Schön HG (1982) Die Bedeutung des austauschbaren Ammoniums in überfluteten Reisböden für die Ertragsbildung von Reis und für die Basis einer Stickstoffdüngerempfehlung. PhD dissertation, University of Gießen, Gießen

Schön HG, Mengel K and De Datta SK (1985) The importance of initial exchangeable ammonium in the nitrogen nutrition of lowland rice soils. Plant Soil 86: 403–413

Simpson JR, Freney JR, Wetselaar R, Muirhead WA, Leuning R and Denmead OT (1984) Transformations and losses of urea nitrogen after application to flooded rice. Aust J Agric Res 35: 189–200

Stangel PJ and De Datta SK (1985) Availability of inorganic fertilizers and their management - a focus on Asia. Paper presented at the International Rice Research Conference, 1–5 June 1985, International Rice Research Institute, Los Banos, Laguna, Philippines.

Tanaka A, Patnaik S and Abichandani CT (1959) Studies on the nutrition of the rice plant. III. Partial efficiency of nitrogen absorbed by rice plant at different stages of growth in relation to yield of rice of different duration. Proc Indian Acad Sci Soc 49(4): 217–226

Vlek PLG and Craswell ET (1981) Ammonia volatilization from flooded soil. Fert Res 2: 227–245

Vlek PLG and Byrnes BH (1986) The efficacy and loss of fertilizer N in lowland rice. In: De Datta SK and Patrick WH Jr (eds) Nitrogen Economy of Flooded Rice Soils, pp 131–147. Martinus Nijhoff, Dordrecht, The Netherlands

Fertilizer Research **42:** 139–148, 1995.
© 1995 *Kluwer Academic Publishers. Printed in the Netherlands.*

A conceptual assessment of the importance of denitrification as a source of soil nitrogen loss in tropical agro-ecosystems

Peter M Groffman

Institute of Ecosystem Studies, Box AB, Millbrook, NY 12545, USA

Key words: denitrification, methods, ^{15}N, nitrate, nitrogen

Abstract

This paper attempts to answer the question: is denitrification a major route of N loss from tropical agro-ecosystems? This question turns out to be very difficult to answer due to a severe shortage of data on this process for tropical agro-ecosystems other than rice. Given this lack of data, I approach this question by analyzing data on denitrification and nitrous oxide flux in tropical native forest and pasture soils and attempt to make some conclusions and pose some hypotheses about the significance of denitrification in tropical agricultural soils. I also briefly review methods for measuring denitrification. The data analysis suggests that denitrification in tropical forest soils is strongly influenced by the nature and amount of soil C and N turnover. Studies to examine differences in denitrification in different tropical agricultural systems should focus on the effects of system management on C and N turnover. The data analysis also suggests that, just as in temperate regions, denitrification may not be a significant route of N loss from most tropical agricultural systems. However, field studies are necessary to determine if this is actually the case.

Introduction

This paper attempts to answer the question: is denitrification a major route of nitrogen (N) loss from tropical agro-ecosystems? This question turns out to be very difficult to answer due to a severe shortage of data on this process for tropical agro-ecosystems other than rice. The lack of denitrification data for these systems is puzzling given the suspicion that this process may be much more important in wet tropical soils than in temperate soils, where it is generally not that important to the overall N economy. Given this lack of data, I will approach this question by drawing from several data sources that are related to this topic. The result is more a series of hypotheses that need to be tested rather than a definitive answer to the original question.

Several comprehensive reviews of controls, rates and methods for studying denitrification have been published (Aulakh *et al.*, 1992; Nieder *et al.*, 1989; Payne, 1981; Tiedje, 1988; Von Rheinbaben, 1990). I will not duplicate these reviews here. In the sections that follow, I will first briefly review the physiology of denitrification and discuss the environmental factors that control this process at different scales of investi-

gation. I will then give a brief review of methods for studying denitrification, and then go on to discuss the various data that provide insights on the significance of this process in tropical agricultural soils. These data include studies of denitrification in native tropical forest soils and studies of nitrous oxide (N_2O) flux in native forest and pasture sites.

Physiology and controls

Denitrification refers to the reduction of the N oxides NO_3^- and NO_2^- to the gases NO, N_2O and N_2 (Fig. 1). The process is generally considered to be carried out by facultative anaerobes – organisms that normally use oxygen O_2 for respiration but in its absence use N oxides as electron acceptors. There is evidence that some organisms can simultaneously respire O_2 and

$$NO_3^- \rightarrow NO_2^- \rightarrow NO \rightarrow N_2O \rightarrow N_2$$

Fig. 1. Denitrification pathway.

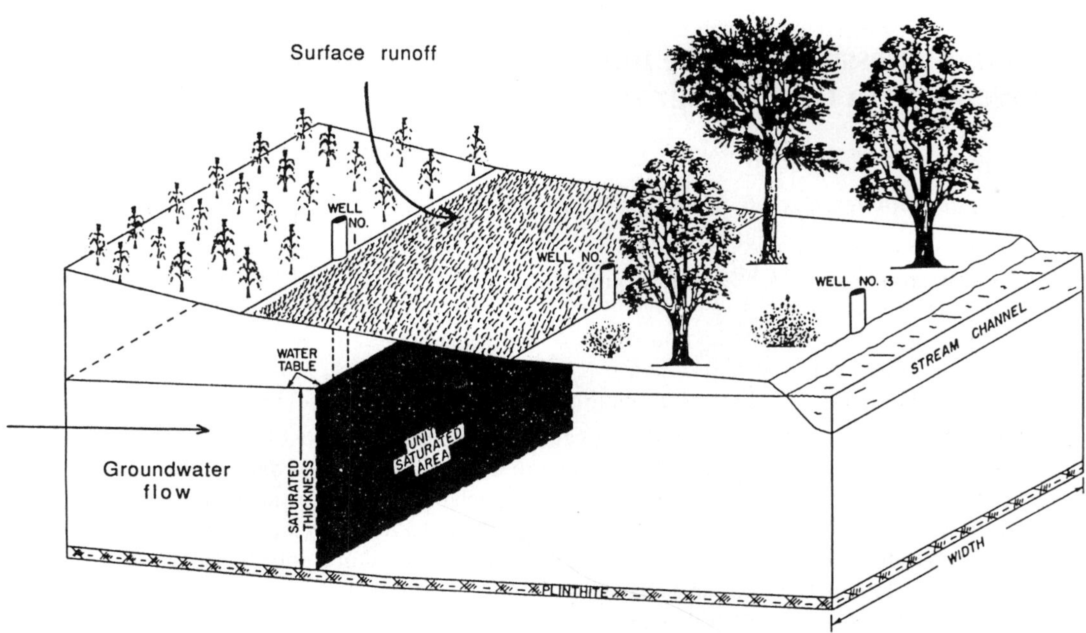

Fig. 2. Conceptual diagram of a landscape containing crops and pasture on upland soils, riparian forest on wetland soils, and a stream. The different components of the landscape are hydrologically linked by surface runoff and groundwater flow. Different components of the landscape, including the groundwater ecosystem, have different potentials to sustain high denitrification N losses under agricultural management or to prevent NO_3^- movement to the stream. Adapted from Lowrance *et al.* (1984).

Table 1. Factors controlling denitrification at different spatial scales. Adapted from Groffman (1991)

Scale	Controlling factors
Organism	O_2, NO_3^-, C
Field	Soil H_2O, NO_3^- supply, C supply
Landscape	Soil type, Plant community type
Region	Geomorphology, Land use
Globe	Biome type, Climate

NO_3^- in what has been termed 'aerobic denitrification' (Lloyd, 1993; Robertson & Kuenen, 1991).

Most denitrification studies have focused on heterotrophic organisms (those that use organic compounds as a source of energy and structural carbon (C), although a variety of chemoautotrophic organisms (S and CH_4 oxidizers) have the ability to denitrify. Detailed studies have analyzed the effects of O_2, NO_3^-, C, pH, temperature and other factors on the activity of denitrifying organisms (see reviews cited above).

It is useful to consider how the factors that control denitrification are expressed at different scales of investigation relevant to specific questions (Table 1, Groffman *et al.*, 1988; Groffman, 1991). For organismal scale studies, we focus on the availability of O_2, NO_3^- and energy and C donors to denitrifying cells. At a slightly larger scale, studies have analyzed denitrification in soil aggregates or the rhizosphere. To evaluate the importance of denitrification as a mechanism of N loss in tropical agro-ecosystems, our interest is at the field or landscape scale. At the field scale, soil moisture controls O_2 availability, NO_3^- supply via nitrification or fertilizer input controls NO_3^- availability, and C supply through residue decomposition or organic inputs controls energy and C availability to denitrifiers. It is important to focus on the appropriate controllers when designing studies and evaluating data at different scales.

There have been many studies to establish relationships between field-scale controllers and denitrification in different temperate agricultural management systems, soil types and climates. These studies have found that denitrification can be an important source of N loss in soils that are excessively wet due either to poor natural drainage, heavy texture, irrigation or high rainfall. The high rainfall and warm temperatures of the humid tropics have led to the suspicion that denitrification may be an important mechanism of N loss from agro-ecosystems in these regions.

At the landscape scale it is useful to focus on soil type (texture, natural drainage) as a controller of variation in soil water, and on plant community type (different agricultural management systems versus native forests versus wetlands versus plantation forests) as a controller of C and NO_3^- availability to denitrifiers. Interest in landscape-scale denitrification studies arises from concern about which soil or cropping systems within a landscape might be subject to high denitrification N loss or from interest in the water quality of the landscape as a whole (Fig. 2). For such whole-landscape assessments, there is a need to determine if particular components of the landscape support high rates of denitrification. There is particular interest in the ability of natural riparian eco-systems to absorb NO_3^- moving in surface runoff and groundwater flow from upland cropped areas, preventing its movement into streams. There is also great interest in the potential for denitrification of NO_3^- that has moved into groundwater.

At landscape, regional and global scales, there is interest in denitrification as a source of N_2O, a 'greenhouse' gas, to the atmosphere (Prinn et al., 1990). There is concern that denitrification of fertilizer N and of excess NO_3^- that leaves agricultural fields in surface runoff or groundwater flow may contribute to the observed increase in atmospheric levels of this gas (Vitousek & Matson, 1993).

Examination of field and landscape scale controls of denitrification suggests that there are inherent relationships between the nature and amount of soil C and N turnover and denitrification. While much field scale work has justifiably concentrated on soil wetness as the prime controller of denitrification, the process cannot proceed without a flow of N and C from plant residues to organic matter, microbial biomass and the specific denitrifying populations. Moreover, focusing on the nature and amount of C and N turnover is a useful vehicle for comparing denitrification under different crop types, residue management systems and nutrient management strategies.

A final field-scale factor that must be considered as a controller of denitrification is the temporal pattern of the cropping system. In many ecosystems, denitrification is not likely to be vigorous in a system with an actively growing plant community. Plants are strong competitors for water and N, and denitrification during the growing season is likely to be limited by high O_2 and low NO_3^- levels in most non-flooded soils. In temperate ecosystems, most denitrification has been observed to occur outside the 'growing season', e.g. before crops are planted and after they are harvested (Goodroad & Keeney, 1984; Groffman & Tiedje, 1989; Myrold, 1988; Schmidt et al., 1988). Outside the growing season, soil water and NO_3^- levels can be relatively high, leading to relatively high rates of denitrification.

The temporal pattern of cropping systems may be a more important controller of denitrification in tropical than in temperate regions. In temperate areas, the non-growing season is usually cold, reducing all microbial activity. In tropical regions, especially in the humid tropics, the potential for denitrification during inter-crop periods may be very high.

Methods for measuring denitrification

Denitrification is a difficult process to measure. Numerous controversies have arisen over which methods are best, and several reviews have been published (Aulakh et al., 1991, 1992; Keeney, 1986; Nieder et al., 1989; Payne, 1991; Tiedje, 1982; Tiedje et al., 1989). Here, I will not extensively review the different methods. Rather, I will make the point that all the different methods have strengths and weaknesses in different circumstances. Investigators must determine if the drawbacks of any particular method are a major problem in the specific situation where it is being applied.

The problem inherent in measuring denitrification is that it is difficult to quantify either the production of the dominant end-product (N_2) or the specific depletion of the substrate (e.g. NO_3^-) due to denitrification. Production of N_2 is difficult to measure due to the high atmospheric background of this gas. Depletion of NO_3^- by denitrification is difficult to quantify due to the fact that this ion is also consumed by plants, heterotrophic microbes, dissimilatory reduction to NH_4^+, leaching and runoff.

Quantification of denitrification is also hindered by high spatial and temporal variability. Rates of denitrification in the field frequently vary over two or three orders of magnitude in both time and space (Burton & Beauchamp, 1985; Folonoruso & Rolston, 1984; Parkin, 1987; Robertson & Tiedje, 1984). High variability hinders quantification of field rates, comparison of treatments and evaluation of different methods.

Acetylene (C_2H_2) inhibition method

The discovery that C_2H_2 inhibits the reduction of N_2O to N_2 was a major development in denitrification research (Balderston *et al.*, 1976; Yoshinari & Knowles, 1976). In the presence of C_2H_2, N_2O is the terminal product of denitrification (Fig. 3). Production of N_2O is relatively easy to quantify due to its low atmospheric background and the availability of sensitive detectors for this gas. Many studies have employed the C_2H_2 inhibition method to quantify denitrification.

There are several problems with the C_2H_2 inhibition method that severely constrain its use in certain circumstances. Perhaps the most important problem is that C_2H_2 inhibits the production of NO_3^- via nitrification (Fig. 3, Hynes & Knowles, 1978; Mosier, 1980; Walter *et al.*, 1979). In soils where NO_3^- concentrations are low and there is a strong coupling between nitrification and denitrification, the C_2H_2 inhibition method will underestimate denitrification rates if incubations are long enough that soil NO_3^- pools become depleted (Ryden & Dawson, 1982). The potential for this underestimation is highest in natural ecosystems where soil NO_3^- levels are usually quite low and especially in wetlands, where coupling between nitrification and denitrification is very strong (Christensen *et al.*, 1989; Kemp *et al.*, 1990; Reddy & Patrick, 1984; Rudolph *et al.*, 1991; Seitzinger *et al.*, 1993).

In addition to inhibiting nitrification, C_2H_2 also inhibits chemoautotrophic oxidation of CH_4, S and possibly Fe (Payne, 1984). These oxidations may be important energy sources for denitrification in groundwater, where heterotrophic activity is limited by a lack of available organic C (Bradley *et al.*, 1992; Korom, 1992; Parkin & Simpkins, 1993; Pedersen *et al.*, 1991; Postma *et al.*, 1991). Acetylene has also been found to inhibit the production of acetate and butyrate via fermentation (Flather & Beauchamp, 1992). Incubations with C_2H_2 may underestimate denitrification in groundwater or other situations where

chemoautotrophic oxidations or fermentation products are important energy sources.

In contrast to problems with inhibition, acetylene can stimulate denitrification by serving as a substrate for C-limited microbes (Topp & Germon, 1986; Yeomans & Beauchamp, 1982). A related problem is that populations of C_2H_2-degrading microbes can develop rapidly (7 days), making it difficult to maintain adequate levels of C_2H_2 for inhibition of N_2O reduction in long incubations (Terry & Duxbury, 1985). A final problem is the presence of contaminants (that can be removed) such as acetone, especially in industrially produced C_2H_2, that can either stimulate or inhibit denitrification (Gross & Bremner, 1992; Hyman & Arp, 1987).

Numerous approaches have been taken to introducing C_2H_2 into soil. In wet, fine-textured soils, penetration of C_2H_2 and release of N_2O to/from the active sites of denitrification can be quite slow (Jury *et al.*, 1982; Parkin *et al.*, 1984; Ryden *et al.*, 1979). The two dominant approaches for introducing C_2H_2 have been 'extracted core' techniques, where soil cores are placed in an enclosed container with an C_2H_2-containing atmosphere, and 'in-field chamber' techniques, where C_2H_2 is introduced into a soil profile in the field and the flux of N_2O from the soil surface is captured by placing a chamber over the soil surface.

Both core and chamber approaches are hindered by the high spatial variability of denitrification. There is an inherent conflict between the desire to use methods that minimize physical disturbance to the soil system (e.g. chambers) versus the desire to use simpler techniques that allow for a large number of replicate samples to be taken (e.g. cores). Several comparison studies have found no consistent differences between the techniques, except when cores were held for long periods of time (several days) before incubation (Aulakh *et al.*, 1991; Burton & Beauchamp, 1984; Ryden *et al.*, 1987).

Direct flux methods

Although it is difficult to directly measure the fluxes of denitrification substrates or products as discussed above, direct flux techniques have application in certain cases. These techniques are particularly useful in situations where the use of C_2H_2 is inappropriate.

Measuring depletion of NO_3^- can be used as a quantification of denitrification when other possible fates of NO_3^- have been either measured or eliminated. Production of N gases has been measured to quantify den-

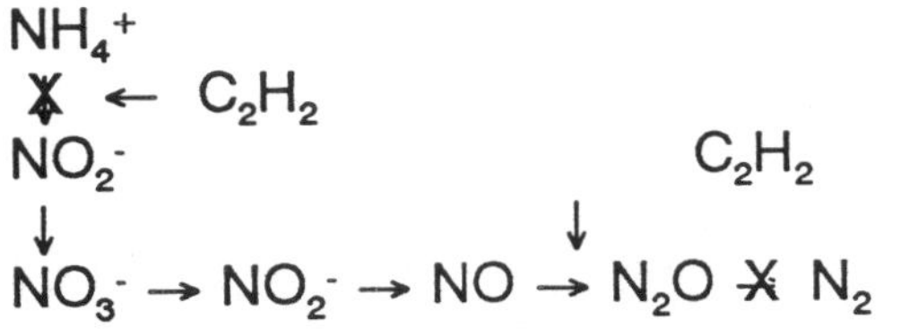

Fig. 3. Effects of acetylene (C_2H_2) on nitrification and denitrification.

itrification in laboratory studies with artificial atmospheres (e.g. without N_2, Monaghan & Barraclough, 1993; Seitzinger *et al.*, 1980). Devol (1991) developed a technique for direct measurement of N_2 production from marine sediments. This technique is based on measuring the accumulation of N_2 gas dissolved in water within a field chamber placed on the sediment surface and may be applicable to flooded soils.

^{15}N balance methods

These methods involve the addition of $^{15}NH_4^+$ or $^{15}NO_3^-$ to soil for a period of time, tracing the movement of ^{15}N into various pools and processes (plants, volatilization, leaching, runoff, soil inorganic and organic pools) and quantifying denitrification as the ^{15}N unaccounted for at the end of the experiment (Mosier *et al.*, 1986; Parkin *et al.*, 1985; Rolston *et al.*, 1979). The accuracy of this estimate of 'unaccounted-for N' is reduced by the accumulated errors associated with estimates of the other pools and processes. The specificity of this estimate as a measure of denitrification is limited by how well other loss processes (leaching, runoff, volatilization) are controlled or quantified.

In addition to accumulated error problems, the validity of the ^{15}N balance technique is limited by how well added ^{15}N simulates the behavior of soil N. While added inorganic ^{15}N is likely a good surrogate for fertilizer N, it may not be a good tracer of N in soil organic matter and microbial biomass. A final concern is that all ^{15}N methods are problematic when soil N contents are low and addition of ^{15}N can significantly enrich the N pools under study, leading to artificially high rates of activity.

$^{15}N_2$ flux techniques

Perhaps the most sophisticated denitrification methods involve quantification of ^{15}N in the gas phase of soil systems treated with ^{15}N. To date, $^{15}N_2$ flux techniques have been limited to situations where soil N levels are relatively high due to the need to highly enrich soil NO_3^- pools with ^{15}N to allow for detection of the label in the gases produced. However, recent improvements in mass spectrometer techniques have made it possible to make measurements with very low, tracer-level additions of ^{15}N (see paper by Barrie, 1995). As with ^{15}N balance approaches, while $^{15}N_2$ flux methods can reliably trace fertilizer-derived fluxes, their ability to depict fluxes of N associated with soil organic matter turnover is less certain.

Perhaps the most important constraint on the use of $^{15}N_2$ flux techniques is that they are expensive and time-consuming to carry out. Costs of ^{15}N and for mass spectrometer analysis are high, and sample preparation and collection techniques are time-consuming. These cost and time constraints limit the number of flux measurements that can be made, which is a serious problem given the high spatial and temporal variability of denitrification.

Methods – conclusions

Many studies have compared different denitrification methods, with few conclusive results (Aulakh *et al.*, 1992; Keeney, 1986; Nieder *et al.*, 1989; Payne, 1991; Tiedje, 1982; Tiedje *et al.*, 1989). In most cases, high variability has made it difficult to determine differences between the different techniques. The different techniques all have drawbacks that will be more or less important in different situations. We can only conclude that it is important for investigators to determine which drawbacks of the different methods are critical in each case, and choose methods accordingly.

Rates, patterns and control of denitrification in tropical soils

There have been no extensive studies of denitrification in upland tropical agricultural soils. Several studies in Australia have documented the effects of fertilizer placement and nitrification inhibition on fertilizer recovery in the soil-plant system in cotton and sugar cane, but no direct denitrification measurements were made (Chapman *et al.*, 1991, 1992; Chen *et al.*, 1994; Freney *et al.*, 1993; Vallis & Keating, 1994). Given the lack of comprehensive studies of denitrification in upland tropical agricultural soils, we need to exploit other sources of data that may provide insights on this topic. In this section, I will examine selected data on denitrification and N_2O flux in native tropical forest and pasture soils and attempt to derive some conclusions, or to propose some hypotheses about the nature and significance of denitrification in upland tropical agricultural soils. I will also briefly review the much more extensive database on denitrification in rice.

Table 2. Mean denitrification rates in humid tropical forest sites at different stages of succession. Data from Robertson & Tiedje (1988)

Forest site	Denitrification ($g\ N\ m^{-2}\ month^{-1}$)
Primary	0.11
Mid-successional	0.03
Early successional	0.16

Table 3. Spearman rank correlations between denitrification potential and variables related to soil C and N turnover in control, annually cleared and continually cleared humid tropical forest soils. Data from Griffiths *et al.* (1993)

Variable	Denitrification
Respiration	0.697**
N mineralization	0.755**
Microbial biomass C	0.771**
Soil organic (total) N	0.635*
Soil organic (total) C	0.639*

*Indicates statistical significance at $p < 0.01$.
**Indicates statistical significance at $p < 0.001$.

Native forest studies

Robertson & Tiedje (1988) quantified denitrification in tropical forest sites of different successional age in Costa Rica (Table 2). Denitrification was much higher in early (sampling began several weeks following disturbance) or late (primary forest that had never been cleared) successional sites than in mid-successional (2–25 years following disturbance) sites. These data have several important potential implications for denitrification in tropical agricultural soils. First, the data suggest that there is a strong link between plant community dynamics and denitrification in tropical soils. A large body of data from temperate regions suggests that in recently disturbed ecosystems (early successional) plant control over water and N cycling is reduced (Bormann & Likens, 1979) and the potential for denitrification is thus increased (Robertson & Tiedje, 1984). In mid-successional forest ecosystems, vegetation is rapidly growing and exerts strong control over water and N availability, reducing the potential for denitrification. In mature (late successional) forests, parts of the ecosystem begin to senesce, increasing water and N availability and the potential for denitrification.

In addition to showing the importance of plant control over denitrification over long time periods (decades), the Robertson & Tiedje (1988) data also illustrate the importance of plant activity over shorter time periods. In one of their early successional sites, denitrification rates declined from 90 ng N $cm^{-2}\ h^{-1}$ shortly after clearing to less than 10 ng N $cm^{-2}\ h^{-1}$ within four months and to less than 2 ng N $cm^{-2}\ h^{-1}$ after one year. These data suggest that, just as in temperate regions, the period after one crop is harvested and before the next is planted is likely to be a key period for denitrification in tropical soils. This period should be a focus in denitrification sampling programmes in tropical agro-ecosystems.

The final important implication of the Robertson & Tiedje (1988) study is that the magnitude of N loss was not that high. The rates of denitrification they reported are not much higher than those measured in temperate forest soils (Bowden, 1986; Groffman & Tiedje, 1989). While it is dangerous to extrapolate from forest to agricultural soils, these data may suggest that denitrification in tropical agricultural soils may not be any higher than it is in temperate forest soils, where it is generally not a major route of N loss (Aulakh *et al.*, 1992; Nieder *et al.*, 1989; Von Rheinbaben, 1990).

Griffiths *et al.* (1993) compared denitrification in humid tropical forest soils subjected to treatments that manipulated the amount of organic matter input to soil (Table 3). Undisturbed primary forest soils had much higher denitrification rates than annually harvested or continually cleared soils that had much lower rates of organic matter input from vegetation. There was a strong correlation between denitrification and various indices of soil C and N turnover such as respirable C, microbial biomass C, and total soil C and N. These data suggest that, as in temperate regions, there is a strong relationship between the nature and amount of C and N turnover and denitrification in tropical soils.

Like the Robertson & Tiedje (1988) study, the data of Griffiths *et al.* (1993) may suggest that upland tropical agricultural soils do not support particularly high denitrification rates. Robertson & Tiedje's (1988) data show that tropical forests do not denitrify at much greater rates than temperate forest soils and the data of Griffiths *et al.* (1993) suggest that when organic matter inputs to tropical soils are decreased, denitrification rates decrease accordingly. Due to harvest and soil disturbance, cultivated agricultural soils (both temperate and tropical) generally have less C and N in organic matter and microbial biomass than forest soils (Paul & Clark, 1989). Tropical agricultural soils should therefore have no more, and may have much less, denitrification than tropical forest soils. Alternatively, Parsons *et al.* (1993) found that the potential for denitrification given NO_3^- and C additions was higher in humid tropical pasture soils than in native forest soils. Denitrification potential may have been higher in the pasture soils due to soil compaction which decreased aeration in these soils. Compaction associated with agricultural activities may stimulate denitrification to a greater extent in humid tropical soils than it does in temperate agricultural soils.

While data on denitrification in humid tropical forest soils is scarce, studies in dry tropical regions are even less common. Abbadie & Lensi (1990) found strong relationships between C and N mineralization rates and denitrification potential in a range of humid savannah ecosystems in West Africa, results that are consistent with Griffiths *et al.* (1993). They also suggested that actual rates of denitrification in their soils were quite low.

N$_2$O flux studies

Studies of N_2O flux in tropical forest soils are much more common than studies of denitrification. Tropical forest soils generally have much higher rates of N_2O emission than temperate forest soils (Keller *et al.*, 1983, 1986), contributing to the suspicion that denitrification might be a more significant route of N loss in tropical agro-ecosystems than in temperate agro-ecosystems. The value of N_2O flux data as an indicator of denitrification is quite limited, however. Two soils can have the same rates of denitrification but much different rates of N_2O production due either to differences in the ratio of N_2O to N_2 produced during denitrification or to N_2O produced by nitrification. The data of Parsons *et al.* (1993) suggest that the ratio of N_2O to N_2 produced during denitrification is much higher in tropical soils than in temperate soils.

Data on N_2O flux from tropical forest soils reinforce the idea that denitrification is strongly linked to the nature and amount of C and N turnover in these soils (Matson & Vitousek, 1987). Matson & Vitousek (1990) reported that N_2O fluxes are much higher in high fertility tropical forest soils than in low fertility soils (Table 4). Higher fertility forests have higher productivity and decomposition rates than low fertility

Table 4. Annual N_2O flux in humid tropical forest soils of different fertility. Data from Matson & Vitousek (1990)

Forest site fertility	N_2O flux (kg N ha^{-1} y^{-1})
High	2.6
Intermediate	1.9
Low	0.47

forests and therefore have a relatively higher flow of C and N to microbial populations, including denitrifiers.

Data on N_2O flux from tropical pastures give further evidence that N gas fluxes in tropical soils are influenced by changes in plant control over water and N availability and by the nature and amount of soil C and N turnover. Matson & Vitousek (1990) observed very high N_2O fluxes from humid tropical forest soils converted to pasture. Keller *et al.* (1993) quantified N_2O fluxes in a range of pasture sites of different age. While recently (0–10 years) converted sites had high N_2O fluxes, older sites had fluxes equal to or less than native forest sites. Exactly what ecosystem changes are controlling these temporal variations is not clear. Measurements of denitrification in tropical pasture sites of different age are clearly needed.

Several studies have quantified N_2O and NO emissions from seasonally dry tropical forests and savannahs (Davidson *et al.*, 1993; Johansson *et al.*, 1988; Sanhueza *et al.*, 1990). While annual fluxes of these gases appear to be lower in these ecosystems than in humid tropical ecosystems, the transition period between dry and wet seasons has very high fluxes (Davidson, 1992; Davidson *et al.*, 1993). Wetting of dry soil kills a portion of the microbial biomass by osmotic stress leading to a pulse of C and N mineralization, nitrification and denitrification (Birch, 1964; Groffman & Tiedje, 1988; Kieft *et al.*, 1987). This period of high flux is brief, however, and should not be a major source of N loss from dry tropical agro-ecosystems.

Riparian forest studies

One recent study suggests that riparian forests may be hotspots of denitrification and NO_3^- removal in humid tropical landscapes, just as they are in temperate landscapes (Bowden *et al.*, 1993; McDowell *et al.*, 1993). These investigators found a high potential for denitrification in humid tropical riparian forests and that this potential was influenced by landscape scale factors such as topography and soil parent material. In their studies of tropical savannahs in West Africa, Abbadie & Lensi (1990) found that riparian ecosystems had the highest denitrification potential of all the sites studied. Interest in riparian zone denitrification as a 'sink' for NO_3^- moving out of upland agricultural soils will likely increase in the future, along with intensification of cropping patterns and fertilizer use. This 'beneficial' denitrification activity will need to be evaluated relative to concerns about N_2O emissions, however.

Rice

Several studies have quantified denitrification in rice (see reviews by Aulakh *et al.*, 1992 and by DeDatta, 1995). Concern about denitrification in rice is high due to the fact that 'unaccounted-for N' in ^{15}N studies in rice is usually greater than 50%.

Nitrification is the key controller of denitrification in rice cropping systems. These systems are dominated by large volumes of anaerobic soil with NO_3^- production restricted to the soil/water interface during flooded conditions and to periods when the systems are not flooded. Fertilizer inputs to rice are almost entirely as either NH_4^+ salts or urea and can therefore not be denitrified unless nitrification occurs first. As in upland ecosystems, managing the interface between cropping systems and between wet and dry seasons will be critical for preventing high denitrification N losses from rice.

Detailed studies of denitrification in rice using direct $^{15}N_2$ flux techniques have found denitrification not to be the major route of N loss from these ecosystems (see reviews cited above). In several studies, directly measured $^{15}N_2$ and $^{15}N_2O$ losses have been much lower than total unaccounted-for ^{15}N, suggesting that other loss processes, most notably NH_3 volatilization, are more important than denitrification. Alternatively, there may be problems with the $^{15}N_2$ flux method arising from ^{15}N-labelled gases remaining 'trapped' in the soil or transmitted through plants.

Conclusions and recommendations

1. There is a clear need for measurements of denitrification in upland tropical agricultural soils. While inferences from tropical forest studies suggest that suspicions that this process is a much more important route of N loss in tropical soils than in temperate soils may be unfounded, these inferences are not a substitute for direct measurements.

2. Denitrification in tropical soils is strongly related to the nature and extent of soil C and N turnover. Denitrification measurement programmes need to focus on treatments that influence C and N turnover such as residue management, organic inputs, intercropping and harvest intensity.

3. Denitrification in tropical soils is strongly influenced by plant control over soil water and N availability. Sampling schemes should focus on the interface between different cropping seasons and between wet and dry seasons.

4. Landscape scale studies should be done to quantify the potential of certain soils to support high denitrification N losses, or to absorb significant amounts of N from other areas in the landscape. The use of 'offsite' denitrification to absorb N must be evaluated relative to concerns about N_2O emissions during this denitrification activity.

References

Abbadie L & Lensi R (1990) Carbon and nitrogen mineralization and denitrification in a humid savanna of West Africa (Lamto, Côte d'Ivoire). Acta Ecol. 11: 717–728

Aulakh M S, Doran J W & Mosier A R (1991) Field evaluation of four methods for measuring denitrification. Soil Sci. Soc. Am. J. 55: 1332–1338

Aulakh M S, Doran J W & Mosier A R (1992) Soil denitrification – significance, measurement, and effects of management. *In*: Stewart B A (ed.) Advances in Soil Science, Vol. 18. Springer-Verlag, New York

Balderston W L, Sherr B & Payne W J (1976) Blockage by acetylene of nitrous oxide reduction in *Pseudomonas perfectomarinus*. Appl. Environ. Microbiol. 31: 504–508

Barrie A (1995) Developments in the use of labelled nitrogen for soil/plant nitrogen research. Fert. Res. (This volume)

Birch H F (1964) Mineralization of plant nitrogen following alternate wet and dry conditions. Plant Soil. 20: 43–49

Bormann F H & Likens G E (1979) Pattern and Process in a Forested Ecosystem. Springer-Verlag, New York

Bowden W B (1986) Gaseous nitrogen emissions from undisturbed terrestrial ecosystems: an assessment of their impacts on local and global nitrogen budgets. Biogeochemistry 2: 249–279

Bowden W B, McDowell W H, Asbury C E & Finley A M (1992) Riparian nitrogen dynamics in two geomorphologically distinct tropical rain forest watersheds: nitrous oxide fluxes. Biogeochemistry 18: 77–99

Bradley P M, Fernandez Jr. M & Chapelle F H (1992) Carbon limitation of denitrification in an anaerobic groundwater system. Environ. Sci. Technol. 12: 2377–2381

Burton D L & Beauchamp E G (1984) Field techniques using the acetylene blockage of nitrous oxide reduction to measure denitrification. Can. J. Soil Sci. 64: 555–562

Burton D L & Beauchamp E G (1985) Denitrification rate relationships with soil parameters in the field. Commun. Soil Sci. Plant Anal. 16: 539–549

Chapman L S, Haysom M B C, Saffigna P G & Freney J R (1991) The effect of placement and irrigation on the efficiency of use of ^{15}N labelled urea by sugar cane. Proc. Aust. Soc. Sugar Cane Technol. 13: 44–52

Chapman L S, Haysom M B C & Saffigna P G (1992) N cycling in cane fields from ^{15}N-labelled trash and residual fertiliser. Proc. Aust. Soc. Sugar Cane Technol. 14: 84–89

Chen D L, Freney J R, Mosier A R & Chalk P M (1994) Reducing denitrification loss with nitrification inhibitors following presowing applications of urea to a cottonfield. Aust. J. Exp. Agric. 34: 75–83

Christensen P B, Nielsen L P, Revsbech N P, & Sørensen J (1989) Microzonation of denitrification activity in stream sediments as studied with a combined oxygen and nitrous oxide microsensor. Appl. Environ. Microbiol. 55: 1234–1241

Davidson E A (1992) Pulses of nitric oxide and nitrous oxide flux following wetting of dry soil: an assessment of probable sources and importance relative to annual fluxes. Ecol. Bull. 42: 149–155

Davidson E A, Matson P A, Vitousek P M, Riley R, Dunkin K, Garcia-Mendez G & Maass J M (1993) Processes regulating soil emissions of NO and N$_2$O in a seasonally dry tropical forest. Ecology 74: 130–139

DeDatta S K (1995) Nitrogen transformation in wetland rice systems. Fert. Res. (This Volume)

Devol A H (1991) Direct measurement of nitrogen gas fluxes from continental shelf sediments. Nature 349: 319–321.

Flather D H & Beauchamp E G (1992) Inhibition of the fermentation process in soil by acetylene. Soil Biol. Biochem. 24: 905–911

Folonoruso O A & Rolston D E (1984) Spatial variability of field-measured denitrification gas fluxes and soil properties. Scil Sci. Soc. Am. J. 49: 1087–1093

Freney J R, Chen D L, Mosier A R, Rochester I J, Constable G A & Chalk P M (1993) Use of nitrification inhibitors to increase fertilizer nitrogen recovery and lint yield in irrigated cotton. Fert. Res. 34: 37–44

Goodroad L L & Keeney D R (1984) Nitrous oxide emissions from soils during thawing. Can. J Soil. Sci. 64: 187–194

Griffiths R P, Caldwell B A & Sollins P (1993) Effects of vegetation regime on denitrification potential in two tropical volcanic soils. Biol. Fertil. Soils 16: 157–162

Groffman P M (1991) Ecology of nitrification and denitrification in soil evaluated at scales relevant to atmospheric chemistry. In: Whitman W B & Rogers J (eds.) Microbial Production and Consumption of Greenhouse Gases: Methane, Nitrogen Oxides and Halomethanes: 201–217. American Society of Microbiology, Washington, DC

Groffman P M & Tiedje J M (1988) Denitrification hysteresis during wetting and drying cycles in soil. Soil Sci. Soc. Am. J. 52: 1626–1629

Groffman P M & Tiedje J M (1989) Denitrification in north temperate forest soils: Relationships between denitrification and environmental factors at the landscape scale. Soil Biol. Biochem. 21: 621–626

Groffman P M, Tiedje J M, Robertson G P & Christensen S (1988) Denitrification at different temporal and geographical scales: Proximal and distal controls. In: Wilson J R (ed.) Advances in Nitrogen Cycling in Agricultural Ecosystems: 174–192. CAB International, Wallingford, UK

Gross P J & Bremner J M (1992) Acetone problem in use of the acetylene blockage method for assessment of denitrifying activity in soil. Commun Soil. Sci. Plant Anal. 23: 1345–1358

Hyman M R & Arp D J (1987) Quantification and removal of some contaminating gases from acetylene used to study gas-utilizing enzymes and microorganisms. Appl. Environ. Microbiol. 53: 298–303

Hynes R K & Knowles R (1978) Inhibition by acetylene of ammonia oxidation in Nitrosomonas europaea. FEMS Microbiol. Lett. 4: 319–321

Johansson C, Rodhe H & Sanhueza E (1988) Emission of NO in a tropical savanna and a cloud forest during the dry season. J. Geophys. Res. 93: 7180–7192

Jury W A, Letey J & Collins T (1982) Analysis of chamber methods used for measuring nitrous oxide production in the field. Soil Sci. Soc. Am. J. 46: 250–255

Keeney D R (1986) Critique of the acetylene blockage technique for field measurement of denitrification. In: Hauck R D & Weaver R W (eds.) Field Measurement of Dinitrogen Fixation and Denitrification: 103–115. Soil Science Society of America, Madison, Wisc.

Keller M, Goreau T J, Wofsy S C, Kaplan W A & McElroy M B (1983) Production of nitrous oxide and consumption of methane by forest soils. Geophys. Res. Lett. 10: 1156–1159

Keller M, Kaplan W A & Wofsy S C (1986) Emissions of N$_2$O, CH$_4$ and CO$_2$ from tropical forest soils. J. Geophys. Res. 91: 11791–11802

Keller M, Veldkamp E, Weitz A M & Reiners W A (1993) Effect of pasture age on soil trace-gas emissions from a deforested area of Costa Rica. Nature 365: 244–246

Kemp W M, Sampou P, Caffrey J, Mayer M, Henriksen K & Boynton W R (1990) Ammonium recycling versus denitrification in Chesapeake Bay sediments. Limnol Oceanogr. 35: 1545–1563

Kieft T L, Soroker E & Firestone M K (1987) Microbial biomass response to a rapid increase in water potential when dry soil is wetted. Soil Biol. Biochem. 19: 119–126

Korom S F (1992) Natural denitrification in the saturated zone: A review. Water Resour. Res. 28: 1657–1668

Lloyd D (1993) Aerobic denitrification in soils and sediments: from fallacies to facts. Trends Ecol. Evol. 8: 352–356

Lowrance R, Todd R, Fail Jr. J, Hendrickson Jr. O, Leonard R & Asmussen L (1984) Riparian forests as nutrient filters in agricultural watersheds. BioScience 34: 374–377

Matson P A & Vitousek P M (1987) Cross-system comparisons of soil nitrogen transformations and nitrous oxide flux in tropical forest ecosystems. Global Biogeochem. Cycles 1: 163–170

Matson P A & Vitousek P M (1990) Ecosystem approach to a global nitrous oxide budget. BioScience 40: 667–672

McDowell W H, Bowden W B & Asbury C E (1993) Riparian nitrogen dynamics in two geomorphologically distinct tropical rain forest watersheds: subsurface solute patterns. Biogeochemistry 18: 53–75

Monaghan R M & Barraclough D (1993) Nitrous oxide and dinitrogen emissions from urine-affected soil under controlled conditions. Plant Soil 151: 127–138

Mosier A R (1980) Acetylene inhibition of ammonium oxidation in soil. Soil Biol. Biochem. 12: 443–444

Mosier A R, Guenzi W D & Schweizer E E (1986) Field denitrification estimation by nitrogen-15 and acetylene inhibition techniques. Soil Sci. Soc. Am. J. 50: 831–833

Myrold D D (1988) Denitrification in ryegrass and winter wheat cropping systems of western Oregon. Soil Sci. Soc. Am. J. 52: 412–415

Nieder R, Schollmayer G & Richter J (1989) Denitrification in the rooting zone of cropped soils with regard to methodology and climate: A review. Biol. Fertil. Soils 8: 219–226

Parkin T B (1987) Soil microsites as a source of denitrification variability. Soil Sci. Soc. Am. J. 51: 1194–1199

Parkin T B, Kaspar H F, Sexstone A J & Tiedje J M (1984) A gas-flow soil core method to measure field denitrification rates. Soil Biol. Biochem. 16: 323–330

Parkin T B, Sexstone A J & Tiedje J M (1985) Comparison of field denitrification rates determined by acetylene-based soil core and nitrogen-15 methods. Soil Sci. Soc. Am. J. 49: 94–99

Parkin T B & Simpkins W W (1993) Methane as an energy source for denitrification in groundwater. Agron. Abstr. 85: 256

Parsons W F J, Mitre M E, Keller M & Reiners W A (1993) Nitrate limitation of N_2O production and denitrification from tropical pasture and rain forest soils. Biogeochemistry 22: 179–193

Paul EA & Clark F E (1989) Soil Microbiology and Biochemistry. Academic Press, New York

Payne W J (1981) Denitrification. John Wiley & Sons, New York

Payne W J (1984) Influence of acetylene on microbial and enzymatic assays. J. Microbiol. Methods 2: 117–133

Payne W J (1991) A review of methods for field measurements of denitrification. For. Ecol. Manage. 44: 5–14

Pederson J K, Bjerg PL & Christensen T H (1991) Correlation of nitrate profiles with groundwater and sediment characteristics in a shallow sandy aquifer. J. Hydrol. 124: 263–277

Postma D, Boesen C, Kristiansen H & Larsen F (1991) Nitrate reduction in an unconfined sandy aquifer: Water chemistry, reduction processes, and geochemical modelling. Water Resour. Res. 27: 2027–2045

Prinn R, Cunnold D, Rasmussen R, Simmonds P, Alyea F, Crawford A, Fraser P & Rosen R (1990) Atmospheric trends and emissions of nitrous oxide deduced from ten years of ALE-GAGE data. J. Geophys. Res. 96: 959–990

Reddy K R & Patrick W H (1984) Nitrogen transformations and loss in flooded soils and sediments. Crit. Rev. Environ. Control 13: 273–309

Robertson G P & Tiedje J M (1984) Denitrification and nitrous oxide production in old growth and successional Michigan forests. Soil Sci. Soc. Am. J. 48: 383–389

Robertson G P & Tiedje J M (1988) Deforestation alters denitrification in a lowland tropical rain forest. Nature 336: 756–759

Robertson L A & Kuenen J G (1991) Physiology of nitrifying and denitrifying bacteria. In: Rogers J E & Whitman W B (eds.) Microbial Production and Consumption of Greenhouse Gases: Methane, Nitrogen Oxides and Halomethanes: 189–199. American Society for Microbiology, Washington DC

Rolston D E, Broadbent F E & Goldhammer D A (1979) Field measurements of denitrification: II. Mass balance and sampling uncertainty. Soil Sci. Soc. Am. J. 43: 703–708

Rudolph J, Frenzel P & Pfennig N (1991) Acetylene inhibition technique underestimates in situ denitrification rates in intact cores of freshwater sediment. FEMS Microbiol. Ecol. 85: 101–106

Ryden J C, Lund L J, Letey J & Focht D D (1979) Direct measurement of denitrification loss from soils: II. Development and application of field methods. Soil Sci. Soc. Am. J. 43: 110–118

Ryden J C & Dawson K P (1982) Evaluation of the acetylene-inhibition technique for the measurement of denitrification in grassland soils. J. Sci. Food Agric. 3: 1197–1206

Ryden J C, Skinner J H & Nixon D J (1987) Soil core incubation system for the field measurement of denitrification using acetylene-inhibition. Soil Biol. Biochem. 19: 753–757

Sanhueza E, Hao W M, Scharffe Donoso D & Crutzen P J (1990) N_2O and NO emissions from soils of the northern part of the Guyana Shield. J. Geophys. Res. 95: 22481–22488

Schmidt J, Seiler W & Conrad R (1988) Emission of nitrous oxide from temperate forest soils into the atmosphere. J. Atmos. Chem. 6: 95–115

Seitzinger S, Nixon S, Pilson M E Q & Burke S (1980) Denitrification and N_2O production in near-shore marine sediments. Geochim. Cosmochim. Acta 44: 1853–1860

Seitzinger S, Nielsen L P, Caffrey J & Christensen P B (1993) Denitrification measurements in aquatic sediments: A comparison of three methods. Biogeochemistry 23: 147–167

Terry R E & Duxbury J M (1985) Acetylene decomposition in soils. Soil Sci. Soc. Am. J. 49: 90–94

Tiedje J M (1982) Denitrification. In: Page A L (ed.) Methods of Soil Analysis, Part II, Chemical and Microbiological Properties: 1011–1026. American Society of Agronomy, Madison, Wisc.

Tiedje J M (1988) Ecology of denitrification and dissimilatory nitrate reduction to ammonia. In: Zehnder A J B (ed.) Biology of Anaerobic Microorganisms: 179–244. John Wiley & Sons, New York

Tiedje J M, Simkins S & Groffman P M (1989) Perspectives on measurement of denitrification in the field including recommended protocols for acetylene-based methods. Plant Soil 115: 261–284

Topp E & Germon J C (1986) Acetylene metabolism and stimulation of denitrification in an agricultural soil. Appl. Environ. Microbiol. 52: 802-806

Vallis I & Keating B A (1994) Uptake and loss of fertiliser and soil nitrogen in sugarcane crops. Proc. Aust. Soc. Sugar Cane Technol. 16: 105–113

Vitousek P M & Matson P A (1993) Agriculture, the global nitrogen cycle and trace gas flux. In: Oremland R S (ed.) Biogeochemistry of Global Change: 193–208. Chapman & Hall, New York

Von Rheinbaben W (1990) Nitrogen losses from agricultural soils through denitrification – a critical evaluation. Z. Pflanzenernähr Bodenkd. 153: 157–166

Walter H M, Keeney D R & Fillery I R (1979) Inhibition of nitrification by acetylene. Soil Sci. Soc. Am. J. 43: 195–196

Yeomans J C & Beauchamp E G (1982) Acetylene as a possible substrate in the denitrification process. Can. J. Soil Sci. 62: 137–144

Yoshinari T & Knowles R (1976) Acetylene inhibition of nitrous oxide reduction by denitrifying bacteria. Biochem. Biophys. Res. Commun. 69: 705–710

Fertilizer Research **42**: 149–158, 1995.
© 1995 *Kluwer Academic Publishers. Printed in the Netherlands.*

Soil, plant and atmospheric conditions as they relate to ammonia volatilization

R.R. Sharpe & L.A. Harper
USDA/ARS, Southern Piedmont Conservation Research Center, P.O. Box 555, Watkinsville GA 30677, USA

Key words: Micrometeorology, N flux, livestock waste, NH_3, ^{15}N

Abstract

Gaseous ammonia (NH_3) transport is an important pathway in the terrestrial N cycle. In the atmosphere NH_3 neutralizes airborne acids and is a major factor determining air quality and acid rain deposition patterns. Redeposition of atmospheric NH_3 plays an important role in the N balance of natural ecosystems and has been implicated in forest decline, plant species change and eutrophication of surface water. Much of the N in soil-plant animal systems can be lost to the atmosphere, particularly with surface applied livestock waste, or urea and anhydrous ammonia fertilizers. Plants can have a significant impact on NH_3 transport because they can both absorb and desorb atmospheric NH_3. Under conditions of low soil N or high atmospheric NH_3 concentrations, plants absorb NH_3. Under conditions of high soil N or low atmospheric NH_3 concentrations, plants volatilize NH_3. This article discusses methods for evaluating NH_3 transport in the filed, the rate of NH_3 volatilized from fertilizer application, and the effects of plants on net NH_3 transport.

Introduction

Ammonia (NH_3) and ammonium (NH_4) are important atmospheric components. Gaseous NH_3 originates in both natural and agricultural systems but the largest fraction comes from livestock waste, fertilizers, and other agricultural sources. Once in the atmosphere, NH_3 is the dominant alkaline gas and is the principal agent for neutralizing airborne acids, such as sulfuric, nitric and hydrochloric acid.

$$2NH_3(g) + H_2SO_4(l) - - \rightarrow (NH_4)SO_4(l, s)$$
$$NH_3(g) + HNO_3(g) \longleftrightarrow NH_4NO_3(s)$$
$$NH_3(g) + HCl(g) \longleftrightarrow NH_4Cl(s)$$

The primary atmospheric reaction is with sulfate (SO_4) because there is a much greater affinity of NH_4 for sulfate than nitrate (NO_3) and little NO_3 will be incorporated into aerosol until the SO_4 is almost completely neutralized [41]. Ammonia thus influences the pH of aerosols and cloudwater and is a major factor in determining air quality and acid rain deposition patterns [28].

From an agricultural viewpoint, NH_3 volatilization is a direct economic loss to the farmer and as much as 40 to 70% of some surface applied N fertilizers can be volatilized [4,6,7,15,35]. In research, if NH_3 losses or gains are not accounted for in N budget studies estimates to evaluate leaching and/or denitrification from agricultural fields can be in serious error [24]. Ammonia released from agricultural systems can also play an important role in the N balance of natural ecosystems.

In northern Europe, it has been estimated that 94% of the NH_3 released from agricultural sources is redeposited into surrounding ecosystems as either NH_3 or NH_4 [1]. Atmospheric NH_3 deposited in natural ecosystems has been implicated in forest decline (12,44) and in plant species change threathening the loss of moors and heathlands [3,40]. Absorption of atmospheric NH_3 by surface water can also accelerate eutrophication. Hutchinson [26] found that surface water can absorb as much as 73 kg NH_3 ha^{-1} y^{-1} from a nearby feedlot.

Comparison of methods

There are inherent difficulties in the measurement of atmospheric NH_3 under both laboratory [36] and field

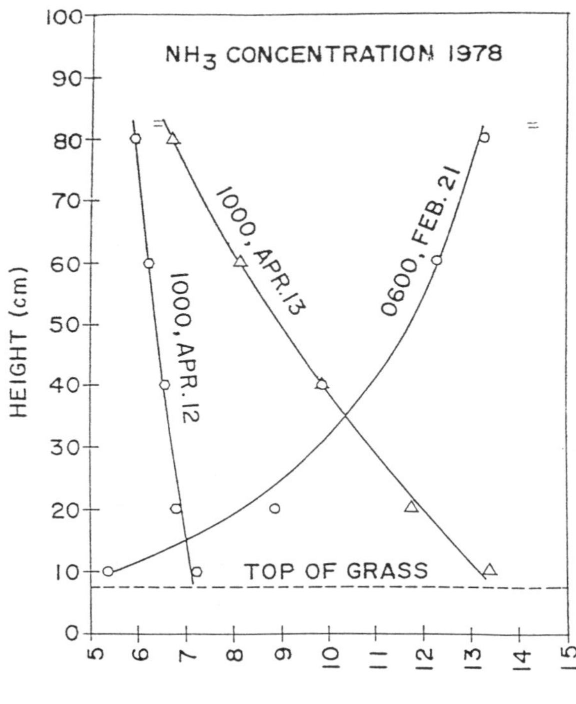

Fig. 1. Ammonia concentration profiles with height obtained over a subtropical pasture.

conditions [24]. Atmospheric NH_3 concentrations are generally low (5 to 20 ppb) [24] and NH_3 forms strong hydrogen bonds and adheres to any unheated surface, leading to memory effects and sample losses [36]. The three possible approaches to measuring NH_3 volatilization in the field are micrometeorological (MM) methods, enclosure (chambers) methods, and isotopic ^{15}N methods [21].

There is general agreement that MM techniques are to be preferred in principle [9,17]. They do not disturb the soil, plant, or environmental processes which influence NH_3 exchange and they provide a measure of average, integrated flux over a large area, thus minimizing problems associated with point to point variations. Micrometeorological techniques are limited to cropping situations where the air mass has blown over a large homogeneous surface so that profiles of gas concentrations in the air are in equilibrium with the local rate of exchange. Other requirements include high measurement sensitivity, relatively expensive instrumentation, and an understanding of the physics of gas transport under field conditions. Examples of typical NH_3 profiles above a grass pasture are shown in Fig. 1.

On April 12 at 1000 h, there was no significant change in atmospheric NH_3 concentrations with height and thus no net transport into or out of the system. Twenty-four hours later on April 13 at 1000 h, there was a decrease in NH_3 concentration with height above the canopy from about 14 to 7 $\mu g\ m^{-3}$ indicating loss from the pasture. On Feb 21 (0600), the NH_3 profile was reversed with an increase in NH_3 concentration with height, indicating absorption of atmospheric NH_3.

Enclosure methods are popular techniques because they are relatively simple, permit replication, are suitable for small experimental plots with different land treatments, and have a lower sensitivity requirement for measuring gas exchange. There are, however, several problems in using enclosures to measure NH_3 fluxes [9]. Ammonia is highly reactive and will readily absorb and desorb from enclosure and pipe walls and any water condensed within the system. Enclosures also modify environmental conditions such as radiation, evaporation, temperature, wind speed and dew formation which can greatly influence NH_3 volatilization. Also, the enclosed surface area in chambers is relatively small and point to point variability of soil gas emissions is often very large. Therefore, accurate measurements of NH_3 fluxes using enclosures are difficult and caution must be taken when interrupting results [9,19,219] although flow-through chambers have shown good results [30,43] when precautions have been taken for NH_3 absorption/desorption.

The ^{15}N balance method estimates NH_3 loss indirectly by measuring all other pathways of N loss from the system and assumes the "unaccounted" for N is lost via NH_3 volatilization. This method allows the soil and/or crop to be open to natural conditions and, with care, the measured amount of NH_3 lost can be determined accurately. Since the ^{15}N method is an indirect measurement, it cannot be used to evaluate NH_3 loss if there are any unaccounted for N losses such as in systems in which denitrification is not measured. Substitution of ^{14}N for ^{15}N in an actively growing canopy may also lead to erroneous results. Plants can absorb and desorb NH_3 [14], and the substitution of ^{14}N from surrounding plants for ^{15}N plants in a microplot can result in an apparent NH_3 loss from the system although the net NH_3 flux from the entire field is zero [24]. This substitution results in a loss of $^{15}NH_3$ from the microplot but no net loss of N from the entire system and thus an overestimation of NH_3 volatilization.

Table 1 gives a listing of accuracies for and comparisons the three methods reported in the literature. Reported accuracies of the MM methods are about 20%

Table 1. Comparison of reported accuracies of microclimate, enclosure and nitrogen isotope methods (after Harper [21]).

Reference	Microclimate (MM)	Enclosure (Encl)	^{15}N	Method
Harper(1971)	±20%			Summation of measurement errors
Denmead *et al.*, (1996)	+20%			Estimate
Lemon & van Houtte (1980)	±20%			Estimate
Hoff *et al.*, (1981)		-52%		Comparison of NH$_3$ loss from plastic sheets
Ferguson *et al.*, (1988)	37%>Encl	37%<MM		Comparison over bare soil
Ferguson *et al.*, (1988)	91%>Encl	91%<MM		Comparison over wheat residue
Harper *et al.*, (1993)	7%^{15}N		7%>MM	Field measurement comparison
Harper & Sharpe (1994)	83%<^{15}N		83%>MM	Comparison of related experiment

based on estimates or summations of probably measurement errors [10,20,29]. Hoff *et al.* [25] compared NH$_3$ volatilization from soil applied swine manure and from manure spread on plastic sheets and calculated a −52% error for the losses measured from and enclosure as from plastic sheets. In a comparison of MM and enclosure methods, Ferguson *et al.* [17] showed considerably greater losses with MM methods than with enclosures over both bare soil (37%) and over wheat residue (91%). Harper *et al.* [22] found a 7% difference in loss of NH$_3$ as measured by MM techniques and by ^{15}N method. This comparison was conducted with a dormant grass sod under dry soil conditions in which denitrification was unlikely and thus no substitutions in the plants of ^{14}N for ^{15}N and NH$_3$. In a study with irrigated corn, Harper and Sharpe [24] reported an NH$_3$ volatilization loss of only 3.6 kg N ha^{-1} for the season. In a nearby field under similar environmental conditions and N levels, Francis (personal communication) estimated by the ^{15}N method about 23 kg N ha^{-1} loss as volatile NH$_3$. Harper and Sharpe [24] attributed the difference in NH$_3$ losses to substitution of atmospheric ^{14}NH$_3$ for ^{15}NH$_3$ emitted by the plants.

Catchpoole *et al.* (6) also showed the affects of actively growing plants on NH$_3$ volatilization as measured by the MM and ^{15}N method (Table 2). During the autumn when there was little plant activity due to heavy grazing and drought conditions, the MM and

Table 2. Percentage of N volatilized as ammonia from a subtropical pasture after fertilization with urea.*

Method	Summer	Autumn	Winter	Spring
^{15}N	29	45	23	20
Flux gradient	12	42	13	9

*Fertilized with 94 kg urea ha^{-1} during each season. After Catchpoole *et al.,* [6]

^{15}N methods measured similar amounts of NH$_3$ losses. During the spring and summer when plants were actively growing, the MM method measured less than 50% of the losses by the ^{15}N method.

Ammonia volatilization from fertilizer and livestock waste

In addition to excellent reviews which have been published concerning NH$_3$ volatilization from fertilizers [4,43], NH$_3$ losses from urea fertilizer is discussed elsewhere in this publication (Freney). Thus this paper will present only a quick overview of the factors affecting NH$_3$ volatilization from fertilizers.

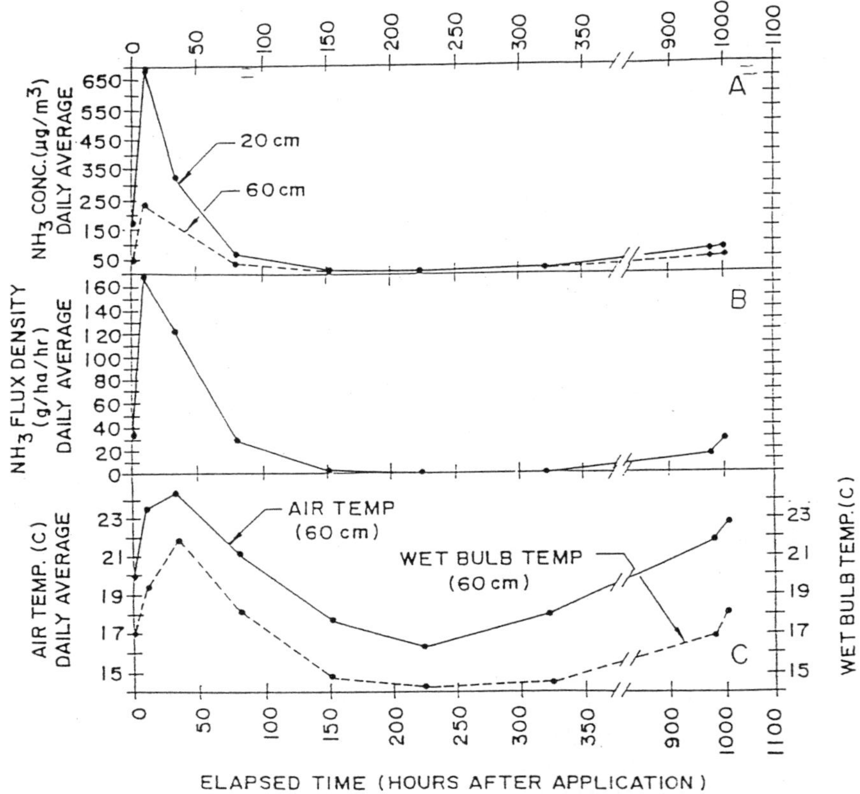

Fig. 2. Aerial NH_3 concentrations and flux densities with average daily air temperatures taken over a subtropical pasture after fertilization with 94 kg urea ha^{-1} in the spring [23].

The quantity of NH_3 volatilized from fertilizers is dependent on the type and timing of fertilizer applied, soil type, and environmental conditions at the time of application. The increase in use of N fertilizer worldwide coupled with a trend towards more extensive use of anhydrous ammonia and urea has increased the potential for NH_3 loss from fertilizers.

Soil factors which effect volatilization include cation exchange capacity (CEC), soil pH and moisture. There is a negative correlation between CEC and NH_3 loss [16,32,38]. It appears that a minimum CEC of 25 meq is required to substantially reduce NH_3 volatilization [16,32]. Amonia losses increase with higher soil pH because of the increased dissociation of NH_4 to NH_3, thus increasing the potential for volatilization [8,33,38].

Environmental factors which effect NH_3 volatilization include air temperature, windspeed, and atmospheric NH_3 concentration. Temperature and windspeed are the climatic factors most directly related to NH_3 loss. Increasing temperature increase the NH_3/NH_4 ratio at a given pH, decreases the solubility of NH_3 in water, and increases the diffusion of NH_3 through the soil [5,15]. If a steady supply of NH_3 is available, increasing windspeed would promote more rapid transport of NH_3 away from the soil surface [11,18].

The effect of temperature on NH_3 flux can been seen by comparing Fig 2 and Fig 3. In the spring, NH_3 efflux reached maximum values of about 170 g ha^{-1} h^{-1} within 24 hours after fertilization and declined to near background concentrations after four days (Fig. 2B). During this period daily average temperatures ranged from 20 to 25 °C (Fig. 2C). During the cooler winter measurement period, temperatures were 9 to 15 °C (Fig. 3C) which contributed to a slower flux rate. A moderate efflux began after 24 hours with maximum values of about 70 g ha^{-1} h^{-1} after four days. Flux rates returned to background levels after about four days.

High background NH_3 concentrations can cause a shift in NH_3 transport from efflux to influx (Fig. 4). A

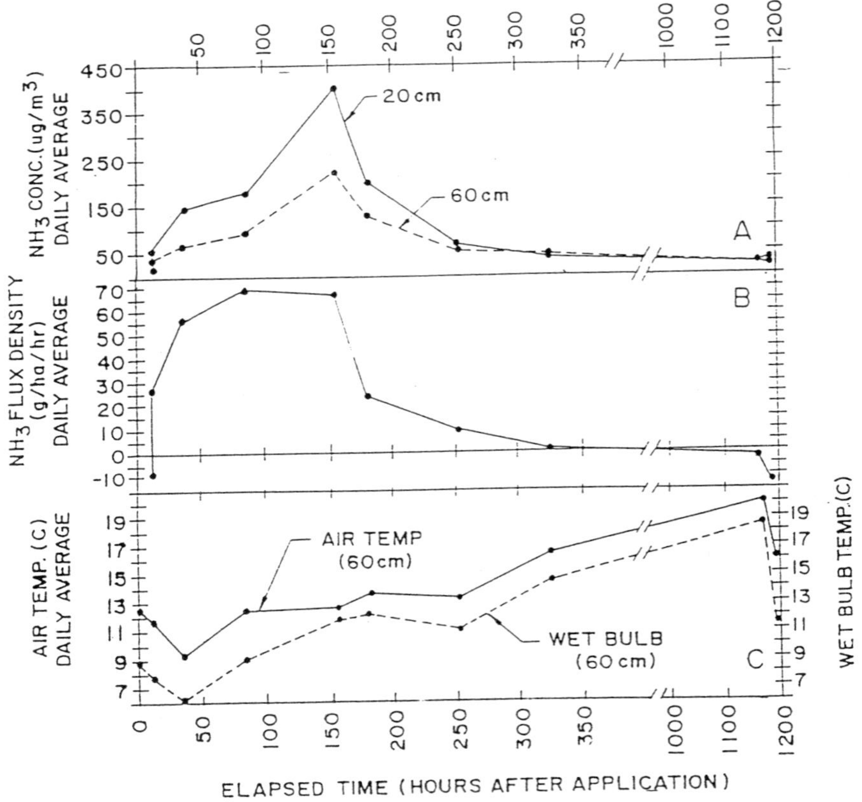

Fig. 3. Aerial NH_3 concentrations and flux densities with average daily air temperatures taken over a subtropical pasture after fertilization with 94 kg urea ha^{-1} in the autumn [23].

study with irrigated corn in central Nebraska showed a net NH_3 efflux from the system except between tasselling and silking (VT to R1 [37]) and the R4 and R5 stages of growth, periods in which NH_3 volatilization would normally be expected [24]. During these periods background NH_3 concentrations increased from 2 to 18 μg m^{-3} causing the corn to shift from volatilization to absorption.

Management practices that decrease the pH of soil or irrigation water, increase CEC, or move the fertilizer deeper into the soil profile (i.e. injecting fertilizer or irrigation after application) would decrease NH_3 loss. Also, applying fertilizer when it is cooler and less windy would tend to decrease volatilization.

As stated previously, livestock waste is a major source of atmospheric NH_3 [34]. In northern Europe it is estimated that volatilization from animal waste constitutes 50% of the total NH_3 emissions [44]. Injecting or incorporating animal slurry greatly reduces volatilization losses in contrast to surface applications which can result in the loss of 32 [45] to 74% [42] of

the applied N (Table 3). Injecting or incorporating the slurry reduced volatilization losses to less than 2% and 7% of the applied N, respectively. Differences in NH_3 losses in the different experiments were attributed to differences in radiation, soil temperature, surface pH, windspeed [45], soil water content [42] or differences in initial composition of the slurry [31].

Volatilization of NH_3 from surface application is very rapid, with most of the loss within the first 24 hours. Ammonia flux rates are highest immediately after application and often decrease 70 to 90% after the first 24 hours [2,42]. Stevens and Logan [39] reported that 24% (37 kg N) of the total N applied to the soil surface was volatilized within seven days, with 76% of this loss coming within 24 hours.

Influence of plants on ammonia flux

Plants play an important role in determining NH_3 flux and a number of studies have shown significant absorp-

154

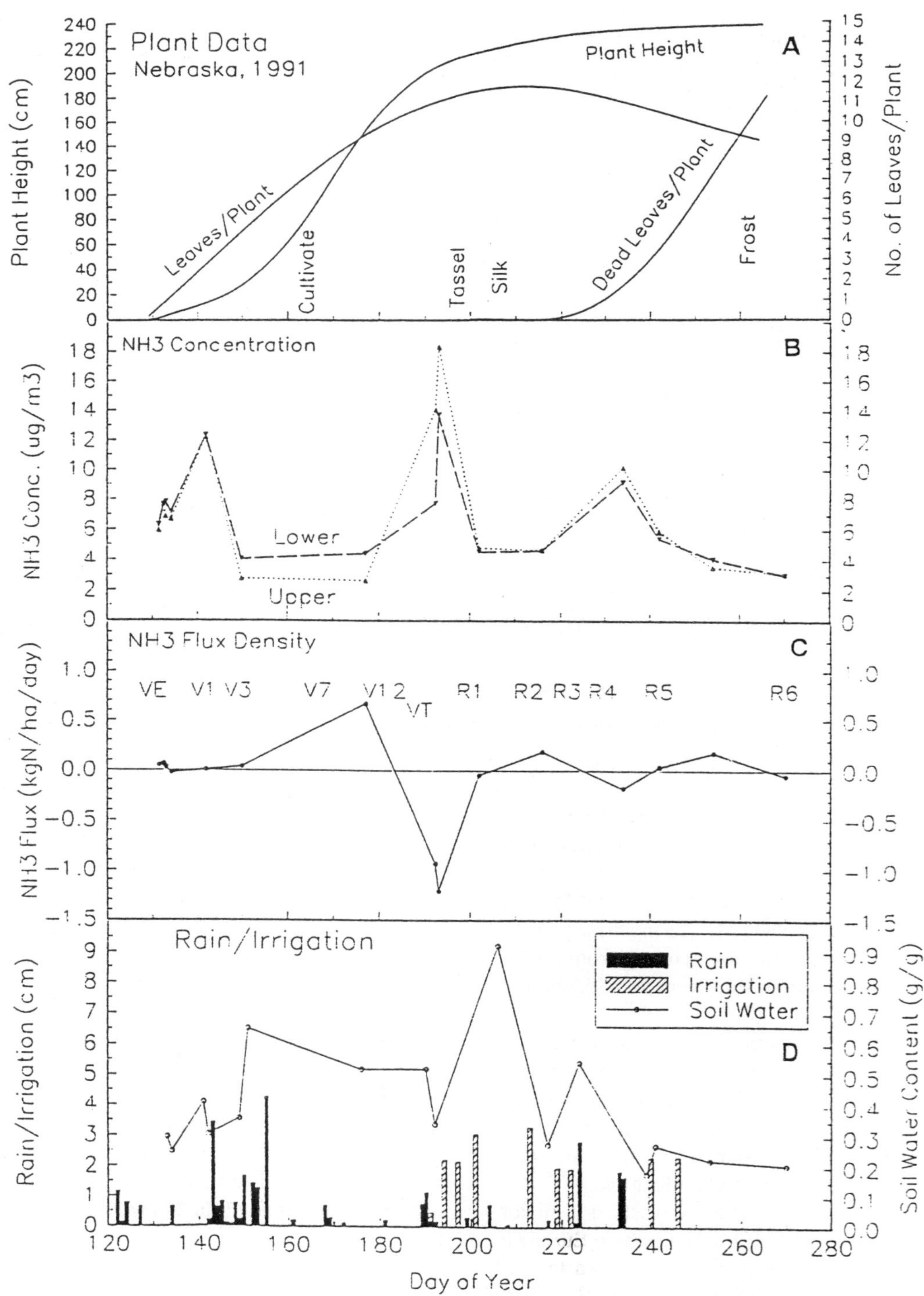

Fig. 4. Average NH₃ concentrations and flux density in corn throughout the measurement season (May–Oct. 1991).

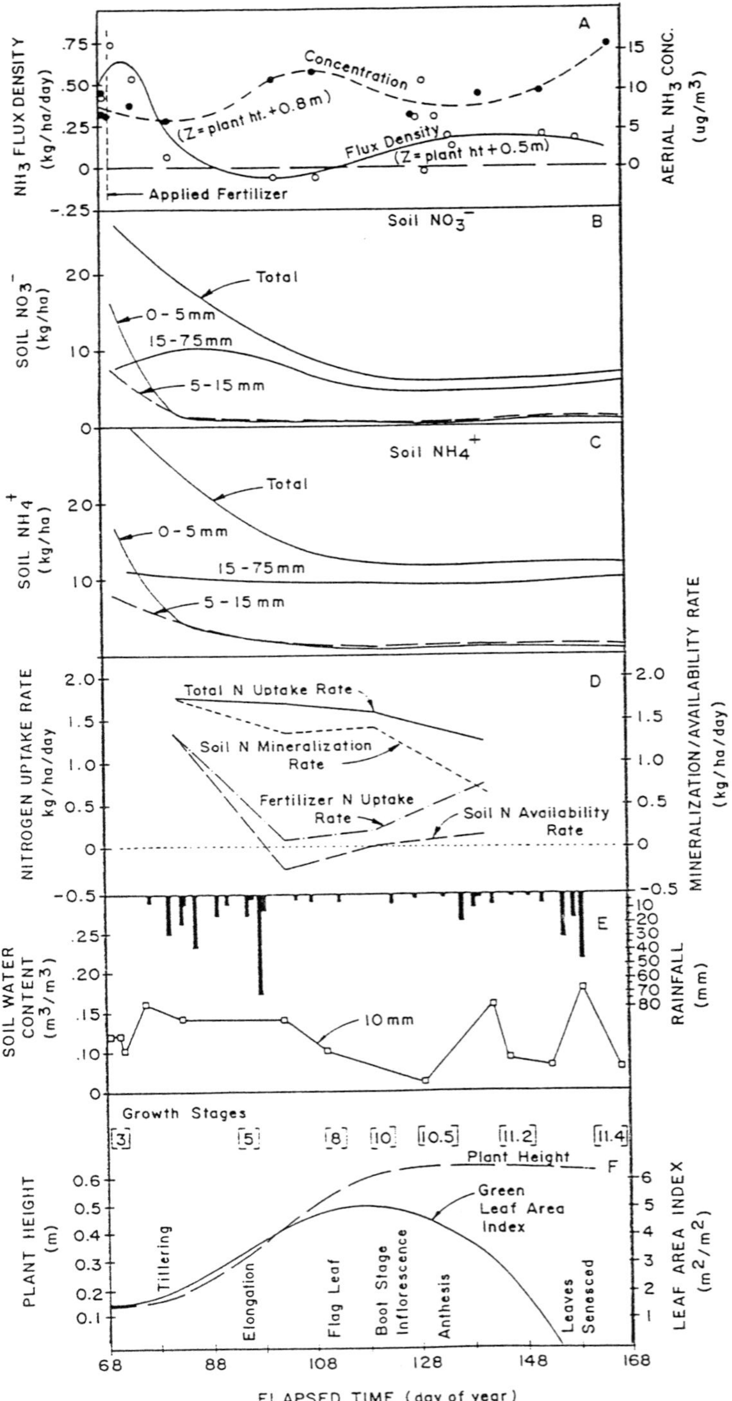

Fig. 5. Comparison of average daily NH_3 concentrations and flux density in wheat with residual soil, plant, and microclimate parameters (March–June, 1983).

156

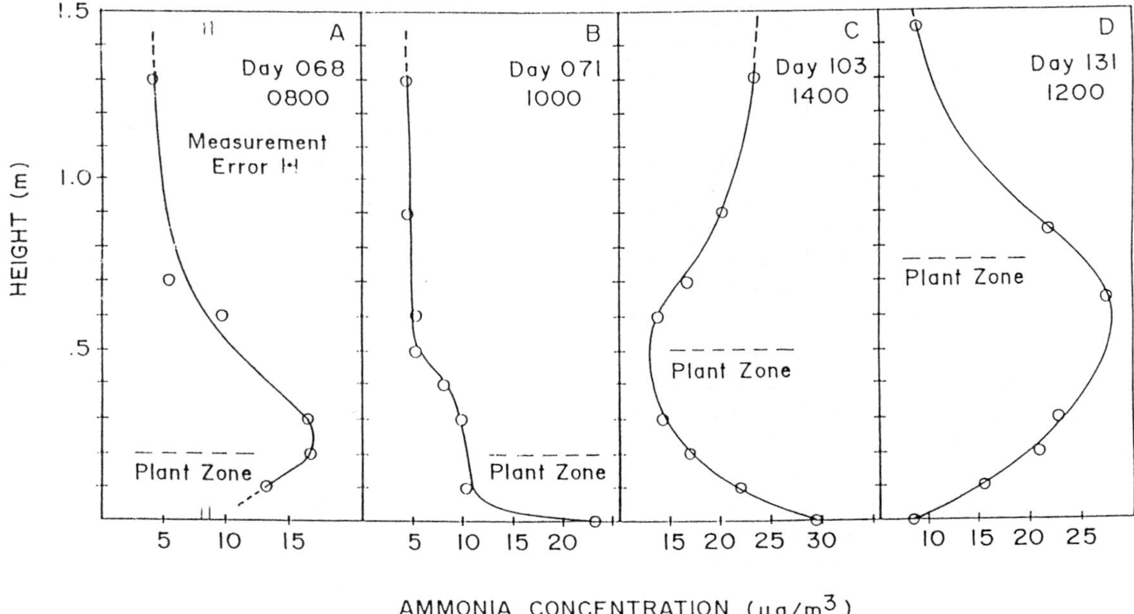

AMMONIA CONCENTRATION (μg/m^3)

Fig. 6. Ammonia concentration profiles representing efflux (A) before fertilization, (B) after fertilization, (C) influx into plant canopy and (D) efflux from plant canopy.

Table 3. Volatilization of ammonia from animal slurry.

Type of slurry	Appl. method	Total loss (hg N ha^{-1})	% Lost	Reference
Cattle	Surface (w)	77.0	74.0	Thompson
	Injected (w)	2.1	1.9	*et al.,* 1987
	Surface (s)	53.0	48.0	
	Injected (s)	1.5	1.3	
Cattle	Surface	27	32	Vander Molen
	Surface	57	67	*et al.,* 1990
	Incorporated	5	6	
	Incorporated	6	7	
Pig	Surface	48.8	78	Lockyer
Poultry	Surface	27.1	49	*et al.,* 1989
Cattle	Surface	10.3	40	

Table 4. Ammonia compensation point (NH$_3$CP) in irrigated corn.

Time	Environ.* Variable	r^2	NH$_3$CP (μg m^{-3})	Highest correlation
All day	All	0.62	7.7	NH$_3$
(0000–2400)	NH$_3$	0.61	6.6	–
Daytime	All	0.86	8.2	NH$_3$
(0800–1000)	NH$_3$	0.85	8.2	–
Nighttime	All	0.64	5.8	T
(0000–2400)	NH$_3$	0.14	6.5	–
Dawn	All	0.99	6.7	Plant N
(0000–2400)	NH$_3$	0.0	6.9	–
Morning	All	0.98	7.3	NH$_3$
(0000–2400)	NH$_3$	0.90	6.6	–
Afternoon	All	0.99	8.5	u
(0000–2400)	NH$_3$	0.23	7.9	
Evening	All	0.97	7.3	Ri
(0000–2400)	NH$_3$	0.39	7.8	
Dusk	All	0.99	3.0	NH$_3$
(0000–2400)	NH$_3$	0.99	3.4	

*All variables include radiation (Ri), temperature (T), windspeed (μ), Plant N and soil water content. After Harper & Sharpe [25].

tion and desorption of NH$_3$ by plants [22,23,27,36]. The NH$_3$ compensation point (NH$_3$CP) is the atmospheric NH$_3$ concentration at which no net exchange of NH$_3$ between the plant and atmosphere occurs. It has been found to vary with plant type, temperature, phenological growth stage, time of day, soil and plant N status, and soil water content [13,23,24,36]. In a study with irrigated corn, Harper and Sharpe [24] found the NH_3 CP to range from 6 to 9 $\mu g\ m^{-3}$ except at dusk when it dropped to about 3 $\mu g\ m^{-3}$ (Table 4). In their study, there appeared to be a sufficient supply of N throughout the season and NH_3 flux was most closely associated with atmospheric NH_3 concentrations than with other measured variables (Table 4). Plants absorbed NH_3 when atmospheric concentrations were greater than 9 $\mu g\ m^{-3}$ and volatilized NH_3 when atmospheric concentrations were less than 6 $\mu g\ m^{-3}$.

Absorption or volatilization of NH_3 is strongly influenced by plant and soil N status. Fig. 5 presents soil, plant and atmospheric N transport for a wheat crop fertilized with NH_4NO_3. Ammonia loss prior to fertilization on day 69 was primarily from the plants since there was little free NH_4 in the soil. The increase in net NH_3 efflux after fertilization was due to higher plant N (and NH_4) concentrations and increased soil NH_4 concentrations at the soil surface (Figs 5A and 5C). When soil NH_4 and NO_3 concentrations in the upper layers decreased to about background levels (Figs 5B and 5C), net NH_3 transport to the atmosphere decreased to zero. Simultaneously, during this early vegetative period (days 78–108), fertilizer availability decreased due to immobilization of fertilizer N (as measured by ^{15}N; Fig. 5D) but plant N content increased. These decreases in soil N and fertilizer uptake resulted in a short term N deficiency during which time the plants absorbed atmospheric NH_3 (days 90–110; Fig. 5A). After day 110, soil N availability increased due to remineralization of fertilizer N and plant growth rate decreased due to the start of senescence (Fig. 5F) and the plants volatilized excess N.

Ammonia concentration profiles above the wheat canopy are shown in Fig. 6. On days 68 (before fertilization) and 131, the greatest NH_3 concentrations were at the top of the canopy indicating that the plants were an NH_3 source causing an efflux out of the system (Fig. 6A and 6D). After fertilization, NH_3 volatilized both from the soil and plants (Fig. 6B). During the period of N deficiency, plants were a sink for atmospheric NH_3 as indicated be the reduced NH_3 concentrations within the plant zone (Fig. 6C).

Agriculture is a major source of atmospheric NH_3 and volatilization of NH_3 from surface applied manure and fertilizer represents a direct economic loss to the farmer.

Micrometeorological techniques are usually the preferred method for measuring NH_3 transport but under certain conditions enclosure methods and ^{15}N isotopic measurement techniques have been used suc-cessfully. Plants can play an important role in net NH_3 transport. Under conditions of low soil N availability or high atmospheric NH_3 concentrations, plants can absorb atmospheric NH_3. Under conditions of high soil N availability or low atmospheric NH_3, plants can volatilize NH_3. In the atmosphere, NH_3 is important in neutralizing acidic species such as H_2SO_4, HNO_3 and HCl. Neutralizatian of acidic species results in the formation of NH_4 salts in cloud droplets and precipitation which can contribute to forest decline, plant species change and eutrophication of surface water.

References

1. Asman WAH & van Jaarsveld HA (1992) A variable-resolution transport model applied for NH_4 in Europe. Atmos Environ 26:445–464
2. Beauchamp EG, Kidd GE &Thurtell G (1978) Ammonia volatilization from liquid dairy cattle manure in the field. Can J Soil Sci 62:11–19
3. Bobbink R, Heil GW & Raessen MBAG (1992) Atmospheric deposition and canopy exchange processes in heathland ecos stems. Environ Poll 75:29–37
4. Bock BR & Kissel DE, eds. (1988) Ammonia volatilization from urea fertilizers. Nat Fert Dev Cent Muscle Shoals, Al
5. Bremner JM & Mulvaney (1978) Urease activity in soils. In: RJ Burns (ed.) Soil Enzymes, pp. 149–196. Academic Press, London
6. Catchpoole VR, Harper LA & Myers RJK (1981) Annual losses of ammonia from a razed asture fertilized with urea. Proc XIV Inter Grassland Cong, Lexington, KY pp. 344–347
7. Catchpoole VR, Oxenham DJ & Harper LA (1983) Transformation and recovery of urea applied to a grass pasture in southeastern Queensland. Aust J Exp Agric Anim Husb 23:80–86
8. Chai HH & Hou TT (1975) Studies on the volatilization gaseous nitrogen from ammonium sulfate, urea and sodium nitrate applied to soils. Soils Fert Taiwan 12–31
9. Denmead OT (1983) Micrometeorolagical methods for measuring gaseous losses of nitrogen in the field. In: JR Freney & JR Simpson (eds.) Gaseous Loss of Nitrogen from Plant-Soil Systems, Martinus Nijhoff/W. Junk Pub., The Hague pp 133–157.
10. Denmead OT, Freney JR & Simpson JR (1976) A closed-ammonia cycle within a plant canopy. Soil Biol Biochem 8:161–164
11. Denmead OT, Freney JR & Simpson JR (1982) Dynamics of ammonia volatilization during furrow irrigation of maize. Soil Sci Soc Am J 46:149–155
12. Duyzer JH, Verhagen HLM & Weststrate JH (1992) Measurement of dry deposition flux of NH_3 on to coniferous forest. Environ Poll 75:3–13
13. Farquhar GD, Firth PM, Wetselaar R & Weir B (1980) On the gaseous exchange of ammonia between leaves and the environment. Determination of the ammonia compensation point. Plant Physiol 66:710–714
14. Farquhar GD, Wetselaar R & Weir B (1983) Gaseous nitrogen losses from plants. In: JR Freney and JR Simpson (eds.) Gaseous loss of nitrogen from plant-soil systems, Martinus Nijhoff/W. Junk Pub. The Hague pp 159–180.

15. Fenn, LB & Kissel DE (1974) Ammonia volatilization from surface applications of ammonium compounds on calcareous soils. II. Effects of temperature and rate of ammonium nitrogen application. Soil Sci Soc Am Proc 38:606–610

16. Fenn, LB & Kissel DE (1976) the influence of cation exchange capacity and depth of incorporation on ammonia volatilization from ammonium compounds applied to calcareous soils. Soil Sci Soc Am J 40:394–398

17. Ferguson RB, Mc Innes KJ, Kissel DE & Kanemasu ET (1988) A comparison of methods of estimating ammonia volatilization in the field. Fert Res 15:55–69

18. Freney JR, Denmead OT, Watanabe I & Craswell ET (1981) Ammonia and nitrous oxide losses following applications of ammonium sulfate to flooded rice. Aust J Agric Res 32:37–45

19. Hargrove WL, Bock BR, Raunikar RA & Urban WJ (1987) Comparison of a forced-draft technique to nitrogen-15 recovery for measuring ammonia volatilization under field conditions. Soil Sci Soc Amer J 51:124–128

20. Harper LA (1971) Mass and energy transfer between the atmosphere and two plant canopy types. Ph.D. Diss. Univ. of Georgia, Athens

21. Harper LA (1988) Comparisons of methods to measure ammonia volatilization in the field. In: BR Bock & DE Kissel (eds.) Ammonia volatilization from urea fertilizers, National Fertilizer Development Center, TVA, Muscle Shoals, Alabama pp 93–109.

22. Harper LA, Catchpoole VR, Davis R & Weier KL (1983) Ammonia volatilization soil, plant, and microclimate effects on diurnal and seasonal fluctuations. Agron J 75: 212–218

23. Harper LA, Sharpe RR, Langdale GW & Giddens JE (1987) Nitrogen cycling in a wheat crop soil, plant and aerial nitrogen transport. Agron J 79:965–973

24. Harper LA, & Sharpe RR (1994) Nitrogen dynamics in irrigated corn: soil-plant nitrogen and atmospheric ammonia transport. Agron J (in press)

25. Hoff KD, Nelson DW & Sutton AL (1981) Ammonia volatilization from liquid swine manure applied to cropland. J Environ Qual 10:90–95

26. Hutchinson GL (1969) Nitrogen enrichment of surface water by absorption of ammonia loss from cattle feedlots. Science 166:514–517

27. Hutchinson GH, Millington RJ & Peters DB (1972) Atmospheric ammonia: absorption by plant leaves. Science 175:771–772

28. Langford AO, Fehsenfeld FC, Zacharaissen J, & Schimel DS (1992) Ammonia fluxes and background concentrations in terrestrial ecosystems. Global Biogeochem Cycles 6:459–483

29. Lemon ER & van Houtte R (1980) Ammonia exchange at the land surface. Agron J 72: 876–883

30. Lockyer DR (1984) A system for the measurement in the field of losses of ammonia through volatilization. J Sci Food Agric 35: 837–848

31. Lockyer DR, Pain BF & Klarenbeek JV (1989) Ammonia emissions from cattle, pig and poultry wastes applied to pasture. Environ Poll 56:19–30

32. Lyster S, Morgan MA & 0'Toole P (1980) Ammonia volatilization from soils fertilized with urea and ammonium nitrate. J Life Sci Royal Dublin Soc 1:167–176

33. More SD & Varade SB (1978) Volatilization losses of ammonia from different nitrogen carriers as affected by soil-moisture, organic matter and method of fertilizer application. J indian Soc Soil Sci 26:112–115

34. Moller D & Schieferdecker H (1985) A relationship between agricultural ammonia emissions and atmospheric sulpher dioxide content over industrial areas. Atmos Environ 9:695–700

35. Nommik H (1973) Assessment of valatilization loss of ammonia from surface applied urea on forest soils by ^{15}N recovery. Plant Soil 38:589–603

36. Parton WJ, Morgan JA, Altenhofen JM & Harper LA (1988) Ammonia volatilization from spring wheat plants. Agron J 80:419–425

37. Ritchie SW, Hanway JJ & Benson GO (1992) How a corn plant develops. pp 21 Iowa State Univ Crop Ext Serv Spec Rept No 48

38. Ryan J, Curtin D & Safi I (1981) Ammonia volatilization as influenced by calcium carbonate particle size and iron oxides. Soil Sci Soc Am J 45:338–341

39. Stevens RJ & Logan HJ (1987) Determination of the volatilization of ammonia from surface-applied cattle slurry by the micrometeorological mass balance method. J Agric Sci Camb 109:205–207

40. Sutton MA, Moncrieff JB & Fowler D (1992) Deposition of atmospheric amrnonia to moorlands. Environ Poll 75:15–24

41. Tang IN (1980) On the equilibrium partial pressures of nitric acid and ammonia in the atmosphere. Atmos Environ 14:819–828

42. Thompson RB, Ryden JC & Lockyer DR (1987) Fate of nitrogen in cattle slurry following surface application or injection to grassland. J Soil Sci 38:689–700

43. Vallis I, Harper LA, Catchpoole VR & Weier KL (1982) Volatilization of ammonia from urine patches in a subtropical pasture. Aust J Agric Res 33:97–107

44. Van der Eerden LJ, Lekkerkerk LJA, Smeulders SM & Jansen AE (1992) Effects of atmospheric; ammonia and ammonium sulphate on Douglas fir (*Pseudotsuge menziesii*). Environ Poll 76:1-9

45. Van der Molen J, Van Fassen HG, Leclerc MY, Vriesema R & Chardon WJ (1990) Ammonia volatilization from arable land after application of cattle slurry. l. Field estimates. Neth J Agric Sci 38:145–158

Fertilizer Research **42**: 159–163, 1995.

© 1995 *Kluwer Academic Publishers. Printed in the Netherlands.*

Nitrous oxide (N$_2$O) emissions from soils in warm climates

Tom Granli & Oluf Chr. Bøckman
Norsk Hydro Research Centre, P.O. Box 2560, N-3901 Porsgrunn, Norway

Key words: nitrous oxide, tropical soils, agriculture, N availability, temperature, soil water content

Abstract

N$_2$O emission rates seem to be higher from soils in warm climates than from soils in temperate climates. Warm and moist conditions promote microbial processes that generate N$_2$O. Clearance of tropical forests enhances N$_2$O formation, but emission measurements from other agricultural operations in the tropics are few. Limiting fertilizer application to recommended rates applied at appropriate times and avoiding fallow land wherever practical serves to limit N$_2$O emissions. More specific advice for agriculture in warm climates requires further studies.

Background

N$_2$O is a natural constituent of the atmosphere, but the concentration is increasing by about 0.25% year^{-1} [17]. The preindustrial atmospheric N$_2$O concentration was probably about 280 ppbv, and the present concentration is 310 ppbv. This may increase to about 340 ppbv in 50 years' time if present emission rates continue unabated.

N$_2$O has an atmospheric lifetime of about 130 years. The natural emission rate of about 9.5 Tg N$_2$O-N year^{-1} (1 Tg = 10^{12} g = 1 million tonnes) has increased to about 14 Tg N$_2$O-N year^{-1}. Thus the increase is about 4.5 Tg N$_2$O-N year^{-1}, or about 50% of the natural background emission [21].

Increasing atmospheric concentration of N$_2$O is a matter of environmental concern:

- N$_2$O is a "greenhouse" gas, and present emissions contribute to about 5% of the anthropogenic greenhouse effect;

- N$_2$O in the atmosphere decomposes slowly in the stratosphere, partly to NO. This NO formation may contribute to a thinning of the ozone layer.

Microbiological processes are the principal source of atmospheric N$_2$O, both natural and anthropogenic. Transport and industrial processes also generate some N$_2$O, but man's major impact on the emission rate seems to be indirect: through use and management of soil and water.

Daily emission rates from soil can vary greatly, and annual emissions are difficult to quantify. However, cultivated soils seem to have an enhanced flux of N$_2$O compared to forest soils. Fertilizer N inputs explain some, but not all of this enhanced emission rate. Other factors including management practices also influence the emission. Bouwman [2] listed estimated fluxes from various land use regimes:

- uncultivated lands and natural ecosystems in temperate regions, mostly below 1 kg N$_2$O-N ha^{-1} year^{-1};

- natural ecosystems in tropical regions, mostly below 2 kg N$_2$O-N ha^{-1} year^{-1};

- cultivated land in temperate regions: fluxes vary greatly, but the median of 36 sets of measurements was about 2.5-3.0 kg N$_2$O-N ha^{-1} year^{-1}.

This paper will give a brief overview of the mechanisms and processes that determine N$_2$O emissions from soil, with a special emphasis on situations typical for warm climates. More details, notably on emissions in temperate climates are found in our recent review [14].

Production of N$_2$O in soil

N$_2$O emission rate from soil depends on complex interactions between soil, climatic and management factors that influence microbial processes. All N_2O derived from microbial processes originates either with:

- nitrification; stepwise oxidation of NH_4^+ to NO_3^-. This process requires the presence of oxygen, it is an aerobic process;
- denitrification; stepwise reduction of NO_3^- to N_2. This process requires that oxygen is in short supply, it is an anaerobic process.

N_2O is an intermediate or a byproduct in these microbial processes.

There are more published field studies on N_2O formation in soils under temperate than under warm climatic conditions, notably from agricultural soils. However, since the processes that generate N_2O are the same, deductions can also be made for tropical regions, though the controlling factors may differ markedly.

Soil is a heterogeneous material, different processes proceed at the same time, and the limiting factor for N_2O production can change, sometimes rapidly, e.g. during rain.

N_2O emission rates from soil depend on:

- the rates of nitrification and denitrification;
- the ratio of N_2O to the other, often main, reaction products, NO_3^- and N_2;
- the ease of diffusion and escape of N_2O from soil to atmosphere.

The main controllers for N_2O production and emission are:

- N availability, or more correctly soil content of NH_4^+ and NO_3^-, the substrates for N_2O producing processes;
- soil content of degradable organic carbon, which serves as energy for most microbial processes;
- soil aeration, which is determined by soil content of air and water and by soil texture.

In addition to these primary controllers, soil temperature and pH are important factors.

Some of the principal factors that influence N_2O emissions are outside the farmers control: soil type, rainfall, season and temperature. Nevertheless, some can be influenced, at least partially. Such factors are:

- soil aeration, affected by tillage;
- water status, controlled by irrigation and drainage;
- time, type, amounts and application method of fertilizers and manure;
- soil pH;
- supply of easily degradable organic material;
- cropping patterns.

Factors affecting N_2O production in soil

The potential of a soil for N_2O emission depends on the amount of N available for microbial transformation and on the rate of these processes (the turnover rate) [24]. Further, NO_3^- is both a substrate for denitrification and also an inhibitor for microbial transformation of N_2O to N_2. Hence, application of mineral fertilizers or manures is usually followed by a rapid increase in N_2O emission rates that lasts for one or a few weeks [6, 29]. Timing and magnitude of this emission depends on other soil and climatic factors.

In temperate climates the direct contribution of fertilizer application to N_2O emissions is given by the N_2O yield:

$$N_2O \ yield = \frac{N_2O\text{–}N_{(f)} - N_2O\text{–}N_{(c)}}{N \ application} \times 100\%$$

where: $N_2O\text{–}N_{(f)}$: emission from fertilized field;
$N_2O\text{–}N_{(c)}$: emission from unfertilized field.

Published N_2O yields vary within wide limits. A regression analysis of published data by Bouwman [2] point to 0.5% as the typical N_2O yield. However, this does not take into account indirect effects of fertilization practices on a soil's ability to form N_2O in the longer term e.g. enhanced fertility. Hence, Mosier [25] suggested that 1% is an appropriate factor for calculation of emission budgets. However, these numbers are derived mainly from measurements in the US, Canada and north-western Europe. In tropical and subtropical climates N efficiency (in crop production) tends to be low, below 60%, as discussed by others in this volume [1, 27, 36]. Hence, more measurements are required before such calculation factors can be used with confidence for fertilized land in warm climates.

The emission depends on the availability of O_2 for microbial processes, and the ease of escape of N_2O by diffusion through soil pores. Soil water content is a controlling factor for both. N_2O emissions are especially favoured when the soil is sufficiently wet to restrict oxygen availability. Such conditions give a high N_2O/NO_3^- product ratio in nitrification, and are also necessary for denitrification. If the soil becomes very wet, nitrification ceases, denitrification proceeds increasingly to N_2 and gas escape from the soil is hindered. Thus, a soil water content where both denitrification and nitrification can proceed, will generally give the maximum emission of N_2O. The range of this soil water content is normally in the range of 45 to 75%

water-filled pore space [9], which is normally close to the soil's field capacity, except for heavy, clay soils.

A strong increase in N_2O emission following irrigation or rainfall is normally observed. In this regard Davidson *et al.* [10] studied emissions in a dry tropical forest. Emissions were higher in the rainy season than in the dry season, and addition of water to dry soil caused rapid formation of NH_4^+ from mineralization and large pulses of N_2O emissions. Further, N_2O emissions are notably high when the soil alternates between dry and wet conditions rather than being constantly wet. This can be due to:

- increased availability of nutrients and stimulation of the microbial activity;
- rapid change of aeration status that furthers N_2O production;
- accumulation of NO_3^-, NO_2^- and dead, degradable organic materials during dry spells.

The effect of water and N inputs depends on soil texture, and vice versa. Clay soils can keep more water for longer periods than can light easily drained sandy soils. Nitrification is usually more rapid in light than in heavy soils, while denitrification is favoured in the latter. N_2O produced at depth can be reduced to N_2 as it moves upwards through the soil profile, especially when diffusion is slow as in heavy textured soils and/or if the soil is very wet. However, N_2O can escape more easily from coarse-textured soil types. Overall, clay soils seem to emit more N_2O than sandy soils, but this tendency can be masked or reversed by other factors.

Ploughing and cultivation increase aeration and ease moisture evaporation. Accessibility of crop residues for soil microbes is enhanced, and this may induce pulses of N_2O emission. However, the general impression is that the rates of denitrification and N_2O losses are lower from ploughed soils than from cultivation without tillage. Further, soil compaction caused by tractor traffic or trampling of cattle alter soil structure and generally enhance both denitrification and N_2O emission rates [12, 15].

The presence of degradable organic matter is necessary for a high level of microbial activity. Pockets of organic matter in an otherwise aerobic soil can cause conditions of local oxygen deficiency. Availability of organic matter can be an important limiting factor for N_2O emissions. N_2O emissions tend to be substantial where organic matter is available in abundance, e.g. from organic soils [11]. The effect depends on N supply. If NO_3^- is sparsely available, denitrification goes mostly to N_2, but if the NO_3^- supply is ample, reduction

of N_2O is impeded and N_2O emissions can be at a high level (e.g. after fertilization). Thus, input of degradable organic materials with manure and crop residues can create conditions favourable for N_2O formation, especially when the C/N ratio in the amendment is low [4] or if it is applied together with mineral N fertilizer [6].

Microbial activity, denitrification and nitrification rates all increase with temperature. The optimum temperature for denitrification is reported to range from 30 to 67 °C. The optimum temperature range for nitrification in soils is usually between 25 and 35 °C, but both nitrifiers and denitrifiers have temperature optima adapted to their climatic regions. Laboratory studies and field measurements show that N_2O production and emission also increase with increasing temperature, at least up to 20-40 °C [6, 7, 13, 26].

Large spatial and temporal variation in N_2O emission is a problem for obtaining reliable estimates from various soils and ecosystems. Diurnal, daily and seasonal variations in N_2O emission in a field can at least partly be explained by variations in temperature [31].

The effect of soil pH on N_2O emission is complex and dependent on nutrient status. Where denitrification is the main source, increasing pH tends to decrease emissions, at least below pH 6 [35]. Where N_2O mainly originates with nitrification, emissions tend to increase as pH increases, at least in the range of pH 6 to 8 [4].

Tropical situations

Several recent publications have pointed to the tropical regions, especially the rain forests, as a dominant source of atmospheric N_2O [3, 34]. Measurements from various natural ecosystems show that emission rates from tropical soils are higher, and sometimes much higher per hectare, than their temperate counterparts [18]. Activity of man enhances these emissions through forest clearing, biomass burning and agricultural/pastoral exploitation.

The principal characteristics that tropical soils–with the exception of high mountain areas–have in common, is high temperature, and limited seasonal temperature variations. There are no special tropical soil processes that produce N_2O. It is warm and frequently moist conditions favouring rapid decomposition of organic matter [28], often with a low C/N ratio,

162

that make the warm regions the principal background sources of N_2O [3].

Emissions from dry areas in the tropics are reported to be low during the dry season, but during the wet season and after irrigation, they can be comparable to those from areas with humid climates [12, 16, 33]. Seasonal frequent and heavy rainfall with intermittent periods of drying of the topsoil increase emissions. While there are a number of studies available for N_2O emissions from tropical forests and grasslands, there is limited data on emissions from fertilized and irrigated fields (excepting the special case of paddy rice). This is unfortunate as there may be an optimum combination of fertilizer application and irrigation practice. Currently, much effort is placed on increasing food production in the developing countries through more extensive use of irrigation. The very large expansion of irrigated cereal production in these parts of the world points to the importance of achieving more emission data from irrigated and fertilized soils in tropical regions.

In the humid tropics, soils are often acidic with pH below 6. Low pH may, to some extent, explain the high emission rates reported from these soils, at least where denitrification can be substantial, as in wet clay soils. Soils in the drier regions have higher pH values with possibilities of NO_2^- accumulation during the dry season. This NO_2^- originates with nitrification and may give high pulses of N_2O in wet periods, at least when the supply of NH_4^+ is adequate [10].

Conversion of tropical forests to pasture and other agricultural land is a major environmental issue [8]. Large amounts of N can be mobilized by such land clearance. This can enhance annual N_2O emission rates, at least as long as decomposable organic matter is present and N is available for transformation, possibly for a decade or two [19, 32, 34]. Thereafter, fluxes from cleared pastures and savannas can be lower than from similar forest sites [20, 30]. The enhanced N_2O emissions following forest clearing and pasture management are more pronounced during wet, rather than dry periods [12, 22]. Fertilization and other management practices (e.g. cultivation of forage legumes) may have potential for extending the period of enhanced N_2O emissions, but this topic has not been investigated for tropical pastures. Pastures merit further investigation as an N_2O source, as productive pastures combine good availability for microbial transformation of N and organic matter, urine patches are known to have high rates of N_2O emissions, and soil compaction by trampling of cattle can also increase N_2O emissions.

Various studies have been made where N_2O emissions from different crops have been compared. No systematic differences are apparent, with a few exceptions. Legumes contribute an N input to soils through biological N fixation. This input like other N inputs increases N_2O formation. Intercropping with legumes is practised and promoted in tropical and subtropical regions where the supply of fertilizers can be difficult.

Some important tropical crops such as banana and sugar cane may merit priority for studies. These crops are heavy fertilized. After harvest of bananas the mass of leaves are cut and left on the surface to decompose under warm, moist conditions. However, there are no reports on measurements of N_2O emissions in banana plantations. On the other hand, sugar cane residues generally decompose under more dry conditions where N_2O production should not be favoured to the same extent. A few published measurements of N_2O emissions from this crop indicate that the fluxes are not remarkable [11, 37].

Another crop more typical for warm than for temperate climates where data are available, is paddy rice. The emission rate is known to be modest while fields are flooded, but Byrnes [5] found that emissions could be as high as 6 to 7 kg N_2O-N ha^{-1} for a growing season. The N_2O originates with NO_3^- accumulated in the soil during the dry season fallow, and denitrified when the rainy season starts. However, it is not known how representative these results are for rice cultivation in general.

Some countries in temperate climates have established codes for good agricultural practice (e.g. [23]). One purpose of such codes is to prevent NO_3^- leaching, e.g. through limiting irrigation and fertilizer application to the amounts necessary to ensure optimal plant growth. Further, NO_3^- accumulation in soil should be prevented by avoiding fallow and keeping the soil under green cover for as much of the year as is practical. These measures should also serve to reduce N_2O emissions. Thus, much advice to farmers in temperate climates on how to minimize N_2O losses from agricultural land is already found in existing codes. This advice may provide a starting point for elaboration of similar information for other regions, provided care is taken to adjust the advice to local farming conditions. Agriculture in the tropics is even more diversified and complex than in temperate regions. There is ample scope for scientific work in the warm climates on increasing plant nutrient efficiency and minimizing environmental impacts of agriculture.

References

1. Agrawal RP (1995) Soil management options for improving nitrogen use. This volume
2. Bouwman AF (1990) Exchange of greenhouse gases between terrestrial ecosystems and the atmosphere. In: Bouwman AF (ed) Soils and the Greenhouse Effect, pp 61–127. Chichester: John Wiley and Sons Ltd
3. Bouwman AF, Fung I, Matthews E and John J (1993) Global analysis of the potential for N_2O production in natural soils. Global Biogeochem Cycles 7: 557–597
4. Bremner JM and Blackmer AM (1981) Terresterial nitrification as a source of atmospheric nitrous oxide. In: Delwiche CC (ed) Denitrification, Nitrification and Atmospheric N_2O, pp 151-170. Chichester: John Wiley and Sons
5. Byrnes BH, Holt LS and Austin ER (1993) The emission of nitrous oxide upon wetting a rice soil following a dry season fallow. J Geophys Res 98: 22,925–22,929
6. Cates RL and Keeney DR (1987) Nitrous oxide production throughout the year from fertilized and manured maize fields. J Environ Qual 16 443–447
7. Conrad R, Seiler W and Bunse G (1983) Factors influencing the loss of fertilizer nitrogen into the atmosphere as nitrous oxide. J Geophys Res 88: 6,709–6,718
8. Cook AG, Janetos AC and Hinds WT (1990) Global effects of tropical deforestation: Towards an integrated perspective. Environ Conserv 17: 201-212
9. Davidson EA (1991) Fluxes of nitrous oxide and nitric oxide from terrestrial ecosystems. In: Rogers JE and Whitman WB (eds) Microbial Production and Consumption of Greenhouse Gases: Methane, Nitrogen Oxides and Halomethanes, pp 219–235. Washington, DC: American Society for Microbiology
10. Davidson EA, Matson PA, Vitousek PM, Rile R, Dunkin K, Garcia-Méndez G and Maass JM (1993) Processes regulating soil emissions of NO and N_2O in a seasonally dry tropical forest. Ecology 74: 30–139
11. Duxbury JM, Bouldin DR, Terry RE and Tate RL (1982) Emissions of nitrous oxide from soils. Nature 298: 462–464
12. García-Méndez G, Maass JM, Matson PA and Vitousek PM (1991) Nitrogen transformations and nitrous oxide flux in a tropical decidious forest in Mexico. Oecologia 88: 362–366
13. Goodroad LL and Keeney DR (1984) Nitrous oxide production in aerobic soils under varying pH, temperature and water-content. Soil Biol Biochem 16: 39–43
14. Granli T and Bøckman OC (1994) Nitrous oxide from agriculture. Norw J Agric Sci, Suppl 12: 1–127
15. Hansen S, Mæhlum JE and Bakken LR (1993) N_2O and CH_4 fluxes in soil influenced by fertilization and tractor traffic. Soil Biol Biochem 25: 621–630
16. Hao WM, Scharffe D, Crutzen PJ and Sanhueza E (1988) Production of nitrous oxide, methane, and carbon dioxide from soils in the tropical savanna during the dry season. J Atmos Chem 7: 93–105
17. Houghton JT, Callander BA and Varney SK (eds) (1992) Climate change. The supplementary report to the IPCC Scientific Assessment. Cambridge: Cambridge University Press
18. Keller M, Kaplan WA and Wofsy SC (1986) Emissions of N_2O, CH_4 and CO_2 from tropical forest soils. J Geophys Res 91: 11,791–11,802
19. Keller M, Jacob DJ, Wofsy SC and Hamss RC (1991) Effects of tropical deforestation on global and regional atmospheric chemistry. Climatic Change 19: 139–158
20. Keller M, Veldkamp E, Weitz AM and Reiners WA (1993) Effect of pasture age on soil trace-gas emissions from a deforested area of Costa Rica. Nature 365: 244–246
21. Khalil MAK and Rasmussen RA (1992) The global sources of nitrous oxide. J Geophys Res 97: 14,651–14,660
22. Luizão F, Matson PA, Livingston G, Luizão R and Vitousek PM (1989) Nitrous oxide flux following tropical land clearing. Global Biogeochem Cycles 3: 281–285
23. MAFF (Ministry of Agriculture, Fisheries and Food) (1991) Code of good agricultural practice for the protection of water. London: Ministry of Agriculture, Fisheries and Food, Welsh Office, Agriculture Department
24. Matson PA and Vitousek PM (1990) Ecosystem approach to a global nitrous oxide budget. Processes that regulate gas emissions vary in predictable ways. Bioscience 40: 667–671
25. Mosier AR (1993) Nitrous oxide emissions from agricultural soils. In: van Amstel AR (ed) Proceedings of International IPCC Workshop: Mehane and nitrous oxide. Methods in National Emissions Inventories and Options for Control, pp 273–285. Amersfoort, The Netherlands, 3-5 February 1993.
26. Nõmmik H (1956) Investigations on denitrification in soil. Acta Agric Scand 6: 195–228
27. Pilbeam CJ and Warren GP (1995) Studies with ^{15}N fertilizer N recovery and N mineralization in semi-arid Kenya. This volume
28. Ross SM (1993) Organic matter in tropical soils: Current conditions, concerns and prospects for conservation. Prog Phys Geogr 17: 265–305
29. Ryden JC (1981) Nitrous oxide exchange between a grassland soil and the atmosphere. Nature 292: 235–237
30. Sanhueza E, Hao WM, Scharffe D, Donoso L and Crutzen PJ (1990) Nitrous oxide and nitric oxide emissions from soils of the northern part of the Guayana Shield, Venezuela. J Geophys Res 95: 22,481–22,488
31. Slemr F, Conrad R and Seiler W (1984) Nitrous oxide emissions from fertilized and unfertilized soils in a subtropical region (Andalusia, Spain). J Atmosph Chem 1: 59–169
32. Steudler PA, Melillo JM, Bowden RD and Castro MS (1991) The effects of natural and human disturbances on soil nitrogen dynamics and trace gas fluxes in a Puerto Rican wet forest. Biotropica 23: 356–363
33. Vitousek PM, Matson PA, Volkmann C, Maass JM and García G (1989) Nitrous oxide flux from dry tropical forests. Global Biogeochem Cycles 3: 375–382
34. Vitousek PM and Matson PA (1992) Tropical forests and trace gases. Potential interactions between tropical biology and the atmospheric sciences. Biotropica 24: 233–239
35. Weier KL and Gilliam JW (1986) Effect of acidity on denitrification and nitrous-oxide evolution from atlantic coastal-plain soils. Soil Sci Soc Am J 50: 1,202–1,205
36. Wild A, Jenkinson D and Barraclough D (1995) The potential problems associated with introducing nitrogen fertilizers into tropical agriculture. This volume
37. Zachariassen J, Matson PA and Vitousek PM (1993) Annual nitrous oxide emissions from intensively managed soils in Maui, Hawaii. Bull Ecol Soc Am 784: 498

Fertilizer Research **42**: 165–174, 1995.
© 1995 *Kluwer Academic Publishers. Printed in the Netherlands.*

The role of *Azolla* in curbing ammonia volatilization from flooded rice systems

Paul L.G. Vlek, Moussa Y. Diakite & Henning Mueller
Institute of Agronomy in the Tropics, Georg August University, Grisebachstr. 6, D-37077 Goettingen, Germany

Key words: Anabaena, cropping systems, fertilizer management, integrated nitrogen management, ^{15}N, nitrogen balance, *Oryza sativa*, urea, wetlands

Abstract

By the year 2020, an additional 300 million tons of rice are needed annually to meet the demands of a growing population. If our natural resource base is to be preserved, intensification strategies should rely on integrated nutrient management, making full use of biological nitrogen fixation. The *Azolla-Anabaena* complex is amongst the most effective systems of fixing nitrogen. In this paper we present evidence from greenhouse studies on the potential of *Azolla* to curb the volatilization of NH_3 following the application of urea to a mixed *Azolla*-rice culture, providing a new incentive for developing ways of integrating *Azolla* in intensive rice cultivation systems.

The results of a series of short term greenhouse experiments show that a full cover of *Azolla* can significantly reduce losses of applied urea-N from 45 and 50% to 20 and 13% for the 30 and 60 kg N ha^{-1} treatments, respectively. About one-quarter of the applied N was tied up in the *Azolla* biomass. The applied N inhibited *Azolla* growth as well as the amount of N fixed. Inoculation with smaller quantities of *Azolla* allowing for more vigorous *Azolla* multiplication was equally effective in reducing NH_3 volatilization and doubled the amount of ^{15}N tied-up by *Azolla*. The reduction in NH_3 volatilization is largely related to the depression by *Azolla* of the floodwater pH, which in its absence may reach values between 9 and 10 as a result of algal activity.

Early rice growth responded positively to urea as well as the large quantities of applied *Azolla* and increased the yield potential of the crop. Smaller quantities of *Azolla* alone were not effective in this regard. The conservation of fertilizer N by *Azolla*, particularly when it fully covered the water, was reflected in a synergistic effect on rice dry matter production, amounting to 9% at the 30 kg N rate and 16% at the 60 kg N rate. In all likelihood this interaction is attributable to the higher efficiency of the applied N. The benefits of *Azolla* in conserving basal urea-N even in small quantities (200–500 kg fresh material ha^{-1}), outweighed competition for the applied N and may be as important as its BNF. The most promising integrated *Azolla*/rice management systems emerging from our studies should be given further attention under field conditions.

Introduction

Nearly 150 million hectares of land are planted annually producing nearly 500 million tons of rice (*Oryza sativa* L.). Ninety percent of it is grown in South and Southeast Asia. Being semi-aquatic, nearly 75% of the rice is produced under flooded or intermittently flooded conditions. Food shortages in Asia have been largely avoided through intensification of rice production. With the advent of higher-yielding varieties and chemical fertilizers, rice yields have moved through the traditional ceiling of 1.5–2 ton ha^{-1} in the mid-sixties and currently stand at an average of around 3.5 ton ha^{-1}. The key ingredient in accomplishing this was undoubtedly chemical N, of which more than 90% is applied in the form of urea.

The efficiency of urea applied to flooded rice is notoriously low, commonly ranging between 30 and 40% (De Datta *et al.*, 1983). It is widely accepted that this low efficiency is due in large part to the susceptibility of urea to NH_3-volatilization, particularly of the early applied N (Vlek and Fillery, 1984). As much as 50% of the applied urea can thus be lost within 2 weeks of application dependent on fertilizer management and the environmental conditions (Fillery and Vlek, 1986). The cost and environmental problems of NH_3-loss associated with the use of urea in flooded rice have been well documented (Vlek and Byrnes, 1986).

NH_3 volatilization is a function of the partial pressure of NH_3 (P_{NH_3}) in the floodwater which in turn is determined by the total (NH_3 + NH_4) - N level and the pH of the floodwater (Vlek and Crasswell, 1981). P_{NH_3} increases ten-fold for every unit increase in pH. NH_3 loss is negligible from floodwater at pH levels below 7.5, but urea-fertilized floodwater will turn alkaline upon hydrolysis due to the alkaline nature of the $(NH_4)_2CO_3$-solution formed. In addition, algae play a key role in the loss of NH_3 from flooded soils through their consumption of CO_2. Diurnal fluctuations in floodwater pH from near neutral to 10 are common in systems where green algae are active (Vlek and Craswell, 1981). Algae blooms, mostly green, rapidly develop upon N fertilization of flooded rice (Watanabe *et al.*, 1977), but die off once light penetration is limited by the rice canopy. This process might be accelerated when *Azolla* grown in association with rice shades out the algae. In this case the *Azolla* serves as a volatilization barrier while the *Anabaena*, being independent of the CO_2 in the floodwater (Tel-Or *et al.*, 1991), fixes nitrogen without driving up the floodwater pH.

The N content of a single standing crop of *Azolla* may range from 20 to 146 kg N ha^{-1}. Under farmers' conditions *Azolla*-N yields may be at the lower end of this spectrum (Watanabe, 1987). In assessing the direct contribution of nitrogen fixed by *Azolla* to rice, Watanabe *et al.* (1981) found that 25% of the applied *Azolla*-N was available to the standing rice plant. Kumarasinghe and Eskew (1993) reported recoveries ranging from 26–37% for *Azolla*-^{15}N when applied at rates from 73–144 kg N ha^{-1}. These recoveries are in line with fertilizer N recovery rates reported for these experiments and elsewhere (Vlek and Byrnes, 1986).

Many experiments document the effectiveness of *Azolla*-N in increasing rice yields. Literature data from over 70 experiments summarized by Steinhauser (1992) covering the seven most frequently tested *Azolla* management systems showed that, on average, green manuring produced 0.45 t ha^{-1} extra rice in comparison with the *Azolla*-free crop. Mixed cropping with or without incorporation yielded an additional 0.5 and 0.6 t ha^{-1}, respectively, whereas an incorporated pre crop gave an extra 0.8 t ha^{-1}. In India, a basal N dressing is recommended if *Azolla* is used as a green manure or in mixed cropping (Singh and Singh, 1986; Manna and Singh, 1989). Literature data confirm that the best responses (1–1.1 t ha^{-1}) are indeed obtained by combining pre or mixed cropping with a basal dose of 30 kg N ha^{-1} (Steinhauser, 1992).

The effect of *Azolla* on the efficiency of applied urea-N has been given scant attention. Field experiments in various countries co-ordinated by the International Atomic Energy Agency (IAEA) (Kumarasinghe and Eskew, 1993) compared the response and ^{15}N recovery by rice for 30 kg N ha^{-1} applied as urea into fields with and without 2 tons of fresh *Azolla* inoculated at transplanting. The ^{15}N uptake data showed that urea N was more efficiently used in the mixed *Azolla*/rice crop when compared to the sole rice crop, increasing the N recovery by 10, 60, and 53% in China, Sri Lanka, and Thailand, respectively. Only in Thailand, the improved urea-N uptake resulted in a significant *Azolla* effect, increasing yield by 52%. This site specificity might reflect differences in NH_3-volatilization as a problem at the three locations.

Vlek *et al.* (1992) showed some preliminary results indicating reduced NH_3 volatilization when *Azolla* is present. In their experiment without ^{15}N, the estimated N losses in the *Azolla*-free system amounted to 80% of the basal broadcast urea-N, irrespecive of the amount of N applied. In the case of deep-placed urea-N losses were between 12 and 25%, in line with losses previously reported for deep-placement of N (Vlek and Fillery, 1984). Deep-placed urea with *Azolla* cover yielded a net increase in N in the soil-plant system of 72 kg N ha^{-1} with little difference among N rates. The sum of this net gain in the deep-placed urea treatments with *Azolla* and the loss of N in the absence of *Azolla* provides an estimate of the average BNF over the three N rates considered (0, 20, 40 kg N ha^{-1}) of 78 kg N ha^{-1}. Based on this, N losses can be estimated to have been reduced from 80% of applied urea in the absence of *Azolla* to 10% and 35% for the 20 and 40 kg N ha^{-1} rate, respectively, when an *Azolla* mat was present.

The objective of the studies we report here was to verify the above results using ^{15}N, allowing a better assessment of the role that *Azolla* plays in preventing NH_3 volatilization from urea, applied at early growth stages, to flooded rice. Various alternative management systems of integrating the application of urea and *Azolla* and its consequences in terms of fertilizer efficiency and biological nitrogen fixation were evaluated.

Materials and methods

Four greenhouse experiments were conducted at the University of Goettingen, Germany, during the summers of 1992, 1993 and 1994. Mean maximum and minimum temperatures in each year were 28°C during the day and 24°C at night. Relative humidities varied from year to year with 62 during the day and 68% at night in 1992. In 1993 the relative humidities were 47% and 70% as compared to 55 and 60% in 1994 for day and night. The light intensity between 06:00 and 20:00 was kept at a minimum of 15–18 Klux with the help of HPL-N lights (Phillips Type 57221, G/74).

Black 41 polypropylene pots were filled with 3 kg low-fertility sandy soil from Solling, Germany, (pH_{CaCl_2} 6.5), giving a soil surface area of 0.025 m². The native soil N content of 0.014% was raised to 0.02% with 1.5 g kg^{-1} organic compost. Soils were screened to < 2 mm, fertilized with mono-calcium phosphate (MCP) at 90 kg P ha^{-1} and flooded with 4 cm standing demineralized water. Following 3 weeks of pre-flooding, soils were fertilized with K_2SO_4 and $MgSO_4$ at rates of 80 kg K ha^{-1} and 40 kg Mg ha^{-1}, respectively. In 1994 the P fertilization was delayed until 7 days before transplanting or transplanting time. An N-free micro nutrient cocktail (Van Hove et al., 1983) was applied in order to avoid possible deficiencies particularly in *Azolla*.

The 1992 experiment comprised a factorial design with 3 levels of urea (0, 30, and 60 kg N ha^{-1}) applied at transplanting and 3 types of floodwater coverage (no cover, complete *Azolla* cover, complete cover with white 3 mm-Styropor balls). In 1993, the factorial design involved 3 urea application treatments (0, 40 kg N ha^{-1} at transplanting and 40 kg N ha^{-1} one week after transplanting) and 5 levels of superimposed *Azolla* inoculation (0, 10, 20, 40, and 60% floodwater coverage) at transplanting. Both experiments had 4 replications. The experiments were laid out in a randomized complete block design. Urea enriched in ^{15}N at an atom excess of 5.308% was broadcast into the floodwater. Two 3-week old (1992) or 2-week old (1993) seedlings (Var. IR50) were transplanted to the centre of the pot immediately prior to Styropor application or *Azolla* inoculation using *Azolla caroliniana*, Adul CA in its second ramification stage (~ 5% N in dry matter). Soils remained flooded with 2.5–4 cm of distilled water throughout the experiment.

In order to ascertain if small quantities of *Azolla* applied prior to transplanting were able to prevent the development of high partial pressures of NH_3 in the floodwater (P_{NH_3}), a parallel experiment was conducted in 1993. A 2% *Azolla* cover (172 kg fresh weight ha^{-1}) was placed on the floodwater at 2 weeks before and at transplanting followed by urea application one week after transplanting. As in the previous two experiments, floodwater pH, temperature and ammoniacal-N concentration were analyzed daily between 12:00 and 13:00 for the first ten days following the application of urea. Temperature and pH were measured *in situ*, whereas ammoniacal N was measured in a 10 mL aliquot using an NH_3 electrode. Ammoniacal N pressures were calculated according to Denmead et al. (1983).

The ^{15}N experiments conducted in 1992 and 1993 were not designed to optimize agronomic differences between treatments. In 1994 a final experiment was conducted in order to verify the synergistic effect of early urea applications in combination with *Azolla* on rice dry matter production. Rice plants were grown in 64 pots as described before. Half the pots received P only and half P combined with *Azolla* at a rate of 1 ton ha^{-1} (10% coverage). These applications were made either one week prior to transplanting or at transplanting time. Each of these four treatments was combined with 4 urea application treatments, 0, or 40 kg N ha^{-1} at one, two or three weeks after transplanting. Seedlings were 3 weeks old at transplanting.

The first symptoms of N deficiency were noted after 3–5 weeks, particularly in the zero-N treatments, while slight damage to the *Azolla* was visible throughout the duration of the experiment in the 60 kg N ha^{-1} treatments (1992). The 1992 and 1994 experiments were terminated after 7 weeks, just before flowering (1992) or at panicle initiation (1994), respectively. Harvesting in 1993 took place after 10 weeks, the time it took to reach panicle initiation due to young seedling age and unfavourable growing conditions in the fall.

In all *Azolla* treatments rice roots grew into the *Azolla* mat whereas in the absence of *Azolla* they were confined to the soil. The *Azolla* was carefully separated from rice roots, harvested and its fresh weight measured. Styropor was washed and discarded. After draining the soil, rice plants were cut at soil level and weighed. Rice roots were harvested separately, washed and weighed. *Azolla* and rice were subsequently dried at 50°C for three days. Soils from each pot were thoroughly mixed, sub-sampled and dried at 105°C for subsequent analysis.

Soil and plant nitrogen from finely ground samples taken at transplanting and at harvesting were analyzed on an elemental analyzer (Carlo Erba NA 1500). Every

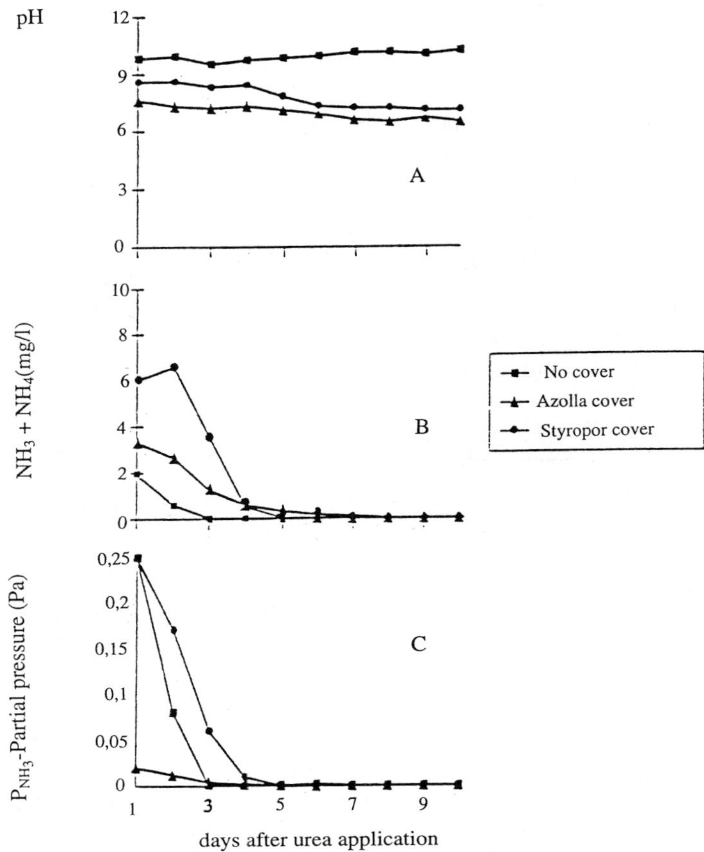

Fig. 1. Dynamics of pH, ammoniacal N and resulting P_{NH_3} in floodwater following the application of 30 kg Urea-N in variously covered floodwater (1992).

tenth sample was also analyzed using the conventional Kjeldahl method (Bremner and Mulvaney, 1982). The ^{15}N content in plant and soil was analyzed in triplicate on an elemental analyzer connected to a mass spectrometer (Reineking *et al.*, 1993) of the central isotope laboratory of the university, providing the ^{15}N-ratio as well as an internal check on the total N values.

Results and discussion

Fate of applied N

In 1992, midday temperatures in the floodwater ranged from 27 to 38 °C with temperatures under Styropor 2–3 °C lower than under *Azolla* or without cover, most likely due to the effective reflection of solar radiation by the white Styropor. The temperature under *Azolla* was slightly higher than when uncovered, a phenomenon observed and explained earlier by Kroeck *et al.* (1988b).

The pH of the floodwater for the 30 kg N application rate ranged from 9 to 10.5 for floodwater without coverage as compared to 7 to 8.5 and 6.5 to 7.5 for Styropor and *Azolla* cover, respectively (Fig. 1a). Thus mid-day pH differentials in the greenhouse experiments reached 1.5–2.5 units. The $NH_3 + NH_4$ - N (AN) levels in the floodwater are presented in Figure 1b. Over the first 5 days, highest levels are maintained under Styropor and lowest levels in the uncovered treatments. The combined effect of the changes in AN and pH on the partial pressure of NH_3 (P_{NH_3}) – the driving force of NH_3 volatilization – as calculated after Denmead *et al.* (1983) can be seen from Figure 1c.

With the low partial pressures observed in the *Azolla*-covered systems (< 0.05 Pa) it is doubtful that any loss of NH_3 could have taken place. On the other hand, NH_3 volatilization in the uncovered systems

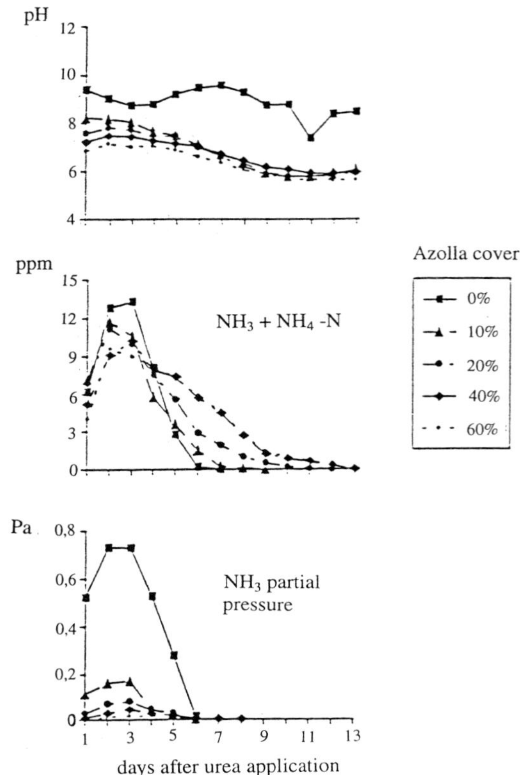

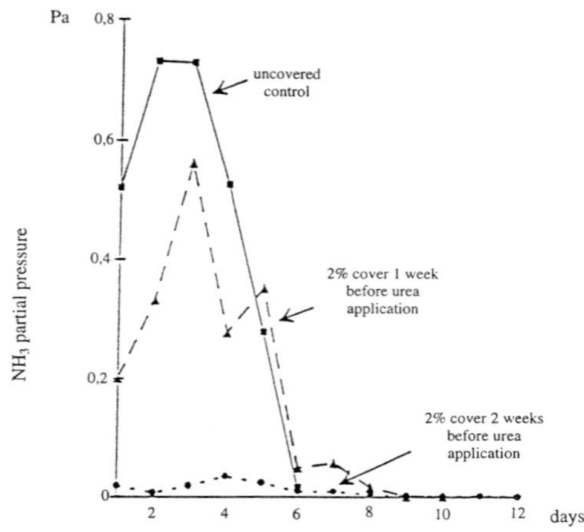

Fig. 3. Partial pressure of ammonia as affected by the timing of urea application with respect to *Azolla* inoculation (1993).

Fig. 2. Floodwater dynamics following urea application to floodwater one week after transplanting and *Azolla* application (1993).

must have been swift during the first 4 to 5 days after N application, and caused a rapid decline of AN in the floodwater. The presence of Styropor reduced the pH, but this was compensated for by the persistence of AN in the floodwater leading to P_{NH_3}-values exceeding those in the uncovered system. The data for the 60 kg N ha^{-1} rate followed essentially the same pattern but gave maximum P_{NH_3} values on day one of 0.55–0.8 Pa.

The results of the 1993 experiment (Fig. 2) confirm this effect of *Azolla*: the driving force of NH_3 volatilization (P_{NH_3}) is reduced with increasing *Azolla* coverage at the time of fertilizer application up to an initial coverage of 40%. As is clear from Figure 2, most of the reduction in the NH_3 volatilization potential is brought about by the reduction in floodwater pH, whereas the ammoniacal N in the floodwater was hardly affected by the levels of applied *Azolla*.

Based on these results it appears that a coverage with *Azolla* of 0.4 m^2 m^{-2} at the time of fertilizer application suffices to minimize the NH_3 volatilization potential, by reducing P_{NH_3} peak levels from around 1

Pa without *Azolla* cover, to 0.6 Pa at 10% coverage, 0.3 Pa at 20%, and below 0.2 Pa with a 40% cover (data not shown). Delaying urea application by one week, a practise widely used by farmers, further reduced the requirement for *Azolla* inoculation to 0.1 m^2 m^{-2} (Fig. 2). Given that a 100% *Azolla* cover requires approximately 10 ton ha^{-1} of fresh material, such a reduction to 1 ton ha^{-1} would substantially increase the chances of adoption of this practice.

The improved performance of advanced *Azolla* inoculation was further demonstrated by varying the timing of application of as little as 172 kg ha^{-1} of fresh *Azolla* (2% coverage). As can be seen from Figure 3, the P_{NH_3}-level in the floodwater was reduced only slightly in comparison with those when no *Azolla* was applied when the *Azolla* inoculation took place only one week prior to urea fertilization. However, when *Azolla* was allowed to grow for two weeks prior to urea application, the P_{NH_3} built-up in the floodwater was effectively prevented even though the *Azolla* mass at the time of urea application was no more than 30% compared to that of the full cover. This result indicates that, given adequate time to grow, relatively small but fast-growing *Azolla* populations may adequately lower floodwater P_{NH_3}-levels.

A combination of mechanisms could be responsible for these results such as eliminating algal growth and uptake of applied NH_3 in combination with possible exudation of protons from the root. An important mechanism might be the respiration of *Azolla* derived CO_2 by *Anabaena* resulting in an increased P_{CO_2} of the floodwater and a reduction of the floodwater pH. The transfer of metabolic carbon from *Azolla* to *Anabae-*

Table 1. Uptake and Loss of basally applied urea-^{15}N after 7 weeks of rice growth (1992)

Coverage	N rate	Uptake rice (%)	Uptake *Azolla* (%)	Soil (%)	Loss[a] (%)
None	30	43.8		6.0	50.2 (2.0)
	60	42.8		7.6	49.7 (2.5)
Styropor	30	50.7		11.6	37.8 (5.9)
	60	55.4		6.6	38.0 (3.2)
Azolla	30	52.1	23.5	3.4	21.0 (0.8)
	60	57.5	23.5	6.1	13.0 (2.0)

[a]Numbers within brackets represent one standard deviation from the mean.

Table 2. The fate of urea-N (40 kg N ha^{-1}) applied at transplanting of rice and *Azolla* (basal) or 7 days after transplanting (delayed) (1993)

Time of N application	*Azolla* cover (%)	*Azolla*	Rice tops	Rice roots	Soil	Loss[a]
				^{15}N recovery (%)		
Basal	0	-	13.3	4.6	22.2	59.9 a
	10	31.2	13.6	4.3	14.4	36.5 c
	20	47.5	16.0	4.5	10.6	21.4 d
	40	51.4	18.8	4.6	6.6	18.7 de
	60	57.3	15.6	4.0	7.3	15.9 def
Delayed	0	-	21.7	6.4	23.9	48.0 b
	10	43.3	27.9	7.3	9.9	11.5 efg
	20	50.2	27.1	7.0	7.1	8.6 fg
	40	53.9	27.4	5.3	8.5	4.9 g
	60	44.6	33.7	7.9	6.1	7.6 fg

[a]Ranking according to Tukey's test.

na is well documented (Tel-Or *et al.*, 1991); however, the fate of this carbon has not been studied. Presumably, the *Azolla-Anabaena* complex could function as a CO_2-pump by the release of respired CO_2 into the floodwater thus countering the rise in floodwater pH normally associated with the hydrolysis of urea and the presence of photosynthesizing algae (Vlek and Craswell, 1991). The latter two mechanisms would explain the observed effectiveness of *Azolla* with an incomplete coverage of the floodwater surface. The relative importance of these mechanisms will be discussed below.

A first indication of whether the observed reductions in P_{NH_3} indeed can eliminate or reduce NH_3 volatilization can be obtained from the construction of an N balance using ^{15}N. In the greenhouse experiment in 1992, we measured the losses of applied N from the plant-soil system indirectly after 7 weeks of vegetative growth. The calculated losses based on the ^{15}N balance for the soil-plant system at harvest are presented in Table 1. Losses of applied N were significantly reduced by the presence of the *Azolla* mat from around 50% to less than 21%. The Styropor, which essentially was meant as a physical barrier to NH_3 volatilization, caused a reduction in N-loss of around 10%, thus confirming that the reduction in NH_3 volatilization is largely related to the pH effect and N uptake of *Azolla*.

The 1993 experiment was conducted to establish the critical mass of *Azolla* necessary to effectively prevent the increase in P_{NH_3} and thus conserve applied N. Results presented in Table 2 show that ^{15}N losses were drastically reduced from approximately 60% in the absence of an *Azolla* cover at the time of transplanting and urea application to 36% with a cover of only one-tenth of the water surface and around 20% with one-fifth or more of the surface area covered at the time of urea application. Delaying the urea application by one week, thus allowing the initial *Azolla* coverage to grow, increased the fraction of the urea-N recovered in the soil-plant system even further. With as little as one-tenth of the water surface initially covered with *Azolla*, losses of applied N were held down to around 10%.

The increased recovery was in large part due to the uptake of urea-N by the actively growing *Azolla*, which contained 31 to 57% of the applied N after ten weeks of growth (Table 2). The ^{15}N recovery rates by *Azolla* averaged 47% and are double those recorded for the 1992 experiment of 24%. The major difference between the experiments was the initial coverage, which was 100% for the 1992 experiment as compared to less than 60% in 1993. As a result, the *Azolla* growth rate in 1993 was clearly higher than in the previous year, leading to a much higher demand for N. In fact, in the 1992 experiment, the application of urea significantly restricted *Azolla* growth from 8.1 g pot^{-1} in the absence of urea to 7.6 g pot^{-1} at 30 kg N ha^{-1} and 6.5 g pot^{-1} at 60 kg N ha^{-1}, respectively.

The effectiveness in N uptake by fast-growing *Azolla* was further confirmed by the ^{15}N recovery when inoculation at as low a rate as 172 kg fresh material ha^{-1}, referred to in Figure 3. In this case, the application of both *Azolla* and urea gave recoveries of 13% after ten weeks of growth. The delayed application doubled this recovery to 25.4% whereas *Azolla* application two weeks prior to urea application raised the recovery level by *Azolla* to 46%. Apparently, the conservation of applied N by incorporation into the *Azolla* biomass can be substantially increased if the time between *Azolla* application and urea application can be extended.

The increasing *Azolla* coverage led to an increase in N recovery by the rice tops and roots (Table 2). The delayed urea application proved particularly advantageous. The rice tops recovered 8% more on the delayed application in the absence of *Azolla* and, on average, 13% more when an *Azolla* cover was in place. In the presence of an actively growing *Azolla* population, the underlying soil was a poor competitor for the applied N as is reflected in the sharp decrease in the recovery of applied N in the soil. On the basis of the 1993 experiment it appears that as little as 1 to 2 ton fresh *Azolla* per ha at transplanting suffices to practically eliminate NH_3 volatilization if urea is applied at least one week after transplanting.

Dry matter production

The application of urea caused a reduction in *Azolla* growth. As a result, the amount of N present in the harvested *Azolla* in 1992 was significantly reduced from 0.30 g pot^{-1} to 0.26 g pot^{-1} (30 kg N ha^{-1}) and 0.22 g pot^{-1} (60 kg N ha^{-1}). This reduction is equivalent to approximately half the N applied in the form of urea. In addition, 0.02 and 0.04 g of the *Azolla* N pot^{-1} was derived from the applied urea at the low and high application rate, respectively. These two factors combined reflect a reduction in nitrogen fixation by *Azolla* offsetting 75% of the applied N. In the 1993 experiment the total N accumulation in *Azolla* was reduced from 0.35 to 0.31 g N pot^{-1} by the application of 40 kg N ha^{-1}, of which 0.05 g was derived from fertilizer N thus off-setting 90% of the applied N.

Despite these poor net gains in nitrogen, the rice yields for the 1992 experiment showed a significant rice dry matter response to urea-N from 7.6 g pot^{-1} for the control to 16.6 g pot^{-1} at 30 kg N ha^{-1} and 22.9 g pot^{-1} at the highest N rate. The Styropor cover did not significantly affect matter production when compared with the uncovered pots, consistent with the P_{NH_3}-levels observed in the floodwater (Fig. 2) and the resulting losses of fertilizer N (Table 1).

Table 3. Dry matter production of rice 7 weeks after transplanting (1992)

N rate	No Azolla	With Azolla	Effect of N × Azolla
(kg ha^{-1})	(g pot^{-1})	(g pot^{-1})	(g pot^{-1})
0	5.97	11.60	0
30	13.30	20.72	1.79
60	18.12	27.68	3.93

The effect of *Azolla* on dry matter production of rice is summarized in Table 3. A significant N-fertilizer effect is evident with and without *Azolla*. The presence of *Azolla* significantly increased the dry matter production of rice, and more so with increasing levels of applied urea. This synergistic effect on rice dry matter production is consistent with the higher increased fertilizer recovery rates of fertilizer N (Table 1) and amounted to 1.79 g pot^{-1} (9.5%) at the 30 kg N rate and 3.93 g pot^{-1} (17%) at the 60 kg N rate (Table 3). This interaction was significant only at a level of $p < 0.1$ and might not be noteworthy if it were not reinforced by the effect of *Azolla* coverage on total nitrogen uptake by rice. Here, the presence of *Azolla* yielded a significant ($p < 0.05$) positive synergistic effect of 19% and 31% of N uptake at the 30 and 60 kg N ha^{-1} rates, respectively.

Whereas in 1992 the effect of *Azolla* without urea significantly increased early rice growth, the 1993 experiment does not show this effect (Fig. 4). Contrary to 1993, the *Azolla* in 1992 fully covered the water surface from the onset. Any further growth most likely caused layering of *Azolla* and dying-off of part of the population. The decaying *Azolla* would thus be releasing N to the standing rice plants. In 1993, however, ample space was available for *Azolla* to expand without layering and to retain its vigour, preventing any transfer of fixed N_2 to the young rice plants.

Having been planted in late summer, the 1993 rice plants grew less vigorous with a control yield of only 57% (tops and roots) of the 1992 experiment. The analysis of variance revealed a strong effect of nitrogen management, with yields averaged for all *Azolla* treatments increasing from 3.0 g pot^{-1} without N to 6.7 g pot^{-1} with basally applied N and 9.2 in the case of delayed application of N (Fig. 4). A weaker but significant effect ($p < 0.05$) was found for the presence of *Azolla*. Given the absence of any yield increase due to

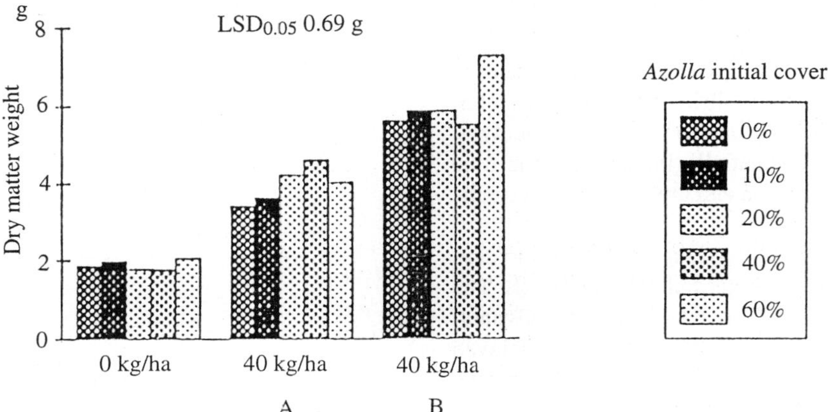

Fig. 4. Dry matter weights of rice shoots after 10 weeks of growth (1993). 40 kg ha^{-1} urea-N = 100 mg N pot^{-1}. A = urea application 1 day after transplanting. B = urea application 1 week after transplanting.

Table 4. Shoot dry weight of rice after 7 weeks of plant growth as affected by the rate and timing of *Azolla* and urea application (1994)

Inoculation	Timing[a]	Rate	Timing	Shoot yield
Azolla (ton ha^{-1})		Urea (kg ha^{-1})		(g pot^{-1})
No *Azolla* + P	0 dat	0	-	0.5 g
		40	7 dat	2.6 e
		40	14 dat	2.3 ef
		40	21 dat	2.8 e
	7 dbt	0	-	0.5 g
		40	7 dat	3.0 de
		40	14 dat	3.0 de
		40	21 dat	2.8 e
Azolla + P	0 dat	0	-	1.6 f
1 ton		40	7 dat	4.4 bc
		40	14 dat	4.3 bc
		40	21 dat	3.8 cd
	7 dbt	0	-	2.1 ef
		40	7 dat	5.6 a
		40	14 dat	5.7 a
		40	21 dat	4.8 ab
LSD				0.2

[a]dbt = days before transplanting
dat = days after transplanting.

Azolla in the absence of applied urea, this *Azolla* effect is likely related to the conservation of applied N.

The dry matter production in the 1994 experiment included the best performing combinations of *Azolla* and urea management surmised from the previous experiments. The analysis of variance of this experiment confirms the advantage of combining urea and *Azolla* (Table 4). Whereas there was the expected significant response to applied N, the timing of N application did not significantly affect dry matter production, with or without *Azolla* present. The presence of *Azolla* also gave a highly significant increase in rice growth. More importantly, the *Azolla* × N interaction was highly significant ($p = 0.0001$) in this experiment, reflecting the greater response to urea-N in the presence of *Azolla* than when *Azolla* was absent. Most likely this synergism is due to the conservation of applied N demonstrated in the earlier experiments. Applying the equivalent of 1 ton ha^{-1} of *Azolla* one week before transplanting allowed the *Azolla* to cover the water surface entirely ($\sim$ 10 ton ha^{-1}) before transplanting and increased its effect on rice growth (Table 4).

Conclusions

There are many practical problems associated with the growth and use of *Azolla* such as disease control, light sensitivity, P fertilization, water management, labour cost and difficulties of incorporation, particularly when rice is randomly planted or direct seeded. These constraints have prevented widespread adoption and are the cause of drastic reduction in the use of *Azolla* in Vietnam and China (Kulasoorya, 1991). Yet, literature data indicate that yield reductions in mixed rice/*Azolla* crops are rare, while mixed crops with a low basal dosage of applied N show a yield advantage of around

1 ton ha^{-1} (Steinhauser, 1992). For *Azolla*/rice culture to survive, the benefits will have to far outweigh the costs and labour inputs. Associated benefits through the introduction of fish in the system which is currently promoted in China may help overcome some of the constraints to *Azolla* adoption (Van Hove, 1989).

Unfortunately, in most countries *Azolla* has been looked upon as a substitute for mineral N only, thus requiring relatively large quantities of *Azolla* to be applied. The potential benefits of *Azolla* in terms of increased efficiency of applied N have only recently been recognized and have so far been poorly studied. Villegas (1985) demonstrated the reduction in the rate of volatilization due to an *Azolla* mat under laboratory conditions. Kroeck *et al.* (1988a, b) later reported the prevention of an algae-induced pH rise and a reduction in water temperatures under *Azolla* cover.

Early plant growth to a large extent determines the yield potential of the rice crop (Vlek *et al.*, 1979). The greenhouse research we reported here confirms that early rice growth can be increased with *Azolla* mixed cropping combined with the application of urea. Even relatively modest quantities of inoculated *Azolla* appear effective in increasing the efficiency of applied urea-N, thus substantially reducing land and labour requirements. However, inoculation with small quantities of *Azolla* alone was not effective as a substitute for a basal application of urea.

The ^{15}N balance studies in our experiments clearly show that the presence of *Azolla* leads to a reduction in N losses and an increased recovery of applied N by rice. The quantities needed for inoculation decrease with increasingly delayed urea application. On the other hand, *Azolla* can lock up a substantial fraction of the applied N, with the amounts dependent on whether a full coverage was established at the time of urea application (24%) or not (48%). Moreover, there is a reduction in nitrogen fixation by *Azolla* when fertilizer N is added. Whether the reduced BNF would cause a yield loss in later stages of rice growth and counteract the early gains in fertilizer efficiency remains to be determined. Further information is also needed on the fate of fertilizer N tied up in *Azolla* as the growing season progresses.

Greenhouse data alone are not adequate to establish the optimum management such as inoculation rates of *Azolla* and the time of urea application to increase the efficiency of the applied N. Given the promising results in these greenhouse experiments, it seems justified to initiate a field research program to confirm the benefits of the integrated management of N in the flooded rice/*Azolla* system. Such a program should attempt to quantify the benefits of this approach while simultaneously searching for means of minimizing the associated cost. Therefore, management systems should be tested that are both socially acceptable and affordable to the farmer by avoiding, to the extent possible, the need for 1) *Azolla* nurseries, 2) the transfer of large quantities of *Azolla*, and 3) *Azolla* incorporation.

References

Bremner J M and Mulvaney C S (1992) Nitrogen-total. *In*: Page A L (ed.), Methods of Soil Analysis, Agronomy No 9, Part 2. Am. Soc. Agron., Madison, Wisconsin

De Datta S K, Fillery I R P and Craswell E T (1983) Results from recent studies on nitrogen fertilizer efficiency in wetland rice. Outlook Agric. 12: 125–134

Denmead O T, Freney J R and Simpson J R (1983) Dynamics of NH$_3$ volatilization during furrow irrigation of maize. Soil Sci. Soc. Am. J. 47: 618

Fillery I R P and Vlek P L G (1986) Reappraisal of the significance of NH$_3$ volatilization as a N loss mechanism in flooded rice fields. Fert. Res. 9: 79–98

Kroeck T, Alkämper J and Watanabe I (1988a) Effect of an *Azolla* cover on the conditions in floodwater. J. Agron. Crop Sci. 161: 185–189

Kroeck T, Alkämper J and Watanabe I (1988b) Temperature regime of *Azolla* under rice. J. Agron. Crop Sci. 161: 316–321

Kulasoorya S A (1991) Constraints for the widespread use of *Azolla* in rice production. *In*: Polsinelli, Matarassi and Vincenzini (eds.) Nitrogen fixation. Kluwer Acad. Publ., Dordrecht, The Netherlands

Kumarasinghe K S and Eskew D L (1993) Isotopic Studies of *Azolla* and Nitrogen Fertilization of Rice. Kluwer Acad. Publ. Dordrecht, The Netherlands

Kumarasinghe K S, Zapata F, Kovacs G, Eskew D L and Danso S K A (1986) Evaluation of the availability of *Azolla* and urea to rice using ^{15}N. Plant Soil 90: 293–299

Manna A B and Singh P K (1989) Rice yields as influenced by *Azolla* N$_2$ fixation and urea-N fertilization. Plant Soil 114: 63–68

Reineking A, Langel R and Schikowski J (1993) ^{15}N, ^{13}C-on-line measurements with an elemental analyser (Carlo-Erba, NA 1500), a modified trapping box and a gas isotope mass spectrometer (Finnigan, MAT 251). Isotopenpraxis Environ Health Stud. 29: 169–174

Singh A L and Singh P K (1986) Comparative studies on different methods of *Azolla* utilization in rice culture. J. Agric. Sci., Camb. 107: 273–278

Steinhauser C (1992) Auswirkungen verschiedener Gründüngungsverfahren mit *Azolla* auf Reiserträge in Asien. (The effect of different green manure systems with *Azolla* on rice-yield levels in Asia). M. Sc. Thesis. Inst. Agron. Tropics Univ. Goettingen, Germany

Tel-Or E, Rozen A, Ofir Y, Kobiler D and Schönfeld M (1991) Metabolic relations and intercellular signals in the *Azolla-Anabaena* association. Isr. J. Bot. 40: 171–181

Van Hove C (1989) *Azolla* and its multiple uses with emphasis on Africa. FAO, Rome, Italy

Van Hove C, Diara H F and Godard P (1983) *Azolla* in West Africa. WARDA Publ., Bouaké, Côte d'Ivoire

Villegas G (1985) Effect of *Azolla* cover on nitrogen in flooded Maahas clay. M. Sc. Thesis, University of the Philippines, Los Banos, Laguna, Philippines

Vlek P L G, Hong C W and Youngdahl L J (1979) An analysis of N nutrition on yield and yield components for the improvement of rice fertilization in Korea. Agron. J. 71: 829–833

Vlek P L G and Craswell E T (1981) NH_3 volatilization from flooded soils. Fert. Res. 2: 247–259

Vlek P L G and Byrnes B H (1986) The efficacy and loss of fertilizer N in lowland rice. Fert. Res. 9: 131–147

Vlek P L G and Fillery I R P (1984) Improving nitrogen efficiency in wetland rice soils. The Fertilizer Society, London

Vlek P L G, Fugger W and Biker U (1992) The fate of fertilizer N under *Azolla* in wetland rice. Proc. 2nd ESA Congress, Warwick Univ. UK

Watanabe I and Liu C C (1992) Improving nitrogen-fixing systems and integrating them into sustainable rice farming. Plant Soil 141: 57–67

Watanabe I (1987) Summary report of the *Azolla* programmes of the International Network on Soil Fertility Evaluation for Rice. *In*: *Azolla* Utilization: 197–205. IRRI, Los Banos, Philippines

Watanabe I, Lee K K, Alimagno V B, Sato M, Del Rosario D C and De Guzman M R (1977) Biological nitrogen fixation in paddy field studies by *in-situ* acetylene-reduction assays. IRRI Res. Pap. Ser. 3: 1–16

Watanabe I, Bai K Z, Berja N S, Espina C R, Ito O and Subidhi R P R (1981) The *Azolla-Anabaena* complex and its use in rice culture. IRRI Res. Pap. Ser. 69:

Fertilizer Research **42**: 175–184, 1995.
© 1995 *Kluwer Academic Publishers. Printed in the Netherlands.*

Downward movement of nitrate and ammonium nitrogen in a flatland ultisol

Fredericka M. Deare[1], N. Ahmad[1] & T.U. Ferguson[2]
[1]*Department of Soil Science, Faculty of Agriculture, The University of the West Indies, St. Augustine, Trinidad and Tobago, West Indies;* [2]*Department of Crop Science, Faculty of Agriculture, The University of the West Indies, St. Augustine, Trinidad and Tobago, West Indies*

Key words: ammonium-N, leaching, mulching, nitrate-N, okra, tillage, urea

Abstract

Three field trials were conducted in the early and late rainy season on a Piarco Series soil (*Aquoxic Tropudults*) in Central Trinidad to monitor downward movement of NH_4^+-N and NO_3^--N under flat-tilled and ridge-tilled conditions as affected by mulch application. The first experiment was carried out in the early rainy season under bare-fallowed conditions, while okra (*Abelmoschus esculentus*) was the test crop used in the two remaining trials, which were conducted during the early and late rainy season periods. The plots were fertilized with urea seven days after crop emergence and 14 days after land preparation for the cropped and bare-fallowed experiments respectively. Soil samples were collected weekly after fertilizer application and analyzed for NH_4^+-N, NO_3^--N and soil water.

Urea application increased NH_4^+-N levels in the soil and NH_4^+-N was the dominant inorganic N form observed for the first four weeks after fertilization. Mulch application decreased NH_4^+-N and NO_3^--N soil levels. Ridging the soil increased downward movement of NH_4^+-N and NO_3^--N. Under bare-fallowed conditions, downward movement of NH_4^+-N was noted in the 30 to 45 cm soil layer at seven days after fertilization, while under cropped conditions its movement was restricted to the 15 to 30 cm layer within the same period. In bare-fallowed soil, increased NO_3^--N and its downward movement was noted after four weeks of fertilization. In the cropped soil, downward movement of NO_3^--N was observed one week after fertilization in the early rainy season and three weeks after fertilization in the late rainy season.

Introduction

Farmers often use systems of seedbed preparation and planting based on their experience and traditions that produce the best results under local conditions. However, these systems may not always be sustainable and may lead to excessive nutrient loss. Under flatland conditions, farmers tend to use "conventional tillage" practices which include ploughing, rotovating or harrowing. The crop is usually planted on the flat land after tillage or the tilled land is laid out in ridges and furrows and the crop planted on the ridges.

Flat-tilled and ridge-tilled systems have inherent differences and produce varying effects on soil and water conservation, nutrient movement and nutrient availability. Flat cultivated soils are less aerated than ridge-tilled soils and may be more compacted and are generally less drained. Therefore, soils are generally ridged to improve drainage. However, ridging may increase soil and overland movement of nutrients (Deare, 1993).

The cultural practice of mulching has been recommended to reduce soil erosion and the resultant nutrient losses. However, mulching increases water infiltration and conductivity through the soil and as a result may lead to increased downward movement of nutrient ions, particularly NO_3^--N, out of the rooting zone. Thus, the studies of Fanning & Carter (1963) showed that cotton bur mulch on a ridge and furrow system resulted in leaching of soluble salts in a clay loam to at least 75 cm under rainfed, temperate conditions, during the first five months. Because of the great potential for N leaching losses in the humid tropics, similar studies need to be carried out.

Most of the studies conducted in the tropics on downward movement of soil N have been restricted to flat-tilled soil (Arora & Juo, 1982; Chesney, 1967; Nkrumah, 1982; Omoti *et al.*, 1983). This study reports on N transformations and movement on a flat and ridged tilled Ultisol under bare and cropped conditions.

Materials and methods

The experimental site was located on a Piarco Series soil (*Aquoxic Tropudults*), at the Sugarcane Feeds Centre, Longdenville in Central Trinidad. The soil is the most pedologically developed terrace soil in Northern Trinidad and is characterized by extreme clay migration from the top 30 cm and accumulation of the clay at lower depths (Ahmad & Gumbs, 1978). This feature leads to restricted internal drainage. Table 1 shows the particle size distribution and chemical features of the soil. Field capacity for the surface soil ranged between 0.264 and 0.147 by mass, permanent wilting point 0.093 and 0.147, and available water 0.134 and 0.171.

The study comprised of three field experiments. The first experiment was carried out under bare-fallowed conditions during the early rainy season (between August and September, 1989). In the two remaining experiments, the plots were cropped with okra (*Abelmoschus esulentus*), variety Clemson Spineless. They were conducted during the early and late rainy season, between July and September, 1990 and October and January, 1989, respectively.

Treatments and experimental design

The field was brush-cut, disc-ploughed and rotovated. Eight randomly assigned plots were ridge-tilled. Ridges were 0.75 m apart with a sideslope gradient of approximately 23%. The remaining plots were levelled. The overall land gradient was approximately two percent.

The following treatments were tested: i. no mulch, flat-tilled; ii. mulched, flat-tilled; iii. no mulch, ridge-tilled and iv. mulched, ridge-tilled. The experiments were arranged in a randomized block and each treatment was replicated four times.

Each plot was fertilized with $(NH_2)_2CO$ (112 kg N ha^{-1}). The fertilizer was banded at the side of the ridge to a depth of 4 to 5 cm. The bagasse mulch (6.0

Table 1. Chemical and physical characteristics of the soil

Soil depth (cm)	pH (H$_2$O)	Chemical properties[a]						CEC (cmol kg^{-1})	Exchangeable cations (cmol kg^{-1})				Particle size distribution				Texture[b]
		Exch.Al (cmol kg^{-1})	O.M. (%)	T-C (%)	T-N (%)	C/N Rato	Av.P. (mg kg^{-1})		Ca	Mg	K	Na	Coarse sand (%)	Fine sand (%)	Silt (%)	Clay (%)	
0–1	4.2	1.38	2.19	1.28	0.13	9.85	29.2	4.30	1.78	0.31	0.13	0.24	4.1	68.0	13.4	14.6	Sl
1–50	3.9	1.85	–	–	0.05	–	8.2	2.58	0.55	0.12	0.11	0.10	1.8	66.4	12.3	19.1	Sl
50–89	4.0	4.00	–	–	0.03	–	5.6	4.59	0.13	0.11	0.05	0.13	1.8	62.1	13.0	23.1	Scl
89–110	4.2	6.64	–	–	0.04	–	8.2	6.31	0.06	0.14	0.07	0.09	1.4	53.5	12.9	32.2	Scl

[a]Exch. - Exchangeable; Av. - Available.

[b]Sl - sandy loam; Scl - sandy clay loam.

t ha^{-1}) was then spread on the relevant plots over the ridges and at an equivalent distance in the flat-tilled plots (Fig. 1). In the bare-fallowed experiment, plots were fertilized 14 days after land preparation. In the late rainy season trial, the crop was fertilized eight days after seedling emergence, while in the early rainy season experiment, fertilizer was added to the crop 14 days after crop emergence.

Sampling and analyses

Soil samples from the various treatments were collected weekly at 15 cm intervals up to depth of 60 cm, using a 2 cm diameter soil auger. Two samples were taken from each of three designated positions in the plot for the individual depth (Fig. 2). These six samples were mixed to form a composite sample representing the depth from which they were removed. Each hole was refilled with soil to prevent free water movement and marked to avoid subsequent sampling. Prior to sampling, the mulch was removed from the point of sampling and replaced after refilling the hole. The samples were quickly transported to the laboratory and frozen at $-13.0\,°C$ until they were analyzed.

Inorganic N was extracted with 1 N KCl in 1:10 soil:KCl ratio and extractable NH_4^+ and NO_3^- were determined by steam distillation (Keeney & Nelson, 1984).

Moisture content of the soil was determined gravimetrically for the four depths by collecting three samples weekly from each treatment and oven drying at 105 °C for 48 hours. Rainfall amount was obtained using a siphon rainfall gauge, located about 100 m from the field trial.

The data was analyzed statistically using analysis of variance (Caribbean Agricultural Research and Development Institute statistical package: ANOVA). Because of the high soil N variabilities between depths and with time, it was assumed that NH_4^+-N and NO_3^--N concentrations were logarithm normally distributed and the significant results are reported as the back-transformed means. This is in keeping with work done by Biggar (1978).

Result and discussion

Bare-fallowed conditions

During the experiment, total rainfall was 103.5 mm and ranged from 0.3 mm to 25.4 mm per day, with

Table 2. Changes in soil water of the flat and ridged plots at different days after fertilizer application

Treatment	Soil water content (g g^{-1})			
	Days after fertilizer application			
	7	14	21	28
Flat	0.13	0.15	0.14	0.16
Ridge	0.09	0.09	0.09	0.10
LP^a	*** ^{b}			
T × LP	*			
No mulch	0.10	0.13	0.14	0.13
Mulch	0.12	0.11	0.10	0.10
MAp	*			
T × MAp	*			

^{a}LP: Land Preparation; MAp: Mulch application; T: Time.
^{b} * - significant ($p<0.05$); ** - significant ($p<0.01$); *** - significant ($p<0.001$); NS - not significant.

maximum rainfall recorded at 18 days after fertilizer application (DAF). Generally, soil water was greater in the flat-tilled soil compared to the ridge-tilled treatment (Table 2). Ridging the soil improved drainage by allowing greater runoff of water from the soil. With increased overland and downward water movement, less water would be retained in the upper layers of the soil. The absence of mulch led to the retention of more water in the soil (Table 2); the high water absorptive capacity of bagasse would account for lower rainfall actually reaching the soil especially at low rainfall intensities. Similar decreases in soil water content were obtained by ap Griffith (1952) when a bagasse mulch was used.

Inorganic-N distribution and movement in the soil
There were higher levels of NH_4^+-N at the initial stage of the experiment (7 DAF), but by 14 DAF there were substantial losses with time. Mulching the flat-tilled soil generally led to lower levels of NH_4^+-N (Fig. 3), particularly in the surface soil (0 to 15 cm) where the soil without mulch gave about 31% higher amounts of NH_4^+-N than the bare soil. This result was probably due to volatilization and immobilization of applied N. Some researchers have showed that volatilization losses increased when urea was applied to soils covered with plant residues (Khan & Rashid, 1971; Meyer et al., 1961). Also, Freney et al. (1992) reported that there was substantial urease enzyme activity in sug-

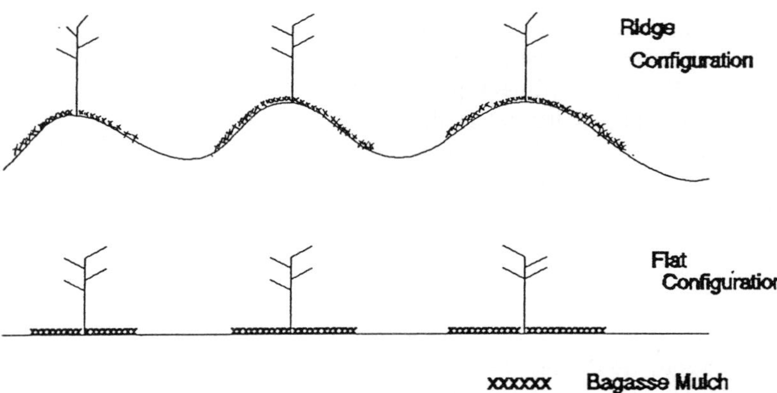

Fig. 1. The position of the bagasse mulch in the plots.

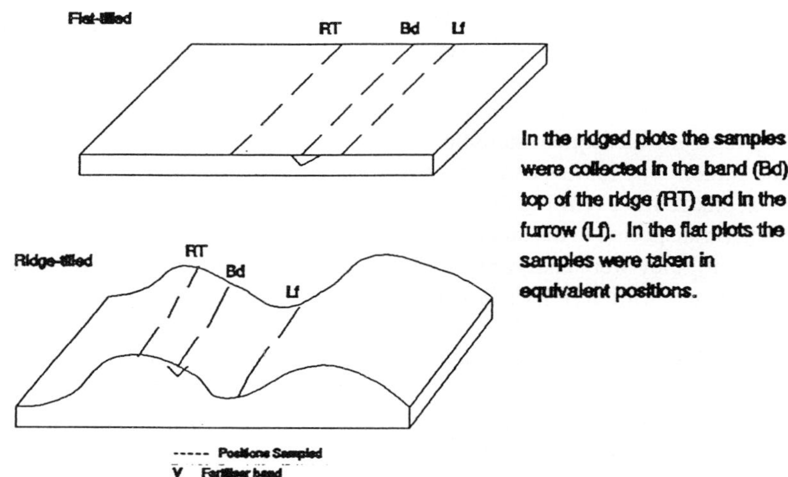

Fig. 2. The schematic representation of the sampling positions on the flat-tilled and ridge-tilled soil

ar cane residue and this would give a higher rate of $(NH_2)_2CO$ hydrolysis and so increased volatilization losses from the Ultisol. Studies have demonstrated that the application of carbonaceous materials can immediately convert inorganic N to organic N (Ahmad *et al.,* 1969). Bagasse, which contains large supplies of carbon sources, between 42 to 47% cellulose and 2 to 3% sucrose (Barnes, 1974), may have led to rapid immobilization of applied N within the first seven days after fertiliser applications.

The soil NH_4^+-N concentrations before and after fertilizer application indicated the NH_4^+-N movement was rapid, as evident by high levels of NH_4^+-N in the 30 to 45 cm soil layer observed seven days after fertilization (DAF) (Fig. 4). It can also be assumed that such movement occurred in response to $(NH_2)_2CO$ application. Most of the NH_4^+-N recovered at 30 to 45 cm had disappeared seven days later (14 DAF), suggesting that further movement had occurred which could have been partly due to the low CEC of the soil.

There was a corresponding lower NH_4^+-N concentration at 0 to 15 cm at 14 DAF and this level was maintained until final sampling (28 DAF) when there was also a significant decrease and corresponding increase in NO_3^--N content. Soil NO_3^--N concentration in the ridge-tilled soil was, on average, lower than the flat-tilled soil (8.4 and 9.2 mg kg^{-1} respectively). This difference in NO_3^--N was more pronounced at 28 DAF at which time, about 45% more NO_3^--N was recovered from the flat-tilled soil than the ridged soil (Fig. 5).

Though increases in NO_3^--N were recorded in the topsoil as early as 7 DAF, high levels (a fourfold increase) were only observed at 28 DAF (Fig. 6). The

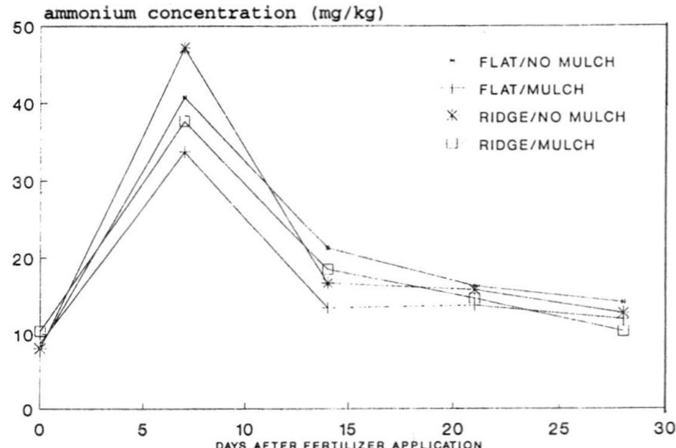

Fig. 3. Changes in soil NH_4^+-N as affected by soil surface management at different times after fertilizer application.

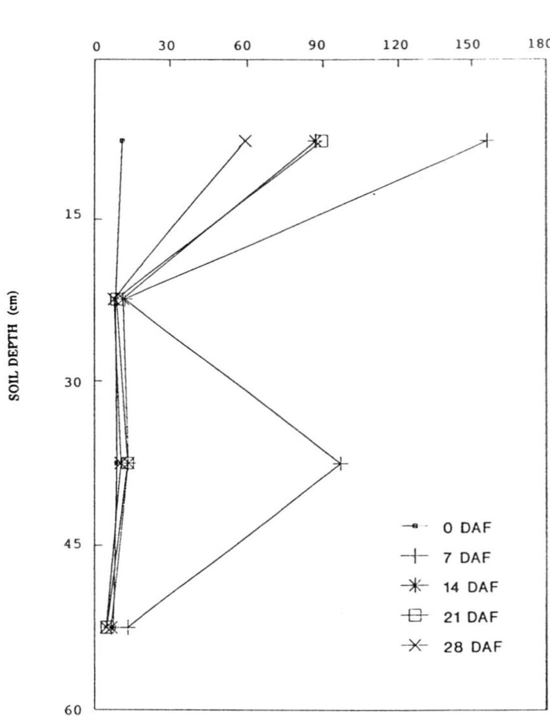

Fig. 4. Soil NH_4^+-N distribution at different times after fertilizer application.

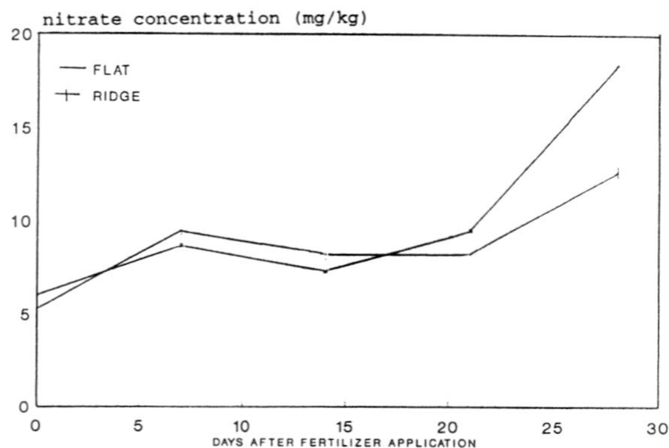

Fig. 5. Changes in soil NO_3^--N as affected by land preparation at different times after fertilizer application.

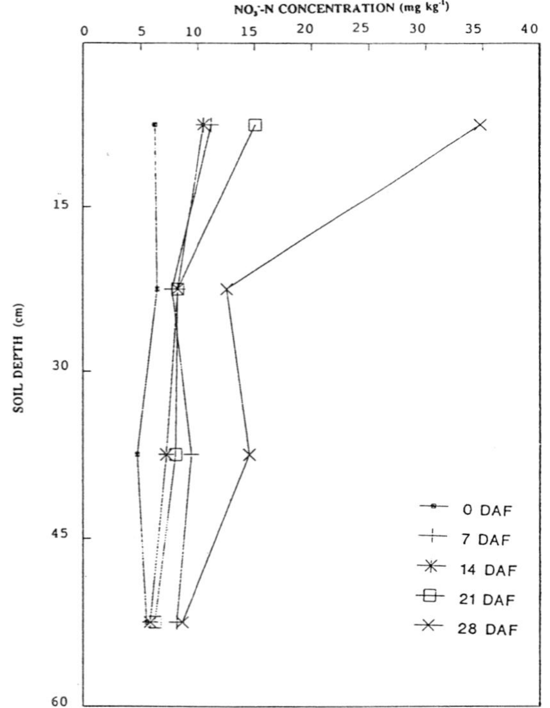

Fig. 6. Soil NO_3^--N distribution at different times after fertilizer application.

delay in the rapid build up of NO_3^--N may be as a result of high pH in the fertilizer band which would adversely affect the nitrifying bacteria and so limit the conversion of NH_4^+-N to NO_3^--N. This increase in NO_3^--N in the surface soil layer also corresponded with increased concentrations of NO_3^--N in the deeper soil layers and this accumulation may be attributed to the leaching of NO_3^--N. This suggests, therefore, that rapid leaching of NO_3^--N would be important for similar Ultisols as soon as rapid nitrification occurs.

Cropped conditions

Precipitation totalled 375.6 mm and 479.4 mm for the late and early rainy season trials, respectively. Average daily rainfall in the late rainy season was 10.4 mm with minimum and maximum values of 0.5 and 25.4 mm, respectively. In early rainy season period, daily rainfall ranged between 0.8 and 25.4 mm, with a mean daily rainfall of 12.1 mm.

During the late rainy season experiment, soil water content with time ranged between 0.186 g g^{-1} and 0.152 g g^{-1}, with maximum soil water noted at initial sampling (7 DAF) and thereafter, reductions were observed with time. In the early rainy season, however, soil water contents fluctuated throughout the period of the investigation, ranging between 0.151 g g^{-1} and 0.188 g g^{-1}, reaching the maximum at 21 DAF. In both experiments, the topsoil (0 to 15 cm) was generally the wettest layer, with the surface layer of the flat-tilled soil having more water than the ridged soil (Table 3). This meant, therefore, that flat-tillage allowed for greater water entry and water retention in the surface layer. While in the late rainy season mulching did not significantly affect the water distribution in the soil, it led to significantly higher water retention in the surface soil in the early rainy season (Table 3).

Inorganic-N distribution and movement in the soil
As in the bare-fallowed soil, NH_4^+-N was the predominant form of inorganic N in the cropped surface soil during the first four weeks after fertilizer application. In both and late and early rainy season trials, increased NH_4^+-N levels were detected at the 15 to 30 cm soil layer at 7 DAF (Fig. 7). However, increased NH_4^+-N levels beyond the 30 cm depth was observed at different times and so in the late rainy season, NH_4^+-N increase was noted 28 DAF, while in the early rainy season, it was detected at 14 DAF. The rainfall patterns during the time of the experiments can explain the differences obtained. In the initial trial (late rainy season), 165.3 mm of rainfall was recorded within the first 28 DAF, with an average of 5.9 mm day^{-1}. However, in the second trial, within the first 14 days after $(NH_2)_2CO$ application 150.1 mm day^{-1} of rain was received, resulting in an average daily rate of 10.7 mm day^{-1}. With this increased amount of rainfall in the early rainy season study, NH_4^+-N movement to the lower layers could have occurred in a shorter time.

In late and early rainy season periods, mulching of the flat-tiled soil led to the lowest levels of NH_4^+-N detected within the first 30 cm soil zone (Table 4). However, in the late rainy period, more NH_4^+-N was retained in the mulched, ridged topsoil (0 to 15 cm) while, in the early rainy season, the no mulch, flat-

tilled plots had the highest NH_4^+-N concentration in the surface soil.

The reduction of NH_4^+-N content in the surface soil with time was accompanied by increases in NO_3^- concentration in the soil. Rapid increases in NO_3^- content were observed one week after fertilizer application in the early rainy season and three weeks after fertilization in the late rainy season (Fig. 8). In the second experiment which was conducted at the beginning of the rainy season, higher NO_3^--N levels were present prior to fertilizer application than in the initial experiment which was carried out in the late rainy season. This observation was in agreement with results obtained by Wild (1972a, b). Mineralization and subsequent nitrification of soil organic matter at the start of the rains may be responsible for the higher levels detected.

In the initial experiment, maximum NO_3^--N concentration in the uppermost soil layer was observed at 21 DAF and resulted in a five-fold increase in NO_3^--N levels compared to the unfertilized soil at 7 DAF. In the second trial, the NO_3^--N concentration reached a maximum 42 DAF, though there were no significant differences from 28 DAF (Fig. 8).

The increases in NO_3^--N level at the greater soil depths with time suggest that this was due to leaching of NO_3^--N. In the late rainy season, while NH_4^+-N was leached to the 15 to 30 cm soil depth within seven days after $(NH_2)_2CO$ application, NO_3^--N movement was detected in the 45 to 60 cm soil layer at the same time. Lower concentrations at seven days later (14 DAF) also suggest movement beyond the measured areas of the experiment, while further movement was observed at 21 DAF. In the early rainy season trial, downward movement of NO_3^--N was not observed until 28 DAF. Compared to the first trial, there were high NO_3^--N levels throughout the profile at the beginning of the experiment. Subsequent reductions indicate possible movement of NO_3^--N beyond the measured limits.

On average, ridging the field led to significantly higher concentrations of NO_3^--N in the surface soil which was probably as a result of improved aeration and lower soil water content (Table 5). In the first trial, the absence of the mulch in the ridged plots resulted in 70% of the NO_3^--N occurring in the first 30 cm of soil. On the other hand, mulching led to a greater build up of NO_3^- in the 30 to 45 cm layer, (86% of the NO_3^--N in the first 45 cm of the soil profile).

In the second experiment, mulch application led to reduced levels of NO_3^--N in the soil profile, giving an average of 15.1 mg kg^{-1} and 13.2 mg kg^{-1} for the no mulch and mulched plots, respectively (Table 5).

Table 3. Soil water distribution in the late and early rainy season experiments as affected by soil surface management

| Soil depth | Soil water content (g g^{-1}) | | | | | |
| | Late rainy season | | Early rainy season | | | |
(cm)	Flat	Ridge	Flat	Ridge	No mulch	Mulch
0–15	0.185	0.166	0.186	0.175	0.176	0.185
15–30	0.165	0.163	0.170	0.173	0.173	0.170
30–45	0.162	0.156	0.167	0.164	0.167	0.164
45–60	0.163	0.162	0.168	0.166	0.170	0.164
Depth (Dp)		*	Dp	*		
LP		*	LP	*		
			MAp	*		

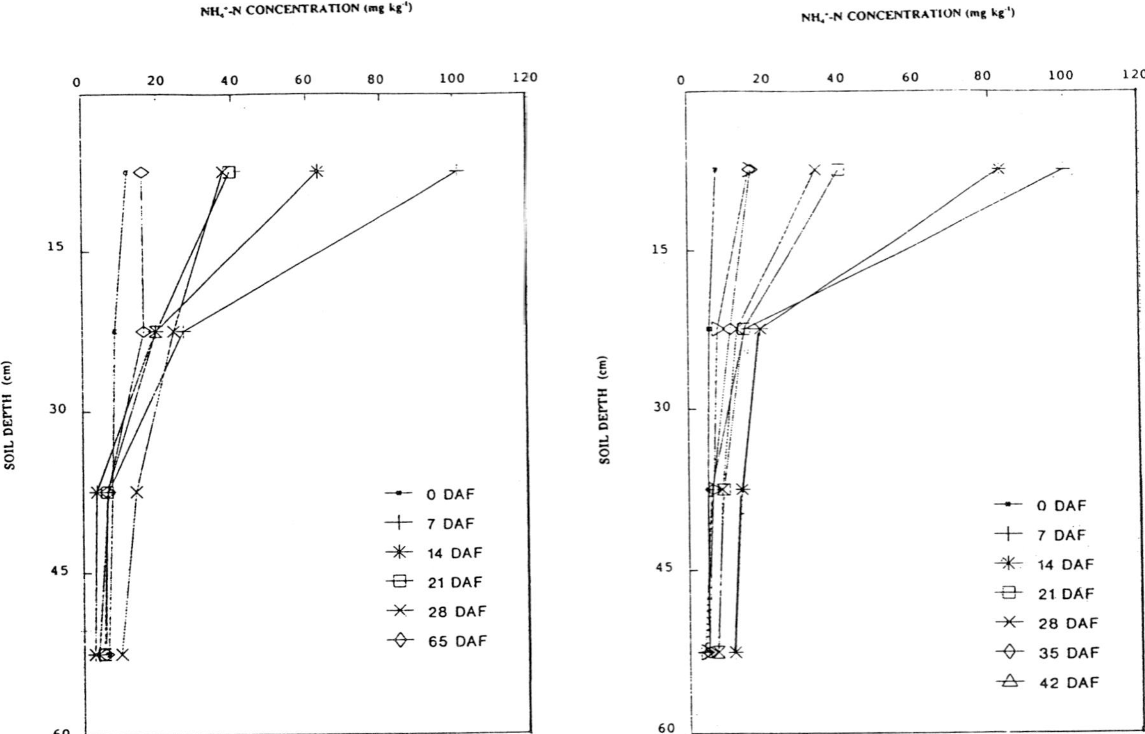

Fig. 7. Soil NH$_4^+$-N distribution at different times after fertilizer application.

About 51% more NO$_3^-$-N was found in the surface soil of the unmulched plots compared to the mulched treatments. The NO$_3^-$-N at 15 to 30 cm soil depth for the mulched plots was about 36 % lower than on the soil surface, suggesting that NO$_3^-$-N may be held within the soil aggregates at the surface of the mulched soil.

In the early and late rainy season trial, N uptake was significantly ($p < 0.05$) affected by land preparation. The mean total uptake of N values for okra grown in the flat-tilled and ridge-tilled treatments were 21.6 kg ha^{-1} and 59.7 kg ha^{-1}, in the late rainy trial and 43.8 kg ha^{-1} and 83.6 kg ha^{-1} in the early rainy season, respectively.

Table 4. Distribution of ammonium-N in the soil profile as affected by land preparation and mulch application

Soil depth (cm)	NH$_4^+$-N concentration (mg kg^{-1})							
	Late rainy season				Early rainy season			
	Flat No mulch	Flat Mulch	Ridge No mulch	Ridge Mulch	Flat No mulch	Flat Mulch	Ridge No mulch	Ridge Mulch
0–15	36.0	31.3	31.3	43.3	41.0	21.0	28.4	33.7
15–30	15.3	16.2	24.7	18.1	13.2	12.1	12.2	8.8
30–45	6.2	7.2	7.5	7.2	8.2	7.4	8.4	6.6
45–60	6.1	4.8	5.5	5.5	5.3	5.6	6.8	6.4
Dp			***		Dp		***	
LP × MAp × Dp			**		LP × MAp × Dp		**	

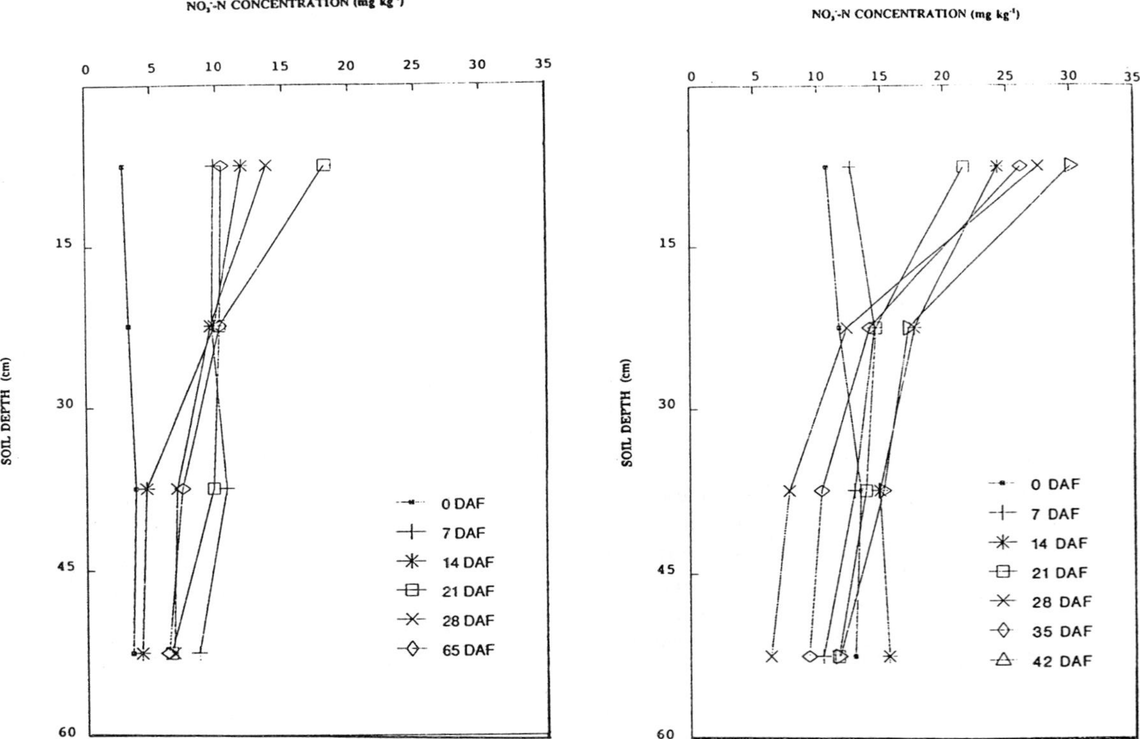

Fig. 8. Soil NO$_3^-$-N distribution at different times after fertilizer application.

Conclusion

The application of $(NH_2)_2CO$ increased NH$_4^+$-N levels in the soil and this was the dominant form of inorganic N for the first four weeks after fertilization. The management practices had varying effect on the quantities of NH$_4^+$-N and NO$_3^-$-N in the soil. Under bare-fallowed conditions, mulch application resulted in lower levels of NH$_4^+$-N, particularly in the flat-tilled soil, and this was probably due to increased hydrolysis and subsequent volatilization and/or immobilization. On the other hand, under cropped conditions, differences in NH$_4^+$-N content depended on the period of crop cultivation. In the early rainy season, tillage did not affect

Table 5. Distribution of nitrate-N in the soil profile as affected by land preparation and mulch application

Soil depth (cm)	NO_3^--N concentration (mg kg^{-1})					
	Late rainy season				Early rainy season	
	Flat No mulch	Flat Mulch	Ridge No mulch	Ridge Mulch	No mulch	Mulch
0–15	7.2	6.3	15.3	13.2	16.7	25.2
15–30	5.4	5.0	14.5	11.8	13.2	16.2
30–45	6.5	3.9	7.6	10.6	12.0	12.6
45–60	6.7	4.7	5.5	5.6	11.3	10.2
	Dp	***			Dp	***
	Dp × LP	*			Dp × LP	NS
	Dp × MAp	NS			Dp × MAp	*
	MAp × LP	**			MAp × LP	NS
	Dp × MAp × LP	***			Dp × MAp × LP	NS

NH_4^+-N levels in the soil, while mulching decreased it, possibly due to increased volatilization. In the late rainy season, NH_4^+-N levels in the ridge-tilled were higher than for the flat-tilled sites, particularly under mulched conditions.

Because of the low cation retention capacity of the soil, downward movement of NH_4^+-N was detected along with NO_3^--N. Under bare-fallowed conditions, substantial leaching of NH_4^+-N was observed beyond the root zone of the crop at 7 DAF (roots were normally located within the first 30 cm of the soil). By contrast, NO_3^--N leaching was detected at 28 DAF, at the time when rapid accumulation of NO_3^--N occurred. These findings suggest that in situations where N fertilizers are applied to such soils prior to planting or immediately after planting, N is likely to be lost in the form of NH_4^+-N before it can be used by the crop. Under cropped conditions, $(NH_2)_2CO$ was added when the soil was partially protected by the vegetative cover and this was associated with less leaching of NH_4^+-N. Its movement was mainly restricted to the 15 to 30 cm soil zone, while nitrate-N leaching, was detected further down the profile at an even earlier period.

References

Ahmad N & Gumbs FA (1978) Soil and Land Use Study of the Sugarcane Feeds Centre Project site, Longdenville. A report (unpublished) sponsored by the University of the West Indies and the Canadian International Development Agency (CIDA)

Ahmad, Kai H & Harada T (1969) Factors affecting immobilization and release of nitrogen in soil and chemical characteristics of the nitrogen newly immobilized. III Effect of carbon sources on immobilization and release of nitrogen in soil. Soil Sci Plant Nutr 15: 252–258

Arora Y & Juo ASR (1982) Leaching of fertilizer ions in a kaolinitic ultisol in the high rainfall tropics: Leaching of nitrate in field plots under cropping and bare fallow. Soil Sci Soc Am J 46: 1212–1218

Barnes AC (1974) The Sugar Cane. Leonard Hill Books, London UK

Biggar JW (1978) Spatial variability of nitrogen in soils. In: Nielson DR & McDonald JG (ed) Nitrogen in the Environment, pp 201–211. Academic Press, New York, USA

Deare FM (1993) Pathways of nitrogen loss as affected by soil surface management of an Ultisol. Ph D Thesis, Dept. Soil Science, The Univ. West Indies, St. Augustine, Trinidad and Tobago. 282p

Chesney HA (1967) Denitrification in some Trinidad soils. M Sc Thesis, Dept. Soil Science, The Univ. West Indies, St. Augustine, Trinidad and Tobago 86p

Fanning CD & Carter DL (1963) The effectiveness of a cotton bur mulch and a ridge-furrow system in reclaiming saline soils by rainfall. Soil Sci Soc Am Proc 27: 703–706

Freney JR, Denmead OT, Wood AW, Saffigna PG, Chapman LS, Ham GJ, Hurney AP & Stewart RL (1992) Factors controlling ammonia loss from trash covered sugarcane fields fertilized with urea. Fert Res 31: 341–349

Griffith G ap (1952) The interception of rain water by a vegetable mulch. Trop Agric (Trin) 21: 50–53

Keeney DR & Nelson DW (1984) Inorganic forms of nitrogen. In: Page AL, Miller RH & Keeney DR (eds) Methods of Soil Analysis, Part 2. Chemical and Microbiological Properties. Monograph 33, pp 643–698. ASA Inc., SSSA Inc. Publ., Madison, Wisconsin, USA

Khan DH & Rashid GH (1971) Losses of nitrogen from urea in some soils of East Pakistan. Exp. Agric 7: 107–112

Meyer RD, Olson RA & Rhoades HF (1961) Ammonia losses from fertilized Nebraska soils. Agron J 53: 241–244

Nkrumah M (1982) Crop utilization of added N, movement of NO_3^--N and transformation of $(NH_2)_2CO$-N over varying seasons. M Sc Thesis, Dept. Soil Science, The Univ. West Indies, St. Augustine, Trinidad and Tobago. 83p

Omoti U, Ataga DO & Isenmila AE (1983) Leaching of losses of
    nutrients in oil palm plantations determined by tension lysimeters.
    Plant Soil 73: 365–376
Wild A (1972a) Mineralization of soil nitrogen at a savanna site in
    Nigeria. Exp Agric 8: 91–97
Wild A (1972b) Nitrate leaching under bare fallow at a site in North-
    ern Nigeria. J Soil Sci 23: 315–324

Fertilizer Research **42**: 185–192, 1995.
© 1995 *Kluwer Academic Publishers. Printed in the Netherlands.*

^{15}N isotope dilution techniques to study soil nitrogen transformations and plant uptake

D. Barraclough
Department of Soil Science, University of Reading, London Road, Reading RG1 5AQ, UK

Key words: ^{15}N, nitrogen, soil

Abstract

The use of ^{15}N as a tracer in soil/plant research is examined. The limitations of the so-called Ndff approach are discussed to show the need to consider not just the fate of the added label but also the path that was followed and the rate of the transformation. The development of ^{15}N isotope dilution techniques to determine gross rates of nitrogen transformation in soil is reviewed with some indications as to the further development of the approach.

Introduction

Labelled compounds (^{15}N, ^{13}C, ^{14}C, ^{32}P) etc. are usually described as tracers. This reflects their most common use; to follow, or trace, the fate of a particular compound in a complex system such as the human body or, in our case, the soil. In the last 25 years, ^{15}N isotope tracer techniques have increased our understanding of the soil/plant system considerably. The so-called Ndff equation (Equation 1) allows the recovery of labelled N to be determined in the plant, in the soil inorganic N pools, in soil organic matter, and, with care, in the gases given off from the soil surface. Equation (1) is robust ; it relies only on the assumption that no significant isotopic discrimination occurs during the biological transformations in the soil/plant nitrogen cycle. At the ^{15}N enrichments normally used in tracer experiments (1–10 at.%), this assumption is safe. And with the advent of easily maintained mass spectrometers and continuous flow isotope ratio techniques (see A.Barrie, this volume), it is also relatively straightforward to obtain the data necessary for Ndff calculations. However, after 25 years of productive use, it is legimitate to examine the Ndff approach critically and ask where can we go from here? What are the drawbacks in the Ndff approach? Could we get more out of ^{15}N isotope experiments?

The basic Ndff Equation

Derivation

The addition of an amount of labelled N (eitheir ammonium or nitrate) to the soil results in the formation of 'sub-pools'. One, comprising the inorganic N originally present in the soil, and any subsequently produced by the mineralization of the soil organic matter (SOM), is described as unlabelled athough it is more accurate to say it is at, or very near, the natural abundance of ^{15}N, 0.3663 at.%. The other sub-pool, the added label, usually has an ^{15}N abundance well above natural abundance. A plant taking up N from the soil can take it up from either the unlabelled or labelled sub-pool. Over a reasonable time, the N atoms taken up from the labelled pool will be a mixture of ^{15}N and ^{14}N and, unless the plant particularly favours one or other isotope, the ratio of ^{15}N to ^{14}N atoms taken up from the labelled pool will be equal to the ^{15}N abundance of that pool. Thus if the labelled pool has a ^{15}N abundance of 10%, one in every ten N atoms taken up from the pool will be ^{15}N. Similarly, of the N atoms taken up from the unlabelled pool, one in every 273 (i.e 0.3663 at.%) will be ^{15}N. The amount of N in the plant, at some time t, will be the sum of the N taken up from the labelled and unlabelled pools plus the N that was in the plant at the beginning of the experiment. Similarly, the amount of ^{15}N in the plant at some time t will be the sum of the labelled atoms recovered from the labelled and unlabelled pools plus the labelled atoms that were in the plant at the beginning of the experiment. We can express these two statements mathematically as follows.

$$N = N_1 + N_s + N_o \qquad (1)$$

where N is the total amount of N in the plant, N_1 is the N recovered from the labelled pool, N_s is the N recovered from the unlabelled (soil) pool, and N_o is the N in the plant at the beginning of the experiment. For the ^{15}N we can write

$$^{15}P \cdot N = {}^{15}L \cdot N_1 + {}^{15}S \cdot N_s + {}^{15}N \cdot N_o \qquad (2)$$

where ^{15}P is the ^{15}N abundance in the plant at time t, ^{15}L and ^{15}S are the ^{15}N abundances in the labelled and unlabelled pools respectively and $^{15}N_o$ is the ^{15}N abundance of the N in the plant at the beginning of the experiment. Equation (2) is simply equating the amount of ^{15}N in the plant with that taken up from the two pools plus what was in the plant already. Normally the ^{15}N abundance in the unlabelled soil pool and in the plant at the beginning of the experiment is 0.3663 at.%, so we can rewrite Equation (2) as

$$^{15}P \cdot N = {}^{15}L \cdot N_1 + 0.3663 \cdot N_s + 0.3663.N_0 \qquad (3)$$

In reality the ^{15}N abundance of soil inorganic N is often slightly greater than 0.366 at.% but the error introduced by taking it as 0.3663 is very small. We can express ^{15}N abundance as the ^{15}N abundance of the pool above natural abundance. This is called the ^{15}N *excess* abundance, $^{15}P^*$ and is given by (in the case of the plant)

$$^{15}P^* = {}^{15}P - 0.3663$$

or rearranging this

$$^{15}P = {}^{15}P^* + 0.3663 \qquad (4)$$

Using this substitution in Equation (3) gives

$$\begin{aligned}
(^{15}P^* + 0.3663) \cdot N = {} & (^{15}L^* + 0.3663)N_1 \\
& + (0 + 0.3663)N_s \\
& + (0 + 0.3663)N_0 \quad (5)
\end{aligned}$$

As $N = N_1 + N_s + N_o$ Equation (5) simplifies to

$$^{15}P^* \cdot N = {}^{15}L^* \cdot N_1$$

With a little rearranging this gives the basic ^{15}N equation

$$N_1 = N \cdot \frac{^{15}P^*}{^{15}L^*} \qquad (6)$$

This is the basic Ndff equation. It allows the calculation of the amount of N recovered from the labelled N added to the soil. It makes no assumptions about the relative exploitation of the labelled and unlabelled pools. The only assumption involved in the derivation is that within a pool, labelled or unlabelled, processes such as plant uptake do not discriminate between ^{15}N and ^{14}N. All the evidence to date suggests that, at the enrichments normally used in tracer experiments (1–10 at.%), this is a safe assumption. Thus Equation (6) is robust, easy to use and is applicable to all ^{15}N experimentation.

What Ndff does not tell us

Given its simplicity and robustness, it is perhaps not surprising that the Ndff approach has some drawbacks. Most centre on the fact that Ndff gives information on the origin of nitrogen in the various soil/plant nitrogen pools, but gives no information on the path taken to reach the pool and only limited information on the rate at which the nitrogen reached the pool. Specifically:

- Equation (6) gives the origin of the nitrogen; it does not give the route taken by the nitrogen in going from pool A to pool B. Thus if ^{15}NH$_4$ is used in an experiment and 30% of the N recovered in the plant is found to have come from the added label this does *not* mean that 30% of the plant N was taken up as ammonium. Nitrification might result in ^{15}N originating in the ammonium pool being taken up as nitrate. Equation (6) doesn't distinguish between this and ^{15}N taken up as ammonium because it only gives the origin of the N, not the route taken to get from A to B.
- The second reservation concerns the amount of information Equation (6) yields on the fate of the unlabelled pool in the soil, the so-called soil N pool. In fact Equation (6) gives no direct information about soil N. It is possible to obtain the recovery of soil N in the plant for example, by subtracting that recovered from the added label from the total nitrogen content of the plant. But for many of the processes we are interested in (mineralization, nitrification, uptake of ammonium and nitrate by plants) we need more information than this. Thus, if we use ^{15}N to determine the immobilization of N in the microbial biomass we need to know the total immobilization, not just that involving the added label.
- The third resevation concerns the lack of any information on the rate of the processes responsible for

transporting the ^{15}N between pools. How fast is nitrification? Does the rate of N_2O evolution from soil (a greenhouse gas) follow the rate of nitrification or the rate of denitrification?

If we are to understand the processes making up the soil/plant nitrogen cycle, we need information in addition to that we acquire via Ndff. We also need to deal in gross, not net, rates of nitrogen transformations such as mineralization and nitrification. Net rates of mineralization, for example, compound two processes, mineralization and immobilization. It appears there is a qualitative relation between the C/N ratio of the organic substrate being decomposed and whether net mineralization or immobilization occurs. If we wish to translate this qualitative relationship into a more useful generally applicable quantitative one, we are faced with a problem. Mineralization appears to occur when C/N ratios are narrow; innmobilization when they are wide. Thus, in relating net mineralization to C/N ratios we are compounding two processes which are affected in opposite directions by the driving variable we are focussing on. Clearly in this case, as in others, we need to be dealing in gross rates of transformation.

^{15}N isotope dilution

In 1954 and 1955 Kirkham and Bartholomew published a set of equations relating the rate at which the ^{15}N abundance in a labelled pool declined over time as a function of the gross rate of either mineralization (in the case of a labelled ammonium pool) or nitrification (in the case of a labelled nitrate pool)(Kirkham & Bartholomew, 1954, 1955). Partly because of their unwieldy formulation and the restricted conditions under which the equations were valid, they were little used. In recent years a number of workers extended and simplified the equations (Barraclough, 1991a; Barraclough *et al.*, 1985; Myrold & Tiedje, 1986; Nisho *et al.*, 1985), removing many of the restrictions in the original formulations. I summarise the basic steps in deriving the equation; a full derivation is available (Barraclough *et al.*, 1985)

The soil/plant nitrogen cycle is a complex web of interacting processes. Label added to one pool does not necessarily follow one path from A to B; there may be many. Thus in the example above, ^{15}N added to the ammonium pool can appear in the plant either by direct take up as ammonium, or following nitrification to nitrate and take up as nitrate. To avoid complica-

tions such as this, it is essential to perform experiments in a way that they are interpretable in terms of the nitrogen transfer between two pools, not between three or four or more. To study the main processes in the soil nitrogen cycle, mineralization, immobilization, nitrification and plant uptake, distinct procedures are followed. For mineralization, only the ammonium pool is labelled. The rate at which the ^{15}N abundance of the pool declines over time is then a function of the rate at which mineralization introduces unlabelled ammonium from the soil organic matter into the pool, and the rate at which ammonium is removed from the pool by other processes. The experiment can only continue as long as the mineralized N remains at natural abundance. To determine nitrification rates, the nitrate pool is labelled. The rate at which the pool abundance declines is now a function of the rate at which unlabelled ammonium enters the pool via nitrification. Again, the experiment can only proceed while the ammonium pool remains unlabelled.

Derivation

Consider a soil ammonium pool to which labelled ammonium has been added at t=0 to give a uniformly labelled pool with a ^{15}N excess abundance of A_o^*. The assumption that the pool is uniformly labelled is important and will be discussed later. If it is satisfied, processes such as immobilization, nitrification and plant uptake will remove ^{14}N and ^{15}N in proportion to the amounts present and will *not* in themselves alter the ^{15}N abundance of the pool. This assumption of proportional exploitation may not hold if isotopic fractionation occurs at or near natural abundance, but where label has been added is probably safe. Under these conditions, mineralization, because it introduces ammonium at natural abundance into the pool, will cause a decline in pool ^{15}N abundance. The greater the rate of mineralization, the more rapid the decline. And, although in themselves they do not alter the pool abundance, the processes removing ammonium from the pool will modify the effect of a given rate of mineralization, because they remove labelled nitrogen that is replaced, via mineralization, with nitrogen at natural abundance. Thus, the greater the rate of removal from the pool, the more rapid the decline in pool abundance for a given rate of mineralization.

If the time interval is short enough to allow all rate processes to be described by zero-order kinetics, these qualitative relationships between the rate of mineralization, the rate of removal from the pool and the decline in pool abundance, can be expressed in quantitative terms as in Equation (7) (Barraclough *et al.*, *1985*.

$$A_t{}^* = \frac{A_o{}^*}{(1 + \frac{\theta t}{C_o})^{(\frac{m}{\theta})}} \tag{7}$$

where $A_o{}^*$ and $A_t{}^*$ are the ^{15}N excess abundances in the ammonium pool at t=0 and t=t respectively, C_o is the size of the ammonium pool at t=0, θ is the rate at which the ammonium pool changes in size and m is the gross rate of mineralizaition. It describes the excess ^{15}N abundance (i.e the abundance in excess of natural abundance, 0.3663 at.%) in an initially labelled ammonium pool at time t, in terms of the initial abundance, the rate at which the pool changes size, the initial size of the pool and the gross rate of mineralization. The equation relies on the following assumptions.

- Zero-order kinetics i.e. over the period of the experiment, all rate processes occur at constant rate. This is not too restrictive an assumption. Over the ideal interval for the approach (2–7 days), most rate processes can be adequately represented by zero-order kinetics.
- Mineralized N is at 0.3663 at.%. This is more restrictive. It assumes that ^{15}N immobilized by the soil microbial biomass is not remineralized during the experiment. When working on a new soil, this should be checked by performing pool dilution experiments over varying time intervals. If remineralization is a significant problem, the gross rate of mineralization, determined from Equation (7) will appear to drop over time. The experimental interval should then be a compromise between the need to allow sufficient time for pool ^{15}N's to have changed significantly, (Davidson *et al.*, 1991) but not long enough for remineralization to become a problem. Our experience is that for temperate soils at temperatures up to 15 °C, 7 days is a reasonable maximum interval; in tropical soils, where temperatures can reach 23 °C, 3–5 days is probably safer.
- The added label must mix uniformly wiith the indigeneous soil N. This is discussed in more detail later.

Given these caveats, the experimental requirements for pool dilution are, in some ways easier than those for an Ndff experiment. What is needed is the size and ^{15}N abundance of the ammonium pool at a minimum of two times following label addition. A typical procedure would be to add the ^{15}N label to the soil and immediately take a series of t=0 samples. At a suitable interval, say 2 or 3 days, a further set of samples are taken. The exchangeable soil ammonium is extracted in 1 or 2 *M* KCl and sufficient extract is subsampled to ensure adequate nitrogen for ^{15}N/^{14}N isotope ratio analysis (20 – 50 μg N on modern mass spectrometers; 100 – 500 μg N on older machines). Using recently developed diffusion methods, (Brooks *et al.*, 1989), the subsample is incubated for 5 days in an Erlenmeyer flask with MgO (preheated to 1000 °C for 2 hours) and the evolved ammonia trapped on a glass microfibre disc acidified with 10 μl 2.5 *M* KH$_2$SO$_4$. The microfibre disc is dried over CaSO$_4$ and introduced into the mass spectrometer either via an elemental analyser (see Barrie, this volume), or after treatment with LiOBr in a conventional mass spectrometer inlet system. The ammonium concentration in the soil extract can be determined by colorimetry (flow injection or continuous flow), or by steam distillation and titration (Bremner, 1965).

Treatment of results

A pool dilution experiment will yield the size and ^{15}N abundance of the ammonium pool at at least two times following label addition. If only two samples have been taken, at t=0 and t=t, the experimental values are inserted into Equation 7, or its rearranged form (Eq.8), to obtain the gross rate of mineralization (Powlson & Barraclough, 1992).

$$m = \theta \cdot \frac{(\log A_o{}^*/A_t{}^*)}{\log(1 + \theta t/C_o)} \tag{8}$$

If more than two samples were taken, the opportunity exists to use non-linear curve-fitting techniques to fit the data to Equation (7). The example in Figure 1 was obtained using a commercially available curve-fitting package in the graphics software SigmaPlot ©.

Examples

Figure 1 shows an example of the pool dilution technique applied to an acidic woodland soil. Soil was incubated for 8 days following labelled ammonium addition (5 μg N g^{-1} as ^{15}NH$_4$NO$_3$ at 5 at.% in the ammonium moiety). The figure shows the final 7 days. The points are the measured ammonium ^{15}N abundance; the line is fitted to the data using Equation (7). Two data sets are shown. One refers to the soil incubated alone; the other included 20 μg g^{-1} of the nitrifrcation inhibitor N-Serve. The rates of mineralization in the two cases are virtually identical, confirming that pool

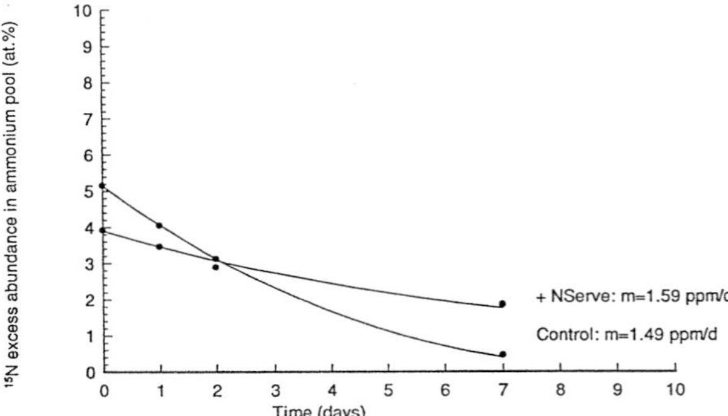

Fig. 1. The use of Equation (7) to determine gross rates of nitrogen mineralization. Symbols are experimental values; lines are best-fit lines calculated from Equation (7).

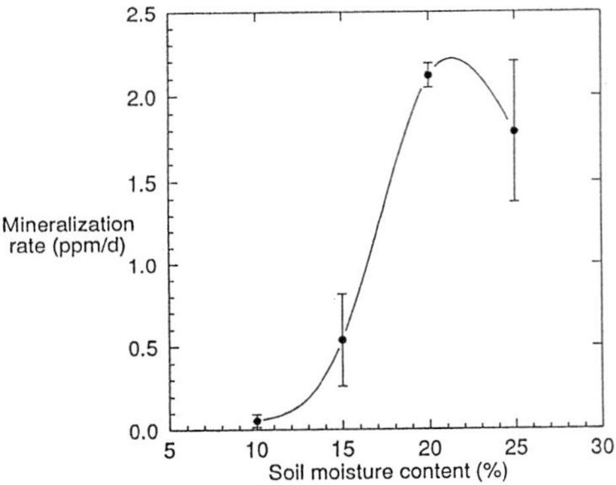

Fig. 2. Gross mineralization rate as a function of moisture content in a Kenyan Orthic Ferralsol.

dilution can reliably isolate constituent processes in the soil/plant nitrogen cycle. Figure 2 gives an example of pool dilution used to determine the relation between soil moisture content and gross nitrogen mineralization rates in a Kenyan Orthic Ferralsol (Pilbeam *et al.*, 1993). The results indicate that during the dry season, when soil moisture contents were in the range 8–10%, nitrogen mineralization rates are very slow. In contrast, during the rainy season, soil moisture contents reach 20% and mineralization rates increase markedly.

Since 1985, a number of papers have been published in which pool dilution techniques have been used to determine gross rates of nitrogen transformations (Barraclough, 1991b; Barraclough *et al.*, 1985; Bjarnason, 1988; Chalk *et al.*, 1990; Davidson *et al.*, 1991; Myrold & Tiedje, 1986; Nishio *et al.*, 1985; Pilbeam *et al.*, 1993). Where the techniques have been used appropriately, they have provided insights into soil nitrogen transformations that could not have been obtained by any other method. Problems have arisen, however,

where the methods have been applied incorrectly or in situations where one of the underlying assumptions is violated. The most restrictive assumption concerns remineralization. If the experimental period is too long, immobilized ¹⁵N will become remineralized causing a progressive apparent decline in the measured rate of mineralization. In the author's experience, departure from zero-order kinetics is less of a problem.

Extensions to pool dilution

These equations have another use. Earlier it was noted that a major drawback to the basic Ndff equation (Equation 6) was that it only gave information on the added label. What we really need is information on the whole ammonium or nitrate pool, not just the added label. To get this we need to abandon the picture of two sub-pools, one labelled, the other unlabelled, that was introduced earlier. We now need to view the whole pool (either ammonium or nitrate), both the labelled and unlabelled fractions. Consider a soil containing 20 kg N ha^{-1} as ammonium and the same amount as nitrate. To the soil is added 100 kg N ha^{-1} of ammonium nitrate with 5.366 at.% in the ammonium moiety. After some interval the crop is harvested and found to contain 2.241 at.% ¹⁵N. Conventional Ndff calculation gives the proportion of crop nitrogen recovered from the ammonium fertilizer as

$$100 \times (2.241 - 0.366)/(5.366 - 0.366)) = 37.5\%$$

So if the crop contained 40 kg N ha^{-1}, 15.0 kg came from the ammonium fertilizer. Assume for a moment that, following the addition of the fertilizer, no mineralization occurred to dilute the label in the ammonium pool. If all the processes that consumed ammonium exploited the ¹⁵N and the ¹⁴N in proportion to their relative amounts (i.e the labelled and unlabelled N were well mixed and there was no preferential exploitation of either isotope), then the fraction of the labelled and unlabelled nitrogen in the soil recovered by the crop would be the same (i.e. if 50% of the labelled-N was recovered in the crop, 50% of the unlabelled would be also). The proportion of the added fertilizer ammonium recovered in the plant was (15.0/50) i.e. 30%, so the amount of unlabelled ammonium recovered in the plant is given by 30 × 20/100 = 6.0 kg N ha^{-1}. The total recovery of nitrogen from the ammonium pool (i.e. fertilizer + soil) is therefore (15.0 + 6.0) = 21.0 kg N ha^{-1}. We can get the same answer by a different route that introduces the idea of a mean pool ¹⁵N. In the example above the ^{15}N abundance of the ammonium pool remains constant because mineraliization has kindly stopped! That pool ^{15}N is given by

$$((20 \times 0.366) + (50 \times 5.366))/70 = 3.9374 \text{ at.\%}$$

Now, if instead of calculating the recovery of crop-N from the added ammonium fertilizer, we consider the recovery from the whole ammonium pool we get, by conventional Ndff calculation

$$40 \times ((2.241 - 0.366)/(3.9374 - 0.366))$$
$$= 21 \, \text{kg N ha}^{-1}.$$

This example illustrates the point that *if* we can identify the right mean pool ^{15}N over the uptake period, we can calculate the recovery of both fertilizer and soil N from that pool. Obviously things are not normally that simple because mineralization will dilute the label in the ammonium pool over time, so the calculation of the mean pool ^{15}N is a little trickier. And the procedure is limited to intervals over which plant uptake (or immobilization) is linear or approximately so. But if these conditions are met we can obtain the mean pool abundance over a time imterval t_1, to t_2 by integrating Equation (7) to obtain

$$[A^*]_{t1-t2} = \frac{\int_{t1}^{t2} A^* \cdot dt}{t_2 - t_1}$$
$$= \frac{1}{t_2 - t_1}\left[A_o^{m/\theta} A_o^* \frac{(A_o + \theta t)^{1-m/\theta}}{\theta - m} \right]_{t1}^{t2} \quad (9)$$

Now instead of determing the recovery of the added label in a, given pool using Equation (6), we can determine the recovery from the whole pool as

$$P \cdot (P^*)/[A^*] \quad (10)$$

where P is the amount of N in the plant, P* is the excess ^{15}N abundance in the plant, and [A*] is the mean ^{15}N excess abundance in the pool to which the label was originally added (obtained from Eq. 9). To avoid double counting nitrogen taken up or immobilized as nitrate but derived, via nitrification, from the ammonium pool, it is better to use equation on the nitrate pool first; the uptake as ammonium follows by difference. Figure 3 shows the results from an experiment in which this approach was used over a 14 day period. It gives the proportion of nitrogen taken up as ammonium by a grass sward receiving 90 kg N ha^{-1} in varying proportions of ammonium:nitrate, from all ammonium (100:0) to all nitrate (0:100) (Barraclough,

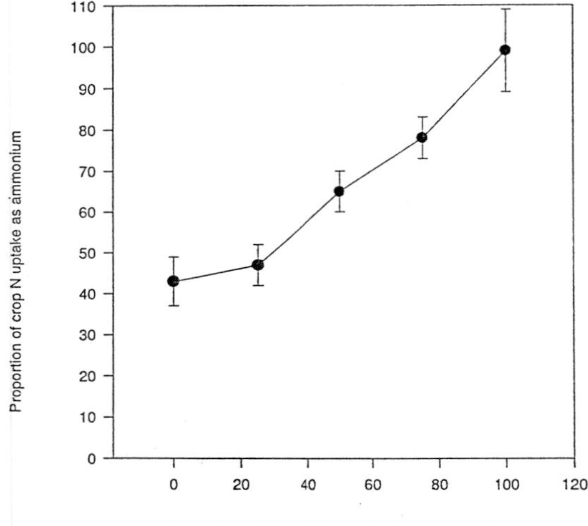

Fig. 3.   Relative uptake of ammonium and nitrate by perennial ryegrass *(Lolium perenne)* at different ratios of ammonium:nitrate applied.

1991b). The results clearly show that at low proportions of ammonium in the fertilizer, there is apparently marked preferential uptake of ammonium. Once again, however, it must be emphasised that this technique is for short-term measurements; some reports in the literature have used long time intervals and encountered serious problems.

Other problems with isotope dilution

The most intractable problem encountered in using isotope dilution concerns the distribution of the added label (assumption 3). To avoid preferential exploitation of either the added label or the indigeneous soil N, the added label must be well mixed in the soil. In laboratory incubation experiments this is not a problem, but in field experiments, with essentially undisturbed soil, it is a major difficulty. Injection of the label with syringes is a common method. If the added label has a random distribution in the soil, the resultant errors in gross rates of mineralization appear to be small; of the order of $5-10\%$. Work in our laboratory, however, has indicated that if the distribution is biased, a more likely event in structured soils, the errors in mineralization rates can reach 40% (R.Monaghan, pers.comm.). In these circumstances, great care is needed to ensure as even a distribution as possible; multiple injections may be required to achieve this.

Further extensions

To determine gross rates of mineralization is a considerable advance. Most work to date has determined net not gross rates of mineralization; such determinations are subject to serious errors and anyway only give the balance between mineralization and immobilization. They are of little use if we want to undertand the processes involved in mineralization immobilization turnover (MIT). It is clearly important to assemble a reliable database on gross rates of mineralization from a range of soils and environments. But we can go further, and hopefully understand more of the processes involved in N mineralization and the differences between soils in different climates. If soil organic matter (SOM) were homogeneous, it would be relatively easy to explain differences between soils in terms of climate. But SOM is not homogeneous. It is a variagated mix of animal and plant residues laid down over a span of time, under different conditions, and differing in composition and decomposability. Radiocarbon dating indicates that different fractions of SOM have 'ages' ranging from 400–2700 years. Many models of SOM turnover explain differences in MIT in terms of the quantity and quality of three main fractions of SOM: fresh plant residues, biomass and humus. Models are tuned to a particular data set by adjusting the amounts and decomposibilty of the fractions to obtain a good fit between simulated and observed values of soil carbon. This is a powerful technique but it has limitations. The main ones are:

- Any fitting procedure involving adjusting the values of a number of parameters to obtain 'fit' runs the risk that any set of best-fit parameters is not unique; another set of values could give an equally good fit.
- The best-fit values obtained by tuning the model to one observed data set may not be applicable to another data set.

It would be more satisfactory to have definite information the relative contribution of the various fractions of SOM to turnover. Thus we need answers to questions such as: "What proportion of the carbon and nitrogen mineralization results from the decomposition of fresh crop residues, biomass turnover and the processing of humic material"? In the Russian tundra, for example, where primary plant productivity is low, we might expect that the humus material is an important source of mineralized C and N. In contrast, in a grassland in the humid tropics, where productivity is high, the decomposition of fresh crop residues would be expected to be an important source of C and N.

As a first hypothesis we can say that the quantity of NH_4 mineralized in a given soil is the sum of the amounts released during decomposition of various fractions of the SOM. But here we hit the first problem. What are the fractions? In a material as complex and heterogeneous as SOM you can virtually find as many fractions as you want, depending on your criteria. Even a cursory glance at some of the models of SOM turnover around will reveal a number of 'pools' postulated. Very often these pools are created to enable the model to fit observations. That is fine. What does pose problems is when the pools are conceptual in too literal a sense; it simply is not possible to identify them experimentally. If we cannot isolate such pools and study their behaviour we can never test the models that gave them birth. The models then become self-fulfilling, untestable cul-de-sacs.

What is needed is to start with identifiable fractions of the SOM that we expect to differ in their contribution to the C and N mineralization flux. For the present, we will start with 4 pools. They are not immutable; if they don't work, we'll change them.If they do, we will try to subdivide them further. The pools are:

- Current crop residues - these are crop residues laid down in the preceding year and those from the current crop.
- The macro-organic matter or light fraction with a specific gravity less than 1.0 g cm^{-3} - this is crop and animal residues more than a year old in a state of partial decomposition.
- The soil microbial biomass itself - as well as processing all the organic matter in soil, the biomass turns over and is processed caniballistically (cryptically).
- The older SOM - the humus material.

Thus, over some interval, we can postulate that the N mineralization flux in the soil is the sum of that released from the processing of the various fractions of SOM i.e.

$$m = m_c + m_l + m_b + m_h \qquad (11)$$

where subscripts c, l, b and h indicate the mineralization flux resulting from the decomposition of current crop residues, the light macro-organic matter fraction, the microbial biomass and humus, respectively. Figure 4 shows this idea in diagrammatic form.

The aim then would be to determine what proportion of the N mineralization flux in different soils comes from the breakdown of these different fractions.

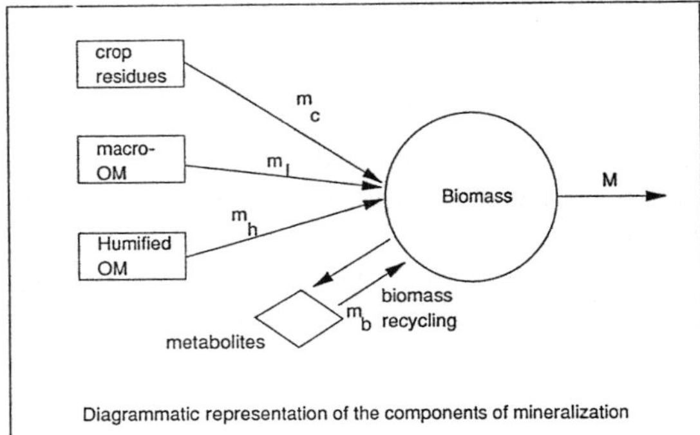

Fig. 4. Diagrammatic representation of the components of mineralization.

Current crop residues and the macro-organic matter fraction are not too difficult. We are currently studying their contribution to mineralization by determining gross rates of mineralization in the presence or absence of these fractions. The soil microbial biomass is more difficult and at present remains a problem. However, this approach, in which the components of mineralizition are broken down, may offer one way of understanding differences in nitrogen turnover rates in different soils and climates.

The techniques involved in isotope dilution are similar to those used in an Ndff experiment. Building isotope dilution into such experiments would require only a little extra effort, but would yield valuable high-grade information.

References

Barraclough D (1991a) The use of mean pool abundances to interpret nitrogen-15 tracer experiments. I. Theory. Plant and Soil 131: 89–96

Barraclough D (1991b) The use of mean pool abundances to interpret nitrogen-15 tracer experiments. II. Application. Plant and Soil 131: 97–105

Barraclough D, Geens EL, Davies GP & Maggs JM (1985) Fate of fertilizer nitrogen. III. The use of single and double labelled ^{15}N ammonium nitrate to study nitrogen uptake by ryegrass. J Soil Sci 36: 593–603

Barrie A (1991) New methodologies in stable isotope analysis. In: Stable Isotopes in Plant Nutrition, Soil Fertility and Environmental Studies. International Atomic Energy Agency, Vienna Austria,

Bjarnason S (1988) Calculation of gross nitrogen immobilization and mineralization in soil. J Soil Sci 39: 393–406

Bremner J (1965) Inorganic forms of nitrogen. In: Black CA (ed) Methods of Soil Analysis, Part 2. Chemical and Microbiological Properties. Agronomy 9, pp 1179–1237. Am Soc Agron, Madison, Wisconsin, USA

Brookes PD, Stark JM McInteer BB & Preston T (1989) Diffusion method to prepare soil extracts for automated nitrogen-15 analysis. Soil Sci Soc Am J 53: 1707 – 1711

Chalk PM, Victoria RL, Muraoka T & Piccolo MC (1990) Effect of a nitrification inhibitor on immobilization and mineralization of soil and fertilizer nitrogen. Soil Biol Biochem 22: 533–538

Davidson EA, Hart SC Shanks CA & Firestone MK (1991) Measuring gross nitrogen mineralization, immobilization, and nitrification by ^{15}N isotopic pool dilution in intact soil cores. J Soil Sci 42: 335–349

Kirkham D & Bartholomew WV (1954) Equations for following nutrient transformations in soil, utilizing tracer data. Soil Sci Soc Am Proc 18:33–34

Kirkham D & Bartholomew WV (1955) Equations for followin nutrient transformations in soil utilizing tracer data. II. Soil Sci Soc Am Proc I9: 189–192

Myrold DD & Tiedje JM (1986) Simultaneous estimation of several nitrogen cycle rates using ^{15}N. Theory and application. Soil Biol Biochem 18: 559–568

Nishio T, Kanamori T & Fujimoto T (1985) Nitrogen tranformations in an aerobic soil as determined by a ^{15}NH$_4$ dilution technique. Soil Biol Biochem 17: 149–154

Pilbeam CJ, Mahapatra BS & Wood M (1993) Soil matric potential effects on gross rates of nitrogen mineralization in an Orthic Ferralsol from Kenya. Soil Biol Biochem 25: 1409–1414

Powlson DS & Barraclough D (1992) Mineralization and assimilation in soil-plant systems. In: Knowles R & Blackburn TH (eds) Nitrogen Isotope Techniques. pp 209–242. Academic Press, London UK

Fertilizer Research **42**: 193–203, 1995.
© 1995 *Kluwer Academic Publishers. Printed in the Netherlands.*

Nitrogen transformations in wetland rice ecosystems

S.K. De Datta
*Crop and Soil Environmental Sciences Department and The Office of International Research and Development,
1060 Litton-Reaves Hall, Virginia Polytechnic Institute and State University, Blacksburg, VA 24061-0334. USA*

Key words: ammonia volatilization, denitrification, green manuring, integrated N management, methane production,
N use efficiency, soil N dynamics

Abstract

In Asia, rice production has increased an average 2.7% annually - due to greater fertilizer use and crop intensification together with varietal improvement and investment in irrigation facilities. Nitrogen efficiency in tropical rice is low. ^{15}N recovery rarely exceeds 30–40% in wetland rice production systems. Ammonia (NH_3) volatilization and denitrification are recognized as major nitrogen loss mechanisms in such systems. Information on the relative importance of the two loss processes is available for few sites in Asia. The greatest losses of N are reported to occur when the fertilizer treatment leads to a high concentration of ammoniacal N in the floodwater. Results from the studies using micrometereological technique suggest that ammonia volatilization may be the most important loss process in wetland rice ecosystems. Directly measuring denitrification in the field proved more difficult than measuring NH_3 volatilization due to difficulty in distinguishing the main end product of denitrification (N_2) against a large background of atmospheric N_2. However, the directly measured ($N_2 + N_2O$) - ^{15}N flux for rice in Indonesia, Thailand and the Philippines rice fields was less than 1% of the applied N. Green manure incorporation in wetland rice fields reduced N losses from mineral N source due to resulting lower floodwater pH and lower partial pressure of NH_3 (pNH_3) than that of urea applied alone. At present, the integrated use of green manure and mineral N is receiving much attention in the hope of meeting farmers' desire to reduce cost of production as well as ecological considerations such as increased methane production which contribute to global climate change. Other promising alternative practices for increasing fertilizer N efficiency include improved timing and application methods, particularly through better incorporation of basal N fertilizer without standing water, deep placement, and use of coated fertilizers.

Introduction

To meet the demand for rice (*Oryza sativa* L.), the world's annual rice production must increase to 556 million tons by 2000 and to 758 million tons by 2020 (IRRI,1989). Of the world's riceland, more than 80% is grown to wetland rice. About 50% of this, and 75% of total rice production are obtained from irrigated wetland rice (De Datta, 1985). This increase in rice production is only possible if soil, water, nutrients and other production inputs are used more efficiently in the future. Nitrogen is the nutrient most limiting rice production worldwide (De Datta, 1981) and it accounts for about 67% of the total amount of fertilizers applied to this crop (De Datta and Nantasomsaran, 1990).

An estimated 24% of the increase in Asian rice production from 1965 to 1980 was attributed to the use of fertilizer, mainly N (Barker *et al.*,1985). Over the past 30 years, the need for N in rice production has been increasingly met through chemical fertilizer use. Total annual fertilizer N consumption in rice has reached 3 million tons, over 90% of which is urea (Vlek and Byrnes,1986).

Despite this impressive growth in world production and use, research over the past 20 to 30 years has shown that crop utilisation of N fertilizers are generally inefficient, with less than 40% of the N applied normally used by the crop (De Datta *et al.*, 1968). In the past two decades, a major research effort to devise efficient management practices and modified forms of urea to improve the level of efficiency has been advocated. Basic N transformation processes to improve N

---

* Paper presented at the International Symposium on Nitrogen Economy in Tropica Soils. Trinidad, West Indies. Jan 9-14, 1991.

194

fertilizer use efficiency in wetland rice were reviewed by Savant and De Datta (1982).

It is important to understand how wetland (lowland) soils differ from dryland (upland) soils in order to fully understand N transformation processes in wetland rice soil ecosystems. In Asia, a wetland rice soil is a soil in a bunded area that is prepared by tillage such as plowing, followed by puddling for transplanted or broadcast-seeded rice culture. The area may be rainfed or irrigated, or a combination of both. The characteristic soil moisture regime varies from near saturation to submergence (flooding) resulting in several centimeters of standing floodwater for most of the crop growth period (De Datta, 1981).

In other areas of the world – the United States, Australia, parts of Europe, and in some Asian countries – rice land is prepared dry and flooded later. Therefore, puddling for transplanting or broadcasting with pregerminated rice is an essential ingredient of a wetland rice soil. Many results discussed in this paper are equally relevant to rice-growing soils that are flooded but not puddled.

Flooding soil at the time of land preparation for wetland rice triggers several physicochemical and microbiological processes. The resultant soil-water atmosphere system is highly complex and heterogeneous in nature (De Datta, 1981; Patrick and Reddy, 1978; Ponnamperuma, 1972; Yamane, 1978).

The behavior of nitrogen in wetland soils is markedly different from its behavior in dry soil conditions. Flooding the soil results in the accumulation of NH_4-N, the instability of NO_3-N and a lowered N requirement for organic matter decomposition. Flooding also causes inefficient utilization of applied N. Ammoniacal N is subject to fixation by clays, loss by volatilization, leaching, runoff, seepage and nitrification followed by loss through denitrification. Significant inorganic N is assimilated into the organic fraction of the wetland soil even though the magnitude of assimilation is lower than in dryland soil (Savant and De Datta, 1982).

Understanding the N transformation processes would greatly facilitate regulating N losses from wetland rice soils and increase the availability of N. Based on this understanding of N transformation processes, alternative management agendas can be developed to minimize fertilizer and soil N losses and increase efficiency of this important nutrient element in these ecosystems.

My paper focuses on N transformation processes in wetland rice ecosystems. Also, it summarizes information primarily on biochemical and physicochemical processes relevant to N transformation processes for improving fertilizer N efficiency in wetland rice production.

Nitrogen dynamics in wetland soils

Ammonium N is the dominant form of mineral N in wetland rice soils, existing in three major fractions:
- Ammonium in soil solution
- Ammonium in exchange sites
- Ammonium in non-exchangeable form
- Ammonium N in soil solution and at exchange sites is readily available to the rice crop, whereas reports on the availability of non-exchangeable NH_4N are conflicting (Mengel *et al.*, 1986).

Relation between initial NH_4^+ on N uptake

Initial NH_4^+ content in the analyzed soils ranged from 12 to 62 kg N ha^{-1}. The relations were substantially improved when the N uptake of the crop was plotted against fertilizer N+ initial exchangeable soil NH_4^+ (Mengel *et al.*, 1986).

Field experiments by Keerthisinghe *et al.* (1985) at three sites in major rice growing areas in the Philippines showed that NH_4^+ fertilizer application was clearly reflected by an increase of exchangeable and non-exchangeable soil NH_4^+, while N uptake resulted in a decrease of both NH_4^+ fractions. In one wetland soil (Maligaya silty clay loam) containing montmorillonite/vermiculite as principle clay minerals (Typic Pellustert), significant increases in exchangeable NH_4^+, however, occurred only in the fertilized plots. Exchangeable NH_4^+ decreased continually with crop growth, reaching the lowest level at 60 days after transplanting IR36 rice. This decrease was reflected by an increased N uptake by rice, particularly with fertilized N treatments (Keerthisinghe *et al.*, 1985). From these studies, the same authors concluded that the exchangeable NH_4^+, and in some soils the non-exchangeable NH_4^+, are the most important soil N fractions easily available to wetland rice.

Net release of non-exchangeable NH_4^+ occurs if the NH_4^+ concentration of the surrounding medium is depleted to a low level. Intensive crop growth associated with a high uptake of NH_4^+ -N may bring about such a depletion of NH_4^+ in the soil solution and consequently favor the release of non-exchangeable NH_4^+. The high soil moisture content of wetland soils promotes the release process due to its favorable effects on NH_4^+ diffusion to plant roots and also by promoting the expansion of 2:1 clay minerals.

From these studies, we concluded that a substantial amount of N taken up by a rice crop may originate from the non-exchangeable fractions of clay bound NH_4^+ (Keerthisinghe *et al.*, 1984).

Nitrogen dynamics in alternating dryland/wetland soils

Wetland rice soils often undergo alternate wet and dry conditions, particularly under rainfed conditions even where continuous wetland rice production is planned. Soils are saturated for at least part of the time during production of rice, but during intervals between rice crops, the soil usually dries and becomes aerated. At this time, either the soil is left fallow or upland crops are grown.

In aerobic dryland soil conditions, ammonium formed from mineralization of organic N or from N fertilizer can be nitrified to NO_3-N, which can accumulate in the soil or be used by plants. When the dryland (aerobic) soils are flooded, soil oxygen is rapidly depleted and soil NO_3-N is prone to loss by denitrification and leaching. In wetland soils the conversion of NH_4-N to NO_3-N is restricted by the limited supply of soil O_2; hence, NH_4-N is the form of mineral N that accumulates. At the end of a wetland rice crop, soil NO_3-N is normally negligible and soil NH_4-N, the dominant form of mineral N is typically low because of N uptake by rice.

Effects of green manure and urea in nitrogen dynamics

Mineralization and immobilization processes occur simultaneously in wetland soils and depend on soil properties and environmental factors. Mineralization of organic N (native or added organic N sources) is the most important process in the N nutrition of wetland rice. Nitrification processes, however, are inhibited under reduced soil conditions (Ponnamperuma, 1972). Thus, mineralization stops at NH_4^+. Temperature, soil moisture regime, soil pH, organic matter content, C/N ratio, amount and kind of organic residues or organic matter added are major environmental factors that affect mineralization/immobilization turnover of N in the soil-plant system.

Soils to which organic matter has been applied are prone to low Eh, because such materials favor the growth and metabolism of anaerobic microorganisms (Wantanabe, 1984). This affects the rate of organic matter breakdown and mineralization of organic N. This N source is of major importance.

Mineralization of soil organic N is the major N source for rice plant nutrition. Fifty to 75% of N uptake by rice is derived from native soil N, even in N fertilized rice fields (Broadbent, 1978).

Experiments with ^{15}N-labeled fertilizer often show that plants given fertilizer N take up more N from the soil than do plants not treated with inorganic or organic fertilizer N. This is often called 'priming effect' or added N interaction (ANI) (Broadbent, 1970; Jenkinson *et al.*, 1985). The origin and nature of the positive priming effect or ANI are controversial. ANIs can be 'real' where fertilizer N induces increased root growth where more soil is explored by roots. 'Apparent' ANIs can be caused by pool substitution or by isotope displacement reactions (Fig. 1). Pool substitution is the process by which added labeled N 'stands Proxy' for native unlabeled soil N that otherwise would have been removed from the pool (Jenkinson *et al.*, 1985).

In a recent experiment Dickmann (1990) monitored the pattern of N release from the incorporated green manure and urea treated wetland soils. Exchangeable NH_4^+-N concentrations peaked at 60 to 120 mg N kg^{-1} at 10 days after flooding, then declined to as low as 4 mg N kg^{-1} dry soil (Fig.2).

Generally, exchangeable NH_4^+-N concentrations after urea applications peaked earlier than did those after green manure applications. These results with non-isotopic methods show that N derived from green manures (e.g. *Sesbania rostrata* and *Aeschynomene afraspera*) is readily available and can be used efficiently by a wetland rice crop. Thus, green manure has the potential to supplement for mineral N (Dickmann *et al.*, 1993).

Nitrogen transformation processes in relation to gaseous nitrogen losses

Various reasons have been cited for the poor efficiency of N fertilizers, including loss of fertilizer N by leaching, run-off or emission of NH_3 and/or N_2 to the atmosphere. In most wetland rice growing areas, leaching of N is not a problem. Rice is primarily grown on heavy clay soils, on sandy soils overlying impervious clay horizons, or on soils puddled to reduce water flow into subsurface horizons. Loss of N by run-off is also controlled in most wetland rice fields, but there are exceptions (De Datta, 1981).

196

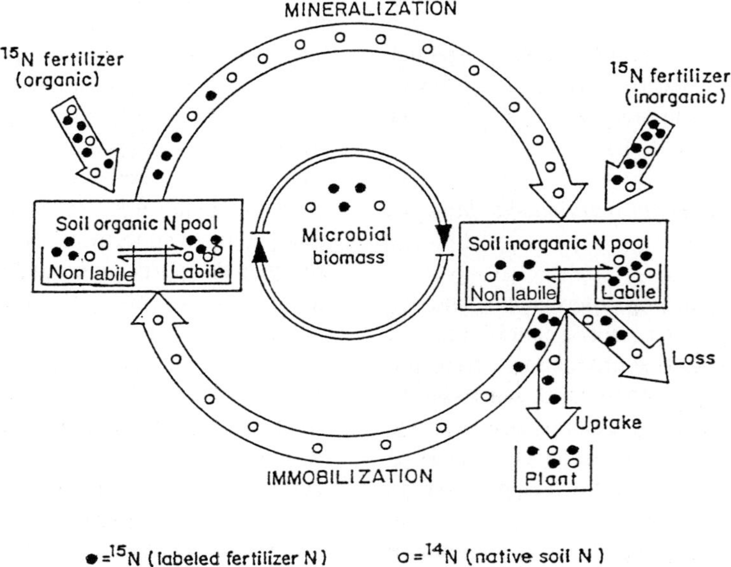

Fig. 1. Schematic presentation of the turn-over of N in relation to pool substitution of fertilizer N. (From KH Dickmann and SK De Datta, 1990, IRRI, unpublished).

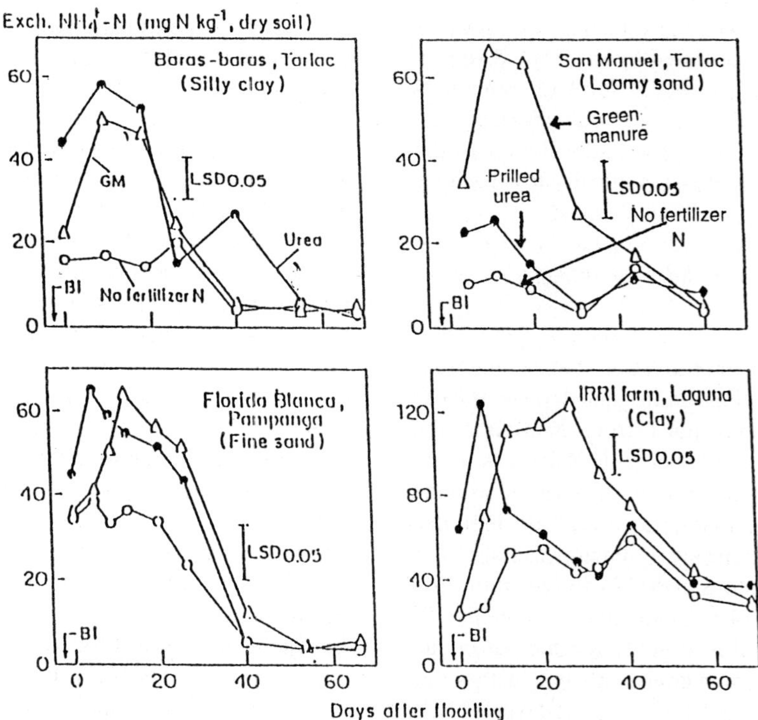

Fig. 2. Kinetics of exchangeable NH_4^+-N in four flooded soils in the presence of rice plants as affected by green manure[a] or urea application. Central Luzon and IRRI, Philippines, 1987 wet season. Urea applied: 60 kg N ha^{-1}. [a] Averaged values for *Sesbaina rostrata* and *Aeschynomene afraspera*. (From KH Dickmann and SK De Datta, 1990, IRRRI, unpublished).

Ammonia volatilization loss

Ammonia volatilization is a major loss mechanism that affects the efficiency of urea and other N fertilizers in irrigated wetland rice. The magnitude of ammonia loss depends on wind speed, temperature, rainfall, ammoniacal-N ($NH_4^+ + NH_3$) concentration, pH and cation exchange capacity (Freney *et al.*, 1983). In wetland rice fields in Asia, the floodwater ammoniacal-N concentrations following N application, high temperatures common in the tropics, and elevated floodwater pH resulting from algal photosynthetic activity create a favorable environment for NH_3 loss (Mikkelsen *et al.*, 1978, Fillery *et al.*, 1986). Vlek and Craswell (1981) explained that aqueous NH_3 content in floodwater increases almost linearly with increasing temperature, which results in nearly a fourfold increase in the range from 10 to 40 °C. In field systems, floodwater pH displays a diurnal pattern seemingly synchronized with cycles of photosynthesis and not respiration or the addition or the depletion and additions of CO_2 to floodwater (Mikkelsen *et al.*, 1978).

Measurement of ammonia volatilization
Because NH_3 volatilization from wetland rice ecosystems is affected by environmental as well as chemical and biological factors (Fillery *et al.*, 1984; Fillery and Vlek, 1986), the emission of NH_3 should be determined by techniques that do not disturb the field environment. While micrometereological techniques meet this requirement (Denmead, 1983), most of them cannot be employed when a large number of small plots are being used to compare different management practices.

A promising technique for comparing NH_3 losses from small plots is the bulk aerodynamic method (Freney *et al.*, 1985). The technique is only applicable in environments with standing water.

Magnitude of ammonium loss
The first direct measurements of NH_3 volatilization loss using the micrometereological technique in tropical irrigated rice fields were made by Freney *et al.*, (1981). Their study was conducted with $(NH_4)_2 SO_4$ applied to a wetland rice field in the Philippines. Ammonia loss accounted for 5% of $(NH_4)_2 SO_4$ broadcast and incorporated (80 kg N ha^{-1}) before transplanting. Subsequent field measurements of NH_3 loss have focused primarily on urea. Earlier studies using micrometereological techniques have shown that NH_3 volatilization can be an important loss mechanism after application of urea-N to floodwater (Fillery *et al.*, 1984). Subsequent studies by Fillery and De Datta (1986) reported NH_3 fluxes accounting for 38 and 36% of N applied as $(NH_4)_2SO_4$ or urea, respectively, in an 8-day period. Some of the differences in the pattern of NH_3 loss from urea and $(NH_4)_2SO_4$ amended fields outlined in the study by Fillery and De Datta (1986), can be understood after considering the ammoniacal-N concentrations and the pNH$_3$ in the floodwater. The delay in the build-up of ammoniacal-N in floodwater after urea application has been noted elsewhere (Fillery *et al.*, 1984), and thus appears to be a characteristic of flooded soil systems with low urease activity in floodwater. Recent micrometereological measurements of NH_3 loss from urea in Asian irrigated wetland rice environments was reported to be from 7 to 54%. In the Philippines where most of the measurements were made, NH_3 loss tended to be greater from urea broadcast into floodwater at 10 to 21 days after transplanting than from urea either broadcast and incorporated before transplanting, or broadcast application 5 to 7 days before panicle initiation of rice. The high measured NH_3 loss from urea broadcast at 10 to 21 days after transplanting, a common farmers' practice in the Philippines and many areas in Asia, highlights the importance of NH_3 volatilization as a loss mechanism.

Earlier studies on NH_3 volatilization at IRRI were limited to transplanted rice (Fillery *et al.*, 1984; De Datta *et al.*, 1988). But with the widespread popularity of broadcast seeding among farmers in the Philippines, Thailand and Malaysia due to increased labor cost in transplanting, quantification was sought of NH_3 volatilization loss in broadcast seeded wetland rice.

Field measurements were made on an Aeric Tropaquent at Calauan, Laguna, Philippines during the 1988 dry season. Computing this cumulative NH_3 loss from transplanting and broadcast seeded wetland rice ecosystems, NH_3 loss was 49% in transplanted and 30% in broadcast seeded wetland conditions (Fig. 3). Comparing the diurnal floodwater temperature and pH, the trend for both planting methods were similar. Wind speed was the most important parameter that affected the differences in NH_3 loss from these planting methods. Wind speed patterns at heights 0.8 to 3 m above the floodwater were similar for both planting methods. However, at heights as low as 0.03 to 0.1 m above the floodwater, wind speed was higher for transplanting than for broadcast seeded wetland systems (De Datta *et al.*, 1988). This explained the higher cumula-

198

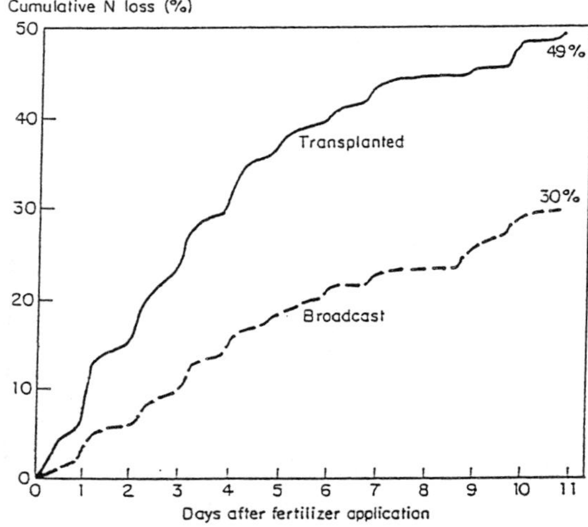

Fig. 3. Cumulative NH$_3$ loss from urea applied to broadcast seeded and transplanted IR66 rice in circular plots. Calauan, Laguna, Philippines, 1988 dry season.

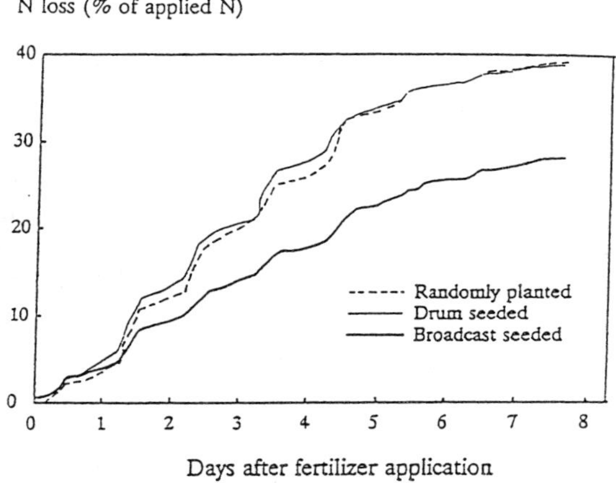

Fig. 4. Cumulative NH$_3$ loss from urea applied to broadcast seeded, row seeded with drum seeder and randomly transplanted IR72 rice in circular plots. Victoria, Laguna, Philippines, 1990 dry season.

tive NH$_3$ loss in transplanted than in broadcast seeded crop. Most farmers in Asia still randomly transplant rice. When subsequent field studies compare randomly transplanted with row seeded (using drum seeder) and broadcast seeded wetland rice, NH$_3$ losses were similar for the randomly transplanted and row seeded rice. Least NH$_3$ loss occurred with broadcast seeded rice, confirming earlier results (Fig. 4). Ammonia fluxes were lowest with broadcast seeding (Fig. 5), which is due to lower wind speed recorded at 0.42 m height with that method (Fig. 6).

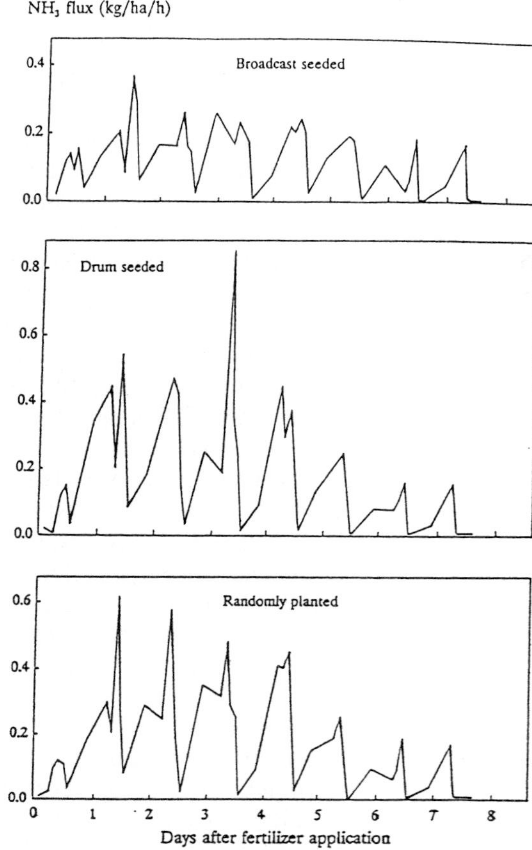

Fig. 5. Ammonia fluxes from urea applied to broadcast seeded, row seeded and randomly transplanted IR72 rice in circular plots. Victoria, Laguna, Philippines, 1990 dry season.

Effect of green manures

Leguminous green manures had important contributions to rice production in its traditional agriculture in China. But with the intensification of agriculture, production of green manure crops decreased. In the Asian tropics, the integrated use of green manure and mineral N has received considerable attention in the recent past in the hopes of meeting the farmers' economic needs as well as for long-term ecological considerations. Thus, it is critical to understand the transformation processes in relation to availability and loss mechanisms with green manure crops which will contribute to developing integrated N management in wetland rice systems. *Sesbania rostrata* and *Aeschynomene afraspera* are tropical legumes which form N$_2$-fixing nodules on both stems and roots, are flood tolerant, and can accumulate large amounts of N within a short period of time,

Wind speed (m/s)

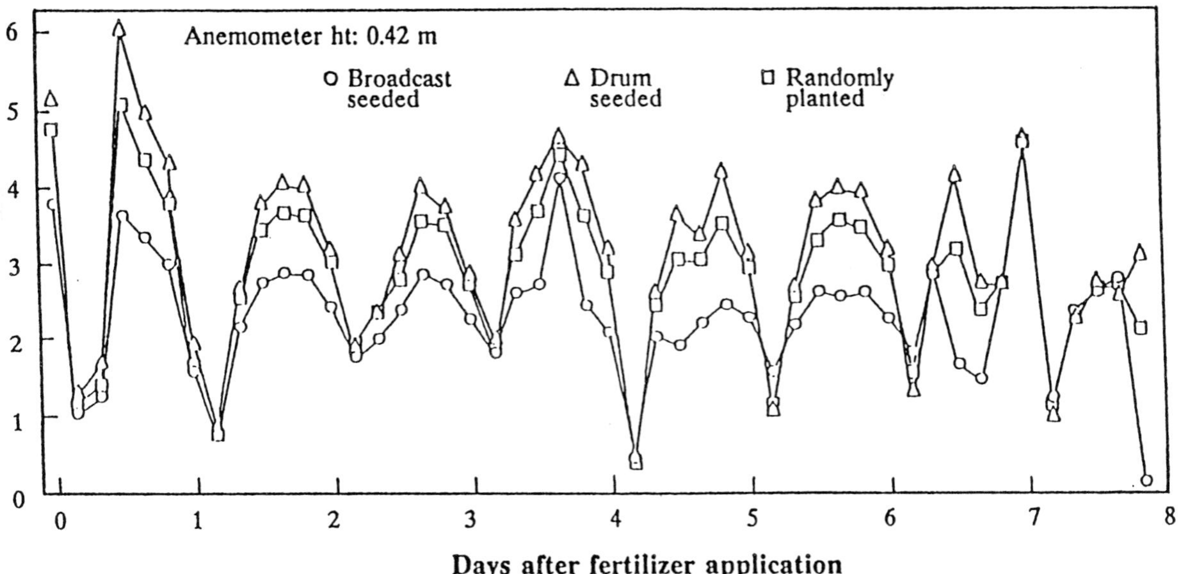

Days after fertilizer application

Fig. 6. Windspeed patterns at 0.42 m in height according to the various methods of planting IR72 rice during the measurement of ammonia fluxes. Victoria, Laguna, Philippines, 1990 dry season.

even under wetland conditions (Rinaudo *et al.*, 1983). In a field experiment Dickmann *et al.* (1993) reported that the substantial amount of N unaccounted for from urea is probably due to losses by NH_3 volatilization and denitrification from floodwater occurring during the first few days after fertilizer application. This suggestion is supported by the results of calculations of the partial pressure of NH_3 in the floodwater (pNH_3) which was significantly lower in green manure or green manure treatments plus urea than urea application alone. Results show that application of green manure with urea may reduce NH_3 losses from urea by lowering the pH in the floodwater. The large recoveries of ^{15}N derived from labeled green manure are presumably due to less N loss resulting from the low ammonium concentration in the floodwater because of slow decomposition of green manure.

In another field study NH_3 fluxes were directly measured from fields amended with urea and green manures using micrometereological techniques. Ammonia fluxes were considerably lower (Fig. 7) and cumulative N losses were also lower with green manure than with urea (Fig. 8). This is due to increased pCO_2 concentration in floodwater with green manure incorporation in the wetland soil (Fig. 9) which lowers the pH of the floodwater thereby reducing NH_3 loss.

However, methane production increases with addition of organic matter such as green manures. Increased methane (CH_4) production from the wetland systems may be higher than 40 Tg year^{-1} (Neue, 1993). This increase in CH_4 production from the wetland rice ecosystems contributes to global climate change.

Denitrification loss

Nitrification in oxidized soil zones and floodwater converts the ammoniacal N formed by ammonification and hydrolysis of urea into NO_3-N. This NO_3-N can thereafter move into reduced soil zones where it is readily denitrifed to dinitrogen (N_2) and nitrous oxide (N_2O) (Reddy and Patrick, 1986). Although denitrification in wetland rice soils has been the subject of laboratory investigation, quantification of denitrification losses in wetland rice soil systems has been hindered by the lack of viable methodology for direct measurement (Buresh and De Datta, 1990).

Measurement of denitrification

The end products of denitrification are N_2 and relatively smaller amounts of N_2O and nitric oxide (NO). Although direct quantitative measurement of all the N gases formed by denitrification is desirable, the direct

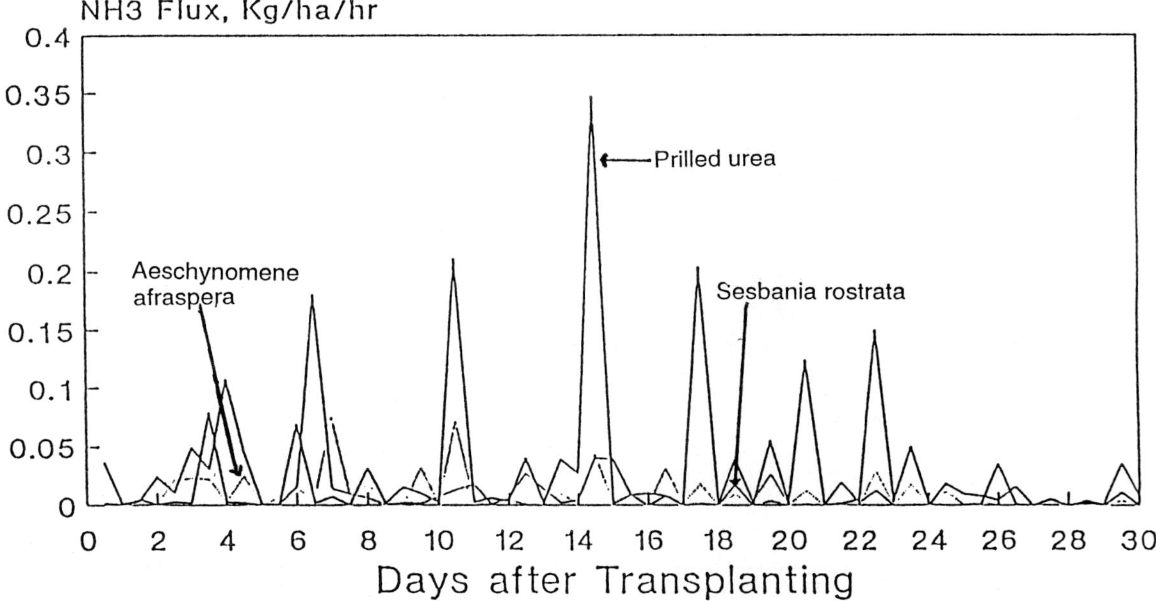

Fig. 7. Ammonia fluxes from urea and two green manure crops. Victoria, Laguna, Philippines 1991 dry season.

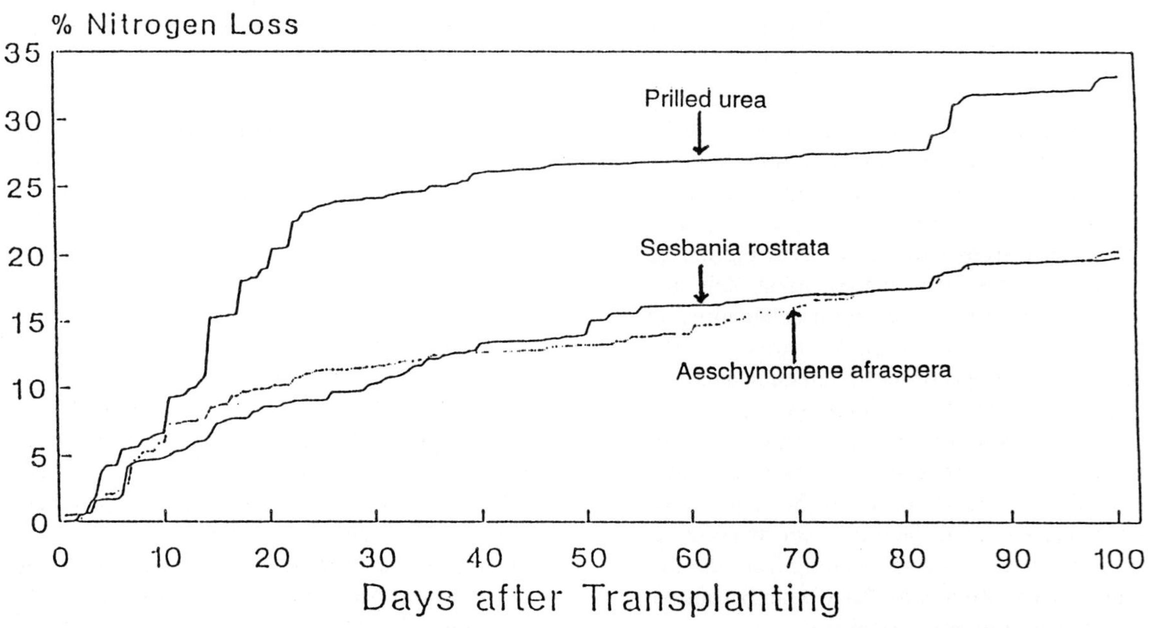

Fig. 8. Cumulative nitrogen losses from urea and green manure crops. Victoria, Laguna, Philippines 1991 dry season.

measurement of N_2 evolved from the soil is difficult because the small quantities of N_2, produced by denitrification are evolved into an atmosphere already containing 78% N_2. Both direct and indirect methods of N_2 and N_2O evolution were attempted to determine N gases.

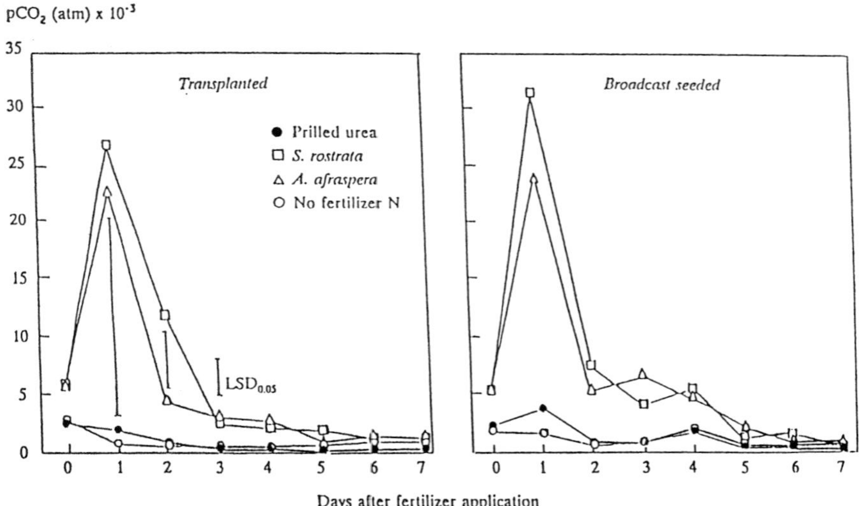

Fig. 9. Partial pressure of carbon dioxide (pCO_2) in floodwater at 1300–1400 h as affected by green manure and urea management in flooded transplanted and broadcast seeded IR72 rice. IRRI, 1990 dry season. N rate = 90 kg ha^{-1}. (B Dekamedhi and SK De Datta, Unpublished)

For direct measurement, Buresh and De Datta (1990) and Buresh *et al.* (1991), used ^{15}N tracer methodology. To measure $(N_2 + N_2O)$ - ^{15}N evolution in wetland soils, a highly ^{15}N-enriched source was applied to microplots bunded by frames pushed into the soil. The ^{15}N gases that evolve from the soil-floodwater systems are then collected in chambers periodically placed either between or over the rice plants.

Directly measured flux of $(N_2 + N_2O)$ - ^{15}N from urea was low in experiments conducted in 1987 and 1988, with the highest of only 39 g N ha^{-1} day^{-1} (Fig. 10).

Use of KNO$_3$, as a reference treatment to check for accuracy of the method showed that the $(N_2 + N_2O)$ - ^{15}N flux was much greater than that obtained from urea in spite of the lower N application rates. Total recoveries of evolved $(N_2 + N_2O)$ - ^{15}N for the same experiment were 37 and 22%, respectively, of the added NO$_3$-N.

The cumulative recovery of $(N_2 + N_2O)$ - ^{15}N over 10- to 20-day measurement periods following the urea application by several different methods was small, ranging from less than 0.1% to 2.2% of the applied ^{15}N (Buresh and De Datta, 1990). Total ^{15}N loss, determined from a ^{15}N - balance in the microplots after the 10- to 20-day measurements, was much larger, ranging from 10 to 56% of the applied ^{15}N. The total ^{15}N loss presumably represented only gaseous N loss by NH$_3$ volatilization and denitrification since microplot borders prevented runoff and since the leaching loss was

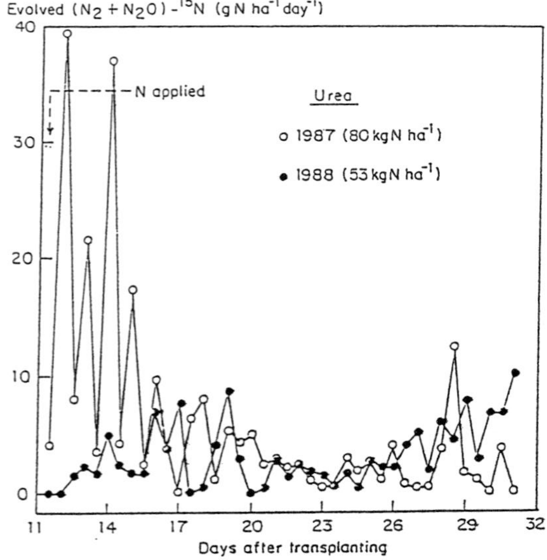

Fig. 10. Evolved (N_2+N_2O)-^{15}N after application of urea into floodwater at 11 days after transplanting. Calauan, Laguna, 1987 and 1988 dry seasons. (From WN Obcemea, JA Real, MI Samson, SK De Datta, and RJ Buresh, 1988, IRRI, unpublished mimeo).

negligible on these wetland soils with a percolation rate of approximately 2.3×10^{-8} m s^{-1} or less (Buresh and De Datta, 1990).

One possibility for low recovery of $(N_2 + N_2O)$ - ^{15}N is the entrapment of $(N_2 + N_2O)$ - ^{15}N in the soil. To substantiate this possibility, Samson *et al.* (1990) determined whether gaseous products of denitrification remain trapped in the soil. Results suggest that ^{15}N level gas remained trapped in the soil even after the evolution of $(N_2 + N_2O)$- ^{15}N had decreased to less than 15 g N ha^{-1} day^{-1}. Quantification of denitrification may then require the measurement of ^{15}N gas trapped in the soil.

Interdependence of ammonia volatilization and denitrification

There is little information on the relative contribution of NH_3 volatilization to the overall N loss from wetland rice system under different management practices. In one study by Freney *et al.* (1990), the relative importance of NH_3 volatilization and denitrification as loss processes following the application of urea to wetland rice was assessed at the four sites in the Philippines. The results show that reducing NH_3 loss by incorporating urea into the soil does not necessarily result in reduced total N loss, and suggests the efficiency of fertilizer N will be improved only when total N-loss is controlled.

Experiments in Thailand, Indonesia and the Philippines suggested that the high denitrification losses investigated by different methods (total ^{15}N loss by ^{15}N - balance minus NH_3 loss measured by bulk aerodynamic method) had not been confirmed by high directly measured loss of N_2 and N_2O (Buresh *et al.*, 1991).

The main objective of the studies on gaseous N losses is to enable the total N loss following fertilizer application to be reduced through the design of simple, cost-effective, management practices that can be adopted widely by resource-poor farmers. If the NH_3 and denitrification losses are independent, then it should be possible to achieve some reduction in total N loss and an increase in fertilizer efficiency. However, if the losses are complementary, then reduction in NH_3 loss through these practices will not lead to a reduction in the total N loss because N conserved would be nitrified, then denitrified (Freney *et al.*, 1990).

Prospect for increasing nitrogen use efficiency

From the large number of research trials throughout Asia we conclude that improved timing and incorporating urea into wetland rice soils offer considerable prospect for increasing agronomic efficiency by reducing N loss and better matching N availability to plant requirements at critical growth stages. Where the gaseous N loss is primarily NH_3 volatilization, it is pos-

sible to substantially reduce the loss by improved water management and through incorporation of fertilizer N into drained soil immediately before transplanting (De Datta *et al.*, 1989). De Datta *et al.* (1988) earlier reported that losses of N fertilizer are less with broadcast seeded wetland rice than in transplanted rice.

Improved incorporation or placement of urea, coatings of urea, use of urease inhibitors would offer considerable potential for increasing the efficiency of urea fertilizer application (Buresh *et al.*, 1988).

Green manures and other organic fertilizers offer a number of apparent agronomic advantages. However, use of green manures increases methane production which pollutes the atmosphere contributing to global climate change (De Datta and Buresh, 1989). Furthermore, the economics of widespread adoption and use of green manures in rice based cropping systems is less clear. Cultural practices should be developed to minimize the emissions of these gases from wetland rice fields.

The nitrogen contribution of legumes in wetland rice - food legume sequences, as in dryland (upland crop) - legume sequences, depend on the quantity of legume N derived from N_2 -fixation, the proportion of legume N mineralized, and the efficiency of use of this mineralized N by the succeeding crop. Losses of NO_3-N, formed from mineralization of soil organic matter and legume N, must be minimized in wetland rice legume sequences in order to maximize the N contribution of legumes to rice and to minimize environmental pollution (Buresh and De Datta, 1991). Because the effectiveness of N management practices can depend on season, soil type, climate, water management, and cropping pattern, there will probably be no universally best management practice. Improvements in N management could involve alternative water management, modifed urea sources, alternative timing and rates, and integrated use of inorganic and organic N sources.

References

Barker R, Herdt RW and Rose B (1985) The Rice Economy in Asia. Resources for the Future. Washington DC. 324p

Broadbent FE (1970) Variables affecting A values as a measure of soil nitrogen availability. Soil Sci Soc Am J 110: 19–23

Broadbent FE (1978) Transformations of soil nitrogen. In: Nitrogen and Rice, pp 108-118. International Rice Research Institute, Philippines

Buresh RJ and De Datta SK (1990) Denitrification losses from puddled rice soils in the tropics. Biol Fert Soils 9: 1–13

Buresh RJ, De Datta SK, Samson MI, Phongpan S, Snitwongse P, Fagi AM and Tejasarwana R (1991) Dinitrogen and nitrous oxide flux from urea basally applied to puddled rice soils. Soil Sci Soc Am J, 55: 268–273

Buresh RJ and De Datta SK (1991) Nitrogen dynamics and management in rice-legume cropping systems. Adv Agron 45: 1–59

Buresh RJ, De Datta SK, Padilla JL and Samson MI (1988) Effects of two urease inhibitors on floodwater ammonia following urea application to lowland rice. Soil Sci Soc Am J 52: 856–861

De Datta SK, Magnaye CP and Moomaw JC (1968) Efficiency of fertilizer nitrogen (^{15}N Labelled) for flooded rice. Trans. 9th Int'l Soil Sci Congr., Adelaide, Australia, IV. pp 67-76

De Datta SK (1981) Principles and Practices of Rice Production. John Wiley and Sons, New York. 618 p

De Datta SK (1985) Availability and management of nitrogen in lowland rice in relation to soil characteristics. In: Proc. Workshop on Characterization, Classification, and Utilization of Wetland Soils, 26 March - 6 April, 1984, pp 247–267. International Rice Research Institute, Philippines

De Datta SK, Buresh RJ, Samson MI and Kai-Rong Wang (1988) Nitrogen use efficiency and nitrogen-15 balances is broadcast - seeded, flooded and transplanted rice. Soil Sci Soc Am J 52: 849–855

De Datta SK and Buresh RJ (1989) Integrated nitrogen management in irrigated rice. Adv Soil Sci 10: 143–169

De Datta SK, Trevitt ACF, Freney JR, Obcemea WN, Real JG and Simpson (1989) Measuring nitrogen losses from lowland rice using bulk aerodynamic and nitrogen-15 balance methods. Soil Sci Soc Am J 53: 1275–1281

De Datta SK and Nantasomsaran P (1991) Status and Prospects of Direct Seeded Flooded Rice in Tropical Asia. In: Direct Seeded Flooded Rice in the Tropics, pp 1–16. International Rice Research Institute, Philippines

Denmead OT (1983) Micrometereological methods for measuring gaseous losses of nitrogen in the field. pp 131–157 In: Freney JR and Simpson JR (eds) Gaseous Loss of Nitrogen from Plant-soil Systems, pp 131–157. Martinus Nijhoff/Dr. W. Junk Publ., The Hague, Netherlands

Dickmann KH, De Datta SK and Ottow JCG (1993) Nitrogen uptake and recovery from urea and green manure in lowland rice measured by ^{15}N- and non-isotope techniques. Plant and Soil 148: 91–99

Fillery IRP, Simpson JR and De Datta SK (1984) Influence of field environment and fertilizer management on ammonia loss from flooded rices. Soil Sci. Soc Am J 48: 914–920

Fillery IRP and De Datta SK (1986) Ammonia volatilization from nitrogen sources applied to rice fields. I. methodology, ammonia fluxes and nitrogen-15 loss. Soil Sci Soc Am J 50: 80–86

Fillery IRP, Simpson JR and De Datta SK (1986) Contribution of ammonia volatilization to total nitrogen loss after applications of urea to wetland rice fields. Fert Res 8: 193–202

Fillery IRP and Vlek PLG (1986) Reappraisal of the significance of ammonia volatilization as a N loss mechanism in flooded rice fields. Fert Res 9: 79–98

Freney JR, Denmead OJ, Watanabe I and Craswell ET (1981) Ammonia and nitrous oxide losses following applications of ammonium sulphate to flooded rice. Aust J Agric Res 32: 37–45

Freney JR, Simpson JR and Denmead OJ (1983) Volatilization of Ammonia. In: Freney JR and Simpson JR (eds) Gaseous loss of nitrogen from plant - soil systems, pp 1-32. Martinus Nijhoff/Dr. W Junk Publisher, The Hague, Netherlands.

Freney JR, Leuning R, Simpson JR, Denmead OJ and Muirhead WA (1985) Estimating ammonia volatilization from flooded rice fields by simplified techniques. Soil Sci Soc Am J 49: 1049–1054

Freney JR Trevitt ACF, De Datta SK, Obcemea WN and Real JG (1990) The Interdependence of ammonia volatilization and denitrification and nitrogen loss processes in flooded rice felds in the Philippines. Biol Fert. Soils 9: 31–36

Jenkinson DS, Fox RH and Rayner JH (1985) Interaction between fertilizer nitrogen and soil nitrogen - The so-called 'priming' effect. J. Soil Sci 36: 425–444

Keerthisinghe G, De Datta SK and Mengel K (1985) Importance of exchangeable and nonexchangeable soil NH_4 in nitrogen nutrition of lowland rice. Soil Sci 140: 194–201

Keerthisinghe G, Mengel K and De Datta SK (1984) The release of nonexchangeable ammonium (^{15}N-Labelled) in wetland rice soils. Soil Sci. Soc Am J 48: 291–294

Mengel K, Schon HG, Keerthisinghe G and De Datta SK (1986) Ammonium dynamics of puddled soils in relation to growth and yield of lowland rice. Fert Res 9: 117–130

Mikkelsen DS, De Datta SK and Obcemea WN (1978) Ammonia volatilization losses from flooded rice soils. Soil Sci Soc Am J 42: 725–730

Neue HU (1993) Methane emission from rice felds. Wetland rice fields may make a major contribution to global warming. Bioscience 43: 466–474

Obcemea WN, Real JA, Samson MI, De Datta SK and Buresh RJ (1988) Direct measurement of ammonia volatilization and denitrification losses in flooded tropical rice. IRRI, Saturday Seminar, 8 Oct. 1988. (*unpublished mimeo*)

Patrick WH Jr and Reddy CN (1978) Chemical changes in rice soils. In: Soils and Rice, pp 361–380. International Rice Research Institute, Philippines

Ponnamperuma FN (1972) The Chemistry of Submerged Soils. Adv Agron 24: 29–96

Reddy KR and Patrick WH Jr (1986) Denitrification losses in flooded rice felds. In: De Datta SK and Patrick WH Jr (eds) Nitrogen Economy of Flooded Rice Soils, pp 99–116. Martinus Nijhoff, Dordrecht, Netherlands

Rinaudo G, Drefus B and Domergues Y (1983) *Sesbania rostrata* green manure and the nitrogen content of rice crop and soil. Soil Biol Boichem 15: 111–113

Samson MI, Buresh RI and De Datta SK (1990) Evolution and soil entrapment of nitrogen gases formed by denitrification in flooded soils. Soil Sci Plant Nutr 36: 299–307

Savant NK and De Datta SK (1982) Nitrogen transformations in wetland rice soils. Adv Agron 35: 241–302

Vlek PLG and Craswell ET (1981) Ammonia volatilization from flooded rice soils. Fert Res 2: 227–245

Vlek PLG and Byrnes BH (1986) The efficiency and loss of fertilizer N in lowland rice. In: De Datta SK and Patrick WH Jr (eds) Nitrogen Economy of Flooded Rice Soils, pp 131–147. Martinus Nijhoff, Dordrecht, The Netherlands

Watanabe I (1984) Anaerobic decomposition of organic matter in flooded rice soils. In: Organic matter and Rice, pp 237–272. International Rice Research Institute, Philippines

Yamane I (1978) Electrochemical changes in rice soils. In: Soils and Rice, pp 381–398. International Rice Research Institute, Philippines

Fertilizer Research **42**: 205–214, 1995.
205
© 1995 *Kluwer Academic Publishers. Printed in the Netherlands.*

Urea-N uptake by dasheen (*Colocasia esculenta* L. Schott) in relation to the fertilizer placement method

Gregory A. Gouveia, Nazeer Ahmad & Selwyn M. Griffith
Department of Soil Science, The University of the West Indies, St. Augustine, Trinidad

Key words: Colocasia esculenta, fertilizer efficiency, fertilizer management, ^{15}N, urea transformations, volatilization

Abstract

Urea has become the most important N carrier in many parts of the world and its reaction when added to soil is unique in many ways. Two field experiments were therefore undertaken using ^{15}N to investigate the uptake efficiency of the added urea-^{15}N which was banded in Experiment I and broadcast in Experiment II. In both experiments the uptake efficiencies were not affected by N-rate and cropping system (Exp. I) or crop residue management (Exp. II) and averaged 17.4 and 16.9% respectively. These low values were supported by evidences of high losses; high pH increases following urea application (volatilization), downward movement of N (leaching), and cycles of waterlogged and well drained conditions in the soil (de-nitrification). Evidence of leaching at least down to 30 cm in the profile was observed in the first experiment where urea was banded but not in experiment II where it was broadcast. The proportion of N in the crop that was derived from added urea (%Ndff) was 57.7% and 36.4% in experiments I and II respectively, suggesting that band application resulted in a higher proportion of the added N in the root zone compared to that for broadcast application. The results indicate the need to investigate other management strategies, such as higher application frequencies and placement closer to the root zone, in order to improve the uptake efficiency of added urea-N in upland rainfed dasheen.

Introduction

Urea transformation in soil remains one of the most important phenomenon associated with the use of this fertilizer in tropical crop production. Many pathways of transformation have been proposed relating primarily to the enzymatically-catalysed hydrolytic reduction of urea to NH_4^+ (Ladd & Jackson, 1982; Mulvaney & Bremner, 1981); however, it is generally accepted that the process occurs in the presence of water and the urease enzyme, first with the formation of bicarbonate and then reduction to ammonium using up free H^+, thereby increasing pH (Ferguson *et al.,* 1984). The rate at which hydrolysis proceeds has been shown to be extremely fast at times, being completed within hours of addition of urea to the soil (Nkrumah *et al.,* 1989; Sankhayan & Shukla, 1977), but generally within a few days (Kumar *et al.,* 1988; Overrein & Moe, 1967).

A major concern relates to the rise in pH following hydrolysis which is directly related to the rate of hydrolysis, urea movement in the soil and atmospheric conditions. The extent to which alkalinity occurs determines the potential for urea-N to be lost to the atmosphere as NH_3 when the fertilizer is added to the soil. Soils with the ability to buffer this tendency for pH to increase through adsorption of NH_4^+ on the exchange sites thereby reducing NH_3 loss, have been found to be associated with a high cation exchange capacity (C.E.C.) (Ferguson *et al.,* 1984; Fullerton, 1990; O'Toole *et al.,* 1985). On agricultural soils with low C.E.C.'s, the use of urea as a N- carrier is a matter of great concern as its fertilizer use efficiency can be very low due to significant NH_3-N loss. High correlations have been reported between NH_3 loss and maximum pH attained following urea application, (Fullerton, 1990; O'Toole *et al.,* 1985) and changes in pH at the soil surface presents indirect evidence of the importance of volatilization. This paper reports on some of the results of two experiments in which the N uptake efficiency in dasheen (known internationally as 'taro') was measured, and aims to determine the importance of N loss mechanisms inferred from variables measured in each situation. Dasheen is an important root crop in the Caribbean and the corms (underground edible portion of the mother plant) are consumed widely; the cormels (corms from sucker plants) are used as animal feed in many of the islands.

Research methodology

Treatments and experimental design

Two experiments were conducted to investigate the efficiency of uptake of urea-N in dasheen (*Colocasia esculenta* L. Schott) production systems on the River Estate Series (*Fluventic Eutropepts*) in Trinidad. The clay content and some chemical properties of the soil series (top 15 cm) are as follows: 33.9% clay, pH of 5.4, C.E.C. of 9.13 cmol kg^{-1}, total N of 0.19%.

In the first experiment, the land was ploughed, roto-vated and ridges and furrows were formed which are a common practice particularly in the rainy season. The ridges were spaced 90 cm apart and were about 45 cm high. Two cropping systems, monocropping and intercropping, and three N rates, 150, 300 and 450 kg N ha^{-1} band applied as urea five cm deep and five cm away from the plants and split equally at two and 18 weeks after planting, were investigated factorial-ly in a randomized block design with three replicates. Dasheen plants about 30 cm long were planted in the furrows of all plots to a depth of 15 cm in the soil and pigeon pea (*Cajanus cajan*) was sown on the ridges of the intercrop plots two weeks later. The intra-row spacing was 50 cm for dasheen and 25 cm for pigeon pea.

In the second experiment, monocrop dasheen was planted after various crops with different types of residue management. The crops and management before the cultivation of dasheen were dasheen (DD), maize (*Zea mays*) (MD), pigeon pea (PD) and pigeon pea with the plant residue returned (PRD). The residue of PRD was chopped and incorporated into the soil, while the crop residues for the other treatments were removed. The plots were roto-tilled to a depth of five cm, followed by manual preparation of ridges about 30 cm high and spaced 180 cm apart. The pigeon pea residue was incorporated in the soil between the ridges

for the PRD-treatment plots following this operation. Dasheen plants were planted in the furrows in rows 90 cm apart and 50 cm within rows to a depth of 15 cm in the soil. Each plot then received urea broadcast and incorporated into the top two cm of soil between the ridges, in equal splits at two and 18 weeks after planting at a rate of 300 kg N ha^{-1}.

The dasheen planting in both experiments was one week apart, in mid-June of 1990, at the beginning of the rainy season. All plots received basal applications of P and K at equal rates of 100 kg ha^{-1} together with the first N application and in the same manner described for N applications in the two experiments. At that time all three nutrient fertilizers were mixed well before application.

Microplots of dimensions 3.6 × 2.0 m and having 16 plants were established in all plots of both exper-iments to which 1.0 atom% urea-^{15}N solutions were applied at the rate and with the method of application specific to the experiment and treatment.

Sampling and analyses

Soil samples from the experimental plots were col-lected to determine changes in the exchangeable NH_4^+ and NO_3^- in the 0–15 cm and 15–30 cm soil layers by direct-distillation techniques (Keeney & Nelson, 1982). Samples were air-dried immediately, ground with a roller to pass a 2 mm sieve and stored in plastic bags.

In the first experiment, these samples were taken just prior to the first application of urea at two weeks after planting (w.a.p.), then at 10, 14, 16,18, 20, 21, 24 and 30 w.a.p., with the sampling at 18 w.a.p. being made just prior to the second application of urea. At each sampling, a composite sample was obtained from three locations per furrow in two randomly selected furrows. The locations in each furrow were one on the fertilizer band, one on the opposite side of the plants and the third one in the middle of the furrow between the plants.

In the second experiment, the samples were taken just before the first application of urea at two w.a.p., then at 9, 11, 14, 16, 18 w.a.p., then at 3 and 10 days, and 4 and 11 weeks after the second application of urea at 18 w.a.p. At each sampling, a composite sam-ple was made with soil obtained from three locations across the furrow in two randomly selected furrows. The locations in the furrow were one within each of the two rows of plants and the third was taken between the rows.

The changes in pH were monitored following the second application of urea. Undisturbed samples were taken by inserting a five cm diameter metal core into the soil to a depth of five cm and excavating around the core. The cores were then quickly taken to the laboratory and the pH to a depth of one cm in the middle of the core was determined with a pH meter after preparing the sample into a paste with water. In both experiments, these core samples were taken from three locations in the furrow similar to that described above. In experiment I the measurements were made six hours, and 2, 3, 6, 13, 22 and 38 days after the second fertilizer application at 18 w.a.p. One replicate for both monocrop and intercrop plots receiving the lowest (150 kg N ha^{-1}) and highest (450 kg N ha^{-1}) levels of urea were measured. In the second experiment, the measurements were made just six hours after the second application of urea, then 1, 3, 6, 9 and 22 days after urea application. For both experiments, since the measurements were made for one replicate, the data were not subjected to statistical analyses.

The four central plants in the microplots were harvested just as senescence was observed (physiological maturity) at 30 w.a.p. and seperated into leaves, petioles, main corms and sucker corms (cormels). These plants were used to determine the fertilizer uptake efficiency of applied urea-^{15}N in the crop. The isotope and N content of the plant tissues were determined on a continuous flow, (Barrie, 1991) automatic N analyser (Carlo Erba NA 1500) interfaced to a mass spectrometer (Europa Scientific Tracermas - Stable Isotope Analyser) (Craswell & Eskew, 1991; Jensen, 1991).

Marketable corm yields in 'yield plots' 7.2 m^2 were measured at 34 w.a.p. when the 'corm heads' were just visible above ground, at which stage senescence was well established and the corms were ready to be harvested. At this stage though, there would have been variable degrees of senescence among plants and translocation of N out of the plant into the soil would have occurred making it an unsuitable time to determine N uptake efficiency. The main corms from the mother plants were weighed separately from the cormels and subsampling was done to determine moisture content.

Results and discussion

Corm yields and ^{15}N uptake

Experiment I

There was no significant ($p > 0.05$) interaction between cropping system and N rate on corm yield nor was there a significant ($p > 0.05$) effect of N rate. However, the main corm yield, corm yield and total corm yield were significantly ($p < 0.05$) higher for the monocrop than for the intercrop dasheen (Table 1). The average corm yield of the intercrop dasheen was about 80% of that for the monocrop. The decrease in corm yields with intercropping has been attributed to shading (Horiuchi, 1985). The yields obtained, however, compared favourably with upland dasheen yields in Hawaii where highest yields were about 23.0 t ha^{-1}.

The uptake efficiency of urea-^{15}N in dasheen was low (Table 2), averaging 17.4%, which is lower than reported efficiencies for lowland rice (*Oryza sativa*) of between 30 and 40% (De Datta *et al.*, 1983). There was no significant ($p > 0.05$) interaction between cropping system and N rate, nor did these treatments significantly ($p > 0.05$) affect the efficiency of urea-N uptake. These low efficiencies indicate that losses of applied N

Table 1. The effect of N rate on the fresh corm yield of dasheen in Experiment I

| | N rate (kg ha^{-1}) | | | | Probability | |
	150	300	450	SEM(10df)	(p)	Mean
Main Corms						
Intercrop	12.7	14.9	10.5	1.09	0.16	12.7
Monocrop	15.8	14.5	14.3	1.09	0.16	14.9
Mean	14.3	14.7	12.4	0.77	0.13	13.8
Cormels						
Intercrop	7.03	7.44	6.43	1.05	>0.5	6.97
Monocrop	8.04	9.55	9.53	1.05	>0.5	9.04
Mean	7.54	8.49	7.98	0.74	>0.5	8.00
Total corms						
Intercrop	19.7	22.3	16.9	1.82	0.39	19.7
Monocrop	23.9	24.0	23.8	1.82	0.39	23.9
Mean	21.8	23.2	20.4	1.29	0.35	21.8

SEM(10df) cropping system; Main corms: 0.63; Cormels: 0.61; Total corms: 1.05.
p (cropping system); Main corms : 0.034; Cormels : 0.035; Total Corms : 0.017.

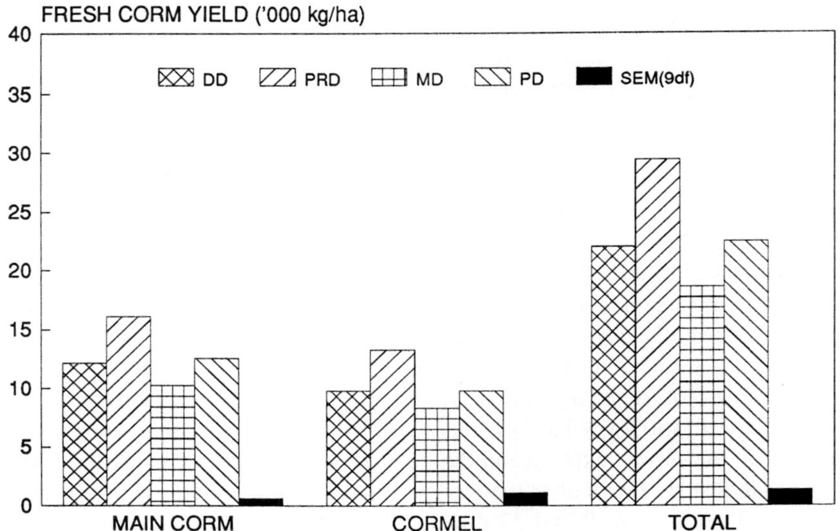

Fig. 1. Fresh corm yields of dasheen (*Colocasia esculenta*) as affected by cropping and residue management, Experiment II. DD = Dasheen, PRD = Pigeon pea (residue added) - Dasheen, MD = Maize-Dasheen, PD = Pigeon pea - dasheen.

Table 2. ^{15}N recovery by *Colocasia esculenta* as affected by cropping system and N rate with band applied urea, Experiment I

| N rate (kg ha^{-1}) | Plant ^{15}N recovery (%) | | |
| | Cropping system | | |
	Intercrop	Monocrop	Mean
150	18.0	18.3	18.2
300	17.7	19.2	18.5
450	14.0	17.0	15.5
Mean	16.6	18.2	

Overall mean =17.4%.
SEM(10df); Cropping system means: 0.7; N rate means: 0.87; Interaction means: 1.23
Probability (p); Cropping system: 0.13; N rate: 0.07; Interaction: >0.5.

Table 3. ^{15}N recovery in *Colocasia esculenta* as affected by the previous crop or residue management with broadcast urea, Experiment II

Cropping system code	Mean	SEM(9df)
DD	15.0	1.38
PRD	18.1	1.38
MD	16.0	1.96
PD	18.5	1.96
Overall mean	16.9	

were probably very high and/or the crop is inefficient in utilizing the available N.

Experiment II

The effect of the previous crop and residue management on the main corm yield was significant ($p < 0.05$) (Fig. 1); however, there was no significant ($p < 0.05$) effect on the cormel yield. The dasheen crop succeeding pigeon pea yielded 18 and 40% more than when the previous crops were dasheen and maize, respectively. Incorporation of the pigeon pea residue resulted in a 30% increase in yield. Similar benefits for upland rice preceded by cowpea have been reported (John et al., 1992). It was also reported that as a result of green manuring or residue incorporation of the cowpea, the fertilizer N requirement for rice was greatly reduced.

The efficiency of urea-^{15}N uptake by dasheen was very low as in the first experiment, averaging 16.9% and was not affected by the previous crop or residue management (Table 3).

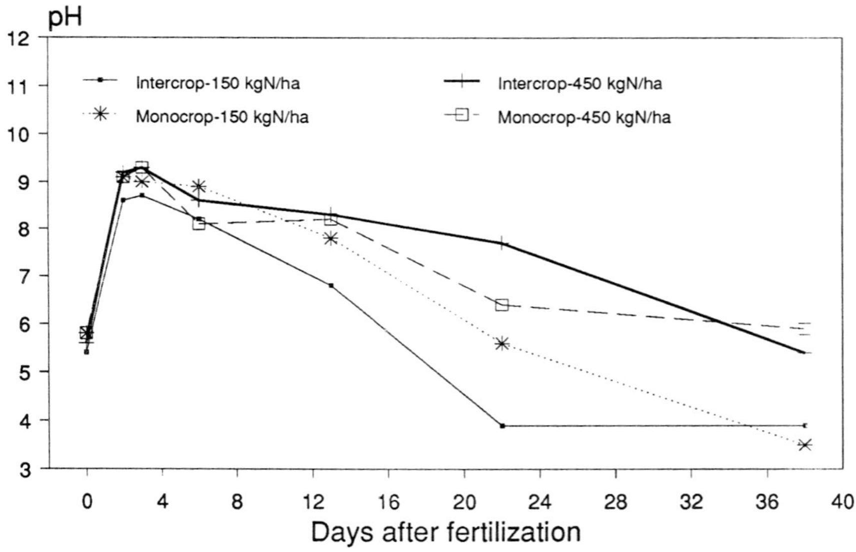

Fig. 2. pH of the top 1 cm soil layer of monocrop and intercrop soil at 150 and 450 kg N ha^{-1} after the second application of N, Experiment I.

Changes in pH following urea application

The pH of the top 1 cm of soil in the fertilized band after the second application of urea to dasheen in the first experiment showed that the highest pH values were attained and sustained longest with the application of 450 kg N ha^{-1} regardless of cropping system (Fig. 2). There were no pH changes observed 10 cm or more away from the band. The acidifying effect of urea application, which results from nitrification of the ammonium formed after hydrolysis, was observed 22 days after fertilizing (d.a.f.) with the lowest N rate of 150 kg N ha^{-1}, but even at 38 d.a.f. the pH had not dropped for the 450 kg N ha^{-1} treatments. At the lowest N rate, intercrop dasheen showed better growth than the monocrop and may have had a greater capacity for N uptake, thereby reducing the accumulation of this nutrient in the soil potentially available for volatilization. This theory is supported by the fact that at the lowest N rate, the pH of soil in the intercrop did not increase to the extent that it did for the monocrop, indicating that volatilization was probably lower in the intercrop. However, at the highest rate, the growth of dasheen was better in the monocrop and the behavior of pH in the soil (lower values for the monocrop) in the two cropping systems was the reverse to what occured at the lowest rate.

In the absence of data on growth and development, the status of soil moisture in the planted furrow for intercrop and monocrop dasheen at the lowest and highest rates of N application gives some indication on the quality of growth in these treatments. Accordingly, for a brief period of 11 days during which the second application of N fertilizer was made, soil tension in the top 15 cm was monitored along with rainfall and evaporation (Fig. 3). At the lowest N rate of 150 kg N ha^{-1} the monocrop remained wetter than the intercrop implying that evapotranspiration was greater in the former due to more vigorous growth. However, at the highest N rate, the reverse was true for dasheen in the two cropping systems. This supports the relationship proposed between pH and dasheen growth referred to in the previous paragraph but also highlights some other possibilities relating to the pH behavior observed. The positive relationship between pH and soil moisture content has been found to be due to increased rates of hydrolysis at higher levels of soil moisture (Campbell *et al.,* 1984). Additionally, movement of urea, carbonate or NH_4^+ to the surface as a result of higher soil moisture could have led to higher pH values in the wetter soils.

An important feature was that the pH increased to about 9 for all treatments by the second day after fertilizing. This suggests that the applied urea diffused to the soil surface. In a previous report, the pH was highest at the depth of the band and remained higher for longer periods than observed in these experiments (Yadvinder - Singh & Beauchamp, 1980). In the second

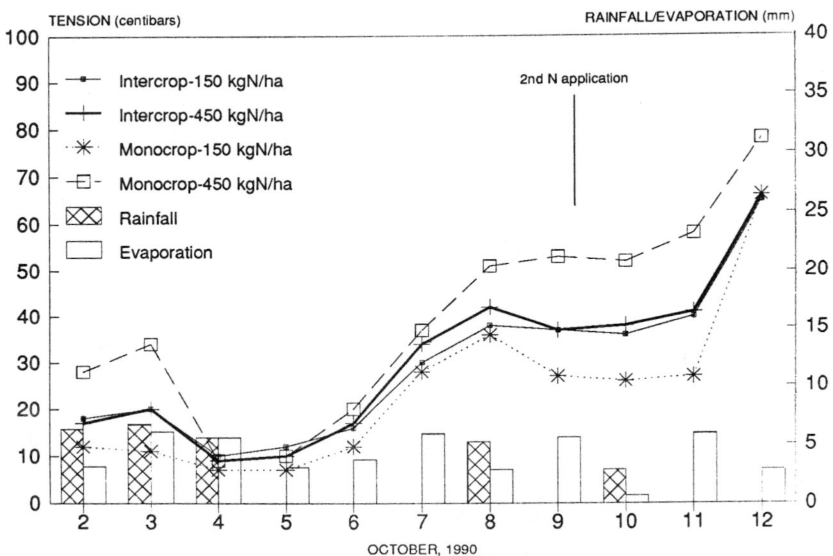

Fig. 3. Variation in soil moisture rentention, rainfall and evaporation around the time of the second application of N, Experiment I.

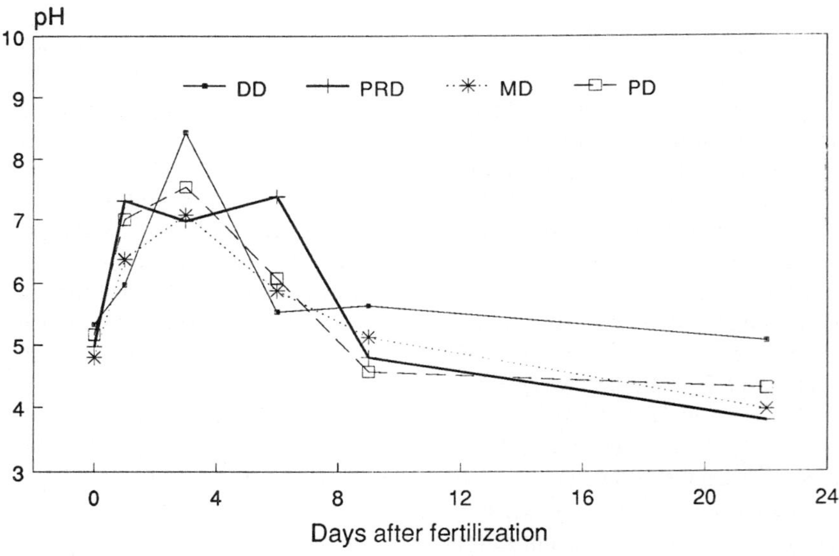

Fig. 4. pH of the top 1 cm of soil following the second application of N, experiment II. DD = dasheen, PRD = Pigeon pea (residue added) - dasheen, MD = Maize - dasheen, PD = Pigeon pea - dasheen.

experiment without banding, the soil pH maximum was lower (ranging from 7.0 to 8.5) at 3 d.a.f. and the pH declined rapidly over the next six days (Fig. 4). The pH of the soil with the pigeon pea residue elevated to a maximum of 7, but persisted 2 days longer than any of the other treatments. This effect may be related to moisture retention changes resulting from the incorporation of the residue in the soil (Kumar *et al.,* 1988).

Distribution of exchangeable NH_4^+ and NO_3^-

Experiment I

In Experiment I, urea banded in the soil hydrolysed quickly, as evidenced by the rapid increase in exchangeable NH_4^+-N within three days of the second application of fertilizer (Fig. 5). Increasing the level of N applied, significantly ($p < 0.05$) increased the exchangeable NH_4^+-N (up to 50 mg kg^{-1} with the

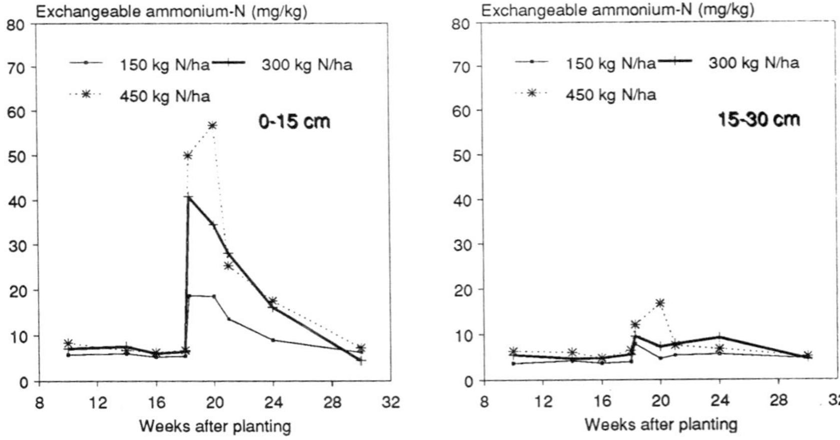

Fig. 5. Exchangeable NH_4^+-N concentrations in the 0–15 cm and 15–30 cm soil layers, Experiment I.

application of 450 kg N ha^{-1}) recovered in the 0–15 cm layer up to six weeks after fertilizing (i.e. 24 w.a.p.). However, at three and six weeks after fertilizing (i.e. 21 and 24 w.a.p.), as the N rate increased from 300 to 450 kg N ha^{-1}, no further increase in the exchangeable NH_4^+-N concentration occurred. By the 12th week after fertilizing, nitrification apparently reduced the exchangeable NH_4^+-N level to its initial value of approximately five mg kg^{-1} regardless of rate or method of urea application.

There was less NH_4^+ in the lower layer (15–30 cm) and the effect of increasing the N rate on the exchangeable NH_4^+-N concentration was not as obvious, but indicates that urea and/or NH_4^+ likely moved to the lower level. The only significant ($p < 0.05$) effect was observed two weeks after fertilizing (20 w.a.p.) when an increase of up to approximately 16 mg kg^{-1} at the greatest N rate occurred. Generally, the maximum concentration of exchangeable NH_4^+-N in this layer for each N rate was about one-third that for the top layer. These results show that movement of NH_4^+ down the profile may have occurred which was consistent with an earlier report (Chesney, 1967), for an even slightly finer-textured, less freely drained soil in Trinidad.

Soon after application of urea, the NO_3^--N concentration increased sharply (Fig. 6), showing that nitrification was very rapid. Even up to 12 weeks after the second application of urea (i.e. 30 w.a.p.), a significant ($p < 0.05$) effect of N rate was observed; the NO_3^--N concentration increasing with N rate. The maximum NO_3^--N concentration in the 0–15 cm layer of about 45 mg kg^{-1} was observed six weeks after the second application of urea (24 w.a.p.) at the highest N rate of

450 kg N ha^{-1}. The trend in the 15–30 cm layer was similar to the top layer and following the second application of urea, the maximum NO_3^--N concentration was achieved within two weeks. Subsequent levels of NO_3^--N generally showed a decline.

The most important observation from a fertilizer management standpoint was that added N, particularly when transformed to the NO_3^-, was lost from the top layer. This appeared to have been more important after the first application of urea when the soil was wetter, although the frequency of sampling around that time was not the same as it was following the second application. Whether it was due to infiltration of the mineral N forms in solution or movement of urea prior to hydrolysis, much of the added N may have become unavailable to the dasheen by being moved below the dasheen rooting zone or denitrified. That the movement was by infiltration of the soil solution in the fertilized zone to the lower depth, is supported in theory (Kirda *et al.*, 1974). However, displacement by as much as five to 10 cm through diffusion in response to an osmotic gradient was observed by other researchers (Passioura & Wetselaar, 1972; Wetselaar *et al.*, 1972). They also found that there was a drastic redistribution of NO_3^- close to the soil surface which could explain the persistence in the 0–15 cm layer of high NO_3^- relative to high exchangeable NH_4^+. The movement of urea by both infiltration and diffusion probably occurred and contributed to the loss of added N, but their importance was difficult to assess.

212

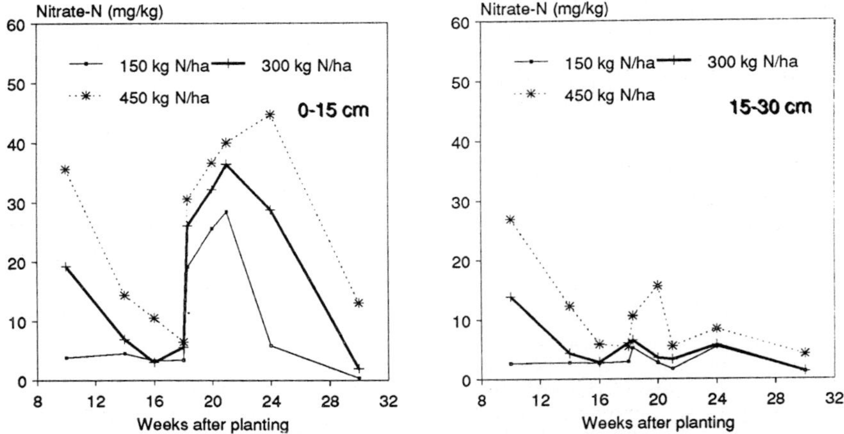

Fig. 6. NO$_3^-$ concentrations in the 0–15 cm and 15–30 cm soil layers, Experiment I.

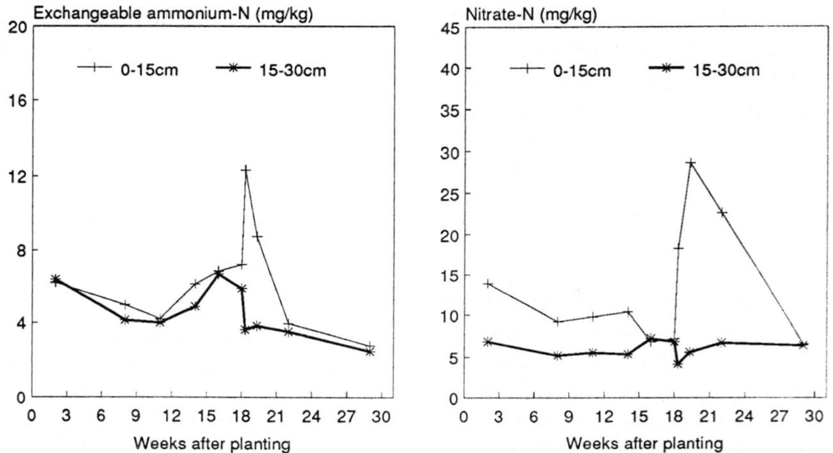

Fig. 7. Exchangeable NH$_4^+$-N and NO$_3^-$-N concentrations in the 0–15 cm and 15–30 cm soil layers, Experiment II.

Experiment II

There was no significant ($p > 0.05$) effect of the previous crop and residue management on the concentration of exchangeable NH$_4^+$-N or NO$_3^-$-N in either of the two soil layers at any sampling time. The changes over time, however, were variable, much more so for the 0–15 cm than for the 15–30 cm layer (Fig. 7). The mean exchangeable NH$_4^+$-N levels in the top layer measured three days after the second application of urea was about 12 mg kg^{-1}, which was twice the amount as before the application. On the other hand, there was no observed increase in exchangeable NH$_4^+$-N in the 15–30 cm layer following the application of the fertilizer. The mean NO$_3^-$-N concentration in the top layer showed a rapid increase from about seven to 18 mg kg^{-1} by the third day after fertilizing and up to 28 mg

kg^{-1} seven days later (Fig. 7). Again, for this N form, there was no observed increase in the concentration in the lower layer after urea was applied.

The NO$_3^-$-N levels in the soil were related to the exchangeable NH$_4^+$-N, except for a few differences probably related to the fact that urea was broadcast on the soil surface in this experiment. Nitrate was generally in greater abundance than exchangeable NH$_4^+$, which indicated that nitrification was rapid and not inhibited by conditions such as high NO$_2^-$ accumulation, which was reported to occur when urea is banded in the soil (Passioura & Wetselaar, 1972; Wetselaar *et al.*, 1972; Yadvinder-Singh & Beauchamp, 1988).

There was no evidence of movement of either NH$_4^+$ or NO$_3^-$ from the top layer down the profile. This was consistent with work done on a similar soil in Trinidad

(Nkrumah *et al.*, 1989). Fixation of NH_4^+ is important in many Trinidad soils (Ahmad *et al.*, 1982; Rodriguez, 1954). This phenomenon may have been important in restricting the movement of NH_4^+ in this experiment, particularly since the urea was broadcast and even with minor infiltration into the soil, the mass of the NH_4^+ per unit mass of soil would have been much lower than where the fertilizer was banded.

Conclusions

The corm yields were comparable to the highest reported on an upland soil; nevertheless, recoveries of added urea-N in the crop averaged only 17.4% and 16.9% for band and broadcast application of the fertilizer respectively. The pH conditions at the soil surface following application of urea either banded (Exp. I) or broadcast (Exp. II) were high enough for volatization to be an important means by which the added N was lost from the soil-plant system. Hydrolysis was not seriously restricted when urea was banded in the soil; however, it appeared that nitrification NH_4^+ to NO_3^- may have been inhibited somewhat by the very high pH attained. Further evidence of this was the higher concentration of NH_4^+-N relative to NO_3^--N for banded urea, relative to when the fertilizer was broadcast. The conditions

in the former suggest an ideal situation for possible accumulation of NO_2^-. When urea was banded, downward movement of NO_3^- and to a lesser extent NH_4^+ (since the C.E.C. of the soil was less than 10 cmol kg^{-1}) was important but was not observed when it was broadcast.

A higher proportion of the N in the dasheen was derived from the fertilizer (%Ndff) for band compared with broadcast application of the urea (Table 4). This meant that the urea was closer to the root system when banded in the soil. However, since there was no observed difference in the plant recoveries in both experiments, it follows that since the uptake of N was greater in the Experiment II, where the urea was broadcast, there was also alot more luxury consumption since yields were not higher. There is therefore need to investigate other methods of application, such as placement closer to the root zone and more frequent smaller doses in order to increase efficiency.

Table 4. The %Ndff in *Colocasia esculenta* as affected by method of fertilizer application

			%Ndff
			Method of application
Band	N rate ($kg\ ha^{-1}$)	150	47.5
		300	58.5
		450	67.1
		SEM(10df)	0.86
		Overall mean	57.7
		p	<0.01
		DD[a]	33.7
		PRD	36.9
Broadcast ($300kg\ N\ ha^{-1}$)		MD	36.7
		PD	38.0
		SEM(9df)	2.18
		Overall mean	36.4
		p	>0.05

[a]DD – Dasheen - dasheen, PRD – Pigeon pea (residue added) - dasheen, MD – Maize - dasheen, PD – pigeon pea - dasheen.

References

Ahmad N, Reid ED, Nkrumah M, Griffith SM & Gabriel L (1982) Crop utilization and fixation of added ammonium in soils of the West Indies. Plant Soil 67: 167–186

Barrie A (1991) New methodologies in stable isotope analysis. In: Stable Isotopes in Plant Nutrition, Soil Fertility and Environmental Studies, pp 3–6. Proceedings of a symposium, Vienna, October 1–5, 1990. Vienna, Austria

Campbell CA, Myers RJK, Catchpoole VR, Vallis I & Weier KL (1984) Laboratory study of transformation and recovery of urea-N in three Queensland soils. Aust J Soil Res 22(4): 433–441

Chesney HAD (1967) Denitrification in some Trinidad soils. M.Sc. Thesis. Dept. of Soil Science, The University of the West Indies, St. Augustine, Trinidad and Tobago

Craswell ET & Eskew DL (1991) Nitrogen and nitrogen-15 analysis using automated mass and emission spectrometers. Soil Sci Soc Am J 55: 750–756

De Datta SK, Fillery IRP & Craswell ET (1983) Recent results on nitrogen efficiency studies on wetland rice. Outlook Agric 12: 125–134

Ferguson RB, Kissel DE, Koelliker JK & Basel W (1984) Ammonia volatilization from surface-applied urea : effect of hydrogen ion buffering capacity. Soil Sci Soc Am J 48(3): 578–582

Fullerton TP (1990) Volatilization of ammonia from urea in contrasting Trinidad soils. Ph.D. Thesis. Dept. of Soil Science, The University of the; West Indies, St. Augustine, Trinidad and Tobago

Horiuchi T (1985) Growth and yields of field crops in intercropping culture. JARQ 19(1): 1–6

Jensen ES (1991) Evaluation of automated analysis of ^{15}N and total N in plant material and soil. Plant Soil 133: 83–92

John PS, Pandey RK, Buresh RJ & Prasad R (1992) Nitrogen contribution of cowpea green manure and residue to upland rice. Plant Soil 142: 53–61

Keeney DR & Nelson DW (1982) Nitrogen - Inorganic forms. In: Page AL *et al.* (eds) Methods of Soil Analysis. Part 2. Chemical and Microbiological Properties pp 643–698. ASA–SSSA, Madison, Wisconsin

Kirda C, Nielson DR & Biggar JW (1974) The combined effects of infiltration and redistribution on leaching. Soil Sci 117: 323–330

Kumar V, Yadav DS & Singh M (1988) Effects of urea rates, farmyard manure, $CaCO_3$ salinity and alkalinity on urea hydrolysis and nitrification in soils. Aust J Soil Res 26: 367–374

Ladd JN & Jackson RB (1982) Biochemistry of ammonification. In: Stevenson FJ (ed) Nitrogen in Agricultural Soils, pp 173–228. ASA–CSSA–SSS, Madison, Wisconsin

Mulvaney RL & Bremner JM (1981) Control of urea transformations in soils. In: Paul EA & Ladd JN (eds) Soil Biochemistry. Vol. 5, pp 153–196. Marcel Dekker, Inc., New York, USA

Nkrumah M, Griffith SM & Ahmad N (1989) Lysimeter and field studies on ^{15}N in a tropical soil. II. Transformation of $(NH_2)_2CO$-^{15}N in a tropical loam in lysimeter and field plots. Plant Soil 114: 13–18

O'Toole P, Mc Garry SJ & Morgan MA (1985) Ammonia volatilization from urea-treated pasture and tillage soils: effects of soil properties. J Soil Sci 36(4): 613–620

Overrein LN & Moe PG (1967) Factors affecting urea hydrolysis and ammonia volatilization in soil. Soil Sci Soc Am Proc 31: 57–61

Passioura JB & Wetselaar R (1972) Consequences of banding nitrogen fertilizers in soil. II. Effects on the growth of wheat roots. Plant Soil 36: 461–473

Rodriguez G (1954) Fixed ammonia in tropical soils. J Soil Sci 5: 264–274

Sankhayan SD & Shukla UC (1977) Rates of urea hydrolysis in five soils of India. Geoderma 16: 171–178

Wetselaar R, Passioura JB Singh BR (1972) Consequences of banding nitrogen fertilizers in soil. I. Effect on nitrification. Plant Soil 36: 159–175

Yadvinder-Singh & Beauchamp EG (1988) Nitrogen transformations near urea in soil with different water potentials. Can J Soil Sci 68: 569–576

Fertilizer Research **42**: 215–230, 1995.
© 1995 *Kluwer Academic Publishers. Printed in the Netherlands.*

Nitrogen fixation by trees in relation to soil nitrogen economy

Y.R. Dommergues
11 rue Maccarani, 06000 Nice, France

Key words: Bradyrhizobium, clonal selection, *Frankia*, inoculation, nitrogen-fixing trees, *Rhizobium*, soil nitrogen economy

Abstract

The N_2-fixing potential (NFP) (i.e. the amount of fixed N_2 in a constraint-free environment) of N_2-fixing trees (NFTs) varies with the genotype. The NFP can be higher than 30-50 g N_2 fixed tree^{-1} year^{-1} in the most active species, be they leguminous trees such as *Albizia lebbeck*, *Gliricidia sepium* and *Leucaena leucocephala*, or actinorhizal trees such as *Casuarina equisetifolia*. The actual amount of nitrogen fixed (ANF) (i.e. the amount of N_2 fixed in the field) is lower than the NFP or even nil because of various constraints, especially drought, nutrient deficiencies, excess of available N and pathogenic nematodes. As tree litters are mineralized, the amount of available N in the soil increases with time, this process leading to the cessation of N_2 fixation in aging plantations. When the mineralization rate is slowed down or inhibited, N_2 fixation can continue. NFTs improve the N status of soils, but the transfer of fixed N to associated plants is not always ensured. Three main approaches are appropriate to increase N_2 fixation: clonal selection of trees combined with vegetative propagation, inoculation with effective rhizobium or *Frankia* strains, and proper fertilization (especially P). In the absence of major environmental constraints, a positive response to inoculation is expected only when specific (non-promiscuous) NFTs are grown in sites where the density of compatible rhizobia is low or nil. The potentialities of NFTs are far from being fully exploited. Further investigations are proposed and the economics of NFT management is briefly discussed.

Introduction

N_2 fixing trees (NFTs) are found in 10 plant families, including nine flowering plant families and one family of gymnosperms (Zamiaceae). According to the symbiotic microorganism associated with the host plant, N_2-fixing plants fall into three categories: Leguminosae and a few Ulmaceae associated with rhizobia (*Rhizobium* or *Bradyrhizobium*); actinorhizal plants associated with *Frankia*; and Cycadales associated with Cyanobacteria. Most N_2-fixing plants belong to the family Leguminosae: 3003 species (including woody and non-woody ones) have been identified as N_2-fixers. The majority are Papilionoideae (2514), followed by Mimosoideae (409) and Caesalpinioideae (80) [10, 31]. Taking into account the fact that only a fraction of the Leguminosae have been examined for nodulation, Brewbaker [10] estimates that 4500 species of Leguminosae (out of a total of 17600 species) would be NFTs. Actinorhizal plants are distributed among eight families. To date, about 194 species belonging to 24 genera have been identified [8]. They are woody shrubs or trees except species belonging to the family Datiscaceae (Table 1).

The most remarkable trait of NFTs is that they are able to meet their own requirements in N, which allows them to colonize and rehabilitate wasted lands. NFTs are also the major source of N in tropical ecosystems: forests, various agrosystems including mixed and sequential cropping systems, agroforestry and silvopastoral systems [111, 112]. Finally NFTs provide timber wood (fuel, timber, pulp), fodder and even human food. Excellent reviews and books have recently been published specifically discussing the establishment and functioning of NFT symbioses [13, 23, 41, 78, 86, 95].

The objective of the present communication is not to repeat these presentations. It will be an attempt (i)

Table 1. Symbiotic microorganisms associated with NFTs [10]

Families	Representative genera	Symbiotic organisms
Leguminosae	*Acacia, Dalbergia, Leucaena, Prosopis, Erythrina, Gliricidia, Sesbania*	*Rhizobium* and/or *Bradyrhizobium*
Ulmaceae	*Parasponia*	*Bradyrhizobium*
Betulaceae	*Alnus*	*Frankia*
Casuarinaceae	*Allocasuarina, Casuarina Gymnostoma*	*Frankia*
Coriariaceae	*Coriaria*	*Frankia*
Elaeagnaceae	*Elaeagnus, Hippophae, Shepherdia*	*Frankia*
Myricaceae	*Comptonia, Myrica*	*Frankia*
Rhamnaceae	*Ceanothus, Colletia, Discaria Trevoa*	*Frankia*
Rosaceae	*Cercocarpus, Dryas, Purshia*	*Frankia*
Zamiaceae	*Macrozamia*	Cyanobacteria

Remark. Plants in symbiosis with *Frankia* are called actinorhizal plants because *Frankia* is an actinomycete. The actinorhizal family Datiscaceae is not included in the list above since it is an herbaceous plant.

to throw light on the specific role of NFTs in the N fluxes through forestry and agroforestry ecosysterns and (ii) to draw a general picture of the different means currently available to maximize the N_2-fixing rate of NFTs in the field.

N_2-fixing potential and actual N_2 fixation

To ensure the development of improved methods for enhancing N_2 fixation and more specifically to properly manage NFTs, it is mandatory to determine (i) the N_2-fixing potential (NFP), i.e. the amount of N_2 fixed in a constraint-free environment; (ii) the actual N_2 fixed (ANF), i.e. the amount of N_2 fixed under field conditions, where various constraints affect the expression of the NFP.

NFP is an intrinsic characteristic of the species (or more exactly of the 'tree genotype × strain combination') under study. A correct evaluation of the NFP requires that the tree under study be grown under the most favorable conditions: adequate inoculation, use of a soil with a low content of available N to avoid the inhibitory effect of this form of N, proper fertilization (including trace elements and major elements except N), carefull irrigation, and appropriate climatic condi-

tions. Obviously these ideal conditions can never be met, but they can be approached close enough to give a fair evaluation of the NFP.

In the example given in Table 2, the NFP of *Casuarina equisetifolia* is shown to be 42.4 g N_2 fixed tree^{-1} year^{-1} (by extrapolation 84.8 kg N_2 fixed ha^{-1} year^{-1} in a plantation with a density of 2,000 trees ha^{-1}). The ANF is much lower than the NFP because the process of N_2 fixation is limited by environmental constraints. At Notto station the ANF was 4.3–6.0 g N_2 fixed tree^{-1} year^{-1}; at Malika it was 29.0. The differences in the ANF measured in the two stations can be explained by the fact that at Malika the soil is more fertile than at Notto and the sea is much closer so that the trees benefit from a complementary source of water (dew and spray from the ocean).

It is clear that the data on N_2 fixation published in the literature refer to systems that have been subject to diverse constraints. Therefore it is not surprising that, for a given tree species, there are large variations in the estimations of the total amount or percentage of N_2 fixed.

At present we have very few data on N_2-fixing rates of trees grown in optimal conditions (i.e. in conditions nearly equivalent to those required to estimate the NFP). However, by interpreting the information

Table 2. N_2-fixing potential (NFP) and actual N_2 fixation (ANF) of *Casuarina equisetifolia*

Exp. n° and experim. site	Method of assess-ment[a]	Age	N_2 fixed per tree (mean cumulative value)	N_2 fixed tree^{-1} year^{-1} (mean)	Ref. [b]
N_2-fixing potential[b]					
1. ORSTOM	TND	24 mo	84.8 g	42.4 g	[92]
station	AV	24 mo	84.4 g	42.2 g	
Dakar					
Actual N_2 fixation (as assessed in the field)					
2. Notto	TND	3 y	13.4 g	4.5 g	[91]
3. Notto	TND	3 y	18.0 g	6.0 g	[61]
	NA	3 y	13.0 g	4.3 g	
4. Malika	Bal.	13 y	377.0 g[c]	29.0 g	[21]

[a] Method of assessment of N_2 fixation: TND, total N difference; AV, A value; Bal., N balance; NA, N^{15} natural abundance.
[b] Trees grown in 1 m^3 containers filled with a sandy N-deficient (total N, 0.01%) soil, carefully irrigated and fertilized (addition of 200 ml of a N-free nutrient solution every week).
[c] Assuming a plantation density of 2000 trees ha^{-1}.

Table 3. Tentative classification of some N_2-fixing trees according to their N_2 fixing potential (NFP)

Species proved to have high or medium NFP (> 30–50 g N_2 fixed tree^{-1} year^{-1}):
Albizia lebbeck, Gliricidia sepium, Leucaena leucocephala, Casuarina equisetifolia

Species proved to have low NFP (< 30 g N_2 fixed tree^{-1} year^{-1}):
A. senegal, Faidherbia albida

Species assumed to have high or medium NFP: *Acacia auriculiformis, A. crassicarpa, A. dealbata, A. mearnsii, A. saligna, A. seyal, Calliandra calothyrsus, Erythrina poeppigiana, Flemingia macrophylla, F. strobilifera, Inga jinicuil, Paraserianthes falcataria, Prosopis tamarugo, P. glandulosa, Robinia pseudoacacia, Sesbania spp., Casuarina glauca, C. cunninghamiana.*

Species whose NFP is still unknown:
Acacia nilotica, A. raddiana, Mimosa scabrella, Prosopis spp., Allocasuarina verticillata, A. decaisneana, Casuarina junghuhniana, C. oligodon, Gymnostoma spp.

Non N_2-fixing species:
Bauhinia sp., Ceratonia siliqua, Parkia biglobosa, Parkinsonia aculeata, Senna (syn.:Cassia) siamea, S. spectabilis, Tamarindus indica

published in different reviews [13, 23, 41, 97], we propose a tentative classification of some NFT based on their NFP (Table 3). Many more investigations are obviously required to place the fifty major NFT species identified by the Nitrogen Fixing Tree Association [10] in the framework of the proposed classification. There are still few data on the amounts of N_2 fixed by NFTs in the field (ANF) because of the great task of destructively harvesting and then analysing aerial parts and roots to evaluate their total N content. This task has not been completed so that the data concern only the percentage of N_2 fixed, which is also designated as N_2 derived from the atmosphere (Ndfa %). This percentage is by no means characteristic of a given species, but allows comparisons between different species grown in similar environmental conditions; it can also facilitate the evaluation of the effect of environmental conditions on the activity of a given N_2-fixing species.

The methodology and problems raised by the accurate assessment of N_2 fixation itself are discussed in other papers in this issue.

Specific features of N_2-fixing trees

The expression of the N_2-fixing activity of NFTs differs somewhat from that of annual plants because of some specific characteristics regarding their nodulation and ability to recycle N.

Nodulation

In addition to transient nodules, a number of NFTs such as *Albizia lebbeck*, some *Acacia* and all species of the Casuarinaceae, bear perennial nodules. The relative advantages of perennial vs. transient nodulation in semi-arid or arid conditions are assumed to be twofold: (i) N_2 fixation would function longer when the rains have stopped, (ii) active lobes from existing nodules would readily develop at the onset of the rainy season, at a time when new infections are still limited by water constraints. Both assumptions require further investigation [74, 97].

The pattern of nodule distribution exhibits wide variations. In many circumstances nodules occur in the litter layer or in upper soil horizons. Some NFTs, especially phreatophytes such as *Prosopis* sp. [32, 53, 94] or *Allocasuarina decaisneana* [74] form nodules at considerable depths (sometimes more than 5–10 m below-ground), which allows N_2 fixation to function throughout the dry season. Deep rooting also increases tolerance to dry conditions by supplying the nodules located in upper horizons with adequate moisture.

Finally one should mention that a few plants form nodules not only on their roots (i.e. underground) but also on their stems (i.e. in the air). One well known example is that of *Sesbania rostrata* [27], another one is that of *Aeschynomene elaphroxylon*, a perennial legume growing in Sahelian lowlands; this woody *Aeschynomene* bears sparse aerial nodules at the base of its trunk [27]. In specific conditions, two species of Casuarinaceae, *Casuarina cunninghamiana* and *C. glauca*, have been reported to bear many actinorhizal nodules on their trunks [27, 71]. Aerial nodulation would probably make the host plant more independant of soil constraints, thus circumventing the inhibitory effect of available N that is accumulated in many aging plantations (see infra).

In favorable situations, the nodule biomass of NFTs can be as high as 500 g tree^{-1}. In *Inga jinicuil*, a NFT used for shade in Mexican coffee plantations, nodule biomass was reported to be approximately 346 g (dry weight) tree^{-1} [77]. In a number of sites nodules cannot be detected even when compatible rhizobia or *Frankia* are present in the soil because (i) nodules may have been formed but decomposed too rapidly or they are too small [54] or (ii) nodulation was impeded by environmental constraints.

Mineralization of litters; N recycling and time course of N_2 fixation in NFT plantations

There is a flow of N from the canopy to the forest floor in litter fall (leaves, branches, flowers and pods), and in throughfall and stemflow of rainwater. There is a similar flow from the root system to the soil in underground litter and root exudates. N is gradually released, and thus made available, from the decomposing biomass (essentially above-ground and below-ground litters) by decomposition mediated by soil animals and microorganisms, the mineralization rate depending very much on species, soil and climate [2, 94]. Part of the N released is taken up by the tree, a process known as N recycling. When the amount of available N in the soil exceeds a certain threshold, this form of soil N is preferred over atmospheric N_2 for meeting the N requirements of the tree; simultaneously, nodulation and N_2 fixation are inhibited.

The first evidence of internal N recycling between a N_2-fixing tree and a non-N_2-fixing understory vegetation was provided by the study of a *Leucaena leucocephala* plantation established in January 1986 at

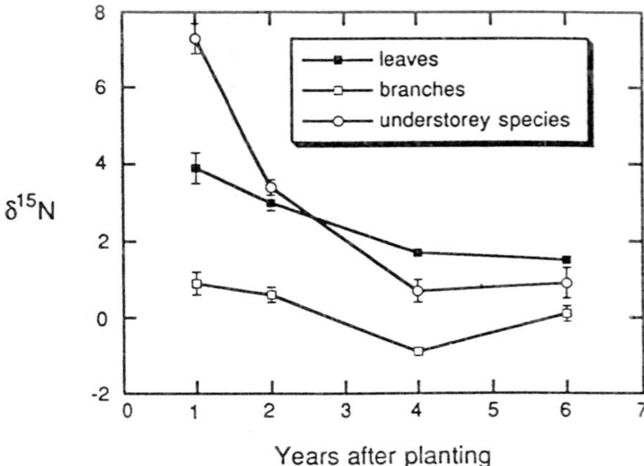

Fig. 1. Changes with time in the mean δ ^{15}N values of *Leucaena leucocephala* leaves and branches and understory species. The *L. leucocephala* plantation was established in January 1986 at an altitude of 320 m on the island of Maui, Hawaii. Annual precipitation: 1875 mm. Soil: clayey, ferritic, isohyperthermic humoxic tropohumult. Inoculation of *L. leucocephala* with a mixture of strains (TAL1145, TAL 82 and TAL 582). Fertilization: 500 kg P ha^{-1} and enough dolomite to increase pH to 6.0. Spacing 1.5×1.5 m [107].

an altitude of 320 m on the island of Maui, Hawaii [107]. The authors measured the changes over time in the δ^{15}N values of *Leucaena leucocephala* leaves and branches and understory species. The δ^{15}N of the *L. leucocephala* leaves was 3.9‰ at 1 year after planting (YAP), 3.2‰ at 2 YAP, 1. 7‰ at 4 YAP and 1.5‰ at final harvest (Fig. 1). During the same period the δ^{15} of the understory species decreased from 7.3‰ at 1 YAP, to 3.4‰ at 2 YAP, O.7‰ at 4 YAP, and 0.9‰ at final harvest. At 6 YAP, the δ^{15}N of *L. leucocephala* leaves and branches and that of the non-N$_2$-fixing understory vegetation were similar. These data indicate that: (i) some of the N$_2$ fixed by *L. leucocephala* had been transferred to soil and associated non-N$_2$-fixing plants, the replenishment of the mineral soil N pool being achieved through net mineralization of *L. leucocephala* litters; (ii) *L. leucocephala* had absorbed some N from the mineral soil N pool, thus recycling a part of the N$_2$ it had fixed; (iii) the N recycling was rapid since the decline in the δ^{15}N value of the understory vegetation was remarkably sharp during the second year.

N cannot be recycled if the litters are not decomposed. Though exceptional, this situation is known to occur. It has been reported in the case of *Prosopis alba* and *P. tamarugo* forests of the rainless Atacama desert, Chile, that the trees, which are typical phreatophytes, obtain their moisture from ground water at depths of 6-8 m or more. The litter is deposited on the forest floor where it remains undecomposed because of the drought and also because the roots are mechanically prevented from growing into the litter by a thick carbonate layer. The litter, which accumulated in a 45 cm-thick layer during a period of 80–122 years, averages 85 kg m^{-2}. The trees derive all the N they require from biological N$_2$ fixation, which persists in mature plantations since it is not inhibited by available soil N. In other words these forests behave as open systems [29].

An intermediate situation was found in *Casuarina equisetifolia* plantations growing on the coastal sand dunes of Senegal where the decomposition of the litter is limited by different factors (drought, lack of P and absence of adapted microfauna). The litter accumulated in a 8 cm-thick layer during a period of 34 years averages 12 kg m^{-2} [60]. In these plantations the amount of N recycled is probably trivial and the available soil N pool is too low to impede nodulation (which is still abundant in old plantations) and consequently N$_2$ fixation.

Considerable evidence indicates that:

(i) Like other trees, NFTs translocate nutrients (including N) from soil horizons well beyond the reach of crops and release them at the soil surface, thus enhancing the nutrient status of surface soil beneath their canopies.

(ii) Internal fluxes of N occur within the tree itself. They result from the immobilization of N as reserves or its remobilization for regrowth. These processes complicate the assessment of N$_2$ fixation by isotope methods, a problem which has already been addressed by a number of authors [13, 20 113].

Transfer of fixed N$_2$ to soil and associated crops

Transfer to soil

Forests

Afforestation with NFT always increases soil N content provided that the ANF is significant. Thus at Banthra station, Lucknow, India, 8 years after planting total N increased about 6-fold in the surface soil of *Prosopis juliflora* and 4-fold in that of *Acacia nilotica*, compared with the initial values [36]. The rate of increase with depth was higher in the *P. juliflora* plantation than in the *A. nilotica* plantation. The amount of N accumulated under *Prosopis* sp. growing in other sites was found to be of the same order [101]. Another example is that of *Acacia mollissima* grown for tannin production in South Africa. The soil total N content was 0.35% in non afforested sites and 0.53% under 30-year-old plantations, which suggests an accumulation of more than 100 kg N ha^{-1} year^{-1} under the forest cover [67].

When the decomposition rate of the litter is slow, N is accumulated mostly as litter. In Senegalese drifting sand dunes, the soil total N content (almost entirely as litter N) increased from 80 kg N ha^{-1} to 309 kg N ha^{-1}, a 4-fold increase 13 years after planting *Casuarina equisetifolia* [21]. More recently Mailly and Margolis [60] estimated that the accumulation of N had reached 1567 kg ha^{-1} in 17–34 year-old *C. equisetifolia* plantations growing in the same region. The latter estimation is much higher than the former one because the plantations studied were older and located in more fertile microsites.

The increment of soil N would be much lower if the trees were felled and the entire aboveground biomass exported. In the case of a 7-year-old plantation of *Acacia mangium* in the Ivory Coast, the total N content of the aboveground components was 832 kg N ha^{-1}, that of the wood component was only 435 kg N ha^{-1} [9]. To reduce N losses only wood should be harvested and leaves, fruit and, when possible, branchlets (60 kg N ha^{-1}) should be left on the spot. Such a procedure would also limit the loss of other nutrients.

Alley cropping
Here crops are grown in alleys formed by hedgerows of NFT. The NFTs are periodically pruned and the biomass thus harvested is used as a green manure for the crops. The net N contribution of the prunings (Ndfa) is calculated as:
Ndfa = pruning biomass×N%×Ndfa%,
N% being the percentage of total N in the prunings and Ndfa % the percentage of pruning total N derived from N$_2$ fixation.

In favorable conditions, the N input to soil can be higher than 100 kg N ha^{-1}. Thus if 10 t of prunings (N% = 3.5 %) are applied ha^{-1}, and if Ndfa % is 45 % (which is an average value in the case of *L. leucocephala* growing in Nigeria), the input originating from the N$_2$ fixation process would be 160 kg N ha^{-1} [55]. However, such a beneficial effect is not observed when the tree biomass is low, the litter quality poor and the cropping sequence inadequate [62].

Prunings from the NFT hedgerows add much needed organic matter to the soil and contribute significantly to the recycling of nutrients other than N, those elements being also essential to ensure sustained crop production [55].

An interesting question is whether or not N$_2$ fixation can compensate the losses of N following the clearing of a forested area for agricultural use. The answer is no, because losses are considerable, especially during the years immediately following deforestation. Thus on a land newly cleared from a secondary forest regrowth of about 5 years in southwestern Nigeria and subsequently grown in rotation with maize (with 120 kg N fertilizer applied at each crop) and cowpea (with no N fertilizer) between alleys of *Gliricidia* or *Leucaena*, the loss of N reached approximately 4000 kg N ha^{-1} during the two first years and 1000 kg N ha^{-1} during the two last years (recalculated from [58]). Another example illustrating this sharp decline of soil N content resulting from deforestation is given in this issue by du Preez (This issue).

Transfer to associated crops

Mixed forests
The field experiment [107] mentioned above clearly shows that the N$_2$ fixed by the NFT (*Leucaena leucocephala*) is transferred to the associated non-N$_2$-fixing plants of the understory. This transfer occurs after mineralization of the aboveground and underground litters, including root exudates. When the mineralization rate of NFT litters is slow (e.g. in *Casuarina equisetifolia* growing in semi-arid conditions) the transfer is very limited.

Alley cropping
The transfer of N from prunings to associated crops represents less than 30–40% of the N yield of the prunings. This low efficiency probably results from the lack of synchronization between crop demand and N release from the prunings, which is rapid since 55% of the pruning N is released within 52 days [110]. In addition, there are losses, through denitrification, volatilization and leaching. Factors affecting N utilization efficiency have been discussed in a recent review [55].

In alley cropping systems based on the use of *Leucaena leucocephala* and *Gliricidium sepium* in the humid tropics, the pruning contribution varies from 5 to 50 kg N ha^{-1} [55].

N is not only released from the prunings through decomposition [110], but also from exudates originating from NFT roots and nodules, the exudation being triggered by defoliation and plant cutting [81].

Improved fallows
Including NFT in fallows reduces the length of the fallow period by accelerating soil restoration through the addition of substantial amounts of N and organic matter to the soil, the recycling of nutrients from deeper horizons and the improvement of soil physical properties. This is perhaps the most realistic management option for sustaining agroforestry systems. It enables periodic replenishment of the nutrient resources provided the cropping phase has not been too long and led to irreversible changes [2, 7]. Promising NFT species include *Acacia senegal* in semi-arid conditions [2], *Leucaena leucocephala* and *Albizia lebbeck* in the humid tropics (R. Oliver, private commun.) and *Acacia mollissima* in tropical highlands. On La Réunion Island the latter NFT has been sucessfully used in rotation with geranium for many years.

Park-like agroforestry systems
Faidherbia albida-based systems. *F. albida* is considered as a highly valuable component of park-like agroforestry systems in the Sahel [12, 105]. It is well known that crops growing under *F. albida* have higher yields than crops growing away from the canopy. Different explanations are proposed to explain this beneficial effect. They include decrease of soil surface temperature early in the cropping season, return of nutrients from deep layers to the surface soil through litter fall, deposition of feces and urine by cattle seeking shade and N_2 fixation [38]. It is now clear that in semi-arid and arid conditions the soil improvement does not result from N_2 fixation since *F. albida* bears few or no nodules [72]. Using the natural abundance method, Schulze et al. [85] have also shown that the N_2-fixing activity of *F. albida* in semi-arid conditions is negligible. By contrast, in moist sites (annual rainfall higher than 1000–1500 mm), nodulation is good to prolific [28, 72], suggesting that the contribution of *F. albida* as a N_2 fixer to the soil N pool is significant. However, the amount of N_2 fixed in these sites is still unknown.

Prosopis cinerea-based systems. They are also recognized for their outstanding role in the rural economy of the north-west region of the Indian subcontinent [87]. Whether higher soil productivity in these systems

partly results from N_2 fixation or exclusively from other processes is still unclear.

Improving N_2 fixation by trees

Inoculation with selected symbiotic N_2-fixing bacteria

Different genera of symbiotic N_2-fixing bacteria
Legumes form a symbiosis with species of the bacterial family Rhizobiaceae which comprises three recognized genera: *Rhizobium*, *Bradyrhizobium* and *Azorhizobium* [88, 97]. Other *Rhizobium* species and genera have been proposed, namely *Photorhizobium* or *Sinorhizobium* [41] and new ones are currently under study [72]. For the sake of simplification, we admit here that rhizobia from tropical NFT fall into two broad groups: the fast-growing, acid-producing rhizobia, designated as *Rhizobium*, and the slow-growing rhizobia, which do not produce acid an yeast-mannitol media, designated as *Bradyrhizobium*.

The proposed classification of *Frankia* into three major host specificity groups (*Alnus*-infecting, *Elaeagnus*-infecting and *Casuarina*-infecting strains) based on preliminary cross-inoculation studies was not confirmed by subsequent infection tests, nor by genetic or phylogenetic means. The situation is complicated because the diversity of the *Frankia* strains is considerably greater than initially suspected. Future taxonomic studies will have to take into account the fact that only a limited number of infective strains from four artinorhizal families (Betulaceae, Myricaceae, Elaeagnaceae and Casuarinaceae) out of eight have been isolated up to now [8, 65, 76, 86].

Selecting strains of rhizobia and Frankia
Screening wild strains. In a first step, a large collection of wild strains should be developed using isolates of rhizobia or *Frankia* obtained from the NFT species under study or other species within the same inoculation group (i.e. having the same specificity for nodulation). Isolates nodulating a given group of host plants are said to be compatible with this group; they are homologous.

In a second step, the strains should be tested under growth room or greenhouse conditions using the same methodology as for annual legumes [102] or pasture legumes [98]. The screening procedure should yield strains compatible with the host plant, effectively fixing N_2 over a range of environmental conditions (including excess of available N), out-competing other strains) having good genetic stability and persisting in the soil. When our knowledge on the diverse roles of the symbiotic bacteria have progressed, it will be possible to take into account the symbiont's ability to affect the host plant physiology by synthesizing phytohormones, siderophores or other subtances directly or indirectly influencing plant growth [18, 41]. In fact, screening procedures currently are left largely to the whims of individual investigators and a standard protocol for screening superior strains in the laboratory is needed [50].

The third step should be to evaluate the response of NFT species to inoculation with the selected strains in the nursery and later in the field after transplanting. Field testing should include periodic evaluation of the following parameters: nodulation and nodule occupancy by introduced strains (competitive ability); tree growth (height, diameter and possibly biomass); amount of N_2 fixed (effectiveness).

Molecular biology methods. The recent developments of molecular biology have made it possible to map the genes involved in N_2 fixation and nodulation, and improve our understanding of the mechanisms of establishment of the symbioses. They also have allowed the identification of strains of rhizobium or *Frankia* by DNA probes or marker genes and clarified their evolutionary relationships [8, 15, 41, 47, 70]. Engineering novel bacterial strains is possible. However, too little is known about the genes which make organisms useful and it will be a long time before new microsymbionts are used in the field [47].

Specificity of the host plant
We have already stressed that rhizobium or *Frankia* specificity for nodulation and effectiveness serves as the basis for selecting the strains suitable for preparing the inoculants.

The question of specificity should also be looked at not only from the microsymbiont point of view but also from that of the host plant. NFTs differ drastically in their rhizobial or frankial requirements. Like non-woody tropical legumes [14, 44], tropical woody legumes can be grouped into three broad categories based on their nodulation and N_2 fixation specificity.

The first group contains species that are nodulated by a large range of genetically diverse strains and fix

N_2 effectively in association with these strains. These species are promiscuous for nodulation and effectiveness, e.g. *Acacia crassicarpa* (SG) (unpublished result), *Acacia auriculiformis* (SG), *Albizia lebbeck* (SG), *Paraserianthes falcataria* (SG) [101].

The second group comprises species that are nodulated by a wide range of genetically diverse strains but fix N_2 with a limited number of specific strains. These species are promiscuous for nodulation and specific for effectiveness, e.g. *Acacia mangium* (SG) (A. Galiana, personal commun.), *Robinia pseudoacacia* (FG), *Acacia mearnsii* (SG) [101].

The third group is made of species that are very specific in their requirements i.e. that nodulate effectively with a narrow range of strains. These species are specific for nodulation and also for effectiveness, e.g. *Leucaena leucocephala* (FG), *Gliricidia sepium*, (FG) and *Calliandra calothyrsus* (FG) [101].

This classification is appropriate whatever type of rhizobium is involved, be it slow-growing (SG) or fast-growing (FG.).

The distinction in promiscuous and non-promiscuous species holds for actinorhizal trees. Specific actinorhizal genera are *Allocasuarina* and *Casuarina*. Promiscuous actinorhizal genera are *Myrica* (Myricaceae) and to a lesser extent *Alnus* (Betulaceae) and *Gymnostoma* (Casuarinaceae). Since molecular and paleontologic approaches indicate (i) that the family Myricaceae would have first diverged followed by the Betulaceae and Casuarinaceae and (ii) that the genus *Gymostoma* would have diverged before *Casuarina* and *Allocasuarina* within the Casuarinaceae, one can assume that evolution had proceeded in the direction of specificity. Therefore promiscuity could be ancestral and specificity could be interpreted as a more specialized feature [59].

Inoculant technology
The production of rhizobial inoculants requires simple technology. We shall not present it again here since it has been very well described many times [41, 56, 64, 79, 104].

The development of *Frankia* inoculants has been delayed for a number of years because isolating *Frankia* is sometimes very difficult and the culture of this microorganism is not easy. Advances have been recently made which allow the production of *Frankia* inoculants of high quality. The most convenient formulation of *Frankia* inoculum is obtained by entrapping the hyphae and sporangia in a polymer gel, generally as alginate beads, with kaolinite added to the gel [18, 19, 33, 42, 91].

Response to inoculation with rhizobia and Frankia
Response to the absence of environmental constraints.
Inoculation can be expected to increase NFT growth only when the density of native compatible effective rhizobium or *Frankia* in the soil is low, that is below 10-50 infecting units per g soil [88, 104]. Except in sterilized nursery soils, desert soils, degraded lands such as overexploited or eroded soils, and mining soils, promiscuous NFT species respond positively to inoculation much less frequently than specific NFT species, because the density of strains compatible with promiscuous species is usually higher than that of strains, compatible with specific species.

Therefore inoculation should generally be focused on specific NFTs and on sites where the densities of compatible rhizobia or *Frankia* are low or nil.

Response in the presence of environmental constraints.
A positive response can be expected only when the constraints can be overcome by proper agronomic methods, a goal which may be difficult to achieve.

Assessment of native population densities. Estimating compatible rhizobium or *Frankia* population densities is mandatory to predict the response of NFTs to inoculation. In the case of legume NFTs, two methods of assessment can be used: the most-probable-number assay, which was shown to be in reasonable agreement with the plate count method [103] and the method of estimation of nodulation potential proposed by Virginia and Jarrell [109]. The latter method can also be used to evaluate the populations of *Frankia*.

NFT nodulating with fast-growing (FG) rhizobia vs NFT nodulating with slow-growing (SG) rhizobia.
NFT species nodulating with FG rhizobia respond more often to inoculation than species nodulating with SG rhizobia [23, 25, 104].

Multi-strain inoculants There are two schools of thought. A first school advocates the use of single-strain inoculants highly specific for a plant genotype. This type of inoculant leads to excellent inoculation response, but lacks flexibility. Using a new plant genotype may require a new strain which may not easily displace the first strain used. The second school advocates the use of multi-strain inoculants, the strains used

being characterized by their broad host spectrum. This second type of inoculant performs well with a large range of plant genotypes [50].

Successes and failures of NFT inoculations in a number of diverse tropical soils have been recently reported [11]; this report clearly illustrates the principles governing the response to inoculation presented above.

Need for inoculation with mycorrhizal fungi
Most NFTs form endomycorhizas. A limited number of NFTs (e.g. *Afzelia* spp., *Acacia* spp., *Casuarina* spp.) are known to form ectomycorhizas or both ecto-and VA endomycorrhizas. These infections occur spontaneously in most soils. But in some particular sites (e.g. degraded lands or waste rockdumps) inoculating the trees with mycorrhizal fungi is mandatory to establish healthy plantations [48, 75] (see also Dobereiner's paper in this issue). The degree of dependence on the mycorrhizas for capture of P, non soluble elements and water is still poorly documented, and further investigations are needed to elucidate the specific NFT requirements in this domain. Attempts to grow VA endomycorrhizal fungi in pure culture have so far been unsuccessful, but large-scale multiplication of these symbionts is achieved by growing them on the roots of living plants [99]. Different techniques of processing VA endomycorrhizal inoculants have been proposed [64]. Most formulations are suited for use in nursery inoculations.

Developing host plants with improved NFP and high tolerance to environmental constraints

Considerable interprovenance and intraspecific variability in the NFP has been found in the following species : *Acacia albida, Acacia mangium, Casuarina equisetifolia, C. cunninghamiana, Gliricidia sepium* and *Leucaena leucocephala* [6, 35, 82, 83, 84, 90, 96]. Large differences have also been shown between provenances in their tolerance to environmental constraints, such as tolerance to salinity in *Casuarina glauca* [30, 43]. Further investigations will probably show that other important traits - such as ability to fix actively N_2 in the presence of significant amounts of available N in soil - vary with the NFT genotype as already shown in annual crops, e.g. soybean [49].

224

Screening NFTs for high NFP
This selection procedure is based on the screening of germplasm (provenances, clones) in the nursery or in the field rather than by breeding, the approach being similar to that of the selection of rhizobia or *Frankia*. Subsequently, the selected clones can be vegetatively propagated either through cuttings or using *in vitro* techniques [35].

One complication derives from possible "host genotype x symbiotic strain" interactions, the question being essentially to know whether the ranking of host genotypes according to their NFP is affected by the symbiotic strain. This question has been addressed by a number of investigators. In an experiment on *Casuarina equisetifolia*, three clones inoculated with three different *Frankia* strains were cultivated on N-free sand for 10 months and irrigated with a N-free nutrient solution. The variance analysis of different N_2 fixation parameters of the nine combinations showed a highly significant effect of the clones ($p<0.01$), a highly significant ($p<0.01$) effect of the strains, with a clone effect definitely more important than the strain effect. There was a significant ($p<0.05$) "*Casuarina equisetifolia* clone×*Frankia* strain" interaction for most of the parameters used including the amount of N_2 fixed, but the ranking of the clones was not affected by the *Frankia* strains [93]. A similar experiment comprising 15 "*Acacia mangium* clone×*Bradyrhizobium* strain" combinations showed that there was no significant "*A. mangium* clone×*Bradyrhizobium* strain" interaction [34]. The conclusion of these experiments was that the ranking of host genotypes according to N_2 fixation traits (nodule weight, N_2 fixed, dry weight of plants grown in the absence of available N) was not affected by strain choice. Consequently, it was suggested that a two-step selection procedure be adopted, the first step regarding the host plant, the second one the *Frankia* strains, the best combination being "the best host genotype×the best symbiotic strain".

However, other studies have shown that such a conclusion cannot be generalized. Thus N_2 fixation appeared to be more influenced by the "host genotype×strain" interaction than by the plant genotype or the symbiotic strain *per se* in the case of the *Gliricidia sepium-Rhizobium* symbiosis. The contribution of the "host genotype×*Rhizobium*" interaction to the total phenotypic variation was then over 50% [6]. A similar "host genotype×*Rhizobium*" interaction in total N was reported in the case of *Leucaena leucocephala*, *L. diversifolia* and their hybrid [89].

More recently a field experiment carried out in the Ivory Coast showed a significant ($p<0.01$) interaction "*Acacia mangium* provenance×*Bradyrhizobium* strain" on tree height (A. Galiana and Y. Prin, private commun.).

Screening NFTs for other desirable traits
Selection for high NFP needs to be accompanied by selection for other desirable traits namely: fast growth and tree shape; tolerance to drought, including ability to capture water through deep rooting, or by foliar water uptake [37]; low requirements in nutrients (especially P); ability to nodulate and fix N_2 in the presence of appreciable amounts of available N; resistance to pests; reduced interspecific competition in mixed cropping systems like alley cropping or mixed plantations. To summarize, NFTs should be considered in the context of the ecosystem and not for their N_2-fixing potential alone [13].

Molecular biology methods
With the discovery of efficient vectors (like *Agrobacterium tumefaciens* or *A. rhizogenes*) or direct techniques (electroporation, liposomes, microinjection, DNA encapsulated microprojectiles) capable of transferring fragments of DNA from one organism to another, it is now possible to engineer almost at will organisms exhibiting a wide variety of new and desirable qualities, which would not be attainable through the use of conventional breeding programs. The first transgenic species of Casuarinaceae, *Allocasuarina verticillata*, was obtained in 1991, using *Agrobacterium rhizogenes* as a vector. In this preliminary attempt, a portion of the *Agrobacterium* DNA, the T-DNA, was transferred into the plant genome. The transformation and subsequent regeneration of transgenic plants was confirmed by the presence of specific opines and by the Southern blot analysis [26, 69]. A similar procedure made possible the transformation of other NFTs, such as *Robinia pseudoacacia* [46] and *Alnus glutinosa* [26].

It appears that further development of genetic engineering depends not only of a reliable and predictable gene transfer system, but also on the availability of genes possessing the desired traits to be introduced in the NFTs. These traits would comprise not only increased N_2 fixation in association with the symbiotic bacterium but also a high level of tolerance to drought and salinity, viral, bacterial, fungal, insect and nematode resistance and control of the inhibition of N_2 fixation by available N.

The present state of our knowledge is not advanced enough to consider using genetic engineering to introduce most of the above mentioned traits at this time [26]; but, due to recent advances in the manipulation of pest resistance genes into the cells of a number of plant species [51, 57], transformed pest resistant NFTs will hopefully be developed in the near future.

Appropriate management practices to alleviating soil constraints

Appropriate management practices can enhance N_2 fixation by reducing the effect of the major physical (high temperatures and drought), chemical (nutrient deficiencies, acidity, aluminium toxicity and excess of available N) and biological (pests and diseases) constraints. These constraints are well known and their influence on N_2 fixation aptly described by a number of authors [41, 74]. They can act at different levels, hindering the growth of the host plant, reducing the growth or survival of the symbiotic bacteria in the soil and affecting the process of infection or the functioning of the nodule, once it has been formed.

Here we shall restrict the discussion to the use of low input management practices to alleviating four major soil constraints: drought, nutrient deficiencies, excess of available N and nematodes.

Drought
A variety of irrigation systems have been developed to ensure normal plant growth and photosynthesis, and subsequently, the establishment and functioning of the symbioses. However, most of them do not guarantee a sustained economic return. Therefore, low input practices are recommended, especially rainwater harvesting systems that have been successfully used to direct infrequent runoff into the rooting zone of trees or to assist groundwater recharge [4]. This approach has been adopted to modernize the management of *Acacia nilotica* pasture lands in Djibouti [5]. Temporary watering may be necessary during the establishment phase of plantations to assist root systems to reach permanently moist soil horizons. This practice is currently applied to ensure the success of *Casuarina equisetifolia* plantations on the Senegalese coast and *C. glauca* windbreaks in Egypt.

Since mycorrhizae are assumed to improve the drought resistance of host plants by increasing water supply and water use efficiency [39], inoculating NFTs with selected mycorrhizal strains would probably enhance the host plant growth and its N_2-fixing activity in soils with low populations of mycorrhizal fungi.

Nutrient deficiencies
Alleviating deficiencies in nutrients such as P and Mo in acid soils or Fe and Zn in soils with high pH is easily achieved by applying the required elements, provided that the deficiencies have been properly diagnosed. In the literature there are a number of examples of the beneficial effect of P application on nodulation and N_2 fixation in NFTs, e.g. *Inga jinicuil* [106]. In sandy Senegalese dune soils, satisfactory growth and N_2 fixation of irrigated *Casuarina equisetifolia* can be obtained only if trace elements are applied regularly (unpublished results).

To allow the successful establishment of trees and subsequent recyling of nutrients on infertile acid soils, application of lime and possibly P is recommended [80]. This recommendation is certainly relevant to NFTs.

In P-deficient soils inoculation with VA endomycorhizal fungi can improve P uptake through increasing the volume of soil explored and subsequently enhance N_2 fixation and growth of NFTs. As already mentioned this positive response can be expected only in soils with very low populations of VA endomycorhizal fungi.

Excess of available N
The inhibition of nodule formation and N_2 fixation in annual crops by excess of available N in soil has been discussed by many investigators [52, 108]. This inhibitory effect occurs also in NFTs, e.g. *Inga jinicuil* [106]. We have already indicated that the accumulation of available N in the soil of a *Leucaena leucocephala* plantation caused the steady decline of the tree N_2-fixing activity, which ceased within 6 years of planting. We have also reported that trees whose litter is decomposed less readily than that of *L. leucocephala* would fix N_2 for longer periods.

With few exceptions, there is a substantial increase in the proportion of N_2 fixed (Ndfa %) in a legume grown in association with a cereal. The increase is attributed to the efficient depletion of available N from the soil by the cereal, thus reducing the inhibitory effect of available N and to a lower competitive ability for soil N by the legume than by the cereal grown in association. This could cause the legume to depend more on atmospheric N_2 than when grown alone [3, 41]. Such a situation probably occurs when NFTs are grown in association with non-N_2 fixing plants.

When the amount of N transferred from NFTs to associated crops is insufficient to meet the N requirements of this crop, N should always be added in a form compatible with N_2 fixation, either as slow-release fertilizer, or as organic amendment, namely manure or compost [22]. Organic amendments not only do not inhibit N_2 fixation but have also been shown to significantly enhance N_2 fixation (F. Ganry, Thesis, Univ. of Nancy, France, 1990).

Parasitic nematodes

Root-knot, cyst and root-lesion nematodes are known to impede nodulation and affect the activity of the nodules of a number of annual legumes [40]. Similar effects would probably be found in NFTs which are infested by nematodes. *Meloidogyne javanica* was reported to infest *Prosopis juiiflora*, *Acacia holosericea* and *Faidherbia albida* and *Meloidogyne incognita* the same species plus *Acacia nilotica* and *A. raddiana*. [73]. Other NFTs are reputed to be susceptible to nematode attacks, namely *Sesbania grandiflora*, *Sesbania sesban*, and *Cajanus cajan*. In addition to their detrimental effect on the growth and N_2-fixing activity of the host tree itself, parasitic nematodes harbored in NFT roots can contaminate the neighboring crops, affecting their growth, vigor and productivity. Therefore it is mandatory to develop reliable methods of control of these dangerous parasites in forestry and agroforestry systems. The following preventive measures are proposed [16]: sterilisation of nursery soils (a measure seldom applied); direct seeding or plantation of disinfected cuttings; plantation of species that are resistant to nematodes or directly or indirectly contributing to their decline, like *Leucaena leucocephala*, as observed in many countries.

Conclusion

Interest in the use of NFTs in forestry, agroforestry and sylvopastoral systems has rapidly increased in recent years with the necessity for rehabilitating deforested or degraded areas and the need for developing low input strategies leading to sustained productivity even in moderately fertile lands.

Let us consider first the case of NFTs planted for land rehabilitation, afforestation or sylvopastoral pur-

poses. Not only can NFTs grow in entirely N-deficient soils, but they also indisputably improve the soil N status. A number of approaches to maximizing their ability to fix N_2 (e.g. clonal selection combined with vegetative propagation; inoculation or P fertilization) have been presented.

If we now consider agroforestry or agro-sylvopastoral systems, the introduction of NFTs in these systems raises on the one hand the question of their N_2-fixing ability and on the other, that of the transfer of fixed N_2 to associated plants (crops, grasses or non N_2-fixing trees). The amount of N transferred varies greatly, depending on factors such as mineralization rate of litter or agroforestry practices, like pruning. Complex interactions (e.g. competition and commensalism) involved in these systems are still poorly understood.

It is obvious that the potentialities of NFTs are far from being fully exploited. Given the present state of our knowledge, what are the strategies to be adopted and what are the future research needs? Further investigations are obviously required at the laboratory level not only to elucidate the mechanisms involved in the establishment and functioning of the symbioses (inhibition by available N is a major problem), to explore the diversity of the microsymbiont populations, to identify the best strains and exploit their genetic capabilities, to improve the methods of plant breeding using methods such as clonal selection and genetic transformation. The results obtained in the laboratory would be worthless if they were not tested in the field, the questions raised being for instance the following:

- Are multi-strain inoculants better than single-strain inoculants?
- How long does the response to inoculation (with selected N_2-fixing symbionts or mycorrhizal fungi) last after transplanting?
- Do vegetatively propagated NFTs behave like seedlings?
- Do NFTs selected for their tolerance to excess of available soil N fix N_2 for a longer period of growth than unselected NFTs?

Other specific field experiments are required to improve the management practices by answering questions such as the following:

- What are the processes involved in the interactions between NFTs and non N_2-fixing plants in a given agroforestry system?
- Is it possible to synchronize N uptake and nutrient release from NFTs?

- What are the most efficient methods of biological control of nematodes?

Addressing all these questions requires that (i) mechanistic studies be initiated instead of phenomenological studies that are usually adopted, (ii) reliable and accurate methods of assessment of N_2 fixation in the field be systematically used (S. Danso, This issue) and (iii) soil heterogeneity within the experimental designs be as reduced as possible.

Studies on the economics of NFT management are not many. The use of inoculants is considered to be beneficial provided that they are conducive to a worthwhile response [64]. When comparing the cost of inoculation on a per ha basis for crops vs trees, it is clear that tree inoculation is much less expensive than crop inoculation since the density of forest plantation is generally ca. 2000 plants ha^{-1} whereas that of crops is in the range of 200 000 to 500 000 plants ha^{-1} [64].

A cost and return analysis carried out in Nigeria indicated that alley cropping with *Leucaena leucocephala* associated with vegetable crops, including low application of fertilizers, was profitable [68]. In this regard, it can be stated that the judicious blending of trees into farming systems can materially benefit crop and livestock production and has important implications for food security [45]. These are encouraging considerations. However, further estimations of costs and labour requirements of specific N_2-fixation practices are necessary, using research data and farmers' experiences [1]. Finally, attempts should be made to increase the benefits that could be obtained from NFTs, for example by promoting the use of species providing human food [63] like some Australian *Acacia* [100], which would reinforce the interest of farmers towards trees, especially NFTs.

References

1. Allison JR (1992) The economics of using legumes in cropping systems. In Mulongoy KM Gueye M and Spencer DSC (eds) Biological Nitrogen Fixation and Sustainability of Tropical Agriculture, pp 323–333. John wiley, Chichester
2. Anderson LS and Sinclair FL (1993) Ecological interactions in agroforestry systems. Agrofor Abstr 6: 57–91
3. Anon (1990) Recently completed co-ordinated research programme. Soils Newsletter Joint FAO-IAEA Division. Vienna 13: 6–8
4. Armitage FB (1985) Irrigated Forestry in Arid and Semi-arid Lands: a Synthesis. International Development Research Centre, Ottawa
5. Audru J, Labonne M, Guérin H and Bilha A (1993) *Acacia nilotica*. Son intérêt fourrager et son exploitation chez les éleveurs Afars de la vallée du Madgoul à Djibouti. Bois For Trop 235: 59–70
6. Awonaike KO and Hardarson G (1992) Biological nitrogen fixation of *Gliricidia/Rhizobium* symbiosis as influenced by plant genotype, bacterial strain and their interactions. Trop Agric (Trinidad) 69: 381–385
7. Balasubramanian V and Sekayange L (1992) Five years of research on improved fallow in the semi-arid highlands of Rwanda. In: Mulongoy K, Gueye M and Spencer DSC (eds) Biological Nitrogen Fixation and Sustainability of Tropical Agriculture, pp 405–422. John Wiley, Chichester
8. Benson DR and Silvester WB (1993) Biology of *Frankia* strains, actinomycete symbionts of actinorhizal plants. Microbiol Rev 57: 293–319
9. Bernhard-Reversat F, Diangana D and Tsatsa M (1993) Biomasse, minéralomasse et productivité en plantation d'*Acacia mangium* et *A. auriculiformis* au Congo. Bois For Trop 238:35–44
10. Brewbaker JL (1990) Nitrogen-fixing trees. In: Werner D and Müller P (eds) Fast-Growing Trees and Nitrogen-Fixing Trees, pp 253–262. Gustav Fisher Verlag Stuttgart
11. Brunck F, Colonna JP, Dommergues YR, Ducousso M, Galiana A, Prin Y, Roederer Y and Sougoufara B (1990) La maîtrise de l'inoculation des arbres avec leurs symbioses racinaires: synthèse d'une sélection d'essais au champ en zone tropicale. Bois For Trop 223: 24–42
12. CTFT (Centre technique forestier tropical) (1988) *Faidherbia albida* (Del.) A. Chev. (Synonyme *Acacia albida* Del.). CTFT, Nogent-sur-Marne.
13. Danso SKA, Bowen GD and Sanginga N (1992) Biological nitrogen fixation in trees in agro-ecosystems. Plant Soil 141: 177–196
14. Date RA (1982) Assessment of rhizobial status of the soil. In: Vincent JM (ed) Nitrogen Fixation in Legumes, pp 85–94. Academic Press, Sydney
15. Dénarié J and Roche P (1992) Rhizobium nodulation signals. In: Verma DPS (ed) Molecular Signals in Plant-microbe Communications, pp 295– 320. CRC Press, Boca Raton
16. D'Hondt-Defrancq M (1993) Nématodes et agroforesterie. L'agroforesterie aujourd'hui 5: 5–9
17. Diem HG and Dommergues YR (1990) Current and potential uses and management of Casuarinaceae in the tropics and subtropics. In: Schwintzer CR and Tjepkema JD (eds) The Biology of *Frankia* and Actinorhizal Plants, pp 317–342. Academic Press, New York
18. Diem HG (1993) Réexamen des critères de sélection des souches de rhizobium utilisées en agriculture. Bull Soc Fr Microbiol 8: 83–87
19. Diem HG, Ben Khalifa K, Neyra M and Dommergues YR (1988) Recent advances in the inoculant technology with special emphasis on plant symbiotic microorganisms. In: Leone U, Rialdi G and Vanore R (eds) International Workshop on Advanced Technologies for Increased Agricultural Production, Santa Margherita Ligure, pp 196–210. Consiglio Nazionale delle Ricerche Italy (CNR), Roma and Universitá degli Studi di Genova (USG),Genova (Italy)
20. Domenach AM and Kurdali F (1989) Influence des réserves azotées sur la formation des feuilles d'*Alnus glutinosa* et ses conséquences dans l'estimation de la fixation de l'azote. Can J Bot 67: 865–871
21. Dommergues Y (1963) Evaluation du taux de fixation de l'azote dans un sol dunaire reboisé en filao (*Casuarina equisetifolia*). Agrochimica 7: 335–340

22. Dommergues YR (1978) Impact on soil management and plant growth. In: Dommergues YR and Krupa SV (eds) Interactions between Non-Pathogenic Soil Microorganisms and Plants, pp 443–458. Elsevier, Amsterdam

23. Dommergues YR (1987) The role of nitrogen fixation in agroforestry. In: Steppler HA and Rarmachandran Nair PK (eds) Agroforestry, a Decade of Development, pp 245–271. ICRAF, Nairobi

24. Dommergues YR (1992) Maximizing nitrogen fixation in *Prosopis* spp. In: Dutton RW (ed) *Prosopis* species. Aspects of their value, Research and Development, pp 207–218. CORD, Durham, UK

25. Dreyfus B and Dommergues YR (1981) Nodulation of *Acacia* species by fast- and slow-growing tropical strains of *Rhizobium*. Appl Environ Microbiol 41: 97–99

26. Duhoux E, Bogusz D and Franche C (1992) Transformations chez les plantes fixatrices d'azote. In: Interactions Plantes-microorganismes, Sénégal Février 1992, pp 462–474. Fondation Internationale pour la Science (IFS), Stockholm

27. Duhoux E, Prin Y and Dommergues YR (1993) Comparison of aerial nodulation of *Casuarina* sp. with legume stem nodulation. In: Subba Rao NS and Rodriguez-Barrueco C (eds) Symbioses in Nitrogen-Fixing Trees, pp 85–93. Oxford and IBH Publishing Co, New Delhi

28. Dupuy N and Dreyfus BL (1992) *Bradyrhizobium* populations occur in deep soil under the leguminous tree *Acacia albida*. Appl Environ Microbiol 58: 2415–2419

29. Ehleringer JR, Mooney HA, Rundel PW, Evans KD, Palma B and Delatorre J (1992) Lack of nitrogen cycling in the Atacama desert. Nature 359: 316–318

30. El-Lakany MH and Luard EJ (1982) Comparative salt tolerance of selected *Casuarina* species. Aust For Res: 13, 11–20

31. Faria SM de, Lewis GP, Sprent JI and Sutherland JM (1989) Occurrence of nodulation in the Leguminosae. New Phytol 111: 607–619

32. Felker P and Clark PR (1982) Position of mesquite (*Prosopis* spp.) nodulation and nitrogen fixation (acetylene reduction) in 3-m long phreatophytically simulated soil columns. Plant Soil 64: 297–305

33. Frioni L, Le Roux C, Dommergues YR and Diem HG (1994) Inoculant made of encapsulated *Frankia*: assessment of *Frankia* growth within alginate beads. World J Microbiol Biotech 10: 118–121.

34. Galiana A, Tibok A and Duhoux E (1991) Nitrogen-fixing potential of micropropagated clones of *Acacia mangium* inoculated with different *Bradyrhizobium* strains. Plant Soil 135: 161–166

35. Galiana A, Athée J, Phelep M and Duhoux E (1992) Improving nitrogen fixation in trees through clonal selection and genetic transformation of the host. In: Dutton RW (ed) *Prosopis* species. Aspects of their value, Research and Development, pp 197–206. CORD, University of Durham, UK

36. Garg VK and Jain RK (1992) Influence of fuelwood trees on sodic soils. Can J For Res 22: 729–735

37. Gates PJ and Brown K (1988) *Acacia tortilis* and *Prosopis cineraria*: Leguminous Trees for Arid Areas. Outlook Agric 17: 61–64

38. Geiger SC, Vandenbeldt RJ and Manu A (1992) Preexisting soil fertility and the variable growth of *Faidherbia albida*. In: Vandenbelt JR (ed) *Faidherbia albida* in the West african Semi-Arid Tropics, pp 121–125. ICRAF, Nairobi

39. Gianinazzi-Pearson V and Diem HG (1982) Endomycorrhizae in the tropics. In: Dommergues YR and Diem HG (eds) Microbiology of Tropical Soils and Plant Productivity, pp 209–251. Nijhoff Junk, The Hague

40. Gibson AH, Dreyfus BL and Dommergues YR (1982) Nitrogen fixation by legumes in the tropics. In: Dommergues YR and Diem HG (eds) Microbiology of Tropical Soils and Plant Productivity, pp 37–71. Nijhoff Junk, The Hague

41. Giller KE and Wilson KJ (1991) Nitrogen Fixation in Tropical Cropping Systems. CAB International, Wallingford Oxon, UK

42. Girgis MGZ and Schwencke J (1993) Differentiation of *Frankia* strains by their electrophoretic patterns of intracellular esterases and aminopeptidases. J Gen Microbiol 139: 2225–2232

43. Girgis MG, Ishac YZ, Diem HG and Dommergues YR (1992) Selection of salt tolerant *Casuarina glauca* and *Frankia*. Acta Oecol 13: 443–451

44. Graham PH and Hubbell DH (1975) Legume-rhizobium relationships. In: Doll EC and Mott GO (eds) Tropical Forages in Livestock Production Systems, Special Publ. 24, pp 9–21. American Society of Agronomy, Madison, Wisc

45. Gregersen H, Draper S and Elz D (1989) People and trees. The World Bank, Washington DC

46. Han K-H, Keathley DE, Davis JM and Gordon MP (1993) Regeneration of a transgenic woody legume (*Robinia pseudoacacia* L., black locust) and morphological alterations induced by *Agrobacterium rhizogenes*-mediated transformation. Plant Sci 88: 149–157

47. Hardarson G (1993) Methods for enhancing symbiotic nitrogen fixation. Plant Soil 152: 1–17

48. Herrera MA, Salamanca P and Barea JM (1993) Mycorrhizal associations and their functions in nodulating nitrogen-fixing trees. In: Subba Rao NS and Rodriguez-Barrueco C (eds) Symbioses in Nitrogen-fixing Trees, pp 141–165. Oxford and IBH Publishing Co, New Delhi

49. Herridge DF and Betts JH (1988) Field evaluation of soybean genotypes selected for enhanced capacity to nodulate and fix nitrogen in the presence of nitrate. Plant Soil 110: 129–235

50. Hubbell DH (1993) Screening for inoculant-quality strains of rhizobia. In: Glick BR and Thompson JE (eds) Methods in Plant Molecular Biology and Biotechnology, pp 321–330. CRC Press, Boca Raton

51. Hubbes M (1993) Impact of molecular biology on forest pathology: a literature review. Eur J For Path 23: 201–217

52. Imsande J (1986) Inhibition of nodule development in soybean by nitrate or reduced nitrogen. J Exp Bot 37: 348–355

53. Jenkins MB, Virginia RA and Jarrell WM (1988) Depth distribution and seasonal populations of mesquite-nodulating rhizobia in warm desert ecosystems. Soil Sci Soc Am J 52: 1644–1650

54. Johnson HB and Mayeux HS (1990) *Prosopis glandulosa* and the nitrogen balance of range lands: extent and occurrence of nodulation. Oecologia 84: 176–185

55. Kang BT and Mulongoy K (1992) Nitrogen contribution of woody legumes in alley cropping systems. In: Mulongoy K, Gueye M and Spencer DSC (eds) Biologicai Nitrogen Fixation and Sustainability of Tropical Agriculture, pp 367–375. John Wiley, Chichester

56. Keyser HH (1992) Inoculating tree legume seeds and seedlings with rhizobia. World J Microbiol 8: 360–361

57. Kumar A, Cooper-Bland S and Powell W (1992) Transfer of disease resistance genes into crop plants: the role of biotechnology. In: Interactions Plantes-microorganismes, Senegal, February 1992, pp 475–490. International Foundation for Science (IFS), Stockholm

58. Lal R (1989) Agroforestry systems and soil surface management of a tropical alfisol. Agrofor Syst 8: 7–29, 97–111, 113–132, 197–215

59. Maggia L and Bouquet J (1994) Molecular phylogeny of the actinorhizal Hamamelidae and relationships with host promiscuity towards *Frankia*. Molecular Ecol 3: 459–467

60. Mailly D and Margolis HK (1992) Forest floor and mineral soil development in *Casuarina equisetifolia* plantations on the coastal sand dunes of Senegal. For Ecol Manage 55: 259–278

61. Mariotti A, Sougoufara B and Dommergues YR (1992) Estimation de la fixation d'azote atmosphérique par le traçage isotopique naturel dans une plantation de *Casuarina equisetifolia* (Forst). Soil Biol Biochem 24: 647–653

62. Matthews RB, Holden ST and Lungu S (1992) The potential of alley cropping in improvement of cultivation systems in the high rainfall areas of Zambia. Agrofor Syst 17: 219–240

63. Maydell von HJ (1987) Agroforestry in the dry zones of Africa: present, past and future. In: Steppler HA and Ramachandran Nair PK (eds) Agroforestry, a Decade of Development, pp 89-116. ICRAF, Nairobi

64. Mulongoy K, Gianinazzi S, Roger PA and Dommergues YR (1992) Biofertilizers: agronomic and environmental impacts and economics. In: DaSilva EJ, Ratledge C and Sasson A (eds) Biotechnology, Economic and Social Aspects, pp 54–69. UNESCO Cambridge University Press, Cambridge

65. Nazaret S, Simonet P and Normand P (1993) Contribution of molecular biology techniques to *Frankia* strain characterization and molecular ecology. In: Subba Rao NS and Rodriguez-Barrueco C (eds) Symbioses in Nitrogen-fixing Trees, pp 213–231. Oxford and IBH Publishing Co., New Delhi

66. Normand P, Fernandez M, Simonet P and Domenach AM (eds) (1992) *Frankia* and actinorhizal plants. Acta Oecol 13: 368–516

67. Orchard ER and Darby GD (1956) Fertility changes under continued wattle culture with special reference to nitrogen fixation and base status of the soil. In: CR. Sixième Congrès International Science du Sol, Paris, 4, 45, pp 305–309 Société Internationale de Science du Sol, Paris

68. Palada MC, Kang BT and Claassen SL (1992) Effect of alley cropping with *Leucaena leucocephala* and fertilizer application on yield of vegetable crops. Agrofor Syst 19: 139–147

69. Phelep M, Petit A, Martin L, Duhoux E and Tempé J (1991) Transformation and regeneration of a nitrogen-fixing tree, *Allocasuarina verticillata* Lam. Bio/Technol 9: 461–466

70. Pillai SD, Josephson KL, Bailey RL and Pepper IL (1992) Specific detection of rhizobia in root nodules and soil using the polymerase chain reaction. Soil Biol Biochem 24: 885–891

71. Prin Y, Duhoux E, Diem HG and Dommergues YR (1991) Aerial nodules in *Casuarina cunninghamiana*. Appl Environ Microbiol 57: 71–874

72. Prin Y, Galiana A, Ducousso M, Dupuy N, Lajudie P de and Neyra M (1993) Les rhizobiums d'*Acacia*. Bois For Trop 238 : 5–20

73. Prot JC (1986) Sensibilité de sept légumineuses arborescentes aux nématodes *Meloidogyne javanica*, *Meloidogyne incognita*, *Scutellonema cavenessi* et *Dolichorhyncus elegans*. Revue Nématol 9: 416–418

74. Reddell P (1993) Soil constraints to the growth of nitrogen-fixing trees in tropical environments. In: Subba Rao NS and Rodriguez-Barrueco C (eds) Symbioses in Nitrogen-Fixing Trees, pp 65–83. Oxford and IBH Publishing Co, New Delhi

75. Reddell P and Milnes AR (1992) Mycorrhizas and other specialized nutrient-acquisition strategies: their occurrence in woodland plants from Kakadu and their role in rehabilitation of waste rock dumps at a local uranium mine. Aust J Bot 40: 223–242

76. Rodriguez-Barrueco C, Cervantes E and Subba Rao NS (1993) Host specificity in *Frankia* symbiosis. In: Subba Rao NS and Rodriguez-Barrueco C (eds) Symbioses in Nitrogen-Fixing Trees, pp 195–209. Oxford and IBH Publishing Co, New Delhi

77. Roskoski JP (1981) Nodulation and N_2 fixation by *Inga jinicuil*, a woody legume in coffee plantations. Plant Soil 59: 201–206

78. Roughley RJ (1987) Acacias and their root-nodule bacteria. In: Turnbull JW (ed) Australian Acacias in Developing Countries, pp 45–49. ACIAR, Canberra

79. Roughley RJ (1988) Legume inoculants: their technology and application. In: Beck DP and Materon LA (eds) Nitrogen Fixation by Legumes in Mediterranean Agriculture, pp 259-268. Martinus Nijhoff, Dordrecht

80. Sanchez PA (1987) Soil productivity and sustainability in agroforestry systems. In: Steppler HA and Ramachandran Nair PK (eds) Agroforestry, a Decade of Development, pp 205–223. ICRAF, Nairobi

81. Sanginga N, Mulongoy K and Ayanaba A (1988) Nitrogen contribution of *Leucaena-Rhizobium* symbiosis to soil and subsequent maize crop. Plant Soil 112: 137–141

82. Sanginga N, Bowen GD and Danso SKA (1990) Assessment of genetic variability for N_2 fixation between and within provenances of *Leucaena leucocephala* and *Acacia albida* estimated by ^{15}N labelling techniques. Plant Soil 127: 169–178

83. Sanginga N, Bowen GD and Danso SKA (1990) Genetic variability in symbiotic nitrogen fixation within and between provenances of two *Casuarina* species using the ^{15}N-labelling methods. Soil Biol Biochem 22: 539–547

84. Sanginga N, Danso SKA and Bowen GD (1992) Variation in growth, sources of nitrogen and N-use efficiency by provenances of *Gliricidia sepium*. Soil Biol Biochem 24: 1021–1026

85. Schulze ED, Gebauer G, Ziegier H and Lange OL (1991) Estimates of nitrogen fixation by trees on an aridity gradient in Namibia. Oecologia 88: 451–459

86. Schwintzer CR and Tjepkema JD (1990) The Biology of *Frankia* and Actinorhizal Plants. Academic Press, San Diego

87. Singh GB (1987) Agroforestry in the Indian subcontinent: past, present and future. In: Steppler HA and Ramachandran Nair PK (eds) Agroforestry, a Decade of Development, pp 117–138. ICRAF, Nairobi

88. Singleton PW, Bohlool BB and Nakao PL (1992) Legume response to rhizobial inoculation in the tropics: myths and realities. In: SSSA Special Publication n° 29 Myths and Science of Soils of the Tropics, pp 135–155. Soil Science Society of America and American Society of Agronomy, Madison, Wisc

89. Somasegaran P and Martin RB (1986) Symbiotic characteristics and *Rhizobium* requirements of *Leucaena leucocephala-Leucaena diversifolia* hybrid and its parental genotypes. Appl Environ Microbiol 52: 1422–1424

90. Sougoufara B, Duhoux E, Corbasson M and Dommergues YR (1987) Improvement of nitrogen fixation by *Casuarina equisetifolia* through clonal selection: A research note. Arid Soil Res Rehab 1: 129–132

91. Sougoufara B, Diem HG and Dommergues R (1989) Response of field-grown *Casuarina equisetifolia* to inoculation with *Frankia* strain ORS 021001 entrapped in alginate beads. Plant Soil 118: 133–137

92. Sougoufara B, Danso SKA, Diem HG and Dommergues YR (1990) Estimating N_2 fixation and N derived from soil by *Casuarina equisetifolia* using labelled ^{15}N fertilizer: some problems and solutions. Soil Biol Biochem 22: 695-7021

93. Sougoufara B, Maggia L, Duhoux E and Dommergues YR (1992) Nodulation and N_2 fixation in nine *Casuarina* clone-*Frankia* strain combinations. Acta Oecol 13: 497–503

94. Sprent JI (1988) The Ecology of the Nitrogen Cycle. Cambridge University Press, Cambridge

95. Subba Rao NS and Rodriguez-Barrueco C (eds) (1993) Symbioses in Nitrogen-Fixing Trees. Oxford and IBH Publishing Co, New Delhi

96. Sun J, Simpson RJ and Sands R (1991) Studies with seedlings of provenances in *Acacia mangium*. In: Richardson AE and Peoples MB (eds) Nitrogen fixation: towards 2000 and beyond. The ninth Australian Nitrogen Fixation Conference Jan 1991. Australian National University, Canberra

97. Sutherland JM and Sprent JI (1993) Nitrogen fixation by legume trees. In: Subba Rao NS and Rodriguez-Barrueco C (eds) Symbioses in Nitrogen-Fixing Trees, pp 32–63. Oxford and IBH Publishing Co, New Delhi

98. Sylvester-Bradley R, Mosquera D and Méndez JE (1988) Selection of rhizobia for inoculation of forage legumes in savanna and rainforest soils of tropical America. In: Beck DP and Materon LA (eds) Nitrogen Fixation by Legumes in Mediterranean Agriculture, pp 225–234. Martinus Nijhoff, Dordrecht

99. Sylvia DM and Jarstfer G (1992) Sheared-root inocula of vesicular-arbuscular mycorrhizal fungi. Appl Environ Microbiol 58: 229–323

100. Thomson L (1992) Australia's subtropical dry-zone *Acacia* species with human food potential In: House APN and Harwood CE (eds) Australian Dry-zone Acacias for Human Food, pp 3–36. Australian Tree Seed Centre, Canberra

101. Tiedemann AR and Klemmedson JO (1973) Effects of mesquite on physical and chemical properties of soil. J Range Manage 29: 1–14

102. Turk D and Keyser HH (1992) Rhizobia that nodulate tree legumes: specificity of the host for nodulation and effectiveness. Can J Microbiol 38: 451–460.

103. Turk D and Keyser HH (1993) Accuracy of most-probable-number estimates of rhizobia for tree legumes. Soil Biol Biochem 25: 69–74

104. Turk D, Keyser HH and Singleton PW (1993) Response of tree legumes to rhizobial inoculation in relation to the population density of indigenous rhizobia. Soil Biol Biochem 25: 75–81

105. Vandenbeldt RJ (ed) (1992) *Faidherbia albida* in the West african Semi-Arid Tropics. ICRAF, Nairobi

106. Van Kessel C and Roskoski JP (1983) Nodulation and N_2 fixation by *Inga jinicuil*, a woody legume in coffee plantations. Plant Soil 72: 95–105

107. Van Kessel C, Farrell RE, Roskoski JP and Keane KM (1994) Recycling of the naturally occurring ^{15}N in an established stand of *Leucaena leucocephala*. Soil Biol Biochem 26: 757–762

108. Vessey JK and Waterer J (1992) In search of the mechanism of nitrate inhibition of nitrogenase activity in legume nodules: recent developments. Physiol Plant 84: 171–176

109. Virginia RA and Jarrell WM (1987) Approaches for studying the function of deep root systems. In: Tenhunen JD *et al.* (eds) Plant Response to Stress, pp 107–127. Springer-Verlag, Berlin

110. Xu ZH, Saffigna PG, Myers RJK and Chapman AL (1993) Nitrogen cycling in leucaena (*Leucaena leucocephala*) alley cropping in semi-arid tropics. Plant Soil 148: 63–72, 73–82

111. Young A (1986) Effect of trees on soils. In: Prinsley RT and Swift MJ (eds) Amelioration of Soil by Trees, pp 10–19. Commonwealth Science Council, London

112. Young A (1986) The potential of agroforestry as a practical means of sustaining soil fertility. In: Prinsley RT and Swift MJ (eds) Amelioration of Soil by Trees, pp 121–144. Commonwealth Science Council, London

113. Zakra N, Weaver RW and Taffin G de (1993) Distribution of ^{15}N from ammonium fertilizer in field-grown *Acacia*. Plant Soil 151: 147–150

Fertilizer Research **42**: 231–239, 1995.
© 1995 *Kluwer Academic Publishers. Printed in the Netherlands.*

Extension of symbiotic biological nitrogen fixation technology in developing countries*

D.H. Hubbell
Soil and Water Science Department, University of Florida, Gainesville, FL 32611, USA

Key words: BNF genetics, extension, inoculation technology, nitrogen economy, tropical soils

Abstract

There is critical involvement of microbial activities in the management of nitrogen in tropical soils. One of the most important of these activities is the provision of nitrogen to the soil via symbiotic biological nitrogen fixation, mediated primarily by the rhizobia-legume association. This symbiotic system commonly produces a significant amount of the nitrogen required by plants growing in a nitrogen-deficient soil. The exploitation of this system is especially important for subsistence farmers in developing countries, who typically are relegated to exist on a few hectares of land of marginal potential for agricultural use. Nitrogen, the nutrient most frequently found limiting to crop growth, is generally not available, economically or logistically, in the form of chemical fertilizers in sufficient amounts to produce a crop. Legumes are utilized throughout the world as sources of food, forage, fuel and cover crops. The efficiency of their use in farming practice depends heavily on the nitrogen-fixing properties of the native rhizobia with which they form symbiotic nitrogen-fixing associations. Native rhizobia are often less effective in N-fixation than are selected strains isolated and screened specifically for high N-fixation. These latter strains are mass produced to form a seed-soil inoculum to be applied at planting, thus assuring a high rate of nodulation and N-fixation in the legume host. This legume inoculation is simple, inexpensive and an invaluable component of the farming systems of agriculture-based cultures throughout the world. This invaluable technology is largely unavailable to subsistence farmers, world-wide, who need it most. The absence or ineffectiveness of appropriate extension programs is discussed as a major contributing factor in this unrealized potential.

Introduction

As part of this Symposium on Nitrogen Economy in Tropical Soils, this paper discusses specifically the microbial involvement in nitrogen economy of tropical soils. We can begin by defining in general terms what is meant by the term 'economy' in the present context. Economy can be formally defined as 1) managing, 2) insuring maintenance or productiveness, 3) thrifty use or administration (of a natural resource), and 4) frugality in expenditure. All of these definitions are appropriate in reference to the consideration of microbial aspects of economy of tropical soils.

Nitrogen in tropical soils leads a tenuous existence. A daunting array of soil and climatic conditions deter-

mine the nitrogen economy of these soils. Some appreciation for the range of variability encountered and its effects on nitrogen economy can be obtained from numerous sources [1, 3, 5, 12, 13, 26, 29, 33].

Any broad discussion of the nitrogen economy of tropical soils must encompass a wide variety of considerations. For reasons of achieving perspective it should be indicated that the term 'tropical soils' is in a sense arbitrary and arguable. There are no clearly defined criteria which separate tropical and temperate soils. Instead, there is a spectrum of soil properties which vary widely in a quantitative sense, and ultimately determine the classification of a soil as 'tropical' or 'temperate'. However, use of this admittedly artificial separation remains valuable as a guide to orientation of thinking and experimentation. For example, a person initiating work on a 'tropical soil' problem should

---

* Florida Agricultural Experiment Station Journal Series No. R-072388

be aware that such traits as low pH, high phosphorus fixation, heavily weathered mineral fraction, etc. are often associated with such soils.

The nitrogen economy of tropical soils, thus broadly defined, is of particular importance to agriculture. Much of the arable land of the world is, climatically and edaphically, tropical. This land is customarily further sub-divided into humid, semi-arid and arid, based on annual rainfall. These climatic and edaphic extremes, encountered in all possible combinations, pose an ever-present challenge to agriculture, and in particular to the nitrogen economy of these soils. It is axiomatic that nitrogen is the nutrient element most frequently found limiting to plant growth. This may be one way of saying that, of all the known plant growth-limiting factors, we deal with nitrogen deficiency least effectively in agricultural systems. Perhaps this is a reflection of the highly individual nature of nitrogen economy of these soils. Each problem situation is new and unique in some way. In each instance, one may begin studies using broad generalizations as a starting point but the unique aspect(s) of each situation soon require the initiation of original research to solve the 'new' problem. The work is on-going and essential. There is no alternative. Populations and food demands continue to increase worldwide, presenting a real challenge to adequate crop production [4, 2]. The problem is exacerbated by degradation (sometimes irreversible) of agricultural soils. Thus, science and technology are under strong pressure to 1) maintain currently productive soils, 2) reclaim abused and unproductive agricultural soils, and 3) convert marginal soils to some significant level of productivity. A key aspect of such efforts is, frequently, teaching farmers to deal with the nitrogen economy of these soils. The vehicle for this is the existence and functioning of an extension system which can translate scientifically generated information into appropriate technology and transfer it to farmers.

The scientific-technical aspects of this work have been comprehensively treated in published proceedings of numerous recent symposia. For that reason I will not attempt an exhaustive review of the assigned topic. Instead, I will (1) discuss recent research trends (state-of the-art) in symbiotic biological nitrogen fixation (SBNF), (2) identify 'cutting edges' in relevant current research, and (3) identify some problems and critically examine current priorities in relation to future professional activity in the area of SBNF.

As implied earlier, the multiple complex interactions to be considered in studying the nitrogen economy of soils dictate an integrated experimental approach. The present volume comprehensively reviews these aspects, including the opposing technologies of chemical nitrogen fertilizers vs biological nitrogen fixation (BNF) as possible solutions to nitrogen economy problems.

The broad topic of the involvement of nitrogen-fixing microbes in the economy of nitrogen in tropical soils encompasses discussion of the involvement of free-living and associative (rhizosphere) microbes (Roger; Boddey and Dobereiner; respectively, this volume). This paper discusses the role of SBNF systems in the nitrogen economy of tropical soils, as exemplified by the extensively studied legume-rhizobia association, long regarded as a classic example of symbiosis and unquestionably the major plant-microbe contributor to the economy of nitrogen in tropical soils.

In order to fully appreciate the current and potential future impact of this system on soil nitrogen economy, and hence on global agriculture, one should be aware of the state of the science and technology of the system, as currently known and practiced. This information is available in a host of comprehensive and authoritative treatments. A partial list of representative sources is given in the references. Many of them relate specifically to tropical soils [8, 15, 20, 23, 27, 30, 31, 32].

Historical role of SBNF in nitrogen economy

It is well-documented that nitrogen is the element most frequently found to be limiting for crop production. Early recognition of this first led to farming practices which alleviated or circumvented this limitation. This was manifested in such practices as application of 'green manures' such as *Azolla* on an annual basis, or crop rotation, in which nitrogen-rich legumes are intentionally grown to maturity and plowed into the soil. The renewal of fertility by the practice of fallowing is generally attributed to transport of mineral nutrients by roots to the plow layer by the fallow vegetation. It is interesting to speculate that part of this fallow effect may also result from a small augmentation of soil nitrogen via nitrogen fixation both 'associative' and by free-living microbes.

Ironically, a myriad of traditional farming practices have been 're-discovered' and are being given formal respect, recognition and attention as valid research subjects. They may even be hailed as new, innovative or cutting edge studies by those who are unacquainted with the history of agricultural practice. Many of these practices utilize SBNF in one form or another to facilitate nitrogen economy. Such practices include use of legumes as green manures, cover crops, companion crops, and alley crops. The use of tree legumes is treated elsewhere in this volume (see Dommergues, this volume).

Present and future role of SBNF in nitrogen economy

The land most amenable to satisfying present and future food demands, in terms of land area and favorable climatic conditions for crop growth, is primarily tropical in nature [15]. The appropriate use of this land in agricultural practice now and in the future is crucial to satisfying projected world food needs [3,18]. These soils, beset as they are with certain properties (ex. low pH, low organic matter, high phosphorus fixation, etc.) which lead to destruction of soil productivity when mis-managed, must be subjected to careful and constant husbandry. The nitrogen economy of these soils is of paramount importance. The critical role of SBNF systems in the nitrogen economy of these soils is no longer arguable. It is a fact [5].

Newly emerging 'cutting edge' developments in biotechnology which relate to SBNF systems demonstrate high potential for significant contribution to the over-all efficacy of SBNF systems in the nitrogen economy of soils. These include the co-inoculation of legumes with rhizobia together with other selected microbes, i.e., a consortium of soil microbes designed to act synergistically in their beneficial effects on plant growth. This includes co-inoculation of rhizobia with arbuscular mycorrhizal fungi to enhance phosphorus as well as nitrogen nutrition of the legume [17]; plant growth promoting rhizobacteria (PGPR) intended to enhance growth-health of legume roots via such mechanisms as production of plant growth hormones [2]; and microbes of various types which show potential for biological control of root pathogens by a variety of mechanisms [19].

Relatively new recognition of the potential importance of 'borderline' SBNF systems, exemplified by such microbes as azospirilli and cyanobacteria (see this volume) has resulted in highly justified intense investigation and review (Skinner *et al.*, [29] Jones [18]). These studies, previously experimentally impossible or difficult to conduct, are proving very amenable to study using readily available ^{15}N methodology (see this volume and [9]).

Studies of the ecology of *Rhizobium* have been greatly expanded by the use of antibiotic resistant mutants [10].

What can be said about the 'cutting edge' of *research* on SBNF? In a word - GENETICS. Identification of the *nif* gene complex and subsequent success in manipulating its inter-organism transfer are events which were inconceivable or perhaps, at best, entertaining science fiction thirty years ago. They are now a reality and hold undisputed title as the cutting edge of research in SBNF. Some realized or imminent accomplishments in this area include the transfer of relevant genetic information (a) between microbes and (b) between microbes and plants, including non-legumes, such as cereals, in addition to the traditional legume host plant. The prospect of nitrogen fixing corn or wheat is no longer a laughing matter [14, 21, 28].

What can be said about the 'cutting edge' of *application* of SBNF, i.e., inoculation technology?

There is abundant testimony to the existence and successful application of this technology at the resource-rich, large-farm level where innovation carries little risk. Some notable examples may be found in Australia, Brazil, Argentina, the USA, and several others nations.

How much does inoculation apply to resource-poor subsistence farming, where innovation is high risk? There is ample evidence that application of inoculation technology at the small farm level is globally minimal [7,16].

The rise and fall of traditional extension

The application of legume inoculation technology is pivotal in the nitrogen economy of nitrogen-deficient agricultural soils, whether temperate or tropical. In the United States this technology application reached its apogee as part of a diverse extension package which was made available to small farm operators in the late 1940s and early 1950s. The vehicle was the USDA Cooperative Extension Service (CES), which relied heavily on the efforts of broadly knowledgeable county extension agents, who traveled the rural by-roads and discussed (and acted on) problems unique to their clientele, the small farmers. The evolution of extension (transfer of technology) originated in the private sector and involved person-to-person discussions between farmers, salesmen, and owners of small, rural agricultural supply stores. Over the years the demand increased for more and newer information to address the new problems which inevitably accompany growth. The land grant universities emerged to address this need through their emphasis on education in the agricultural sciences. This progressed naturally to increased effort in research on agricultural problems. The CES subsequently emerged as a highly effective means of transferring this information from the laboratory to the farm [22]. Extension agents mediated the translation of scientific findings into usable farm practice and carried it in person to the users. This effort was highly successful initially but relatively ephemeral. The advent of basic changes in agricultural practice, such as emphasis on monoculture of high value crops, sophisticated mechanization and greatly expanded use of agricultural inputs, have made this personalized, broadly-based extension increasingly anachronistic. Farming became less of a vehicle for low profit but highly sustainable family operation and more of a large, profit-oriented business. Farming has faded significantly as a way of family life and has largely become a commercial enterprise. The factors which engendered this transformation led to new and complex problems which demanded the application of increasingly sophisticated and expensive biotechnological solutions. The generalized knowledge of the county extension agent was less adequate to the task and less needed. As a result of increased size and complexity of problems there are fewer small farms with a diversity of 'small' problems, addressable by the traditional extension agent, to more large, consolidated farms with fewer but larger problems of potentially high economic impact, which require the attention of high technology, research-based solutions requiring the intervention of an extension agent SPECIALIST. In the U.S., the extension agent is still the right person, in the right place, at the right time to deal with extant agriculture-related problems. For extension agents, the transition from 'generalized' to 'specialized' was a natural sequence of events. Paradoxically, the advanced biotechnology which engendered and fostered this transition is creating new problems. Paramount among these are the practices of (1) monoculture and the potential for development of plant disease problems which are more devastating than those encountered in diverse cropping systems and (2) increased use of chemical fertilizers and pesticides which, inappropriately used, can lead to serious degradation of the quality and quantity of our diminishing soil and water resources.

The role of Cooperative Extension Service AGENTS changed in the U.S. because the nature of the problems it was designed to solve eventually changed completely. The nature of the newly evolved problems require and are receiving the attention of *specialized extension agents*. The technology transfer (extension) efforts exerted by U.S. scientists in developing countries have been equivocal at best. In hindsight, the reason is obvious. Current biotechnology, as transferred by extension agent specialists in the U.S., was not transferable to the subsistence farmers of developing countries who were at a stage of agricultural evolution where they were most receptive to the highly individualized attention of *extension agent generalists* who could identify individual farmer problems and provide feasible low-input solutions to those problems. Fencing masters went abroad to lead a bayonet charge. Predictably, it did not work. Tragically, it resulted in great disillusionment in the future of traditional agricultural extension methods as a viable solution to the problems of subsistence agriculture [35]. An agricultural development specialist from a U.S. land-grant college of agriculture recommended that agricultural output be expanded substantially by investing more in commodity related research, with the additional advice: 'Do not invest in extension.' (Cited by York [35]). Policy and decision makers failed to identify the nature of the problem. When agent specialists failed to do what agent generalists had done exceedingly well in the U.S., the door was closed on traditional extension as a proven, effective mechanism for transfer of adaptive-adoptive technology to subsistence farmers in developing countries. This change in attitude was reflected in changed priorities in the nature of agricultural research considered 'fundable' by higher administration. We have been cautioned [22] to remember the evolutionary process of our extension system, considering 'the time and sequence of events', before we attempt to transfer our model extension package to developing countries.

Some developing countries have no extension service; in other countries it may exist but is largely ineffective. There are many identifiable reasons. Some of the difficulties facing extension workers in these countries have been enumerated [22], as follows:

1. "Because of the shortage of research programs, they often have little to extend."

2. "They are usually men and they tend to bypass women, who often do much of the farming."

3. "Much of their time may be spent on administrative work for programs that are not production-oriented."

4. "They often find it difficult to reach farmers because of primitive or nonexistent transportation and communication networks. Shortages of vehicles and fuel compound the problem."

Effective transfer has many requirements. These requirements, once recognized and addressed, are satisfied in due course, leading to a time-place-situation which is optimum for effective use of the extension system. The U.S. Cooperative Extension Service is a joint county-state-national government function. Its strength is still in its grassroots support at the county level. Certainly, a primary requirement of successful transfer of the U.S. extension system to a developing country presupposes the full approval and support of the national, state, and local government of the developing country. The country must be fully committed to extension as a top priority of their agricultural program.

Several cogent approaches to improving existing extension programs in developing countries have been suggested [22], including:

1. relieve extension agents of non-production responsibilities such as regulatory enforcement,

2. strengthen training and technical backup of extension workers,

3. focus research and extension on improved technologies for priority crops and animals,

4. increase dialogue between extension workers and researchers, including participation in on-farm research and feedback to the researchers from the farmers,

5. coordinate public sector extension with mass media approaches and private sector activities,

6. increase the use of female extension workers and direct attention to the needs of women and other low resource farmers.

Again, these are issues, considerations, responsibilities, etc. which can and should be effectively addressed by national governments. External scientific, technical and financial support should, by invitation, be made available to these countries from the international community.

This invitation will only come from the governments of those developing countries which have the vision to recognize that technology, such as legume inoculation, can be a critical component of a sustainable, long-term solution to the problem of inadequate agricultural productivity.

The ascent of research

Let us talk about problems and priorities in SBNF research. With the ever-increasing shift from small-scale farming to large-scale farming there was a shift in the nature of production-oriented problems. There were fewer small farms and hence less demand for research on small farm problems. Concomitantly, there was an increased demand for research related to large farm problems. Research money flows in the direction of perceived need. Research priorities-funds shifted from practical small farm problems to sophisticated high technology solutions to large farm problems. An aggravating factor is the truism that one must 'publish or perish'; it is firmly established in fact. The initiation and continuance of extramural financial support for research depends strongly on tangible evidence of productivity - i.e. publications. In the U.S., such support from *private* or *state* sources is generally minimal. At the *national government* level the support has plateaued or declined. Competition for grant support is severe as a result of shrinking available funds and ever-increasing quality and quantity of grant applications. A carefully planned high technology laboratory research project may generate several solid 'hard science' papers in a year. The acceptability of field experiments for publication is not great until the work has been repeated at least once, and preferably twice. Publications are fewer and slower in arriving. If you were beginning your career, would you opt for laboratory or field research? Most choose the former as a matter of professional survival; others, perhaps as a function of opportunity, choose combined lab and field studies - certainly the best option whenever possible. How does this relate to SBNF research? Developed countries are amassing a wealth of valuable knowledge of basic aspects relevant to SBNF. But they are losing sight of an important objective. The critical importance of doing science for the sake of science, i.e., gaining new knowledge, cannot be denied. However, increasingly, developed countries seem to be doing basic research on SBNF because that is where the funds, publication, prestige, professional advancement, etc. lead. They are, in effect, prostituting themselves. Production-oriented field research, resulting in technology which can/would be transferable to subsistence farmers by extension agents, is the baby that is being thrown out with the bath water.

There needs to be a serious re-assessment of priorities in the area of SBNF research. There must be some agreement on the identity of the clientele. Developed

236

countries need to identify a major part of the clientele as the small farmers throughout the world who stand to benefit from field application of the existing technology. A global survey [7] indicates that 'inoculant technology has had little impact on food production at the small-farm level'. The crux of the problem is not the shortage of scientific information. The 'cutting edge' of inoculation technology does not lie in its *development* but in its *transfer*. It exists. It has been available commercially in appropriate form for small-farm application for eighty years [18]. It is not being transferred to subsistence farmers in developing countries.

This gap in information transfer can be filled by a user-friendly extension service.

This requires traditional extension agents who know farmer problems and how to adopt-adapt current technology to solutions to those problems, often in very localized situations. Authoritative description of an extension methodology class, appropriate to the training of extension agents in international agriculture, is available [25]. This facet of an agricultural program has a low profile in national politics. It is not generally associated with power, prestige or money. This relegates it to low priority with highly placed administrators (policy and decision makers) who ultimately determine national programs, funding priorities, etc.

Why do I belabor the issue of agricultural extension? I do so because this is the mechanism whereby legume inoculation technology is (can be; should be) made availablte to subsistence farmers world-wide. It is one of the most important components in the technology package of extension agents in the majority of developing countries.

Millions of subsistence farmers now barely survive on the margin of unnecessarily low crop productivity. Given access to the services of agricultural extension agents, these farmers can achieve crop productivity levels in significant excess to what is required for personal needs. This excess production is then available to the urban, non-farming population. This should serve to delay predicted mass hunger and starvation until such time as society agrees to abate the current rampant population increase which may otherwise prove inimical to the survival of mankind [4, 24].

For example, the benefits to be derived from application of inoculant technology are indicated in Table 1. There is direct evidence that this technology can be successfully tranferred at both large-scale and small-scale farm levels [6] in developing countries. A detailed

scenario for potential future successes of this nature is available [26].

Extension programs which enjoy government support show outstanding results. Thailand is an excellent example of what can be accomplished [6]. Given appropriate government priority, this success can be duplicated throughout the world. It is then a problem of *educating* highly placed policy-and-decision makers concerning the importance of establishing a well-trained corps of extension agents [15]. Whether we relish the role or not - it is clearly our responsibility to work on this education process. Until we become involved and become part of the *solution* we will continue to be part of the *problem*.

Resurrection of traditional extension

The real-life cutting edge in SBNF is not in *research*. It is in *extension*. In this statement I address specifically the unmet needs of subsistence farmers for usable, low level (appropriate) technology of all kinds, but in particular reference to the need for access to legume inoculation technology. There are many valid reasons for the widespread lack of adoption of agricultural technology by subsistence farmers. They include 'lack of credit, limited access to information, aversion to risk, inadequate farm size, inadequate incentives associated with farmer tenure arrangements, insufficient human capital, absence of equiprnent to relieve labor shortages (thus preventing timeliness of operations), chaotic supply of complementary inputs (such as seed, chemicals, and water) and inappropriate transportation infrastructure'[11] . In regard to these reasons, one in particular stands out in specific reference to the adoption of legume inoculation technology. That reason is 'limited access to information'. The traditional unadorned technology is simple, inexpensive and highly successful (increased yield) when properly practiced. This technology has been available and practiced in the U.S. and other developed countries for several decades. It remains largely unpracticed in developing countries because it is unknown to subsistence farmers, or practiced ineffectively due to lack of information, initial supervision, etc. There is now no dearth of information on how, why, when and where to inoculate.

In years past the local availability of high quality legume inoculant was a severe limitation to adoption. The problem was often exacerbated by the production, distribution and use of low quality inoculants which led inevitably to inoculant failures. Bad news travels

Table 1. Impact of NifTAL training, technical assistance, and transfer of inoculation technology on marginal economic benefit and economic and environmental opportunity costs of N fertilizers. (reprinted by permission from NifTAL Director's Annual Report to Agency for International Development, 1993)

Country	Training	Technical assistance	NifTAL technology	Inoculant produced tons	Inoculant value $M	Value of yield response to inoculation $M	Fertilizer N saved kg N (M)	Foreign exchange value of fertilizer $M	N loss to environment avoided kg N (M)
Thailand 1984–1992	Extension specialists, inoculant production, Rhizobium technology	Facilities design public and private sector, protocols, research, germplasm	NifTAL designed fermentors, production methods, germplasm	701	1.40	162.98	102.89	12.35	68.94
Indonesia 1989–1992	Extension specialists, Rhizobium technology, graduate, inoculant production	Facilities design (private sector), equipment specification, protocols	NifTAL designed fermentor, production methods, germplasm	90	1.290	27.00	12.11	1.450	8.11
Zambia 1984–1992	Inoculant production, quality control extension specialists, graduate	Materials testing, facilities design, production methods, research design	NifTAL designed fermentor, production methods, germplasm	67	.311	17.88	3.56	2.02	2.38
Uganda 1988–1993	Extension specialists-PVO/NGO, inoculant production, quality control	Facilities-lab design, materials testing and processing, production methods	NifTAL designed fermentors, germplasm	3	.002	1.00	.028	.011	.017

Fertilizer use efficiency assumed to be 40%; N fertilizer equivalent for inoculant equal to inoculant marginal yield increase/.8 (N harvest index)/.40 (fertilizer use efficiency); Value of N fertilizer $400 metric ton; Responses from inoculation estimated from field trials or in the case of Indonesia field trials and soil sampling techniques, area of responsive legumes inoculated at current recommended rates of inoculation per unit area; Value of crop estimated at local prices and current exchange rates, no allowance is made for changes in exchange over time; marginal yield response value for Uganda uses U.S. $0.21/kg since local prices reflect second country barter arrangements for subsidized products. Value of inoculant produced is actual or set at $2.00 ha^{-1} if currently subsidized below market prices.

fast and subsistence farmers have a long memory for failed innovations which have caused losses that they cannot afford. The situation is much improved. Reliable producers - distributors of quality inoculants are found world-wide. The production aspect has been significantly improved by the development of streamlined inoculant productiorn technology by NifTAL [30]. The system is characterized by simplicity, low cost and efficiency. It effectively removes quality inoculant availability (production) as a limitation to adoption of the technology in developing countries (Table 1).

In summary, the availability of information, resources and on-site assistance related to transfer of technology to developing countries, as demonstrated by the NifTAL Project, is unprecedented. There can be no valid excuse for the limitation of legume crop productivity anywhere in the world due to the non-existence of legume inoculation technology. The technology exists. I beleive that the limitation to adoption of this technology is attributable primarily to the absence or ineffectiveness of an extension system in most developing countries, where such systems are accorded little or no government priority as a national need.

References

Andrew CS and Kamprath EJ (eds) (1978) Mineral Nutrition of Legumes in Tropical and Subtropical Soils. Proceedings of a Workshop Held at CSIRO Cunningham Laboratory, Brisbane, Australia, January 16-21, 1978. CSIRO, Melbourne, Australia. 415p.

Arshad M and Frankenberger WT (1993) Microbial production of plant growth regulators. In: Metting FB (ed) Soil Microbial Ecology, p646 Marcel Dekker Inc., New York. 646p.

Beck DP and Materon LA (eds) (1987) Nitrogen Fixation by Legumes in Mediterranean Agriculture. Proceedings of a Workshop on Biological Nitrogen Fixation on Mediterranean-type Agriculture Held at the International Center for Agricultural Research in the Dry Areas, Aleppo, Syria, April 14-17, 1986. Kluwer Academic Publishers, Boston. 379p.

Bongaarts J (1994) Can the growing human population feed itself? Sci Am, March 1994.

Bornemiszsa E and Alvorado A (eds) (1975) Soil Management in Tropical America. Proceedings of a Seminar Held at CIAT, Cali, Colombia, February 10-14, 1974. Soil Science Department, North Carolina State University, Raleigh. 565p.

Chanaseni C and Kongngoen S (1992) Extension programs to promote rhizobial inoculants for soybean and groundnut in Thailand. Can J Microbiol 38 :594–597

Eaglesham ARJ (1989) Global importance of *Rhizobium* as an inoculant. In: Campbell D and MacDonald RM (eds) Microbial Inoculation of Crop Plants. Vol. 25, Special Publications of the Society for General Microbiology. IRL Press, New York.

Evans HJ, Bottomley PJ and Newton WE (eds) (1985) Nitrogen Fixation Research Progress. Proceedings of the 6th International Symposium on Nitrogen Fixation Held at Corvallis, Oregon, August 4-10, 1985. Martinus Nijhoff Publishers, Boston. 731p.

FAO/IAEA Programme (1990a) Use of ^{15}N Methodology to Quantify Biological Nitrogen Fixation. A selection of nineteen publications compiled by G. Hardarson. FAO/IAEA, Seibersdorf Laboratory, Austria

FAO/IAEA Programme (1990b) Use of Antibiotic Resistant Mutants to Study *Rhizobium* Ecology. A selection of eleven publications compiled by G. Hardarson. FAO/IAEA, Seibersdorf Laboratory, Austria

Feder G, Just RE, and Zelberman D (1985) Adoption of agricultural innovations in developing countries–a survey. Econ Dev Cult Change. 33:255–298

Graham PH and Harris SC (eds) 1982 Biological Nitrogen Fixation Technology for Tropical Agriculture: Papers Presented at a Workshop Held at CIAT, March 9-13, 1981. Centro Internacional de Agricultura Tropical, Cali, Colombia. 768p.

Hardarson G (1993) Methods for enhancing symbiotic nitrogen fixation. Plant Soil 152:1-17

Hennecke H and Verma DPS (eds) (1991) Advances in Molecular Genetics of Plant-Microbe Interactions. Vol. 1. Kluwer Academic Publishers, Boston.481p.

Halliday J (1981) Agrotechnologies Based on Symbiotic Systems that Fix Nitrogen. Issue Paper Presented at an International Workshop on Biological Nitrogen Fixation Technology for Tropical Agriculture Held at CIAT, March 9-13, 1981. Cali, Colombia.

Hubbell DH (1988) Extension/transfer of BNF technology In: Beck DP and Materon LA (eds) Nitrogen Fixation by Legumes in Mediterranean Agriculture, P 379.Martinus Nijhoff Publishers, Boston.

Jarstfer AG and Sylvia DM (1993) Inoculum production and inoculation strategies for vesicular-arbuscular mycorrhizal fungi. In: Metting FB (ed) Soil Microbial Ecology. Marcel Dekker, Inc. New York. 646p.

Jones DG (1991) Symbiotic nitrogen fixation-exploitation and unachieved potential. Ann Appl Biol 118:249–259

Kloepper JW (1993) Plant growth-promoting rhizobacteria as biological control agents. In: Metting FB (ed) Soil Microbial Ecology. Marcel Dekker, Inc., New York. 646p.

Lopez-Real JM and Hodges (eds) (1986) The Role of Microorganisms in a Sustainable Agriculture. Selected Papers from the Second International Conference on Biological Agriculture Held at University of London, Wye College, Wye, Kent, United Kingdom, September 3-7, 1984. AB Academic Publishers, Berkhamstead, Herts, UK. 246p.

McCardell AJ, Cregan PB and Sadowsky MJ (1993) Genetics and improvement of biological nitrogen fixation In: Metting FB (ed) Soil Microbial Ecology. Marcel Dekker, Inc. New York. 646p.

McPherson MP (985) Extension: do we really have anything useful or new to offer the developing countries? Opening Address at Conference on the International Role of Extension: Future Directions. Held at Michigan State University, March 31, 1985. pp 50-55

Mulongoy K, Gueye M and Spencer DSC (eds) (1992) Biological Nitrogen Fixation and Sustainability of Tropical Agriculture. Proceedings of the Fourth International Conference of the African Association for Biological Nitrogen Fixation Held at the International Institute of Tropical Agriculture, Ibadan, Nigeria, September 24-28, 1990. John Wiley and Sons, New York. 488p.

Pirnentel D, Hannan R, Pacenza M, Pecars J and Pimentel M (1994) Natural resources and an optimum human population. Popul Enviro 15:347–369

Shroyer MG, Shroyer JP and Hicks DR (1992) An extension methodology class: A component of an international agricultural development project. J Nat Resour Life Sci Educ 21: 175–177

Singleton P (1989) Employing BNF technology in Indonesia: Assessment of infrastructure, manpower and research needs. A report to the Secondary Food Crops Development and Communication for Technology Transfer in Agriculture Projects. NifTAL Project, Paia, Hawaii

Singleton PW, Bohlool BB and Nakao PL (1992) Legume response to rhizobial inoculation in the tropics-myths and realities. Soil Sci Soc Am Special Publication no. 29, pp 135–155

Singleton PW, Keyser HH, Zachariassen J and Smith RS (1994) The importance of legume-based BNF in world agriculture-an examination of the major commercial and environmental issues In: Lee RG (ed) Nitric Acid-Based Fertilizers and the Environment. SP-21, International Fertilizer Development Center, Muscle Shoals, Alabama *(In press)*

Siqueira JO, Naix MG, Hammerschmidt R and Safir GR (1991) Significance of Phenolic Compounds in Plant-Soil-Microbial Systems. Crit Rev Plant Sci 10: 63–126

Skinner FA, Boddey RM and Fendrik I (eds) (1987) Nitrogen Fixation with Non-Legumes. Proceedings of the Fourth International Symposium on Nitrogen Fixation with Non-Legumes Held at

Rio de Janeiro, Brasil, August 23-28,1987. Kluwer Academic Publishers, Boston. 336p

Somasegaran P and Hoben HJ (1985) Methods in Legume-*Rhizobium* Technology. NifTAL Project and MIRCEN, University of Hawaii, Paia, Maui, Hawaii. 367p.

Thompson JA (ed) (1991) Expert Consultation on Legume Inoculant Production and Quality Control. Proceedings of a Meeting Held at FAO Headquarters, March 19-21 Rome, Italy. 145p.

Vincent JM (ed) (1982) Nitrogen Fixation in Legumes. Based on Proceedings of an International Seminar Sponsored by the Austrailian Development Assistance Bureau and the University of Sydney. Academic Press, New York. 288p.

Vincent JM, Whitney AS and Bose (eds) (1977) Exploiting the Legume- Rhizobium Symbiosis in Tropical Agriculture. Proceedings of a Workshop Held at Kahului, Maui, Hawaii, August 23-28, 1976. College of Agriculture Misc. Pub 145, Department of Agronomy and Soil Science, University of Hawaii, Honolulu. 469p.

York ET (1985) International technology development and transfer. Keynote Address at Conference on the International Role of Extension: Future Directions, Held at Michigan State University, April 1, 1985. pp 56-66

Fertilizer Research **42**: 241–250, 1995.
© 1995 *Kluwer Academic Publishers. Printed in the Netherlands.*

Nitrogen fixation associated with grasses and cereals: Recent progress and perspectives for the future

Robert M. Boddey & Johanna Dobereiner
EMBRAPA - Centro Nacional de Pesquisa de Agrobiologia, Km 47, Seropédica, Itaguaí, 23581–970, Rio de Janeiro, Brazil

Key words: Acetobacter diazotrophicus, Azospirillum spp., cereals, grasses, *Herbaspirillum* spp., nitrogen fixation, sugar cane

Abstract

Over the last 20 years many new species of N_2-fixing bacteria have been discovered in association with grasses, cereals and other non-nodulating crops. Virtually all of these bacteria are microaerophylic, fixing N_2 only in the presence of low partial pressures of oxygen. Until a few years ago much attention was focussed on members the genus *Azospirillum* and it was assumed that N_2 fixation was restricted to the rhizosphere or rhizoplane of the host plants. Through the use of N balance and ^{15}N techniques it has been shown that in the case of lowland rice, several tropical pasture grasses and especially sugar cane, the contributions of biological N_2 fixation (BNF) are of agronomic significance.

More detailed study of the N_2-fixing bacteria associated with sugar cane (*Acetobacter diazotrophicus* and *Herbaspirillum* spp.) has shown that they occur in high numbers not only in roots of this crop but also in the stems, leaves and trash but are rarely found in the soil. Some of these endophytic diazotrophs have now also been found in forage grasses, cereals, sweet potato and cassava, although evidence of significant BNF contributions is still lacking.

The identification of these endophytic diazotrophs as the organisms probably responsible for the high contributions of N_2 fixation observed in sugar cane suggests that it may be possible to attain significant BNF contributions in some other gramineae and perhaps root crops.

Introduction

Until about 25 years ago, apart from the rhizobium group of bacteria, the known heterotrophic N_2-fixing bacteria (eg. *Azotobacter, Beijerinckia, Derxia* spp., etc) had all been isolated on N-free agar plates using traditional bacteriological techniques. They were able to grow on plates in the absence of fixed N as they possess oxygen-protection mechanisms such as production of copious quantities of gum which hinder O_2 diffusion and/or they 'burn off' O_2 by a decoupled respiratory pathway [40, 41]. Gum production or rapid respiration consume large amounts of carbon substrate and thus N_2 fixation in these conditions has a far higher carbon cost than in cultures grown at low pO_2.

After the development of the acetylene reduction (AR) technique [24, 76] considerable nitrogenase activity was detected associated with roots of wetland rice [72, 85, 86], maize [82], and various tropical grasses and sugar cane [26], which could not be accounted for by the presence of the known species of N_2-fixing bacteria. The root-associated nitrogenase activity was found to be maximal at low oxygen partial pressures and Dobereiner and Day [25] developed semi-solid culture media to isolate N_2-fixing fixing bacteria from the roots and rhizosphere of these grasses and cereals. In this medium, the diffusion of O_2 to the site of bacterial growth is in equilibrium with the respiration rate of the growing bacteria so that as the bacteria multiply the organisms migrate gradually towards the surface of the medium producing a veil-like pellicle at

242

the point where O_2 diffusion matches O_2 uptake and consequently maintains a pO_2 low enough for nitrogenase to function. This simple, but very powerful technique, quickly led to the discovery of two species of diazotrophs of the new genus *Azospirillum* [25, 79] and subsequently three other species of this genus; *A. amazonense* [54], *A. halopraeferans* [68] and *A. irakense* [49]. The microaerophilic N_2-fixers, *Campylobacter nitrofigilis* [56], *Bacillus azotofixans* [77] and *Herbaspirillum seropedicae* [7] were also originally isolated using this technique. All these diazotrophs were found in association with the roots of species of grasses or cereals.

Associations of *Azospirillum* with grasses and cereals

To obtain maximum yields of the most important cereals such as maize, wheat and rice, the largest fertilizer requirement is N, and the discovery of *Azospirillum*, and high AR (nitrogenase) activity associated with the roots of cereals, immediately attracted much interest among agronomists and soil microbiologists. Many studies were performed on the inoculation of these crops with *Azospirillum* spp, particularly the *Azospirillum brasilense* type strain Sp7 originally isolated from the rhizosphere soil of *Digitaria decumbens* (and the closely related strain Cd), but in only a few studies were positive crop yield responses reported (see reviews of [12, 15, 16, 36]). These largely negative results, along with criticisms of the use of the acetylene reduction technique, particularly with excised roots [13], led to the consensus that biological N_2 fixation (BNF) associated with grasses and cereals had very little agronomic potential [37].

In the early 1980s, results from our laboratory showed that there was a certain degree of specificity in the infection of cereal roots by *Azospirillum* spp. [4, 5, 6, 8]. The evidence for this specificity was obtained by surface sterilizing the cereal roots and only when roots were treated in this manner was the predominance of certain strains evident. These data suggested that the interior of the root could be colonized by azospirilla, and data to support this were obtained from microscopic examination of tetrazolium stained roots [64, 55]. Although the infection of the interior of cereal and grass roots by azospirilla, and strain specificity in the infection process, was hotly contested [14], infection of the interior of Bermuda grass roots (*Cynodon dacty-*

lon) by *A. brasilense* was demonstrated using the fluorescent antibody technique by Schank et al. [75].

The results obtained in our laboratory showed that azospirillum strains isolated from surface sterilized roots of a certain cereal showed a greater aptitude to reinfect the same cereal and to promote responses in crop yield and/or N accumulation when these 'homologous' strains were inoculated. In this respect several studies showed that the *A. brasilense* strain Sp 245, isolated from surface-sterilized wheat roots repeatedly increased grain yield and N accumulation of wheat in both field and pot experiments where Sp 7 and/or Cd promoted little or no plant response [6, 8, 9]. Similar results have been obtained with maize, wheat and pearl millet by other authors [33, 71, 84].

Only a few studies have been made on the mechanism of crop response to azospirillum inoculation. In two studies where the ^{15}N isotope dilution technique was employed, positive responses of wheat dry matter and N accumulation to inoculation were observed but the ^{15}N data showed that extra N accumulated by the plants was derived from soil and fertilizer N and not from N_2 fixation [46, 18]. Several mechanisms for the observed plant growth responses have been proposed, including the production of plant-growth promoting substances such as auxins [44, 88, 31], or the improvement of uptake of mineral nutrients by other mechanisms [53, 47]. Data from studies performed by at EMBRAPA-CNPAB suggested also that there may be a role of bacterial nitrate reductase in increasing soil N assimilation [18, 32]. In these studies the dry matter and N accumulation wheat responded to inoculation with *A. brasilense*, but failed to respond when they were inoculated with mutants of the strains which were deficient in nitrate reductase. These various mechanisms, and their possible importance in affecting plant growth, have been discussed in several recent reviews [12, 30, 61].

The possibility that different plant genotypes may respond differentially to azospirillum inoculation was indicated from results of experiments on pearl millet [84] and wheat [57]. Recent experiments performed in Argentina with maize showed a decisive effect of plant genotype on the response of plant growth to azospirillum inoculation [35]. In two field experiments, significant and consistent increases as well as decreases of grain yield and plant N accumulation were observed due to the inoculation with a mixture of Brazilian and Argentinian strains of *A. brasilense* and *A. lipoferum*, isolated from surface sterilized maize roots. The highest yielding genotype (Dekalb 4F37) showed yield

Table 1. Effect of inoculation of a mixture of seven strains[a] of *Azospirillum* spp. on grain yield and N accumulation of seven maize varieties in two consecutive field experiments. Values are means of three replicates. After Garcia de Salamone [35]

| Maize genotype | 1989/90 | | | | 1990/91 | | | |
| | Grain yield | | N accumulation | | Grain yield | | N accumulation | |
	non-inoc. (kg ha^{-1})	inoc. % change	non-inoc. (kg ha^{-1})	inoc. % change	non-inoc. (kg ha^{-1})	inoc. % change	non-inoc. (kg ha^{-1})	inoc. % change
Dekalb 47–37	7344	−32*	134	−23*	4647	−39*	96	−42*
Morgan 251	7135	−09	100	−07	3484	+42*	59	+28
Morgan 318	3745	+40*	68	+39*	6496	+27*	98	+28*
Morgan 315	6921	−07	124	+02	4421	−21	79	−21
Dekalb 2F-11	5643	+05	86	+11	3154	+12	64	+06
Delalb 4D-70	4664	+45*	53	+82*	3775	+24	58	+32
CMS 22	2934	+37*	92	+38*	3307	+11	67	+02
HSD [b](Tukey)	580		16		1415		26	

[a]Mixture of four strains of *A. brasilense* isolated from surface sterilized maize roots in Agentina (7M, 10M, 40M and 42M) and three strains of *A. lipoferum* isolated in Brazil (BSL 505, Sp 242 and 582).
[b]Honest significant difference.
*Response to *Azospirillum* inoculation significant at $p=0.05$ (Tukey).

Table 2. Total N accumulation, ^{15}N enrichment and estimates of contributions of biological nitrogen fixation to maize variety Dekalb 4D-70 inoculated with different strains of *Azospirillum* spp. [a]Means of four replicates. After Garcia de Salamone [35]

Azospirillum strain	N accumulation g plant^{-1}	Atom % ^{15}N excess	% N derived from BNF[b]
Sterilized inoculant	0.98b	0.1368a	-
A.brasilense			
42M	2.35a	0.0512c	62
42M NR$^-$	1.51ab	0.1155ab	15
A. lipoferum			
Sp 242	1.01b	0.1169ab	15
Sp 242 NR$^-$	1.04b	0.0752bc	45
Mixture[c]	2.48a	0.0793bc	42
Mixture NR^{-}[d]	2.27a	0.0873abc	36
Coefficient of variation (%)	61.0	34.0	-

[a]Plants (2 per pot) grown in pots, maintained under conditions, containing 25 kg of soil amended with ^{15}N labelled organic matter 17 months before planting.
[b]Estimated using ^{15}N enrichment of control inoculated with sterilized inoculant.
[c]Mixture of strains 42M, 40M, Sp 242 and BSL 505.
[d]Mixture of nitrate reductase-negative mutants of strains 42M, 40M, Sp 242 and BSL 505.

decreases between 30 to 40%, whereas four lower yielding genotypes showed yield increases between 10 and 45% due to inoculation (Table 1). A subsequent pot experiment with ^{15}N labelled soil, using one of the maize genotypes which responded positively to *Azospirillum* inoculation, suggested that up to 62% of the plant N could be obtained from BNF. In this experiment nitrate reductase-negative mutants obtained from these strains showed variable effects on yield and ^{15}N enrichment (Table 2).

Contributions of biological N_2 fixation

While there is only limited evidence that azospirillum can contribute significant quantities of biologically fixed N to cereals or grasses, in recent years evidence has been obtained that several graminaceous crops can benefit from plant-associated BNF although which diazotrophs are responsible for these contributions is unknown.

Long term studies on field-grown wetland rice have consistently shown positive N balances, suggesting BNF inputs of between 30 and 60 kg N ha^{-1} crop^{-1}, [1, 50, 74, 83] but uncertainty exists as to the proportion of these contributions derived from plant-associated heterotrophic bacteria or from cyanobacteria, and in some cases, rainfall or irrigation water. More reliable estimates of BNF contributions to wetland rice have been obtained from N balance experiments performed in pots. In these studies cyanobacteria were controlled by shading of the soil surface and in the first study contributions of plant associated BNF were found to be equivalent to 18% of total plant N [2]. In a subsequent experiment the N balance associated with 76 commercial cultivars and two wild varieties of wetland rice was found to vary from the equivalent of 16 to 70 kg N ha^{-1} crop^{-1}, when the balances were extrapolated on the basis of 250,000 plants ha^{-1} [3].

That heterotrophic N_2-fixing bacteria can contribute significant quantities of N to wetland rice has been firmly established by several $^{15}N_2$ gas incorporation experiments [29, 43, 87]. Similar studies on pasture grasses also confirmed small but significant BNF contributions to various pasture grasses in Texas [60], *Digitaria decumbens* [23] and sugar cane [73] in Brazil, and sorghum seedlings at ICRISAT in India [37]. However, because such experiments must be performed in sealed enclosures, it is not possible with this technique to evaluate the BNF contributions to a crop over the whole growing season under field conditions.

At our Centre several studies have been conducted using the ^{15}N isotope dilution technique to quantify the contribution of BNF to tropical pasture grasses grown in pots or cylinders maintained under field conditions. In two different studies on four *Brachiaria* species and 11 ecotypes of *Panicum maximum* grown in concrete cylinders (60 cm diameter) containing approximately 140 kg of labelled soil, it was found that *B. humidicola* and *B. decumbens* and several of the *P. maximum* ecotypes showed consistently higher N yields and lower ^{15}N enrichments than the control crop (*B. radicans*). From these studies we concluded that BNF contributions could be as high as 7 to 10 kg N ha^{-1} month^{-1} during the warm wet summer months [17, 58]. A further study performed in pots of ^{15}N-labelled soil on 25 ecotypes of *P. maximum* using the same control crops confirmed that BNF contributions varied widely between the different ecotypes and up to 39% of plant N could be obtained from this source [59].

In a ^{15}N-aided N balance experiment on 4 sugar cane varieties grown in large pots containing 64 kg soil, it was found that the variety CB 47–89 accumulated over 34 g of N in a 22 month period (plant crop and 1 ratoon) of which only 8.5 g could be accounted for as being derived from the soil and added ^{15}N-labelled urea [52]. This large positive balance, equivalent to over 150 kg N ha^{-1} year^{-1}, was significantly higher than the N balance associated with the other three varieties

and the ^{15}N enrichment data suggested that over 50% of plant N was derived from BNF.

In a subsequent study a large tank ($20 \times 6 \times 0.8$ m) was filled to a depth of 60 cm with ^{15}N-labelled soil fertilized with P, K and micronutrients to estimate the contribution of BNF to 7 commercial sugar cane varieties and three other species of *Saccharum* [81]. In this study the cane varieties CB 45–3 and SP 70–1143 maintained yields over a 3-year period equivalent to more than 200 tons cane ha^{-1} year^{-1} and accumulated over 230 kg N ha^{-1} year^{-1} with no decrease in N accumulation, even though all plant material was removed at each annual harvest and no N fertilizer or other N amendment was made during the experiment. Similar results were obtained with the *Saccharum spontaneum* variety, Krakatau but most other cane varieties showed decreasing yields with time.

Over the whole 3 years the weighted mean ^{15}N enrichments of all of the sugar cane varieties were much lower than of the *Brachiaria arrecta* control plant, indicating large contributions of plant associated BNF (Table 3). It was clear that the ^{15}N enrichment of the soil mineral N decreased during the study 3 particularly during the first year. At the second and third annual harvests there were only small differences in the ^{15}N enrichments between different varieties and that of the control crop which would suggest lower BNF inputs at this time. However, the data showed that this was due to carry over of labelled N from one harvest to the next in the stem bases and roots of cane varieties, which did not occur in the case of the *B. arrecta*.

The interpretation of the ^{15}N data was further complicated by the fact that the uptake of soil N by the *B. arrecta* was almost certainly restricted towards the end of the growing season due to shading of this crop by the tall sugar cane plants, and this probably resulted in a somewhat higher ^{15}N enrichment in the control crop than otherwise would have occurred. These difficulties are fully discussed in the original paper [81] and because of them it was decided to perform a total N balance on the whole tank by careful analysis of the N content of soil samples taken at plant emergence in the first year in comparison with samples taken at the final harvest.

These data show that there were significantly (p <0.05) positive N balances associated with the varieties CB 45–3, SP 70–1143 and Krakatau, and that there was a good agreement between the ^{15}N dilution and the total N balance estimates (Table 3). The data again indicated that over 60% of the plant N was derived

Table 3. ^{15}N enrichment and total nitrogen accumulation of 10 sugar cane varieties and *Brachiaria arrecta* control, and estimates of nitrogen derived from BNF using N balance and ^{15}N isotope dilution techniques (g N m^{-2}). Means of 4 replicates. After Urquiaga *et al.* [81]

Variety/ Species	Weighted mean atom% ^{15}N excess	Final N content of soil	N accum. whole plant 3 years (g N m^{-2})	Estimates of BNF contribution			
				Annual mean			
				N balance		^{15}N	
				(g N m^{-2})	(%)	(g N m^{-2})	(%)
CB 47–89	0.191bcd	835	61.4bc	13.2	65	11.6c	57
CB 45–3	0.166cde	864	84.3ab	20.9	74	17.5b	62
NA 56–79	0.198bc	884	57.8c	12.0	62	10.9c	56
IAC 52–150	0.188bcd	924	59.6bc	12.6	64	11.3c	57
SP 70–1143	0.146de	852	77.5bc	18.6	72	17.3b	67
SP 71–799	0.183bcd	860	56.9c	11.7	62	11.1c	58
SP 79–2312	0.198bc	845	63.6c	14.0	66	11.8c	56
Chunee	0.227b	826	33.0d	3.8	34	5.6d	51
Caiana	0.190bcd	857	11.6d	−3.4	–	2.2d	58
Krakatau	0.133e	857	102.8a	27.0	79	23.9a	70
B.arrecta	0.443a	830	24.9	1.1	4	–	–
CV (%)	13.6	5.1	25.0	29.2		29.2	

[a]N balance estimate of BNF contribution = total N accumulated by crop + mean total N content of soil in tank at final harvest – mean total N content of soil in tank at emergence of the sugar cane. Mean change in soil N content from emergence until final harvest = 27.1 g N m^{-2} with a standard error of the difference between the means of 22.0 g N m^{-2}. N balances greater than 37.7 g N m^{-2} (12.4 g N m^{-2} year^{-1}) were significantly greater than zero ($p=0.05$, Student t test).
[b]^{15}N isotope dilution estimate of BNF contribution = (total N accumulated by the crop) × (1 - (weighted mean atom % ^{15}N excess of sugar cane) weighted mean atom % ^{15}N excess of *B. arrecta*)).
[c]% of plant N derived from BNF (%Ndfa).

from BNF (the equivalent of more than 170 kg N ha^{-1} year^{-1}) in these three cane varieties.

Endophytic N$_2$-fixing bacteria

Acetobacter diazotrophicus

The discovery of very large contributions of BNF to certain Brazilian varieties of sugar cane immediately led our team to investigate more closely the occurrence of N$_2$-fixing bacteria associated with this crop. Instead of restricting the search to the roots and rhizosphere, all plant organs were sampled and initially attempts to isolate diazotrophs were made using a semi-solid medium based on sugar cane juice [21]. Immediately, a new species of N$_2$-fixing bacteria was found to occur in large numbers in the roots and stems of cane plants taken from many areas in Brazil which has now been classified as *Acetobacter diazotrophicus* [39].

This bacteria is a small, Gram-negative, aerobic rod, showing pellicle formation in N-free semi-solid medium with 100 g l^{-1} sucrose but without cane juice, forming a thick surface pellicle after 7 to 10 days. Best growth occurs with high sucrose or glucose concentrations (100 g l^{-1}) and strong acid production results in a final pH of 3.0 or less. Growth and N$_2$ fixation (more than 100 n moles C$_2$H$_2$ ml^{-1} h^{-1}) continues at this pH for several days [78]. Ethanol is also used as a C source for growth and is oxidized to CO$_2$ and H$_2$O. Dark brown colonies form on potato agar with 100 g l^{-1} sucrose, and dark orange colonies on N-poor (0.02 g l^{-1} yeast extract) mineral agar medium with 100 g l^{-1} sucrose and bromothymol blue. The bacterium possesses no nitrate reductase and N$_2$ fixation is not affected by high levels (25 mM) of NO$_3^-$. Also NH$_4^+$

causes only partial inhibition of nitrogenase, especially when grown on 100 g l^{-1} sucrose [80, 19].

Another interesting aspect is that *A. diazotrophicus* growing in 10% sucrose showed an optimum dissolved oxygen concentration for acetylene reduction in equilibrium with 0.2 kPa O_2 in the atmosphere, but continued to fix N_2 up to 4.0 kPa, showing a much higher O_2 tolerance than *Azospirillum* spp. [70].

Experiments on mixed cultures of *A. diazotrophicus* with an amylolytic yeast (*Lipomyces kononenkoae*), used as a model system for plant/bacteria interactions, showed that over 40% of the total N_2 fixed by the bacteria was transferred to the yeast, starting right from the beginning of the culture [22]. These results are important in that until now the lack of evidence for efficient transfer of fixed N from diazotrophs to plants has been a source of scepticism that such associations could be of agronomic importance.

This bacterium has been found in many sugar cane varieties in several regions of Brazil as well as from Mexico, Cuba and Australia [51, 34] with numbers ranging from 10^3 to 10^7 in roots, basal and apical stems, leaves and in sugar cane trash [27]. It was not found in soil between rows of sugar cane plants or roots from 12 different weed species taken from cane fields. It was also not found in grain or sugar sorghum, but was isolated from a few samples of washed roots and aerial parts of *Pennisetum purpureum* cv Cameroon, and from sweet potato [27, 28, 65].

Sterile micropropagated sugar cane seedlings were not infected by *A. diazotrophicus* by traditional root inoculation methods; generally infection of cane plants by this bacterium is rare except when inoculated *in vitro*. However, under these conditions *A. diazotrophicus* was found to colonize extensively the exterior and interior of the shoot and root [45]. This study was performed using immuno-gold labelling with both optical and electron microscopic techniques.

The difficulty of infection of plants grown in soil or vermiculite can be overcome by co-inoculation with VA mycorrhizal fungi, especially originating from fungal spores infected with the bacteria [66]. This technique of introduction of a N_2-fixing bacteria into sugar cane plants may be important for introducing selected or genetically improved strains into plants for further propagation in the field via stem cuttings.

Herbaspirillum spp.

Bacterial taxonomists working in Belgium found that the bacteria known as *Pseudomonas rubrisubalbicans*, a sugar cane endophyte which causes mottled stripe disease in some varieties from the USA and other countries, but not in Brazilian varieties, was closely related genetically to a N_2-fixing bacterium called *Herbaspirillum seropedicae* [38]. As mentioned before, *Herbaspirillum* was first isolated from the roots of maize and other cereals at our Centre [7]. Most of the isolates of *P. rubrisubalbicans* were found to be able to fix N_2 and were identical in most other respects to *Herbaspirillum* [67]. Recently results from DNA/rRNA hybridization and computer-assisted auxanographic tests have established that this generically-misnamed plant endophyte, '*Pseudomonas*' *rubrisubalbicans*, must now be included in the genus *Herbaspirillum* [38]. Subsequently a more specific culture medium (JNFb) for this organism has been developed and $^{15}N_2$ gas incorporation confirmed, not only in strains of the original *H. seropedicae*, but also in isolates from different culture collections of *H. rubrisubalbicans*, identified as the causitive organism of mottled stripe disease [10]. *Herbaspirillum* spp. have been isolated from sugar cane leaves, stems and roots and do not survive well in the soil, but only within plants.

When non-sterile soil was inoculated with 10^8 cells g^{-1} of either species of *Herbaspirillum*, the number of viable cells decreased until the bacteria was undetectable after 21 days with *H. rubrisubalbicans* and 28 days with *H. seropedicae* [62]. However, 50 days after *Herbaspirillum* spp. were undetectable, surface-sterilized sorghum seeds were planted in these pots and the bacteria could be reisolated from the roots when the plants were 30 days old.

In both monoxenic sugarcane and sorghum plants inoculated with *Herbaspirillum* spp., the bacteria have been localized, using the immuno-gold technique and both electron and optical microscopy, within the meta and protoxylem (Olivares *et al.*, in prep.). In the case of a sugar cane variety (B-4362), susceptible to mottled stripe disease, *H. rubrisubalbicans* was found to completely block some of the xylem vessels, whereas in a resistant variety, the bacteria were encapsulated by membranes, probably of plant origin.

Other endophytic diazotrophs

While it has been shown that *Azospirillum* spp. are able to infect the roots and even stems of various graminaceous crops, members of this genus cannot be described at true endophytes because they survive in high numbers in the soil, especially in the tropics.

Recently, N_2-fixing bacteria found within the roots of Kallar grass (*Leptochloa fusca*) in Pakistan have been classified in the new genus *Azoarcus* [69]. The evidence available suggests that this bacterium may also be an endophyte but data on its survival in soil are not yet available [42].

Perspectives for the future

It seems logical that N_2-fixing bacteria located within plant tissue (endophytic diazotrophs) will be able to fix N_2 and transfer fixed N to the host plant with much greater efficiency than diazotrophs occurring in the plant rhizosphere. In the rhizosphere the supply of carbon substrates to the bacteria and fixed N to the plant is likely to be inefficient. The N_2-fixing organisms have to compete with other rhizosphere microorganisms and the cycles of wetter and drier soil will cause variations in the diffusion of O_2 to, and hence the pO_2 in, this zone. The fact that some Brazilian sugar cane varieties have been shown to obtain very large BNF contributions and at the same time high numbers of endophytic diazotrophs have been isolated from their roots, leaves and stems suggests that these endophytes are responsible for the high BNF contributions observed.

Acetobacter diazotrophicus has been found in *Pennisetum purpureum* (cv. Cameroon) and sweet potato, both plants which accumulate sugar or starch and are propagated vegetatively [27, 65]. As this diazotroph has been shown not to survive in soil and to infect plants only with difficulty, it seems that it is transferred from crop to crop in the stem pieces used for replanting. *Herbaspirillum* spp. has been found to occur in parts of many graminaceous crops including the roots and aerial tissue of rice [11, 63] and *P. purpureum* [28], and in the roots of various forage grasses, maize and many weeds in sugar cane and maize fields [28]; however, the propagation of this organism has not yet been elucidated. *Azoarcus* spp. have been found so far to occur naturally only in Kallar grass but Hurek et al. [42] reported that it was able to colonize the interior of rice roots. None of these crops has been found to obtain such large BNF contributions as sugar cane, but

the fact that these microorganisms are present, often at high numbers, suggests that further research on these associations, and especially the sugar cane system may lead to strategies to increase BNF in many important cereal crops.

The complete absence of *A. diazotrophicus* in soil and the restricted occurrence of *Herbaspirillum* spp. and other endophytic diazotrophs, suggests that once selected (or even genetically manipulated) strains of these bacteria are established in plants in the field, the chances are slight that wild type strains will contaminate the plants to compete with them. In the case of sugar cane for phytosanitary reasons the use of direct planting of monoxenic micropropagated cane plantlets is now being tested at several cane plantations in São Paulo State. This may soon offer an economically viable opportunity to propagate cane plants infected by improved, or genetically manipulated, strains of endophytic diazotrophs.

The various recent findings discussed in this paper have established several new possibilities for the extension of BNF to non-leguminous crops and certainly seem more promising for the near future than the transfer of *nif* genes to plants, or through attempts to nodulate non-legumes [48].

References

1. App A, Santiago T, Daez C, Menguito G, Ventura W, Tirol A, Po J, Watanabe I, De Datta SK and Roger P (1984) Estimation of the nitrogen balance for irrigated rice and the contribution of phototrophic nitrogen fixation. Field Crops Res 9: 17–27
2. App AA, Watanabe I, Alexander M, Ventura W, Daez G, Santiago T and De Datta SK (1980) Nonsymbiotic nitrogen fixation associated with the rice plant in flooded soils. Soil Sci 130: 283–289
3. App AA, Watanabe I, Ventura TS, Bravo M and Jurey CD (1986) The effect of cultivated and wild rice varieties on the nitrogen balance of flooded soil. Soil Sci 141: 448–452
4. Baldani VLD and Döbereiner J (1980) Host-plant specificity in the infection of cereals with *Azospirilum* spp. Soil Biol Biochem 12: 433–439
5. Baldani JI, Pereira PAA, Rocha REM da and Döbereiner J (1981) Especificidade na infecção de raízes por *Azospirillum* spp. em plantas com via fotossintética C3 e C4. Pesq Agropec Bras 16: 325–330
6. Baldani VLD, Baldani JI and Döbereiner J (1983) Effects of *Azospirillum* inoculation on root infection and nitrogen incorporation in wheat. Can J Microbiol 29: 924–929
7. Baldani JI, Baldani VLD, Seldin L and Döbereiner J (1986) Characterization of *Herbaspirillum seropedicae* gen. nov., sp. nov., a root-associated nitrogen-fixing bacterium. Int J Syst Bacteriol 36: 86–93
8. Baldani VLD, Alvarez MA de B, Baldani JI and Döbereiner J (1986) Establishment of inoculated *Azospirillum* spp. in the

rhizosphere and in roots of field grown wheat and sorghum. Plant Soil 90: 35–46

9. Baldani VLD, Baldani JI and Döbereiner J (1987) Inoculation of field-grown wheat (*Triticum aestivum*) with *Azospirillum* spp. in Brazil. Biol Fertil Soils 4: 37–40

10. Baldani VLD, Baldani JI, Olivares FL and Döbereiner J (1992) Identification and ecology of *Herbaspirillum seropedicae* and the closely related *Pseudomonas rubrisubalbicans*. Symbiosis 13: 65–73

11. Baldani VLD, James E, Baldani JI and Döbereiner J (1992) Localization of the N_2-fixing bacteria *Herbaspirillum seropedicae* within root cells of rice. An Acad Bras Cienc 64: 431

12. Bashan Y and Levanony H (1990) Current status of *Azospirillum* inoculation technology: *Azospirillum* as a challenge for agriculture. Can J Microbiol. 36: 591–608

13. Berkum P van and Bohlool BB (1980) Evaluation of nitrogen fixation by bacteria in association with roots of tropical grasses. Microbiol Rev 44: 491–517

14. Berkum P van, Mc Clung CR and Sloger C (1982) Some pertinent remarks on N_2 fixation associated with the roots of grasses. *In*: Graham PH and Harris SC (ed) Biological Nitrogen Fixation Technology for Tropical Agriculture, pp 513–525. Centro International de Agricultura Tropical (CIAT), Cali, Colômbia

15. Boddey RM and Döbereiner J (1982) Association of *Azospirillum* and other diazotrophs with tropical graminae. *In*: Non-Symbiotic Nitrogen Fixation and Organic Matter in the Tropics, pp 28–47. Indian Society of Soil Science, New Delhi

16. Boddey RM and Döbereiner J (1988) Nitrogen fixation associated with grasses and cereals: recent results and perspectives for future research. Plant Soil 108: 53–65

17. Boddey RM and Victoria RL (1986) Estimation of biological nitrogen fixation associated with *Brachiaria* and *Paspalum* grasses using ^{15}N labelled organic matter and fertilizer. Plant Soil 90: 265–292

18. Boddey RM, Baldani VLD, Baldani JI and Döbereiner J (1986) Effect of inoculation of *Azospirillum* spp. on the nitrogen assimilation of field grown wheat. Plant Soil 95: 109–121

19. Boddey RM, Urquiaga S, Reis V and Döbereiner J (1991) Biological nitrogen fixation associated with sugar cane. Plant Soil 137: 111–117

20. Boureau M (1977) Application de la chromatographie en phase gazeuse à l'étude de l'exudation racinaire du riz. Cah ORSTOM sér Biol 12: 75–81

21. Cavalcante VA and Döbereiner J (1988) A new acid-tolerant nitrogen-fixing bacterium associated with sugarcane. Plant Soil 108: 23–31

22. Cojho EH, Reis VM, Schenberg ACG and Döbereiner J (1993) Interactions of *Acetobacter diazotrophicus* with an amylolytic yeast in nitrogen-free batch culture. FEMS Microbiol Lett 106: 341–346

23. De Polli H, Matsui E, Döbereiner J and Salati E (1977) Confirmation of nitrogen fixation in two tropical grasses by $^{15}N_2$ incorporation. Soil Biol Biochem 9: 119–123

24. Dilworth MJ (1966) Acetylene reduction by nitrogen fixing preparations of *Clostridium pasteurianum*. Biochem Biophys Acta 127: 285–294

25. Döbereiner J and Day JM (1975) Associative symbioses in tropical grasses: characterization of microorganisms and nitrogen-fixing sites. *In*: Newton WE and Nyman CJ (ed) Proceedings of the 1st International Symposium on Nitrogen Fixation, pp 518–538. Washington State University Press, Pullman

26. Döbereiner J, Day JM and Dart PJ (1972) Nitrogenase activity in the rhizosphere of sugar cane and some other tropical grasses. Plant Soil 37: 191–196

27. Döbereiner J, Reis V and Lazarine AC (1988) A new N_2 fixing bacteria in association with cereals and sugarcane. *In*: Bothe H, De Bruijn FJ de and Newtan WE (ed) Nitrogen Fixation: Hundred years after, pp 717–722. Gustav Fischer, Stuttgart

28. Döbereiner J, Reis VM, Paula MA and Olivares F de (1993) Endophytic diazotrophs in sugar cane, cereals and tuber plants. *In*: Palacios R, Mora J and Newton WR (ed) New Horizons in Nitrogen Fixation, pp 671–676. CRC Press, Dordrecht, Netherlands

29. Eskew DL, Eaglesham ARJ and App AA (1981) Heterotrophic N_2 fixation and distribution of newly fixed nitrogen in a rice-flooded soil system. Plant Physiol 68: 48–52

30. Fages J (1994) *Azospirillum* inoculants and field experiments. *In*: Okon Y (ed) *Azospirillum*/Plant Associations, pp 87–109. CRC Press, Boca Raton, Florida

31. Fallik E, Okon Y and Fischer M (1988) The effect of *Azospirillum brasilense* inoculation on metabolic enzyme activity in maize root seedlings. Symbiosis 6: 17–28

32. Ferreira MCB, Fernandes MS and Döbereiner J (1987) Role of *Azospirillum brasilense* nitrate reductase in nitrate assimilation by wheat plants. Biol Fertil Soils 4: 47–53

33. Freitas JLM de, Rocha REM da, Pereira PAA and Döbereiner J (1982) Matéria orgânica e inoculação com *Azospirillum* na incorporação de N pelo milho. Pesq Agropec Bras 17: 1423–1432

34. Fuentes-Ramirez LE, Jiminez-Salgado T, Abarca-Ocampo IR and Caballero Mellado J (1993) *Acetobacter diazotrophicus*, an indolacetic acid-producing bacterium isolated from sugarcane cultivars in Mexico. Plant Soil 154: 145–150

35. Garcia de Salamone IE (1993) Influencia de bacterias del genero *Azospirillum* sobre el rendimento y nutricion nitrogenada del cultivo de maiz (*Zea mays* L.). MSc thesis, University of Buenos Aires. 172p

36. Gaskins MH, Albrecht SL and Hubbell DH (1985) Rhizosphere bacteria and their use to increase plant productivity. Agric Ecosys Environ 12: 99–116

37. Giller KE and Day JM (1985) Nitrogen fixation in the rhizosphere: significance in natural and agricultural systems. *In*: Fitter AH (ed) Ecological Interactions in Soil, pp 127–147. Blackwell Scientific Publication, Oxford

38. Gillis M, Döbereiner J, Pot B, Goor M, Falsen E, Hoste B, Reinhold B and Kersters K (1991) Taxonomic relationships between [*Pseudomonas*] *rubrisubalbicans*, some clinical isolates (EF group 1), *Herbaspirillum seropedicae* and [*Aquaspirillum*] *autotrophicum*. *In*: Polsinelli M, Materassi R and Vincenzini M (ed) Nitrogen Fixation, pp 292–294. Kluwer Academic Publ., Dordrecht, Netherlands

39. Gillis M, Kerters B, Hoste DJ, Kroppenstedt RM, Stephan MP, Teixeira KRS, Döbereiner J and De Ley J (1989) *Acetobacter diazotrophicus* sp. nov. a nitrogen fixing acetic acid bacterium associated with sugar cane. Int J Syst Bacteriol 39: 361–364

40. Hill S (1971) Influence of oxygen concentration on the colony type of *Derxia gummosa* grown on nitrogen-free media. J Gen Microbiol 67: 77–83

41. Hill S, Drozd JW and Postgate JR (1972) Environmental effects on the growth of nitrogen-fixing bacteria. J Appl Chem Biotechnol 22: 541–558

42. Hurek T, Reinhold-Hurek B, Van Montagu M and Kellenberger E (1991) Infection of intact roots of Kallar grass and rice seedlings by 'Azoarcus' Dev. Plant Soil Sci 48: 235–242

43. Ito O, Cabrera D and Watanabe I (1980) Fixation of dinitrogen-15 associated with rice plants. Appl Environ Microbiol 39: 554–558

44. Jain DK and Patriquin DG (1985) Characterization of a substance produced by *Azospirillum* which causes branching of root hairs. Can J Microbiol 31: 206–210

45. James EK, Reis VM, Olivares FL, Baldani JI and Döbereiner J (1994) Infection of sugar cane by the nitrogen-fixing bacterium *Acetobacter diazotrophicus*. J Exp Bot 45: 757–766

46. Kapulnik Y, Feldman M, Okon Y and Henis Y (1985) Contribution of nitrogen fixed by *Azospirillum* to the N nutrition of spring wheat in Israel. Soil Biol Biochem 17: 509–515

47. Kapulnik Y, Okon Y and Henis Y (1985) Changes in root morphology of wheat caused by *Azospirillum* inoculation. Can J Microbiol 31: 881–887

48. Kennedy IR and Tchan YT (1992) Biological nitrogen fixation in non-leguminous field crops: Recent advances. Plant Soil 141: 93–118

49. Khammas KM, Ageron E, Grimont PAD and Kaiser P (1989) *Azospirillum irakense* sp. nov., a nitrogen-fixing bacterium associated with rice roots and rhizosphere soil. Res Microbiol 140: 679–693

50. Koyama T and App AA (1979) Nitrogen balance in flooded rice soils. *In*: Nitrogen and Rice, pp 95–104. IRRI, Manila, Philippines

51. Li R and MacRae IC (1992) Specific identification and enumeration of *Acetobacter diazotrophicus* in sugarcane. Soil Biol Biochem 24: 413–419

52. Lima E, Boddey RM and Döbereiner J (1987) Quantification of biological nitrogen fixation associated with sugar cane using a ^{15}N aided nitrogen balance. Soil Biol Biochem 19: 165–170

53. Lin W, Okon Y and Hardy RWF (1983) Enhanced mineral uptake by *Zea mays* and *Sorghum bicolor* roots inoculated with *Azospirillum brasilense*. Appl Environ Microbiol 45: 1775–1779

54. Magalhães FM, Baldani JI, Souto SM, Kuykendall JR and Döbereiner J (1983) A new acid-tolerant *Azospirillum* species. An Acad Bras Cienc 55: 417–430

55. Magalhães FMM, Patriquin D and Döbereiner J (1979) Infection of field grown maize with *Azospirillum* spp. R Bras Biol 39: 587–596

56. McClung CR, Patriquin DG and Davis RE (1983) *Campylobacter nitrofigilis* sp. nov., a nitrogen-fixing bacterium associated with roots of *Spartina alterniflora* loisel. Int J Syst Bacteriol 33: 605–612

57. Millet E, Avivi Y and Feldman M (1984) Yield response of various wheat genotypes to inoculation with *Azospirillum brasilense*. Plant Soil 80: 261–266

58. Miranda CHB and Boddey RM (1987) Estimation of biological nitrogen fixation associated with 11 ecotypes of *Panicum maximum* grown in nitrogen-15-labeled soil. Agron J 79: 558–563

59. Miranda CHB, Urquiaga S and Boddey RM (1990) Selection of ecotypes of *Panicum maximum* for associated biological nitrogen fixation using the ^{15}N isotope dilution technique. Soil Biol Biochem 22: 657–663

60. Morris DR, Zuberer DA and Weaver RW (1985) Nitrogen fixation by intact grass-soil cores using ^{15}N$_2$ and acetylene reduction. Soil Biol Biochem 17: 87–91

61. Okon Y and Kapulnik Y (1986) Development and function of *Azospirillum*-inoculated roots. Plant Soil 90: 3–16

62. Olivares FL, Baldani VLD, Baldani JI Döbereiner J (1993) Ecology of *Herbaspirillum* spp. and ways of infection and colonization of cereals with these endophytic diazotrophs. *In*: Poster at 6th Int. Symp. Nitrogen Fixation with Nonlegumes. Ismailia, Egypt. 6–10 Sept. Program and Abstracts 118p

63. Oliveira E (1992) Estudo da associação entre bactérias diazotróficas e arroz. MSc Thesis, Universidade Federal Rural do Rio de Janeiro, Itaguaí, RJ

64. Patriquin DG and Döbereiner J (1978) Light microscopy observations of tetrazolium-reducing bacteria in the endorhizosphere of maize and other grasses in Brazil. Can J Microbiol 24: 734–742

65. Paula MA de, Döbereiner J and Siqueira JO (1989) Efeito da inoculação com fungo micorrízico VA e bactérias diazotróficas no crescimento e produção de batata-doce. *In*: 22° Congresso Brasileiro de Ciência do Solo, pp 109. Recife: Sociedade Brasileira de Ciência do Solo. Programa e resumos

66. Paula MA de, Reis VM and Döbereiner J (1991) Interactions of *Glomus clarum* with *Acetobacter diazotrophicus* in infection of sweet potato (*Ipomoea batatas*), sugarcane (*Saccharum* spp.) and sweet sorghum (*Sorghum vulgare*). Biol Fertil Soils 11: 111–115

67. Pimentel JP, Olivares F, Pitard RM, Urquiaga S, Akiba F and Döbereiner J (1991) Dinitrogen fixation and infection of grass leaves by *Pseudomonas rubrisubalbicans* and *Herbaspirillum seropedicae*. Plant Soil 137; 61–65

68. Reinhold B, Hurek T, Fendrik I, Pot B, Gillis M, Kersters K, Thielemans S and De Ley J (1987) *Azospirillum halopraeferens* sp. nov., a nitrogen-fixing organism associated with roots of kallar grass (*Leptochloa fusca* L. Kunth). Int J Syst Bacteriol 37: 43–51

69. Reinhold Hurek B, Hurek T, Gillis M, Hoste B, Vancanneyt M Kersters K and De Ley J (1993) *Azoarcus* gen. nov., a nitrogen fixing Proteobacteria associated with roots of Kallar grass (*Leptochloa fusca* (L.) Kunth), and description of two species *Azoarcus indigens* sp. nov. and *Azoarcus communis* sp. nov. Int J Syst. Bacteriol 43: 574–588

70. Reis VM, Zang Y and Burris RH (1990) Regulation of nitrogenase activity by ammonium and oxygen in *Acetobacter diazotrophicus*. An Acad Bras Cienc 62: 317

71. Reynders L and Vlassak K (1982) Use of *Azospirillum brasilense* as biofertilizer in intensive wheat cropping. Plant Soil 66: 217–223

72. Rinaudo G and Dommergues Y (1971) Validité de l'estimation de la fixation biologique de l'azote dans la rhizosphère par la methode de reduction de l'acetylene. Ann Inst Pasteur 121: 93–99

73. Ruschel AP, Henis Y and Salati E (1975) Nitrogen-15 tracing of N-fixation with soil-grown sugar cane seedlings. Soil Biol Biochem 7: 181–182

74. Santiago Ventura T, Bravo M, Daez G, Ventura V, Watanabe I and App AA (1986) Effects of N-fertilizers, straw, and dry fallow on the nitrogen balance of a flooded soil planted with rice. Plant Soil 93: 405–411

75. Schank SC, Smith RL, Weiser GC, Zuberer DA, Bouton JH, Quesenberry KH, Tyler ME, Milam JR and Littell RC (1979) Fluorescent antibody technique to identify *Azospirillum brasilense* associated with roots of grasses. Soil Biol Biochem 11: 287–295

76. Schöllhorn R and Burris RH (1966) Study of intermediates in nitrogen fixation. Fed Proc 24: 710

77. Seldin L, Van Elsas JD and Penido EGC (1984) *Bacillus azotofixans* sp. nov., a nitrogen-fixing species from Brazilian soils and grass roots. Int J Syst Bacteriol 34: 451–456

78. Stephan MP, Oliveira M, Teixeira KRS, Martinez Dretz G and Döbereiner J (1991) Physiology and dinitrogen fixation of *Acetobacter diazotrophicus*. FEMS Microbiol Lett 77: 67–72

79. Tarrand JJ, Krieg NR and Döbereiner J (1978) A taxonomic study of the *Spirillum lipoferum* group, with descriptions

of a new genus, *Azospirillum* gen. nov. and two species, *Azospirillum lipoferum* (Beijerinck) comb. nov. and *Azospirillum brasilense* sp. nov. Can J Bot 24: 967–980

80. Teixeira KRS, Stephan MP and Döbereiner J (1987) Physiological studies of *Saccharobacter nitrocaptans* a new acid tolerant N₂-fixing bacterium. *In*: Poster presented at Fourth International Symposium on Nitrogen Fixation with Non-Legumes, Rio de Janeiro, Final program abstracts

81. Urquiaga S, Cruz KHS and Boddey RM (1992) Contribution of nitrogen fixation to sugar cane: Nitrogen-15 and nitrogen balance estimates. Soil Sci Soc Am J 56: 105–114

82. Von Bülow JFW and Döbereiner J (1975) Potential for nitrogen fixation in maize genotypes in Brazil. Proc Nat Acad Sci USA 72: 2389–2393

83. Walcott JJ, Chauviroj M, Chinchest A, Choticheuy P, Ferraris R and Norman BW (1977) Long-term productivity of intensive rice cropping systems on the central plain of Thailand. Exp Agric 13: 305–316

84. Wani SP, Chandrapalaiah S and Dart RJ (1985) Response of pearl millet cultivars to inoculation with nitrogen-fixing bacteria. Exp Agric 21: 175–182

85. Yoshida T and Ancajas RR (1971) Nitrogen fixation by bacteria in the root zone of rice. Soil Sci Soc Am Proc 35: 156–157

86. Yoshida T and Ancajas RR (1973) Nitrogen-fixing activity in upland and flooded rice fields. Soil Sci Soc Am Proc 37: 42–46

87. Yoshida T and Yoneyama T (1980) Atmospheric dinitrogen fixation in the flooded rice rhizosphere as determined by the N-15 isotope technique. Soil Sci Plant Nutr 26: 551–560

88. Zimmer W, Roeben K and Bothe H (1988) An alternative explanation for plant growth promotion by bacteria of the genus *Azospirillum*. Planta 176: 333–342

Fertilizer Research **42**: 251–259, 1995.

© 1995 *Kluwer Academic Publishers. Printed in the Netherlands.*

Recent developments on the use of urease inhibitors in the tropics

B.H. Byrnes[1] & J.R. Freney[2]
[1]*Research and Development Division, International Fertilizer Development Center, P.O. Box 2040, Muscle Shoals, AL 35662, USA;* [2]*Division of Plant Industry, CSIRO, G.P.O. Box 1600, Canberra, A.C.T. 2601, Australia*

Key words: ammonia volatilization, N efficiency, phenyl phosphorodiamidate, phosphoric triamides, urea

Abstract

Urea has become the most widely used form of N fertilizer in the world, particularly in the tropics. Its efficiency, however, is decreased by losses of N through ammonia volatilization when the urea is not incorporated into the soil. High temperatures and high biological activity at the soil surface promote rapid hydrolysis of urea to ammonia and carbonate species by the soil enzyme urease, leading to large ammonia losses. These conditions have generated interest in materials that can inhibit the urease enzyme, slowing urea hydrolysis and allowing the urea to move away from the soil surface to where it is not as susceptible to ammonia loss. The phosphoryl di- and triamides, which are structural analogs of urea, meet the requirements for effective soil urease inhibition to varying degrees depending on the conditions of their use. Until the discovery of these compounds, there was little hope that urease inhibition could be achieved either economically or in an environmentally acceptable way. Included in this group is N-(n-butyl) thiophosphoric triamide (NBTPT), which is that most widely tested proinhibitor or precursor of the actual inhibitor N-(n-butyl) phosphoric triamide. Recent research in tropical rice systems indicates that urease inhibitors such as N-(n-butyl) phosphoric triamide and cyclohexylphosphoric triamide can play an important role in increasing urea efficiency. In some experiments where urease inhibition was only partially successful, better results were obtained when the phosphoroamides were used in conjunction with an algicide, to restrict ammonia loss, and nitrification inhibitors, to reduce loss of N by denitrification. Further research on tropical soils in different environments is required to determine the most suitable combination of inhibitors to reduce N loss and increase the efficiency of fertilizer N use.

Introduction

Urea has become the most used N fertilizer in the world, accounting for approximately 40% of the total synthesized N supply, or about 30 million tons of N [5]. Its market share is increasing because it is the least expensive form of solid N fertilizer available and its high nutrient content (46% N) offers transportation advantages over other sources. In tropical agriculture, it accounts for about 79% of the total fertilizer N use. Although urea has cost advantages over other products, it also has an important disadvantage in that considerable losses of N can occur if the urea is not incorporated into the soil soon after application. The loss occurs through ammonia volatilization after the

urea is hydrolyzed at the soil surface by reaction with the enzyme urease.

Soil pH, moisture, cation exchange capacity, type and quantity of cations present, and wind speed also affect the extent of ammonia volatilization [30]. Carbonate or bicarbonate species are also formed as a result of urea hydrolysis, providing buffering capacity at a high pH at the site where NH_3 is released, further facilitating the loss of NH_3 and increasing NH_3 toxicity to germinating seeds. To reduce ammonia volatilization losses and thereby increase the efficiency of the urea applications, compounds have been tested for their ability to inhibit the urease enzyme. Decreasing ammonia toxicity to germinating seeds and increasing nitrate accumulation are also benefits of urease inhibition [42]. This paper reviews recent developments

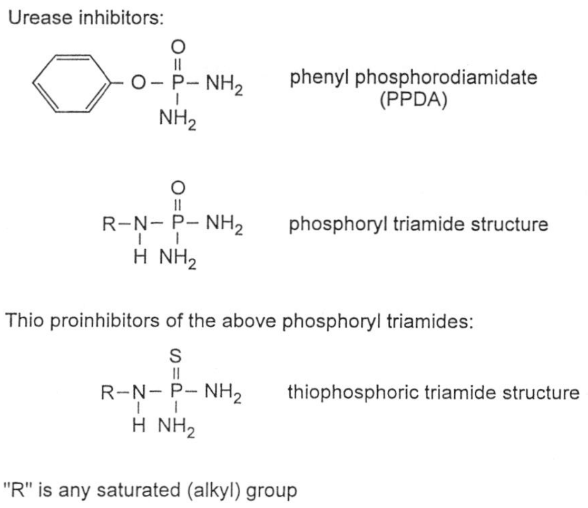

Urease inhibitors:

phenyl phosphorodiamidate
(PPDA)

phosphoryl triamide structure

Thio proinhibitors of the above phosphoryl triamides:

thiophosphoric triamide structure

"R" is any saturated (alkyl) group

Fig. 1. Urease inhibitors and proinhibitors which are, or produce, structural analogs of urea.

in urease inhibitors to improve the efficiency of urea fertilizer applications, particularly regarding their use in tropical agriculture.

Urease is an extremely powerful enzyme produced by practically all microbial and plant species, which may indicate that it is extremely advantageous for organisms to be able to utilize N excreted from animals as urea. Considering that primitive life forms predate higher forms of life, it may indicate that the gene to produce urease originated with the development of life in a primordial sea, in which urea may have been a major N-containing compound. Urease activity of soils is highly related to biological activity, which depends on organic substrates or other energy sources (including sunlight), temperature, and moisture. Urease is apparently protected from protease enzymes by its association with organic-mineral complexes in soils and therefore can persist in the abiotic soil fractions. Although urease activity can be temporarily increased by transitory occurrences that affect biological activity, its inherent (minimum) soil activity is dependent on the amount and types of clays and organic material present in surface soils [4, 43].

Because of the wide distribution and high specific activity of urease, urea hydrolysis in soils normally occurs at a faster rate than do microbially mediated N transformations; often hydrolysis occurs within 3–7 days of urea application to soils unless moisture is limiting. The distribution of the urea can greatly affect the hydrolysis rate in that the greater the access of urea to urease, the more rapid the hydrolysis.

Urease inhibitors are expected to be most beneficial on soils when (i) incorporation of urea is difficult, (ii) there is little opportunity for the urea to move into the soil with infiltrating water, and (iii) the soil surface has a high urease activity due to the lack of cultivation or the accumulation of organic material.

Given the ubiquitous and the high specific activity of soil urease, and the ability of soils to precipitate and adsorb, urease inhibition is difficult to achieve. In addition, a urease inhibitor must meet other requirements as well:

1. The inhibitor must move with the urea, which is not absorbed to any extent by soils, or the urea will still be hydrolyzed near the soil surface as it moves out of the zone of inhibition.

2. It must be reasonably stable in soils, but be ultimately degraded to benign residues.

3. It must be safe for workers exposed to the compound and be environmentally acceptable.

4. It must be inexpensive enough to compete with alternative N management options and must cost less than the fertilizer N conserved.

5. For ease of application, it must be compatible with urea production methods and be stable in admixtures with urea.

6. The demand for the inhibitor must be large enough to warrant development, testing, production, and marketing of the material.

7. The N saved from ammonia volatilization must not be subject to loss by other mechanisms to such a degree that increased plant uptake is not achieved.

Research on urease inhibition dates back more than 50 years [21], and a large number of compounds with a wide range of characteristics have been tested and patented as urease inhibitors [31, 42]. Some inhibitors apparently coordinate with the sulfhydryl group at the active site of the enzyme, whereas others chelate with the essential nickel atom [40]. Despite the considerable effort to identify urease inhibitors for application with fertilizer urea, most of the compounds tested were ineffective for use in soils. The only compounds that have proved to be viable for soil application are the structural analogs of urea (the phosphoryl di- and triamides), and possibly hydroquinone and 2,5-dimethyl p-benzoquinone [52], which must be used at much higher concentrations than the phosphoryl amides. This paper concentrates on the recent research results with the structural analogs of urea.

Phosphoramides

The bond angles and lengths of the amide groups of the phosphoryl di- and triamides (Fig. 1) are similar to those of urea [40], and yet these amides are not substrates of urease. This characteristic apparently enables the inhibitor to react with the active site of the urease and occupy the site for long periods. Whether it degrades, leaving a residue attached to the active site, has not been established. The first structural analog found to be a potent urease inhibitor was phenyl phosphorodiamidate (PPDA), which was identified by East German scientists through screening over 14,000 compounds [40]. An electron-withdrawing group, such as phenol or trichloroethyl [6], is necessary for the diamide to achieve inhibition. Substitutions, such as Cl on the phenol ring, can improve inhibition if they do not cause stearic hindrance by being located adjacent to the ester group [40]. In contrast, a triamide must contain a saturated or alkyl group to have appreciable inhibition. The more commonly tested compounds have contained either n-butyl or cyclohexyl as the 'R' group (Fig. 1). Compounds with other alkyl groups show the same urease inhibitory characteristics, although the solubility and mobility of the compound can be reduced as the group becomes larger (IFDC, unpublished reports). Phosphoryl triamide itself (where 'R' is a proton rather than an alkyl group) is also a urease inhibitor, but it is not as potent as the other compounds [35, 48].

The most widely tested triamide, N-(n-butyl)*thio*phosphoric triamide (NBTPT), is not a urease inhibitor, nor are the other thiophosphoric triamides. Rather they are precursors or proinhibitors of their oxygen analogs, which are the actual inhibitors. Numerous *in vitro* tests of the thio compounds in a purified state have shown their total ineffectiveness. Their effectiveness as inhibitors is dependent on their conversion to an oxygen analog upon contact with soil or other material. This has been shown conclusively by separate groups using various means [22, 39].

The factors that affect the rate of NBTPT conversion to its oxygen analog N-(n-butyl) phosphoric triamide (NBPTO) have not been elucidated. In unpublished work at IFDC, the pure oxygen analog was identified after reaction of an aqueous solution of NBTPT with activated charcoal. Separation of the analogs was achieved with a Sephadex chromatographic column, and the isolated and dried oxygen analog was identified by high-performance liquid chromotography.

Formation of the oxygen analog can be accurately determined in soil extracts or in floodwater using an

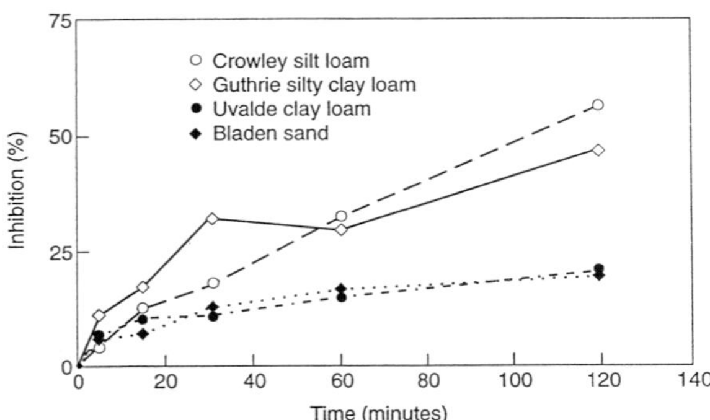

Fig. 2. Development of urease inhibition from perfusing 25 ml of 9.21 mM solution of NBTPT through 15 g of soil; 500 μl aliquots of perfusate used in *in vitro* system (IFDC, unpubl.).

in vitro Jack bean urease assay in a pH-stat system [13]. Briefly, this method requires an accurately reproducible amount of Jack bean urease, which is reacted in solution with an aliquot containing the active inhibitor. An aliquot of urea solution is then added to the partially inhibited urease solution. Following the urea addition, urea hydrolysis produces ammonia at a constant rate; the ammonia is titrated with acid to maintain the pH of the solution at 6.7. If the quantities of urease and inhibitor correspond to achieve partial, but not complete, inhibition of the urease, the rate of acid addition with the pH-stat system is inversely proportional to the quantity of active inhibitor added. Standard curves can be developed with PPDA or the oxygen analogs of the triamides.

Using this system, we found that detectable amounts of the oxygen analog developed within a few minutes of NBTPT addition to the four aerobic soils tested (Fig. 2). Formation of the oxygen analog was also rapid when NBTPT was added to pure bentonite clay. We also tested the development of the active form in the floodwater of rice (*Oryza sativa* L.) soils from the United States, Malaysia, Indonesia, Thailand, and Egypt. The U.S. soils developed appreciable quantities of the active inhibitor within 2 h; however, it took 2 days to achieve the same concentration of active inhibitor in the floodwater of the tropical soils. These soils were preincubated under flooded conditions for 3 weeks, and algal growth was vigorous in all soils. The reasons for the differences between these groups of soils have not been elucidated.

Phenyl phosphorodiamidate has potency similar to that of N-(n-butyl) phosphoric triamide *in vitro* with Jack bean urease (IFDC, unpublished reports). How-

Table 1. Effects of urease inhibitors on total plant recovery, N loss, and grain yield[a]

Treatment	Total plant recovery		Loss		Grain yield (g/pot)	
	TUM	IFDC	TUM	IFDC	TUM	IFDC
	(% of applied ^{15}N)					
Control	-	-	-	-	25.5 C	15.1 D
Urea alone	29.7 D[b]	28.0 C	49.9 D	49.2 C	34.8 B	37.0 C
PPDA	45.3 C	38.1 B	27.8 C	30.8 B	40.4 AB	41.6 ABC
PPDA + algicide	54.0 B	41.8 B	20.9 B	30.2 B	47.3 A	43.5 AB
NBTPT	65.5 A	52.5 A	9.6 A	17.2 A	48.6 A	46.9 A

[a]Greenhouse experiments at the Technical University of Munich (TUM) [14] and IFDC [11].
[b]Data within a column followed by the same letter are not significantly different at the 5% level by Duncan's multiple range test.

ever, PPDA is rapidly degraded in the floodwaters of soils, mainly by base hydrolysis of the ester bond [16]. Degradation in aerobic soils is even faster than in solution and occurs mainly through the mechanism of base hydrolysis, as evidenced by the almost quantitative extraction of the degradation product, phenol, from the soil. Half-lives of PPDA in both sterilized and unsterilized soils were from 2 to 6 h [11]. Soil pH had little effect, whereas the presence of clays and organic matter generally increased degradation rates [11, 38]. The potency of PPDA is attested by the fact that it functions in soil despite its rapid degradation (for example, see [38]).

Urease inhibition in flooded soils

The phosphoroamides PPDA and NBTPT have shown promise for limiting the hydrolysis of urea in laboratory experiments [3, 12, 17, 38, 48] and for increasing N uptake and grain yield in greenhouse studies [14, 15, 54].

In greenhouse experiments at the Technical University of Munich and IFDC, the inhibitors were added at 1% of the weight of urea; electric fans circulated the air above the floodwater to promote ammonia volatilization. Improvement in recovery of applied N was due solely to increased plant uptake, which had to be enhanced by 14% for the increase to be significant (p <0.05 [Table 1]).

In field experiments, however, application of these compounds seldom resulted in significantly (p < 0.05) increased grain yields [8, 9, 17, 25, 28, 44, 45, 51] even though plant uptake of applied N was often

Table 2. Effect of urease inhibitors on grain yield of flooded rice field experiments[a]

Inhibitor/ Rate[b]	Grain yield (t ha^{-1})		Significance[c]	Reference
	Without inhibitor	With inhibitor		
PPDA/2%	5.3	5.2	n.s.	[25]
	4.2	4.0	n.s.	
PPDA/1%	4.7	5.2	*	
	6.0	6.0	n.s.	
PPDA/2%	6.2	6.7	*	[10]
	5.5	5.8	n.s.	
	6.0	6.2	n.s.	
	6.0	6.3	n.s.	
NBTPT/0.9%	6.2	6.8	*	
	5.5	5.6	n.s.	
NBTPT/1%	4.7	5.3	*	[28]
NBTPT/1%	No N response			[45]
NBTPT/1%	4.2	4.7	n.s.	[44]

[a]All experiments are with delayed applications of urea, 15–25 days after transplanting.
[b]Rate of inhibitor on basis of percentage weight of the urea.
[c]n.s. - not significant; * - significant at 95% confidence interval.

increased to the same extent as in the greenhouse studies (Table 2). The promising results with NBTPT in greenhouse experiments were not duplicated in field trials in southeast Asia (IFDC, unpublished reports). In 20 field studies with delayed applications of urea, addition of NBTPT (0.5% of the weight of urea) slowed urea hydrolysis (based on the rate of disappearance of urea from the floodwater) in several experiments and increased uptake of applied N, but did not increase grain yield. When urea is applied 15–25 days after transplanting, as in these experiments, the benefits of a urease inhibitor should have been maximized; the rice plants are well developed at this stage, and delaying ammonia loss by a few days would have resulted in increased uptake of applied N by the rice.

Numerous reasons have been given for the lack of success with these two compounds including (i) microbial decomposition, (ii) degradation of PPDA due to high pH or temperature [11, 16, 38], (iii) the requirement for NBTPT to be converted to the oxygen analog before it is effective [13, 20, 22, 23, 39], (iv) volatilization or nitrification and denitrification of the conserved N [17, 29, 44, 51], and (v) the fact that the saving of 9% to 15% of the applied N is often not sufficient to significantly ($p < 0.05$) increase grain yield in field experiments.

Direct measurements of ammonia volatilization in field studies showed that PPDA slowed ammonia loss for only 2 to 4 days; after that time, ammonia loss was quite high [24, 51]. Delaying ammonia loss for such a short time may not be beneficial because the relative loss from the control and inhibited systems will depend on the pH and temperature of the floodwater and wind speed when the ammoniacal-N concentration of the floodwater is high [51].

The main reason for the lack of success of PPDA in flooded soils seems to be its rapid hydrolysis [16] under the alkaline conditions generated in the floodwater by photosynthetic algae [41]. Floodwater pH values in excess of 10 have been recorded in rice fields in Australia, China, and Malaysia [50]. High floodwater temperatures seem to be less important than pH in the decomposition of PPDA, because PPDA was found to be active for at least 7 days when the pH of the floodwater was controlled by the addition of the algicide terbutryn [26] or simazine [11]. In the presence of algicide, PPDA significantly ($p < 0.05$) reduced ammonia loss, increased N uptake by rice, and increased grain yield in greenhouse experiments (Table 1). Likewise, in a flooded rice field in Thailand, addition of terbutryn with PPDA significantly ($p < 0.05$) slowed the hydrolysis of urea for more than 7 days, reduced ammonia loss from 15% to 7.3% of the applied N, and increased grain yield from 3.6 to 4.03 t ha^{-1} [46].

The success or failure of NBTPT in flooded soils seems to be related to the urease activity of the soil and the capacity of the soil to convert NBTPT to the oxygen analog NBPTO. Thus, in soils with high urease activity, most of the urea is hydrolyzed before appreciable conversion of NBTPT occurs, and in strictly anaerobic soil, NBTPT fails to inhibit urease activity (IFDC, unpublished reports; [35, 36, 37, 55]). Inhibition of urease activity was more successful when NBTPT was added to soils of low urease activity, but the hydrolysis was merely slowed and not completely stopped.

In laboratory experiments, the capacity of NBTPT to inhibit urease activity was enhanced by aerobic conditions (in contrast to anaerobic conditions [36]) created by bubbling oxygen into the floodwater [37] or by the growth of photosynthetic algae, which add oxygen to the flooded system [35]. Complete inhibition of urease activity was not achieved in any of these experiments. More effective inhibition was achieved by adding PPDA and NBTPT in combination [37]. It appeared that PPDA initially inhibited urease activity and during this period part of the NBTPT was converted to the active oxygen analog; then as the concentration of PPDA declined, the concentration of NBPTO became high enough to inhibit urea hydrolysis.

The use of both inhibitors together also worked successfully on flooded rice in Thailand when terbutryn was applied to prolong the effectiveness of PPDA [46]. The combined treatment decreased ammonia loss from 15% to 3% of the applied N and increased grain yield from 3.6 to 4.1 t ha^{-1} [46]. Obviously, it would have been more satisfactory to apply a phosphoric triamide rather than the mixture of inhibitors; however, at the time of conducting the experiment, a phosphoric triamide was not available. It appears that inclusion of an inhibitor with immediate inhibitory activity rather than use of NBTPT alone will be essential to achieving the necessary effect in tropical flooded rice systems.

In greenhouse experiments with flooded soils, the NBTPO inhibited urease activity to a greater extent than did NBTPT in 6 of 10 temperate U.S. soils when the urea and phosphoroamides were incorporated into the soils. However, both analogs performed well when they were applied to the floodwater without mixing (IFDC, unpubl. data).

Application of NBPTO to a flooded rice field in Thailand resulted in much lower ammoniacal-N concentrations in the floodwater than in the control treatment following the application of urea [27]. Ammonia loss was significantly ($p < 0.05$) decreased from 14.4% to 7.6% of the applied N, but grain yield was not increased by this treatment alone. Increased grain yield was not obtained unless an algicide or nitrification inhibitor was added to decrease N loss by ammonia volatilization and denitrification [27]. These results confirm the observations of Phongpan and Byrnes [44] and Byrnes and Amberger [12] that denitrification of conserved N was responsible for the lack of inhibitor response on uptake or yield when it was demonstrated that urea hydrolysis had been slowed.

Keerthisinghe and Freney [35], in laboratory studies, found that cyclohexylphosphoric triamide (CHPT) was even more effective as a urease inhibitor than was NBPTO, and this was confirmed in a field study in Thailand [27]. CHPT maintained the ammoniacal-N concentration of the floodwater at a lower level than did NBPTO (< 2 compared to 5 g N m^{-3}), decreased ammonia loss to a greater extent (50% versus 90%), and resulted in greater uptake of applied N by rice (15.7% versus 10.5%). Again significant increases in grain yield were only obtained in the presence of algicide or algicide plus nitrification inhibitor to further conserve applied N [27].

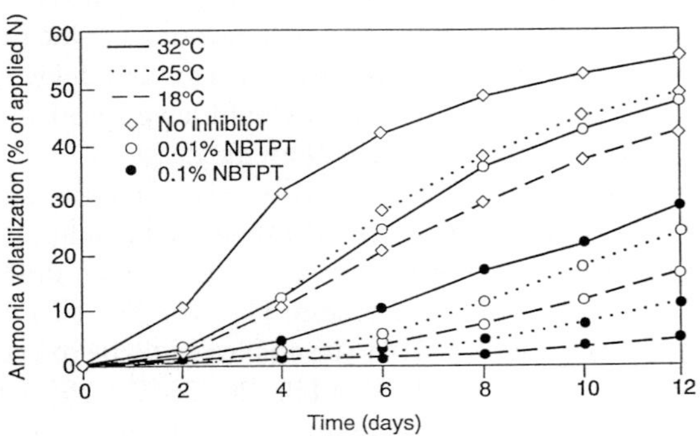

Fig. 3. Effect of temperature and NBTPT concentration on cumulative ammonia volatilization losses from urea applied on a soil surface (adapted from [18]).

Urease inhibitors in upland soil systems

The loss of N from surface-applied urea can be an important loss mechanism, particularly on plantation and long season crops that allow development of high urease activities at the soil surface. Warm, moist tropical soils, particularly those that receive additions of organic matter from leaf litter or support algal growth on the soil surface, develop very high urease activities, and ammonia loss can be extremely high and variable. Soil moisture, urease activity, pH, and cation exchange capacity, rainfall, wind speed, temperature, and humidity affect the magnitude of loss [30, 47]. Despite the high potential for ammonia loss following application of urea, few studies have evaluated urease inhibitors in upland systems in the tropics. The growing importance of no-tillage or limited-tillage production practices in maize (*Zea mays* L.) has increased the interest in the use of urease inhibitors. Most of the work in the United States has been with surface applications of urea and urea ammonium nitrate (UAN) solutions (approximately 50% of the N in the form of urea).

Laboratory testing under aerobic soil conditions indicated the superiority of the triamides, both thio and oxygen analogs, over PPDA [2, 3, 7, 49]. The promising results with NBTPT have prompted numerous field studies. Tests were conducted with NBTPT rather than other analogs because the performance of NBTPT on aerobic soils was generally better than that of other compounds. This has been attributed to the greater stability of NBTPT, particularly in acidic soils [23, 34]. In 21 field experiments from 1989, mostly in the Midwest of the United States, NBTPT increased average grain yields of maize by 750 kg ha^{-1} (9%)

at an average fertilization rate of 100 kg N ha^{-1} [32, 33]. The yields with NBTPT were equivalent to those resulting from 80 kg N ha^{-1} of additional urea-N without NBTPT. The average increase for 78 trials over 5 years at all rates of N was 270 kg of grain per hectare (4%). In southern Illinois, maize yields were increased by an average of 525 kg ha^{-1} (9%) for 13 experiments with broadcast urea and 750 kg ha^{-1} (14%) for 9 experiments in which urea was band placed on the soil surface [53]. Use of NBTPT on ryegrass pasture in Northern Ireland improved urea performance by about 12%, making it comparable to calcium ammonium nitrate [56].

In most of these field studies, NBTPT was applied at 0.25% of the weight of urea. Urea hydrolysis and ammonia volatilization are sensitive to temperature, as is the effectiveness of NBTPT (Fig. 3). These results indicate that a higher concentration of NBTPT will be needed for the warmer tropical soils than for the temperate soils to achieve similar effects on N losses [18].

The marketing of NBTPT in admixture with urea in the United States, principally for use with maize, is tentatively to be at the rate of approximately 0.15% of the weight of urea, and the product is to be sold under the trade name of Agrotain. A second product will likely be marketed, which will also contain a nitrification inhibitor, dicyandiamide (DCD), with the NBTPT (A. Sutton, IMC-Agrico, pers. commun.). Market availability of the products, slated for 1995, will greatly assist the widespread evaluation of the materials for their economic and environmental benefits.

Additional considerations

In aerobic soils, both the phosphoric and the thiio-phosphoric analogs containing the cyclohexyl moiety performed slightly better than did the n-butyl compounds at low (0.01% by weight) concentrations [20]. The oxygen analogs of the triamides are readily synthesized; because of this capacity for immediate inhibition, it is logical that mixtures of the oxygen and thio compounds may find a niche in markets with flooded rice or on soils that convert the thio compound to its active form too slowly to be effective. However, there may be economic or technical constraints that make the use of the oxygen analog or mixtures less viable than NBTPT.

Several years ago, IFDC researchers became concerned about the stability of NBTPT in combination with urea because the effectiveness of urea-NBTPT mixtures used for research decreased with time. They found that such mixtures produced by melt prilling and by spraying a methanol solution of NBTPT onto urea can degrade fairly rapidly if not stored at cold. temperatures (-4 °C). Degradation occurs through acid hydrolysis (deamination) followed by a much slower secondary step, which evolves H_2S. The main findings from this work were as follows:

1. Larger crystals of NBTPT degraded more rapidly than did smaller crystals which suggests a self-catalysis in that any initial degradation gives rise to an acidic residue that causes more rapid acid hydrolysis.
2. Removal of NH_3 from the product increased the degradation rate. Therefore, the top layers of open piles tended to degrade more rapidly than did those lower in the pile.
3. Moisture in the urea and purity of the urea or NBTPT had little effect on stability. The dark color of the urea-NBTPT admixture is not an indication of degradation of the NBTPT.
4. During storage, NBTPT can recrystallize out of urea in white, needle-shaped crystals.
5. Laboratory-scale preparations of urea-NBTPT showed highly variable degradation rates and different patterns of degradation, even when they were prepared in the same way. The major factor(s) affecting degradation of NBTPT in urea were never determined.
6. Despite the problems observed in laboratory-scale preparations, the materials produced by melt granulation in a commercial-sized plant were extremely stable. Thus, the problem of NBTPT degradation is

apparently of more concern to researchers intending to repeat experiments over seasons than for commercialization.

The conversion of NBTPT (or other thio phosphoryl triamides) to the active oxygen analog will likely be dependent on a number of as yet undetermined biotic and abiotic soil properties. The rate of conversion along with the relative absorption and degradation rates of the two compounds will determine the compound's effectiveness. The experience with flooded soils indicates that these rates can be affected not only by intrinsic but also by transient soil factors, such as algal growth or soil mixing. This area of research will be important in further attempts to use the thio proinhibitors in flooded soil systems.

These very active triamide inhibitors and proinhibitors have mobility very similar to that of urea in soils [19]. Thus, they facilitate movement of urea away from the soil surface prior to hydrolysis, slow nitrification, and cause less increase in pH at the microsite of the fertilizer granule. Despite a recent assertion [1], these compounds, as far as is currently known, have no toxic or detrimental effects, nor do their degradation products (A. Sutton, IMC-Agrico, pers. commun.). Prior to their use in tropical systems, further research on the forms and amounts needed will be required. Commercialization of NBTPT in the United States may lead to its availability in tropical countries as the requirements and economic considerations are worked through.

References

1. Al-Kanani T, MacKenzie AF, Fyles JW, Ghazala S and O'Halloran IP (1994) Ammonia volatilization from urea amended with lignosulfonate and phosphoroamide. Soil Sci Soc Am J 58: 244–248
2. Beyrouty CA, Sommers LE and Nelson DW (1988) Ammonia volatilization from surface-applied urea as affected by several phosphoroamide compounds. Soil Sci Soc Am J 52: 1173–1178
3. Bremner JM and Chai HS (1986) Evaluation of N-butyl phosphorothioic triamide for retardation of urea hydrolysis in soil. Commun Soil Sci Plant Anal 17: 337–351
4. Bremner JM and Mulvaney RL (1978) Urease activity in soils. In: Burns RG (ed) Soil Enzymes, pp 149–196. Academic Press, London
5. British Sulphur (1993) Statistical supplement: Nitrogen fertilizer statistics 1991. Nitrogen 203: 37–40
6. Broadbent FE, Nakashima T and Chang GY (1985) Performance of some urease inhibitors in field trials with corn. Soil Sci Soc Am J 49: 348–351
7. Bronson KF, Touchton JT, Hiltbold AE and Hendrickson LL (1989) Control of ammonia volatilization with N-(n-butyl)

thiophosphoric triamide in loamy sands. Commun Soil Sci Plant Anal 20: 1439–1451

8. Buresh RJ, De Datta SK, Padilla JL and Chua TT (1988) Potential of inhibitors for increasing response of lowland rice to urea fertilization. Agron J 80: 947–952

9. Buresh RJ, De Datta SK, Padilla JL and Samson MI (1988) Field evaluation of two urease inhibitors with transplanted lowland rice. Agron J 80: 763–768

10. Buresh RJ, De Datta SK, Padilla JL and Samson MI (1988) Effect of two urease inhibitors on floodwater ammonia following urea application to lowland rice. Soil Sci Soc Am J 52: 856–861

11. Byrnes BH (1988) The degradation of the urease inhibitor phenyl phosphorodiamidate in soil systems and the performance of N-(n-butyl) thiophosphoric triamide in flooded rice culture. Ph.D. thesis, Technical University of Munich, Weihenstephan, Germany

12. Byrnes BH and Amberger A (1989) Fate of broadcast urea in a flooded soil when treated with N-(n-butyl) thiophosphoric triamide, a urease inhibitor. Fert Res 18: 221–231

13. Byrnes BH and Christianson CB (1988) Development of a urease inhibitor from N-(n-butyl) thiophosphoric triamide. Agron Abs 212

14. Byrnes B, Gutser R and Amberger A (1989) Greenhouse study of the urease inhibitors phenyl phosphorodiamidate and N-(n-butyl) thiophosphoric triamide on the efficiency of urea applied to flooded rice. Z Pflanzenernähr Bodenkd 152: 67–72

15. Byrnes BH, Savant NK and Craswell NT (1983) Effect of a urease inhibitor phenyl phosphorodiamidate (PPD) on the efficiency of urea applied to rice. Soil Sci Soc Am J 47: 270–274

16. Byrnes B, Vilsmeier K, Austin E and Amberger A (1989) Degradation of the urease inhibitor phenyl phosphorodiamidate in solutions and floodwaters. J Agric Food Chem 37: 473–477

17. Cai GX, Freney JR, Muirhead WA, Simpson JR, Chen DL and Trevitt ACF (1989) The evaluation of urease inhibitors to improve the efficiency of urea as a N-source for flooded rice. Soil Biol Biochem 21: 137–145

18. Carmona G, Christianson CB and Byrnes BH (1990) Temperature and low concentration effects of the urease inhibitor N-(n-butyl) thiophosphoric triamide (nBTPT) on ammonia volatilization from urea. Soil Biol Biochem 22: 933–937

19. Christianson CB, Baethgen WE, Carmona G and Howard RG (1993) Microsite reactions of urea-nBTPT fertilizer on the soil surface. Soil Biol Biochem 25: 1107–1117

20. Christianson CB, Byrnes BH and Carmlona G (1990) A comparison of the sulfur and oxygen analogs of phosphoric triamide urease inhibitors in reducing urea hydrolysis and ammonia volatilization. Fert Res 26: 21–27

21. Conrad JP (1940) The nature of the catalyst causing the hydrolysis of urea in soils. Soil Sci 50: 119–134

22. Creason GL, Schmitt MR, Douglass EA and Hendrickson LL (1990) Urease inhibitory activity associated with N-(n-butyl) thio-phosphoric triamide is due to formation of its oxon analog. Soil Biol Biochem 22: 209–211

23. Douglass EA and Hendrickson LL (1989) Urease inhibition of N-(n-butyl) thiophosphoric triamide and its oxon analog in diverse soils. Agron Abs 214

24. Fillery IRP and De Datta SK (1986) Ammonia volatilization from nitrogen sources applied to rice fields: I. Methodology, ammonia fluxes, and nitrogen-15 loss. Soil Sci Soc Am J 50: 80–86

25. Fillery IRP, De Datta SK and Craswell ET (1986) Effect of phenyl phosphorodiamidate on the fate of urea applied to wetland rice fields. Fert Res 9: 251–263

26. Freney JR, Keerthisinghe DG, Chaiwanakupt P and Phongpan S (1993) Use of urease inhibitors to reduce ammonia loss following application of urea to flooded rice fields. Plant and Soil 155/156: 371–373

27. Freney JR, Keerthisinghe DG, Phongpan S, Chaiwanakupt P and Harrington K (1994) Effect of urease, nitrification and algal inhibitors on ammonia loss, and grain yield of flooded rice in Thailand. Fert Res 40: 225–233

28. Freney JR, Simpson JR, Zhu ZL and Aziz B (1989) Gaseous nitrogen loss from urea fertilizers in Asian cropping systems. Australian Centre for Int Agric Res, Canberra, Australia. 17p

29. Freney JR, Trevitt ACF, De Datta SK, Obcemea WN and Real JG (1990) The interdependence of ammonia volatilization and denitrification as nitrogen loss processes in flooded rice field in the Philippines. Biol Fertil Soils 9: 31–36

30. Hargrove WL (1988) Soil, environmental, and management factors influencing ammonia volatilization under field conditions. In: Bock BR and Kissel DE (eds) Ammonia Volatilization From Urea Fertilizers, Bull. Y-206, pp 17–36. National Fertilizer Development Center, TVA, Muscle Shoals, AL

31. Hauck RD (1984) Technological approaches to improving the efficiency of nitrogen fertilizer use by crop plants. In: Hauck RD (ed) Nitrogen in Crop Production, pp 551–560. Am Soc Agron., Madison, Wisc

32. Hendrickson LL (1992) Corn yield response to the urease inhibitor NBPT: Five-year summary. J Prod Agric 5(1): 131–137

33. Hendrickson LL (1990) Corn yield response to the urease inhibitor N-(n-butyl) thiophosphoric triamide (NBPT) when applied with urea. Presented at the North-Central Extension-Industry Soil Fertility Conf., St. Louis, MO, Nov. 14 and 15, 1990

34. Hendrickson LL and Douglass EA (1993) Metabolism of the urease inhibitor N-(n-butyl) thiophosphoric triamide (NBPT) in soils. Soil Biol Biochem 25: 1613–1618

35. Keerthisinghe DG and Freney JR (1994) Inhibition of urease activity in flooded soils: effect of thiophosphorictriamides and phosphorictriamides. Soil Biol Biochem (*In press*)

36. Lu W, Lindau CW, Pardue JH, Patrick WH Jr, Reddy KR and Khind CS (1989) Potential of phenylphosphorodiamidate and N-(n-butyl) thiophosphoric triamide for inhibiting urea hydrolysis in simulated oxidized and reduced soils. Commun Soil Sci Plant Anal 20: 775–788

37. Luo Qui-xiang, Freney JR, Keerthisinghe DG and Peoples MB (1994) Inhibition of urease activity in flooded soils by phenylphosphorodiamidate and N-(n-butyl) thiophosphorictriamide. Soil Biol Biochem (*In press*)

38. Martens DA and Bremner JM (1984) Urea hydrolysis in soils: Factors influencing the effectiveness of phenyl phosphorodiamidate as a retardant. Soil Biol Biochem 16: 515–519

39. McCarty GW, Bremner JM and Chai HS (1989) Effect of N-(n-butyl) thiophosphoric triamide on hydrolysis of urea by plant, microbial, and soil urease. Biol Fertil Soils 8: 123–127

40. Medina R and Radel RJ (1988) Mechanisms of urease inhibition. In: Bock BR and Kissel DE (eds) Ammonia Volatilization From Urea Fertilizers, Bull. Y-206, pp 137–174. National Fertilizer Development Center, TVA, Muscle Shoals, AL

41. Mikkelsen DS, De Datta SK and Obcemea WN (1978) Ammonia volatilization losses from flooded rice soils. Soil Sci Soc Am J 42: 725–730

42. Mulvaney RL and Bremner JM (1981) Control of urea transformations in soils. In: Paul EA and Ladd JN (eds) Soil Biochemistry, Vol. 5, pp 153–159. Marcel Dekker, Inc., NY

43. Pettit NM, Smith HRJ, Freedman RB and Burns RG (1976) Soil urease: activity, stability and kinetic properties. Soil Biol Biochem 8: 479–484

44. Phongpan S and Byrnes BH (1993) Effect of methods of application on the efficiency of urea broadcast onto lowland rice (*Oryza sativa* L.). Biol Fertil Soils 15: 235–240

45. Phongpan S and Byrnes BH (1990) The effect of the urease inhibitor N-(n-butyl) thiophosphoric triamide on the efficiency of urea application in a flooded rice field trial in Thailand. Fert Res 25: 145–151

46. Phongpan S, Freney JR, Keerthisinghe DG and Chaiwanakupt P (1995) Use of phenylphosphorodiamidate and N-(n-butyl) thiophosphorictriamide to reduce ammonia loss and increase grain yield following application of urea to flooded rice. Fert Res 41: 59–66

47. Popp T (1988) Nitrogen losses from urea-techniques and means for their control. In: Pushparajah E, Hussin A, Bachik AT (eds) Int Symp on Urea Technol and Utilization, pp 321–334. Malaysian Society of Soil Science, Kuala Lumpur

48. Radel RJ, Gautney J and Peters GE (1988) Urease inhibitor developments. In: Bock BR and Kissel DE (eds) Ammonia Volatilization From Urea Fertilizers. Bull. Y-206, pp 111–136. National Fertilizer Development Center, Muscle Shoals, AL

49. Schlegel AJ, Nelson DW and Sommers LE (1986) Field evaluation of urease inhibitors for corn production. Agron J 78: 1007–1012

50. Simpson JR and Freney JR (1988) Interacting processes in gaseous nitrogen loss from urea applied to flooded rice fields. In: Pushparajah E, Husin A, Bachik AT (eds) Int Symp on Urea Technol and Utilization, pp 281–290. Malaysian Society of Soil Science, Kuala Lumpur

51. Simpson JR, Freney JR, Muirhead WA and Leuning R (1985) Effects of phenyl phosphorodiamidate and dicyandiamide on nitrogen loss from flooded rice. Soil Sci Soc Am J 49: 1426–1431

52. Tomlinson TE (1970) Urea: agronomic applications. Proc Fert Soc 113: 1–76

53. Varsa EC, Jemison JM, Osborn MW, Leis AK, Hnetkovsky SW and Jan N (1993) Effect of NBPT-amended urea and UAN on no-till corn in southern Illinois. Presented at the 23rd North Central Extension Industry Soil Fertility Conf., St. Louis, MO, October 27 and 28, 1993

54. Vlek PLG, Stumpe JM and Byrnes BH (1980) Urease activity and inhibition in flooded soil systems. Fert Res 1: 191–202

55. Wang ZP, Van Cleemput O, Demeyer P and Baert L (1991) Effect of urease inhibitors on urea hydrolysis and ammonia volatilization. Biol Fertil Soils 11: 43–47

56. Watson CJ, Steven RJ and Laughlin RJ (1990) Effectiveness of the urease inhibitor NBPT (N-(n-butyl) thiophosphoric triamide) for improving the efficiency of urea for ryegrass production. Fert Res 24: 11–15

Fertilizer Research **42**: 261–276, 1995.
© 1995 *Kluwer Academic Publishers. Printed in the Netherlands.*

Biological N_2-fixation and its management in wetland rice cultivation

P.A. Roger
Laboratoire de Microbiologie Orstom, Université de Provence, Case 87, 3, Place Victor Hugo,
F-13331-Marseille Cedex 3, France

Key words: Azolla, BNF, cyanobacteria, green manure, heterotrophic bacteria, legume, rice

Abstract

The review summarizes the current status of the utilization of N_2-fixing organisms as biofertilizer in rice cultivation. Heterotrophic bacteria, free-living cyanobacteria, *Azolla*, and legume green manures are considered with regard to their potential for increasing yield, their current use and the prospects for their use with regard to the identified limiting factors.

Biological N_2 fixation has been the most effective system for sustaining production in low-input traditional rice cultivation. On the other hand, the utilisation of N_2-fixing organisms in intensified rice production encounters serious limitations. The utilization of free-living bacteria and cyanobacteria is refrained by their modest potential and the non establishment of inoculated strains. *Azolla* and legumes used as green manures have a high potential as N source, but their utilization is severely limited by socio-economic factors.

Introduction

More than half the world population is dependent upon rice, which provides 20% of global human per capita energy and 15% of per capita protein. In 1991, rice occupied $148 \cdot 10^6$ ha-10% of world's arable land – for a global production of $420 \cdot 10^6$ t [31].

Rice grows in flooded conditions during part or all the cropping period in about 88% of rice land. Flooding changes the chemistry, microbiological properties, and nutrient supplying capacity of soil. It leads to the differentiation of a range of macro- and micro-environments differing by their redox, physical properties, light status, and nutrient sources for the microflora. As a result, all kinds of N_2-fixing organisms can find conditions favorable for their growth in ricefields. They include (1) indigenous organisms: heterotrophic aerobic, microaerophilic and anaerobic bacteria – in soil and associated with rice –, photosynthetic bacteria, and cyanobacteria; and (2) introduced green manures: *Azolla* and legumes. Traditional wetland rice cultivation has been extremely sustainable because biological N_2 fixation (BNF) has permitted a moderate but stable yield to be maintained for thousands of years without

N fertilizer addition and without deterioration of the environment [9].

Reviews on BNF in ricefields deal with its estimation and contribution to N balance [66], agronomic use of N_2-fixing biofertilizers [70], and the microbial management of wetland ricefields [71]. Specific reviews on N_2-fixing organisms in ricefields deal with heterotrophs [101], BNF associated with straw [34], rice varietal differences in stimulating BNF [36], cyanobacteria [62], *Azolla* [89], and legume green manures [33, 37].

This paper summarizes quantitative data on BNF estimates in wetland ricefields and considers for each of the major groups of N_2-fixing organisms: the N_2-fixing potential, the potential to increase rice yield, the current status of their utilization, and the prospects with regards to identified limiting factors.

Assessment of biological N_2-fixation in ricefields

Methods used to estimate BNF by during a rice crop cycle include:

- Balance studies in long-term fertility experiments, or in pots.
- Acetylene reducing activity (ARA) measurements performed at intervals during a crop. This method, despite recognized limitations, was used in about 2/3 of the 38 quantitative BNF studies related to rice published since 1985 [66].
- The determination of the maximum biomass of the N_2-fixing system and the % N derived from the air (Ndfa) of this biomass. The ^{15}N dilution method was used to estimate Ndfa of the rice plant and legume green manures. Difference in natural ^{15}N abundance ($\partial^{15}N$) was used to estimate Ndfa in *Azolla*.

N-balance is currently the only method that provides an estimate of total BNF in ricefields, but the values are underestimated because losses are usually not taken into account. Balance studies in the field encounter additional difficulties, as compared with pot experiments, because of sampling errors, unaccounted subsoil contribution, and losses by leaching. Therefore, after the early measurements in long-term field experiments [25]), there has been an increased interest in pot studies [4, 75, 77, 83].

In a bibliographic compilation of 211 N-balance estimates [66] values ranged from -102 to $+171$ kg N ha^{-1} crop cycle^{-1} and averaged 24 kg N ha^{-1} crop cycle^{-1} (Table 1). The balance was influenced by N-fertilizer application, the presence of rice, and light availability. An average balance of 30 kg N ha^{-1} crop cycle^{-1} was obtained when no N-fertilizer was used. This shows that the average potential of BNF in nonfertilized fields can ensure, on a long-term basis, a yield of about 1.5 t ha^{-1} (assuming that all N_2 fixed is absorbed by the rice plant and 50 kg grain is produced per kg N absorbed). Balance was negligible in the presence of N-fertilizer (4 kg N ha^{-1} crop cycle^{-1}). This results from the two known processes of BNF inhibition by N-fertilizer and N losses by NH_3 volatilization and denitrification [63]. Mean balance values estimated in the presence of light (31 kg N ha^{-1} crop cycle^{-1}) and in the absence of light (13 kg N ha^{-1} crop cycle^{-1}) indicates that, on an average, photodependent BNF contributes 2/3 of the balance.

Heterotrophic N_2-fixation

The presence of N_2-fixing bacteria in rice rhizosphere was reported as early as 1929 [76], but the study of the potential of N_2-fixing heterotrophs started only in 1971, when it was demonstrated that some BNF is associated with wetland rice roots [61, 100]. Early inoculation trials were conducted with *Beijerinckia* [17] and *Azotobacter* [80]. Most reports on bacterial inoculation of rice (40 of 44), however, have been published since 1976. Research on BNF in the rice rhizosphere has also revealed differences in the ability of rice genotypes to stimulate associative BNF and N uptake [36]. This suggests that N utilization by rice can be improved by selection and breeding of varieties that can stimulate the development of a more efficient associated microflora.

Estimations of heterotrophic BNF

Total heterotrophic BNF
Estimates from N balance in unfertilized planted pots covered with black cloth averaged 7 kg N ha^{-1} [3]. Similar trials showed balances negatively correlated with the amount of N applied [83]. Extrapolated values averaged 19 kg N ha^{-1} crop^{-1} with 65 kg N ha^{-1}, -0.3 with 112 kg N, and -14 with 146 kg N. Using available N of a stabilized ^{15}N-labelled soil as control, it was estimated that, when no N-fertilizer was applied and photodependent BNF was restrained, heterotrophic BNF contributed 16–21% of rice N, or 11–16 kg N ha^{-1} crop^{-1} [103].

BNF associated with rice rhizosphere
Reported ARA values in rice rhizosphere range from 0.3 to 2 μmol C_2H_4 plant^{-1} h^{-1}. They are usually highest at or near heading stage. Extrapolated values range from 0.8 to 6 kg N ha^{-1} crop cycle^{-1} [70]. The theoretical maximum associative BNF can be calculated by assuming that all rhizospheric bacteria are N_2 fixers and they use all C flux in rhizosphere (1 t ha^{-1} crop cycle^{-1}) with a high efficiency of 40 mg N g^{-1} C. This would be equivalent to 40 kg N ha^{-1} crop^{-1}. But bacterial enumerations in rice rhizosphere often show a ratio higher than 10 between N_2-fixing and total bacteria [66].

BNF associated with straw
Early estimates of BNF after straw incorporation range from 0.1 to 7 (mean 2.1) mg N g^{-1} straw added, in 30 days [70]. Most data originate from laboratory incubations in darkness of soil enriched with 1 to 100% straw (average 22%) which simulates composting rather than the field situation. Moreover, dark incubation allows heterotrophic BNF only, whereas straw

Table 1. Bibliographic study of N-balance estimates in wetland ricefields* (adapted from [66])

Major statistic of the set of data analyzed

Number of data: 211	Unit: kg N ha^{-1} crop cycle^{-1}
Minimum: -102	Maximum: 171
Mean: 24.2	Median: 27.0
Standard deviation: 33.1	Coefficient of variation: 136%

Histogram of the data

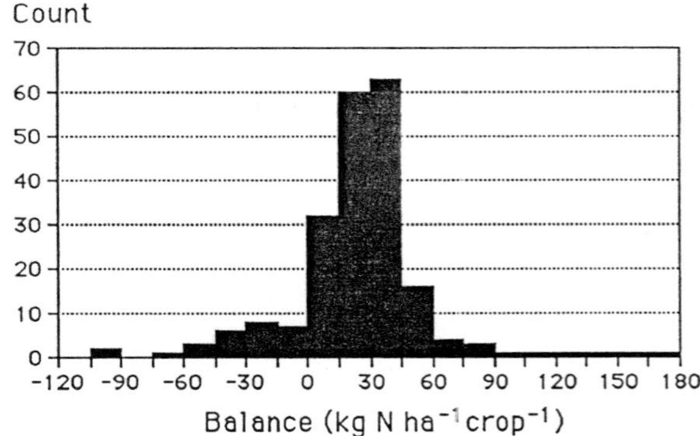

Effect of various factors on N-balance

Factor	Number of data	Mean (kgN ha^{-1} crop cycle^{-1})	Standard error	Level of significance of the difference
N-fertilizer application				
−	166	29.7	25.4	1%
+	45	4.0	47.6	
Planted versus unplanted				
+	193	26.5	30.7	1%
−	18	-0.5	46.2	
Effect of soil exposure to light (treatments where no N-fertilizer was applied)				
+	152	31.2	25.7	1%
−	14	13.2	13.8	

*The set of data is temptatively exhaustive. Data originate from pot and field experiments. Data from pot experiments are extrapolated in kg N ha^{-1} crop cycle^{-1} based on the surface of the pots.

incorporation also may increase populations and N$_2$-fixing activity of photosynthetic organisms [34]. A few semi-quantitative data and laboratory data suggest that straw might increase BNF by 2–4 kg N t^{-1} applied [34].

These data suggest that the N-potential of heterotrophic BNF is the lowest among the N$_2$-fixing agents discussed in this review.

Potential of heterotrophic BNF for agronomic utilization

No method adoptable by farmers has been yet designed to enhance on purpose heterotrophic BNF in ricefields. Research has mostly aimed at promoting associative BNF by inoculating selected strains of N$_2$-fixing bacteria. More recently it was found that there is a potential for selecting/breeding rice varieties more efficient in stimulating associative BNF [36].

Rice inoculation with N$_2$-fixing heterotrophs
Genera of N$_2$-fixing bacteria isolated from rice rhizosphere include *Agromonas, Alcaligenes, Aquaspirillum, Azospirillum, Beijerinckia, Citrobacter, Enterobarter, Flavobacterium, Klebsiella* and *Pseudomonas* [70]. Strains most frequently isolated using exudates of rice seedlings as C source were *Enterobacteriaceae, Azospirillum* spp. and *Pseudomonas paucimobilis* [5, 50, 81].

Bacterial inoculation of rice has been studied by dipping seeds in bacterial cultures or coating them with various carriers, dipping seedlings in bacterial cultures, and inoculating nursery soil and/or the field. The analysis of data from 23 articles reporting 210 trials shows an average increase in yield of 19.8%, but the response varied from -33% to +125% [71]. Average increase was higher in pots (27.6%) than in the field (14%). Average yield increase in field experiments (+14%) (Table 2) was close to the minimum detectable difference (14.5%) which can be expected from the experimental design most commonly used (16-m^2 plots, with 4 replicates) [23]. Thus, experiments with no statistical analysis should be interpreted with caution. The distribution of the differences in yield between controls and inoculated plots was very asymmetrical and the histogram, which exhibited an abrupt rise of the first class of positive values, strongly suggested a bias (Table 2). At least, it indicated that unsuccessful trials were often not reported.

The beneficial effect of bacterial inoculation can be attributed to a combination of (1) increased associative BNF, (2) production of plant growth regulators (PGR) that favor rice growth and nutrient utilization, (3) increased nutrient availability through solubilization of immobilized nutrients by inoculated bacteria, and (4) competition of inoculated strains with pathogens in the rhizosphere. The relative importance of these components has not yet been determined. Current estimates of BNF in rice rhizosphere are insufficient to explain the average 0.5 t ha^{-1} increase in yield reported in field experiments. Assuming that all N fixed is absorbed by the plant, such a yield increase would at least require an increase in BNF of 10 kg N ha^{-1} crop^{-1}; but no data demonstrate a marked and durable increase of BNF in inoculated rice. The hypothesis that PGR production by inoculated bacteria increases nutrient absorption does not agree with the absence of significant difference in N fertilizer efficiency between control plots (18.7 kg grain per kg N applied) and inoculated plots (19.1 kg) (Table 2). Similar harvest index values (grain yield/straw yield) in the controls and inoculated plots (0.57) may indicate that the effect of inoculation probably takes place early in the crop cycle.

Little information is available on the establishment of inoculated strains. In most cases, variations in population densities were too small to be significant [71]. The inoculation of a mutant of *Azospirillum lipoferum* resistant to two antibiotics showed survival of the strain for 50–70 days but no establishment [48]. Inoculated *Azospirillum* was about 500 times less abundant than putative indigenous populations of *Azospirillum*, but inoculation increased the dry weight and total N of the plant.

Trials have been conducted to select the most efficient combination of a N$_2$-fixing strain and a specific rice cultivar, by using a two-step process in which bacterial strains were first isolated from the rhizosphere of actively N$_2$-fixing rice plants [26]. Strains were then tested with rice cultivars using a gnotobiotic system known as the spermosphere model [81] in which an axenic seedling is grown in darkness in a Pankurst tube on a medium without C and N source. This approach has produced both erratic increases in yield [12] and significant increases (6 to 21%) which were higher at the highest level of N-fertilizer (76–96 kg N/ha^{-1}) [50]. If the validity of the method is confirmed, its potential for practical utilization will strongly depend upon the degree of specificity required to select an efficient bacteria for given agro-ecological conditions.

Utilization of varietal differences in promoting associative BNF
The existence of varietal differences in the ability to support associative BNF was demonstrated by N balance studies [4], ARA measurements [82], and N isotope ratios [95]. The plant traits associated with associative BNF were, by decreasing importance: dry weight of roots and submerged portions of the plant at heading, dry weight of shoots at heading, N uptake at heading, and N uptake at maturity [36]. A ranking of 21 rice genotypes for BNF and N utilization, established from the above plant traits and ARA, was fairly reproducible in two dry season trials [35]. Nothing is known, however, about the physiological basis of varietal differences. The idea of breeding varieties with higher N$_2$-fixing potential is attractive because it would enhance BNF without additional cultural practices. A prerequisite is the availability of a rapid screening technique. Even short-term ARA assays are time-consuming and

Table 2. Bibliographic study of the field experiments on the effect of bacterial inoculation on rice yield (adapted from [71])

| | Grain yield | | | Straw yield difference | Harvest index (grain/straw) | | N efficiency (kg grain kg^{-1} N) | |
	Control (t ha^{-1})	Difference (t ha^{-1})	(%)	(%)	Control	Inoc.	Control	Inoc.
Mean	3.99	0.52	14.4	15.1	0.57	0.57	18.7	19.1
Stdev	1.64	0.52	14.1	14.9	−0.17	0.17	17.1	13.8
Maxi	11.09	2.35	59.6	64.8	0.86	0.88	78.0	54.7
Mini	1.00	−1.50	−25.0	−7.5	0.25	0.25	−20.0	−12.0
Count	121	121	123	51	51	51	60	59

Histogram and statistics of yield differences between inoculated and noninoculated treatments

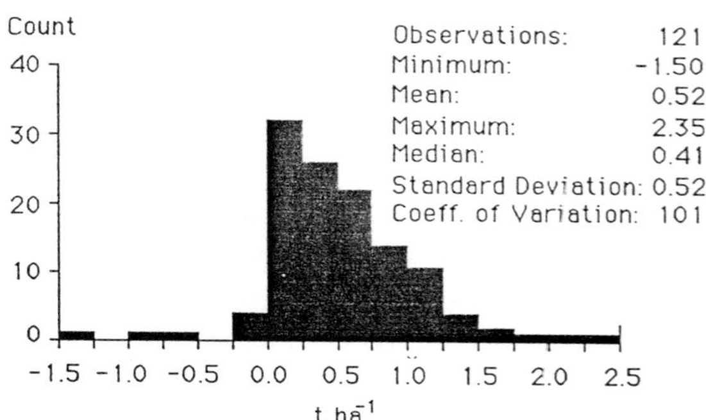

do not allow the screening of a large number of genotypes [82]. ^{15}N dilution can be used for screening and genetic studies, but reference varieties with low BNF stimulation ability must first be identified.

Prospects

So far, the results of bacterial inoculation experiments are inconsistent. Reported increases in yield have not been related with an increase in BNF. Usually inoculated strains did not clearly establish. Therefore reasons for yield increases are still unclear. The potential of methods aiming at selecting the most efficient combination of a N$_2$-fixing bacterial strain and a specific rice cultivar needs (1) further confirmation and (2) the determination of the degree of specificity required, which will determine the feasibility for practical use. Current knowledge is insufficient to establish inoculation methods that can be used by rice farmers.

The existence of varietal differences in promoting associative BNF offers a promising way of taking advantage of heterotrophic BNF. This potential can be better utilized if screening takes into account both the ability to stimulate BNF and to utilize soil N [36].

Free-living cyanobacteria

The agronomic potential of N$_2$-fixing cyanobacteria was recognized in 1939 by De [14], who attributed the natural fertility of wetland ricefields to BNF by these organisms. Research on cyanobacterial inoculation of ricefields was initiated in Japan in 1951 [88] and then continued in India [84]. Research on cyanobacteria agroecology in ricefields developed during the last decade [62].

Potential of cyanobacteria as a biofertilizer for rice

N$_2$ fixation by cyanobacteria has been almost exclusively estimated from ARA. Estimates published before 1980 ranged from a few to 80 kg N ha^{-1} crop^{-1} (mean 27 kg) [64]. About 180 crop cycle measurements in experimental plots at IRRI [67] showed extrapolated values ranging from 0.2 to 50 kg N ha^{-1} crop^{-1} and averaging 20 kg in no-N control plots, 8 kg in plots with broadcast urea, and 12 kg in plots where N was deep-placed.

Biomass measurements provide a rough estimate of the N$_2$-fixing potential of cyanobacteria because they bloom only when the photic zone is depleted of N and most of their N can be assumed to originate from BNF. Cyanobacteria in ^{15}N-labelled plots had about 90% Ndfa [27]. A visible growth of cyanobacteria usually corresponds to less than 10 kg N ha^{-1}, a dense bloom may correspond to 10–20 kg N ha^{-1}; larger biomasses (20–45 kg N ha^{-1}) are recorded only in experimental microplots or in inoculum production plots [28, 68, 69]. The theoretical maximum BNF by cyanobacteria can be calculated by assuming that the photosynthetic aquatic biomass is composed exclusively of N$_2$-fixing cyanobacteria (C:N = 7) and primary production is 0.5 t C ha^{-1} crop^{-1}. This would be equivalent to 70 kg N ha^{-1} crop^{-1}.

Possible beneficial effects of cyanobacteria other than provision of N include (1) competition with weeds, (2) increased soil organic matter content and aggregation, (3) excretion of organic acids that increase P availability to rice, (4) decrease of sulphide injury in sulfate reduction-prone soils by increased O$_2$ content and plant resistance to sulfide, and (5) production of PGR that enhance rice growth [64]. But this last aspect still needs to be demontrated because cyanobacteria extracts may also negatively affect rice germination [53], and despite the numerous reports on algal PGR effects, none shows the isolation and characterization of a microalgal PGR [46].

Algal inoculation technology and its current status

Experimental cyanobacteria inoculation (algalization) of ricefields initiated in Japan [88] was subsequently abandoned there. Applied research on algalization has been conducted mostly in India and, to a lesser extent, in Burma, Egypt, and China. A similar technique of inoculum production in shallow open-air ponds is used in India, Egypt, and Burma [84]. A multi-strain starter inoculum produced from laboratory cultures is propagated, on the spot, in trays or microplots with 5–15 cm water, about 4 kg soil m^{-2}, 100 g superphosphate m^{-2}, and insecticide. If necessary, lime is added to adjust soil pH to 7.0–7.5. In 1–3 weeks, an algal mat develops which is then allowed to dry. Algal flakes are collected for further use at 10 kg ha^{-1}.

Table 3 presents the analysis of 634 field experiments of algalization. The difference in yield between inoculated and noninoculated plots was very variable (C.V. > 100%). Because of the asymmetrical data distribution, the median grain yield (257 kg ha^{-1}) was considered a better index of the average effect of inoculation than the mean (337 kg ha^{-1}) [62]. While the difference in average yield between inoculated and noninoculated plots was significant at $p < 0.01$, only 17% of the 634 individual observed differences were statistically significant. This indicates a small and variable response of yield to inoculation and also an experimental error frequently larger than the response. When interpreting data from the literature, it should also be kept in mind that unsuccessful trials have often not been reported. When they were mentioned, it was usually without quantitative data that could explain the possible reason for failure. For example, a report of a multilocation trial [54] indicates that notwithstanding the 22 sets of data presented, "the results from many other locations in South India, Deccan, and the Konkan region were not received because of the failure of multiplying cyanobacteria at these locations."

Cyanobacteria inoculation is currently used on a trial-and-error basis. Methods to estimate the chance of success of inoculation in a given agroecosystem are unavailable because the factors underlying yield increases associated with successful algal inoculation are not clearly understood or quantified. No published study reporting a significant increase in yield after cyanobacterial inoculation includes estimation of inoculum quality, BNF measurement, or biomass estimates.

Reports on the adoption of algal inoculation are somewhat controversial, but even with the most optimistic evaluations, adoption seems to be restricted to a very limited area in a few Indian states, in Egypt, and possibly in Burma [71]. Farmers' limited acceptance of algalization probably reflects the low and erratic increases in yield obtained.

Prospects

Methods for utilizing cyanobacteria in rice cultivation need to be reconsidered in view of the following results of the agroecological studies of the last decade:
(1) N$_2$-fixing cyanobacteria are ubiquitous in ricefields.
(2) The study of the ratio of indigenous heterocystous cyanobacteria in 102 soils to those contained in the recommended dose of 22 soil-based inocula (10 kg

Table 3. Bibliographic study of the effect of cyanobacterial inoculation on rice yield* (adapted from [62])

1. Major statistics of the data

	Difference between control and inoculated plots	
	Absolute (k ha^{-1})	Relative (%)
Number of observations	634	634
Maximum	3700	168.2
Minimum	−1280	−19.3
Average	337	11.3
Median	257	7.9
Standard deviation	398	16.0
Coefficient of variation	118	141

2. Histogram of the data

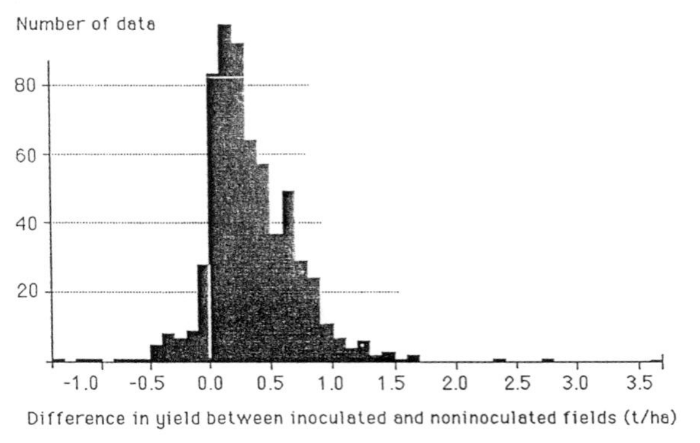

*Data compiled from 41 references listed in [62].

ha^{-1}) showed that in 90% of the cases, indigenous cyanobacteria were more abundant than those in the inoculum [68].

(3) Results also show the infrequent establishment of nonindigenous strains inoculated in various soils, even when grazers were controlled [24, 58, 59]. While cyanobacteria inoculated in five soils persisted for at least 1 month, their growth was rare (1 out of 10 cases). Blooms developed on all soils when grazers were controlled, but were mostly of indigenous strains [58].

This suggests that attention should be paid to practices that enhance the growth of indigenous strains already adapted to the environment. Their growth is most commonly limited by low pH, P deficiency, grazing, and broadcasting of N-fertilizer. Cultural practices that alleviate limiting factors (liming, P application, grazer control, and N-fertilizer deep-placement) favor photodependent BNF and cyanobacteria growth, but their economic feasability is often low.

Liming is rarely economically feasible. The efficiency of P fertilizer (kg N fixed per kg P applied) is usually low (2.3 g N g^{-1} P) [13]. Split P application is more efficient than basal application [28]. Grazer control can be achieved with conventional pesticides [24] but their cost is prohibitive for the result achieved. Pesticides of plant origin might be more economical. Broadcast application of N-fertilizer, which is widely practiced by farmers, inhibits photodependent BNF and causes N-losses by ammonia volatilization [21, 65]. In contrast, the deep-placement of N-fertilizer decreases its inhibitory effect on cyanobacteria and reduces N-losses by volatilization. Delaying N-fertilizer application could also possibly allow the growth of a N$_2$-fixing algal bloom at the early stages of the crop [99], but the resulting effects on N- losses from fertilizer applied in an algal-rich water are unknown.

The ubiquity of N$_2$-fixing cyanobacteria in rice soils does mean that inoculation is unnecessary. Inoculating fields with indigenous strains which are able to establish might be useful because P accumulation by the inoculum (produced with high levels of P) gives it an initial advantage over the indigenous propagules, which are usually P-deficient [69]. Since spore germination is photodependent [57], propagules applied on the soil surface might germinate better than the indigenous ones mixed with the soil. Inoculation with indigenous strains is likely to be useful after an upland crop grown before rice or after a long dry fallow because the low density of natural population density may lead to a lag of several weeks before N$_2$-fixation becomes significant. In fact, the positive effects of cyanobacteria inoculation observed with the method recommended in India could be due to indigenous strains, because when inoculum is multiplied on the spot in shallow trays or plots, it is probable that local strains present in the soil may outgrow the original isolates even before inoculum is added to the field.

In the absence of knowledge on factors that allows foreign strains to establish in a field, the agronomic potential of cyanobacterial inoculation is probably limited to indigenous strains used in agroecosystems favorable to cyanobacterial growth, when inoculation allows to accelerate the formation of an N$_2$-fixing bloom early in the crop cycle. Soil properties, climatic conditions and cultural practices needed for such conditions to occur probably limit the usefulness of cyanobacterial inoculation to a small percentage of the world's ricefields.

Azolla

Azolla is an aquatic fern which harbors the symbiotic N$_2$-fixing cyanobacteria *Anabaena azollae*. *Azolla* usually needs to be inoculated and grown when used as green manure. Its use dates back to the 11th century in Vietnam and the 14th century in China [44]. The N$_2$-fixing symbiont of *Azolla* was identified by Strasburger in 1873 but progress in *Azolla* biotechnology (i.e. recombination and sexual hybridization), is recent [38, 39, 97].

Potential as a biofertilizer for rice

BNF by *Azolla* has usually been estimated from biomass measurement and the assumption that most of *Azolla* N originates from BNF. Recent measurements show an average Ndfa of 75% in *Azolla* [96]. The maximum N potential of *Azolla*, calculated from biomass recorded in experimental plots and assuming that two crops of *Azolla* with a Ndfa of 80% are grown per rice crop is 224 kg N^{-1} crop cycle^{-1}. But, in a field trial at 37 sites in 10 countries, productivity was lower than in experimental plots [91]. Biomass was 5–25 t fresh weight ha^{-1} (10–50 kg N ha^{-1}) for *Azolla* grown before or after transplanting (average 15 t ha^{-1} or 30 kg N). Comparisons with inorganic fertilizer showed that one *Azolla* crop incorporated before or after transplanting was equivalent to the application of 30 kg N ha^{-1} as urea.

Besides providing N to the rice crop, *Azolla*, has several other advantages. Because of its lower K absorption threshold in floodwater than rice, *Azolla* becomes a source of K for rice when incorporated [41]. *Azolla* also enhances the utilization of P fertilizer [74], decreases weed incidence [15, 44], reduces water evaporation [16], and improves soil structure, which is important where rice is grown sequentially with an upland crop [73].

Current usage

Estimates of the extent of *Azolla* use show a marked decrease during the 1980s in China and Vietnam, where it had been a traditional technology. Estimates in China are 6.5 million ha before 1978 [19], 1.34 million ha in 1979 [40], 0.7 million ha in 1982 [44] and a decrease in use was still reported in 1987 [41]. This has been attributed to the availability of cheap sources of urea and potash, and changing governmental economic policies leading to the disbanding of many agricultural communes and the reallocation of labor [73]. In Vietnam *Azolla* was used in about 500,000 ha in 1980 [70]; since then, its use has also continuously decreased.

During the 1980s, *Azolla* was tested for adoption in Brazil, India, Pakistan, the Philippines, Senegal, Sri Lanka, and Thailand. The Philippines is the only country were adoption was sufficient to be quantified during the 1986 *Azolla* Workshop [29]. In this country, farmers adopted *Azolla* on 5,000 ha in South Cotabato in 1981 [32]; success was due mainly to a high level of available P in the soils and a short dry season. *Azolla* utilization extended to 26,000 ha in 1983, and 84,000 ha in 1985 [45]. Since then, *Azolla* use has not progressed in the Philippines and has probably decreased.

Limiting factors for Azolla use and possible remedies

Water control and maintenance of inoculum
Azolla cannot withstand desiccation and requires standing water throughout its growth cycle. Because *Azolla* is propagated vegetatively, inoculum must be maintained in nurseries year-round and multiplied for distribution before field cultivation. Therefore an irrigation network and a network for inoculum conservation, production, and distribution are prerequisites for *Azolla* use. This also implies that *Azolla* adoption by farmers first depends on a government policy to establish such networks. Problems in inoculum conservation, multiplication, and transport could be solved if *Azolla* could be propagated from spores. A method for utilizing sporocarps for inoculum conservation has been developed in China but the growth of sporophytes was too slow to meet the inoculum requirement in the field; 160 kg fresh weight ha^{-1} sporocarps yielded 16 to 21 t fresh weight ha^{-1} of *Azolla* in 52 days [42]. Conditions for sporocarp formation and germination are incompletely understood [98].

Need for P-fertilizer
Reported threshold values of P deficiency are 0.4% dry weight in *Azolla* and 20 ppm available Olsen P in soil [2]. Such P-rich soils are uncommon. A growth test on 972 Philippine soils showed that only 13% of the samples were highly suitable for *Azolla* growth and P fertilization would be required in most soils [11]. This was confirmed by the observation that 80% of field-grown *Azolla* were P deficient ($p < 0.4\%$). To be economically feasible, P-fertilization requires a ratio of N fixed to P applied that is greater than the ratio of the prices of the corresponding fertilizers (about 3 for most countries). Basal P application has a low efficiency and may be economically infeasible, while split application has an efficiency of 5–10 g N_2 fixed per g P applied [93]. P-fertilization limited to the inoculum production plot permits the P-enriched *Azolla* to multiply 6 to 7 times without P application in the main field and ensures a high efficiency of P applied [93].

Pests
Although commercial pesticides effectively control *Azolla* pests, their application is not economically feasible in the field [28] and should be limited to inoculum production [47]. Pesticides of plant origin might be economically feasible for field use. Alternate drainage and irrigation, cultivation in wet fields or with a thin water layer, and reduced application of organic manures, may help controlling pests [102].

Temperature requirement
The optimum temperature for most *Azolla* species (20–30 °C) is below the average temperature in the tropics [43]. Cool weather is a key to successful *Azolla* cultivation in Vietnam and China. Temperature limitations can be reduced by selecting cold- or heat-tolerant strains. Among strains tested at IRRI, *A. microphylla* #418 was most tolerant of high temperature (37 °C day/29 °C night) [94].

Economics
Technologies used in Vietnam and China are labor intensive and therefore have economic limitations. In South Cotabato (Philippines) where *Azolla* spread spontaneously and no P-fertilizer and little labor were needed, economic return from *Azolla* adoption, including cost savings in chemical fertilizers and weed control, was more than US$ 35 ha^{-1} at 1981 prices [32]. However, conditions in the study area were exceptionally favorable and should be viewed realistically. When conditions for *Azolla* growth were not favored by an exceptionally high level of available P, economics were not in favor of *Azolla* use [72]. The economic potential of *Azolla* is greatest where the opportunity cost of labor is low, and labor cost becomes critical where wage rates approach US$ 2 day^{-1} [32]. It is clear that a case study in the Philippines is not enough to allow definite conclusions regarding *Azolla* economics, which may vary according to socio-agricultural systems.

Prospects

Azolla has a N-potential similar to that of legumes but is easier to incorporate and grows well with rice in flooded conditions. Environmental, technological, and economic factors limit its use. Problems in inoculum conservation, multiplication, and transport could be solved if *Azolla* could be propagated from spores. Temperature limitations and requirements for P can be reduced by selecting cold- or heat-resistant strains with low P requirements, and by split application of P-fertilizer, limited or not to inoculum production.

Azolla strains exhibit a wide range of behavior with regard to environmental factors, P requirement, N_2 fixation, and productivity, The ability to combine favorable characters such as resistance to high temperature and pests, low P requirement, and erect growth (permitting higher productivity) would allow strains to be designed for specific conditions. For this purpose, recombination of algal and plant symbionts [38, 39] and sexual hybridization [97] between *Azolla* species proved feasible. The IRRI biofertilizer germplasm contains 23 hybrid strains obtained by algal transfer and 85 obtained by sexual hybridization, some of which exhibit improved characters [94]. However, hybrid formation requires that macro- and micro-sporocarps be obtained and no method has been yet designed to induce sporulation at will. This is a major limiting factor for *Azolla* hybridization.

The key economic costs in *Azolla* use are those of P application, labor, and pest control. Economic limitations are important and need further evaluation but calculations should also consider the long-term benefits of *Azolla* on soil fertility. *Azolla* has potential not only as a green manure but as a multipurpose biofertilizer that can be a weed suppressor, a mineral scavenger – especially for K –, an animal feed, a primary producer in rice-fish-*Azolla* culture [20] and a depollutant. This potential may renew interest in *Azolla* use [41].

Legume green manures

A broad range of legume green manures (LGM) has been used in rice cultivation (Table 4). Potentialities of LGM for rice were early recognized. In 1936 the International Institute of Agriculture reported that: "application of green manure may involve great progress in rice growing by ensuring yields higher than those at present attained". Similar statements were recorded from the proceedings of symposia in 1952, 1953 and 1954 [51]. Afterwards, it seems that less attention was paid to LGM in rice production. They were mentioned in only a few paragraphs of the proceeding of the symposium on "Nitrogen and Rice" held in 1979 at IRRI. The discovery of stem-nodulating legumes [18,1] able to grow, fix N_2, and develop large biomasses under waterlogged conditions; and the concern for agricultural sustainability, have renewed the interest of the scientific community in LGM for wetland rice. It

is significant that a symposium at IRRI in 1987 was devoted to green manure in rice farming [30].

Potential to increase rice yield

BNF by legume green manure (LGM) used for rice has usually been estimated from total N measurement and the assumption that 50–80% of accumulated N is Ndfa. Values of N accumulated in traditional LGM crop [70] average 114 kg ha^{-1} (Table 3). Those values were most often for a crop grown until maturity, which is rarely done with a LGM. More recent values (Table 4) show that in 30–45 days a LGM accumulates 7 to 143 kg N ha^{-1} (mean 63). Values published after 1985 average 133 kg N ha^{-1} [37]. Ranges in kg N ha^{-1} are 40–225 for stem nodulating legumes, 33–115 for grain legumes, and 24–39 for perennial trees. Assuming 50–80% Ndfa, one LGM crop can fix an average 1.0–1.6 kg N ha^{-1} d^{-1} or 60–100 kg N ha^{-1} in 50–60 d.

Beside N and other nutrient provision to the crop, the beneficial effects of LGM incorporation that have been reported include:
- improvement of soil properties, especially (1) total N and organic matter content, (2) available Zn, (3) water holding capacity, and (4) soil aggregation [7, 8];
- control of some rice pests, in particular nematodes [56]; and
- immobilization of nitrogen nitrified during dry fallows, that would otherwise be lost by denitrification when soils are reflooded.

Estimates of yield increase due to traditional LGM range from 30 to 100 kg grain per ton of LGM incorporated [70].

Current utilization

Despite their potential, a setback in use of leguminous green manures has been observed during the last decade. China is the only country where LGM are still noticeably used, but data show a decrease from 10 million ha in 1974 to less than 4 million ha in 1990 [79]. In other countries, usage of LGM seems to have become incidental [70]. In several rice growing areas of India 20-30% of soils were planted to LGM at mid century [78], then green manuring exhibited a strong setback with the increase in cropping intensity and the low cost and ready availability of fertilizer [85].

Table 4. N accumulated by legumes used as green manure in rice cultivation

1. Crops grown until maturity (reproduced from [70])

Species	Nitrogen accumulated	
	(kg ha^{-1})	(% fresh weight)
Astragalus sinicus	108–123	0.35–0.47
Canavalia ensiformis	98	0.47
Cassia mimosoides	97	0.44
Crotalaria anagyroides	98	0.33
Crotalaria juncea	105–129	0.30
Crotalaria quinquefolia	88	0.19
Dolichos biforus	89	0.58
Gycine koidzumii	71	0.42
Phaseolus sp.	-	0.28
Phaseolus lathyroides	90	-
Phaseolus calcaratus	42	0.22
Sesbania aculeata	96–122	0.32–0.36
Sesbania rostrata	267	-
Sesbania sesban	100–202	0.39
Sesbania microcarpa	87	0.50
Sesbania sirececa	146	-
Average	114	0.37

2. N accumulate in crops grown for definite time (calculated from data tabulated in [10])

	Dry weight	N accumulated	
	(t ha^{-1})	(kg ha^{-1})	(kg ha^{-1} day^{-1})
110 crops of 16 species grown for 30 to 178 days (average 52 days)			
Mean	4.3	99	1.9
Maximum	13.3	267	5.1
Minimum	0.2	7	0.2
32 crops of 11 species grown for 30 to 45 days (average 40 days)			
Mean	2.5	63	1.6
Maximum	6.7	143	3.2
Minimum	0.2	7	0.2

Limiting factors

Reasons for decline and constraints associated with LGM use appear in several reviews [7, 22, 70]. Some detrimental effects have been reported, mainly in temperate condition [90]. One important limiting factor is the bulkiness of the LGM and the resulting problems for incorporation. Nitrogen content in legumes varies from 0.2 to 0.6% therefore the fresh weight corresponding to 50 kg N varies from 10 to 26 tons. Incorporation of such a biomass requires draft and/or man power, which are often not available. There are also incidental reasons. For example, in certain areas of India indiscriminate cattle grazing due to inadequate social control explain farmer reluctance to raise a green manure crop [85].

However, major limitations are socio-economic. First, it is clear that LGM are not appealing because they do not yield food or cash directly. Then there are many economical limitations. In situations where

N fertilizer is available, assuming an average yield of 15 kg grain per kg N applied, the cost of inorganic N fertilizer relative to the price of rice is very favorable. Furthermore, many governments have adopted a fertilizer subsidy policy and made cheap credit available for farmers to buy N fertilizer. On the other hand, the costs of green manure seed and land preparation are not favorable. If there is residual moisture in the soil, after the harvest of the rice crop, the economic advantage is very often in favor of growing a cash crop of legumes, groundnut, maize, millet, onion etc. and to resort to application of inorganic fertilizers for the next rice crop [85].

The situation where N fertilizer is not available is most frequently that of subsistence farmers, with small land holdings, who cannot afford to release land used for food or forage crops to LGM cultivation and therefore prefer to grow a catch crop. In some areas where no N fertilizer is available and organic manure was traditionally applied to rice, green manure is now applied preferentially to vegetable cash crops rather than to the rice crop.

Furthermore, the increased availability of inorganic fertilizers and low emphasis on GM by research and extension have contributed to the decline in LGM use.

Prospects

The sustainability issues together with the discovery of stem-nodulating legumes have revived scientific interest in LGM. The formation of stem nodules has been reported in 25 species of the genera *Sesbania* and *Aeschynomene*. Stem-nodulating legumes exhibit adaptation to waterlogged or water-saturated conditions of growth. Data on N accumulated in 40–55 days show a higher N potential of stem-nodulating LGM than of traditional LGM. Several studies have shown the very high N_2-fixing potential of *Sesbania rostrata* and *Aeschynomene afraspera* [6, 37, 49, 52, 60,]. ^{15}N dilution and $\partial\ ^{15}N$ studies showed that Ndfa of 45–55 day-old *S. rostrata* was about 70% and increased to 90% at 65 days. Field experiments show that stem-nodulating LGM offer a better N potential for wetland rice than traditional LGM. In experimental plots, a gain of 267 kg N ha^{-1} was reported after incorporating a 52-d crop [60]. In a 45-d *Sesbania*-rice (WS)/55-d *Sesbania*-rice (DS) sequence, *Sesbania* fixed 303 kg N ha^{-1} year^{-1} when uninoculated, and 383 kg N when inoculated with *Azorhizobium* [37]. Incorporating one 40–60 day old culture of *Sesbania* or *Aeschynomene*

may increase yield by 1–3 t ha^{-1} [86, 87], which is equivalent to the application of 50 to 100 kg fertilizer N ha^{-1} [7, 37, 87].

Photoperiod sensitivity limits the utilization of *S. rostrata* while *A. afraspera* seems to be considerably less photoperiod sensitive [6]. The non availability of seeds of stem-nodulating legumes is currently the major limiting factor for their adoption by farmers. But many of the socioeconomic factors that limit the use of traditional LGM use also will limit that of stern-nodulating LGM.

Conditions for economical use of LGM are that (1) there is no alternative for a more profitable crop, (2) legume establishment does not require soil preparation and is inexpensive, (3) LGM productivity is stable in time, (4) manpower and adequate implements are available for incorporation, and (5) there is no concurrence with manpower needed for rice transplanting [22]. In particular, N_2 fixation by a LGM (and conservation of NO_3 mineralized during the dry season) may be an economically viable proposition if production costs can be kept low, and if the LGM does not compete with marketable or subsistence crops.

Summary and conclusion

Nitrogen is usually the limiting factor to high yields in rice fields. Therefore, the use of BNF as an alternative or supplementary source of N for rice has been the major approach in microbiological management of wetland rice. Whereas N_2-fixing green manures have been used for centuries in some rice-growing areas, research on cyanobacterial and bacterial inoculants for wetland rice is relatively recent, being initiated in the early 1950s for free-living cyanobacteria and in the 1960s for rhizosphere bacteria.

Currently, most bacterial strains tested for inoculation have been N_2-fixing forms, but ARA, ^{15}N, and N balance studies have not provided clear evidence that the promotion of rice growth and N uptake was due to increased BNF. Therefore, several authors refer to the production of PGRs to explain the beneficial effect of bacterial inoculation. No experiment has yet supported this hypothesis. If it is verified that the ability of inoculated strains to produce PGRs is more important than their N_2-fixing ability, it is clear that the screening of bacterial strains for inoculation should not be limited to N_2-fixing strains. The few data available on strain establishment showed that, in most cases, inoculated strains disappeared or established themselves for vari-

273

ous periods of time, but did not multiply. That would limit the effect of inoculation to the earlier growth stage of the plant, a hypothesis that agrees with the absence of an inoculation effect on harvest index. Given the current knowledge, no definite conclusion regarding the potential of bacterial inoculation of rice can be drawn.

The selection and breeding of rice according to the variety's ability to stimulate an associative microflora that promotes BNF and soil N utilization is still limited by the absence of an efficient screening method. The relatively low N_2 fixation potential of associative BNF is not a hindrance to this promising approach, whose major advantage is that the N potential is inherent to the plant and thus requires no additional cultural practice by the farmer.

Free-living cyanobacteria have a moderate potential of about 30 kg N ha^{-1} crop $cycle^{-1}$ which may translate to a yield increase of 250–350 kg ha^{-1}. However, the technology for cyanobacterial inoculation has not progressed beyond the experimental stage of large-scale field testing. As long as inoculation is applied on a trial-and-error basis, it will have little chance of success. Recent developments indicate that foreign strains usually do not establish and more attention should be paid to promoting indigenous strains, which are ubiquitous. Cultural practices to enhance their growth are known but environments where those practices can be efficient and economically viable are probably limited. In the long term, genetic engineering may contribute to cyanobacteria management in wetland rice fields, but it is not yet known if and how engineered strains can establish and/or compete with the indigenous microflora. In-depth agroecological research is required before cyanobacteria technology can be substantially improved.

Azolla has proved useful as a N biofertilizer in some rice-growing countries. Like legumes it has a high N potential, but it is easier to incorporate and grows well with rice under flooded conditions. Environmental, technological, and economic factors limit *Azolla* use. Recent progress in strain hybridization and recombination has opened new ways to alleviate some environmental and nutritional limitations. Socioeconomic limitations are important and are probably increasing, as shown by the setback of *Azolla* in China and Vietnam where it was traditionally used. However, recent studies have shown that *Azolla* has potential not only as a green manure but as a multipurpose biofertilizer. These limiting factors and the potential of *Azolla* as a multipurpose crop, which may revive interest in its use, will decide the extent of its future utilization.

Leguminous green manures have traditionally been used in many rice-growing countries. Estimates of N accumulated in a traditional prerice LGM crop range from 42 to 202 kg N ha^{-1}. Despite this potential, a strong setback of LGM use has been observed during the last decades. In countries other than China, the use of LGM seems to have become incidental, mostly because of socioeconomic limitations. LGM are not appealing because they do not yield food or cash directly. Where N fertilizer is available, green manuring is usually more expensive than inorganic N. Situations where N fertilizer is not available concern mostly subsistence farmers, with small holdings, who cannot afford to release land used for food or forage crops to LGM and prefer to grow a cash crop.

During the last decade the discovery of the high N_2-fixing potential of some flood-resistant stem-nodulating legumes has revived the interest of rice scientists in LGM. Field experiments show that stem-nodulating LGM offer a better N potential for wetland rice than traditional LGM do. However, many of the socioeconomic factors that limit the use of traditional LGM use also will limit that of stem-nodulating LGM. N_2 fixation by a LGM (and conservation of NO_3 mineralized during the dry season) may be an economically viable proposition if production costs can be kept low, and if the green manure does not compete with marketable or subsistence crops.

BNF in rice fields has been the most effective system for sustaining production in low-input traditional cultivation. The general impression when considering the current management of N_2-fixing organisms in ricefields is that 40 years after the first inoculation experiments, the agronomic potential of BNF is underutilized and its intentional use is decreasing.

Considering that rice obtains most of its N from the soil, regardless of the amount of chemical N fertilizer applied, concerns in recent high-input, intensive rice cultivation are (1) sustainability of high yields and (2) the possible environmental impacts of intensive management on soil fertility. Knowledge on this aspect is still limited, but the key roles of the rhizosphere, the photosynthetic aquatic biomass, and their N_2-fixing components in maintaining the fertility of rice soils under intensive cultivation have been recognized and need further study [92].

An additional 300 million tonnes of rice will be needed in 2020 to meet the need of a fast-growing human population. This requires a 65% production increase within 30 years without much expansion of actual cultivated area [31]. But increased rice production should not be at the expense of future generations and should fulfill the concept of sustainability. A major challenge is managing pests and nutrients in ways that reduce agrochemical use. Increased use of inorganic fertilizer is inescapable, but, as pointed out by Postgate (1989) [55], a parallel return to greater exploitation of BNF, still responsible for providing 60–70% of the new N in the biosphere, seems common sense. Currently, BNF is intentionally used in only a few percent of the global rice-growing area. Designing economically viable methods for utilizing N_2-fixing organisms in rice cultivation still remains a major scientific challenge.

References

1. Alazard D and Duhoux E (1987) Nitrogen-fixing stem nodules on *Aeschynomene afraspera*. Biol Fert Soils 4: 61–66
2. Ali S and Watanabe I (1986) Response of *Azolla* to P, K, Zn in different wetland rice soils in relation to chemistry of floodwater. Soil Sci Plant Nutr 32: 239–254
3. App AA, Watanabe I, Alexander M, Ventura W, Daez C, Santaguio T and De Datta SK (1980) Nonsymbiotic nitrogen fixation associated with the rice plant in flooded soils. Soil Sci 130: 283–289
4. App AA, Watanabe I, Santiago-Ventura T, Bravo M and Daez-Jurey C (1986) The effect of cultivated and wild rice varieties on the nitrogen balance of flooded soil. Soil Sci 141: 448–452
5. Bally R, Thomas-Bauzon D, Heulin T and Balandreau J (1983) Determination of the most frequent N_2-fixing bacteria in a rice rhizosphere. Can J Microbiol 29: 881–887
6. Becker M (1990) Potential use of the stem-nodulating legumes *S. rostrata* and *A. afraspera* as green manure for lowland rice. Ph.D. thesis. Justus-Liebig University, Giessen, FRG
7. Becker M, Ladh JK and Ottow JC (1988) Stem-nodulating legumes as green manure for lowland rice. Phillip J Crop Sci 13 (3): 121–127
8. Bouldin DR (1988) Effect of green manure on soil organic matter content and nitrogen availability. In: Sustainable agriculture: Green manure in rice farming, pp 152–163. IRRI PO Box 933, Manila, Philippines
9. Bray F (1986) The Rice Economics. Technology and Development in Asian Societies. UK. Basil Blackwell Inc., 254 p
10. Buresh RJ and De Datta SK (1991) Nitrogen dynamics and management in rice-legume cropping systems. Adv Agron 45: 1–59
11. Callo DP, Dilag RT, Necesario RS, Wood DM, Tagalog FC, Gapasin DP, Ramirez CM and Watanabe I (1985) A 14-day *Azolla* adaptability test on farmer's fields. Phillip J Crop Sci 10 (3): 129–133
12. Charyulu PBBN, Fourcassie F, Barbouche AK, Rondro Harisoa L, Omar AMN, Weinhard P, Marie R and Balandreau J (1985) Field inoculation of rice using *in vitro* selected bacterial and plant genotypes. In: Klingmüller W (ed) *Azospirillum* III: Genetics, Physiology, Ecology, pp 163–179. Springer Verlag, Berlin
13. Cholitkul W, Tangcham B, Sangtong P and Watanabe I (1980) Effect of phosphorus on N_2-fixation measured by field acetylene reduction technique in Thailand long term fertility plots. Soil Sci Plant Nutr 26: 291–299
14. De PK (1939) The role of cyanobacteria in nitrogen fixation in rice fields. Proc R Soc Lond 127 B: 121–139
15. Diara H F, Van Brandt H, Diop AM and Van Hove C (1987) *Azolla* and its use in rice culture in West Africa. In: *Azolla* Utilization 147–152. IRRI PO Box 933, Manila, Philippines
16. Diara HF and Van Hoove C (1984) *Azolla*, a water saver in irrigated rice fields? In: Silver WS Shröder EL (eds) Practical Application of *Azolla* for Rice Production, pp 115–118. M Nijhoff/W Junk, Den Haag
17. Dobereiner J and Ruschel AP (1962) Inoculation of rice with N_2-fixing genus *Beijerinckia* Derx. Rev Brasil Biol 21: 397–402
18. Dreyfus BL and Dommergues YR (1981) Nitrogen-fixing nodules induced by *Rhizobium* on the stem of the tropical legume *Sesbania rostrata*. FEMS Microbiol Lett 10: 313–317
19. FAO (1978) *Azolla* propagation and small-scale biogas technology. FAO Soils Bull. 41
20. FAO (1988) The Rice-*Azolla*-Fish system. RAPA Bulletin Vol 4: 1988. Food and Agriculture Organization of the United Nations, Bangkok, Thailand. 37p
21. Fillery IRP, Roger PA and De Datta SK (1986) Ammonia volatilization from nitrogen sources applied to rice fields: II. Floodwater properties and submerged photosynthetic biomass. Soil Sci Soc Am J 50: 86–91
22. Garrity DP and Flinn JC (1988) Farm-level management systems for green manure crops in Asian rice environments. In: Green Manure in Rice Farming, pp 111–129. IRRI PO Box 933, Manila, Philippines
23. Gomez KA (1972) Techniques for field experiments with rice. IRRI PO Box 933, Manila, Philippines. 46 p
24. Grant IF, Roger PA and Watanabe I (1985) Effect of grazer regulation and algal inoculation on photodependent nitrogen fixation in a wetland rice field. Biol Fert Soils 1: 61–72
25. Greenland DJ and Watanabe I (1982) The continuing nitrogen enigma. Trans Int Congr Soil Sci, New Delhi 5: 123–137
26. Heulin T, Rahman M, Omar AMN, Rafidison Z, Pierrat JC and Balandreau J (1989) Experimental and mathematical procedures for comparing N_2-fixing efficiencies of rhizosphere diazotrophs. J Microbiol Meth 9: 163–173
27. Inubushi K and Watanabe I (1986) Dynamics of available N in paddy soils. II. Mineralized N of chloroform-fumigated soil as a nutrient source for rice. Soil Sci Plant Nutr 32: 561–577
28. IRRI (1986) Annual report for 1985. PO Box 933, Manila, Philippines
29. IRRI (1987) *Azolla* utilization, PO Box 933, Manila, Philippines. 296 p
30. IRRI (1988) Green manure in rice farming. PO Box 933, Manila, Philippines. 377 p
31. IRRI (1993) IRRI Rice Almanac 1993–1995. PO Box 933, Manila, Philippines. 142 p
32. Kikuchi M, Watanabe I and Haws LD (1984) Economic evaluation of *Azolla* in rice production. In: Organic Matter and Rice, pp 569–592. IRRI, PO Box 933, Manila, Philippines
33. Ladha JK, Pareek RP and Becker M (1992) Stem-nodulationg Legume-*Rhizobium* symbiosis and its agronomic use in lowland rice. Adv Soil Sci 20: 148–192

34. Ladha JK and Boonkerd N (1988a) Biological nitrogen fixation by heterotrophic and phototrophic bacteria in association with straw. In: Proc 1st Int Symp on Paddy Soil Fertility, Chiangmai, Thailand Dec 6–13, 1988 pp 173–187. Wien ISSS

35. Ladha JK, Tirol A, Punzalan GC and Watanabe I (1987) N_2-fixing (C_2H_2-reducing) activity and plant growth characters of 16 wetland rice varieties. Soil Sci Plant Nutr 33: 187–200

36. Ladha JK, Tirol-Padre A, Punzalan GC, Watanabe I and De Datta SK (1988c) Ability of wetland rice to stimulate biological nitrogen fixation and utilize soil nitrogen. In: Bothe H, de Bruijn FJ and Newton WE (eds) Nitrogen Fixation, Hundred Years After, pp 747–752. Gustav Fischer, Stuttgart

37. Ladha JK, Watanabe I and Saono S (1988b) Nitrogen fixation by leguminous green manure and practices for its enhancement in tropical lowland rice. In: Sustainable Agriculture: Green Manure in Rice Farming, pp 165–183. IRRI, PO Box 933, Manila, Philippines

38. Lin Chang, Liu Chung-Chu, Zeng De-Yin, Tang Long-Fei and Watanabe I (1988) Reestablishment of symbiosis to *Anabaena*-free *Azolla*. Zhungguo Kexue 30B (7): 700–708

39. Lin Chang and Watanabe I (1988) A new method for obtaining *Anabaena*-free *Azolla*. New Phytol 108: 341–344

40. Liu ChungChu (1979) Use of *Azolla* in rice production in China. In: Nitrogen and Rice, pp 375–394. IRRI, PO Box 933, Manila, Philippines

41. Liu ChungChu (1987) Reevaluation of *Azolla* utilization in agricultural production. In: *Azolla* Utilization, pp 67–76. IRRI, PO Box 933, Manila, Philippines

42. Lu Shu-Ying (1987) Method for using *Azolla filiculoides* sporocarp to culture sporophytes in the field. In: *Azolla* Utilization, pp 26–32. IRRI, PO Box 933, Manila, Philippines

43. Lumpkin TA (1987) Environmental requirements for successful *Azolla* growth. In: *Azolla* Utilization, pp 89–97. IRRI, PO Box 933, Manila, Philippines

44. Lumpkin TA and Plucknett DL (1982) *Azolla* as a green manure. Westview Tropical Agriculture Series, Westview Press, Boulder Co, USA. 230 p

45. Mabbayad BB (1987) The *Azolla* program of the Philippines. In: *Azolla* Utilization, pp 101–108. IRRI, PO Box 933, Manila, Philippines

46. Metting B and Pyne JW (1986) Biologically active compounds from microalgae. Enzyme Microbial Technol 8 (7) 385–394

47. Mochida O (1987) Pests of *Azolla* and control practices. Text book for the training course on *Azolla* use. Fujian Acad Agric Sci and IRRI, Fuzhou, PRC, June 1987. 60 p

48. Nayak DN, Ladha JK and Watanabe I (1986) The fate of marker *Azospirillum lipoferum* inoculated into rice and its effect on growth, yield and N_2 fixation of plants studied by acetylene reduction, $^{15}N_2$ feeding and ^{15}N dilution techniques. Biol Fert Soils 2: 7–14

49. Ndoye I and Dreyfus B (1988) N_2 fixation by *Sesbania rostrata* and *Sesbania sesban* estimated using ^{15}N and total N difference methods. Soil Biol Biochem 20: 209–213

50. Omar AMN, Heulin T, Weinhard P, Alaa El Din MN and Balandreau J (1989) Field inoculation of rice with *in vitro* selected plant growth promoting rhizobacteria. Agronomie 9: 803–808

51. Pandey RK and Morris RA (1983) Effects of leguminous green manuring on crop yields in rice-based croppping systems. Paper presented at the Internat Rice Res Conf April 23–28. IRRI, PO Box 933, Manila, Philippines

52. Pareek RP, Ladha JK and Watanabe I (1990) Estimation of N_2 fixation by *Sesbania rostrata* and *S. cannabina* in lowland rice soil by ^{15}N dilution method. Biol Fert Soils 10: 77–88

53. Pedurand P and Reynaud PA (1987) Do Cyanobacteria enhance germination and growth of rice ? Plant Soil 101: 235–240

54. Pillai KG (1980) Biofertilizers in rice culture. Problems and prospects for large scale adoption. AICRIP Publication no. 196

55. Postgate J (1990) Fixing the nitrogen fixers. New Sci Feb. 3, 57–61

56. Prot JC and Rahman ML (1994) Nematode ecology and management in rice ecosystems in South and Southeast Asia. In: PS Teng, KL Heong and K Moody (eds). Rice Pest Science and Management pp 129–146. IRRI, PO Box 933, Manila, Philippines

57. Reddy PM (1983) Role of chromatic lights in germination of the spores of blue-green algae. Arch Hydrobiol Suppl 67: 299–304

58. Reddy PM and Roger PA (1988) Dynamics of algal populations and acetylene reducing activity in five soils inoculated with cyanobacteria. Biol Fert Soils 6: 14–21

59. Reynaud PA and Metting B (1988) Colonization potential of Cyanobacteria on temperate irrigated soils of Washington State, USA. Biol Agric Hortic 5: 197–208

60. Rinaudo G, Alazard D and Moudiongui A (1988) Stem-nodulating legumes as green manure for rice in West Africa. In: Green Manure in Rice Farming, pp 97–109. IRRI, PO Box 933, Manila, Philippines

61. Rinaudo G and Dommergues YD (1971) Validité de l'estimation de la fixation biologique de I'azote dans la rhizosphère par la méthode de réduction de I'acetylène. Ann Inst Pasteur 121: 93–99

62. Roger PA (1991) Reconsidering the utilization of cyanobacteria in wetland rice cultivation. In: Dutta SK and Sloger C (eds) Bological N_2 Fixation Associated with Rice Production, pp 119–141. Oxford and IBH, New Delhi

63. Roger PA, Grant IF, Reddy PM and Watanabe I (1987a) The photosynthetic aquatic biomass in wetland rice fields and its effects on nitrogen dynamics. In: Efficiency of Nitrogen Fertilizers for Rice, pp 43–68. IRRI, PO Box 933, Manila, Philippines

64. Roger PA and Kulasooriya SA (1980) Cyanobacteria and rice. IRRI, PO Box 933, Manila, Philippines. 112 p

65. Roger PA, Kulasooriya SA, Tirol AC and Craswell ET (1980) Deep placement: A method of nitrogen fertilizer application compatible with algal nitrogen fixation in wetland rice soils. Plant Soil 57: 137–142

66. Roger PA and Ladha JK (1992) Biological nitrogen fixation in wetland ricefields: estimation and contribution to nitrogen balance. Plant Soil 141: 41–55

67. Roger PA, Reddy PM and Remulla-Jimenez R (1988) Photodependent acetylene reducing activity (ARA) in ricefields under various fertilizer and biofertilizer management. In: Bothe H, de Bruijn FJ and Newton WE (eds) Nitrogen Fixation: Hundred Years After, p 827. Gustav Fischer, Stuttgart

68. Roger PA, Santiago S, Reddy PM and Watanabe I (1987b) The abundance of heterocystous cyanobacteria in rice soils and inocula used for application in ricefields. Biol Fert Soils 5: 98–105

69. Roger PA, Tirol A, Ardales S and Watanabe I (1986) Chemical composition of cultures and natural samples of N_2-fixing blue-green algae from ricefields. Biol Fert Soils 2: 131–146

70. Roger PA and Watanabe I (1986) Technologies for utilizing biological nitrogen fixation in wetland rice: potentialities, current usage, and limiting factors. Fert Res 9: 39–77

71. Roger PA, Zimmerman WJ and Lumpkin T (1993) Microbiological management of wetland rice fields. In: Metting B (ed) Soil Microbial Technologies, pp 417–455. Marcel Dekker, New York

72. Rosegrant MW, Roumasset JA and Balisacan A (1985) Biological technology and agricultural policy: an assessment of *Azolla* in Philippine rice production. Am J Agric Econ 67: 726–732

73. Roychoudhury P, Pillai GR, Pandey SL, Krishna Murti GSR and Venkataraman GS (1983) Effect of blue-green algae on aggregate stability and rice yield under different irrigation and nitrogen levels. Soil Tillage Res 3: 61–66

74. Sampaio MJA, Fiore MF and Ruschel AP (1984) Utilization of radioactive phosphorus (^{32}P) by *Azolla-Anabaena* and its transfer to rice plants. In: Silver WS and Shrüder EC (eds) Practical Application of *Azolla* for Rice Production, pp 163–167. M Nijhoff/W Junk, Den Haag

75. Santiago-Ventura T, Bravo M, Daez C, Ventura W, Watanabe I and App AA (1986) Effect of Fertilizers, straw, and dry fallow on the N-balance of a flooded soil planted with rice. Plant Soil 93: 405–411

76. Sen MA (1929) Is bacteria association a factor in nitrogen assimilation by rice plant ? Agric J India 24: 229

77. Singh AL and Singh PK (1987) Nitrogen fixation and balance studies of rice soil. Biol Fert Soils 4: 15–19

78. Singh NT (1984) Green manures as a source of nutrients in rice production. In: Organic Matter and Rice, pp 217–228. IRRI, PO Box 933, Manila, Philippines

79. Stone B (1990) Evolution and diffusion of agricultural technology in China. In: Kotk NG (ed) Sharing Innovation, Global Perspectives of Food, Agriculture and Rural Development, pp 35–93. IRRI PO Box 933, Manila, Philippines

80. Sundara WVB, Mann HS, Paul NB and Mathur SP (1962) Bacterial inoculation experiments with special reference to *Azotobacter*. Indian J Agric Sci 33: 279–290

81. Thomas-Bauzon D, Weinhard P, Villecourt P and Balandreau J (1982) The spermosphere model. Its use in growing, counting, and isolating N$_2$-fixing bacteria from the rhizosphere of rice. Can J Microbiol 28: 922–228

82. Tirol-Padre A, Ladha JK, Punzalan G and Watanabe I (1988) A plant sampling procedure for acetylene reduction assay to detect rice varietal differences in ability to stimulate N$_2$ fixation. Soil Biol Biochem 20: 175–183

83. Troldenier G (1987) Estimation of associative nitrogen fixation in relation to water regime and plant nutrition in a long-term pot experiment with rice. Biol Fert Soils 5: 133–140

84. Venkataraman GS (1981) Cyanobacteria for rice production. A manual for its promotion. FAO Soil Bull. n° 46. 102 p

85. Venkataraman GS (1984) Development of of organic matter-based agricultural systems in South Asia. In: Organic Matter and Rice, pp 57–70. IRRI, PO Box 933, Manila, Philippines

86. Ventura W, Mascarinia GB, Furoc RE and Watanabe I (1987) *Azolla* and *Sesbania* as biofertilizers for lowland rice. Philipp J Crop Sci 12: 61–69

87. Ventura W and Watanabe I (1993) Green manure production of *Azolla microphylla* and *Sesbania rostrata* and their long-term effects on rice yields and soil fertility. Biol Fert Soils 15: 241–248

88. Watanabe A, Nishigaki S and Konishi C (1951) Effect of nitrogen fixing blue green algae on the growth of rice plant. Nature (London) 168: 748–749

89. Watanabe I (1982) *Azolla-Anabaena* symbiosis, its physiology and use in tropical agriculture In: Dommergues Y and Diem H (eds) Microbiology of Tropical Soils, pp 169–185. M Nijhoff, Den Haag

90. Watanabe I (1984) Use of green manures in Northeast Asia. In: Organic Matter and Rice, pp 229–234. IRRI PO Box 933, Manila, Philippines

91. Watanabe I (1987) Summary report of the *Azolla* program of the International Network on Soil Fertility and Fertilizer Evaluation for Rice. In: *Azolla* Utilization, pp 197–205. IRRI, PO Box 933, Manila, Philippines.

92. Watanabe I, De Datta SK and Roger PA (1988) Nitrogen cycling in wetland rice soils. In: Wilson JR (ed) Advances in Nitrogen Cycling in Agricultural Ecosystems, pp 239–256. C.A.B. International Wallingford, UK

93. Watanabe I, Lapis M, Oliveros R and Ventura W (1988b) Improvement of phosphate fertilizer application to *Azolla*. Soil Sci Plant Nutr 34 (4): 557–569

94. Watanabe I, Roger PA, Ladha JK and Van Hove C (1992) Biofertilizer germplasm collections at IRRI. IRRI, PO Box 933, Manila, Philippines. 66 p

95. Watanabe I, Yoneyama T, Padre B and Ladha JK (1987) Difference in natural abundance of ^{15}N in several rice varieties: application for evaluating N$_2$ fixation. Soil Sci Plant Nutr 33: 407–415

96. Watanabe I, Yoneyama T, Talukdar H and Ventura W (1991) The contribution of atmospheric N$_2$ to *Azolla* spp. growth in flooded soils. Soil Sci Plant Nutr 37: 101–109

97. Wei Wen-Xiong, Jin Gui-Ying and Zhang Ning (1986) Preliminary report on *Azolla* hybridization studies. Bull Fujian Acad Agric Sci 1: 73–79 (*In Chinese*)

98. Xiao QingYuan, Shi YanRu, Yang GuangLi and Peng KeLin (1987) Germination of *Azolla filiculoides* Lam. sporocarps and factors affecting their growth. In: *Azolla* Utilization, pp 33–38. IRRI, PO Box 933, Manila, Philippines

99. Yanni YG (1991) Potential of indigenous cyanobacteria to contribute to rice performance under different schedules of nitrogen application. World J Microbiol Biotechnol 7: 48–52

100. Yoshida T and Ancajas RR (1971) Nitrogen fixation by bacteria in the root zone of rice. Soil Sci Soc Am Proc 35: 156–157

101. Yoshida T and Rinaudo G (1982) Heterotrophic N$_2$ fixation in paddy soils. In: Dommergues YR and Diem HG (eds) Microbiology of Tropical Soils and Plant Productivity, pp 75–107. M Nijhoff and W Junk, Den Haag

102. Zhang ZhuangTa, Ke YuSi, Ling DeQuan, Duan BingYuan and Liu XiLian (1987) Utilization of *Azolla* in agricultural production in Guangdong Province, China. In: *Azolla* Utilization, pp 141–145. IRRI, PO Box 933, Manila, Philippines

103. Zhu ZL, Chen DL, Zhang SL and Xu YH (1986) Heterotrophic nitrogen fixation in ricefield soil (in Chinese) Soils (China) 18: 225–229

Fertilizer Research **42:** 277–296, 1995.
© 1995 *Kluwer Academic Publishers. Printed in the Netherlands.*

Managing nitrogen for sustainable crop production

C.A. Campbell[1], R.J.K. Myers[2] & D. Curtin[1]
[1]*Agriculture and Agri-Food Canada, Research Station, P.O. Box 1030, Swift Current, Saskatchewan, S9H 3X2, Canada and* [2]*IBSRAM, P.O. Box 9–109, Bangkken, Bangkok, Thailand*

Key words: nutrient cycling, socioeconomic constraints, sustainable agriculture, temperate/boreal ecosystems, tropical ecosystem

Abstract

This paper discusses the influence of N resources (fertilizer, legume, soil) on sustainable agriculture in temperate/boreal ecosystems (exemplified by the Canadian prairies), and in the humid, subhumid and semi-arid tropic (exemplified by southeast Asia and central and south America). A sustainable agricultural system is one that is economically viable, provides safe, nutritious food, and. conserves or enhances the environment. Consequently, we discuss the impact of N on crop yields, nitrogen use efficiency (NUE), food quality, environmental quality and on socioeconomic factors. Considerably more long-term research has been conducted in the temperate regions, consequently this was where most information was available. However, the principles governing the behaviour of N are very similar in all ecosystems. It is mainly the rates of nutrient cycling and the socioeconomic constraints that differ. Legumes and N fertilizers, used in a responsible manner, will increase crop production, provide quality food, increase net returns, reduce risk of monetary loss, improve soil quality, and reduce N loss via leaching and gaseous means. The key to sustainable management of N is to synchronize N supply with N use by the crop. Because societies in most temperate ecosystems are more affluent they are better positioned to encourage adoption of management techniques that promote sustainability. In contrast, most producers in the tropics are, subsistence farmers; consequently, their immediate goal is economic survival, not preservation of the environment.

Introduction

Sustainable land management, sustainable agriculture and sustainable cropping systems all command considerable attention in all developed and in many developing countries. Here we discuss a microcosm of this large problem, namely, how N (sometimes in combination with P) influences sustainable crop management, concentrating on the general principles applicable to most ecosystems. We present this information in two parts: (a) agroecosystems in developed countries in temperate/boreal climates where large commercialized production predominates [e.g. Canada, USA, Europe and most of the Organization of Economic Cooperation and Development (OECD) member states], and (b) developing countries, mainly located in humid, sub-humid, and semi-arid. tropics and subtropics, where agriculture is predominantly conducted on small, subsistence-type farming systems (e.g. Africa, Southeast Asia, South America, Caribbean). We use the Canadian prairies for (a), and primarily Southeast Asia for (b).

Temperate/boreal ecosystems with large commercialized production systems - the Canadian prairies

Background

Only 7% (67 million ha) of Canada's land area is arable while another 6% is suitable for grazing [12]. Of the arable land, 79% (53 million ha) is located in the three prairie provinces of Alberta, Saskatchewan, and Manitoba (Fig. 1). Long, cold winters; short, hot summers; low, variable precipitation; characterize the climate

278

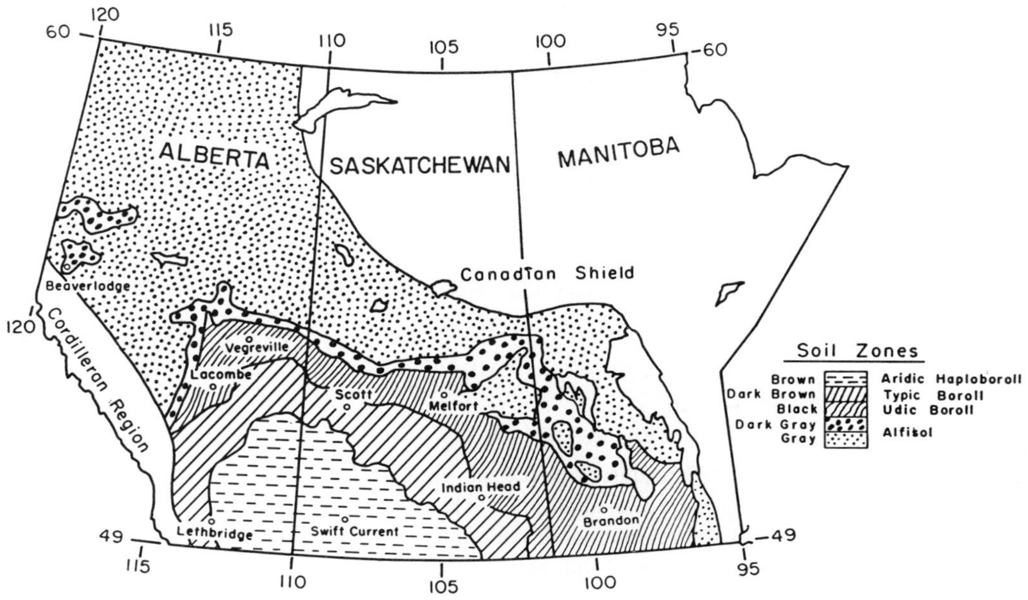

Fig. 1. Prairie Provinces of Canada, showing major soil zones.

and dictate cropping practices. Prairie soils are young and inherently fertile. Thus, crops mainly require N and P fertilizer, in limited cases S, but rarely K. Farms are large (most exceed 4.5 km^2) and highly mechanized to allow timely operations in the 4-month growing season (May–September). Cereals [mainly hard red spring wheat (*Triticum aestivum* L.)) predominate, but mixed farming (grain and livestock) is common on the more humid Gray and Dark Gray Luvisolic soil zones (Alfisols) in the northerly areas, while oilseeds and pulses are significant components of crop rotations on Black Chernozems (Udic Borolls). Monoculture wheat with varying degree of summerfallow (land left bare to conserve water, control weeds, or reduce incidence of disease) predominates in the more arid southern areas where Brown and Dark Brown Chernozems (Aridic and Typic Borolls) predominate. Summerfallow is significant even in the more humid areas. Low precipitation causes low yields (most < 3.5 t ha^{-1} and many only 1.75 t ha^{-1}) requiring low fertilizer inputs. Severe climate helps to control some pests, thereby reducing pesticide requirements. Irrigation is not common except in Alberta where just over 0.4 million ha of land is irrigated. Cereals occupy 50% and hay occupies 25% of the irrigated land [12].

Sustainable agriculture production - The problem

According to Agriculture and Agri-Food Canada, Sustainable Agriculture refers to "agrifood systems that are economically viable, provide safe nutritious food, and allows conservation or enhancement of soil, air and water." On the Canadian prairies, low fertilizer and pesticide use plus low precipitation minimize concerns for air and water pollution. However, soil degradation (soil organic matter loss, salinization, erosion) and especially economics, are major concerns.

The native grassland soils of the prairies were very stable but, when cultivated, lower plant biomass inputs, higher rates of organic matter decomposition, and susceptibility to erosion, made loss of organic matter and fertility inevitable. Furthermore, frequent summerfallowing to conserve water and reduce risk of crop failure hastened soil degradation and increased salinity. Thus, use of improper summerfallow techniques resulted in severe erosion in the dry, windy 1930s ('Dirty Thirties'). Since then, research has led to implementation of several conservation techniques (e.g. stubble mulch tillage, strip cropping, contour farming, use of tree shelterbelts, grass waterways, use of proper crop rotations) and these have helped to abate the problem. However, prairie soils which seldom required N fertilizer in the 1940s and 1950s, now often require this nutrient even after a year of summerfallow because the inherent fertility of the soil has been substantially depleted [66],

Despite the obvious positive contributions of fertilizers in maintaining food production (and as shown later, maintaining soil quality), 75% of Canadians surveyed in 1989 said that they were concerned about fertilizer use [12]. Thus, there is an urgent need to critically assess the true impact of fertilizers (in this case N) on the components of sustainable production systems.

Impact of N on yields and economics

Many studies have been conducted on the Canadian prairies that examine the influence of fertilizers on yield, nutritional quality and economics. Most of these studies were summarized in two recent reviews [23, 66]; thus, only a few pertinent highlights will be discussed in this paper. Studies typically show positive yield responses of all crops to N in all soil zones if moisture is non-limiting and if available N is not already over-abundant [66].

Nitrogen use efficiency

Sustainable crop production requires improved nitrogen use efficiency (NUE). Many factors influence NUE, with perhaps the main factor, weather, being unpredictable. However, historical weather records allow us to estimate the probabilities of obtaining average, or above- or below- average weather conditions, and we can use these to guide our management decisions. In the semiarid climate of southwestern Saskatchewan, yield, and therefore NUE response to N, was shown to be directly influenced by available water (Fig. 2). In this 9 year continuous no-till spring wheat study, the N-supplying capacity of the soil was improved by the combined effect of fertilizing, increasing cropping frequency, and reducing tillage, so that NUE was directly related to soil available N, but it was inversely related to rate of fertilizer N [27]. Factors that will affect NUE are placement of N (banding superior to broadcasting), timing of application (spring application near seeding time superior to fall application), and N source (ammonium nitrate generally superior to urea which may volatilize easily) [27,75]. But, with all these options, the farmer is required to balance the improved NUE against the extra cost (e.g., for banding > broadcasting; ammonium nitrate more costly than urea; fertilizer cheaper when bought in fall than in

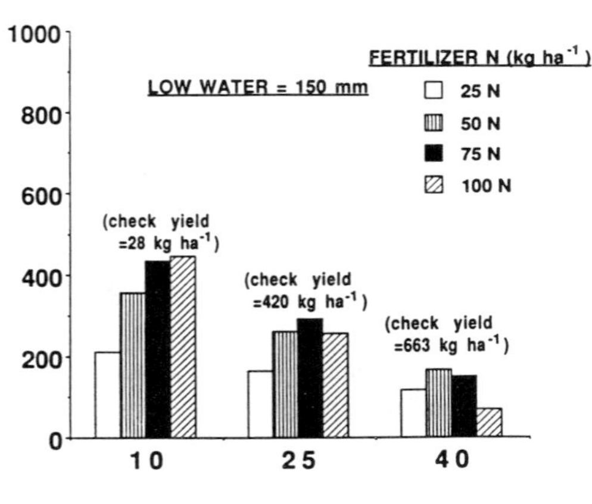

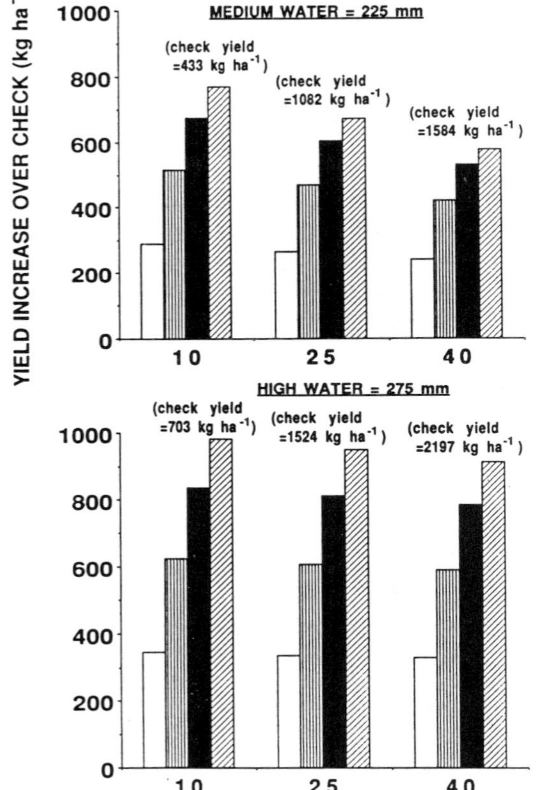

Fig. 2. Yield increase over zero N yields of spring wheat as estimated for some soil water, soil and fertilizer N situations in a 9-year no-till continuous wheat experiment conducted on a Brown Chernozem at Swift Current, Saskatchewan [27].

spring), and they must consider the convenience in terms of distribution of labour [85].

Long-term yield trends

A crop rotation study on a thin Black Chernozem at Indian Head, Saskatchewan [16], demonstrates how fertilizers may influence long-term yield trends. Treatments compared were stubble-crop wheat in fertilized (N and P) and unfertilized fallow-wheat-wheat and continuous wheat systems, and unfertilized sweetclover (*Melilotus officinalis* L.) green manure-wheat-wheat and fallow-wheat-wheat-hay-hay-hay systems. Hay was a mixture of alfalfa (*Medicago sativa* L.) and bromegrass (*Bromus inermis* Leyss). In the first 20 years of this study (1958–1977), N and P fertilizers were low because they were applied based on general recommendations for the crop and soil, but from 1978 a soil test was used as the criterion for fertilizer requirements and the N rates were tripled (on average) compared to rates in the first 20 years.

The benefit of N and P fertilizer in improving yields was readily apparent from the beginning of this study and, after the N rate was increased in 1978, the yield response to fertilizer was even more apparent (Fig. 3). The average rate of increase in yield of fertilized monoculture wheat grown on stubble was 30–33 kg ha^{-1} yr^{-1} (Table 1). The widening difference between fertilized and unfertilized systems over time was partly due to the higher rates of N applied after 1977, and partly because yields on the unfertilized plots had gradually declined, reflecting a decline in fertility of the soil (Table 1).

The rate of decline in yields of the unfertilized treatments was much greater in the fallow-wheat-wheat system (-28 kg ha^{-1} yr^{-1}) than in continuous wheat (-11 kg ha^{-1} yr^{-1}), probably reflecting greater erosion in the fallow-containing system [45]. Further evidence of the decline in fertility of the unfertilized soil was seen in the lower soil organic matter content and N-supplying power of the fallow-containing system compared to the continuous wheat (Table 1).

Sweetclover green manure increased yields of wheat grown on stubble compared to unfertilized fallow-wheat-wheat (Fig. 3), probably a result of N fixation, as evidenced by a 57% higher N-supplying power in the soil under green manure (Table 1). Even so, the yield of wheat grown on stubble in the green manure system has declined by 26 kg ha^{-1} yr^{-1}, much like that of the unfertilized fallow-wheat-wheat. This negative yield trend in the green manure system was partly because legumes do not supply P (Table 1). These results suggest that, on a yield basis, the fertilized system is more sustainable than the green manure system.

Yields of wheat grown on stubble in the 6-yr hay-containing system was much higher than those of stubble-wheat in the fertilized fallow-wheat-wheat system in the first 20 years of study (Fig. 3). This was partly due to less erosion [45], and partly because of the high N-supplying power of soil in the 6-yr rotation (Table 1). However, because the hay crop provides no P (Table 1), and because hay is harvested two in every six years, the 6-yr rotation also did not maintain wheat yields at as high a level as those obtained for the fertilized fallow-wheat-wheat system after the fertilizer rates were based on soil tests.

Although the benefits of N cannot be separated from that of P in this study, the results, clearly show that proper fertilization is required if we are to achieve sustainable crop production. Further, where legumes are used, adequate P must also be applied.

Economics

A prime consideration in achieving sustainability is having an economically viable operation. Much research has been conducted to assess the influence of fertilizers and legumes on the economic viability of crop rotation systems on the Canadian prairies [23]. The optimal system will depend, *inter alia*, entirely on the cost of fertilizer and price obtained for produce and often on Government policies and subsidies.

Under arid conditions (e.g., Brown soil zone), use of N and P fertilizers is profitable for continuous wheat in most economic situations, and fertilizer use often minimized losses (Fig. 4) [84]. However, for fallow-containing wheat rotations, it is only when wheat prices are high, or fertilizer costs low, that applying both N and P fertilizers results in higher net incomes than for rotations to which only N or P is applied.

In the sub-humid, thin Black Chernozemic soil at Indian Head, Saskatchewan, the economic benefit from fertilization is generally significant, particularly for the more intensive cropping systems [84]. Proper fertilization amplifies net income in years of favourable weather, and reduces economic losses in years of unfavourable growing conditions. Prior to 1977, when the fertilizer rates used in the Indian Head study were lower than economic optimums, net returns for legume-containing systems were superior to those of fertilized or unfertilized fallow-wheat-wheat (Table 2). But, after fertilizer rates were increased, the differential between fertilized and unfertilized monocul-

Table 1. Average annual rate of change in yield trends over 34 yr and organic N, available P and potential nitrogen-supplying power of soil at Indian Head, Saskatchewan in 1987 [16]

Rotation and fertilizer treatment[a]	Average yield increase[b] (kg ha^{-1} yr^{-1})	Organic N (t ha^{-1})	Potential nitrogen-supplying power[c] (kg ha^{-1} wk^{-1})	Available-P in 0–1.2 m[d] (kg ha^{-1})
F-W-(W)	-28	3.00	28	37
F-W-(W) (N+P)	+30	3.24	38	55
Cont (W)	-11	3.17	36	31
Cont (W) (N+P)	+33	3.46	55	42
GM-W-(W)	-26	3.37	44	26
F-W-(W)-H-H-H	-5	3.58	48	24
LSD(P<0.10)	–	0.31	8	13

[a]F = fallow, W = wheat, Cont = continuous, GM = sweetclover green manure, H = alfalfa-bromegrass hay and () denotes rotation phase sampled.
[b]Average slope of yield trends lines over time.
[c]The initial potential rate of N mineralization (i.e. $N_0 \times K$).
[d]Olsen-P.

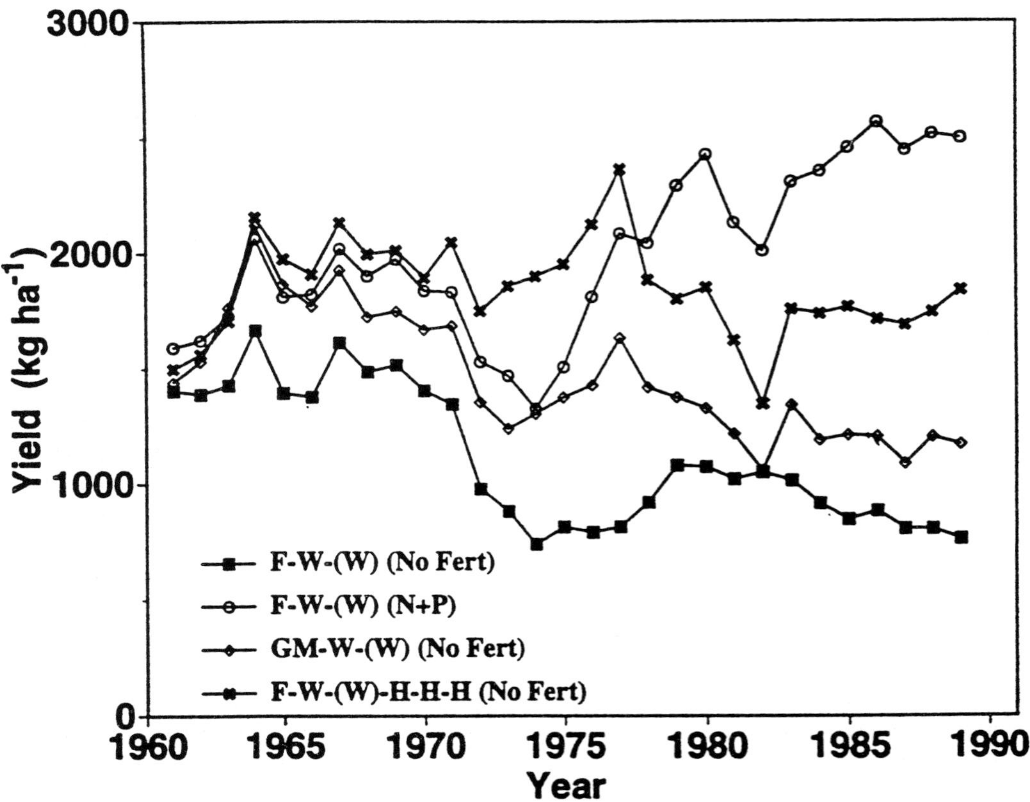

Fig. 3. Yield trends of hard red spring wheat grown on stubble in a fallow-wheat-wheat rotation on a thin Black Chernozem at Indian Head, Saskatchewan, showing the influence of sweetclover green manure, legume-grass hay crops, and of fertilizer. All points are 5-year running means [16].

Table 2. Annual net incomes[a] for rotations during two fertility periods[b] [24]

Rotation Sequence	N and P Fertilized	1960–77			1978–84		
		Mean	Min	Max	Mean	Min	Max
		(-$ ha^{-1})					
F-W	No	10	−100	122	−12	−80	103
F-W	Yes	17	−110	122	3	−80	122
F-W-W	No	−5	−128	155	−25	−115	85
F-W-W	Yes	15	−128	155	17	−73	142
GM-W-W	No	42	−128	180	22	−45	97
F-W-W-H-H-H	No	22	−110	152	32	−45	103
Cont. W	No	−57	−160	175	−99	−248	75
Cont. W	Yes	−10	−178	243	17	−240	233

[a]Base assumptions: Wheat = \$147/t; hay = \$67/t; fertilizer N = \$0.57/kg; P_2O_5 = \$0.60/t; labour = \$10/hr; interest = 11%.
[b]Fertilized according to general recommendations for the region in period 1960–77, but according to soil tests in 1978–84; the latter period required much higher rates of N.

ture wheat widened in favour of the fertilized systems while net returns decreased relative to the first period. The hay-containing system still performed well in this second period. Thus, these economic trends appeared to mimic the long-term yield trends discussed earlier (Table 1 and Fig. 3).

Where soils are very fertile, such as on the thick Black Chernozem at Melfort, Saskatchewan, the economic benefit from fertilization may be much lower than in a less fertile soil where the driving variables are equally good (e.g., Indian Head) (Fig. 4). Thus, the influence of fertilizer on net returns depends on the likelihood of gaining significant enough crop response to overcome added cost of inputs, and this will depend on the inherent fertility of the soil, the weather conditions, and the rates of fertilizer used.

Impact of N on environmental quality

Earlier we showed that adequate fertilization is required to maintain the fertility of soil, otherwise it will degrade, thereby leading to lower crop production. Even the inclusion of legume systems, though they reduce the rate of degradation, may not be sufficient to maintain crop production at a level equivalent to when proper fertilization practices are employed. Let us now consider how some soil characteristics may be influenced by fertilizers and legumes over time.

Impact on soil quality

Although prairie soils have experienced considerable degradation due to poor soil management, most soils still have adequate quantities of organic matter to sustain acceptable crop production [22]. The rate of loss of organic matter from prairie soils was rapid in the first 20 years after the grassland was broken, but then decreased and organic matter levels have begun to approach a steady state in reasonably managed soils [22, 59].

What influence have fertilizers had on the quantity of organic matter in soils? Long-term studies carried out in the various soil zones in western Canada have shown that, without fertilization, soil organic matter levels will not increase and will likely continue to decline, particularly where summerfallow occupies a significant component of the rotation [20]. However, proper fertilization can either reduce the rate of loss in organic matter or, where summerfallow is kept to a minimum, it may increase the soil organic matter content modestly (Table 1 and [20]). If the organic matter level is already high, however, it is unlikely that fertilizers will increase the level very much [14].

Like fertilizers, legumes, whether used as green manure, hay, or pulse crops, can increase the soil organic matter content if used for a long enough period (Table 1 and [20]). In fact, any management practice that results in an increase in the amount of crop residues (especially roots) returned to the land will have a positive influence on soil organic matter content, unless organic matter is already high [13,14].

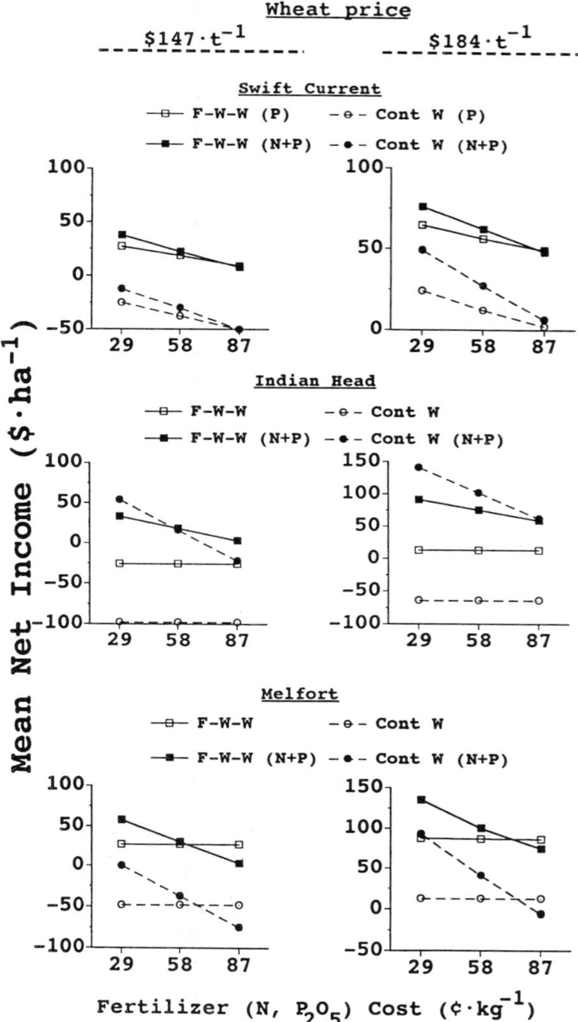

Fig. 4. Effect of wheat price and fertilizer cost on net income from fertilization for long-term crop rotations at Swift Current (Brown Chernozem), at Indian Head and Melfort (Black Chernozems), Saskatchewan. (F - fallow, Cont - continuous cropping, W - spring wheat) [84].

Because crop residue production is directly related to crop production, there will generally be a direct relationship between the maintenance of soil organic matter levels and economic performance. The export of grain or other produce from the land takes with it large amounts of N, P, K and S [11]. If these nutrients are not replaced (e.g., by additions of fertilizers) soil organic matter and the fertility of the soil will gradually decline. This has been well documented for N in a long-term rotation study at Swift Current (Fig. 5).

Table 3. Influence of grain legume on trend in fertilizer N requirements for spring wheat in two continuously cropped systems receiving N and P based on soil tests over a 12-yr period at Swift Current, Saskatchewan [26]

	N fertilizer applied to rotation phase indicated by () (kg ha⁻¹)		
Year	Cont (W)	W-(Len)[b]	Cont(W) minus W-(Len)
1979	39	50	−11
1980	61	61	0
1981	33	28	5
1982	27	27	0
1983	20	21	−1
1984	44	11	33
1985	11	5	6
1986	11	5	6
1987	41	3	38
1988	56	4	52
1989	9	2	7
1990	17	0	17

[a]Prior to commencement of this experiment this land was cropped continuously to spring wheat fertilized with N and P for 12 yr.
[b]The comparison was made between Cont W and the lentil phase of W-Len because the N requirements of the lentil was dictated by the wheat phase of the rotation.

Not only is the amount of organic matter influenced by fertilizers, legumes, or reduced inputs, so too is the quality of the soil organic matter. For example, the number of microbes and their activity in the soil are directly related to fertilizer rates and to the amount of legume inputs into soil [26,54]. Here, too, the impact of fertilizer interacts with cropping frequency (Table 1).

Because fertilizers (and legumes) increase the N-supplying power of soils, their regular use, year after year, can lead to significant improvement in the fertility of the soil. This may then reduce the fertilizer requirements of such land in future years (Table 3).

As stated earlier, fertilization will have a positive effect on crop residues [18]; in turn, this will have a direct effect on soil aggregation (Fig. 6). The latter will influence soil structure, erodibility, soil workability, and water infiltration. Thus, the adoption of proper fertilization practices is imperative for maintaining soil in good physical condition for crop production. Legumes (green manure, and particularly grass-alfalfa hay crops) are very effective in promoting good soil aggregation [18].

284

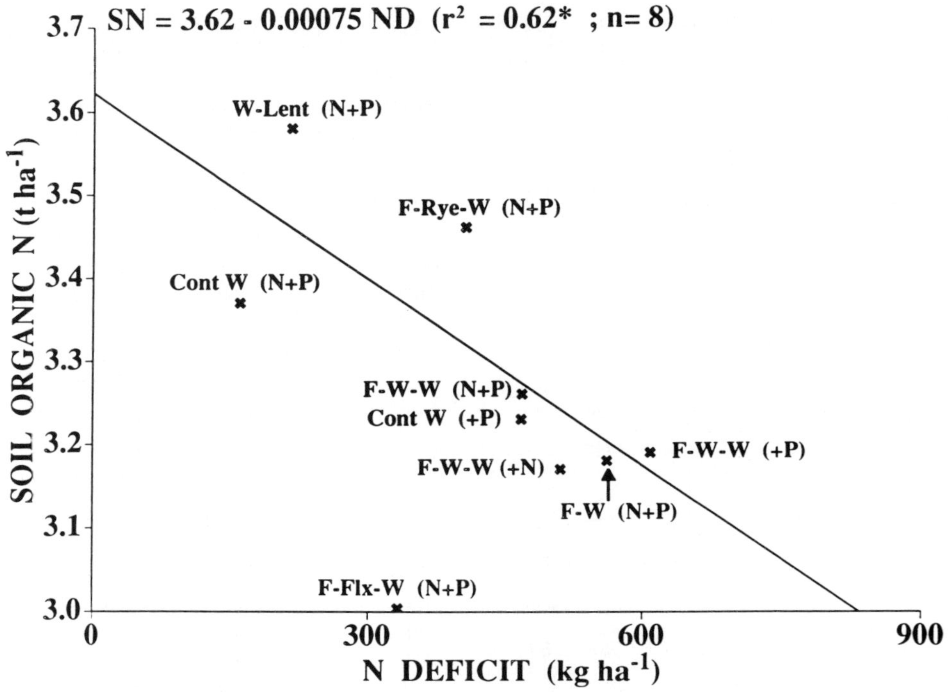

Fig. 5. Relationship between soil organic N in the surface 15 cm of soil and apparent N deficit (i.e. N exported in grain minus N applied as fertilizer) in a 25-yr crop rotation experiment at Swift Current, Saskatchewan. F = fallow; CF = chemical fallow; W - spring wheat, WW - winter wheat, Rye - fall rye, Flx - flax, Lent - grain lentil, Cont - continuous [21].

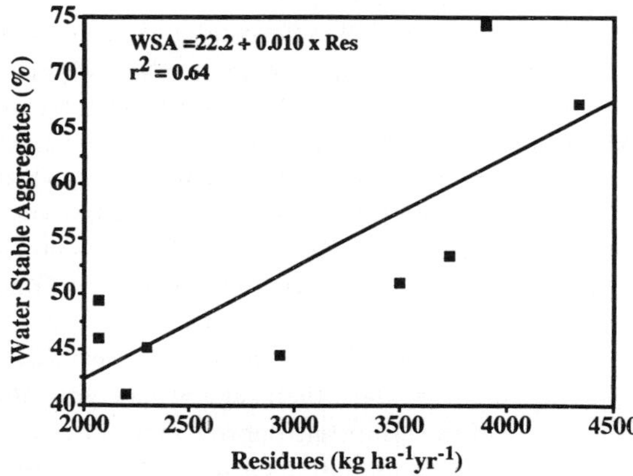

Fig. 6. Relationship between water stable aggregates and crop residues for various treatments in a long-term crop rotation experiment at Indian Head, Saskatchewan [18].

Impact on water quality

High nitrate content in drinking water can be a health hazard, causing methaemoglobinaemia. Although the climate of the Canadian prairies is semiarid, thus resulting in severe moisture deficits on a regular basis, nitrate leaching can take place, especially in the more humid regions, such as in the Black and Grey Luvisolic soil zones [17,30]. This is particularly true where summerfallow constitutes a significant component of the crop rotation, because this allows storage of water, thereby priming the soil for 'leakage'. Further, in the summerfallow period mineralization and nitrification predominates and, without much plant uptake of N, nitrate can be easily leached during early, wet springs (Fig. 7, and [6]).

If fertilizers are applied based on soil test criteria, then nitrate leaching is minimized, especially when continuous cropping is also practiced (Fig. 7) [17]. In contrast, over-fertilization, which may result from use of general fertilization criteria for a crop and soil (Fig. 8), as well as under-fertilization [25], can sometimes lead to large amounts of nitrate being leached. Nor is the use of legumes less likely to result in nitrate leaching than will mineral fertilizers (Fig. 9). This is because legumes increase the N-supplying power of soils (Table 1) and, when they are plowed down, the mineralized/nitrified N can be easily leached, especially if summerfallowing is a component of the rotation. Thus, the key to reducing nitrate leaching is to reduce

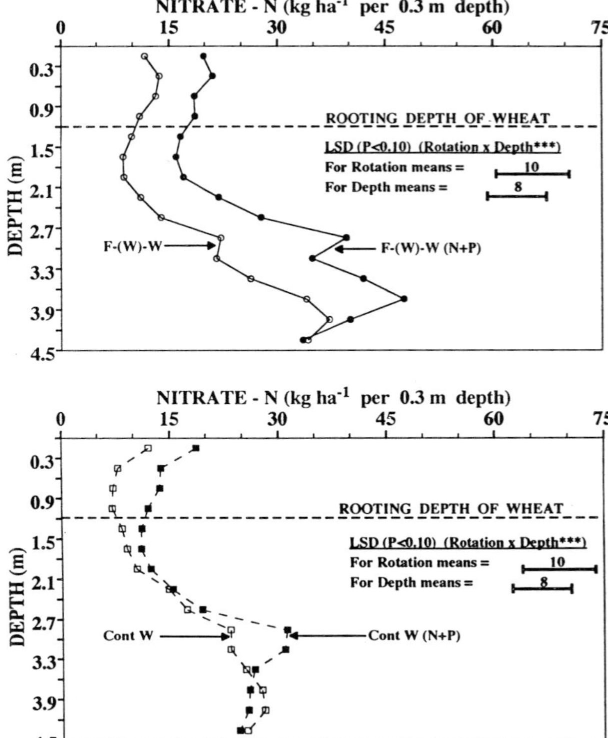

Fig. 7. Effect of fertilizer on nitrate-leaching in two monoculture wheat rotations after 34 years at Indian Head, Saskatchewan. The rotation phase in parenthesis was the one sampled [17].

fallow frequency and to apply fertilizer based on soil tests. Too much, or too little fertilizer, may lead to nitrate leaching beyond the root zone of annual crops [15,25]. Nonetheless, to date there has been no evidence to suggest that the use of fertilizers or legumes have resulted in nitrate pollution of ground waters on the Canadian prairies [47,48]. In the few cases where high concentrations of nitrates have been detected in ground water on the prairies, these were associated with farmsteads, or were located in the vicinity of towns and villages where human and animal wastes were concentrated. [30].

Impact on air quality

When we think of air quality with regard to the impact of N fertilizers, "greenhouse gases", such as CO_2 and the oxides of nitrogen, come to mind. Although the influence of N fertilizers and legumes on evolution of CO_2 from soil (C mineralization or respiration) has been much studied on disturbed soil in the laborato-

ry, few such studies have been conducted *in situ* on the Canadian prairies. Nitrogen fertilization has been shown to reduce CO_2 respiration [56,57], probably due to the negative impact of fertilizers on microbial activity resulting from the lowered pH accompanying N addition [56]. However, incubation studies with disturbed soil from long-term rotations in Saskatchewan, usually show increased CO_2 respiration from soils having a history of regular fertilizer addition (as was observed for N mineralization as well) [13].

The main N-containing greenhouse N gas is nitrous oxide (N_2O). The chief processes responsible for N_2O production in soils are denitrification and nitrification, with the former being the main mechanism involved [68]. Under anaerobic conditions, nitrate or other oxides of N serve as alternative terminal electron acceptors to O_2 for denitrifying bacteria. Thus, the presence of oxides of N is a prerequisite for N_2O production by this process. It is not surprising therefore that the application of N fertilizers can result in dramatic increases in N_2O emissions in the field [31,38]. Annual losses of N as N_2O from fertilized field soil can be as high as 40 kg ha⁻¹ [67] compared to that from unfertilized soils being < 2 kg ha⁻¹ [68]. Based on a summary of 104 field experiments conducted in temperate climates, it was reported that all but one showed evidence of fertilizer-derived N_2O emissions [38]. Further, N_2O emissions generally increased with increasing rate of N. Anhydrous ammonia appeared to cause considerably higher N_2O emissions than other common fertilizers, though the reason for this is uncertain [38]. Most of the N_2O emission occurred soon after the fertilizer was applied. Little information of this type has been published in western Canada, although overwinter loss of N from fall-applied fertilizer has been attributed to denitrification [58]. Denitrification was observed to increase in a Blaine Lake soil in Saskatchewan after fertilization, but the extent of the N emission was not specified [79].

Legumes, like N fertilizers, have been shown to contribute to increased N_2O emissions [38]. Denitrification was increased significantly by clover plowdown at two sites in Saskatchewan [2]; however, the added easily- decomposable organic matter was felt to be a factor in the enhanced denitrification.

In Saskatchewan, significant emission of N_2O was found to accompany the nitrification of ammonia fertilizers under aerobic conditions [3]. However, the amount of N_2O emitted by this mechanism is much less than that lost via denitrification. Further, several recent reports suggest that NO, not N_2O, is the dom-

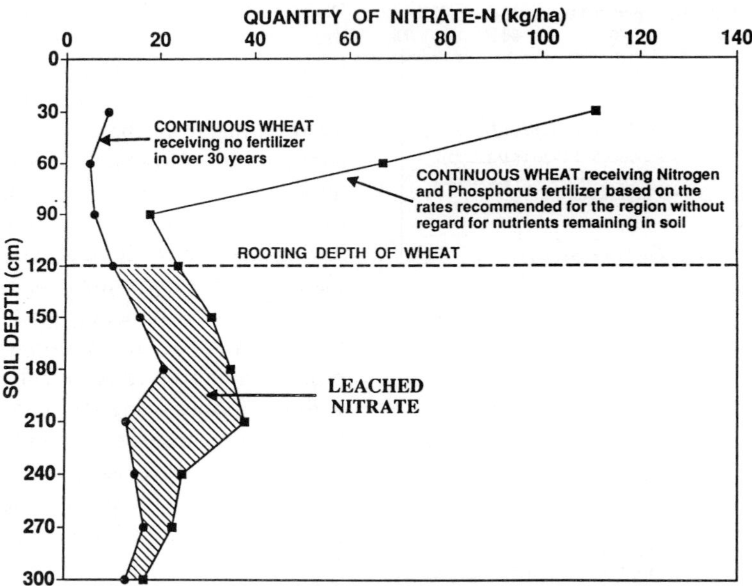

Fig. 8. Application of N fertilizer based on general recommendations for the crop and soil can lead to overfertilization resulting in excessive nitrates in surface and subsoil even in a continuous wheat system, as seen at Melfolt, Saskatchewan. (Campbell, unpubl. data from Melfort crop rotation experiment).

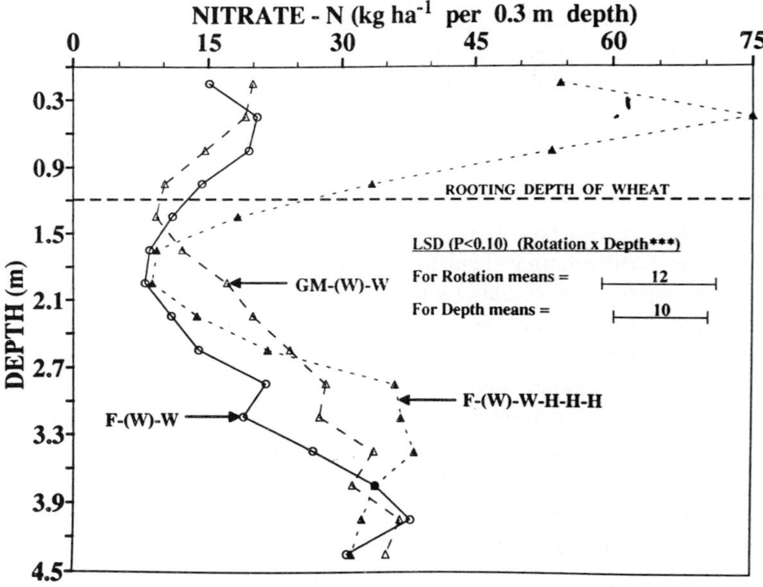

Fig. 9. Effect of legume green manure and legume-grass hay crops on nitrate-leaching in rotations after 34 years at Indian Head, Saskatchewan. The rotation phase in parenthesis was the one sampled [17].

inant N gas produced during nitrification in aerated soils [78].

Nitric oxide, which is the dominant component of the so-called NO_X, has no direct effect on the earth's radiation balance, but it is very active chemically and plays a critical role in its interaction with such oxidants as ozone [31]. Because NO_X is eventually oxidized to nitric acid it will also contribute to acid precipitation.

Nitrification is believed to be the major source of NO in soils with moisture content in the available range [49]. In such soils, NO evolution was an order of magnitude greater than N_2O evolution. Because nitrification is a common occurrence in prairie soils whenever ammonium fertilizers are applied, one can assume that most ammonia-based fertilizers do contribute to evolution of NO_X to some extent. However, the amount of N fertilizers used on the Canadian prairies is small compared to the more humid regions of Canada, the USA, or Europe; therefore, our contribution to the evolution of these noxious gases is, presumably, relatively small.

Tropics - Predominantly subsistence farming

Background

Nitrogen fertilizer use in the tropics has increased more slowly than in more developed countries in Europe and North America. Nitrogen is mainly used on lowland rice (*Oryza sativa*) and on the high value crops, such as sugarcane (*Saccharum officinalis* L.), cotton (*Gossypium hirsutum*) and irrigated wheat. Historically, fertilizer use was encouraged by artificially low cost of fertilizer, but subsidies have tended to decrease with time. Because of the prevalence of subsistence farming, economics is the main concern and environmental questions are incidental. Because of limited resources, there has been minimal long-term research on N cycling in soils of the tropics [46]. In fact, tropical agriculture has mainly relied on results of research from temperate regions. This seems reasonable since the principles involved apply in all climates, varying only in degree as dictated by climate, especially precipitation and temperature [4,5].

Sustainable agriculture production - The problem

The definition of sustainable agriculture used earlier for the developed countries is applicable to the tropics. As stated earlier, economics is the main concern and, as shown later, many farmers cannot afford to buy fertilizers, while wastes such as cow manure may be more valuable as an energy source or for industrial purposes (India) than for building soil organic matter. The second part of this paper will deal more with nitrogen use efficiency than on environmental impact of N.

Impact of N on yields and economics

Many experiments have been conducted on N management of crops in the tropics. Most have been aimed at evaluating N fertilizers and have been directed to lowland rice, the most important crop. In recent years, there has been an increase in the research effort directed at the use of green manures and other organic inputs for both lowland rice and upland crops. Generally, the responses to N have paralleled those obtained in temperate climates. Field experiments in Puerto Rico, Brazil and Ghana showed that the response of maize (*Zea mays* L.) to N was remarkably similar to that obtained on similar soils in the USA [46]. Further, apparent N recovery in the above-ground crop (measured by difference method) was almost identical for tropical and temperate locations, with mean recovery ranging from 60–120 kg ha^{-1} being 56% [46]. On the Canadian prairies mean recoveries of fertilizer N for cereals is 53% [29].

Nitrogen use efficiency

Lowland Rice
Asia uses 39% of the world's N fertilizer and about 60% of this is used on lowland rice [33]; thus, most research has been conducted on this crop. Research, together with favourable fertilizer prices has ensured widespread use of N fertilizer on lowland rice. However, whereas farmers have been quick to adopt practices aimed at optimizing yields, there is still a problem with respect to NUE by rice. Generally, less than 40% of N applied to rice is used by the crop [33]. Loss has been mainly through ammonia volatilization, although leaching and denitrification also contribute to inefficient utilization. Because of the nature of these processes, attempts to reduce losses and improve NUE have focused on water management. In the case of ammonia volatilization, loss is closely related to concentration of ammonium in the floodwater. Estimates of total losses of N as ammonia are between 36–44% [40]. This loss was reduced to 22% by use of urease inhibitors [40], but the cost is prohibitive [10]. Two approaches to reducing loss by reducing floodwater concentration of ammonia have been (a) to drain the floodwater and apply the fertilizer to the soil surface, and (b) to place the fertilizer below the soil surface. It was confirmed that losses were less when fertilizer was applied to the drained soil surface than when it was broadcast on the floodwater [32]. It was shown that uti-

lization efficiency could be raised to more than 60% by placing the fertilizer below the surface of the soil (deep point placement) [34]. Banding it was also effective, but combinations of various split applications, including that favoured by the farmers, gave the poorest efficiency. The latter observation was confirmed by others [50,72]. For example, lowland rice took up 25–34% of urea applied as split application, but 50–61% from deep placed urea supergranules [72].

Reviewing a series of experiments in Indonesia, Myers [62] concluded that there were technical solutions to increasing NUE of lowland rice. Various split applications for example, resulted in improved NUE (Table 4). Placement of fertilizer below the soil surface also consistently gave high yield and NUE. However, generally the improved technology merely improved N concentration in the grain rather than grain yield. In the absence of any price incentives for grain protein, farmers are justifiably unwilling to adopt these technologies. One byproduct of this research is a clear indication of the need for accompanying study of economic and social factors. In the Philippines, similar research led to revised recommendations on N placement which were similarly rejected by farmers.

Upland crops

Because most farmers in the tropics are subsistence farmers, many cannot afford to buy fertilizer; consequently, upland crops receive less fertilizer than lowland crops. Fertilizers are mainly used on higher value crops, such as sugarcane and cotton, or in special situations, such as on irrigated wheat in Zambia; little is used on crops such as corn. Response of these crops to N is widespread (46,55); however, because precipitation is uncertain, farmers need to be aware of the risk of receiving no response from inputs.

Upland crops in the humid tropics are grown on substantial areas where the land is either acidic or often sloping or steep. The high precipitation results in considerable loss of N by denitrification, leaching and runoff. The major crops include cassava, rice, corn and some pulse; such as peanut. Nitrogen deficiency is general, especially in the aridic soils, and deficiencies of nutrients such as P, K, Ca, Mg, B and others is common. Thus, responses to N may be limited unless these other deficiencies are also addressed. Although NUE here, as in the temperate climates, is directly related to crop production (N uptake) and thus precipitation, excessive precipitation will reduce N availability. Thus, NUE was found to be lowest where rainfall was

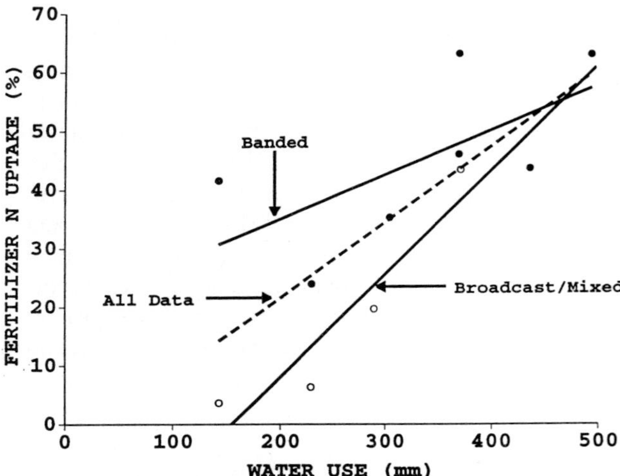

Fig. 10. Fertilizer N uptake in relation to available water and method of application (●banded, ○broadcast/mixed) (From Myers and Hibberd, unpubl. data).

highest [74]. For upland rice grown in Sumatra in the rainy season (900–1300 mm) NUE was 9–18% while for corn grown in the dry season (410–840 mm) NUE was 32–40%.

In the subhumid and semiarid tropics where there are substantial areas of Vertisols and soils are less acid, restriction to upland crop response to N is mainly due to water deficiency. Here we find interactions between water and available soil N and crop response to fertilizer N similar to that reported for temperate climates [27,63,76,77]. As soil fertility declines and available water increases, the frequency and magnitude of response of cereals to N usually increases. Because extremes of water supply are part of the normal climatic variation in subtropical and tropical regions, this demonstrates the risk of using N fertilizer. Under rainfed conditions in India, sorghum took up 29–56% of N applied as urea on a Vertisol where the seasonal rainfall was 695 mm [60]. In an Alfisol, 47–64% crop recovery of N was observed where 830 mm rain was received, compared with 54-67% when 610 mm was received [61]. With these weather-related risks with which to contend, it is not surprising that subsistence farmers hesitate to spend scarce cash on fertilizers. Myers and Hibbard (unpubl. data) found that crop recovery of applied N by sorghum on a Vertisol ranged from 0–60% depending on the amount of water available for use by the crop (Fig. 10).

Sugarcane and cotton, two major upland crops of the subhumid and semiarid tropics, are usually grown

Table 4. Percentage utilization and net uptake of fertilizer applied to flooded rice at different times and different rates determined by ^{15}N [1]

Time of application	Fertilizer applied (kg N ha^{-1})					
	25		50		75	
	Uptake (kg N ha^{-1})	Recovery (%)	Uptake (kg N ha^{-1})	Recovery (%)	Uptake (kg N ha^{-1})	Recovery (%)
Transplanting	6.2	25	13.0	26	–	–
21 d.a.t.	5.1	21	9.4	19	13.5	18
btn 21 d.a.t., PI	5.3	21	11.8	24	17.5	23
7 d.b.PI	8.5	34	–	–	–	–
PI	–	–	20.8	42	29.2	39
FL	–	–	–	–	22.7	30

d.a.t. - days after transplanting, PI - primordial initiation, d.b. - days before, FL - flag leaf emergence.

under irrigation and both crops generally make poor use of N. In the case of sugarcane, N is often lost by volatilization when urea is broadcast onto the trash on the soil surface [41]. In this environment the Australian sugarcane grower is prepared to accept these losses and apply additional N rather than undergo increased expenses in placing the N below the soil surface. However, if irrigation can be scheduled efficiently, it should be possible to increase N uptake and minimize N losses [19]. Care must be taken when using furrow irrigation to ensure that N is not pushed into the tops of ridges where it will be unavailable when the soil dries out [81]. This problem can be circumvented by placing the N deeper in the soil.

In summary, crop response to N, and N use efficiency, are very similar for similar types of crops in tropics and temperate climates [46]. Further, the same factors influence efficiency. Thus. greatest efficiency of N use is obtained when N is placed (e.g., in a band) rather than broadcast, when N is applied to the plant near to seeding time (not long before it's required), and when N availability is in synchrony with N requirements by the crop. The influence of N source on N use efficiency is less clear because workers in Puerto Rico and Brazil reported that, in contrast to expectations, maize used urea-N more efficiently than sulfur-coated urea or lime-coated ammonium nitrate in well-drained acid soils [46].

Green manures and crop residues

The use of organic amendments in tropical soils can provide several benefits, the main one being to sup-

ply nutrients. However, they can also be effective in assisting in weed control, reducing erosion, reducing soil temperature, conserving water and increasing soil aggregation and thereby improving soil structure. The major nutrient supplied is N and the effectiveness of this role is variable. In the developing countries where economic margins are so nebulous at times, producers will not apply any new technique unless they are sure it is going to lead to almost immediate direct economic benefits; that is, they will rarely institute some new management in order to reap some possible future advantage *per se.*

In Australia, rapid acceptance of trash retention for sugarcane production was due to reduction in production costs associated with reduced tillage, not due to perceived improvements in the soil organic matter or N cycling. In the 1920s and 1930s, the West Indies Sugar Co. at Frome, in Westmoreland, Jamaica, started applying organic manures such as bagasse, bagassed-under-compost, filterpress mud, and fly-pen manure (bedding from pens of cows that transport the cane), to the cane fields. Previously only N, P and K fertilizers had been applied. The organic manures (plus N, P, K as determined by tissue testing), resulted in tremendous yield increases and markedly prolonged the productive life of the ratoon cane fields (Fig. 11). At the same time, this allowed the sugar factory to recapture space which had been used to store these byproducts and it reduced the practice of contaminating rivers with toxic dunder. In Indonesia, farmer acceptance of green manuring was partly due to the perceived N inputs, but also because farmers found that seeds of the green manure crop could be used for food (R. Sutanto and

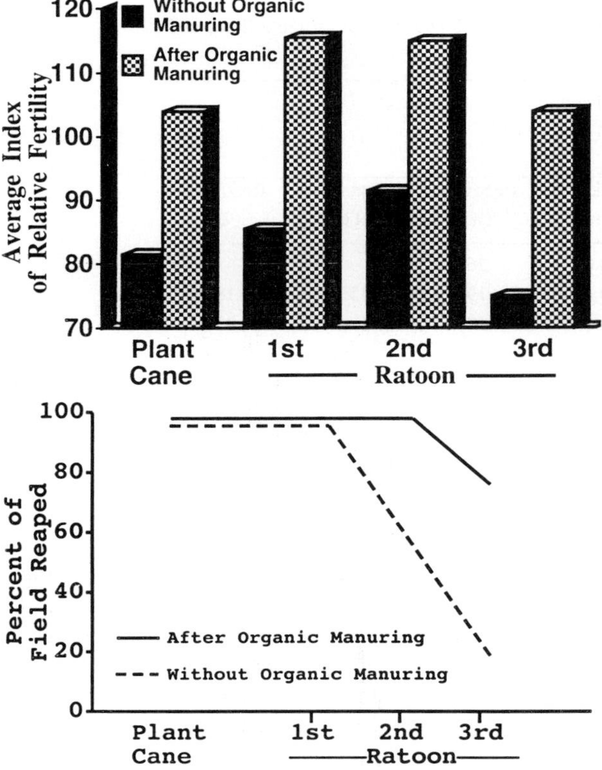

Fig. 11. Influence of organic manuring with sugarcane waste products on yields (top) and on ratooning powers (bottom) of sugarcane fields at Frome Estate, West Indies Sugar Co., Jamaica [51].

A: Supriyo, personal communication). In Vietnam and the Philippines, farmers accept leguminous hedgerows because they provide soil erosion control rather than for the N inputs that farmers cannot see. Nevertheless, there are increasing numbers of examples of substantial fertilizer substitutions provided by leguminous inputs in tropical environments.

Several workers have demonstrated that, in the humid tropics, green manures can be a very effective fertilizer substitute [36,65]. For example, 47 to 49% of the N in *Sesbania rostrata* or *Aeschynomene afraspera* was taken up by lowland rice in the wet season, compared with 23–71% uptake from urea fertilizer [36]. In the dry season, 40–42% was taken up from legume N, compared with 31–54% from urea. For upland crops grown on an Ultisol in Sumatra, N in cowpea residues was found to be more available than N applied as urea [74]. For upland rice, green manure was found to be equivalent to 66 kg N ha^{-1} of fertilizer, and residues left after harvest were equivalent to 70 kg N ha^{-1} [53]. Also, for upland crops there is

interest in the degree to which prunings from hedgerow species in alley-cropping systems can substitute for N fertilizer. Maize was found to only obtain 9% of the N contained in Leucaena prunings whereas 43% of fertilizer N was taken up [82,83]. Applying Leucaena prunings containing 300 kg N ha^{-1} was found to be equivalent to applying 64 kg N ha^{-1} of fertilizer [71]. Variable results with these materials have prompted an interest in gaining an understanding of the processes and this has led to the idea that effectiveness of organic inputs depends on weather-related factors (moisture and temperature) and quality-related factors (chemical and physical). Thus, rapid release of N from organic inputs has occurred mostly in humid conditions. Research into the importance of quality factors has centred on three components – N concentration (for C:N ratio), lignin concentration and polyphenol content.

There is concern that some high quality crop residues or green manures may release N too rapidly so that losses of N may occur. The concept of synchrony of supply of N with the crop's demand has resulted. Thus, an organic material of the right quality may release N at approximately the same rate as required by the crop (Fig. 12). If this happened, there should be less opportunity for N loss by denitrification or leaching. The principle of synchrony has been described in detail and evaluated recently [64]. Some other organic inputs which would supply some N to crops are of potential importance in the tropics. Included are the waste products from milling of palm oil,, municipal wastes, etc. So far, these products have been little researched.

Socioeconomic factors

In the tropics, most farmers either do not have the cash or the access to credit to invest in fertilizer; or, the prices for farm products are too low to make fertilizer use attractive. In some countries there are, or have been, subsidies on certain fertilizer products which have encouraged the use, and occasionally overuse, of fertilizer. As these subsidies have been removed, there has been a reduction in the profitability of fertilizing. There is a contrast between farmer attitude to N fertilizing in developing versus developed countries. For example, sugarcane growers in Thailand tend to apply less than the recommended rates of fertilizer whereas their Australian counterparts tend to apply more than recommended rates (P Pramanee, pers. commun.). This is probably due to a different risk factor – the Australian farmer is managing the risk of losing yield

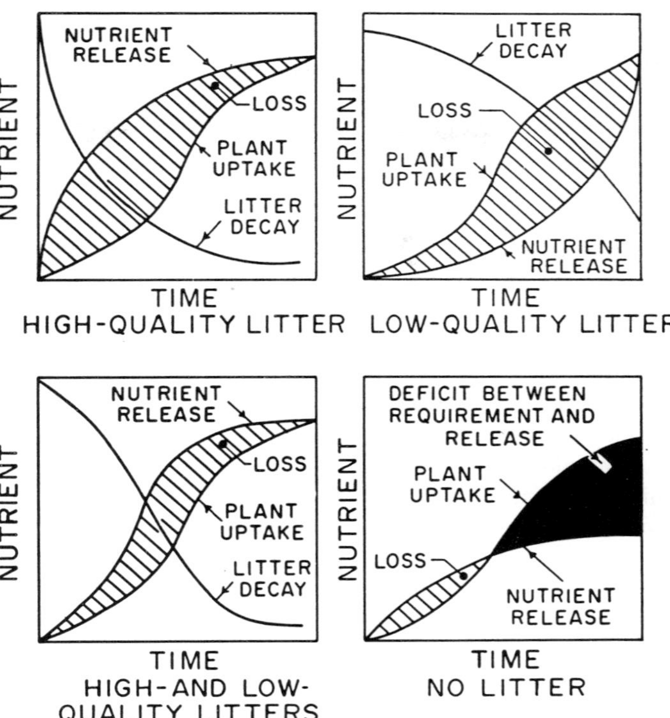

Fig. 12. Hypothetical patterns of nutrient availability in four treatments of an experiment to test the synchrony principle [64].

because of higher than usual N losses, whereas the Thai farmer is managing the risk of not getting the maximum return per unit of investment. Such attitudes create different opportunities for environmental impact. For example, the Australian approach would increase the likelihood of negative environmental impacts through N moving from the field into waters, whereas the Thai farmer's system, which is more conservative, could lead to soil degradation. There are other ways that economics may force farmers to use strategies with potentially poor environmental impact. Where deep point placement of N for lowland rice was recommended to farmers in the Philippines the recommendation was sound in agronomic and environmental terms, because the N was used more effectively by the rice than was N applied according to farmers' practice; however, farmers refused to adopt the improved practice because of the labour cost and inconvenience [42]. They preferred to add a larger amount of N and accept the poorer utilization efficiency. Fujisaka *et al.*, [42] concluded that the farmers were correct in rejecting the official recommendation. There are examples of legislation affecting nutrient management, as in the case of rubber replanting schemes in Malaysia. Here, farmers must sow legume ground covers between the rows of trees to control erosion and to supply N to the system. Unfortunately, such legislation does not ensure success of the legume ground cover, consequently there are many cases where the cost of seed and sowing is wasted because of careless management. Of course, similar situations occur in developed countries as well [12].

Impact of N on environmental quality

There are few long-term trials in the tropics (except some on lowland systems at the International Rice Research Institute) to allow systematic monitoring of the influence of N on environmental quality.

Impact on soil quality

Although organic matter and organic N have been measured on numerous occasions, there appears to be lack of agreement on the amounts of organic matter in tropical soils compared to temperate soils. For example, while some [5] suggest that organic matter is low in tropical soils, others (9) say that many

292

well-drained tropical soils have greater organic matter content than do comparable temperate soils. Sixteen randomly selected soils among four major soil orders in the USA, Brazil and Zaire showed no difference in organic matter [69,70]. In a review paper, Grave [46] reported that organic N in Oxisols and Ultisols in Brazil, Puerto Rico and Zaire ranged from 0.07 to 0.3% in the plow layer while in temperate soils the range was 0.08 to 0.4%.

Some lowland rice systems have been in existence for so long that it might be assumed that the organic matter is in equilibrium so that fertilizer either has no impact on soil quality or that it merely helps to maintain quality. In a 7-yr fertilizer study conducted in Brazil, it was reported that the concentration of soil organic nitrogen in the surface soil of unfertilized plots did not change significantly during the experiment, even though 384 kg of N was removed in above-ground dry matter [46]. Apparently, the amount of organic nitrogen mineralized was sufficiently small, relative to experimental error, that changes in soil organic N were undetectable. Annual fertilization did not influence soil organic C either. Mean concentration of soil organic N (0.124%) in fertilized treatments (sampled in 1975 and 1976) was similar to that in check plots.

As in temperate soils, moisture and temperature are the main factors influencing N mineralization. Because temperature is well above freezing all year, N mineralization can occur all year. Consequently, if water is available, the amount of N mineralized annually should be greater in a tropical soil than in a temperate soil [52]. As well, when a soil is disturbed so that the organic matter level changes towards a new steady state, this rate of change should be more rapid in the tropical system [46]. It is known that N mineralization can occur even in dry soils and nitrate will accumulate in dry seasons [73], then at the onset of rainy reasons, large flushes of N mineralization can occur [7,8]. Such flushes are much greater under conditions of alternate wetting and drying than when moisture is steady [7]. This was also observed in temperate climates [11]. Thus, minor droughts during rainy seasons will enhance mineralization. In the 7-yr study conducted in Puerto Rico, Brazil and Ghana, net N mineralization during a growing season for maize was found to be very similar to that produced in New York and Nebraska, being typically about 60 to 80 kg N ha^{-1} [46].

Ammonium-containing fertilizer is known to acidify soils. However, in flooded rice where highest N rates are used persistently, the system seems to be buffered against acid conditions, provided it is maintained in the flooded condition where N is primarily in the ammonium form. In upland systems, N fertilizer use is generally low in the tropics thus soil acidification from fertilizer should be minimal, except in certain high input systems such as sugarcane and cotton., Similarly, legumes in the tropics may not be fixing sufficient N to provide a positive N balance,, consequently their impact on soil acidification and maintenance of soil organic matter may be less dramatic than in temperate climates.

Impact on water quality

Conditions in well-drained tropical soils are conducive to N loss through leaching, which occurs most rapidly in soils of coarse, sandy texture during the rainy season [39], when rainfall often exceeds evapotranspiration. High correlations between rainfall and N movement have been reported in tropical soils of Australia [80], and in West African Ultisols [28]. Leaching losses of 65 percent of applied N have also been reported in Costa Rican alluvial soils planted to maize [43].

In the tropics, as in other environments, the key to reducing the possible negative impact of N on the environment is to synchronize N availability with crop requirements. Nitrate will move with water thereby entering ground water and streams. The potential for perturbation of watersheds from nitrate addition is particularly significant in the tropics. The N/P ratios in Central American lakes were observed to be lower than in temperate zones by a factor as great as 5 [35]. This suggests that nitrogen deficiency may be general in the tropics and that eutrophication may result from increased nitrogen addition. Thus, proper timing and level of fertilization will reduce potential environmental damage.

Because of its mode of culture, lowland rice must impact on water quality downstream. As well, water quality could be important in respect of waterways downstream of heavily fertilized sugarcane. In the tropics this could impact on reefs. However, only the more wealthy amongst the farmer community can be expected to modify their system to address this problem.

Impact on air quality

The significance of N loss by denitrification in tropical soils is not well documented. Denitrification losses from cultivated, grassland, and forest soils was studied in Ghana and N losses found to be insignificant at field capacity [44]. Significant losses occurred only in waterlogged conditions. Significant denitrification losses were reported for waterlogged surface soils in Puerto Rico, but here glucose had been added to provide adequate supply of energy [37]. Losses from samples of surface and subsurface soils at field capacity were found to be insignificant [37]. Thus, it appears that denitrification loss in humid tropical soils is not likely to be large under field conditions.

It is known that the rice system is a major producer of methane. It must also be responsible for production of some N_2O. While this must be a matter of concern, there can be no expectation that low-income farmers will modify their practices to reduce environmental problems unless there is some other rewarding reason for doing so.

Conclusions

We have clearly demonstrated that, when fertilizers are used responsibly, they will increase crop production, contribute to economic survival, and maintain or improve the physical, biochemical, and biological quality of soils in the temperate/boreal climates. Under such circumstances, fertilizers will not have significant detrimental effect on soil, water, or air quality. On the contrary, failure to apply adequate fertilizer may be uneconomical and may eventually lead to soil degradation. The inclusion of legumes, such as green manure, and hay crops, in cereal- based crop rotations will improve yields, economics, and soil quality for some time; however, because legumes do not supply P, cereal yields in such systems may gradually decline and such systems may eventually become unsustainable.

In tropical climates, most of the same principles observed for the behaviour of N in temperate climates are applicable. The main differences are that much less N fertilizer is used because most producers are subsistence farmers who can rarely afford to buy fertilizer and only lowland rice and higher value upland crops such as sugarcane, cotton and irrigated wheat are likely to receive much fertilizer. Legumes are rarely used just to provide extra N, they must provide other benefits such as erosion control. Even when used, legumes often provide less N than they remove. Despite low use of fertilizer N for upland crops in the tropics, the humid conditions, year-round high temperatures, steep terrain in many instances, and flushes in N mineralization that

occur when dry soils are wetted, will likely result in sizeable N losses through leaching, runoff and denitrification, and eventually cause soil degradation. In contrast to temperate climates, however, only limited research on these factors has been conducted because of limited resources of most tropical countries.

While governments in developed countries, located mainly in temperate and boreal climates, are able to use incentives and/or deterrents to entice farmers to adopt and implement measures that will promote sustainable crop management, in most tropical countries subsistence farming predominates and economics will be the main motivator, thus compliance regarding environmental issues will depend mainly on the incentives that are offered.

References

1. Abdullah N, Sisworo WH & Prawirasamudra A (1980) Pengaruh cara-cara pemberian pupuk nitrogen terhadap hasil dan komponen hasil serta distribusi N-pupuk di daiam tanaman padi pelita I/1. Majalah BATAN 13: 20
2. Aulakh MS, Rennie DA & Paul EA (1983) The effect of various clover management practices on gaseous N losses and mineral N accumulation. Can J Soil Sci 63: 593–605
3. Aulakh MS, Rennie DA & Paul EA (1984) Acetylene and N-serve effects upon N_2O emissions for NH_4^+ and NO_3^- treated soils under aerobic and anaerobic conditions. Soil Biol Biochem 16: 351–356
4. Bartholomew WV (1972) Soil nitrogen and organic matter. In: Drosdoff M (ed) Soils of the Humid Tropics, pp 63–81. Natl Acad Sci, Washington, DC, USA
5. Bartholomew WV (1973) Soil nitrogen in the tropics. In: Sanchez PA (ed) A Review of Soils Research in Tropical Latin America, pp 68–89. N.C. State Univ Agr Exp Sta. Tech Bull. 219
6. Bauder JW, Sinclair KN & Lund RE (1993) Physiographic and land use characteristics associated with nitrate nitrogen in Montana groundwater. J Environ Qual 22: 255–262
7. Birch HF (1958) The effect of soil drying on humus decomposition and nitrogen availability. Plant Soil 10: 9–31
8. Birch HF (1959) Further observations on humus decomposition and nitrification. Plant Soil 11: 262–286
9. Buol SW (1973) Soil genesis, morphology, and classification. In: Sanchez PA (ed) A Review of Soils Research in Tropical Latin America. pp 1–38. N. C. State Univ Agr Exp Sta Tech Bull 219
10. Buresh RJ, DeDatta SK, Padilla JL & Samson MI (1988) Effect of two urease inhibitors on floodwater ammonia following urea application to lowland rice. Soil Sci Soc Am J 52: 856–861
11. Campbell CA (1978) Soil organic matter, nitrogen and fertility. In: Schnitzer M Khan SU (eds) Soil Organic Matter Developments Soil Sci 8, pp 173–272. Elsevier Scientific Publ. Co., Amsterdam, the Netherlands
12. Campbell CA (1992) Adoption of sustainable agriculture in cereal production on the Canadian prairies. Paper presented at Organization for Economic Cooperation and Development

294

(OECD) Workshop on Sustainable Agriculture, Paris, France, Feb. 1992. 11 p (Mimeo)

13. Campbell CA, Biederbeck VO, Zentner RP & Lafond GP (1991) Effect of crop rotations and cultural practices on soil organic matter, microbial biomass and respiration in a thin Black Chernozem. Can J Soil Sci 71: 363–376

14. Campbell CA, Bowren KE, Schnitzer M, Zentner RP & Townley-Smith L (1991) Effect of crop rotations and fertilization on soil organic matter and some biochemical. properties of a thick Black Chernozem. Can J Soil Sci 71: 377–387

15. Campbell CA, DeJong R & Zentner RP (1984) Effect of cropping, summerfallow and fertilizer nitrogen on nitrate-nitrogen lost by leaching on a Brown Chernozemic loam. Can J Soil Sci 64: 61–74

16. Campbell CA, Lafond GP & Zentner RP (1993). Spring wheat yield trends as influenced by fertilizer and legumes. J Prod Agric 6: 564–568

17. Campbell CA, Lafond GP, & Zentner RP (1994) Nitrate leaching in a Udic Haploboroll as influenced by fertilizer and legumes. J Environ Qual 23: 195—201

18. Campbell CA, Moulin AP, Curtin D, Lafond GP & Townley-Smith L (1993) Soil aggregation as influenced by cultural practices in Saskatchewan. I. Black Chernozemic soils. Can J Soil Sci 73: 579–595

19. Campbell CA & Paul EA 1978 Effect of fertilizer N and soil moisture on mineralization, N recovery, A-values under spring wheat grown in small lysimeters. Can J Soil Sci 58: 39–51

20. Campbell CA, Schnitzer M, Lafond GP, Zentner RP & Knipfel JE (1991). Thirty-year crop rotations and management practices effects on soil and amino nitrogen. Soil Sci Soc Am J 55: 739–745

21. Campbell CA & Zentner RP (1993) Soil organic matter as influenced by crop rotations and fertilization. Soil Sci Soc Am J 57: 1034–1040

22. Campbell CA, Zentner RP, Dormaar JF & Voroney RP (1986) Land quality, trends and wheat production in western Canada. In: Slinkard AE & Fowler DB (eds) Wheat Production in Canada - A Review, pp 318–353. Proc. of the Canadian Wheat Production Symposium, Saskatoon, SK, Canada.

23. Campbell CA, Zentner RP, Janzen HH & Bowren KE (1990) Crop rotation studies on the Canadian prairies. Canadian Gov't. Publ. Centre, Supply Services Canada, Hull, P.Q. Publ. No. 1841/E. 133 p

24. Campbell CA, Zentner RP & Lafond GP (1991) Legumes in cereal-based rotations: Influence on profitability and soil quality. Better Crops with Plant Food. Spring, 1991. pp 28–30

25. Campbell CA, Zentner RP, Selles F & Akinremi OO (1993) Nitrate leaching as influenced by fertilization in the Brown soil zone. Can J Soil Sci 73: 387–397

26. Campbell CA, Zentner RP, Selles F, Biederbeck VO and Leyshon AJ (1992) Comparative effects of grain lentil-wheat and monoculture wheat on crop production, N economy and N fertility in a Brown Chernozem. Can J Plant Sci 72: 1091–1107

27. Campbell CA, Zentner RP, Selles F, McConkey BG & Dyck FB (1993) Nitrogen management for spring wheat grown annually on zero-tillage: yields and nitrogen use: efficiency. Agron J 85: 107–114

28. Charreau C (1972) Problems poses par l'utilization agricole des sols tropicaux par des cultures annuelles. Agron Trop (France) 27: 905–929

29. Cowell LE &Doyle PJ (1993) Nitrogen use efficiency. In: Rennie DA, Campbell CA & Roberts TL (eds) Impact of Macronutrients on Crop Responses and Environmental Sustainability on the Canadian Prairies - A Review, pp 49–109. Can Soc Soil Sci Ottawa, ON, Canada

30. Cowell LE & Doyle PJ (1993) Crop nutrients and the environment. In: Rennie DA, Campbell CA & Roberts TL (eds) Impact of Macronutrients on Crop Responses and Environmental Sustainability on the Canadian Praires - A Review, pp 474–527 Can Soc Soil Sci, Ottawa, ON, Canada

31. Curtin D, Selles F, Campbell CA & Biederbeck VO (1993) Role of agriculture as a source and sink of the greenhouse gases carbon dioxide and nitrous oxide. Final report, Greenhouse Gases Initiative Project. Research Branch, Agriculture Canada, Research Station, Swift Current, SK. Publ. No. 379M0082. 72 p

32. DeDatta SK, Fillery IRP, Obcemea WN & Evangelista RC (1987) Floodwater properties, nitrogen utilization, and nitrogen-15 balance in a calcareous lowland rice soil. Soil Sci Soc Am J 51: 1355–1362

33. DeDatta SK & Buresh RJ (1989) Integrated nitrogen management in irrigated rice. Adv Soil Sci 10: 143–169

34. DeDatta SK, Trevitt ACF, Freney JR, Obcemea WN, Real JG & Simpson JR (1989) Measuring nitrogen losses from lowland rice using bulk aerodynamic and nitrogen-15 balance methods. Soil Sci Soc Am J 53: 1275–1281

35. Deevey ES (1957) Limnologic studies in middle America with a chapter on Aztomnology, Conn Acad Arts Sci Trans 39: 213–328

36. Diekmann KH, DeDatta SK & Ottow JCG (1993) Nitrogen uptake and recovery from urea and green manure in lowland rice measured by ^{15}N and non-isotope techniques. Plant Soil 148: 91–99

37. Dubey HD & Fox RH (1974) Denitrification losses from humid tropical soils of Puerto Rico. Soil Sci Soc Am Proc 38: 917–920

38. Eichner MJ (1990) Nitrous oxide emissions from fertilized soils: Summary of available data. J Environ Qual 19: 272–280

39. Engelstad OP & Russel DA (1975) Fertilizers for use under tropic;al conditions. Adv Agron 27: 175–208

40. Fillery IRP & DeDatta SK (1989) Ammonia volatilization from nitrogen source applied to rice fields. I. Methodology, ammonia fluxes and nitrogen-15 loss. Soil Sci Soc Am J 50: 80–86

41. Freney JR, Denmead OT, Wood AW, Saffigna PG, Chapman LS, Ham GJ, Hurney AP & Stewart RL 1992 Factors controlling ammonia loss from trash covered sugarcane fields fertilized with urea. Fert Res 31: 341–349

42. Fujisaka S Guino R & Obusan L 1993 Could farmers recommending a recommendation be right? The question of nitrogen at transplanting for irrigated rice. ASIALAND workshop on interpreting data from field experiments on upland soils. IBSRAM Technical Notes No. 6 (*In press*)

43. Gamboa J, Paez G & Blasco M 1971 Unmodelo para describir los procesos de retencian y lixiviacio en los suelos. Turrialba 21: L 312–316

44. Greenland DJ (1962) Denitrification in some tropical soils. J Sci 58: 227–33

45. Greer KJ (1989) Evaluating the soil quality of long-term crop rotations at Indian Head, M.Sc. Thesis, Univ. of Saskatchewan, Saskatoon, SK.

46. Grove TL (1980) Nitrogen fertility in Oxisols and Ultisols of the humid tropics. Cornell Int Agric Bull 36: 3–27

47. Hedlin RA & Cho CM (1974) Fertilizer use and other soil management practices in relation to contamination of ground and surface water with nitrogen and phosphorus. In: Allocative Conflicts in Water Resource Management, pp 303–323. Agassiz Centre for water studies, Univ. of Manitoba, Winnipeg, Man

48. Henry JL & Meneley WA (1992) Nitrate in the Environment. In: Ed. by John L. Havlin, Kansas State Univ., Manhatan, KS pp 1–14 Great Plains Soil Fertility Conference Proc., Denver, Colorado. March 3–4, 1992. pp 1–14

49. Hutchinson GL, Guenzi WD & Livingston GP (1993) Soil water controls on aerobic soil emission of gaseous nitrogen oxides. Soil Biol Biochem 25: 1–9

50. Ingram KT, Dingkuhn M, Novero RP & Wijangco EJ (1991) Growth and CO_2 assimilation of lowland rice in response to timing and method of N fertilization, Plant Soil 132: 113–125

51. Jamaica Agricultural Society (1954) The Farmer's Guide. The Univ. Press, Glasgow. Scotland

52. Jenny H (1949) Comparative study of decomposition rates of organic matter in temperature and tropical regions. Soil Sci 68: 419–432

53. John PS, Buresh RJ, Pande RK, Prasad R & Chua TT (1992) Nitrogen-15 balances for urea and neem coated urea applied to lowland rice following two cowpea cropping; systems. Plant Soil 120: 233–241

54. Juma NG (1993) The role of fertilizer nutrients in rebuilding soil organic matter. In: Rennie DA, Campbell CA & Roberts TL (eds) Impact of Macronutrients on Crop Responses and Environmental Sustainability on the Canadian Prairies - A Review, pp 363–387. Can Soc Soil Sci, Ottawa, ON

55. Keating BA, Godwin DC & Watiki JM (1991) Optimising nitrogen inputs in response to climatic risk. In: Muchow RC & Bellamy JA (eds) Climatic Risk in Crop Production: Models and Management for the Semiarid Tropics and subtropics, pp 329–358. CAB International, Wallingford, UK

56. Kowalenko CG, Ivarson KV & Cameron DR (1978) Effect of moisture content, temperature and nitrogen fertilization on carbon dioxide evolution from field soils. Soil Biol Biochem 10: 417–423

57. Lueken H, Hutcheon WL & Paul EA (1962) The influence of nitrogen on the decomposition of crop residues in the soil. Can J Soil Sci 42: 276–288

58. Malhi SS & Nyborg M (1983) Field study of the fate of fall-applied ^{15}N-labelled fertilizers in three Alberta soils. Agron J 75: 71–74

59. McGill WB, Campbell CA, Dormaar JF, Paul EA & Anderson DW (1981) Soil organic; matter losses. In: Agriculture land: Our disappearing heritage. A. symposium, pp 72–133. Proc. Ann. Alberta Soil Sci. Workshop, Edmonton. 24–25 Feb. 1981 Alberta Soil and Feed Testing Lab., Edmonton, AB

60. Moraghan JT, Rego TJ, Buresh RJ, Vlek PLG, Burford JR, Singh S & Sahrawat KL, (1984) Labeled nitrogen fertilizer research with urea in the semi-arid tropics. II. Field studies on a Vertisol. Plant Soil 80: 21–33

61. Moraghan JT, Rego TJ & Buresh RJ (1984) Labeled nitrogen fertilizer research with urea in the semi-arid tropics. III. Field studies on Alfisol. Plant Soil 82: 193–203

62. Myers RJK (1992) Management of nitrogen in cropping systems in Indonesia - a review of studies using [15]N. Soil Manage Abstr 4: 109–137

63. Myers RJK & Hibberd DH (1986) Effect of water supply and fertilizer placement on use of nitrogenous fertilizer by grain sorghum on cracking clay soils. Proc. Aust. Sorghum Conf 5. 82–5.92

64. Myers RJK, Palm CA, Cuevas E, Gunatilleke IUN & Brossard M (1994) The synchronisation of nutrient mineralisation and plant nutrient demand. p 81–116. In: Woomer PL and Swift MJ (eds) The management of Tropical Soil Biology and Fertility, pp 81–116. John Wiley and Sons

65. Palm O, Weerakoon WL, De Silva MA & Rosswall T (1988) Nitrogen mineralization of *Sesbania sesban* used as green manure for lowland rice in Sri Lanka. Plant Soil 108: 201–209

66. Rennie DA, Campbell CA & Roberts TL (1993) Impact of macronutrients on crop responses and environmental sustainability on the Canadian prairies - a review. Can Soc Soil Sci, Ottawa, ON, Canada. 527 p

67. Ryden JC, & Lund LJ (1980) Nature and extent of directly measured denitrification losses from some irrigated vegetable crop production units. Soil Sci Soc Am J 44: 505–511

68. Sahrawat KL & Keeney DR (1986) Nitrous oxide emission from soils.
Adv Soil Sci 4: 143–148

69. Sanchez PA (1976) Properties and Management of Soils in the Tropics. John Wiley and Sons, New York

70. Sanchez PA & Buol SW (1975) Soils of the tropics and the world food crisis. Science 188: 598–603

71. Sanginga N, Mulongoy K & Ayanaba A (1988) Nitrogen contribution of *Leucaena/Rhizobium* symbiosis to soil and a subsequent maize crop. Plant Soil 112: 137–141

72. Savant NK, De Datta SK & Craswell ET (1982) Distribution patterns of ammonium nitrogen and ^{15}N uptake by rice after deep placement of urea supergranules in wetland soil. Soil Sci Soc Am J 46: 567–573

73. Semb G & Robinson JBD (1969) The natural nitrogen flush in different arable soils and climates in East Africa. East Afr Agric For J 34: 350–370

74. Sisworo WH, Mitrosuhardjo MM, Rasid H & Myers RJK (1990) The relative roles of N fixation, fertilizer, crop residues and soil in supplying N in multiple cropping systems in a humid, tropical upland cropping system. Plant Soil 121: 73–82

75. Stanford G (1973) Rationale for optimum nitrogen fertilization in corn production. J Environ Qual 2: 159–166

76. Thompson CH (1974) Fertilizing dryland grain sorghum on upland soils in the 20- to 26- inch rainfall area in Kansas. Kansas State University Agricultural Experiment Station Bulletin 579

77. Thompson CH (1975) Fertilizing dryland wheat on upland soils in the 20–26-inch rainfall area in Kansas. Kansas State University Agricultural Experiment Station Bulletin 590

78. Tortoso AC & Hutchinson GL 1990 Contributions of autotrophic and heterotopic nitrifiers to soil NO and N_2O emissions. Appl Environ Microbiol 56: 1799–1805

79. van Kessel C, Pennock DJ & Farrell RE (1992) Temporal denitrification at the landscape scale in a Black soil. In: Proc Soils and Crops Workshop, pp 280–289. Published by Univ. of Saskatchewan, Saskatoon, SK

80. Wetselaar R (1962) Nitrate distribution in tropical soils. Plant Soil 16: 19–31

81. Wright GC & Catchpoole VR (1985) Fate of urea nitrogen applied at planting to grain sorghum grown under sprinkler and furrow irrigation on a cracking clay soil. Aust Agric Res 36: 677–684

82. Xu ZH, Saffigna PG, Myers RJK & Chapman AL (1993) Nitrogen cycling in Leucaena (*Leucaena leucocephala*) alley cropping in semi-arid tropics. I. Mineralization of nitrogen from Leucaena residues. Plant Soil 148: 63–72

83. Xu ZH, Myers RJK, Saffigna PG & Chapman AL (1993) Nitrogen cycling in Leucaena (*Leucaena leucocephala*) alley cropping in semi-arid tropics. II. Response of maize growth to addition of nitrogen fertilizer and plant residues. Plant Soil 148: 73–82

84. Zentner RP, Campbell CA, Brandt SA, Bowren KE & Spratt ED (1986) Economics of crop rotations in western Canada. In: Slinkard AE & Fowler DB (eds) Wheat Production in Canada - A Review, pp 254–317. Proc Can Wheat Prod Symp 3–5 March 1986, Saskatoon, SK

85. Zentner RP, Ukrainetz H & Campbell CA (1989) The economics of fertilizing bromegrass in Saskatchewan. Can J Plant Sci 69: 841–859

Fertilizer Research **42**: 297–314, 1995.

© 1995 *Kluwer Academic Publishers. Printed in the Netherlands.*

Searching for criteria for the selection of efficient tree species for fallow improvement, with special reference to carbon and nitrogen

Götz Schroth, Dorothee Kolbe, Balle Pity[1] & Wolfgang Zech
Institute of Soil Science and Soil Geography, University of Bayreuth, D-95440 Bayreuth, Germany; [1]*Institut des Forêts, Département Foresterie, 08 BP 33 Abidjan 08, Côte d'Ivoire*

Key words: agroforestry, bioassay, fallow improvement, litterfall, nitrogen mineralization, root carbon, root nitrogen, soil respiration.

Abstract

Planted fallows are a potential means for regenerating soil fertility in the humid tropics. N-fixing leguminous trees are often recommended for this purpose, but precise criteria for efficient fallow trees do not exist. A combined field and laboratory study was undertaken, including 9 tree species of five years age, of which 6 were N-fixing, and a spontaneous fallow dominated by *Chromolaena odorata* on a *Ferralic Cambisol* in the Central Côte d'Ivoire. Although the trees differed widely in growth, litterfall and root mass, differences in total soil C and N were small. Soil respiration and N-mineralization were measured by incubating topsoil in the laboratory. The initial CO_2-flush after rewetting differed significantly between species, but non-significant differences afterwards indicated comparable stability and size of the labile C pools of the soils. Total N-mineralization differed significantly between species, with highest values for the control. There was no relationship between the ability of the plants to fix N and N-mineralization or total N-accumulation in the soils. To test the effect of the trees on a subsequent crop, rice was grown in pots filled with topsoil from all plots. Total above-ground dry matter of the rice differed significantly between species and was significantly correlated with N-mineralization in the soil and N-concentration in the rice leaves. N-mineralization was closely related to the amount of litterfall for 7 of the 9 tree species, and to root-mass for the 6 N-fixing tree species. Litterfall and root mass are identified as potential criteria for fallow tree efficiency, but the long-term effect of high N-mineralization rates on available soil-N pools needs further investigation. The importance of soil biological measurements in comparison to chemical soil tests for the assessment of fallow effects is stressed.

Introduction

Agroforestry may play an important role in tropical land use in the future, both by its productive functions and by increasing the sustainability of land use systems [37]. Agroforestry techniques can be divided into permanent tree-crop associations and improved rotational systems involving a woody component. Permanent associations may be better suited than rotational systems for providing permanent ground cover and perennial root systems, thus reducing potential erosion hazards and promoting soil biological activity [28], lowering soil organic matter decomposition and capturing leachable nutrients during phases when crop roots are absent or of low intensity. On the other hand tree-crop associations may suffer from competition between the systems' components [31, 38], and this may hinder the utilization of tree species of high productivity if these are too competitive and oblige the user to conduct rigorous management schemes, like frequent and timely hedge pruning in alley cropping, or root trenching and crown thinning in tree planting along field boundaries.

Improved fallow systems offer the advantage that highly productive tree species can be used irrespective of their competitiveness versus crops. Also, after the establishment phase, during which the trees may be associated with a crop, tree fallows do not require much further input of labour until their reconversion for cropping and may thus be more compatible with the working capacity of smallholders. In deforested areas, planted tree fallows may help to reconstitute a forest microclimate. They may contribute to the regeneration of degraded sites by soil improvement and increase their species diversity [25]; however, when replacing heterogeneous natural fallows, the regrowth of indigenous trees and shrubs will be reduced [33]. Locally, fallow improvement techniques are already practised by farmers [5, 20, 22].

While rapid growth and the production of high quality wood and other tree products are obvious economic advantages for planted fallow trees, there is little information available about their desirable characteristics from a soil fertility perspective. Nutrient and organic matter accumulation in the above ground biomass of a fallow vegetation may be of little importance for subsequent crops if the biomass is exported from the plot for use as fuelwood and fodder. With increasing population density, this will increasingly be the case. Thus, changes in soil fertility during a fallow phase and nutrient storage in the tree root systems may be processes of highest importance for fallow efficiency.

The aim of the following study was to relate characteristics of different tree species to their effects on soil fertility parameters and crop yields when planted as fallow vegetation in order to identify criteria for the selection of efficient fallow trees. Such criteria should be valuable for choosing potentially interesting tree species for screening trials. In this study, most of the emphasis has been placed on carbon and nitrogen because of their prominent role in the fertility of many tropical soils and the sustainability of management systems [39].

Study site and methods

Study site

The field work was carried out from June 1990 to September 1991 on the research station Sangoué of the "Institut des Forêts" at Oumé, Central Côte d'Ivoire (6° 17′ N, 5° 13′ W, altitude 200 m). Mean annual precipitation is 1285 mm (1956–1992) in two rainy seasons, with rainfall maxima from March to June and from September to October, and December and January as the driest months. The year 1990, during which most of the field work was done, was relatively wet with 1481 mm of rainfall. The original vegetation was a semi-deciduous rainforest of the association *Celtis-Triplochiton* [14]. The soils at the study site were *Ferralic Cambisols* according to FAO/Unesco [10], (Table 1). The high pH-values in the topsoil (Fig. 1) may be explained with repeated burning of the vegetation and the deposition of alkaline dust by the harmattan winds during the dry season. The lower part of the profile (Bsq) was hardened and could only be dug with a pickaxe. Only few roots penetrated this horizon.

In 1981, the secondary forest vegetation of the site was removed together with parts of the topsoil by a bulldozer for a planned afforestation. This plan was abandonned, and the site was invaded by a spontaneous fallow vegetation, mainly *Chromolaena odorata* (L.) King & Robinson (formerly: *Eupatorium odoratum* L.). Due to the topsoil truncation, there was little interest in using this site until May 1985, when the vegetation was cut and burned for the establishment of a productivity trial with several multipurpose tree species. The earlier removal of the topsoil was considered favourable for the present study, as it ascertained a high contribution of the trees to the soil organic matter that had formed at the site.

Tree species selection

The tree species for the productivity trial had been chosen and the seeds collected in their area of origin in Central America by the Oxford Forestry Institute [35]. The trial was part of an international study of site adaptability and productivity of mostly little known legume tree species with potential value for forestry and agroforestry. The trees were planted in the field as bagged plants in 1985 after three months in the nursery, where the seeds had been inoculated with a mixture of *Rhizobium* strains. The trial was set up as a randomized block design with four replications. Two of the blocks were incomplete due to mortality of tree seedlings before planting. Each plot consisted of $6 \times 6 = 36$ trees of the same species, planted at 2×2 m (2500 trees per ha). Only the area between the central 4×4 trees was used in this study.

The original trial consisted of 18 tree species. In addition, some indigenous leguminous tree species had been included, but either their growth and survival rate had been very low or the plots were unreplicated, thus the African species could not be included in this study. From the tree species present, nine were chosen for this study, which differed as much as possible in certain characteristics with potential relevance for C- and

Table 1. Analytic data from a soil profile on the experimental site. The soil was classified as Ferralic Cambisol [10].

Horizon depth [cm]	Ah	B1	B2	Bsq
	0–18	18–32	32–58	58–100
Munsell-colour	10YR3/3	7.5YR4/6	5YR4/6	Mottled
Stone-content (> 2 mm)	7%	31%	26%	n.d.
Sand (0.063–2 mm)	56%	59%	51%	n.d.
Silt (0.002–0.063 mm)	12%	7%	12%	n.d.
Clay (< 0.002 mm)	32%	34%	38%	n.d.
Bulk density [kg dm^{-3}]	1.38	1.71	1.57	2.09
pH (0.01 M CaCl$_2$)[a]	7.4	7.1	7.0	6.7
CEC at pH 7[cmol (+) kg^{-1}]	14.1	7.3	8.4	n.d.
CEC at pH 7 [cmol (+) kg^{-1} clay]	44.3	22.7	22.4	n.d.
$C_{org.}$ [g kg^{-1}]	21	6	5	2
$N_{tot.}$ [g kg^{-1}]	2.3	0.8	0.6	0.2
P_{Bray} [mg kg^{-1}]	3.7	1.5	1.1	0.4

n.d. = not determined
[a]see also Fig. 1.

N-cycling, like growth, soil cover by the litter and ability to fix atmospheric N (Table 2). Tree plots which seemed strongly influenced by litter or shade from neighboring plots were ignored. *Caesalpinia coriaria* formed a bush of only 2–3 m height, while the others were trees with one or several stems. A detailed description of the species including their potential uses (fodder, wood etc.) and biomass production is included in several recent publications [18, 35, 36].

In the original trial, control plots with spontaneous vegetation had not been included. Thus, one control plot was chosen at the beginning of this study in the immediate vicinity of each block of tree plots. The four control plots were placed on the four sides of the trial area. They had undergone the same land preparation as the tree plots until 1985 but had never been planted, or the trees had died during the first year (one case). The vegetation of the control plots was dominated by *Chromolaena odorata*, a partly lignified perennial of American origin, which has been introduced to the Côte d'Ivoire in 1952 [11] and is now a typical component of fallows and the understorey of plantations in the forest zone. Between 1985 and 1990, the vegetation in the control plots had either been cut at irregular intervals from about once a year (two cases) to several times a year (one case), or had not been modified at all (one case). When cutting the vegetation, the biomass had been left in the plots. When the field work was carried out, the vegetation of three control plots was about 3 m high and could only be penetrated with the help of a cutlass, and in one plot it had recently been cut and was less than 1 m high.

Stand and litter characteristics

The girth at 1.3 m of the trees was measured in August 1990 and the basal area per tree calculated. For multi-stemmed trees, all stems or branches were included in the measurement. Missing, dead or poorly developed plants of less than 1.3 m height were ignored. The litter layer was sampled at five randomly chosen points per plot, each of 20 × 20 cm. The litter was separated from adhering mineral soil by floatation in water, dried at 70 °C for 48 h and weighed. Branches were only included if < 5 mm in diameter. The litter of the *L. leucocephala* plots contained large amounts of seeds of this tree which were collected after the floatation from the sediment, dried and weighed separately. The percentage soil cover of the litter layer was estimated in increments of 10 %.

The litterfall was measured in all tree plots between 24 July 1990 and 8 February 1991, when the measurement had to be discontinued. This period included the short and most of the long dry season, but not the long rainy season. Two collectors of 1 m^2 were placed in the central part of each plot at 50 cm height. The collectors consisted of a wooden frame and a mosquito net of 1.5 mm mesh size. The litter was collected every two weeks. Branches of > 5 mm diameter and litter from neighboring plots were discarded and the resid-

Table 2. Tree species included in the study, basal area per tree at 1.3 m, ability to fix nitrogen (from [35]) and number of replications

Species	Basal area [cm^2]	N-fixation reported	Replications
Mimosaceae			
Albizia guachapele (Kunth) Dugand.	159	yes	4
Albizia niopoides[a] (Spruce ex Benth.) Burkart	58	yes	3
Leucaena collinsii[b] Britton & Rose			
ssp. zacapana C.E.Hughes	89	yes	2
Leucaena leucocephala (Lam.) de Wit			
ssp. glabrata (Rose) S.Zárate	78	yes	4
Fabaceae			
Ateleia herbert-smithii Pittier	81	yes	4
Gliricidia sepium (Jacq.) Walp.	80	yes	4
Caesalpiniaceae			
Caesalpinia coriaria (Jacq.) Willd.	39	no	4
Caesalpinia eriostachys Benth.	55	no	3
Caesalpinia velutina (Britt. & Rose) Standl.	53	no	3

[a]formerly : *A. caribaea* (Urb.) Britton & Rose.
[b]Previously thought to be a ssp. of *L. diversifolia.*

ual litter was dried at 70 °C for 48 h and weighed. For the three species with very fine leaflets, *C. coriaria*, *A. niopoides* and *L. collinsii*, some of the litter was washed through the mosquito net during rainfalls. Correction factors were obtained by placing buckets covered with identical mosquito net in the plots and dividing the dry weight of the litter in the buckets by that of the litter on the mosquito net. From the results of 25 such measurements each of 14 days duration, an average correction factor of 1.31 was calculated for the three species and applied to the litterfall data. The litterfall in the control plots was not measured as collectors could not be installed without disturbing the vegetation.

Root measurements and soil sampling

The root mass in the plots was quantified blockwise between 27 July and 12 September 1990. Soil cores of 8 cm diameter were taken from 0–10, 10–20, 20–40 and 40–60 cm depth at five randomly chosen points in each plot. The samples of the same depth in a plot were combined and homogenized. A subsample of 500–700 g was taken for chemical analyses and quantification of stone percentage (> 2 mm). A second subsample of 400–500 g (300–350 g for the depth 0–10 cm) was tak-

en for root extraction, and two subsamples of 200–300 g for measuring soil moisture by drying at 105 °C until constant weight. The total sample was also weighed. The volume of the subsample for root extraction was calculated from its weight and the weight and volume of the total sample [30]. The soil bulk density was calculated from weight and volume of the total sample and its water content. The soil samples for root extraction were watered and stored in the refrigerator for 1–2 days. The roots were extracted by washing with tap water above a 0.5 mm sieve and hand sorting under 10 × magnification, if necessary. Live and dead roots were not separated. Root diameter classes 0–1 mm and 1–5 mm were collected separately. Diameter classes > 5 mm gave very erratic results and were thus not considered. The root samples were dried at 70 °C for 48 h. After weighing with 0.1 mg precision they were ground and their C and N contents measured gas-chromatographically with a CN-analyzer. Root weight was converted to root-C to avoid systematic errors due to differences in adhering mineral soil between roots of varying diameter, vigour and surface properties.

Soil analyses

After air-drying and sieving to 2 mm, the following soil analyses were conducted on soil samples from the four depths of all plots and the soil pit: pH by glass-electrode in a 1:2.5 suspension of soil in 0.01 M $CaCl_2$, total C by dry combustion, and total N gaschromatographically with a CN-analyzer. Plant-available P was extracted from topsoil samples of all plots and all depths of the soil pit with 0.03 N NH_4F and 0.025 N HCl (Bray 1) [24] and measured photometrically with the molybdenum-blue method. The following analyses were only conducted on the samples from the soil pit: Cation exchange capacity by washing the soil with 1 M NH_4OAc at pH 7 followed by ethanol leaching and exchanging the ammonium with 1 M KCl [10], measuring the NH_4-N by redox titration; particle-size distribution with the pipet method after removing soil organic matter with H_2O_2 and dispersion of the soil in sodium hexametaphosphate (50 g l^{-1}) an sodium carbonate (7 g l^{-1}) [12]. C and N contents per soil layer were calculated from their concentration in the fine earth, stone content and bulk density. It was assumed that the density of the fraction > 2 mm, which consisted of iron-cemented concretions, was the same as that of the heavily iron and silica cemented subsoil horizon (Bsq, Table 1).

Soil respiration and N mineralization

Soil samples were again taken from 0–10 cm depth from all plots between 16 and 19 September 1990. To minimize microbiological processes during the drying, the samples were first kept in a drying-oven at 38 °C for 1 day, followed by air-drying for 2 days. The samples were then sieved to 2 mm. From every plot, 80 g samples of air-dry soil were weighed into four 1 l glass-containers and moistened to 60% water holding capacity with distilled water. The soil was incubated at 23 ± 1 °C in the dark. CO_2 forming in the containers was collected in a cup with 0.1 N NaOH which was placed on a wire support in the containers [4]. The CO_2 was quantified by two-point titration with 0.05 N HCl to pH 8.4 and pH 4.3 ([19], modified). Soil respiration was measured on day 1, 5, 19, 33 to 35, 54 to 60 and 75 to 81 after the addition of water to the soil. When no measurement was conducted, the lids of the containers were left slightly open to avoid depletion of oxygen. Evaporated water was replaced every 1–2 weeks as necessary.

N-mineralization was measured in the same soil samples. Mineral N was extracted from the air-dry soil at the beginning of the trial and from samples taken from the containers on day 15, 50 and 86 after the addition of water. The extraction was made by end-over-end shaking with 2 N KCl at a 1:5 ratio for 1 h followed by filtration and acidifying the extracts with concentrated H_2SO_4 (4 ml l^{-1}) for conservation. Ammonium and nitrate were measured photometrically with a rapidly flow analyzer. The water content of the soils as obtained on subsamples by drying at 105 °C to constant weight was used to convert the results to soil dry weight. At all three extractions, more than 95% of the N_{min} was present in the samples as nitrate, indicating a sufficient oxygen availability in the incubated soils.

Air-drying and rewetting as well as mixing of soil are known to increase the N-mineralization rates in soils [29]. However, we estimate that these treatments of the soil samples reflect reasonably the processes at the onset of the rains after the dry season and the beginning of a cropping cycle following a fallow period.

Bioassay of soil fertility

Because of other experiments in the tree plots, it was not possible to remove the vegetation and quantify the fallow effects by growing a test crop in the plots at the end of this study. Instead, a pot experiment was conducted. Soil was collected from 10 points per plot from 0–10 cm depth after removing the litter layer. The soil was homogenized and larger roots (> 5 mm diameter) were removed. With the soil from every sampling point, a plastic bag of 26 cm diameter, 30 cm height and a volume of approximately 15 l was filled, giving 10 bags per plot. The bags were placed in an open area in a randomized block design. Rice (*Oryza sativa* L., variety IAC 165) was used as test crop, as it seemed to be the only non-leguminous crop of local importance which could be conveniently grown in the relatively small containers. Fifteen rice grains were sown in each bag. After 11 days, the germinated plants were counted and their number reduced to 5 seedlings per bag. The bags were regularly watered with the same amount of water to avoid any drought stress and weeded. Several insecticide treatments were necessary against crickets. At mid-tillering, leaf samples were taken for foliar analysis. From every bag, the "y-leaf" of a representative plant was sampled and the leaves from all 10 bags from the same field plot were combined. The samples were dried at 70 °C for 48 h. In the ground samples, total N was measured gas-chromatographically with a CN-analyzer. For the other elements, subsamples were digested under pressure in concentrated HNO_3 for 14 h at 170 °C [15]. P was measured photometrically with the molybdenum-blue method, and the nutrient cations were measured by atomic absorption spectrometry. At rice harvest, only bags containing all 5 plants were considered. The plants were cut at 3 cm above the soil. Grain and straw weight were measured after drying at 105°C to constant weight.

Statistical analysis

The results were analyzed by ANOVA followed by F-test. Replicate measurements from the same plots were averaged before the analysis. The values for the missing plots were estimated after Steel and Torrie [34] and the sums of squares corrected according to Yates [40]. In case of significance of the F-test at $p < 5\%$, treatment means were compared by Duncan's multiple range test at the same level of probability. Spearman correlations between different species means were calculated with the program CSS/Statistica. In most cases, correlations were computed between treatment means in order to identify species effects. If values of individual plots were used, this is explicitly stated in the text.

Results and discussion

Litter layer and litterfall

The dry weight of the litter layer varied by a factor of 4.9 between species (Table 3). The highest values for *L. leucocephala* were due to on average 38.4% of seeds in the litter layer. After subtracting these, *L. leucocephala* ranked behind *A. guachapele* and *C. velutina*. The structure of the litter layer differed markedly between species, with the small leaflets of *L. collinsii*, *A. niopoides* and *C. eriostachys* forming a more compact layer than the larger leaflets of *A. herbert-smithii* and *G. sepium*. Three species covered the soil incompletely with their litter, which would be a disadvantage for soil protection.

The litterfall during $6\frac{1}{2}$ months varied by a factor of 4.5 between the highest (*L. leucocephala*) and the lowest values (*C. coriaria*, Table 3). It was at the lower end of yearly litterfall data reported by Szott *et al.* [37] and Bernhard-Reversat [6], possibly because significant litterfall from trees can also occur during

the wet season [17, 21] which was not included in our measurements.

The quotient of the dry weights of litterfall and litter layer (O_L in Table 3) gives a relative estimate of the rate of litter decomposition, although litterfall was not measured for the whole year, and the single quantification of the litter layer cannot be assumed to reflect a steady state situation. *G. sepium* and *C. eriostachys* seemed to have the fastest litter decomposition, thus explaining the incomplete soil cover by the litter, whereas *C. velutina* litter was the most recalcitrant. The similarity of the quotients for the other six species was surprising, as they included species like *C. coriaria* below which the soil was almost covered with worm casts indicating the highest earthworm activity from all species, and *A. niopoides* with an abrupt boundary between litter layer and mineral soil, indicating a biologically rather inactive litter. The quotients of the three *Caesalpinia* species gave no indication that their inability to fix N had an influence on the rate of litter decomposition.

Root systems

Root nodules were found on the roots of all tree species except the three *Caesalpinia* species, in agreement with previous reports [35]. Some nodules were also recorded in plots of *C. coriaria* and in control plots, apparently from undergrowth legumes. Most of the nodules in all plots seemed to be inactive and were often partly decomposed, possibly because the weather was dry during July and the second half of August, when part of the root samples were taken.

The total amount of root-C in 0–60 cm differed significantly between species (Table 4). Assuming an average C concentration in the roots of 45 %, the dry weight of roots < 5 mm ranged from 2.3 Mg ha^{-1} 60 cm^{-1} for *G. sepium* to 7.3 Mg ha^{-1} 60 cm^{-1} for *C. velutina*. The C accumulation in roots < 1 mm (very fine roots = VFR) was lowest in the control plots and highest under *C. eriostachys*. On the average, 67% of the roots < 5 mm were VFR. The root masses in this study were higher than most values cited by Szott *et al.* [37] from agroforestry systems with perennial crops and from annual cropping systems, but lower than most values from natural forests and savannas in the humid tropics.

The root-N contents of the top 60 cm of soil were lowest for the control and highest for *A. niopoides* (Table 4). The latter species had the lowest C/N-ratio in the VFR of all species (Table 4), combined with an intermediate root mass. The lowest N-concentrations

Table 3. Litter layer, litterfall and quotient of the dry weights of litterfall and litter layer (Q_L) by treatment means. Values with the same letter are not different at $p < 0.05$ (Duncan's multiple range test)

Species	Litter layer		Litterfall	Q_L
	Dry matter [kg ha^{-1}]	Soil cover [%]	Dry matter [kg ha^{-1}]	
A. guachapele	5350b	100	5269a	0.98
A. niopoides	3557c	100	2888de	0.81
L. collinsii	3902bc	100	3940bc	1.01
L. leucocephala	6621a	100	5911a	0.89
A. herbert-smithii	2805cd	100	2120f	0.76
G. sepium	1699d	50	3458cd	2.04
C. coriaria	1348d	30	1301g	0.97
C. eriostachys	2208d	80	4243b	1.92
C. velutina	4917b	100	2257ef	0.46
F-test	***		**	

$^{**} = p < 0.01$.
$^{***} = p < 0.001$.

were found in the coarse roots of the four non-N-fixing species (control and *Caesalpinia spp.*). On the average of all species, 48.6 % of the root-C and 52.9 % of the root-N in 0–60 cm were concentrated in the top 10 cm of soil.

Chemical soil properties

The pH-values in the plots (Fig. 1) give evidence of a considerable heterogeneity of soil conditions at the site. The differences between species were only significant in 10–20 cm depth ($p < 0.05$). The two non-N-fixing tree species *C. velutina* and *C. eriostachys* had particularly high pH-values in the subsoil, whereas the two well-nodulated *Leucaena* species and the control plots had relatively low subsoil pH-values. One reason for this was probably higher cation uptake by the *Leucaena* stands, but the risk of increased nitrate and cation leaching which may be associated with N-fixation also needs consideration when using N-fixing fallow species [33]. The *Chromolaena* plots combined particularly high N-mineralization rates (see below) with a subsoil (40–60 cm) root mass of only 28 kg ha^{-1} 10 cm^{-1} of VFR-C in comparison 61 kg ha^{-1} 10 cm^{-1} (*G. sepium*) and 159 kg ha^{-1} 10 cm^{-1} (*L. collinsii*) for the tree plots. Both factors are also known to favour cation leaching and thus soil acidification [26].

The C and N stocks of the soil in 0–60 cm depth (Table 5) were highest for *A. guachapele* and lowest for

the control, but the differences were not significant. On the average, the subsoil (10–60 cm) contributed 59% to the C and 60 % to the N in 0–60 cm. *A. guachapele* had the highest N-contents in the subsoil (5.68 Mg ha^{-1} in 10–60 cm). In 10–20 cm, significant differences were found to all other species except *C. eriostachys* and *A. niopoides*, and in 20–40 cm to all except *A. niopoides*. The control plots showed the lowest N-accumulation in the subsoil with 3.13 Mg ha^{-1} in 10–60 cm.

Soil-N was neither significantly correlated with root-N, nor with the ability of a species to fix atmospheric N. Both, the highest (*A. guachapele*) and the lowest (*L. collinsii*) soil-N reserves from the nine tree species were found under N-fixing species, whereas the *Caesalpinia* species occupied intermediate positions. This does not exclude important differences in N-accumulation in the biomass, but rather shows that the fixed N was either not released to the soil or rapidly lost by leaching beyond 60 cm soil depth.

The concentrations of available phosphorus in the topsoil increased with soil C contents ($r^2 = 0.404^{***}$, n = 35). With 1.9 to 4.6 mg kg^{-1} they fell into the very low to low range [24]. There were no significant differences between species.

In view of the pronounced differences between the fallow species in growth, litter production and root mass and the partial removal of the old topsoil before the beginning of the trial, the differences which were

Table 4. Carbon and nitrogen accumulation in root systems in 0–60 cm soil depth by treatment means, and C/N ratio in two size classes of roots. Values with the same letter are not different at $p < 0.05$ (Duncan's multiple range test)

Species	C in roots		N in roots		C/N	
	0–1mm	0–5mm	0–1 mm	0–5 mm	0–1 mm	1–5mm
	[kg ha^{-1}]	[kg ha^{-1}]	[kg ha^{-1}]	[kg ha^{-1}]		
A. guachepele	1115cd	1588cd	66.8bc	83.8bc	16.7ef	26.9c
A. niopoides	1443bc	2122abcd	107.0a	133.0a	13.6f	24.3c
L. collinsii	2029ab	2730abc	82.0ab	100.9ab	24.8c	36.3bc
L. leucocephala	1282c	1877bcd	65.7bcd	84.2bc	19.6cde	32.4c
A.h.-smithii	761d	1297cd	43.0cde	68.3bc	17.7def	20.9c
G. sepium	731d	1049d	39.7de	59.5bc	18.2def	16.6c
C. coriaria	2013a	2832ab	50.8bcde	66.5bc	39.7a	54.1a
C. eriostachys	2144a	2746ab	64.0bcd	75.8bc	33.4b	53.1ab
C. velutina	1886ab	3304a	63.3bcd	86.1abc	30.1b	63.0a
Control	701d	1202d	32.5e	41.5c	21.6c	60.1a
F-test	**	**	**	*	**	**

$* = p < 0.05.$
$** = p < 0.01.$

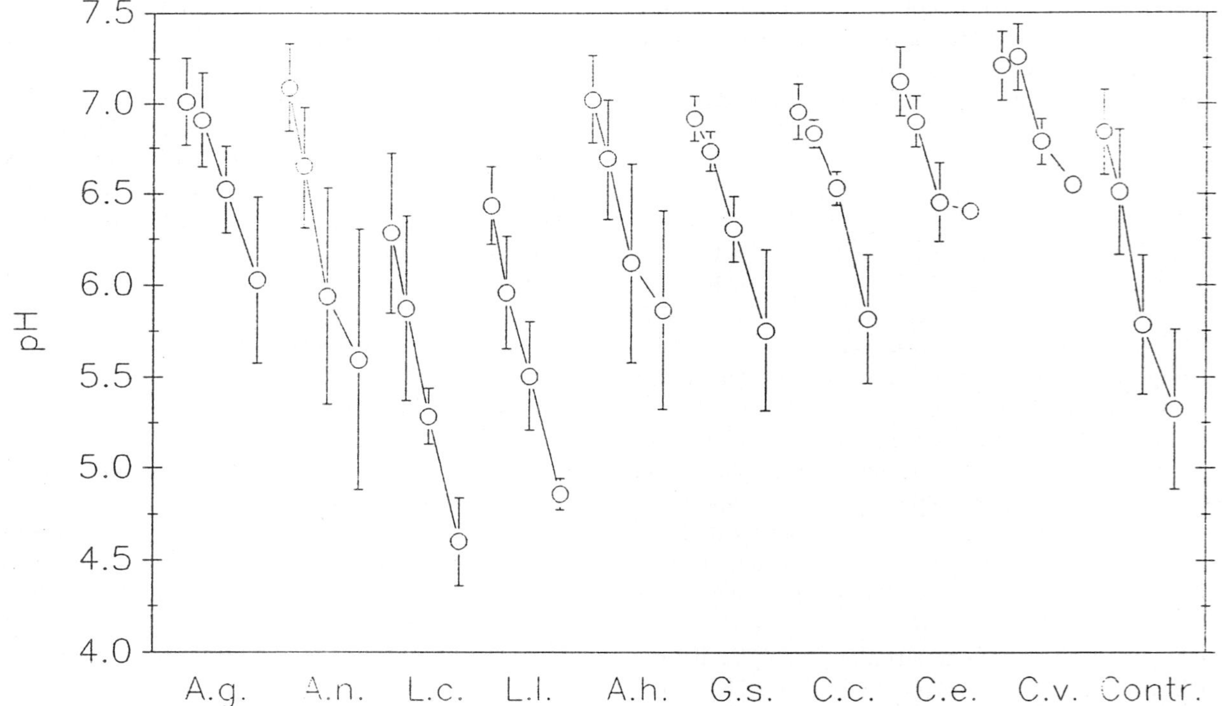

Fig. 1. pH in 0.01 M CaCl$_2$ in (from left to right) 0–10, 10–20, 20–40 and 40–60 cm soil depth (means and S.E.).

detected between the species by routine chemical soil analyses were thus surprisingly small.

Soil respiration

The CO$_2$-development from the incubated soils during $2\frac{1}{2}$ months showed a mineralization flush after rewet-

Table 5. Soil carbon and nitrogen accumulation in 0–10 cm and 0–60 cm soil depth by treatment means

Species	Soil-C		Soil-N	
	0–10 cm [Mg ha^{-1}]	0–60 cm [Mg ha^{-1}]	0–10 cm [Mg ha^{-1}]	0–60 cm [Mg ha^{-1}]
A. guachepele	26.2	69.1	2.93	8.61
A. niopoides	25.9	65.7	2.92	7.90
L. collinsii	21.4	50.7	2.40	5.62
L. leucocephala	22.2	56.1	2.45	6.43
A.h.-smithii	23.3	57.1	2.55	6.56
G. sepium	22.8	59.0	2.64	6.84
C. coriaria	24.5	57.6	2.86	6.90
C. eriostachys	28.3	60.5	3.01	7.10
C. velutina	23.9	57.5	2.55	6.35
Control	21.5	47.5	2.35	5.48
F-test	n.s.	n.s.	n.s.	n.s.

n.s. = not significant.

ting and much lower and continuously decreasing rates in all samples thereafter (Fig. 2). The time curves could very accurately be approximated with functions that assume the existence of two substrate fractions of differing stability:

$$y = a_1 * \exp(-a_2 * t) + a_3 * \exp(-a_4 * t), \quad (1)$$

with y = CO_2-development per day, t = time in days, a_1 to a_4 = variables describing the contribution of the two fractions to the total CO_2-development and its decrease during the incubation. The differences between the approximated parameters of the ten species were not significant.

While the initial mineralization flush after drying and rewetting reflects the decomposition of microbial biomass and very labile soil-C [27, 32], later soil respiration indicates the size and stability of a still relatively labile fraction of soil organic matter against microbial oxidation [3]. A significant species effect was only recorded on the first day of the incubation, with higher mean values for the control than for the tree plots (Table 6). Apparently, *Chromolaena* maintained more microbial biomass or other very labile C compounds in the soil than the trees. Soil respiration during the first day, but not thereafter, was significantly (R = 0.833 *) correlated with litterfall for all tree species except *C. coriaria*. This may indicate that litter was the origin of a highly labile C-fraction, in agreement with previous work [6]. Root-C in the topsoil was not related to soil respiration during any of the two phases.

An initial hypothesis of this study, that the stability of the accumulated soil organic matter might differ between fallow species and that these differences in stability could be measured as soil respiration, could thus not be verified. Neither soil carbon accumulation nor soil respiration after the initial mineralization flush reflected the marked differences in carbon inputs through litterfall and presumably root turnover between species. The most sensitive indicator of species differences was the size of the CO_2-flush of the first day. This has also been observed by Powlson and Jenkinson [27] after $CHCl_3$-fumigation of soils with differing management history.

N mineralization

N-mineralization was also two-phasic (Fig. 3). In most samples more than half of the N_{min} that had formed during the incubation was already mineralized within the first two weeks after rewetting. The highest mineralization-flush was observed in soils from *A. guachapele* and the control. However, the differences were only significant at $p < 0.1$ (Table 6). The N_{min}-flush was closely correlated with the respiration rate of the first day (R = 0.867**) and was probably of equally short duration. The initial N flush also increased significantly with the N-content of the topsoils for the nine tree species (R = 0.767*), but not the control which had the lowest soil-N content of all species, but the second highest N-mineralization.

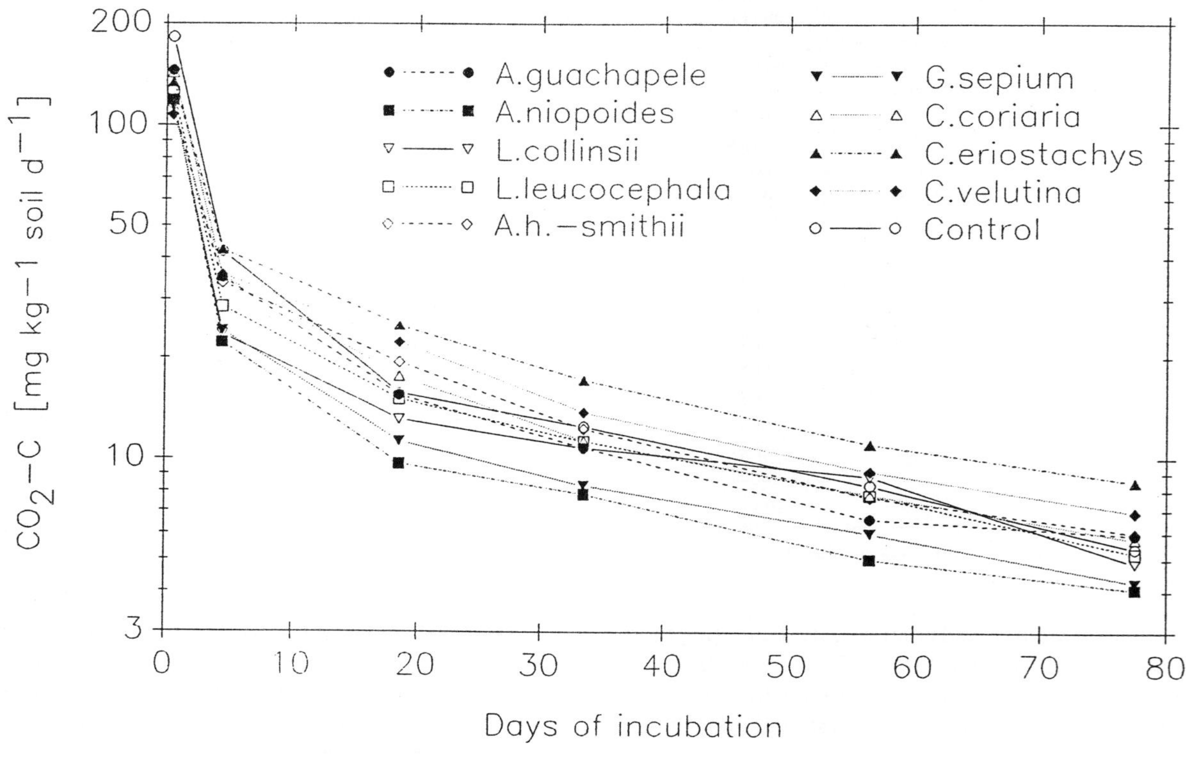

Fig. 2. Time course of daily CO_2-evolution from incubated soils (logarithmic scale).

Table 6. Soil respiration and nitrogen mineralization rates as measured in the incubation trial. Data with the same letter are not significantly different at $p < 0.05$ (Duncan's multiple range test)

Species	Soil-resp. at day 1 [mg CO_2-C/ g soil-C]	N-mineraliz. from day 1 to 15 [mg kg^{-1} d^{-1}]	N-mineraliz. from day 16 to 86 [mg kg^{-1} d^{-1}]	C/N-mineraliz. ratio from day 16 to 78
A. guachapele	6.22ab	4.58	0.618bc	15.8
A. niopoides	5.32bc	3.74	0.539bc	12.1
L. collinsii	5.20bc	2.99	0.869a	10.6
L. leucocephala	6.26ab	3.48	0.715ab	13.5
A. herbert-smithii	4.96c	3.39	0.507c	21.4
G. sepium	5.21bc	3.36	0.572bc	12.6
C. coriaria	5.43bc	3.86	0.597bc	18.5
C. eriostachys	4.43c	3.65	0.575bc	28.6
C. velutina	4.86c	3.04	0.537bc	23.1
Control	7.03a	4.37	0.841a	12.2
F-test	**	n.s.	**	n.s.

** $= p < 0.01$.
n.s. = not significant.

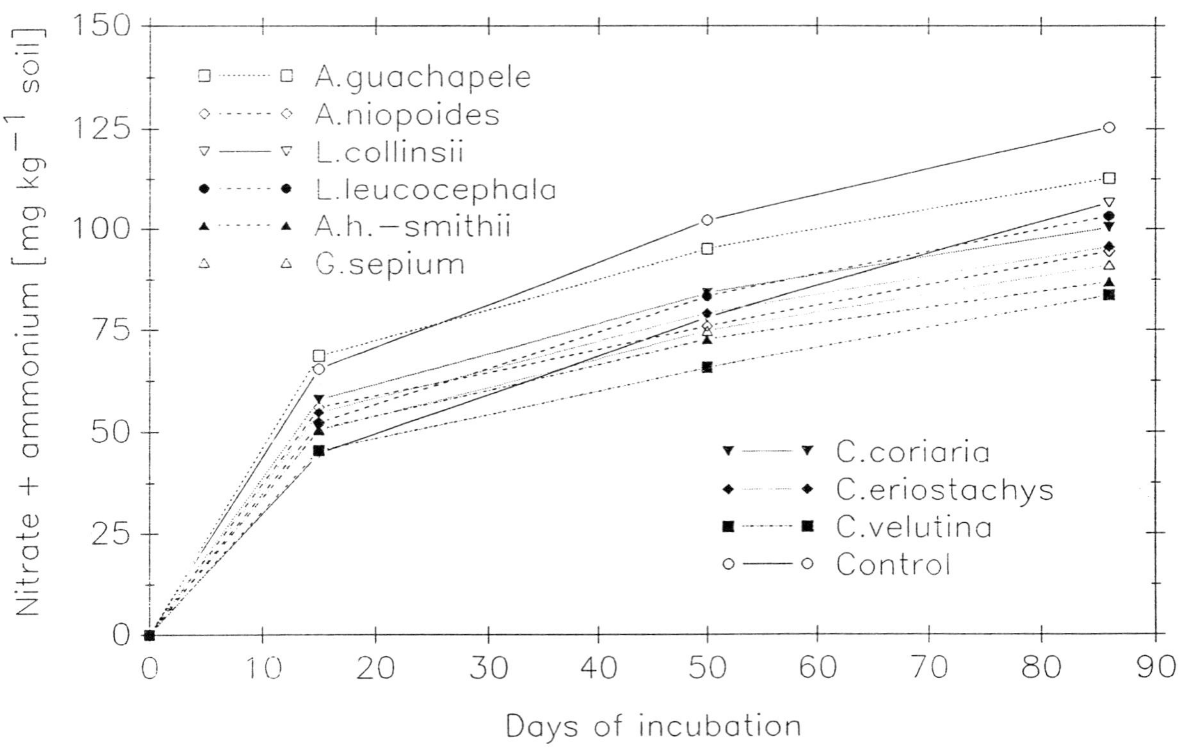

Fig. 3. Mineralization of soil N, cumulative values of NH_4-N + NO_3-N, in incubated soils.

During the subsequent quasi-linear phase, the mean daily N-mineralization differed significantly between species (Table 6). This shows that despite statistically similar values for total N in the topsoils, the stability of the accumulated N against microbial attack differed between species. Highest mineralization rates were found for *L. collinsii* and the control, followed by *L. leucocephala*. The other species differed little. Under field conditions, with a bulk density of 1.25 kg dm^{-3} stone-free soil, the measured values would have corresponded to a daily N-mineralization between 1.1 kg ha^{-1} 10 cm^{-1} for *L. collinsii* and 0.63 kg ha^{-1} 10 cm^{-1} for *A. herbert-smithii*. N-mineralization during this phase was not significantly correlated with the N-content of the topsoils. Also, there was no significant correlation between C- and N-mineralization in the soil from the different species during this phase, suggesting that N was more concentrated in the mineralizable soil organic matter of some species than of others, resulting in differences in net N-mineralization at equal C-mineralization rates.

As a consequence of high mineralization rates in both phases, the concentration of mineral N was high-est in the soil from the control plot at the end of the incubation period (Fig. 3). Here, an equivalent to 156 kg N ha^{-1} 10 cm^{-1} had been mineralized. Second was *A. guachapele*, due to a high mineralization flush, but a lower rate than the control in the second phase. *L. collinsii* had the highest mineralization rate in the second phase after a low mineralization flush. With low rates in both phases, *C. velutina* had the lowest accumulation of mineral N after 86 days, with an equivalent to 105 kg N ha^{-1} 10 cm^{-1}.

N-mineralization increased with litterfall for seven of the nine tree species (Fig. 4), either because of increasing N inputs into the soil, or increasing C availability for N-mineralization, or both. For the six N-fixing tree species, N-mineralization also increased significantly with root mass (Fig. 5). The soil from the three *Caesalpinia* species showed a lower N-mineralization, and the soil from the control plots showed a higher N-mineralization than would have been expected from this relationship. The relationship between N-mineralization and root-C for the N-fixing species can only partly be explained with concomitantly higher contents of root-N in the soils, as root-

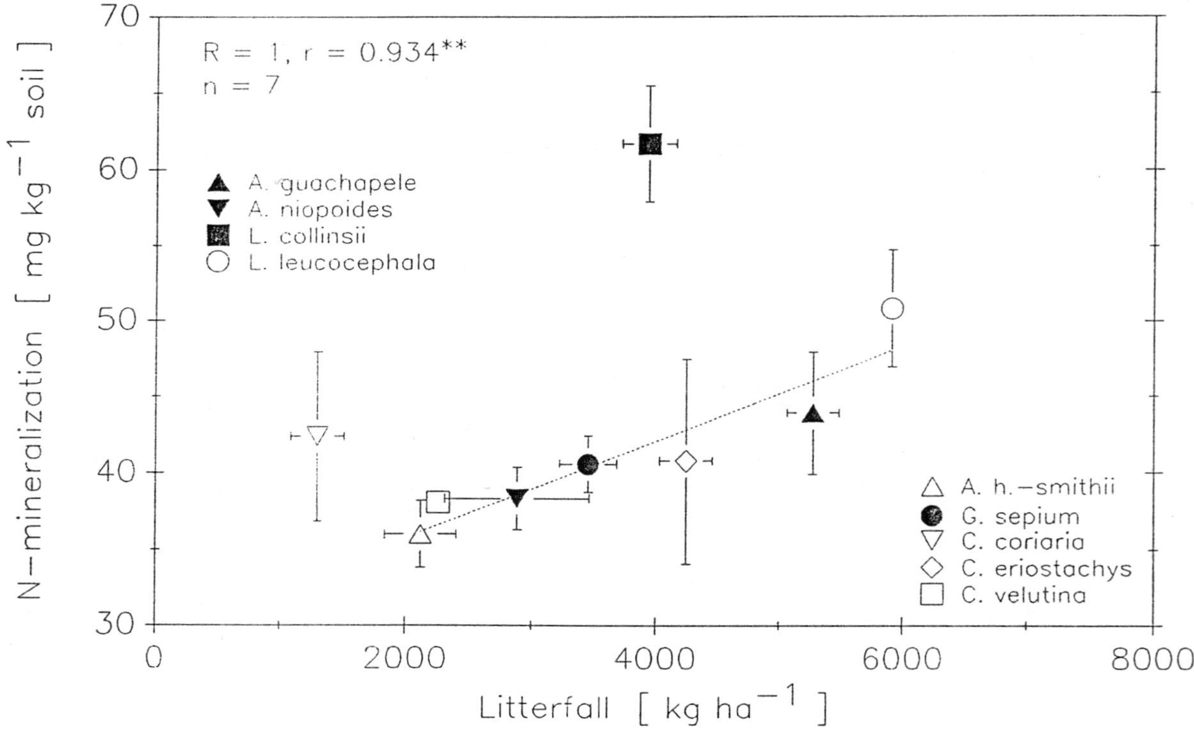

Fig. 4. Relationship between litterfall during $6\frac{1}{2}$ months and N-mineralization in the incubated soils between day 16 and 86 (means and S.E.). Spearman (R) and Pearson (r) correlations were calculated for all species except *L. collinsii* and *C. coriaria*.

N was less closely correlated with N-mineralization than root-C. Increased C-availability for mineralization, increased N-mineralization in the rhizospheric soil [1], and rhizodeposition of N-compounds [13] are probable explanations. In the soil of the non-N-fixing *Caesalpinia*-species, net N-mineralization may have been lower because of N-immobilization in decomposing root particles of wide C/N-ratio (Table 4). An additional explanation for the correlation between root-C and N-mineralization could be the stabilizing effect of roots on soil aggregates, thus physically protecting labile soil organic matter [9]. The disruption of the soil structure by sampling and sieving the soil, or by soil tillage, should then lead to increasing mineralization rates with increasing root density.

Lower N-concentrations in the labile soil organic matter and the immobilization of mineral N in root particles may also explain the higher mean ratio of mineralized C to mineralized N during the quasi-linear phase of the soil from the *Caesalpinia*-species, but not the control, compared to the N-fixing species (Table 6). These ratios were calculated after integrat-

ing equation (1). Although the differences were not significant, they could partly explain why the C/N-ratios of the soil organic matter under the non-N-fixing species were not systematically higher than under the N-fixing species, despite the higher N-concentration in the below-ground and certainly also the above-ground litter of the N-fixing species.

It cannot be said at present why the control soils showed so much higher N-mineralization in relation to root mass than the N-fixing species. The involvement of a particularly abundant and active soil microflora as indicated by the high CO_2-flush at rewetting seems likely. This may have partly depended on the high organic matter inputs when the fallow vegetation was cut and decomposed in the plots (see Methods section).

Rice bioassay

The germination rate of the rice varied between 74 and 80% for all species without significant differences, giving no indications for allelopathic effects. In contrast,

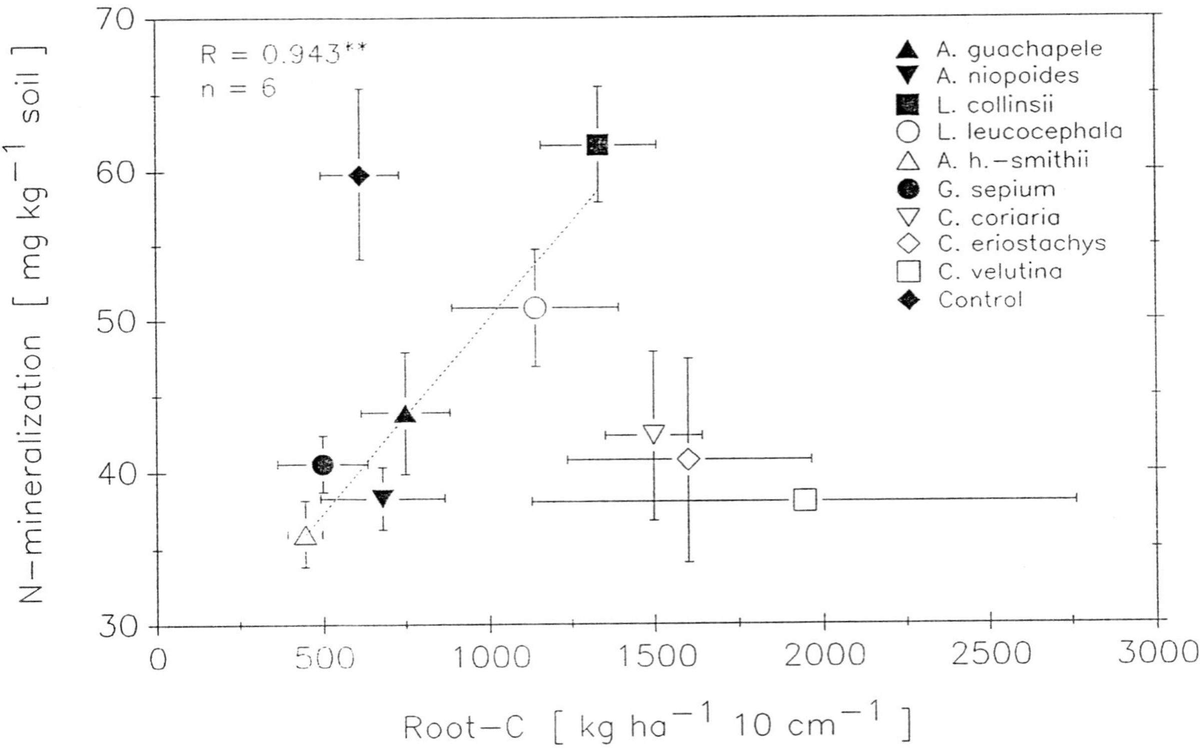

Fig. 5. Relationship between root-carbon (0–5 mm diameter) in 0–10 cm soil depth and N-mineralization in the incubated soils between day 16 and 86 (means and S.E.). The Spearman correlation was calculated for the N-fixing species, that is without the *Caesalpinia* species and the control.

the total rice dry matter per bag differed significantly between species (Table 7). The highest yield was obtained on soil from *L. collinsii* plots, followed by L. leucocephala and the control. *A. niopoides* and *G. sepium* soil gave the lowest yields. The grain yields did not differ significantly between species, but were closely correlated with biomass yields ($R = 0.881^{***}$).

The concentrations of N, P and K in the rice leaves were sufficiently high when compared to deficiency thresholds [7]. However, using the data from all plots (n = 35), significant correlations were obtained between total biomass per bag and foliar N ($R = 0.585^{***}$), P (0.425^*) and Mg (0.522^{**}) levels. Correlations of these nutrients with grain yield were also significant, indicating that rice yields were controlled by N, Mg and P availability. The foliar Mg levels were significantly higher in the rice from the control plots than from all tree plots, probably as an effect of higher Mg immobilization in the biomass of the trees. The effect of the fallow type on N and P levels was only significant at *p* < 0.1 (Table 7).

Using treatment means (n = 10), the correlations of leaf N concentrations with total biomass ($R = 0.890^{***}$) and grain yields ($R = 0.855^{**}$) were both highly significant, whereas all other nutrients gave non-significant relationships to yield parameters. Foliar N levels were significantly correlated with N-mineralization in the soil during the quasi-linear phase ($R = 0.791^{**}$), resulting in a significant correlation of N-mineralization with biomass (Fig. 6) and grain yield ($R = 0.644^*$). Thus, *N was the only nutrient whose availability was controlled by the fallow species in a way that rice productivity was affected.* The N mineralized during the initial flush was not correlated with foliar nutrient levels or rice yield. This was apparently due to the fact that the soil of the bioassay had been collected in the field several weeks after the beginning of the rainy season and had not been dried, thus the rice could not have taken up N released during the mineralization flush after rewetting.

The correlations with rice yield of the variables related to N-mineralization were not always significant. Root-C was significantly correlated with rice grain yield for the N-fixing species ($R = 0.829^*$), while

Table 7. Rice yields and foliar nutrient levels from the bioassay of soil fertility. Values with the same letter are not significantly different at $p < 0.05$ (Duncan's multiple range test)

Species	Total dry matter [g per bag]	Grain dry matter [g per bag]	Foliar nutrient concentrations[a]				
			N	P	K	Mg	Ca
			—[g kg^{-1}]—				
A. guachapele	57.0bc	25.8	40.7	1.34	23.3	2.07b	3.13
A. niopoides	50.1c	23.5	38.7	1.21	22.9	2.05b	2.87
L. collinsii	72.0a	33.4	42.4	1.38	23.4	2.26b	3.01
L. leucocephala	64.9ab	29.0	41.8	1.32	23.2	2.20b	2.97
A.h.-smithii	55.9bc	24.7	40.0	1.47	23.2	2.26b	3.09
G. sepium	54.3bc	25.5	40.6	1.34	23.3	2.12b	3.02
C. coriaria	56.8bc	26.2	41.0	1.55	24.6	2.22b	2.45
C. eriostachys	61.9abc	29.6	41.0	1.59	23.7	2.13b	3.02
C. velutina	56.5bc	26.5	41.0	1.45	24.1	2.16b	3.02
Control	63.2ab	26.5	43.6	1.67	24.4	2.66a	3.03
F-test	*	n.s.	n.s.	n.s.	n.s.	*	n.s.

$*$ = $p < 0.05$; n.s. = not significant;
[a]Micronutrients (Zn, Cu, Mn) did not differ significantly between species and were uncorrelated with yield data.

the correlation to total biomass production was only significant at $p < 0.1$. The correlation of litterfall with total rice yield was significant at $p < 0.1$ (n = 7 as in Fig. 4)

Implications for fallow species selection

The investigated tree species differed in several characteristics with the potential to influence the development of soil fertility under a tree fallow, like growth, ability to fix atmospheric N, litter production and root mass. Among the chemical soil parameters, N-availability showed the most pronounced effect of fallow species, and it was also the most important soil factor controlling rice production. According to the presented results, N-availability for subsequent crops could be increased by choosing fallow trees with high litter production and high root mass.

A problem of this approach may be that in this study, the two species with the highest mineralization rates during the quasi-linear phase, *L. collinsii* and the *Chromolaena*-fallow, had the lowest soil-N contents both in 0–10 cm and in 0–60 cm soil depth, suggesting that the high N-mineralization rates prevented them from accumulating soil-N reserves. For *Chromolaena*, this seems logical as the species is not N-fixing and cannot replace the mineralized N in the soil. For *L. collinsii*, this tendency indicates that the conversion of fixed atmospheric N into soil-N was not sufficient to compensate for the high mineralization losses. Both species seem thus to be less suitable for N enrichment of the soil than for the acceleration of N cycling in a system. Such species may be called 'N-mobilizing' species. For the accumulation of N during fallow phases, other species with lower N-mineralization rates in the soil may be preferable, like *A. guachapele* or *A. niopoides*, for instance. These 'N-accumulating' species were particularly efficient in N-enrichment of the subsoil. However, the accumulated N may not be readily available as indicated by the results of the incubation trial, and crop yields after the fallow may be negatively affected.

When choosing between N-mobilizing and N-accumulating fallow species, a crucial question is for how much time the soil will continue to supply high rates of N after a fallow of N-mobilizing species. If the N-mineralization rates decline after some time as an effect of the more rapid depletion of the mobile soil-N fraction, the cropping phase would have to be shortened and this may offset a short-term yield advantage. A solution may be to grow N-accumulating and N-mobilizing species in association during the fallow phase, to remove the N-accumulating species for the cropping phase, and to grow the crops in association with the N-mobilizing species until the next fallow becomes necessary.

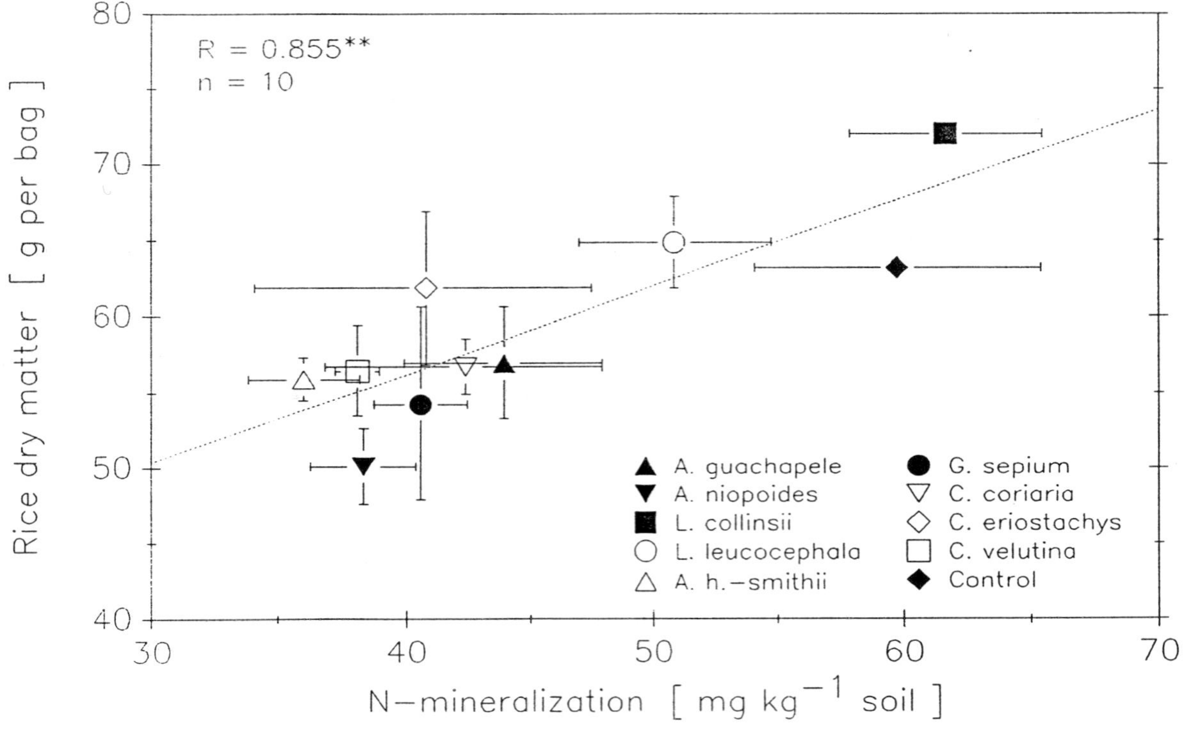

Fig. 6. Relationship between N-mineralization and total rice yield per bag in the incubated soils between day 16 and 86 (means and S.E.).

However, this system may create another problem owing to the observed positive correlation between the N-mobilizing characteristic of a species and its root mass (Fig. 5), and thus possibly root competition with the crops. In this study, the tree species with low root mass in the topsoil had a dense understorey (*G. sepium, A. guachapele*), whereas species with high root mass tolerated little herbaceous vegetation (*C. eriostachys, C. coriaria, L. collinsii*). The initially mentioned approach to the reduction of competition problems in agroforestry, the utilization of competitive, root-intensive tree species in rotational systems and of tree species of low root density in tree-crop associations, would tend to confine the N-mobilizing species to the fallows and the N-accumulating species to the associations, certainly not a desirable situation in terms of rational nutrient management. These questions need to be studied in long-term experiments.

For the efficient utilization of soil-N by a crop, the timing of N-mineralization in relation to nutrient uptake is of great importance. The observed mineralization flush after drying and rewetting the soil is a well-known phenomenon in tropical agriculture and may ensure a considerable part of the N demand of a crop if this is sown early enough at the beginning of the rainy season. However, if sowing is delayed because of the erratic nature of the first rains, the mineralization flush may give rise to high leaching losses. In this study, the size of the N-flush was positively correlated with the N-content of the soil. Thus, at the onset of the rains, leaching losses of N with their soil acidifying effect would be most likely to occur after a fallow of an N-accumulating tree species. A reduction of the mineralization flush at rewetting of the soil, followed by continuously high N-mineralization rates would be desirable and should lead to an efficient utilization of the mineralized N by the crop after a fallow. In this study, the soil from *L. collinsii* came closest to this ideal.

A potentially significant difference between fallow species is their ability to accumulate N in the subsoil. The mineralization of this N may contribute to crop nutrition in dry years, when N-mineralization and root growth in the topsoil are reduced. In this respect, *A. guachapele* was particularly efficient, while the spontaneous *Chromolaena*-fallow was of least value. It should however be mentioned that N mineralized in the subsoil is more likely to be leached than N mineralized in the topsoil.

An often mentioned criterion for a good fallow plant is the ability to fix atmospheric N. There was no relationship between this ability and either the N-accumulation in the soil or N-mineralization in this study. Thus, if only the soil-N is considered, a crop growing after a fallow of an N-fixing legume tree would not have been better supplied with N than a crop growing after a non-N-fixing fallow species. This may have partly been due to low N-fixation rates of the trees, as the soils were severely P-deficient. On the other hand, the N-fixing tree species had much higher N-concentrations in the root biomass than the non-N-fixing species, suggesting that their N-nutrition was effectively improved by N-fixation. Thus, the inclusion of N-fixing legumes in fallow rotations may only be worthwhile if at least the leafy biomass of the fallow vegetation is retained in the plots at their reconversion for cropping so that the crops can profit from the accumulated N. If this biomass is either exported as fodder or burned, the residual effect of a legume fallow on the soil may not differ much from that of a nonleguminous fallow vegetation. Also, other nutrients may then be rapidly depleted by exportation in the tree biomass.

No difference between the soils from the ten plant species was found in this study in rice germination rate, although allelopathic effects on germination and growth of other plants are known from *L. leucocephala* [8] and *G. sepium* [2], and have been suggested for *C. odorata* [16]. It should be stressed that the allelopathic suppression of weeds would be a favourable characteristic of a fallow species, as long as the responsible chemicals are rapidly de-activated in the soil after the reconversion of the fallow for cropping, or sufficiently specific in their effect to avoid negative impacts on subsequent crops.

Conclusions and outlook

It has frequently been attempted to evaluate the efficiency of managed and natural fallow types by conventional soil analyses, like total C and N content or P availability. It seems likely that important fallow characteristics cannot be measured with these methods, but require the quantification of soil-C and N pools and soil biological parameters. Incubation techniques and bioassays with local crop species seem to be useful methods for the identification of differences in chemical and biological soil fertility and permit to monitor their development under a fallow vegetation. The applied methods are, however, insensitive to soil physical characteristics and, if carried out with topsoil only, also to subsoil properties, which may both be of paramount importance for the overall fallow effect. They can thus not replace the integrative quantification of all soil effects by cropping the plots after removing the fallow vegetation, but help to isolate specific fallow processes.

By influencing N-availability and thus crop yields, litter production and root mass were important factors of fallow quality in this trial. However, long-term studies on different N-pools in soils and their changes during fallow and cropping periods as influenced by different fallow species, their biological characteristics and management techniques are necessary. Especially root systems can play an important role in the soil by influencing its physical, chemical and biological properties. The skillful employment of plant root systems, as well as soil fauna, may become one of the most powerful tools of soil fertility management once the interactions between roots and soil processes are sufficiently understood. Plant root characteristics may then become important criteria for the selection of fallow species.

It is clear that the introduction of planted instead of spontaneous fallows is mainly an economic question for the land user. In this trial, none of the tree species gave significantly higher rice yields than the *Chromolaena* fallow, in agreement with previous work [23]. The efficiency of *Chromolaena* to mobilize N is possibly a reason why Ivorian farmers select sites with vigorous *Chromolaena* stands for the establishment of new fields. The question if the wood and fodder production of the tree plots in comparison with the economically worthless *Chromolaena* justifies the effort of tree planting depends on the local availability and the market price of these resources. Another question is the long-term sustainability of different crop-fallow rotations. Can this sustainability be ascertained by a species like *Chromolaena odorata* with a high N-mobilizing potential, but without the ability to enrich the soil with biologically fixed N? The future fertility of extensive areas in West Africa which are invaded by this plant after forest removal could depend in part on this question.

Acknowledgements

This work has been funded by the Deutsche Forschungsgemeinschaft (Ze 154/12–1) and the European Community Commission in a project coordinated by the CIRAD-Forêts (DG XII STD 2 11/45). We gratefully acknowledge the hospitality and the assistance of the personnel of the Institut des Forêts, Oumé, and the generous technical support of the Scierie Jacob during this work. Michel Boli Bi, Nestor N'guessan and Eloi Koupo have committedly assisted in the field work, and Tanja Gonter has conducted the foliar analyses. H. Kolbe wrote the curve-fitting program for the respiration data. Significant improvements of the text have been provided by I. Kögel-Knabner and, especially, Y. Dommergues.

References

1. Abbadie L and Lensi R (1990) Carbon and nitrogen mineralization and denitrification in a humid savanna of West Africa (Lamto, Côte d'Ivoire). Acta Oecol 11: 717–728
2. Alán E and Barrantes U (1988) Efecto alelopático del madero negro (Gliricidia sepium) en la germinación y cricimiento inicial de algunas malezas tropicales. Turrialba 38: 271–278
3. Alef K (1991) Methodenhandbuch Bodenmikrobiologie. Landsberg: Ecomed
4. Anderson J P E (1982) Soil respiration. In: Page A L, Miller R H and Keeney D Z (eds) Methods of Soil Analysis, part 2, pp 831–872. Madison: American Society of Agronomy
5. Berenschot L M, Filius B M and Hardjosoediro S (1988) Factors determining the occurrence of the agroforestry system with Acacia mearnsii in Central Java. Agroforestry Systems 6: 119–135
6. Bernhard-Reversat F (1993) Dynamics of litter and organic matter at the soil-litter interface in fast-growing tree plantations on sandy ferrallitic soils (Congo). Acta Oecol 14: 179–197
7. Chabalier P-F (1987) Rice. In: Martin-Prével P, Gagnard J and Gautier P (eds) Plant Analysis as a Guide to the Nutrient Requirement of Temperate and Tropical Crops, pp 526–530. New York: Lavoisier Publishing Inc
8. Chou C-H and Kuo Y-L (1986) Allelopathic research of subtropical vegetation in Taiwan. III. Allelopathic exclusion of understory by Leucaena leucocephala (Lam.) de Wit. J Chem Ecol 12: 1431–1442
9. Duxbury J M, Smith M S and Doran J W (1989) Soil organic matter as a source and a sink of plant nutrients. In: Coleman D C, Oades J M and Uehara G (eds) Dynamics of Soil Organic Matter in Tropical Ecosystems, pp 33–67. Honolulu: NifTAL-Project, Dept of Agronomy, University of Hawaii
10. FAO/UNESCO (1990) Soil Map of the World, revised legend. Rome:
11. Gautier L (1992) Taxonomy and distribution of a tropical weed: Chromolaena odorata (L.).R. King & H. Robinson. Candollea 47: 645–662
12. Gee G W and Bauder J W (1986) Particle-size analysis. In: Klute A (ed) Methods of Soil Analysis, part 1, pp 383–411. Madison: American Society of Agronomy
13. Giller K E and Wilson K J (1991) Nitrogen fixation in tropical cropping systems. Wallingford: CAB International
14. Guillaumet J-L and Adjanohoun E (1971) La végétation de la Côte d'Ivoire. Mémoires ORSTOM 50: 157–263
15. Heinrichs H, Brunsack H-J, Loftfield N and König N (1986) Verbessertes Druckaufschlußsystem für biologische und anorganische Materialien. Z Pflanzenernähr Bodenk 149: 350–353
16. Holm L G, Plucknett D L, Pancho J V and Herberger J P (1977) The World's Worst Weeds. Honolulu: University Press of Hawaii
17. Hopkins B (1966) Vegetation of the Olokemeji forest reserve, Nigeria IV. The litter and soil with special reference to their seasonal changes. J Ecol 54: 687–703
18. Hughes C E and Styles B T (1984) Exploration and seed collection of multiple-purpose dry zone trees in Central America. Intern Tree Crops J 3: 1–31
19. Jenkinson D S and Powlson D S (1976) The effects of biocidal treatments on metabolism in soil. V. A method for measuring soil biomass. Soil Biol Biochem 8: 209–213
20. Kass D C L, Foletti C, Szott L T, Landaverde R and Nolasco R (1993) Traditional fallow systems of the Americas. Agroforestry Systems 23: 207–218 207-218
21. Khiewtam R S and Ramakrishnan P S (1993) Litter and fine root dynamiics of a relict sacred grove forest at Cherrapunji in north-eastern India. For Ecol Manage 60: 327–344
22. MacDicken K G (1991) Impacts of Leucaena leucocephala as a fallow improvement crop in shifting cultivation on the island of Mindoro, the Philippines. For Ecol Manage 45: 185–192
23. Ngiumbo K A B and Balasubramanian V (1992) Effect of fallow and residue management practices on biomass production, weed suppression and soil productivity. In: Mulongoy K, Gueye M and Spencer D S C (eds) Biological Nitrogen Fixation and Sustainability of Tropical Agriculture, pp 463–473. Ibadan: International Institute of Tropical Agriculture
24. Olsen S R and Sommers L E (1982) Phosphorus. In: Page A L, Miller R H and Keeney D R (eds) Methods of Soil Analysis, part 2, pp 403–430. Madison: American Society of Agronomy
25. Parrotta J A (1992) The role of plantation forests in rehabilitating degraded tropical ecosystems. Agric Ecosys Environ 41: 115–133
26. Pieri C. (1989) Fertilité des terres de savanes. Paris: Ministère de la Coopération et du Dèveloppement and CIRAD-IRAT
27. Powlson D S and Jenkinson (1976) The effects of biocidal treatments on metabolism in soil. II Gamma irradiation, autoclaving, air-drying and fumigation. Soil Biol Biochem 8: 179–188
28. Sanginga N, Mulongoy K and Swift M J (1992) Contribution of soil organisms to the sustainability and productivity of cropping systems in the tropics. Agric Ecosys Environ 41: 135–152
29. Scherer H W, Werner W and Rossbach J (1992) Effects of pretreatment of soil samples on N mineralization in incubation experiments. Biol Fertil Soils 14: 135–139
30. Schroth G and Kolbe D (1994) A method of processing soil core samples for root studies by subsampling. Biol Fertil Soils 18: 60–62
31. Schroth G, Poidy N, Morshäuser T and Zech W (1995) Effects of different methods of soil tillage and biomass application on crop yields and soil properties in agroforestry with high tree competition. Agric Ecosys Environ 52: 129-140
32. Sparling G P and Ross D J (1988) Microbial contributions to the increased nitrogen mineralization after air-drying of soils. Plant Soil 105: 163–167

314

33. Staver C (1989) Shortened bush fallow rotations with relay-cropped Inga edulis and Desmodium ovalifolium in wet central Amazonian Peru. Agroforestry Systems 8: 173–196

34. Steel R G D and Torrie J H (1980) Principles and Procedures of Statistics. New York: McGraw-Hill

35. Stewart J L and Dunsdon A J (1994) Performance of 25 Central American dry zone hardwoods in a pantropical series of species elimination trials. For Ecol Manange 65: 183–193

36. Stewart J L, Dunsdon A J, Hellin J J and Hughes C E (1992) Wood biomass estimation of Central American dry zone species. Tropical Forestry Paper 26. Oxford: Oxford Forestry Institute

37. Szott L T, Fernandes E C M and Sanchez P A (1991) Soil-plant interactions in agroforestry systems. For Ecol Manage 45: 127–152

38. Szott L T Palm C A and Sanchez P A (1991) Agroforestry in acid soils of the humid tropics. Adv Agron 45: 275–301

39. Wambeke A van (1992) Soils of the Tropics. New York: McGraw-Hill, Inc.

40. Yates F (1933) The analysis of replicated experiments when the field results are incomplete. Emp J Exp Agric 1: 129–142

Fertilizer Research **42**: 315–319, 1995.
© 1995 *Kluwer Academic Publishers. Printed in the Netherlands.*

Nitrogen balance and root behavior in four pigeonpea-based intercropping systems[*]

K. Katayama[1], O. Ito[2], R. Matsunaga[2], J.J. Adu-Gyamfi[3], T.P. Rao[3], M.M. Anders[3] &
K.K. Lee[3]
[1]*National Agriculture Research Center, 3-1-1, Kannondai, Tsukuba, Ibaraki, 305 Japan;* [2]*Japan International
Research Center for Agriculture Sciences 1-2 Ohwashi, Tsukuba, Ibaraki, 305 Japan;* [3]*International Crops
Research Institute for the Semi-Arid Tropics, Patancheru, Andhra Pradesh 502 324 India*

Key words: biological N_2 fixation, *Cajanus cajan* L. Millsp. intercropping, minirhizotron, natural ^{15}N abundance,
pigeonpea

Introduction

A medium-duration pigeonpea (*Cajanus cajan* L. Millsp.) is usually grown as intercrop. A wide range of
crop combination in pigeonpea-based intercropping
systems is found in India and eastern Africa (Ofroi
and Stern, 1987; Rao and Willey, 1980; Venkateswarlu
and Subramanian, 1990). Although much information
is available on the production efficiency and monetary advantage of intercropping, very little is known
about the nitrogen (N) economy and root behavior.
The study was carried out to examine how the nitrogen balance sheet and root development of pigeonpea
could be altered by companion crops.

Materials and method

The experiment was conducted during the 1992 rainy
season on a shallow Alfisols at ICRISAT Center, near
Hyderabad, India. Medium duration pigeonpea (cv.
ICP 1–6), hybrid grain sorghum (*Sorghum bicolor* L.
Moench cv. CSH 5), pearl millet (*Pennisetum glau-
cum* L. R.Br. cv. ICMH 89988), groundnut (*Arachis
hypogaea* L. cv. ICGS 11) and cowpea (*Vigna unguiculata* Walp. cv. Russian Giant) were sown on the broad
bed furrow on 25 June 1992. Spacings for the sole
crops were pigeonpea (75 × 20 cm), sorghum (50 ×
15 cm), pearl millet (50 × 15 cm), groundnut (37.5
× 10 cm), and cowpea (50 × 10 cm). The ratio of

the row arrangements in the intercrops was 2:1 for all
combinations except the groundnut/pigeonpea which
was 4:1. The spacing arrangements gave an identical
number of plants on area basis for each component
crop in sole and intercropping system with the exception of pigeonpea. All plots received uniform basal
application of 14.4 kg N ha^{-1} as urea and 36.9 kg P
ha^{-1} as single superphosphate prior to sowing. The
experimental layout was a randomized complete block
design with three replications. The size of each plot
was 6 m × 12 m, consisting of 8 broad beds with 75
cm width.

The crops were harvested at maturity, 83, 105, 110,
130 and 209 days after sowing (DAS) for pearl millet,
groundnut, sorghum, cowpea and pigeonpea, respectively. Land equivalent ratio (LER) for yield was used
to evaluate the biological efficiency of the intercropping system relative to sole cropping (Mead and Willey, 1980). The dry weight of samples was determined
after oven-dried at 70 °C, and a portion of the ground
material used for N analysis. Total N was determined
by indophenol color formation (Chykin, 1969). For
analysis of ^{15}N natural abundance, N_2 gas from digested samples was introduced into a mass spectrometer
(Finnigan Mat 251). Detailed procedure for ^{15}N natural abundance analysis and the estimation of N derived
from air (N_{dfa}) have been described elsewhere (Tobita
et al., 1994).

Root length was measured with the minirhizotron
method (CIRCON MV9011 agriculture system with
MV9390 color CCD microvideo camera). The trans-
parent plastic minirhizotron tubes (58 mm in diameter

---

[*] ICRISAT Journal Article No. 1653

Table 1. Shoot dry matter (SDM) at harvest, harvest index (HI) and grain yield of sole crop and intercrop

Crop	Companion crop	SDM (kg ha^{-1})	HI	Yield (kg ha^{-1})
Pigeonpea	None	8755	0.203	1777
	Sorghum	5771	0.289	1668
	Pearl millet	6150	0.279	1716
	Groundnut	5893	0.303	1786
	Cowpea	2220	0.242	537
	SE ($\pm$)	1470**[a]	0.012**	332**
	CV (%)	25.5	4.6	22.2
Sorghum	None	10292	0.274	2820
	Pigeonpea	9191	0.288	2647
	SE ($\pm$)	183*	0.013[NS]	86.4[NS]
	CV (%)	1.9	4.7	3.2
Pearl millet	None	3950	0.389	1536
	Pigeonpea	4052	0.376	1524
	SE ($\pm$)	207[NS]	0.055[NS]	202[NS]
	CV (%)	5.2	14.2	13.2
Groundnut	None	4219	0.330	1392
	Pigeonpea	3150	0.244	769
	SE ($\pm$)	71.5**	0.046[NS]	149*
	CV (%)	1.9	16.1	13.7
Cowpea	None	3862	0.242	935
	Pigeonpea	3824	0.227	868
	SE ($\pm$)	718[NS]	0.018[NS]	233[NS]
	CV (%)	18.7	7.5	25.8

[a] *: $p < 0.05$, ** $p < 0.01$, NS: not significant.

Table 2. Yield and land equivalent ratio (LER) for yield of four intercrops

Crop combination	Yield (kg ha^{-1})	LER[a]
Pigeonpea/Sorghum	4315	1.88
Pigeonpea/Pearl millet	3240	1.96
Pigeonpea/Groundnut	2555	1.56
Pigeonpea/Cowpea	1405	1.23
SE ($\pm$)	347**[b]	0.26*
CV (%)	12.0	15.1

[a] LER = (Int Y_p/Sole Y_p) + (Int Y_0/Sole Y_0) where Y_p and Y_0: Yield of pigeonpea and other crops
[b] *: $p < 0.05$, **: $p < 0.01$.

and 100 cm in length) were installed at a 45 degree angle between rows of component crops before sowing. Root length at 10 cm intervals up to 70 cm depth was calculated from the number of roots observed on a video display (Upchurch and Ritchie, 1983).

Results

Growth and yield parameters

Shoot dry matter (SDM) and yield of pigeonpea intercropped with cowpea was significantly lower than sole pigeonpea and other intercropped pigeonpea (Table 1). Harvest index (HI) of sole pigeonpea was significantly lower than intercropped pigeonpea. The SDM and yield of groundnut intercropped with pigeonpea was significantly lower than sole groundnut. These yield reductions could be closely associated with the reduction in SDM. There was no significant difference in yield of sorghum, pearl millet and cowpea of sole and intercrops.

The combined yield of pigeonpea/sorghum was the highest, followed by pigeonpea/pearl millet, pigeonpea/groundnut and pigeonpea/cowpea (Table 2). The LER values for grain yield were greater than the unity in all intercrops. Pigeonpea intercropped with cereals and groundnut recorded a significantly higher LER than pigeonpea with cowpea.

Nitrogen balance sheet

Nitrogen yield of pigeonpea intercropped with cowpea was significantly lower than sole pigeonpea and pigeonpea intercropped with other component crops (Table 3). N yield of groundnut intercropped with pigeonpea was significantly lower than sole groundnut. There was no significant difference in N yield of sorghum, pearl millet and cowpea in sole and intercrops. The proportion of nitrogen derived from air (% N_{dfa}) was significantly higher in pigeonpea intercropped with cereals than pigeonpea intercropped with legumes. There was no significant difference in %N_{dfa} between sole and intercrop of groundnut and cowpea. Although the %N_{dfa} of pigeonpea intercropped with cereals was significantly higher than with legumes and sole pigeonpea, the amount of N derived from air (N_{dfa}) was not significant among pigeonpea treatments except pigeonpea intercropped with cowpea. The N_{dfa}

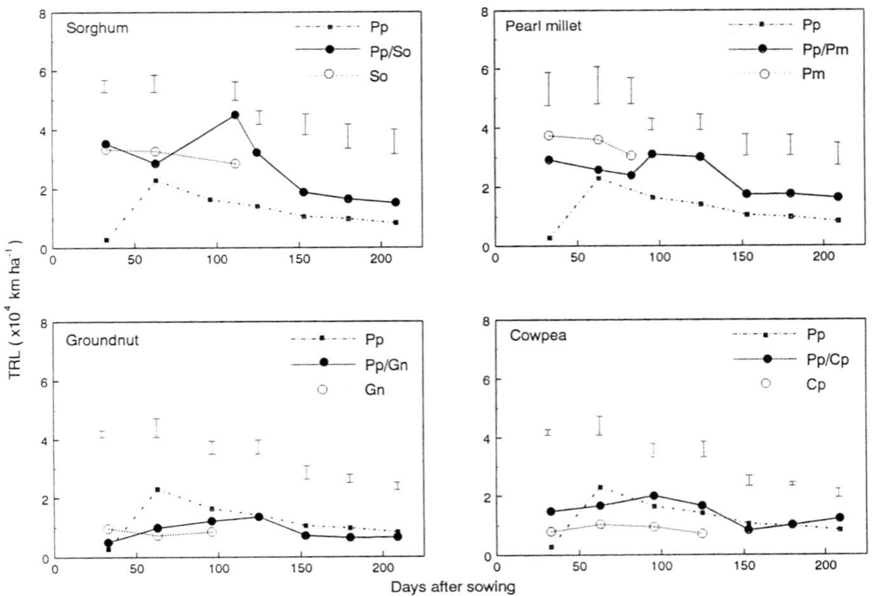

Fig. 1. Seasonal changes in total root length (TRL) of pigeonpea and companion crops in sole and intercropping system. Vertical bars indicate SE's at each sampling time.

for groundnut intercropped with pigeonpea was significantly lower than that of sole groundnut.

Root system development and nitrogen uptake

Total root length (TRL) of sole pigeonpea was significantly lower than sole cereals and pigeonpea/cereals combinations at 33 DAS, and this trend was steady until the end of growth season (Fig. 1). However, there was no significant differences in TRL between legumes in sole crop and legumes intercropped with pigeonpea. The TRL of cereals was significantly higher than that of legumes. The amount of nitrogen derived from soil and fertilizer (N_{dffs}) was calculated using total-N in the plants and $\%N_{dfa}$ in case of legumes in intercropping. Then ratio of N_{dffs} for component crop over pigeonpea was positively correlated with $\%N_{dfa}$ of pigeonpea grain shown in Table 3 (Fig. 2). This suggests that a higher N consumption of companion crop relative to pigeonpea may have increased the dependency on biological nitrogen fixation.

Discussion

Shoot dry matter of pigeonpea intercropped with cowpea was significantly lower than other combinations, though, HI was not affected (Table 1). This reduction in SDM is explained by competition for light between pigeonpea and cowpea. Maximum leaf area index of pigeonpea intercropped with cowpea was 0.41 which was only a third or a fourth compared to those of pigeonpea with other crops (data not shown). Since crop growth is correlated with solar radiation intercepted by leaf, irrespective of crop (Monteith, 1977), pigeonpea with large leaf area would have an advantage in intercepting light over pigeonpea with smaller leaf area. The indeterminate, spreading and climbing nature of the cowpea used in this experiment may be a key factor in suppressing pigeonpea growth. Indeterminate cowpea is reported to reduce the yield of pigeonpea compared to the determinate ones (Rao and Willey, 1980).

Pigeonpea/cereal combinations were more beneficial than pigeonpea/legume combinations as indicated by their respective LERs and grain yield (Table 2). In this regard, the overall yield of pigeonpea/cereal combinations has been reported to be higher than pigeonpea/legume combinations (Rao and Willey, 1980).

A significantly higher $\%N_{dfa}$ of pigeonpea was observed with cereal companion crops than with legumes (Table 3), suggesting that pigeonpea increases its dependency on biological nitrogen fixation (BNF) only when intercropped with cereals. Such an increase in $\%N_{dfa}$ has been reported for fababeans intercropped with barley (Danso *et al.*, 1987) and ricebean with

Table 3. Nitrogen yield in grain (NY), proportion of nitrogen derived from air (% N_{dfa}) and amount of nitrogen derived from air (N_{dfa}) of sole crop and intercrop

Crop	Companion Crop	NY (Kg ha^{-1})	%N_{dfa}	N_{dfa} (kg ha^{-1})
Pigeonpea	None	58.3	63.7	37.1
	Sorghum	50.7	82.6	41.9
	Pearl millet	55.7	84.9	47.3
	Groundnut	55.7	64.8	36.1
	Cowpea	16.0	70.1	11.2
	SE ($\pm$)	12.4**	5.8**	8.7**
	CV (%)	26.2	8.0	25.3
Sorghum	None	28.4	–	–
	Pigeonpea	33.0	–	–
	SE ($\pm$)	7.0^{NS}		
	CV (%)	23.0		
Pearl millet	None	17.7	–	–
	Pigeonpea	18.4	–	–
	SE ($\pm$)	3.5^{NS}		
	CV (%)	19.4		
Groundnut	None	44.5	55.1	24.5
	Pigeonpea	24.2	51.6	12.5
	SE ($\pm$)	1.6**	6.6^{NS}	2.4*
	CV (%)	4.6	12.4	13.0
Cowpea	None	34.6	61.7	21.3
	Pigeonpea	32.2	64.7	20.8
	SE ($\pm$)	9.1^{NS}	2.1^{NS}	5.4^{NS}
	CV (%)	27.3	3.3	25.5

[a] *: p <0.05, **: p <0.01, NS: not significant.

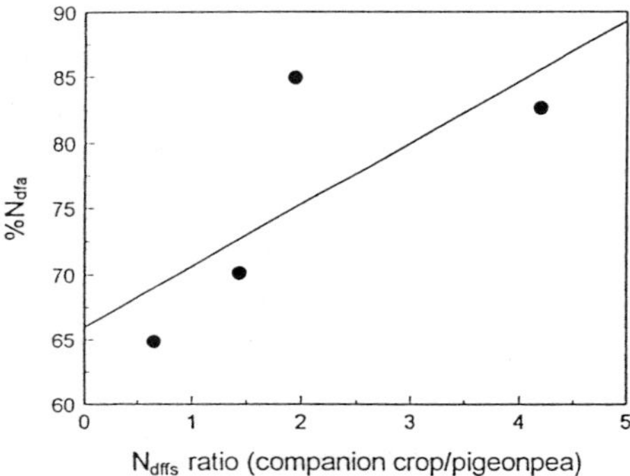

Fig. 2. Relationship between N_{dffs} (nitrogen derived from fertilizer and soil) ratio of companion crop over pigeonpea and %N_{dfa} (nitrogen derived from air) of pigeonpea in intercropping.

maize (Rerkasem *et al.*, 1988). The reduction in soil N level due to the high uptake by cereals, especially at the early stage, may cause the legumes to depend more on BNF (Herridge and Brockwell, 1988)

The TRL of cereals was three to six times higher than that of legumes, suggesting that cereals may have an advantage in exploiting N from soils over legumes. The TRL of the intercrops in pigeonpea/cereal combinations was higher than that of sole pigeonpea even after the harvest of the cereals (Fig. 1). Since cereals developed the ratoon after harvest of above-ground parts, their roots would be kept alive in soils, resulting in an overestimation of the TRL in intercropping. The ratoon roots was unable to be separated out due to difficulties in recognizing roots of two component crops by

the minirhizotron observation. In contrast, the TRL of the intercrops in pigeonpea/legume combinations was not higher than that of sole pigeonpea. Considering that dry matter production of above-ground parts was enhanced by intercropping with legumes as indicated with LER (Table 2), there would be an allelopatic inhibition of root growth between two legume crops.

Assuming that below-ground competition for N from soil and fertilizer is a limiting factor for the growth of each component crop (Fig. 2), cereals may have an advantage over legumes to meet their N requirement, particularly during the active growing stage, due to their more extensive root proliferation in the surface soil. Consequently, cereals would deplete soil N more rapidly than legumes. It is reported that nitrate, which is a main form of N under upland conditions, disappears completely from soils in sole and intercropped sorghum fields at 45 DAS, whilst an appreciable amount still remained unutilized in sole pigeonpea (Ito *et al.*, 1994)

Pigeonpea is reported to develop a deeper root system than soybean and maize on Alfisols (Arihara *et al.* 1991). Therefore intercropped pigeonpea could utilize N in the deeper soil, even though soil N in the surface becomes depleted. When pigeonpea is grown on a shallow Alfisol with a hard stony layer below 30 cm depth, like in the present study, it is unable to exhibit its deep rooting characteristics (Ito *et al.*, 1992). Pigeonpea intercropped with cereals must have been forced to grow under limited soil N from the early growth stage. It is concluded from this experiment that the increase in %N_{dta} of pigeonpea could be attributed to the higher utilization of soil and fertilizer N by the cereal companion crops.

References

Arihara J, Ae N and Okada K (1991) Root development of pigeonpea and chickpea and its significance in different cropping systems. *In*: Johansen C, Lee KK and Sahrawat KL (eds) Phosphorus Nutrition of Grain Legumes in the Semi-arid Tropics, pp 183–194. Patancheru, A.P. 502 324, ICRISAT, India

Chykin S (1969) Assay of nicotinamide deaminase. Determination of ammonia by the indophenol reaction. Anal Biochem 32: 375–382

Danso SKA, Zapata F and Hardarson G (1987) Nitrogen fixation in fababeans as affected by plant population density in sole or intercropped systems with barley. Soil Biol Biochem 19: 411–415

Herridge DF and Brockwell J (1988) Contributions of fixed nitrogen and soil nitrate to the nitrogen economy of irrigated soybean. Soil Biol Biochem 20: 711–717

Ito O, Matsunaga R, Tobita S and Rao TP (1992) Rooting behavior of intercropped pigeonpea (*Cajanus cajan* (L.) Millspaugh) and sorghum (*Sorghum bicolor* (L.) Moench). *In*: Kutschera L, Hübl E, Lichtenegger E, Persson H, Sobotik M (eds) Root Ecology and its Pactical Application: 3.ISRR Symposium, Wien, University Bodenkultur, 1991, pp 419–422. Verein für Wurzeelforschung, A-9020 Klagenfurt, Austria

Ito O, Matsunaga R, Katayama K, Tobita S, Adu-Gyamfi JJ, Rao TP and Gayatri Devi (1994) Nitrate in soil solution under sorghum and pigeonpea plants in mono and intercrops. In: Proceedings of 15th World Congress of Soil Science Vol 5b: 48–49. Acapulco, Mexico

Ofroi F and Stern WR (1987) Cereal-legume intercropping systems. Adv Agron 41: 41–90

Mead R and Willey RW (1980) The concept of a 'Land Equivalent Ratio' and advantages in yields from intercropping. Exp Agric 16: 217–228

Monteith JL (1977) Climate and the efficiency of crop production in Britain. Phil Trans R Soc Lond B 81: 277–294

Rerkasem B, Rerkasem K, Peoples M B, Herridge DF and Bergersen FJ (1988) Measurement of N_2 fixation in maize (*Zea mays* L.)-ricebean (*Vigna umbellata* [Thunb.] Ohwi and Ohashi) intercrops. Plant and Soil 108: 125–135

Rao MR and Willey RW (1980) Preliminary studies of intercropping combinations based on pigeonpea or sorghum. Exp Agric: 16: 29–39

Tobita S, Ito O, Matsunaga R, Rao TP, Rego TJ, Johansen C and Yoneyama T (1994) Field devaluation of nitrogen fixation and use of nitrogen fertilizer by sorghum/pigeonpea intercropping on an Alfisol in the Indian semi-arid tropics. Biol Fertil Soils 17: 241–248

Upchurch DR and Ritchie JT (1983) Root observations using a video recording system in minirhizotrons. Agron J 75: 1009–1015

Venkateswarlu S and Subramanian VB (1990) Productivity of some rainfed crops in sole and intercrop systems.Indian J Agric Sci 60: 106–109

Fertilizer Research **42**: 321–329, 1995.

321

© 1995 *Kluwer Academic Publishers. Printed in the Netherlands.*

Recent research on problems in the use of urea as a nitrogen fertilizer

J. M. Bremner
Department of Agronomy, Iowa State University, Ames, IA 50011, USA

Key words: urea fertilizer, phytotoxicity, ammonia volatilization, nitrite accumulation, urea hydrolysis, urease inhibitors

Abstract

Recent research on the NH_3 volatilization, NO_2^- accumulation, and phytotoxicity problems encountered in the use of urea fertilizer is reviewed. This research has shown that the adverse effects of urea fertilizers on seed germination and seedling growth in soil are due to NH_3 produced through hydrolysis of urea by soil urease and can be eliminated by addition of a urease inhibitor to these fertilizers. It also has shown that the leaf burn commonly observed after foliar fertilization of soybean with urea results from accumulation of toxic amounts of urea in soybean leaves rather than formation of toxic amounts of NH_3 through hydrolysis of urea by leaf urease. It further showed that this leaf burn is accordingly increased rather than decreased by addition of a urease inhibitor to the urea fertilizer applied. N-(*n*-butyl)thiophosphoric triamide (NBPT) is the most effective compound currently available for retarding hydrolysis of urea fertilizer in soil, decreasing NH_3 volatilization and NO_2^- accumulation in soils treated with urea, and eliminating the adverse effects of urea fertilizer on seed germination and seedling growth in soil. NBPT is a poor inhibitor of plant or microbial urease, but it decomposes quite rapidly in soil with formation of its oxon analog N-(*n*-butyl) phosphoric triamide, which is a potent inhibitor of urease activity. It is not as effective as phenylphosphorodiamidate (PPD) for retarding urea hydrolysis and ammonia volatilization in soils under waterlogged conditions, presumably because these conditions retard formation of its oxon analog. PPD is a potent inhibitor of urease activity but it decomposes quite rapidly in soils with formation of phenol, which is a relatively weak inhibitor of urease activity. Recent studies of the effects of pesticides on transformations of urea N in soil indicate that fungicides have greater potential than herbicides or insecticides for retarding hydrolysis of urea and nitrification of urea N in soil.

Introduction

The use of urea as a N fertilizer has increased dramatically during the past 25 years and urea is now the most important N fertilizer in world agriculture. There is a clear and urgent need, therefore, to find methods of reducing problems encountered in use of urea fertilizer, which include damage to seeds, seedlings and young plants, NO_2^- toxicity, phytotoxicity of foliar-applied urea, and volatilization of urea N as NH_3. Gaseous loss of urea fertilizer N as NH_3 is of particular concern because it can exceed 50% of the N applied [87]. The purpose of this article is to review and summarize recent work to account for some of these problems and to find methods of reducing them.

Approaches to reduction of problems

Several approaches have been explored for reduction of the problems encountered in the use of urea fertilizer. They include (a) addition of a urease inhibitor to the fertilizer; (b) coating of the fertilizer with sulfur or other materials to slow its rate of dissolution; (c) acidulation of the fertilizer with inorganic acids such as phosphoric or nitric acid; (d) treatment of the fertilizer with inorganic salts such as $CaCl_2$ or KCl; (e) use of urea supergranules. These approaches have been discussed in numerous articles during the past decade [36,42,47,76,80,89,91,92,101]. All have proved useful under some circumstances, but the urease inhibitor approach has received the most attention during the past 10 years and it has shown the most promise for reduction of NH_3 volatilization and other problems encountered in the use of urea fertilizer.

Ammonia volatilization

It is generally accepted that the NH_3 volatilization problem results largely, if not entirely, from the normally rapid enzymatic hydrolysis of urea to NH_3 and CO_2 by soil urease ($NH_2CONH_2 + H_2O \rightarrow 2NH_3 + CO_2$) and the resulting rise in pH and accumulation of NH_4^+. To minimize this problem and the adverse effects of urea fertilizers on seed germination and seedling growth, farmers have been advised not to surface-apply urea fertilizer without incorporating it into the soil soon after application and not to band-place urea with or near seed. As noted by Hauck [47], however, fertilizer and crop management systems now being adopted by farmers make such restrictions on urea use impractical or inconvenient. The risk of gaseous loss of urea N as NH_3 is particularly high in reduced tillage agriculture in which urea is often surface-applied without incorporation.

The approach currently favored for reduction of the NH_3 volatilization problem is to find compounds that will inhibit soil urease activity and thereby retard urea hydrolysis when applied to soils in conjunction with urea fertilizer. This approach has received considerable attention during the past 25 years, and numerous compounds have been patented or proposed as inhibitors of urea hydrolysis. Early work at Iowa State University to identify promising soil urease inhibitors showed that dihydric phenols and quinones such as hydroquinone, catechol, and p-benzoquinone were the most effective of more than 100 compounds tested and indicated that hydroquinone had special promise because of its low cost [7,8,15,18,72,73]. Continuation of this work, however, showed that several phosphoroamides, notably N-(n-butyl) thiophosphoric triamide (NBPT) and phenylphosphorodiamidate (PPD), are more effective than hydroquinone or p-benzoquinone for retarding hydrolysis of urea fertilizer in soil [5,6,28,60]. It also showed that NBPT is markedly superior to PPD for retarding urea hydrolysis and decreasing NH_3 volatilization and NO_2^- accumulation in upland soils treated with urea [5,6,28]. In contrast to PPD, which is a potent inhibitor of jackbean urease activity but decomposes quite rapidly in soil with formation of phenol [12], which is a poor inhibitor of urease activity, NBPT has very little effect on jackbean urease activity but decomposes quite rapidly in soil with formation of its oxon analog [N-(n-butyl) phosphoric triamide], which is a potent inhibitor of urease activity [65]. It is noteworthy that thiophosphoryl triamide (TPT), which has been under consideration by the Tennessee Valley Authority (TVA) as a dual purpose urease/nitrification inhibitor [76,77], is also a poor inhibitor of jackbean urease activity, but decomposes quite rapidly in soil with formation of its oxon analog, phosphoryl triamide, which is a potent inhibitor of urease activity [66]. Other workers have confirmed that the inhibitory effect of NBPT on urea hydrolysis in soil is due to its oxon analog [32] and have shown that, although NBPT is superior to PPD for retarding urea hydrolysis and NH_3 volatilization in soils under aerobic conditions, PPD is better than NBPT for retarding these processes in soils under waterlogged conditions [24,25,57,89], presumably because aerobic conditions are required for the conversion of NBPT to its oxon analog.

Studies of the effects of phosphoroamides and other urease inhibitors on seed germination and N transformations in soils other than urea hydrolysis have been reported [11,13,20,93,94,100]. They have shown that NBPT and PPD do not inhibit nitrification or denitrification, do not retard mineralization of organic N when applied at the rate of 10 μg g^{-1} soil [13], and do not affect seed germination even when applied at the rate of 250 μg g^{-1} soil [11]. Bremner *et al.* [14] recently studied the persistence of the inhibitory effects of NBPT, PPD, and TPT on urea hydrolysis in soils by measuring the ability of four soils to hydrolyze urea after they had been treated with 5μg phosphoroamide g^{-1} soil and incubated at 15 °C or 30 °C for 0, 3, 7, 14, or 28 days. The soils used differed markedly in pH, texture, and organic-matter content. The data obtained showed that the persistence of the effects of the phosphoroamides studied decreased with increase in soil temperature from 15 °C to 30 °C and that whereas the effect of PPD decreased with increase in the time of incubation, the effects of NBPT and TPT sometimes increased before decreasing with increased time of incubation. These observations are in harmony with the finding that, although PPD is a potent inhibitor of urease activity, it decomposes in soils with formation of phenol, which is relatively weak inhibitor of urease activity, whereas NBPT and TPT have little or no effect on urease activity but decompose in soil with formation of their oxon analogs, which are potent inhibitors of urease activity. The inhibitory effect of NBPT on urea hydrolysis was considerably more persistent than that of PPD or TPT and was significant even after incubation of NBPT-treated soil at 15 °C or 30 °C for 28 days.

Most of the information currently available concerning the potential value of NBPT, PPD and other phosphoroamides for reduction of NH_3 volatilization and other problems encountered in the use of urea fertilizer has emerged from laboratory investigations, but a significant number of field and greenhouse studies have been performed to evaluate these compounds for urea fertilization of corn, wheat, rice, and grasses [3,4,21,22,23,24,48,50,51,55,78,79,82]. The most extensive field studies have been with corn [48].

Although most recent research on urease inhibitors has been focused on phosphoroamides such as NBPT and PPD, hydroquinone (HQ) has received considerable attention during the past decade, especially in China, where it has been formulated with urea. Several workers have reported that application of HQ with urea can increase crop yield and urea efficiency [26,56,79,104]. Recent work has indicated that the use of HQ to retard hydrolysis of urea fertilizer will not have adverse effects on human health or the environment and that the increase in crop yield and urea efficiency observed on application of HQ with urea is due to a beneficial physiological effect of HQ on the plant as well as the inhibitory effect of HQ on urea hydrolysis [97,102,103].

Nitrite accumulation

Soils fertilized with urea tend to accumulate NO_2^- and this can lead to NO_2^- toxicity problems [29,30,31]. There also has been concern that NO_2^- accumulation following urea fertilization may lead to significant gaseous loss of urea N via chemo-denitrification (chemical decomposition of NO_2^-) and to NO_2^- pollution of ground waters.

The NO_2^- accumulation observed following application of urea to soils is believed to be due to the increase in soil pH and accumulation of NH_3 resulting from hydrolysis of urea by soil urease, because there is evidence that the *Nitrobacter* bacteria responsible for oxidation of NO_2^- to NO_3^- in soils are considerably more sensitive to NH_4^+ salts under alkaline conditions than are the bacteria responsible for oxidation of NH_4^+ to NO_2^- [1,2].

Bundy and Bremner [19] found that nitrification inhibitors such as N-Serve (nitrapyrin) eliminated or greatly reduced NO_2^- accumulation and greatly retarded NO_3^- formation in soils that accumulated consid-

erable amounts of NO_2^- when treated with urea. They also found that addition of nitrification inhibitors to soils that accumulated substantial amounts of NO_2^- after treatment with urea had little, if any, effect on the recovery of urea N as (urea + exchangeable NH_4^+ + NO_2^- + NO_3^- + NH_3)-N after various times, which indicated that the 'N deficits' observed in studies of urea N transformations in soils may not be largely due to gaseous loss of urea N through chemo-denitrification and are at least partly due to volatilization and fixation of the NH_4^+ formed by urea hydrolysis in soils. Their work also indicated that although N-Serve and other nitrification inhibitors may prove useful for reduction of the NO_2^- toxicity problem associated with the use of urea as a fertilizer, application of such inhibitors in conjunction with surface-applied urea fertilizer may promote gaseous loss of urea N as NH_3. These observations concerning the effect of N-Serve on NO_2^- accumulation and NH_3 volatilization in soils treated with urea have been confirmed in recent work by Magalhaes and Chalk [59].

Since NO_2^- accumulation in soils treated with urea is believed to result from the rapid hydrolysis of urea by soil urease, it should be reduced by application of a urease inhibitor to the urea. This expectation was confirmed by a study of the effects of four urease inhibitors, namely, NBPT, PPD, DPCA (N-(diaminophosphinyl)-cyclohexylamine], and HQ (hydroquinone), on urea hydrolysis, NH_3 volatilization, and NO_2^- accumulation in five soils treated with urea [6]. This study showed that the ability of these urease inhibitors to retard urea hydrolysis, NH_3 volatilization, and NO_2^- accumulation in the five soils studied decreased in the order NBPT > DPCA >> PPD > HQ. On average, the gaseous loss of urea N as NH_3 and the accumulation of urea N as NO_2^- were decreased from 53% to 5% and from 11% to 1%, respectively, by addition of NBPT at the rate of 10 μg g^{-1} soil (0.47 parts of NBPT per 100 parts of urea). There seems little doubt, therefore, that the NO_2^- accumulation problem in use of urea fertilizer could be eliminated or greatly reduced by amending the fertilizer with NBPT.

Loss of urea N by denitrification

Recent work in Australia has indicated that NH_3 volatilization, leaching, and run off are not important mechanisms for N loss from red-brown earths when urea is applied in irrigation water or broadcast onto the soil and washed in by sprinkler irrigation and that N loss from these soils is due principally to denitrification [84,85]. Support for these conclusions has been provided by studies showing that N loss from these soils *via* denitrification can be reduced, and that the recovery of urea N by irrigated wheat or cotton can be markedly increased, through application of nitrification inhibitors such as wax-coated CaC_2 or 2-ethynylpyridine to retard formation of nitrate by nitrification of urea N [37,38,86].

Adverse effects of urea on seed germination, seedling growth, and early plant growth

Numerous investigations have shown that urea fertilizers can have adverse effects on seed germination, seedling growth and early plant growth in soil [33,40,43,58,88]. Studies to account for these effects have given divergent results, and several explanations have been advanced for each effect [27,30,31,41,47,88]. For example, studies to account for the adverse effect of urea fertilizer on seed germination have suggested that it is due to urea fertilizer impurities such as biuret [40,88,96], to cyanate formed by isomerization of urea in aqueous solutions [81], to the high pH or high concentration of NH_4^+ ions resulting from hydrolysis of urea fertilizer by soil urease [31,32,95], to the NH_3 produced by urea hydrolysis [31,32], or to NO_2^- produced through nitrification of urea N by soil microorganisms [29,30,31]. But recent work [9,10] leaves very little doubt that the adverse effects of urea fertilizer on seed germination and seedling growth in soil are due to the NH_3 formed through hydrolysis of urea by soil urease and are not due to urea itself, to urea fertilizer impurities such as biuret or cyanuric acid, or to NO_2^- formed by nitrification of urea N. Support for this conclusion was obtained from (a) comparison of the effects on seed germination in soil of purified urea, urea fertilizers, urea fertilizer impurities, and compounds formed by enzymatic and microbial transformations of urea in soil (Tables 1 and 2); (b) studies showing that NH_3 volatilized from soils treated with urea completely inhibited germination of seeds close to, but not in contact with, these soils; and (c) experiments showing that the adverse effect of urea fertilizer on seed germination in soil was completely eliminated when the soil was autoclaved to destroy urease or was treated with PPD or NBPT to inhibit soil urease activity before treatment with urea fertilizer (Table 3).

Table 1. Comparison of effects of purified urea, urea fertilizers, and urea fertilizer impurities on germination of seeds in Dickinson soil. After [10].

Substance added to soil	% Germination of seeds in 7 days			
(2.5 mg g^{-1} soil)	Barley	Corn	Rye	Wheat
None (control)	94	86	95	92
Purified urea	0	0	0	0
Urea fertilizers	0	0	0	0
Biuret	95	74	93	36
Cyanuric acid	97	86	94	92

Table 2. Comparison of effects of different forms of N on seed germination in Dickinson soil. After [10].

Form of N added to soil	% Germination of seeds in 7 days			
(1.00 mg N g^{-1} soil)	Barley	Corn	Rye	Wheat
None (control)	96	85	94	93
Urea	0	0	0	0
Ammonium as $(NH_4)_2SO_4$	94	84	96	91
Ammonium as $(NH_4)_2CO_3$	27	75	6	10
Ammonium as NH_4OH	0	0	0	0
Nitrite as KNO_2	0	0	0	0
Nitrate as KNO_3	94	85	92	94

Phytotoxicity of foliar-applied urea

Foliar application of N and other plant nutrients has potential advantages over soil application for fertilization of crops in that it may increase the efficiency of fertilizer use and allow relief of physiological stress [41,45]. Interest in foliar fertilization of soybean was

Table 3. Effect of PPD on germination of seeds in Dickinson soil treated with purified and fertilizer urea. After [10].

	% Germination of seeds in 7 days			
Soil treatment[a]	Barley	Corn	Rye	Wheat
None(control)	96	84	95	92
Purified urea	0	0	0	0
Purified urea + PPD	93	85	92	94
Fertilizer urea	0	0	0	0
Fertilizer urea + PPD	96	87	93	95

[a]Urea added at rate of 2.5 mg g^{-1} soil, PPD at rate of 25 μg^{-1} soil.

greatly stimulated by work by Garcia and Hanway [39] indicating that foliar fertilization of this crop during seed development could lead to substantial increases in the yield. But most studies of foliar fertilization of soybean during seed development have given disappointing results and have indicated that such fertilization usually leads to a decrease in yield and to some degree of leaf burn (leaf-tip necrosis) [44,45].

It is generally believed that leaf burn is at least partly responsible for the reduced yields observed after foliar fertilization [74] and that it is increased by low humidity and high temperature and by use of a too concentrated fertilizer solution [39]. It is also believed that the burn observed depends upon the form of N fertilizer used and that urea is less likely to cause foliage burn than other N fertilizers because it has a lower salt index and is more rapidly absorbed into the leaf [39,75]. But leaf burn has often been observed after foliar fertilization of plants with urea, and it has been reported that the leaf-burn observed with urea increases with leaf urease activity and is due to the NH_3 produced from urea by this activity [46,49,90].

Because the leaf-tip necrosis often observed after foliar fertilization of plants with urea is usually attributed to NH_3 formed through hydrolysis of urea by plant urease, Krogmeier *et al.* [53] studied the possibility that this necrosis could be eliminated or reduced by adding a urease inhibitor (phenylphosphorodiamidate) to the urea fertilizer. They found that although addition of this urease inhibitor to foliar-applied urea increased the urea content and decreased the NH_3 content of soybean leaves fertilized with urea, it increased the leaf-tip necrosis observed after fertilization. They concluded that this necrosis resulted from accumulation of toxic amounts of urea rather than accumulation of toxic amounts of NH_3. This conclusion was supported by their finding that the necrotic areas of soybean leaves treated with urea or with urea and phenylphosphorodiamidate contained much higher concentrations of urea than did the nonnecrotic areas.

Nickel (Ni) is an essential component of urease [34,35], and recent work has shown that plants deficient in Ni tend to be necrotic. For example, Brown *et al.* [17] observed that barley, wheat, and oat plants grown in Ni-deficient media were necrotic, and Shimada and Ando [83] noted that tomato and soybean plants having a low Ni content accumulated urea and developed necrotic leaf-tips when grown with urea as the sole source of N. These observations led Krogmeier *et al.* [54] to study the effect of Ni deficiency in soybean plants on the phytotoxicity of foliar-applied urea,

their rationale being that if this phytotoxicity is due to urea itself, it should be increased if the plants foliar-applied with urea have a low urease activity as a result of Ni deficiency. They measured the urea content, urease activity, and leaf-tip necrosis of leaves of soybean plants treated with urea after growth of the plants in nutrient solutions containing different amounts of Ni. They found that the urease activity of these leaves decreased, and that their urea content and leaf-tip necrosis increased, with decrease in the Ni content of the nutrient solution. Besides supporting the conclusion that the leaf-tip necrosis observed after foliar fertilization of soybeans with urea is due to accumulation of toxic amounts of urea in the soybean leaves, these observations indicate that Ni-deficient plants may be more susceptible to leaf burn when foliar-fertilized with urea than plants that are not deficient in Ni.

Joo *et al.* [50,51] have reported laboratory and field experiments indicating that, by reducing ammonia volatilization, NBPT and PPD have the potential to increase the N use efficiency of urea N applied as an aqueous fertilizer to Kentucky bluegrass turf. Their work indicated that NBPT is more effective than PPD for increasing the efficiency of urea N applied to turfgrass but that it can increase tip burn on fertilization with urea, presumably because its inhibitory effect on urease activity can lead to accumulation of toxic amounts of urea in the grass [52,53].

Effects of pesticides on transformations of urea N in soil

The use of pesticides in conjunction with urea fertilizer has increased dramatically during the past 25 years. There is a clear need, therefore, for information concerning the effects of pesticides on transformations of urea N in soils. To meet this need, Martens and Bremner [61–64] studied the effects of 46 herbicides, 17 fungicides, and 15 insecticides (Table 4) on urea hydrolysis and nitrification of urea N in two fine-textured and two coarse-textured soils. They found that when the 46 herbicides tested were applied at the rate of $5\ \mu g\ g^{-1}$ soil, none of them retarded hydrolysis of urea in the four soils used or retarded nitrification of urea N in the two fine-textured soils, but ten of them (amitrole, 2,4-D amine, chlorpropham, dinoseb, propanil, propham, acifluorfen, diclopfop methyl, fenoxaprop ethyl and tridiphane) retarded nitrification of urea N in the two coarse-textured soils. When the 15 insecticides tested were applied at the rate of $5\ \mu g\ g^{-1}$ soil none

326

Table 4. Pesticides studied

Herbicides	Herbicides	Insecticides	Fungicides
Acifluorfen	Ethalfuralin	Aldicarb	Anilazine
Alachlor	Fenoxaprop ethyl	Carbaryl	Benomyl
Amitrole	Fluazifop butyl	Carbofuran	Captan
Atrazine	Glyphosate	Chlorpyrifos	Chloranil
Bentazon	Haloxyfop methyl	Diazinon	Chloroneb
Bifenox	Linuron	Ethoprop	Chlorothalonil
Bromoxynil	Mefluidide	Fenitrothion	Fenaminosulf
Butylate	Metolachlor	Fenvalerate	Fenarimol
Chloramben	Methoxydim	Fonfos	Folpet
Chloropropham	Monuron	Isofenfos	Iprodione
Cinmethylin	Oryzalin	Lindane	Mancozeb
Cyanazine	Paraquat	Malathion	Maneb
2,4-D amine	Pendimethalin	Phorate	Metalaxyl
2,4-D ester	Picloram	Terbufos	Metham-sodium
Dalapon	Prometryn	Trimethacarb	PCNB
Dicamba	Propachlor		Terrazole
Diclofop methyl	Propanil		Thiram
Dimethazone	Propham		
Dinitramine	Siduron		
Dinoseb	Simazine		
Diuron	Tridiphane		
DPX-602	Trifluralin		
EPTC	Vernolate		

of them affected urea hydrolysis in the four soils used or nitrification of urea N in the two fine-textured soils, but five of them (carbaryl, lindane, trimethacarb, diazinon and fenitrothion) retarded nitrification of urea N in the two coarse-textured soils. When the 17 fungicides studied were applied at the rate of 1 μg g^{-1} soil, seven of them (anilazine, benomyl, chloranil, captan, maneb, mancozeb and thiram) retarded hydrolysis of urea in the two coarse-textured soils and one (maneb) retarded urea hydrolysis in all four of the soils used. All of the fungicides except benomyl, chloroneb, chlorothalonil, fenarimol and iprodione retarded nitrification of urea N in one of the coarse-textured soils when they were applied at the rate of 1 μg g^{-1} soil, and terrazole retarded nitrification of urea N in all four of the soils used when applied at this rate. These studies indicate that fungicides have a greater potential than herbicides or insecticides for retarding hydrolysis of urea and nitrification of urea nitrogen in soil.

Since the NO$_3^-$-N formed by nitrification of urea N in soil is susceptible to loss by denitrification, it is noteworthy that several workers have reported that deni-trification in soils is retarded by very small amounts of certain pesticides and nitrification inhibitors. For example, there are reports that denitrification is retarded by as little as 1 μg g^{-1} soil of atrazine [69,70], simazine [69,70], nitrapyrin [67,68] and etridiazole [71]. These reports could not be confirmed, however, in studies of the effects of 33 pesticides and 30 nitrification inhibitors on denitrification in soil [16,98,99].

Conclusions

The work reviewed indicates that NBPT has considerable potential as a fertilizer amendment for reduction of NH$_3$ volatilization, NO$_2^-$ accumulation, and other problems encountered in the use of urea as a fertilizer. It is not yet available commercially, but should be available in 1996. A recent summary by Hendrickson [48] of the results of 78 field trials with NBPT conducted over 5 years showed that addition of NBPT to urea fertilizer applied to corn increased grain yields by an average of 280 kg ha^{-1} and gave grain yields equivalent to those obtained with ca. 80 kg ha^{-1} of additional unamended urea-N.

As noted by Hauck [47], there is a clear need for a urea-based fertilizer that can be drilled directly with wheat and other cereal grain seeds. There seems little doubt that this need could be met by amending urea fertilizers with NBPT because studies using seeds of wheat, barley, oats, and rye have demonstrated that the adverse effects of urea fertilizer on seed germination, seedling growth, and early plant growth in soil can be eliminated or greatly reduced by amending the fertilizer with as little as 0.01% (wt/wt) of NBPT (9).

References

1. Aleem MIH and Alexander M (1960) Nutrition and physiology of *Nitrobacter agilis*. Appl Microbiol 8: 80–84
2. Alexander M (1965) Nitrification. In: Bartholomew WV and Clark FE (eds) Soil Nitrogen, pp 307–343. Am Soc Agron, Madison, WI
3. Beyrouty CA, Sommers LE and Nelson DW (1988) Ammonia volatilization from surface-applied urea as affected by several phosphoroamide compounds. Soil Sci Soc Am J 52: 1173–1178
4. Beyrouty, CA, Nelson DW and Sommers LE (1988) Effectiveness of phosphoroamides in retarding hydrolysis of urea surface-applied to soils with various pH and residue cover. Soil Sci 145: 345–352
5. Bremner JM and Chai HS (1986) Evaluation of *N*-butyl phosphorothioic triamide for retardation of urea hydrolysis in soil. Commun Soil Sci Plant Anal 17: 337–351
6. Bremner JM and Chai HS (1989) Effects of phosphoroamides on ammonia volatilization and nitrite accumulation in soils treated with urea. Biol Fertil Soils 8: 227–230
7. Bremner JM and Douglas LA (1971) Inhibition of urease activity in soils. Soil Biol Biochem 3: 297–307
8. Bremner JM and Douglas LA (1973) Effects of some urease inhibitors on urea hydrolysis in soils. Proc Soil Sci Soc Am 37: 225–226
9. Bremner JM and Krogmeier MJ (1988) Elimination of the adverse effects of urea fertilizer on seed germination, seedling growth, and early plant growth in soil. Proc Natl Acad Sci USA 85: 4601–4604
10. Bremner JM and Krogmeier MJ (1989) Evidence that the adverse effect of urea fertilizer on seed germination in soil is due to ammonia formed through hydrolysis of urea by soil urease. Proc Natl Acad Sci USA 86: 8185–8188
11. Bremner JM and Krogmeier MJ (1990) Effects of urease inhibitors on germination of seeds in soil. Commun Soil Sci Plant Anal 21: 311–321
12. Bremner JM and Martens DA (1987) Decomposition of phenylphosphorodiamidate in soil. Agron Abstr 79
13. Bremner JM, McCarty GW, Yeomans JC and Chai HS (1986) Effects of phosphoroamides on nitrification, denitrification, and mineralization of organic nitrogen in soil. Commun Soil Sci Plant Anal 17: 369–384
14. Bremner JM, McCarty GW and Higuchi T (1991) Persistence of the inhibitory effects of phosphoroamides on urea hydrolysis in soils. Commun Soil Sci Plant Anal 22: 1519–1526
15. Bremner JM and Mulvaney RL (1978) Urease activity. In: Burns RG (ed) Soil Enzymes pp 149–196. Academic Press, New York
16. Bremner JM and Yeomans JC (1986) Effects of nitrification inhibitors on denitrification of nitrate in soil. Biol Fert Soils 2: 173–179
17. Brown PH, Welch RM and Cary EE (1987) Nickel: A micronutrient essential for higher plants. Plant Physiol 85: 801–803
18. Bundy LG and Bremner JM (1973). Effects of substituted *p*-benzoquinones on urease activity in soils. Soil Biol Biochem 5: 847–853
19. Bundy LG and Bremner JM (1974) Effects of nitrification inhibitors on transformations of urea nitrogen in soils. Soil Biol Biochem 6: 369–376
20. Bundy LG and Bremner JM (1974) Effects of urease inhibitors on nitrification in soils. Soil Biol Biochem 6: 27–30
21. Buresh RJ, De Datta SK, Padilla JL and Samson MI (1988) Effect of two urease inhibitors on floodwater following urea application to lowland rice. Soil Sci Soc Am J 52: 856–861
22. Byrnes BH, Savant NK and Craswell ET (1983) Effect of a urease inhibitor phenyl phosphorodiamidate on the efficiency of urea applied to rice. Soil Sci Soc Am J 47: 270–274
23. Byrnes BH and Amberger A (1989) Fate of broadcast urea in a flooded soil when treated with N-(*n*-butyl)thiophosphoric triamide, a urease inhibitor. Fert Res 18: 221–231
24. Byrnes BH, Gutser R and Amberger A (1989) Greenhouse study on the effects of the urease inhibitors phenyl phosphorodiamidate and N-(*n*-butyl)thiophosphoric triamide on the efficiency of urea applied to flooded rice. Z Pflanzenernahr Bodenkd 152: 67–72
25. Cai GX, Freney JR, Muirhead WA, Simpson JR, Chen DL and Trevitt ACF (1989) The evaluation of urease inhibitors to improve the efficiency of urea as a N-source for flooded rice. Soil Biol Biochem 21: 137–145
26. Calancea L and Kiss S (1984) Effect of hydroquinone on the uptake of nitrogen from soil reserve and urea by *Lolium multiflorum*. In: Fifth Symposium on Soil Biology, pp 65–68. Romanian Society of Soil Science, Bucharest
27. Catchpoole VR (1975) Pathways for losses of fertilizer nitrogen from Rhodes grass pasture in south-eastern Queensland. Aust J Agric Res 26: 259–268
28. Chai HS and Bremner JM (1987) Evaluation of some phosphoroamides as soil urease inhibitors. Biol Fertil Soils 3: 189–194
29. Court MN, Stephen RC and Waid JS (1962) Nitrite toxicity arising from the use of urea as a fertilizer. Nature 194: 1263–1265
30. Court MN, Stephen RC and Waid JS (1964) Toxicity as a cause of the inefficiency of urea as a fertilizer. I. Review. J Soil Sci 15: 42–48
31. Court MN, Stephen RC and Waid JS (1964) Toxicity as a cause of the inefficiency of urea as a fertilizer. II. Experimental. J Soil Sci 15: 49–65
32. Creason GL, Schmitt MR, Douglass EA and Hendrickson LL (1990) Urease inhibitory activity associated with N-(*n*-butyl)thiophosphoric triamide is due to formation of its oxon analog. Soil Biol Biochem 22: 209–211
33. Cummins DG and Parks WL (1961) The germination of corn and wheat as affected by various fertilizer salts at different soil temperatures. Soil Sci Soc Am J 25: 47–49
34. Dixon NE, Gazzola C, Blakeley RL and Zerner B (1975) Jack-bean urease: (E.C.3.5.1.5.3). A metalloenzyme. A simple biological role for nickel. J Am Chem Soc 97: 4131–4133

35. Dixon NE, Blakeley RL and Zerner Burt (1980) Jack-bean urease: (E.C.3.5.1.5.3). A metalloenzyme. A simple biological role for nickel. J Am Chem Soc 97: 4131–4133

36. Fenn LB and Hossner LR (1985) Ammonia volatilization from ammonium or ammonium-forming nitrogen fertilizers. Adv Soil Sci 1: 123–169

37. Freney JR, Smith CJ and Mosier AR (1992) Effect of a new nitrification inhibitor (wax coated calcium carbide) on transformations and recovery of fertilizer nitrogen by irrigated wheat. Fert Res 32: 1–11

38. Freney JR, Chen DL, Mosier AR, Rochester LJ, Constable GA and Chalk PM (1993) Use of nitrification inhibitors to increase fertilizer nitrogen recovery and lint yield in irrigated cotton. Fert Res 34: 37–44

39. Garcia LR and Hanway JJ (1976) Foliar fertilization of soybeans during the seed-filling period. Agron J 68: 653–657

40. Gasser JKR (1964) Urea as a fertilizer. Soils Fert 27: 175–180

41. Gooding MJ and Davies WP (1992) Foliar fertilization of cereals: a review. Fert Res 32: 209–222

42. Gould, WD, Hagedorn C and McCready RGL (1986) Urea transformations and fertilizer efficiency in soil. Adv Agron 40: 209–238

43. Goyal SA and Huffaker RC (1984) Nitrogen toxicity in plants. In: Hauck RD (ed) Nitrogen in Crop Production, pp 97–118. Am Soc Agron, Madison, WI

44. Gray RC (1977) In: Situation 77. Bull Y-115, pp 54–58. National Fertilizer Development Center, Muscle Shoals, AL

45. Gray RC and Akin GW (1984) Foliar fertilization. In: Hauck RD (ed) Nitrogen in Crop Production, pp 579–584. Am Soc Agron, Madison, WI

46. Harper JE (1984) Uptake of organic nitrogen forms by roots and leaves. In: Hauck RD (ed) Nitrogen in Crop Production, pp 165–182. Am Soc Agron, Madison, WI

47. Hauck RD (1984) Technological approaches to improving the efficiency of nitrogen fertilizer use by crop plants. In: Hauck RD (ed) Nitrogen in Crop Production, pp 551–560. Am Soc Agron, Madison, WI

48. Hendrickson LL (1992) Corn yield response to the urease inhibitor N-(n-butyl)thiophosphoric triamide: 5 year summary. J Prod Agric 5: 131–137

49. Hinsvark ON, Wittwer SH and Tukey HB (1953) The metabolism of foliar-applied urea. I. Relative rates of $C^{14}O_2$ production by certain vegetable plants treated with labeled urea. Plant Physiol 28: 70–76

50. Joo YK, Christians NE and Bremner JM (1987) Effect of N-(n-butyl) thiophosphoric triamide (NBPT) on growth response and ammonia volatilization following fertilization of Kentucky bluegrass (Poa pratensis L.) with urea. J. Fertil. Issues 4: 98–102

51. Joo YK, Christians NE Spear GT and Bremner JM (1992) Evaluation of urease inhibitors as urea amendments for use on Kentucky bluegrass turf Crop Sci 32: 1397–1401

52. Krogmeier MJ, McCarty GW and Bremner JM (1989) Potential phytotoxicity associated with the use of urease inhibitors. Proc Natl Acad Sci USA 86: 1110–1112

53. Krogmeier MJ, McCarty GW and Bremner JM (1989) Phytotoxicity of foliar-applied urea. Proc Natl Acad Sci USA 86: 8189–8191

54. Krogmeier MJ, McCarty GW, Shogren DR and Bremner JM (1991) Effect of nickel deficiency in soybeans on the phytotoxicity of foliar-applied urea. Plant and Soil 135: 283–286

55. Liantie L, Wang ZP, Van Cleemput O and Baert L (1993) Urea N uptake efficiency of ryegrass (Lolium perenne L.) in the presence of urease inhibitors. Biol Fertil Soils 15: 225–228

56. Li RH, Zhao XY and Zhou LK (1991) Application of hydroquinone-containing urea in China. In: The Collections of the the 7th National Conference on Soil Science pp 131–132. Chinese National Society of Soil Science, Changsha

57. Lu W, Lindau CW, Pardue JH, Patrick WH, Reddy KK and Khind CS (1989) Potential of phenylphosphorodiamidate and N-(n-butyl)thiophosphoric triamide for inhibiting urea hydrolysis in simulated oxidized and reduced soils. Commun Soil Sci Plant Anal 20: 775–788.

58. Low AJ and Piper FJ (1961) Urea as a fertilizer. J Agric Sci 57: 249–255

59. Magalhaes AMT and Chalk PM (1987) Nitrogen transformations during hydrolysis and nitrification of urea. 2. Effect of fertilizer concentration and nitrification inhibitors. Fert Res 11: 173–184

60. Martens DA and Bremner JM (1984) Effectiveness of phosphoroamides for retardation of urea hydrolysis in soils. Soil Sci Soc Am J 48: 302–305

61. Martens DA and Bremner JM (1984) Effects of fungicides on transformations of urea nitrogen in soil. Agron Abstr 189

62. Martens DA and Bremner JM (1985) Effects of grass herbicides on transformations of urea nitrogen in soil. Agron Abstr 160

63. Martens DA and Bremner JM (1986) Effects of insecticides on transformations of urea nitrogen in soil. Agron Abstr 183

64. Martens DA and Bremner JM (1993) Influence of herbicides on transformations of urea nitrogen in soils. J Environ Sci Health 28: 377–393

65. McCarty GW, Bremner JM and Chai HS (1989) Effect of N-(n-butyl)thiophosphoric triamide on hydrolysis of urea by plant, microbial, and soil urease. Biol Fertil Soils 8: 123–127

66. McCarty GW and Bremner JM (1989) Formation of phosphoryl triamide by decomposition of thiophosphoryl triamide in soil. Biol Fertil Soils 8: 290–292

67. McElhannon WS and Mills HA (1981) Inhibition of denitrification by nitrapyrin with field-grown sweet corn. J Am Soc Hort Sci 106: 673–677

68. McElhannon WS and Mills HA (1981) Suppression of denitrification with nitrapyrin. Hort Science 16: 530–531

69. McElhannon WS, Mills HA and Bush PB (1984) Simazine and atrazine-suppression of denitrification. HortScience 19: 218–219

70. Mills HA (1984) A procedure for rapidly evaluating the effect of chemicals on denitrification. Commun Soil Sci Plant Anal 15: 1007–1016

71. Mills HA and McElhannon WS (1984) Terrazole suppression of denitrification. HortScience 19: 54–55

72. Mulvaney RL and Bremner JM (1978) Use of p-benzoquinone and hydroquinone for retardation of urea hydrolysis in soils. Soil Biol Biochem 10: 297–302

73. Mulvaney RL and Bremner JM (1981) Control of urea transformations in soils. In: Paul EA and Ladd JN (eds) Soil Biochemistry, Vol. 5, pp 153–196. Marcel Dekker, New York

74. Poole WD, Randall GW and Ham GE (1983) Foliar fertilization of soybeans. II Effect of biuret and application time of day. Agron J 75: 201–203

75. Rader LF, White LH and Whittaker CW (1943) The salt index - A measure of the effect of fertilizers on the concentration of the soil solution. Soil Sci 55: 201–218

76. Radel RJ, Gautney J and Peters GE (1988) Urease inhibitor developments. In Bock B and Kissel D (eds) Ammonia

Volatilization from Urea Fertilizers, pp 111–136. Tennessee Valley Authority, Muscle Shoals, AL

77. Radel RJ, Randle AA, Gautney J, Bock BR and Williams HM (1992) Thiophosphoryl triamide: a dual purpose urease/nitrification inhibitor. Fert Res 31: 275–280

78. Rao DLN and Ghai SK (1985) Effect of phenylphosphorodiamidate on urea hydrolysis, ammonia volatilization and rice growth in an alkali soil. Plant and Soil 94: 313–320

79. Rao DLN and Ghai SK (1986) Urease inhibitors: effect on wheat growth in an alkali soil. Soil Biol Biochem 18: 255–258

80. Rappaport BD and Axley JH (1984) Potassium chloride for improved urea fertilizer efficiency. Soil Sci Soc Am J 48: 399–401

81. Rotini OT (1956) La transformazione enzimatica dell'-urea nell terreno. Ann Fac Agric Univ Pisa 17: 1–25

82. Schlegel AJ, Nelson DW and Sommers LE (1987) Use of urease inhibitors and urea fertilizers on winter wheat. Fert Res 11: 97–111

83. Shimado N and Ando T (1980) Role of nickel in plant nutrition. II. Effect of nickel on the assimilation of urea by plants. Nippon Dojo Hiryogaku Zasshi 51: 493–496

84. Smith CJ, Freney JR, Chalk PM, Galbally IE, McKenney DJ and Cai GX (1988) Fate of urea nitrogen applied in solution in furrows to sunflowers growing on a red-brown earth: transformations, losses and plant uptake. Aust J Agric Res 39: 793–806

85. Smith CJ, Freney JR, Chapman SL and Galbally IE (1989) Fate of urea nitrogen applied to irrigated wheat at heading. Aust J Agric Res 40: 951–963.

86. Smith CJ, Freney JR and Mosier AR (1993) Effect of acetylene provided by wax-coated calcium carbide on transformations of urea nitrogen applied to an irrigated wheat crop. Biol Fert Soils 16: 86–92

87. Terman GL (1979) Volatilization losses of nitrogen as ammonia from surface-applied fertilizers, organic amendments, and crop residues. Adv Agron 31: 189–223

88. Tomlinson TE (1970) Urea: Agronomic applications. Proc Fert Soc 113: 1–76

89. Van Cleemput O and Wang Zhengping (1991) Urea transformations and urease inhibitors. Trends Soil Sci 1: 45–52

90. Vasilas BL, Legg JO and Wolf DC (1980) Foliar fertilization of soybeans: Adsorption and translocation of ^{15}N-labeled urea. Agron J 72: 271–275

91. Vlek P and Fillery I (1984) Improving nitrogen efficiency in wetland rice soils. Proc Fert Soc No. 230

92. Voss RD (1984) Potential for use of urease inhibitors. In: Hauck RD (ed) Nitrogen in Crop Production, pp 571–577. Am Soc Agron, Madison, WI

93. Wang Zhengping, Van Cleemput O and Baert L (1989) Effect of urease inhibitors on nitrification in soil. Soil Use Manage 6: 41–43

94. Wang Zhengping, Liante L, Van Cleemput O and Baert L (1989) Effect of urease inhibitors on denitrification in soil. Soil Use Manage 7: 230–233

95. Widdowson, FV Penney A and Cooke GW (1960) The value of calcium nitrate and urea for main-crop potatoes and kale. J Agric Sci 55: 1–10

96. Wilkinson SR and Ohlrogge AJ (1960) Influence of biuret and urea fertilizers containing biuret on corn plant growth and development. Agron J 52: 560–562

97. Xiaoyan, Z, Likai Z, Ronghua L, Guiru A and Bo Z (1993) Effect of hydroquinone on maize yield and urea efficiency. Soil Biol Biochem 25: 147–148

98. Yeomans JC and Bremner JM (1985a) Denitrification in soil: Effects of herbicides. Soil Biol Biochem 17: 447–452

99. Yeomans JC and Bremner JM (1985b) Denitrification in soil: Effects of insecticides and fungicides. Soil Biol Biochem 17: 453–456

100. Yeomans JC and Bremner JM (1986) Effects of urease inhibitors on denitrification in soil. Commun Soil Sci Plant Anal 17: 63–73

101. Youngdahl L, Lupin MS and Craswell ET (1986) New developments in nitrogen fertilizer for rice. Fert Res 9: 149–160

102. Zhao XY, Zhou LK and Wu GY (1988) Environmental impact assessment of hydroquinone used as a urease inhibitor (1). AgroEnviron Protect 7: 18–20

103. Zhao XY, Zhou LK and Xie CG (1989) Environmental impact assessment of hydroquinone used as a urease inhibitor (2). AgronEnviron Protect 8: 28–30

104. Zhou LK and Wu GY (1988) Effect of urease inhibitor hydroquinone on the efficiency of urea fertilizer. Acta Pedol 25: 191–198

Fertilizer Research **42**: 331–338, 1995.

© 1995 *Kluwer Academic Publishers. Printed in the Netherlands.*

Nitrate contamination of groundwater: Measurement and prediction

M.J. Goss & D. Goorahoo
Centre for Land and Water Stewardship, University of Guelph, Guelph, Ontario N1G 2W1, Canada

Key words: groundwater quality, nitrate contamination, nitrogen budgets, Ontario, well survey

Abstract

Agriculture makes a significant contribution to the diffuse source contamination of surface and groundwater resources, particularly contributing to the NO_3^- contamination of groundwater. Two approaches were adopted to evaluate management practices (within the context of the whole farming system) for their impacts on the environment : (1) measurement of the quality of groundwater under different farming systems, and (2) comparison of predictions of the impact of farming systems on water quality, obtained using whole farm N budgets, with measured values.

The Ontario Farm Groundwater Quality Survey evaluated the rural groundwater quality in Ontario, with respect to common contaminants including NO_3^-. Approximately 1300 domestic farm wells were sampled, and wells were drilled in some fields of farms involved in the study. NO_3^- was present at concentrations above the maximum acceptable for drinking water (10 mg N 1^{-1}) in 14% of wells, including 7% of wells that also had unacceptable concentrations of coliform bacteria. Significant levels of NO_3^- contamination were observed under most agricultural land use practices investigated.

Calculation of N budgets was simplified by assuming that there was no net change in the N content of farm assets. The N inputs to agricultural systems considered were: purchases from off-farm suppliers, N_2 fixation and atmospheric deposition. Symbiotic N_2 fixation was estimated from empirical relationships between crop yield and N_2 fixed. The N outputs were in sales of plant and animal produce, gaseous and leaching losses. Gaseous loss was assumed to result only from volatilization of ammonia, estimated to be 39% of total manure N.

We have identified one cash crop farming system where there was a true balance. The rotation included corn soybeans and wheat, with two years of soybean always being grown before corn. Many livestock farms, including two organic farms, gave large imbalances of N which might indicate that these operations were not in equilibrium.

The relationship between measured and predicted values of NO_3^--N expected in the groundwater under the different management systems showed that the simplified N budget overestimated the NO_3^--N concentration by about one third. However, the budget approach appeared to identify farms where contamination was likely even if the actual amount was over estimated. Simplified budgets could therefore be used to compare the potential of different farming systems for causing environmental contamination.

Introduction

There is increasing awareness that water resources are vulnerable to contamination from point and diffuse sources within agricultural watersheds (Byrnes, 1990; Follett and Walker, 1989). In industrial areas much emphasis has been placed on preventing point source contamination from manufacturing sites. However, diffuse source contamination can be much more difficult to deal with. Agricultural activity is recognised as making a significant contribution to diffuse source contamination of surface and groundwater resources (Addiscott *et al.*, 1991). In particular, agriculture is considered to be the major contributor to the NO_3^- contamination of groundwater (Juergens-Gschwind, 1989; Power and Schepers, 1989).

Much effort has been aimed at optimizing application rates of manures and mineral fertilizers to meet crop yield goals. The current goal is to provide sufficient food and to protect water quality. This requires the evaluation of management practices within the context of the whole farming system, and assessment of impact on the environment. One way to study the effectiveness of current practices is to measure the quality of groundwater under different farming systems. Another approach is to compare predictions of the impact of farming systems on water quality, obtained from empirical or mechanistic stimulation, with measured values. Limitations in our understanding of soil and plant processes can then be identified, and possibilities for extending recommendations on fertilizer practices to other soils and climates can be increased. Where agriculture has resulted in contamination, new practices need to be developed or existing systems modified to improve water quality. A potentially useful method for predicting losses of NO_3^- from agriculture to groundwater is to calculate the N balance for a whole farm, taking account of animals and crops. Excess N in the farming system can be equated with losses to the environment. Causes of excess nutrients can be identified by the sensitivity of the balance to different aspects of management.

This paper considers the use of a groundwater survey and a simplified N budget for estimating the impact of agriculture on the contamination of water resources with NO_3^-, including an assessment for farms where organic manures only were used.

Methods

Measurement of NO_3^- contamination

In 1991–92, the Federal Government of Canada, commissioned a major survey of groundwater quality in Ontario "The Ontario Farm Groundwater Quality Survey" (Rudolph and Goss, 1993). The aim was to evaluate the rural groundwater quality in Ontario at a provincial scale. This was approached through two separate sampling programmes of approximately 1300 domestic farm wells. Wells were tested for NO_3^-, several common herbicides, and for total and faecal coliform bacteria. In addition a limited number of monitoring wells were specially installed in fields, and were sampled for the same contaminants. The participating farm families also completed questionnaires about the practices and activities carried out in the vicinity of the well, and on the land-use management. This multifaceted

approach allowed the impact of farming practices on drinking water quality to be studied.

On a number of other farms, solution samplers were used to sample the water draining from the rooting zone in fields representative of the rotation. The samplers, were installed at an angle of 45° to a depth of 0.8 m. Each sampler consisted of a porous ceramic cup, 50 mm long × 20 mm OD, attached to a length of PVC pipe. Two pieces of narrow-bore nylon tubing were sealed into the porous cup, one to apply a pressure of −80 kPa and draw water into the cup, the other for collecting the sample using a replaceable evacuated tube sealed with a septum. Attempts were made to collect samples in spring and fall following major rainfall events.

Prediction of NO_3^- contamination

The basic relationships for the N budget of a farm can be summarized as:

N in inputs = N in output + change in the N contents of the soil, livestock and other components

Calculation of an N budget can be simplified by assuming that there is no net change in the N content of farm assets. Thus for an arable farm it is assumed that soil organic matter content, and consequently soil N content, remain constant on a yearly basis for monoculture systems or over the course of a rotation when a sequence of crops are grown (Fried *et al.*, 1976). For a livestock operation it is assumed further that the number of animals and their demography remain constant.

Components of a simplified N budget

The main components of the N inputs to agricultural systems are derived from purchases from off-farm suppliers (e.g. animals, seed, fertilizers and feed), and from natural processes that occur during the growth of crops (e.g. symbiotic N_2 fixation) or from natural processes that are influenced by anthropogenic activity (e.g. atmospheric deposition).

Fertilizer inputs were obtained from farmer records. N contents of animal manures were taken from published tables (Fraser, 1985), as were the N content of seeds used in crop production (McBride, 1987). Feed contents were determined in the same way. The N in animals bought in were estimated from average weights of cattle, pigs or poultry typically purchased for fattening, breeding or milking, and assuming typical values for protein content (Ensminger and Olentine, 1978).

Natural inputs through symbiotic N_2 fixation were estimated from empirical relationships between crop yields and N_2 fixed (Barry *et al.*, 1993).

A single value of 18.4 kg N ha^{-1} was used for atmospheric deposition (Barry *et al.*, 1993).

The main outputs of N from agricultural systems are in the form of direct sales of plant and animal materials, and from gaseous and leaching losses. Sales were calculated from the crude protein content or N concentration of materials, and the weight of the material. Gaseous loss was assumed to result only from volatilization of ammonia, estimated to be 39% of total manure N (Barry *et al.*, 1993).

The excess N on the farm at the end of a crop cycle was assumed to be susceptible to leaching in the total through drainage. This was estimated to average 160 mm in Ontario (Barry *et al.*, 1993). The concentration of NO_3^--N predicted was compared with that in the drinking water well, in the water collected from the well installed during the groundwater quality survey, or in the water collected from the solution samplers.

Field investigations of variables used in the simplified N budgets

A number of the values used in calculating the N budgets have not been validated. They were best estimates based on information obtained from the literature. A full validation of individual values would require considerable investment in resources, but check measurements were carried out on a number of farms.

Atmospheric deposition

Wet deposition of N was calculated using the monthly precipitation and the average concentration of NH_4^+ and NO_3^- in the rainwater. On three farms, rainfall was collected in plastic-wedge-style rain gauges over the period 1st May to 1st October 1992. The rainwater was transferred to 1 l amber, high density polyethylene bottles for storage in a refrigerator until collected for analysis at the end of each month. Determination of the N concentration in precipitation was restricted to the summer months because the rain gauges and collection bottles were unsuitable for winter conditions, or for snow collection.

Bulk deposition of N was collected for the months of August and September for two of the farms, and for September only in the case of the third farm. Samples were collected using Teflon macrofiltration screen material, mesh size 150 μm, secured over the top of a polypropylene funnel of 180 mm diameter that was connected to a glass bottle. The screen was used to trap any deposition of aerosol particles or particulate matter that settled by gravity. It also prevented the entry of insects into the bottle. At the end of each month, the inner sides of the funnel and the teflon mesh were washed with deionised water, and the total volume of water, rainwater and wash water, in the collection bottle was recorded. A subsample of this water was taken back to the laboratory for analysis. Only on one farm did the N measured in bulk deposition exceed that in the rain gauge. On a further eight farms, rainwater was collected for the months of May and August in gauges considered to be large enough to act as bulk deposition samplers.

The ammoniacal and NO_3^- components of all the water samples collected were determined simultaneously using a TRAACS-800 autoanalyzer (Tel and Heseltine, 1990).

Ammonia volatilization from spread manure

The technique of Schjoerring *et al.* (1992) was adapted to estimate the amount of NH_3 volatilized from surface applied composted liquid dairy cattle manure (CLCM). Passive flux samplers were used to measure the concentration of NH_3 in the air flowing from the source to the surroundings, and vice versa. Each sampler consisted of two glass tubes, 100 mm in length and 7 mm ID, connected by a small piece of silicon tubing. The inner surface of the tubes were coated with oxalic acid. A stainless steel disc, 0.5 mm in thickness and with a 1 mm diameter hole drilled in the centre, was glued to the end of one of the two tubes. The stainless steel disc acted to decrease the air speed inside the tubes in order to achieve a low friction resistance and a high NH_3 collection efficiency. The experimental area was a circular plot of radius 15 m (area 707 m^2) in a cultivated field on a dairy farm. CLCM equivalent to 45,000 l ha^{-1} (140 kg total N ha^{-1} and 45 kg NH_4^+-N ha^{-1}) was applied to the plot by broadcasting to one side of a tractor-drawn, top-loading tank spreader. Masts were quickly erected at right angles to each other on the circumference of the plot. Pairs of samplers, one with the steel disc directed towards the centre, the other with the disc away from the plot, were mounted at heights of 0.50, 1.0, 1.5 and 2.0 m on each mast. At the end of 20 hours exposure the flux samplers were taken down from the masts and closed immediately with parafilm.

Samples of the CLCM were taken from the tanker for analysis of total Kjeldahl N, NH_4^+-N, NO_3^--N and dry matter content. The oxalic acid coating was washed from the passive sampler tubes using deionised water, and the NH_4^+-N determined (Tel and Goorahoo, 1993).

N_2 fixation by soybeans

Soybeans (*Glycine max* cv. Maple Donovan) were grown on three fields of one farm in 1992. The seeds were inoculated with *Bradyrhizobium japonicum* strain 532C prior to sowing in rows 18 cm apart at 108 kg ha^{-1}. Microplots 2×2 m were sown with the non-nodulating variety 'Evans' at the same density. Yield of the nodulating crop was assessed on two randomly selected plots 1×1 m, and yield of the non-nodulating variety was determined in the 2×2 m microplots. The above ground portions of the plants were sampled, and the pods were separated and hulled to obtain the grain. In addition, the above-ground portion and the roots of four plants were collected from each plot. The roots were washed under running tap water to remove adhering soil, and then examined for the presence of pink nodules. The plants were dried at 60 °C, ground and analyzed for N content. The grain yield from each plot was calculated at a moisture content of 14%.

The amount of N_2 fixed by soybeans on each field was determined from the difference in the N content between nodulating and non-nodulating soybeans.

Results

The general quality of rural groundwater

A key finding of the Ontario Farm Groundwater Quality Survey (Rudolph and Goss, 1993) was that NO_3^- in excess of the maximum acceptable concentration (10 mg N l^{-1}) was found in 14% of wells (Table 1), including 7% of wells that also had unacceptable concentrations of coliform bacteria.

Water quality and cropping systems

Farms with clearly identifiable cropping or grazing systems were divided into different land use classes. About 20 per cent of wells in each major class contained NO_3^- in excess of 10 mg N l^{-1} (Table 1).

Table 1. Number of water wells that exceeded 10 mg nitrate-N l^{-1} for farms in Ontario using various cropping systems (Rudolph and Goss, 1993)

Land use	Wells tested	Exceeds Ontario objective
	(no.)	(%)
>90% row crop	183	12.8 ± 5.0[a]
30–90% row crop	385	15.7 ± 3.7
>85% small grains	28	5.4 ± 8.4
>50% hay	158	8.3 ± 4.4
>50% pasture	78	12.9 ± 7.6
Berries	6	16.7 ± 15.2
Field vegetables	18	30.6 ± 21.6

[a] ± 95% confidence interval.

There were no significant differences between classes as diverse as row cropping (continuous cropping with row crops such as soybeans and maize), grain cropping (turf and small grain production) and berry cropping (soft fruit production) (Rudolph and Goss, 1993).

Simplified N budgets

Validation of literature values

Atmospheric deposition

The atmospheric deposition of N determined from bulk samplers ranged between 11.1 and 46 kg N ha^{-1}, and the average value was 18.4 kg N ha^{-1}. As nitric oxide deposition was not likely to have been captured, these values were likely to have been about 3 kg N ha^{-1} less than that actually adding to the local N burden (Ro et al., 1988). Nonetheless, given the range of values the results give good support to the average value of 18.4 kg N ha^{-1} estimated from the literature (Barry et al., 1993).

Ammonia volatilization from spread manure

The horizontal net flux of NH_3 from the plot treated with composted liquid cattle manure, decreased with height (Fig. 1), and could be represented by the following quadratic equation:

$$\sum F_h = 0.019z^2 - 9.22z - 105.4$$

where F_h is the horizontal flux and z is height in metres.

The vertical net flux of NH_3 from the experimental plot was calculated as 29.10 μg NH_3-N $m^{-2}s^{-1}$ or

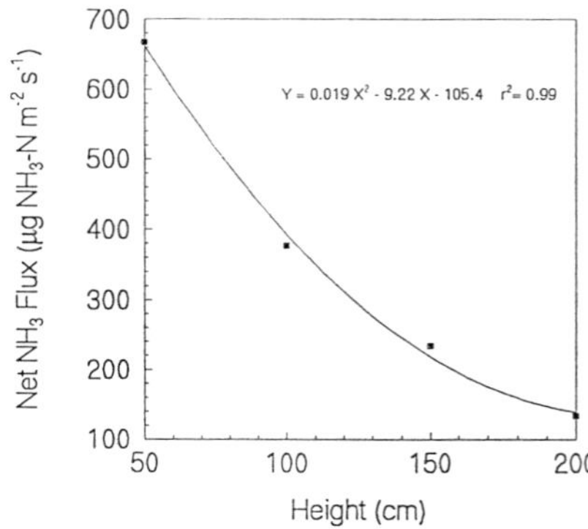

Fig. 1. Relation between net horizontal flux of NH₄ from composted liquid cattle manure and height above the soil surface.

1.05 kg NH$_3$-N ha^{-1}h^{-1}. The NH$_3$ volatilized was therefore 21 kg NH$_3$-N ha^{-1} over the 20 hour period. This amount would represent 15% of the total N and 46% of the ammoniacal N applied.

The vertical net flux of 1.05 kg NH$_3$-N ha^{-1}h^{-1} is comparable to maximum fluxes during a general diurnal pattern observed by Beauchamp *et al.* (1982). In that study, the maximum ammonia volatilization from liquid dairy cattle manure occurred after midday at times corresponding to maximum daily temperatures.

The 46% loss of applied ammoniacal N determined in this study lies within the wide range of volatilization losses reported for various manures (Gordon *et al.*, 1988; Lauer *et al.*, 1976; Thompson *et al.*, 1990). This loss over 20 hours after application is higher than the 24–33% reported by Beauchamp et al. (1982), and the 44–60% reported by Thompson *et al.* (1990), for losses that occurred within six days of application. However, losses as high as 62% of NH$_4^+$-N in pig or cattle slurry have been observed, in which 24–39% of the total NH$_4^+$-N loss occurred during the first hour following application and up to 85% within 12 hours (Pain *et al.*, 1989).

Losses of N from manure in animal housing and during storage appear to be less influenced by management practices than losses during application. For farms in Ontario, the loss from manure during collection and storage (liquid or solid) was about 28% of total N excreted (Beauchamp and Burton, 1985). Similar losses have been measured when manure was composted (Kirchmann, 1985). The loss measured in the field was therefore consistent with an average loss of N from manure, totalling 39% of excreted N over the period of excretion to plant utilization.

N$_2$ fixation by soybeans

The average soybean yields obtained for the three fields were 0.85, 1.33 and 1.55 t ha^{-1} were appreciably less than the 1987–1990 average of 2.3 t ha^{-1} for the farm because adverse weather conditions, particularly the cool temperature, delayed crop maturation. The cumulative crop heat units (an index based on day and night air temperatures, and the relationship between development rate and temperature for warm season crops Brown, 1985) recorded between May and October, 1992 were 9351 1-350 less than average (K. Reid, 1993, Ontario Ministry Agric Food, pers. comm.).

The average N$_2$ fixed was 94 kg N ha^{-1}, and ranged from 53 to 108 kg N ha^{-1}. These values could be underestimations of the actual amount of N$_2$ fixed because fallen leaves were not collected. The values obtained might also have been less than average because of the weather in 1992.

This value of N$_2$ fixation by soybeans was somewhat less than the 108 to 151 kg N ha^{-1} reported from other field experiments with the same variety conducted in Ontario from 1988 to 1989 (Ravuri, 1992). The amount of N$_2$ fixed by soybean showed no correlation with grain yield (Fig. 2). The soybeans had not reached physiological maturity when the plants were harvested in December 1992, and the harvest index was likely to be considerably smaller than normal. As N$_2$ fixation in soybean is virtually completed by the pod filling stage (Deibert *et al.*, 1979), this would account for the measured values being much greater than expected from those predicted by the equation derived by Barry et al. (1993). The proposed method of prediction of N$_2$-fixation still requires validation. Nonetheless, for typical grain yields of soybeans in Ontario of about 2.5 t ha^{-1}, the predicted N$_2$-fixation is close to 100 kg N ha^{-1}, a value close to the mean reported here.

Excess N on farms in Ontario

The main assumption made in the simplified budgetary approach was that the farming practices had reached a state of equilibrium. An established orchard would be expected to be close to equilibrium, and hence be a good. test of the budgetary approach. Close agreement was found between the concentration of NO$_3^-$ predicted to be in the groundwater under an apple orchard farm (13.5 mg N l^{-1}), and the value measured in the drinking water well (13. 6 mg N l^{-1}). (Goss *et al.*, 1993).

In many parts of southern Ontario the most common rotation is based on corn, soybeans and wheat.

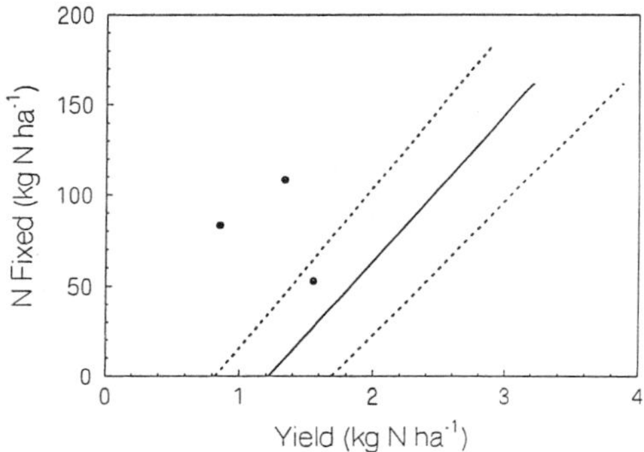

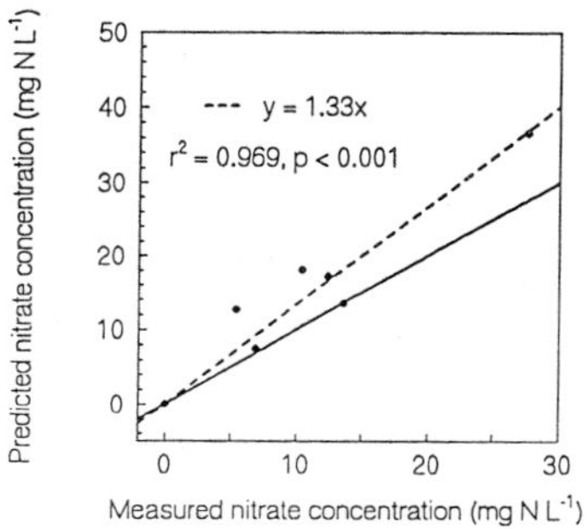

Fig. 2. Relationship between grain yield and the mass of the atmospheric nitrogen fixed by soybeans. The solid line is the regression used for predicting N fixation in the simplified N budget calculation, and the dotted lines indicate the associated 95% confidence intervals. The points are the values determined for three fields in 1992.

Fig. 3. Relationship between measured values of nitrate-N in or moving to groundwater, and values predicted using the simplified N budget for the farms in this study. The solid line is the 1:1 relationship, and the dotted line is the fitted regression.

The simplified N budget approach for farm 1 having this rotation indicated NO_3^- contamination would be approximately 7.5 mg N l^{-1} (Table 2). The concentration of NO_3^- in water from the solution samplers averaged over all the fields was 6.9 mg N l^{-1}.

A modification of this rotation was the introduction of a second year of soybeans immediately after the first. The N budget for a farm (farm 2) with this rotation showed almost a balance between inputs and outputs of N (Table 2). Under these conditions the simplified N budget approach indicated no measurable NO_3^- in the drainage water. The NO_3^- content of the farm well water was below the 0.2 mg N l^{-1} limit of detection (Table 2).

Even where mineral fertilizers constitute the main input of N, the predicted value is not always close to the measured value. On a tobacco farm (farm 3) the concentration of NO_3 predicted to be in the groundwater was 18.1 mg N l^{-1}, whereas the value measured in the well water was only 7.4 mg N l^{-1} (Table 2). However, rye was used as a cover crop to minimize soil erosion, and it is possible that this would affect the quality of the soil organic matter. If so this could influence the rate and timing of mineralization of organic matter.

Many livestock farms appear to give large imbalances of N which could be due to the fact that these operations were not in equilibrium. For example under a dairy farm (farm 4), the predicted concentration of NO_3^- in the groundwater was 36.6, whereas the value in the farm well was 10.1 mg N l^{-1}, but that in the monitoring well was 27.6 mg N l^{-1} (Table 2). In this case the farm well was probably sampling groundwater not connected to that under the worked land.

Inorganic N fertilizers were not used on any of the three farms where components of the simplified budget were tested. One was a cash crop farm (farm 5, Table 2). The prediction from the simplified budget was that there was an excess of N that would lead to NO_3^- moving to groundwater at a concentration of 13.7 mg N l^{-1} (Table 2). No solution samples could be extracted from the soil during the period of the study in 1992. The predicted excess of N for the dairy farm (farm 6) would have resulted in a concentration of 17.2 mg N l^{-1} in the groundwater. The measured value was 12.4 mg N l^{-1} (Table 2). The third farm was a hog farm (farm 7). Again an excess of N was predicted and this was expected to result in 12.7 mg N l^{-1} of soil solution, whereas the measured value was only 4.9 mg N l^{-1} (Table 2). Potential denitrification measured for the different fields on the hog farm averaged 14.5 kg N ha^{-1} y^{-1} (Goorahoo, 1993). When this value was included as an additional output in the budget (farm 7a, Table 2), the predicted value of NO_3^- moving to the groundwater was 3.6 mg N l^{-1} (Table 2).

The relationship between the measured and predicted values of NO_3^--N expected in the groundwater under these different management systems (Fig, 3) showed that on average the simplified N budget over-estimated the NO_3^- concentration by about one third.

Conclusions

The conclusion from the comparisons so far made suggests that the budgetary approach can predict contamination of groundwater with NO_3^- for farming systems that are close to equilibrium. Even in systems where there was poor agreement between the predict-

Table 2. Whole-farm nitrogen budgets for 7 farming systems in Ontario, Canada

	Farm							
	1	2	3	4	5	6	7	7a
Crop rotation (−) or Livestock system (.)	C-Sb-W[a]	W-Sb- Sb-C- Sb-Sb-C	Tb-R	Dairy C,O,Sb, Af	Organic cash-crop. C-O-Ws- Sb-O-B- H-H	Organic dairy H-H-H- Ws-C- mix grn-R for, H	Organic swine. Sb-W-mix grn-H-H- P	Organic swine Sb -Denit induded
N-inputs (kg N ha^{-1}y^{-1})								
Seed	2.7	5.4	0.6	0.7	0.9	1.6	3.5	3.5
Feed	0	0	0	32.5	0	0.1	140.8	140.8
Fertilizer	66.3	50.2	37.4	79.0	0	0	0	0
Manure	0	0	0	22.0	0	0	0	0
Animals	0	0	0	0	0	0	1.4	1.4
N$_2$-fixation								
Non-symbiotic*	5.0	5.0	5.0	5.0	5.0	5.0	5.0	5.0
Symbiotic	51.7	83.7	0	20.9	66.9	35.5	0	0
Atmosphere	18.4*	18.4*	18.4*	18.4*	17.0	45.3	26.0	26.0
TOTAL INPUTS (I$_T$)	**144.1**	**162.7**	**61.4**	**178.5**	**105.5**	**87.5**	**243.6**	**243.6**
N-outputs (kg N ha^{-1}y^{-1})								
Plant produce	132.2	163.0	32.4	54.5	83.6	18.2	56.0	56.0
Animal produce	0	0	0	32.1	0	22.9	37.5	37.5
Manure	0	0	0	0	0	0	84.3	84.3
Volatilization	0	0	0	33.3	0	18.9	45.5	45.5
Denitrification	−	−	−	−	−	−	−	14.5
TOTAL OUTPUTS (O$_T$)	**132.2**	**163.0**	**32.4**	**119.9**	**83.6**	**60.0**	**223.3**	**237.8**
I$_T$ − O$_T$(kg N ha^{-1}y^{-1})	11.9	−0.3	29.0	58.6	21.9	27.5	20.3	5.8
Groundwater recharge (mm y^{-1})	160	160	160	160	160	160	160	160
N in leachate								
Predicted	7.4	0	10.1	36.6	13.7	17.2	12.7	3.6
Measured in well	−	<0.2	7.4	10.1	−	−	−	−
Measured in soil soln	6.9	−	−	27.6	−	12.4	4.9	4.9

[a] Af-alfalfa, B-barley, C-maize, H - alfafa and grass hay, mix grn - mixed grains (barley, oats and peas), O- oats, P - pasture, R - rye, Sb- soybeans, W-wheat, Ws - spelt.
* Values generally applicable in Southern Ontario as determined by Barry *et al.* (1993).

ed and measured concentration in the farm well, the approach has indicated that there will be contamination of groundwater even if the actual amount is over estimated. They can therefore be used to compare the potential of different farming systems to cause contamination of the environment. The survey approach did not serve to discriminate between different farming systems.

This simplified N-budget approach has been used to estimate the potential for N leaching from individual fields, whole farms or from small regions, primarily under intensive management. Many soils in the tropics receive little N-fertilizer, and this might be one reason why appropriate measurements are largely unavailable. Some of the farms investigated in this study had received no mineral-N fertilizer, and the approach was satisfactory for assessing these farming practices. For some regions in India and in Central and South America, fertilizer use is similar to that in some temperate regions, and leaching can be significant during the rainy season. Thus, there are good reasons for investigating the impact of the farming practices on the potability of groundwater. It is hoped that this paper will provide a stimulus for those working in the tropics to have regards for the quality of groundwater, and provide a guide for the study of agricultural practices in this regard.

References

Addiscott TM, Whitmore AP an Powlson DS (1991) Farming, Fertilizers and the Nitrate Problem. CAB International, UK

Barry DAJ, Goorahoo D and Goss MJ (1993) Estimation of nitrate concentrations in groundwater using a whole farm nitrogen budget. J Environ Qual 22(4): 767–775

Beauchamp EG and Burton DL (1985) Ammonia losses from manures. Ontario Ministry Agric. Food Factsheet. Agdex 538. Order no 85–071. Toronto, ON, Canada

Beauchamp EG, Kidd GE and Thurtell G (1982) Ammonia volatilization from liquid dairy cattle manure in the field. Can J Soil Sci 62: 11–19

Brown DM (1985) Heat units for corn in southern Ontario. Ontario Ministry Agric. Food Factsheet. Agdex no. 111/31. Toronto, ON, Canada

Byrnes BH (1990) Environmental Effects of N Fertilizer Use- An Overview. Fert Res 26: 209–215

Deibert EJ, Bijeriego M and Olson JA (1979) Utilization of ^{15}N fertilizer by nodulating and non- nodulating soybean isolines. Agron J 71: 717–723

Ensminger ME and Olentine CG (1978) Feeds and Nutrition - Complete, 1st ed. The Ensminger Publishing Company, USA

Fraser H (1985) Manure Characteristics. Ontario Ministry Agric. Food Factsheet. Agdex 538. Order no 85–109. Toronto, ON, Canada

Fried M, Tanji KK and Van de Pol RM (1976) Simplified long term concept for evaluating leaching of nitrogen from agricultural land. J Environ Qual 3: 391–396

Follett RF and Walker DJ (1989) Ground water quality concerns about nitrogen. In: Follett RF (ed) Nitrogen Management and Ground Water Protection, Ch 1. Elsevier Science Publishers, Amsterdam, Netherlands

Goorahoo D (1993) The use of whole farm nitrogen budgets to estimate nitrate concentrations in groundwater for three organic farms in Bruce County. M.Sc. thesis, University of Guelph, Guelph, Canada. 345p

Gordon R, Leclerc M, Schuepp P and Brunke R (1988) Field estimates of ammonia volatilization from swine manure by a simple micrometeorological technique. Can J Soil Sci 68: 369–380

Goss MJ, Barry DAJ and Goorahoo D (1993) Evaluating nitrogen budgets for assessment of the effectiveness of farming systems in limiting nitrate contamination of groundwater. Agricultural Research to Protect Water Quality Conference Proceedings-February 21–24, 1993. pp 233–235. Conference sponsored by Soil and Water Conservation Society, Minneapolis, Minnesota, USA

Juergens-Gschwind S (1989) Ground water nitrates in other developed countries (Europe) relationships to land use patterns. In: Follett RF (ed) Nitrogen Management and Ground Water Protection, pp 75–138. Elsevier Science Publishers, Amsterdam, Netherlands

Kirchmann H (1985) Nitrogen balance studies in cattle manure during the biological decomposition in composts. Acta Agric Scan, Suppl 24, Ch. 3

Lauer DA, Bouldin DR and Klausner SD (1976) Ammonia volatilization from dairy manure spread on the soil surface. J Environ Qual 5: 134–141

McBride G (1987) Average composition of Ontario feeds by region (5 year summary). Ontario Ministry Agric. Food. Information for industry personnel. Agdex no. 400/60. Univ. of Guelph, Guelph, ON, Canada

Pain BF, Phillips VR, Clarkson CR and Klarenbeek JV (1989) Loss of nitrogen through ammonia volatilization during and following the application of pig or cattle slurry to grassland. J Sci Food Agric 47: 1–12

Power JF and Schepers JS (1989) Nitrate contamination of groundwater in North.America. Agric Ecosys Environ 26: 165–187

Ravuri V (1992) Soybean (*Glycine max* L. Merrill) dinitrogen fixation and residual nitrogen effects on following corn (*Zea mays* L.) crops. Ph.D thesis. Univ. of Guelph, Guelph, ON, Canada

Rudolph D and Goss MJ (1993) Ontario farm groundwater quality survey. Summer 1992. Report prepared for Agriculture Canada under the Federal-Provincial Environmental Sustainability Initiative, 1992

Ro CU, Tang AJS, Chan WS, Kirk RW, Reid NW and Lusis MA (1988) Wet and dry deposition of sulphur and nitrogen compounds in Ontario. Atmos Environ 22: 2763–2772

Schjoerring JK, Sommer SG and Ferm M (1992) A simple passive sampler for measuring ammonia emission in the field. Water Air Soil Pollut 62: 13–24

Tel DA and Heseltine C (1990) The analysis of KCL soil extracts for nitrate, nitrite and ammonium using a TRAACS 800 analyzer. Commun Soil Sci Plant Anal 21: 1681–1688

Tel DA and Goorahoo D (1993) Ammonium determination in oxalic acid sorbent using a TRAACS 800 autoanalyzer. Method presented at the International Symposium on Soil Testing and Plant Analysis: Precision Nutrient Management. August 14–19, 1993, Olympia, Washington. Paper in preparation for publication in Commun Soil Sci Plant Anal

Thompson RB, Pain BF and Lockyer DR (1990) Ammonia volatilization from cattle slurry following surface application to grassland. II Influence of application rate, windspeed, and applying slurry in narrow bands. Plant Soil 125: 119–128

Fertilizer Research **42**: 339–346, 1995.

© 1995 *Kluwer Academic Publishers. Printed in the Netherlands.*

Alternatives for nitrogen nutrition of crops in tropical agriculture

Johanna Dobereiner, Segundo Urquiaga & Robert M. Boddey
EMBRAPA – Centro Nacional de Pesquisa de Agrobiologia, Km 47, Seropédica, Itaguaí, 23581–970, Rio de Janeiro, Brazil

Key words: Biofuel production, biological nitrogen fixation, french bean, green manures, soil organic matter, soybean, sustainable crop production

Abstract

The development of sustainable agricultural systems for the tropics requires among other technologies, alternatives for nitrogen fertilizers which are often limited in availability for financial reasons and also represent a major source of groundwater and air pollution. There are many new alternatives for the development of agricultural systems which make use of biological processes in soil. Biological nitrogen fixation (BNF), that is, the biological conversion of atmospheric dinitrogen into mineral N, is the most important alternative among them. Examples are given of the impact of various technologies used in Brazil. Soybean, introduced into the country 30 years ago, is now the second most important export crop, reaching 24 million tons annually with no N fertilizer application. Consequently, Brazil today is the country in the world which uses the lowest amounts of nitrogen fertilizers in relation to phosphate. Alternatives for crop rotations and pastures are also discussed. Possibilities of expanding BNF to cereals and other non-legume crops are gaining new credibility due to the identification of endophytic associations with diazotropic bacteria. The definite proof of substantial BNF in sugar cane with N balance and ^{15}N methods in certain genotypes selected under low N fertilizer applications opens up new alternatives for sustainable agriculture and will be the key to viable bio-fuel programmes.

Introduction

Over the last 30 years cereal grain yields in the tropical and sub-tropical regions of the developing world have increased immensely with the application of large quantities of fertilizers and pesticides on high yielding crop varieties developed by international and national agricultural research institutes. Countries such as India, with large populations and limited land resources, have become self-sufficient in grain production and even today the application of this 'Green Revolution' technology continues to increase crop yields in many developing countries (Plucknett and Smith, 1993).

As far as fertilizer inputs are concerned, the greatest annual investment for high yielding cereals is for N. The manufacture of N fertilizer requires large energy inputs (more than 6 times that needed for P or K fertilizers), requiring 18.5 Mcal of fossil energy per kg N. The manufacturing process is fueled principally by natural gas, the price of which is tied closely to that of crude oil. With the inevitable price rises of fossil fuels (not to mention proposed carbon taxes) that must occur over the next few decades due to depleting reserves of petroleum and increased production costs of other fuels, the continued use of high N fertilizer inputs may become economically inviable. A further disadvantage of the use of high N applications is that excessive or badly timed additions may result in leaching of excess mineral N into water reserves with resulting risks of $NO_3{}^-$–N pollution of drinking water and, in severe cases, can contribute to the eutrophication of rivers or lakes. This problem has become widespread in areas of intensive farming in Europe and North America.

Nitrogen plays a key role in the maintenance of soil organic matter. Because of millenia of heavy leaching and profound weathering in the humid and sub-humid areas of the tropics, in most of the soils of these regions the clay fraction is predominantly sesquioxides, and clays of the 1:1 lattice type. For this reason, these soils have low cation exchange (CEC) and water retention capacities. It is the organic fraction which enables them to retain nutrients and water for crop growth and the

340

maintenance of this fraction is critical in the preservation of soil fertility. This problem is less serious in temperate regions not only because of soil organic matter mineralization rates are lower in the cooler climate, but also because in the glaciated regions of the north of Europe, Asia and America, the soils are more recent in age and many are higher in 2:1 lattice clays with higher CEC and water retention capacities. It is for this reason that the European model of farming with frequent ploughing and high fertilizer inputs is often completely inappropriate for many tropical regions and has frequently been responsible for causing severe soil loosening and soil erosion.

In the tropics, in the clearing of bush or forest for planting crops or establishing pastures, burning is usually employed, which results in large losses of nitrogen by volatilization and the reduction of other elements to mineral forms in the ash. These elements are available for crop growth but rainfall can cause considerable losses of K through leaching before the root systems of the crops are developed, although P and Ca are generally better conserved due to their lower solubilities. Subsequent losses of soil organic matter are stimulated by mechanical cultivation of the soil and again N is the element most susceptible as it can be lost not only by leaching, but also by NH_3 volatilization and denitrification. While the correction of soil acidity and P deficiency are essential for crop growth in many areas (Urquiaga and Boddey, 1994), once corrected their availability in the soil declines only slowly with time due to the aforementioned low solubility of P and Ca. However, this is not the case with N. High soil mineral N levels are transient, because what is not taken up by the crop or immobilized in the organic matter is quickly lost and therefore cropping systems require continuous or regular additions of N.

It is clear, therefore, that there is a great need for alternative sources of N supply in tropical cropping systems. It is fortunate that it is N, but only N, that can be derived from the air for crop growth through the associations of bacteria which are able to assimilate gaseous N_2 (N_2-fixing bacteria) with plants. The bacteria obtain their carbon requirements from the plant photosynthates and export N to the plants. The best known examples of such associations are the legume symbioses. Extension of biological N_2 fixation (BNF) to cereals and other non-leguminous crops represents today a major challenge for the development of sustainable agriculture.

N_2 fixation in legumes

Grain and forage legumes play important roles in tropical agriculture. Soybean gives an impressive example: after coffee it represents the most important export crop of Brazil today occupying ever-increasing areas of the central highland savannas (Cerrados). Since 1970, Brazil's soybean production increased from less than two million tons to the 24 million expected this year. This was only possible due to plant breeding for N_2 fixation and to the development of new *Rhizobium* inoculants which had to be adapted to the specific microbial dis-equilibrium prevailing in these soils after they have been ploughed, limed and fertilized for cropping. These *Rhizobium* strains are now however, in generalized use in the country. Even though they produce large nodule masses and fix much N_2 they are rather inefficient in terms of transference of the fixed N to the grain. Recently, new strains were identified which can increase N harvest indices from 56 to 69% (Neves *et al.*, 1985). Grain yield increases in the field with these 'super strains' can reach 45% (Table 1), if they can be established. In the 'cerrado' soil, however, they can not yet be used, owing to competition problems with the already established, less efficient strains. Various strategies to make these strains more competitive are under study.

Brazil and Argentina are the only countries in the world where soybean is grown without any N fertilizer. This crop therefore contributes to soil fertility, without hazards to the environment. Biological nitrogen fixation in Brazil contributes an estimated 2.5 million tons of N representing an economy of 1.8 billion US dollars per year.

In the case of the *Phaseolus vulgaris* bean crop, an important protein source for the human population of the tropics, especially Latin America, N_2 fixation has an additional problem related to the genetic instability of the *nod* and *fix* genes in the traditional *Rhizobium phaseoli* strains which is related to the frequent genetic rearrangement in these strains (Martinez *et al.*, 1990). These authors observed two groups of *Rhizobium* capable of nodulating beans, group I being specific for *Phaseolus vulgaris* and genetically unstable in relation to symbiotic performance, and group II which, besides beans, also nodulates *Leucaena leucocaephala* and other tree legumes. Strains of group II have the NOD and FIX genes localized on two large plasmids, show no or little rearrangement and are therefore much more stable. Recently group II was proposed to

Table 1. Rhizobium strain effects on nitrogen fixation and yields of field grown soybeans (Neves *et al.*, 1985).

Strain	N_2 fixed[1] (kg N ha^{-1})	Nodule efficiency[2] (mg N_2 fixed g^{-1} nodule)	Seed yield (kg ha^{-1})	N harvest index[3]
29W	343b[4]	426b[4]	1863c[4]	59b[4]
DF 395	214b	543b	1768cd	52b
SM_1b	204c	491b	1398de	56b
965	278a	899a	2898a	69a
CB 1809	239b	1005a	2682a	69a
DF 383	242b	851a	2284b	68a
Control	0	–	948f	49bc

[1] At maximum plant nitrogen content (109 d), calculated by the difference in total plant nitrogen between inoculated and control plants.
[2] At maximum nodule weight (74 d).
[3] (N seed)/(total N)
[4] Means followed by the same letter are not significantly different at $p = 0.05$ (Tukey).

Table 2. Nodulation and N_2 fixation at high soil temperatures of beans[1] inoculated with *Rhizobium* sp. strains isolated from tree legumes (Hungria *et al.*, 1993).

Origin of strains	Strain No.	Nodule No./plant	Total N (mg/plant)
Gliricidia sp.	BR 8801	8c[2]	15d[2]
	8802	20cd	22cd
	8803	12c	14d
Leucaena sp.	BR 814	46b	41ab
	816	40c	41ab
	817	68a	52a
Lonchocarpus sp.	BR 6009	14de	28c
	6010	60ab	48ab
	6011	20cde	22cd
Phaseolus vulgaris	CNPAF 146	2c	12d
P. vulgaris + 90 mg N/plant		0	35bc

[1] Plants were grown in Leonard jars in a water bath at 40 °C for 8 h per day. Values are means of 4 replicate jars.
[2] Means followed by the same letter are not significantly different at $p = 0.05$ (Tukey).

be a new species of *Rhizobium, R. tropici* (Martinez-Romero *et al.*, 1991).

Under tropical conditions the instability of bean *Rhizobium* strains is aggravated by excessive soil temperatures. Two heat shocks of only 8 h at 38 °C over a 10 day period resulted in complete inhibition of N_2 fixation (Hungria *et al.*, 1993). However, nodulation and N_2 fixation of certain legume tree species (e.g. *Leucaena* or *Prosopis*) was unaffected by an regime of 38 °C for 8 h per day during the whole growth cycle (C.O. Cunha & A.A. Franco, pers. com.). Selection among more then 100 *Rhizobium* strains isolated from 18 tree legume species yielded 14 strains (isolated from *Leucaena, Gliricidia* or *Lonchocarpus* spp.) which were able to nodulate beans effectively and fix N_2. When these strains were used to inoculate beans grown at 40 °C soil temperature for 8 h per day, some of them were able to nodulate and fix N_2, yielding plants equal or superior in productivity to those fertilized with N (Table 2).

In grain legume crops where BNF can contribute more than 70% of total plant N accumulation, the net

Table 3. Effect of crop rotation on maize yield (kg/ha) and profitability over 5 years.[1]

	Maize yield 1982/83	Maize yield 1984/85	Profit all crops after 5 years (US$/ha)
Maize in monoculture	4480	1855	1178.51
Maize in crop rotation[2]	3696	2703	1869.35
With rock phosphate	4808	2671	1780.38
With rock phosphate and green manure[3]	5283	3023	1575.24

[1] After F.F. Duque and G.G. Pessanha, in preparation.
[2] The crop rotation from 1981 to 1985 was *Phaseolus* beans - maize - peanuts - cassava intercropped with cowpea - maize.
[3] Green manure was *Styzolobium*.

gain to, or loss from the soil/plant system will depend on the N harvest index. This means that in order to make a net contribution of N to the system, grain legumes must fix a proportion of N at least as large as the proportion of the N they remove from the soil. In the case of soybean, one of the most efficient N_2-fixing crops which also has a high harvest index (Table 1), the overall N balance of the cropping system can be negative. It follows, therefore, that in a system of crop rotation where non-N_2-fixing (cereal) crops constitute much of the rotation, grain legume crops do not necessarily compensate for the significant drain on soil N. This topic was recently discussed by Giller *et al.* (1994), and they pointed out that some grain legumes such as groundnut, cowpea and pigeon pea, which have rather low harvest indices, should be more suitable than soybean for incorporation into crop rotations. As it is probably unacceptable to farmers to plant grain legumes selected for lower harvest indices, where large inputs from BNF are required, the best solution should be to incorporate green manure legumes into the rotation.

Legumes for green manure

A large variety of tropical legumes are available which can be planted between crops either as intercrops or after the crop is harvested. Several of them can fix large amounts of N_2 and some assimilate P from rock phosphates which are of low availability for grain crops. It has been shown (Silva *et al.*, 1985) that the legume *Stizolobium aterrimum* can obtain as much P from rock phosphate as from soluble phosphates. When such green manure cover crops are incorporated into the soil, not only the organic matter, but also P and N are incorporated in slow release organic forms (Table 3). These features can contribute substantially to make green manuring economically viable.

The main problem here is that farmers are infrequently willing to substitute the planting of a food crop with a green manure crop which gives no immediate financial or grain yield. Here the intercropping of a grain crop with an undersown leguminous green manure may be viable or the use of drought-resistant green manure crops which will grow in the dry season where grain crops are unlikely to produce significant yields. There is an urgent need for more research into the selection of such legumes and the evaluation of the effects of their incorporation into cropping systems.

Pasture legumes

Large areas of the tropics, especially in South America, are dedicated to pastures. These pastures are almost all of pure grass, either native, or exotic species often of African origin. However, while pasture productivity can respond dramatically to N fertilization, if accompanied by liming and additions of P, this practice is rarely viable economically, at least in the case of pastures for extensive beef production. Several years after establishment, the plant, and consequently animal, productivity of these pastures decline and they are invaded by non-palatable species. Again it is the lack of N for plant nutrition which is thought to be the principal cause of this pasture decline (Graham *et al.*, 1985; Steele and Vallis, 1988). As perennial pastures are never ploughed and hence they have permanent rooting systems and mineralization of soil organic matter is not stimulated by mechanical disturbance, it may be thought that N losses are minimal. There are, however, considerable N losses by NH_3 volatilization and N leaching from urine and dung deposited by the grazing animals (Denmead *et al.*, 1974; Denmead *et al.*, 1976; Ball and Keeney, 1983). Soil organic matter is preserved in such systems, and may even accumulate, due to the high dry matter production potential of the C4 grasses, but the problem is that the C:N ratio of the deposited plant material (litter) is so high that release of N by mineralization eventually becomes extremely limited, resulting in pasture decline (Robbins *et al.*, 1989; Boddey *et al.*, 1995).

Conventionally, farmers renovate their pastures by burning them (sometimes annually), or, at intervals of several years, they may plough up the pasture and either replant pasture grasses or put in one or more arable crops, later replanting the grasses. The problem here is that in burning or ploughing considerable losses of N occur, and to a lesser extent other nutrients. After many cycles of renovation, soil organic matter levels can be depleted to the extent that soil fertility is lowered and erosion may occur. While it may be economically viable to add P and K fertilizers and lime, as mentioned before, N fertilizer additions are usually too costly. It is here that the introduction of N_2-fixing forage legumes is likely to be a useful strategy to replace N losses and to produce plant residues of lower C: N ratio.

There are two possible ways of introducing these legumes, either by forming a mixed legume/grass pasture, or by the introduction of pasture or other legumes in the rotation or in the process pasture renovation. The first option requires that forage legumes are selected which are adapted to the usually low fertility conditions of these soils and are able to compete with the existing grasses and survive. Animal management in this respect is critical, over- or under- grazing can cause the legume to be out-competed by the grass. While considerable success in the selection of appropriate legumes and the management of such mixed pastures has been achieved in tropical Australia, insufficient appropriate research has been conducted to develop such pastures for other edapho-climatic regions such as the edaphic savannas of South America.

Forest legumes

Brazilian reafforestation projects, until recently, did not consider one of the important characteristics of so many native legume trees, which is their ability to fix N_2. The most precious hardwood species and many native fast-growing trees are legumes, but little is known about their capacity to nodulate or fix N_2. Surveys in the North Eastern dry regions, in the Amazon rain forests and in South East Brazil revealed many

economically important N_2 fixing trees not known as such before (Table 4).

Mesquite (*Prosopis juliflora*) called algaroba in Brazil, is being planted in large government projects in the North East dry regions and since 1982 is inoculated with commercially available inoculants developed by our Centre. The planting in the field of sabia (*Mimosa caesalpinifolia*), and acacias (*A. mangium, A. auriculiformes*) grown in nurseries in media containing 50% rock phosphate and inoculated with selected mycorrhiza and rhizobia has shown great success in the recuperation of completely eroded soils or subsoils (Franco *et al.*, 1992).

The development of agroforestry systems which include in the crop rotation ten-year periods of legume forests which supply the farm with energy and emergency fodder during dry years and recover eroded soils, building up organic matter from the large amounts of protein-rich leaves which fall on the ground, seems another prospect as yet almost unexplored.

Cereals and grasses

The extension of biological N_2 fixation to cereals and grasses has been a major research challenge in the last two decades. Because the gramineae, in common with other eucaryotes, cannot use molecular N_2, the most promising approach appears to be the search for more or less symbiotic associations of bacteria which are able to fix N_2 with cereals, which can be improved by modern technologies. The transfer of N_2 fixation genes into plant cells seems a more pretentious alternative, which if successful, could eventually become the best solution. Unfortunately, progress in this field is very slow, while many new alternatives have become available during the last 15 years for improving already-identified naturally-occurring associations of cereals with N_2-fixing bacteria.

Significant, although somewhat variable amounts of N_2 fixation have been demonstrated with ^{15}N techniques in association with wetland rice (Yoshida and Yoneyama, 1980; Eskew *et al.*, 1981) and forage grasses (De Polli *et al.*, 1977; Boddey and Victoria, 1986; Miranda and Boddey, 1987). Recent experiments with sugar cane, combining N balance and ^{15}N dilution techniques have brought unequivocal proof that in certain Brazilian varieties more than 50% of plant N can be derived from the air (Lima *et al.*, 1987).

A further ^{15}N dilution and N balance study in a large tank (Urquiaga *et al.*, 1992) indicated that sugar cane fertilized with P and K alone could obtain most

Table 4. Nodulation of Brazilian forest legumes.[1]

	Subfamilies			
	Mimosoideae	Papilionoideae	Caesalpinoideae	Total
No. of species verified	102	119	100	321
No. of species with nodules	82	90	15	187
No. of species found for the first time with nodules	39	55	13	107
No. of genera found for the first time with nodules	2	6	2	10
No. of *Rhizobium* strains isolated	389	387	60	836

[1] Adapted from (Faria *et al.*, 1987).

of the N needed for yields equivalent to 200 t of cane ha^{-1} from BNF. There were pronounced differences between plant genotypes showing that the plant plays an important role and that plant breeding will be an important approach for increasing BNF.

Now that it is known that amounts of N of economic importance can be fixed in association with cereals and other Gramineae, the understanding of the physiology is essential in order to start to manipulate and increase their efficiency. Many different N$_2$-fixing bacteria have been isolated from the rhizosphere and from roots of cereals, but only where plant-bacteria interactions exist can one be convinced of an association. Pathogenic plant–bacteria associations have been known for a long time, the effects of microorganisms being visible as damage to the plant tissue. Characteristic of these associations is the specificity that can be on a strain or species level. Plant breeding for resistance to specific pathogens is one of the major objectives in agricultural research. Breeding for improved associations of plants with N$_2$ fixing bacteria will have to select for opposite characteristics.

Several of the new N$_2$ fixing bacteria as *Acetobacter diazotrophicus* or *Herbaspirillum* spp have not been isolated from soil and therefore can rather be considered plant endophytes or even mild plant pathogens. The infection of sugar cane, sweet potato or sorghum with these bacteria has been shown to be dependent on VA mycorrhizae (Paula *et al.*, 1991). This demonstrates an entirely new type of associaton of N$_2$-fixing bacteria with graminaceous plants which contrasts with the concept of rhizosphere associations known so far. Obviously bacteria which live within plant tissues, especially in sugar-rich plants, can benefit the plants more easily than rhizosphere bacteria which compete for nutrients with the entire soil microflora.

The replacement of N fertilizers by biological N$_2$ fixation (BNF) in the Brazilian sugar cane crop is already very substantial in comparison with countries like Cuba, México, Venezuela and Hawaii, where commonly more than 200 kg N ha^{-1} (three times the Brazilian average applications of N) are applied to sugar cane. The recent results on BNF in certain sugar cane genotypes (Urquiaga *et al.*, 1992; Boddey *et al.*, 1994) indicate that the Brazilian sugar cane yields could be more than doubled without N fertilizer if irrigation is used when necessary and applications of P, K and Mo are increased.

At present the Brazilian sugar cane crop consumes 240 000 t of N per year at a price of US$ 150 million. If this sum were to be used to improve the sugar cane growing technology and yields are doubled with little additional costs, more than half of the Brazilian oil imports could be replaced without increasing the area planted with sugar cane. The actual alcohol production of Brazil of 11 billion litres per year is equivalent to 210 000 barrels of petrol per day.

Biological nitrogen fixation in sugar cane and the energy balance

Of even higher priority for any bio-fuel program, is the energy balance; that is the energy gain/input ratio for growing the crop. Such a program with high energy balance is the only alternative to fossil fuels in the long run, when actual costs will become secondary in priority. In the Brazilian alcohol program virtually all cane is cut by hand and all energy for cane processing and alcohol production is derived from burning the crushed cane stems (bagasse). With this system and the mean application of 60 kg N ha^{-1}, and yields of 65 t cane ha^{-1}, the energy balance is around 4.5 (Silva *et al.*, 1978). The elimination of N fertilizers will increase it to more than 5 and increased yields due to improved tecnologies with little additional energy costs can reach an energy balance close to 6. In the alternative production system, such as that of the USA with more than 200 kg N ha^{-1} applied, and the use of mechanical harvesting, the energy gain/input ratio would be close to 1, making any bio-fuel program senseless (Boddey, 1993). Based on these facts, the contribution of BNF to the Brazilian sugar cane crop can be considered of decisive importance to the energy balance, and with this, for the viability of the 'Pro-alcool' program.

From the ecological point of view, criticisms of the alcohol program, often pronouced by poorly informed persons, have little scientific basis. The sugar cane crop improves soil fertility, prevents erosion, and takes out more CO_2 from the atmosphere than it liberates, even if combustion of alcohol and bagasse is considered. Alternatives for harvesting sugar cane without prior trash burning are under investigation and the results so far indicate that besides increased demand for manual labour in the sugar cane growing areas, which is desirable, this practice can be economically profitable and certainly makes additional contributions to soil fertility and environmental protection (Oliveira *et al.*, 1994).

References

Ball PR and Keeney DR (1983) Nitrogen losses from urine-affected areas of a New Zealand pasture under contrasting seasonal conditions. In: Proceedings of the XIV International Grassland Congress, pp 342–344. Lexington, KY

Boddey RM (1993) 'Green' energy from sugar cane. Chem Ind No. 10, 17 May, pp 355–358

Boddey RM, Oliveira OC de, Urquiaga S, Reis VM, Olivares FL de, Baldani VLD and Dobereiner J (1994) Biological nitrogen fixation associated with sugar cane and rice: contributions and prospects for improvement. In: Trans 15th World Congr Soil Sci Vol 4a pp 273–292. Acapulco, Mexico

Boddey RM and Victoria RL (1986) Estimation of biological nitrogen fixation associated with *Brachiaria* and *Paspalum* grasses using^{15}N-labelled organic matter and fertilizer. Plant Soil 90: 265–292

Boddey RM, Resende C de P, Pereira JM, Cantarutti RB, Alves BJR, Ferreira E, Richter M, Cadisch G and Urquiaga S (1995) The nitrogen cycle in pure grass and grass/legume pastures: Evaluation of pasture sustainability. In: Nuclear Techniques in Soil-Plant Studies for Sustainable Agriculture and Environmental Preservation. pp 307–319, FAO/IAEA, Vienna, Austria.

Denmead OT, Simpson JR and Freney JR (1974) Ammonia flux into the atmosphere from a grazed pasture. Science 185: 609–610

Denmead OT, Simpson JR and Freney JR (1976) A closed ammonia cycle within a plant canopy. Soil Biol Biochem 8: 161–164

De Polli H, Matsui E, Döbereiner J and Salati E (1977) Confirmation of nitrogen fixation in two tropical grasses by $^{15}N_2$ incorporation. Soil Biol Biochem 9: 119–123

Eskew DL, Eaglesham ARJ and App AA (1981) Heterotrophic $^{15}N_2$ fixation and distribution of newly fixed nitrogen in a rice-flooded soil system. Plant Physiol 68: 48–52

Faria SM de, Lima HC de, Franco AA, Mucci ESF and Sprent JI (1987) Nodulation of legume trees from South East Brazil. Plant Soil 99: 347–356

Franco AA, Campello EF, Silva EMR da and Faria SM de (1992) Revegetação de solos degradados. Itaguaí, RJ: EMBRAPA-CNPBS (Comunicado Técnico, 9)

Giller KE, McDonagh JF and Cadisch G (1994) Can biological nitrogen fixation sustain agriculture in the tropics. In: Syers JK and Rimmer BL (ed) Soil Science and Sustainable Land Management in the Tropics. pp 173–191 Wallingford, UK: CAB International.

Graham TWG, Myers RJK, Doran JW, Catchpoole VR and Robbins GB (1985). Pasture renovation: The effect of cultivation on the productivity and nitrogen cycling of a Buffel grass (*Cenchrus ciliaris*) pasture. In: Proceedings of the XV International Grassland Congress, pp 640–642. Kyoto, Japan

Hungria M, Franco AA and Sprent J (1993) New sources of high-temperature tolerant rhizobia for *Phaseolus vulgaris* L. Plant Soil 149: 103–109

Lima E, Boddey RM and Döbereiner J (1987) Quantification of biological nitrogen fixation associated with sugar cane using a ^{15}N aided nitrogen balance. Soil Biol Biochem 19: 165–170

Martinez E, Pardo MA, Martins F, Graham P, Franco AA, Palacios R and Segovia L (1990) Genetic relatedness and taxonomic considerations of *Rhizobium* strains that nodulate *Phaseolus vulgaris* (L.). In: Gresshoff PM, Roth LE, Stacey G and Newton WE (ed) Nitrogen Fixation: Achievements and Objectives, pp 831. Chapman and Hall New York

Martinez-Romero E, Segovia L, Mercante FM, Franco AA, Graham P and Pardo MA (1991) *Rhizobium tropici*, a novel species nodulating *Phaseolus vulgaris* L. beans and *Leucaena* sp. trees. Int J Syst Bacteriol 41: 417–426

Miranda CHB and Boddey RM (1987) Estimation of biological nitrogen fixation associated with 11 ecotypes of *Panicum maximum* grown in nitrogen-15-labeled soil. Agron J 79: 558–563

Neves MCP, Didonet AD, Duque FF and Döbereiner J (1985) *Rhizobium* strain effects on nitrogen transport and distribution in soybeans. J Exp Bot 36: 1179–1192

Oliveira OC de, Urquiaga S and Boddey RM (1994) Burning cane: the long term effects. Int Sugar J 96 272–275

Paula MA, Reis VM and Döbereiner J (1991) Interactions of *Glomus clarum* with *Acetobacter diazotrophicus* in infection of sweet potato (*Ipomoea batatas*), sugarcane (*Saccharum* spp.), and sweet sorghum (*Sorghum vulgare*). Biol Fertil Soils 11: 111–115

Plucknett DL and Smith NJH (1993) Green revolution remains in force. Forum Appl Res Public Pol 8: 65–69

Robbins GB, Bushell JJ and McKeon PK (1989) Nitrogen immobilization in decomposing litter contributes to productivity decline in ageing pastures of green panic (*Panicum maximum* var. trichoglume). J Agric Sci 113: 401–406

Silva EMR da, Almeida DL de, Franco AA and Döbereiner J (1985) Adubação verde no aproveitamento de fosfato em solo ácido. R Bras Ci Solo 9: 85–88

Silva JG, Serra GE, Moreira JR, Conçalves JC and Goldemberg J (1978) Energy balance for ethyl alcohol production from crops. Science 201: 903–906

Steele KW and Vallis I (1988) The nitrogen cycle in pastures. In: Wilson J (ed.) Advances in Nitrogen Cycling in Agricultural Ecosystems, pp 274–291. Wallingford, CAB International, Oxon, UK

Urquiaga S and Boddey RM (1994) Crop production in deleterious soils with special emphasis on acid soils. In: Trans 15th World Congr Soil Sci Vol 5a pp 489–508. Acapulco, Mexico

Urquiaga S, Cruz KHS and Boddey RM (1992) Contribution of nitrogen fixation to sugar cane: Nitrogen-15 and nitrogen balance estimates. Soil Sci Soc Am J 56: 105–114

Yoshida T and Yoneyama T (1980) Atmospheric dinitrogen fixation in the flooded rice rhizosphere as determined by the N-15 isotope technique. Soil Sci Plant Nutr 26: 551–560

N. Ahmad (ed.), Nitrogen Economy in Tropical Soils, 347–353.
© 1996 *Kluwer Academic Publishers. Printed in the Netherlands.*

Behaviour of urea and ammonium sulfate fertilizers and their N uptake relationships in calcareous soils

L.J. Cajuste, E. Sánchez-A & R.J. Laird
Graduate School Edaphology Center, Chapingo-Montecillo, México

Key words: ammonia volatilization, ammonium sulfate, Mexican soils, sorghum, urea hydrolysis

Abstract

Urea hydrolysis, nitrification of ammonium and NH_3 volatilization losses from urea and ammonium sulfate applied to one slightly acid and two calcareous soils were studied in the laboratory. Rates of N applications were 0, 150 and 300 μ g^{-1} soil. Additionally, response of sorghum (*Sorghum bicolor L.*) to 0, 75 and 150 mg N kg^{-1} soil of both fertilizers in a greenhouse experiment was examined. Results showed that rate of urea hydrolysis, expressed as recovered NH_4^+, was proportional to the applied urea doses and varied slightly with soil characteristics.

In two soils with substantial $CaCO_3$ content, urea hydrolysis produced a peak NH_4^+ concentration at 48 h after the fertilizer application; maximum recovered NH_4^+, however, was found at 24 h in an acidic soil. Ammonia volatilization loss from AS application was not detected in the Lirios soil. In contrast, total NH_3 volatilization in the most alkaline soil amounted to more than 50% with the application of 150 μg N g^{-1} regardless of N source; in this soil a smaller percentage volatilization loss of NH_3 was observed with the application of 300 μg N g^{-1} soil, suggesting that variation in soil water content might be responsible for this phenomenon. Amounts of recovered NO_3^- were not always directly related to applied N doses. Dry matter, in general, increased significantly with N doses, although there was no significant difference between 75 and 150 μg N g^{-1} rates in some cases. With 150 μg N g^{-1} of urea application, sorghum yield remained unchanged or decreased significantly, presumably due to high soil pH, nitrite toxicity and NH_3 volatilization. Soil N availability estimates (NH_4^+, NO_3^-, and $NH_4^+ + NO_3^-$), after a 14-day incubation period were moderately associated with N uptake but influenced very significantly plant N content.

Introduction

Broadcast application of urea is becoming increasingly important among Mexican farmers, because of a limited supply of ammonium sulfate, as the industrial production is insufficient to satisfy the nationwide demand. In most agricultural soils, cereal grain yield has increased with increasing nitrogen (N) fertilizer applications (Bandel et al., 1980; Howard & Tyler, 1989; Stecker et al., 1993); however, many crops show low yield response to urea in calcareous soils (Fenn & Kissel, 1975; Fenn & Richards, 1986; Gasser, 1964). This may be attributed to NH_3 volatilization following urea hydrolysis, on the application of urea to these soils (Fan & Mackenzie, 1993; Fenn & Hossner, 1985; Stecker et al., 1993). The objective of this study was to examine the behaviour of urea and ammonium sul-

fate fertilizers and their N uptake relationships in one slightly acid and two calcareous soils.

Materials and methods

Surface soil samples (0-15 cm depth) were collected from three agricultural sites (two Argidic Torrerts and one Typic Calciorthids soils) in the Lirios-Marín county of the Mexican state of Nuevo León. They were incubated in the laboratory with several doses of urea (UR) and ammonium sulfate (AS) to study the fate of N following the application of both fertilizers to the soils. Some selected physical and chemical characteristics of the soils are presented in Table 1.

Urea hydrolysis reaction, nitrification of NH_4^+ and NH_3 volatilization were studied in laboratory experiments. Rates of N applications were 0, 150 and 300 μg

Table 1. Selected physical and chemical characteristics of the soils

Soils	Clay (%)	pH$_w^a$ 1:2	CaCO$_3$	Organic Matter (%)	CEC (cmol kg^{-1})	Classification
Lirios (S-1)	42	6.3	1.2	2.9	19.62	Argidic torrerts
Marin (S-2)	65	7.6	10.9	2.0	27.48	Argidic torrerts
Ancon (S-3)	26	8.0	13.6	1.1	11.70	Typic Calciorthids

[a] w = Water suspension

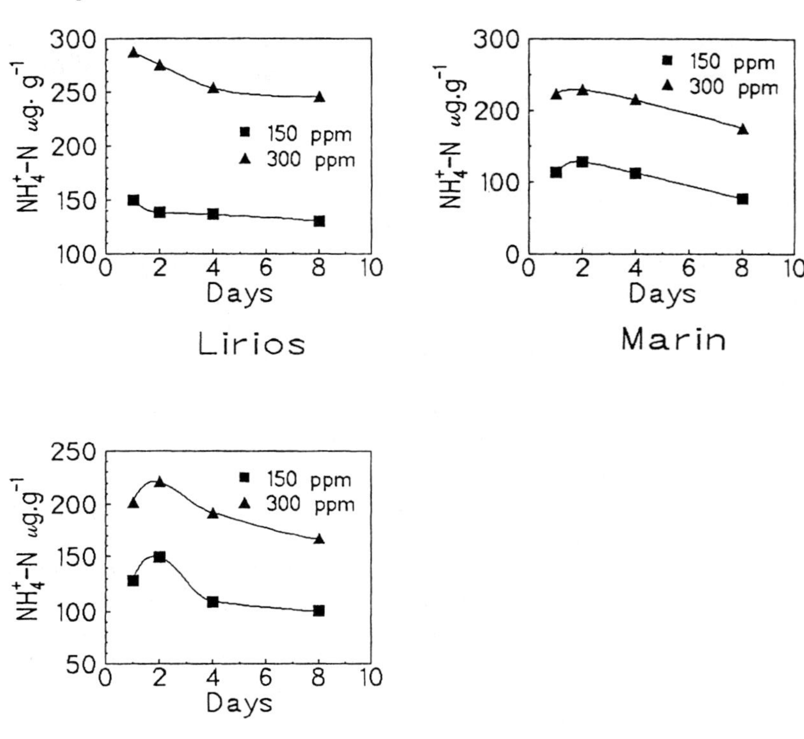

Fig. 1. Urea hydrolysis in the soils as influenced by incubation time

g^{-1} soil. Chemical reagent grade of both N fertilizers was used; the fertilizers were applied in solution to the soil samples. The following procedures were carried out: Urea hydrolysis and NH$_4^+$ − N nitrification experiments, according to Ayanaba and Kang methodology (1976). Briefly, 5 g of air-dried soil were placed in a 100 ml polypropylene centrifuge tube, treated with 1 ml of the appropriate N dose and wetted to 80% field capacity. After one, two, four, and eight days (urea hydrolysis experiment) and 2 and 4 weeks (nitrification experiment) of incubation at 35 °C, the moist soil was extracted with 1 *M* KCl and analyzed for NH$_4^+$ and NO$_3^-$ (Bremner & Kenney, 1966).

Ammonia volatilization was measured in a forced-draft system using covered jars swept with air (a slight-ly modified Fleisher and Hagin procedure) (Fleisher and Hagin, 1981). Ten grams of air-dried soil were placed in a 30 ml jar and sealed with a rubber stopper which was covered with a parafilm sheet. The soils in the jars were moistened with the appropriate N solution and incubated at 22 °C at field capacity for one, two and four weeks. Following the application of fertilizers, jars were covered immediately and connected to the air train. Incoming air passed through 2 l of 1 *M* H$_2$SO$_4$ to remove ambient NH$_3$ and then through 15 l of distilled water to humidify the air before entering the jar. Distilled water was added once a week to replenish evaporated soil water, as the air flow was maintained at 0.75 l min^{-1} jar^{-1} throughout the incubation period. Air coming from sample jars passed into 30 ml of a 20

Table 2. Ammonia volatilization loss (%) from ammonium sulfate (AS) and urea (UR) applications to the soil during the three incubation periods (I, II & III)

Soil	Incubation period	(AS) 150	Rate 300	(UR) 150	Rate 300
		$(\mu g\ g^{-1})$			
S-1	I	nd	nd	14.5	13.6
	II	nd	nd	5.7	4.7
	III	nd	nd	3.2	4.2
	Sum	-	-	23.4	22.5
S-2	I	12.4	15.5	15.1	23.0
	II	8.6	5.3	8.0	5.7
	III	4.3	1.6	5.0	3.3
	Sum	25.3	22.4	28.1	32.0
S-3	I	33.5	20.8	41.5	25.6
	II	6.8	9.8	8.9	8.1
	III	10.7	4.6	6.3	5.0
	Sum	51.0	35.2	56.7	38.7

g kg^{-1} H$_3$BO$_3$ solution which was titrated with 0.005 M H$_2$SO$_4$ to determine the NH$_3$ loss through volatilization.

Amounts of NH$_4^+$ from urea hydrolysis, NH$_3$ loss through volatilization and NO$_3^-$ produced by nitrification were determined as the difference between recovered NH$_4^+$, NH$_3$ and NO$_3^-$ from the corresponding experiments and the amounts of NH$_4^+$, NH$_3$ and NO$_3^-$ present in check samples. All experiments included three replicates of each treatment in a completely randomized design.

A short term greenhouse experiment was carried out with sorghum (*Sorghum bicolor* L.). The plants were cultivated in 250 g soil/pot. N doses of both fertilizer sources were equivalent to 0, 75 and 150 mg kg^{-1}. Treatments were arranged according to a randomized complete block design. The fertilizers were added to the pots and the soils were allowed to incubate for 14 days; soil subsamples were then collected for NH$_4^+$ and NO$_3^-$ analyses (Bremner & Keeney, 1966). Next, each pot was seeded to obtain five sorghum plants, and a nutrient solution as recommended by Waugh and Fitts (1966), but without N, was added. Throughout the experiment, the pots were watered to near field capacity. The aerial portions of plants, harvested after a 21-day growing period, were dried, ground and analyzed for total N according to a semi-micro Kjeldahl procedure (Benton Jones et al., 1991).

Results and discussion

Urea hydrolysis

The effect of incubation time on urea hydrolysis, expressed as the amount of recovered NH$_4^+$, is shown in Figure 1. Recovered NH$_4^+$ was, to a great extent, proportional to the applied urea dose and varied slightly with soil characteristics. For example, recovered NH$_4^+$ was higher in the Lirios soil, regardless of incubation time. Rate of urea hydrolysis followed a similar pattern in the two soils with significant amounts of CaCO$_3$ where NH$_4^+$ reached a peak concentration at about 48 h, after which it decreased gradually. The Lirios acidic soil, on the other hand, showed a maximum rate of urea hydrolysis 24 hours after the application of the fertilizer. This behaviour suggests the existence of better environmental conditions for urease enzyme activity in the acidic soil than in the others (Bouwmeester et al., 1985; Faurie and Bardin, 1979).

Ammonia volatilization losses

There was a reduction in the rate of NH$_3$ volatilization with incubation time in all soils. In the Ancon soil, the NH$_3$ losses during the first week and during the third and fourth weeks with the application of 150 μg N g^{-1} were 33.5 and 10.7% with AS, and 41.5 and 6.3% with UR, respectively (Table 2).

Table 3. Amount of $NO_3^- - N$ recovered from ammonium sulfate and urea applications to the soils after two and four weeks incubation periods (I & II)

Soil	N Rate $\mu g\ g^{-1}$	Ammonium	Sulfate I	II	UREA I	II
				(%)		
S-1	150		26	82	19	44
	300		-	20	5	6
S-2	150		41	49	27	34
	300		14	63	2	3
S-3	150		28	31	31	31
	300		25	22	9	17

Table 4. Dry matter weight and N content of sorghum (*Sorghum bicolor* L.) as influenced by N rate from two fertilizer sources applied to the soils

Soil	N Rate $(\mu g\ g-1)$	Dry Matter (g/pot) AS	UR	N Content (%) AS	UR
S-1	0	0.087a	0.090a	2.05a	2.07a
	75	0.133b	0.120b	2.76b	4.27b
	150	0.133b	0.090a	4.03c	5.37c
	LSD (0.05)	0.0239	0.0250	0.0528	0.2973
S-2	0	0.090a	0.086a	1.47a	1.27a
	75	0.143b	0.140b	3.57b	3.09b
	150	0.149b	0.120b	4.14c	4.28c
	LSD (0.05)	0.0167	0.0167	0.1448	0.1213
S-3	0	0.173a	0.176a	2.14a	2.28a
	75	0.260b	0.273c	2.96b	4.36b
	150	0.290c	0.240b	4.36c	4.60c
	LSD (0.05)	0.0220	0.0236	0.0508	0.2136

In the Lirios soil, no NH_3 volatilization loss was detected from the AS application, irrespective of N rate and incubation time. Total ammonia volatilization in the Ancon soil, however, reached more than 50% with the application of 150 μg N g^{-1}, regardless of N source. These results are not surprising, as the acidic reaction of the Lirios soil clearly reduced the NH_3 loss (Mills et al., 1974), whereas the $CaCO_3$ and organic matter contents of the Ancon soil, 136 g kg^{-1} and 11 g kg^{-1} respectively, tended to increase NH_3 volatilization.

Surprisingly, the percentage loss of NH_3 due to volatilization was smaller in the Ancon soil with the application of 300 μg N g^{-1} as urea. It is unlikely that the higher soil pH would be responsible for this phenomenon; the observed reduction in NH_3 loss was more likely due to a decrease in soil water content, thus lowering the NH_3 vapor pressure of the soil solution. Similar results have been reported by Fox and Hoffman (1981), with the application of ammonium nitrate urea (UAN).

Ammonium nitrification

The percentage of applied N in the form of $NO_3^- - N$, following the application of 300 μg N g^{-1}, was less than that observed with the lower N dose in most cases (Table 3); this may indicate that nitrification was, to

Table 5. Dry matter production and N content of sorghum (*Sorghum bicolor* L.) as influenced by N source applied to the soils[1]/

Soil	N Source	N Rate (μg g^{-1})	Dry Matter (g/pot)	N Content (%)
S-1	UR	0	0.090a	2.07a
	AS	0	0.087a	2.50a
	UR	75	0.120a	4.27a
	AS	75	0.133a	3.76b
	UR	150	0.090a	3.37a
	AS	150	0.133b	4.03b
S-2	UR	0	0.086a	1.27a
	AS	0	0.090a	1.46a
	UR	75	0.140a	3.09a
	AS	75	0.143a	3.57b
	UR	150	0.120a	4.28a
	AS	150	0.140b	4.14b
S-3	UR	0	1076a	2.28a
	AS	0	0.173a	2.24a
	UR	75	0.273a	4.36a
	AS	75	0.260a	2.96b
	UR	150	0.240a	4.60a
	AS	150	0.290b	4.36b

Means in a column within one N rate followed by the same letter are not significantly different at p = 0.05 as indicated by LSD values

Table 6. Correlation coefficients (r) between plant growth parameters (*Sorghum bicolor* L.) and soil N availability estimates with applied ammonium sulfate and urea

Plant Parameters	UREA			Ammonium Sulfate		
	$NH_4^+ - N$	$NO_3^- - N$	$(NH_4^+ - N + NO_3^- - N)$	$NH_4^+ - N$	$NO_3^- - N$	$NH_4^+ - N + NO_3^- - N$
Dry matter	0.146	0.144	0.133	−0.058	0.290	0.077
N content	0.886***	0.952***	0.977***	0.589*	0.911***	0.779***
N uptake	0.580*	0.568*	0.603**	0.176	0.586*	0.360

some extent, inhibited by a high NH_4^+ concentration in the soil solution which was detrimental to the growth of nitrifying bacteria (Smith, 1964). Effect of incubation time on the amount of recovered $NO_3^- - N$ was not consistent. For instance, with the 150 μg N g^{-1} of both fertilizers there was a $NO_3^- - N$ increase with time in most cases, but little or no change in half of the comparisons with the 300 μg N g^{-1} rate of fertilization.

Yield and N uptake

Dry matter production increased significantly with the application of AS in all soils, although there was no

Table 7. Regression of plant parameters on N availability estimates for two nitrogen fertilizer sources

Equations	R^2
1 - Ammonium Sulfate	
$Y_1 = 0.129 - 0.00069X_1 + 0.00160X_2$	0.180
$Y_2 = 1.608 + 0.00101X_1 + 0.03915X_2$	0.956***
$Y_3 = 0.233 - 0.00310X_1 + 0.01326X_2$	0.403*
2 - UREA	
$Y_1 = 0.118 + 0.00056X_1 - 0.00002X_2$	0.021
$Y_2 = 1.092 + 0.03620X_1 + 0.00382X_2$	0.956***
$Y_3 = 0.129 + 0.00776X_1 + 0.00283X_2$	0.370*

Y_1 = Dry matter, Y_2 = Plant N content, Y_3 = N uptake, $X_1 = NH_4$, $X_2 = NO_3$

significant difference between the 75 and 150 μg N g^{-1} rates, except in the Ancon soil (S-3) where the yield with 150 μg N g^{-1} was significantly higher than with 75 μg N g^{-1} (Table 4). With the urea application, the 75 μg N g^{-1} rate increased dry matter production in all soils; with 150 μg N g^{-1} the yield remained unchanged in the Marin soil, and decreased significantly in the other soils (compared to the 75 μg N g^{-1} treatment). The decrease in dry matter is presumably due to the high NH_4^+ concentration, following urea hydrolysis of the 300 μg N g^{-1}, which created an unfavorable environment (high pH, nitrite toxicity, NH_3 volatilization) for plant growth, (Bremner & Doublas, 1971; Fenn & Hossner, 1985; Freney et al., 1983; Shaviv & Mikkelsen, 1993).

N percentage in sorghum increased significantly with N rates, regardless of N sources (Table 4). N content was significantly different with N sources at both 75 and 150 μg N g^{-1} soil; however, dry matter production was affected by N source only at the 150 μg N g^{-1} soil (Table 5).

Plant growth parameters and soil N availability

Amounts of NH_4^+ and $NO_3^- - N$ recovered from both fertilizers after a 14 day incubation of the soils did not correlate with dry matter yields of sorghum (Table 6). N uptake correlated significantly with NH_4^+ or NO_3^- and very significantly with the pool of both inorganic forms of N released from urea, but was directly associated only with $NO_3^- - N$ of the ammonium sulfate, indicating that the urea was a better source than ammonium sulfate for N absorption by sorghum plants. The dependence of N uptake on soil inorganic N from

both fertilizers was moderate (Table 7). However, 83 and 96% of the variation in plant N content could be explained by differences in the amounts of NH_4^+ and $NO_3^- - N$ extracted from the soils fertilized with AS and UR, respectively, after a two-week incubation preceding sorghum growth.

Conclusions

Rate of urea hydrolysis, the volatile loss of ammonia and the amount of ammonium nitrification varied widely with soil characteristics. Dry matter production of sorghum, in general, increased with ammonium sulfate application, but decreased with the application of more than 75 μg N g^{-1} as urea, presumably because of nitrite toxicity. Plant N content was influenced significantly by N source; however, dry matter was affected by source only at the highest N rate. Soil N availability estimates ($NH_4^+ - N$ and $NO_3^- - N$) did not correlate with dry matter, but were closely associated with plant N content and moderately with N uptake.

Acknowledgements

The authors wish to express their sincere gratitude to Mrs. Laura Santamaría for typing the manuscript.

References

Ayanaba A & Kang BT (1976) Urea transformation in some tropical soils. Soil Biol Biochem 8:313-316

Bandel VA, Dziena S & Stanford G (1980) Comparision of N fertilizers for no-till corn. Agron J 72:337-341

Benton Jones Jr. J, Wolf B & Mills HA (1991) Plant Analysis handbook. Micro-Macro Publishing, Inc. pp 197-199

Bouwmeester RJB, Vlek PLG & Stumpe JM (1985) Effect of environmental factors on ammonia volatilization from a urea-fertilizer soil. Soil Sci Soc Am J 49:376-381

Bremner JM & Doublas LA (1971) Decomposition of urea phosphate in soils. Soil Sci Soc Am Proc 35:575-578

Bremner JM & Keeney DR (1966) Determination and isotope-ratio analysis of different forms of nitrogen in soils. 3. Exchangeable ammonium, nitrate and nitrite by extraction distillation methods. Soil Sci Soc Am Proc 30:577-582

Fan MX & Mackenzie AF (1993) Urea and phosphate interactions in fertilizers microsites: Ammonia volatilization and pH changes. Soil Sci Soc Am J 57:839-845

Faurie G & Bardin R (1979) La volatilization de l'ammoniaque I Influence de la nature du sol et des composés azotes. Ann Agron

Fenn LB & Kissel DE (1975) Ammonia volatilization from surface application of ammonium compounds on calcareous soils: 4. Effect of calcium carbonate content. Soil Sci Am Proc 39:631-633

Fenn LB & Richards J (1986) Ammonia loss from surface applied urea-acid products. Fert Res 9:265-275

Fenn LB & Hossner LR (1985) Ammonia volatilization from ammonium or ammonium forming nitrogen fertilizers. Adv Soil Sci 1:123-169

Fleisher Z & Hagin J (1981) Lowering ammonia volatilization losses from urea application by activation of nitrification process. Fert Res 2:101-107

Fox RH & Hoffman LD (1981) The effect of N fertilizer source on grain yield, N uptake, soil pH and lime requirement in no-till corn. Agron J 73:891-895

Freney JR, Simpson JR & Denmead. OT (1983) Volatilization of ammonia. In: Freney JR & Simpson. JR (eds) Gaseous Loss of Nitrogen from Plant-Soil Systems, pp 1-32. Martinus Nijhoff Publishers, Dordrecht, the Netherlands

Gasser JKR (1964) Some factors affecting losses of ammonia from urea and ammonium sulfate applied to soil. J Soil Sci 15:258-272

Howard DD & Tyler DD (1989) Nitrogen source rate and application method for no-tillage corn. Soil Sci Soc Am J 53:1573-1577

Mills HA, Barker AV & Maynard DN (1974) Ammonia volatilization from soils. Agron J 66:355-358

S.A.S. Institute (1985) SAS user's guide: Statistics 5th Ed. SAS Inst., Cary, NC, USA

Smith JH (1964) Relationships between soil cation-exchange capacity and the toxicity of ammonia to the nitrification process. Soil Sci Soc Am Proc 28:640-644

Shaviv A & Mikkelsen RL (1993) Controlled-released fertilizers to increase efficiency of nutrient use and minimize environmental degradation - A review. Fert Res 35(1-2):1-12

Stecker JA, Buchholz DD, Hanson RG, Wollenhaupt NC & McVay KA (1993) Broadcast nitrogen sources for no-till continuous corn and corn following soybean. Agron J 85:893-897

Waugh DL & Fitts JW (1966) Soil test interpretation studies: Laboratory and potted plant, Technical bulletin No. 3. International Soil Testing, Raleigh, North Carolina, USA pp 18-19

N. Ahmad (ed.), Nitrogen Economy in Tropical Soils, 355–361.
© 1996 *Kluwer Academic Publishers. Printed in the Netherlands.*

Simple field testing sites to determine the extent of nitrogen leaching from agricultural areas

P. Cepuder & M. Tuller
*University of Agriculture, Forestry and Renewable Resources, Institute for Hydraulics and Rural Water
Management, Nussdorfer Laende 11 A-1190, Vienna, Austria*

Abstract

Thirty percent of Austria's potable water is taken from groundwater and groundwater pollution with nitrate caused
by agriculture is one of the greatest problems in many areas. To ensure a high quality of drinking water in future,
the protection of the groundwater will be very important. The Institute for Rural Water Resources Management has
established experimental sites to measure the extent of nitrogen pollution, to be able to give recommendations to
reduce nitrogen leaching from areas used for agriculture. To measure the quantity and the quality of the percolating
soil water, modified small lysimeters for quantity assessments combined with suction cups for quality assessments
and gypsum blocks are used. The required suction is matched to the tension of the soil with the help of the gypsum
blocks.

Kurzfassung

Die Trinkwasserversorgung Österreichs wird zu ca. 30% durch Grundwasser gedeckt. Um der Bevölkerung auch
in Zukunft ein qualitativ gutes und hygienisch einwandfreies Grundwasser zur Verfügung stellen zu können, ist ein
umfassender Schutz der großen Grundwasservorkommen in den Tal- und Beckenlandschaften von großer Bedeutung. Die Landwirtschaft ist in letzter Zeit immer mehr ins Kreuzfeuer der Kritik geraten, maßgebend für die hohen
Nitratbelastungen der Grundwasservorkommen verantwortlich zu sein. Um das Ausmaß dieser Verschmutzung
genauer erfassen zu können, wurden vom o.a. Institut einfache Feldmeßstellen entwickelt. Es werden modifizierte
Unterdrucklysimeter in Kombination mit Saugkerzen und Gipsblöcken zur qualitativen und quantitativen Bestim-
mung der Nitratbelastung verwendet. Durch die Anpassung des Unterdruckes an die Saugspannungsverhältnisse
im Boden kann die Perkolation gemessen werden. Die qualitative Nitratbestimmung erfolgt anhand der, durch die
im ungestörten Boden eingebauten Saugkerzen gewonnenen, Wasserproben.

Introduction

Nitrogen in the form of nitrate is one of the fundamental
nutritive substances for plants. High rates of mineralization beyond the growing season and exaggerated use
of fertilizers increase the capacity for nitrogen leaching down the soil profile. In this way nitrogen is lost
from agricultural areas and may, moreover, pollute the
groundwater with nitrate. For some years the Institute
for Rural Water Management at the University of Agriculture in Vienna has been discussing the problem of
nitrogen leaching in areas used for agriculture. The current research is done with respect to shallow and deep
soils with different degrees of cultivation and different N fertilizer carriers (e.g. mineral fertilizers, liquid
manure, sewage sludge). Simple field testing sites and
testing installations have been built up at various locations where samples of soil water are taken at different
depths to determine the extent of nitrogen leaching. In
this paper the simple field measuring devices and their
use are described.

The simple field measuring device

The introduced field measuring device is primarily
designed to determine the extent of the nitrogen loss in
agricultural areas.

To eliminate disturbing influences e.g. continuous fences, windbreak hedges from the boundaries of the field, the measuring devices should be installed at an adequate distance from the edge of the field, normally about 7 m to 8 m, depending on the working width of the agricultural machines used. The distance from wind protection appliances should be at least 50 m.

A field measuring device consists of the following parts:
  * small lysimeter
  * two suction cups
  * two gypsum blocks
  * three bottles for collecting samples
  * a vacuum container
  * vacuum tubes and wires for the gypsum blocks (Figures 1 to 3)

For the installation of the appliance parts, a cavity must be dug. The soil is cleared level by level and stored separately. The excavation of the hole should be done extremely carefully, in order to avoid unnecessary disturbance of the soil structure. The excavation of each level of soil should be done manually, however, with adequate care, it can also be done mechanically. In such cases, appliances with low ground pressure are to be preferred.

After the completion of the cavity, two approximately 20 cm deep holes are dug with an auger in a depth of 105 cm on the front side. The holes have to be wide enough to be filled up with a suction cup and a gypsum block which is designed to test the degree of moisture. Afterwards the holes are filled up to the middle with super-saturated soil and the suction cup or the moisture detector is pressed in; the remaining cavity is then closed up. In the center of the cavity a flat surface is made and the lysimeter is installed. The lysimeter consists of a ceramic suction plate (diameter 29 cm, Figure 1) in a PVC-tube of diameter 30 cm and height 40 cm closed at the bottom. In addition, a suction cup and a moisture detector are installed directly above the suction plate. Above the ceramic suction plate, *material of fine sand* should be used, so that coarse soil fragments cannot cause any mechanical damage (Figure 4). The lysimeter container should be filled up with water to the middle in order to accelerate the consolidation of the installed soil. Afterwards the lysimeter and the cavity are refilled according to the natural structure of the soil. To determine the content of soil moisture, TDR-detectors or gypsum blocks can be used as measuring detectors.

From the suction plate, the suction cup and the gypsum block tubes and wires are laid in a depth of 50

cm to a small measuring station at the edge of the field. (Figure 3).

For the sampling of the soil water, two small and two large bottles are needed. In the small bottles the water of the suction cups is collected and in one of the large ones the water of the lysimeter. The second large bottle is used as an overflow container for the three other bottles. These should be placed in a container beneath the ground surface (Figure 3). By means of this system, extensive fluctuations of temperature can be avoided. An attached steel container with a volume of at least 50 litres is designed to ensure the necessary vacuum. Each suction unit (suction cup or suction plate) is handled separately by means of a valve. It is therefore possible to turn off defective suction appliances which are meant to serve the undisturbed collection of water samples with the properly working suction appliances. Each of the bottles is closed up with a rubber stopper, and connected with the suction cup (plate) and the vacuum container by tubes.

Method and analysis of the samples

Most of the agricultural plants have a utilizable root - zone of approximately 90 cm provided that the soil is adequately deep. It is assumed for this method of measurement that the water which percolates down to more than 90 cm contributes to the ground water. The water that is collected by the lysimeter in a depth of 105 cm is therefore seen as the depth of new ground water. Because of the disturbance of the structure of the soil that is caused by the installation of the lysimeter, the qualitative analysis of the seepage water is done with respect to the water collected by the suction cups in the undisturbed area of soil (Figure 3). The degree of the nitrogen emission within the respective area can be calculated on the basis of this quantitative and qualitative assessments. The suction cup in the lysimeter serves comparative measurements. The moisture detectors show the water conditions inside as well as outside the lysimeter, so that the adequate suction can be adjusted. Normally, the field capacity has a suction from 0.1 to 0.3 bar. Sandy soils usually have 0.1 bar, whereas clayey and loamy soils have a suction of 0.3 bar. For additional or different kinds of soil, the results have to be modified by interpolation. The difference of elevation between the suction plate and the collecting bottle is 1 m; for that reason the conveyor height has to be added to the necessary suction.

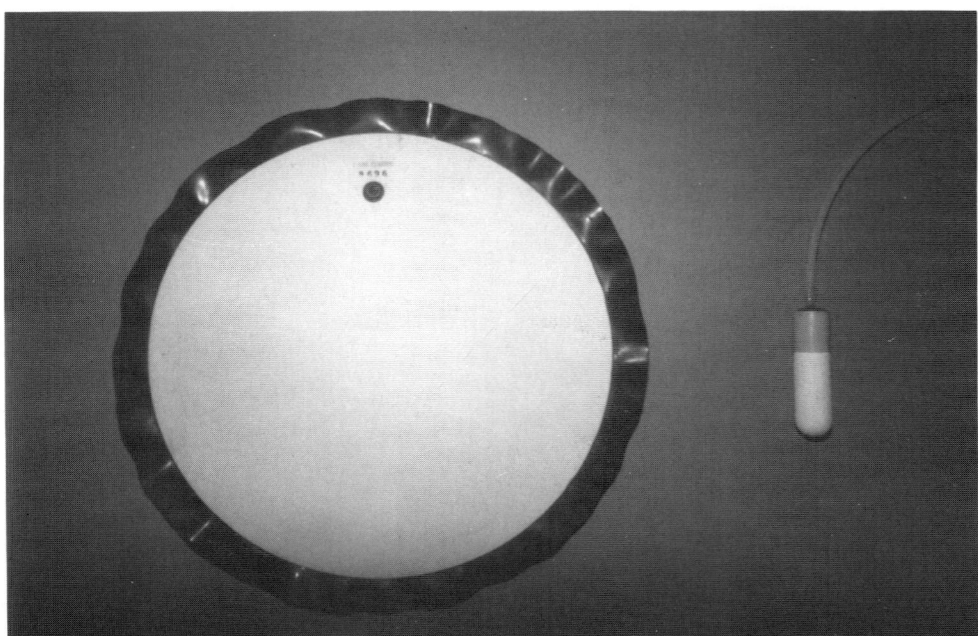

Fig. 1. Suction plate and suction cup of a small lysimeter

Therefore if the field capacity has 0.3 bar, the vacuum container is tuned in on a vacuum of 0.4 bar. The degree of vacuum must be diminished, however, when a front of seepage water percolates through the soil profile. During periods of aridity, the degree of vacuum can be increased up to a maximum of 0.8 bar. Because of the drying of the ceramic cups and ceramic plates, air may get into the measuring system. During those periods, the instruments can be turned off, for no samples of water can be taken. The whole appliance will be turned on again when the degree of ground water in the area of the lysimeter increases, which will be indicated by the moisture detectors. Sandy soils are more adequate for representative results because of their faster consolidation. Clayey and loamy soils will cause a longer preparation time as well as more frequent defects of the suction cups and plates. The samples of the soil water are taken from the immediate surrounding of the suction cup and from the container of the lysimeter. The collected samples of water should be taken out once a week and stored in vacuum-closed plastic bottles in a cool place. On the basis of the continuous work of the field measuring device, fronts of seepage water and their degree of nitrate can be registered. By means of that information, the temporal process of the concentration of nitrate in the different levels of soil can be recorded and the spread of nitrate observed. Dependent on the kind of soil, the scheme of the profile, the temperature and the precipitation, the plants as well as the cultivation (fertilizing, tillage), the concentration of nitrate in the ground water alternates.

Results of two testing plants

In a typical agriculturally used area in the south-east of Austria, testing sites were installed in two plots with very different soils. The first soil type was a deep soil (silty loam) with 15% sand, 65% silt and 20% clay. The second one was a shallow loamy sand (70% sand, 20% silt, 10% clay) with 60% of gravel. Each of the installations consists of two small-lysimeters, two suction cups and two gypsum blocks in a depth of 1.05 m.

Both plots were cropped with corn from April to October. Due to the danger of nitrate leaching, bare soil was covered with grass during the winter period. Every year 180 kg ha^{-1} nitrogen was applied to corn.

On a long term basis the average air temperature was 8.9 °C and the average precipitation was 924 mm. In 1991 the amount of precipitation was nearly 1000 mm. In 1992 and 1993 the amount was approximately 750 mm.

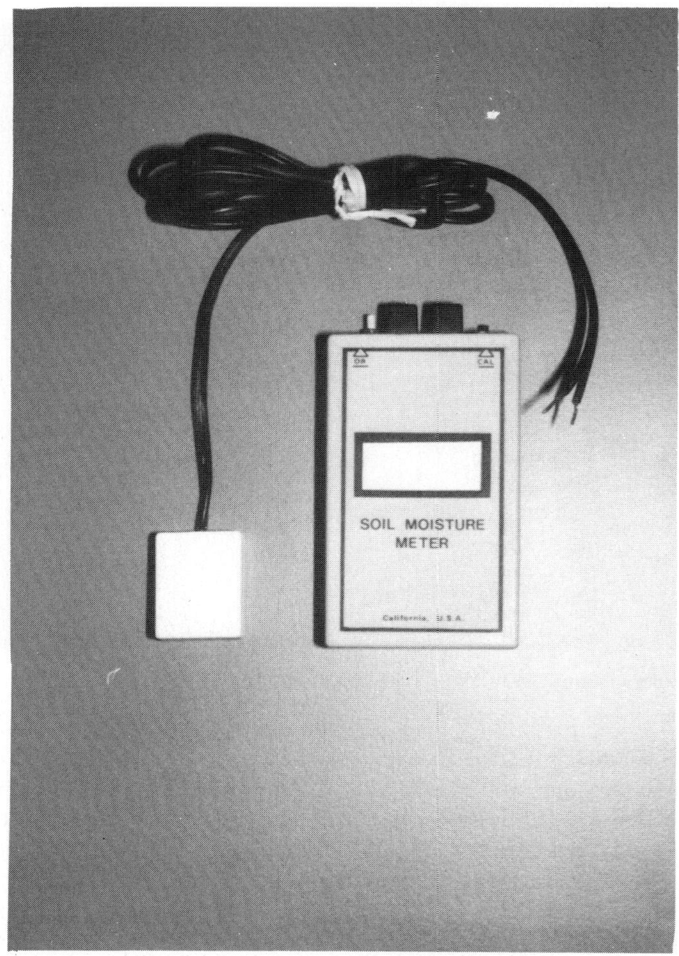

Fig. 2. Gypsum block and the appropriate measuring instrument

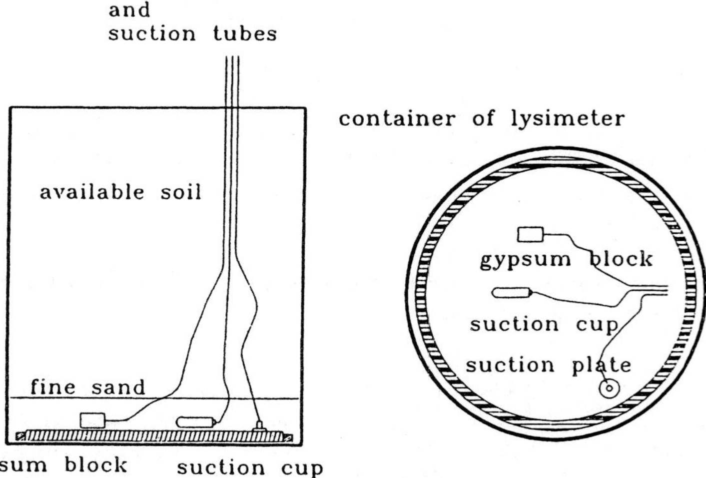

Fig. 3. Design of a simple field measuring device

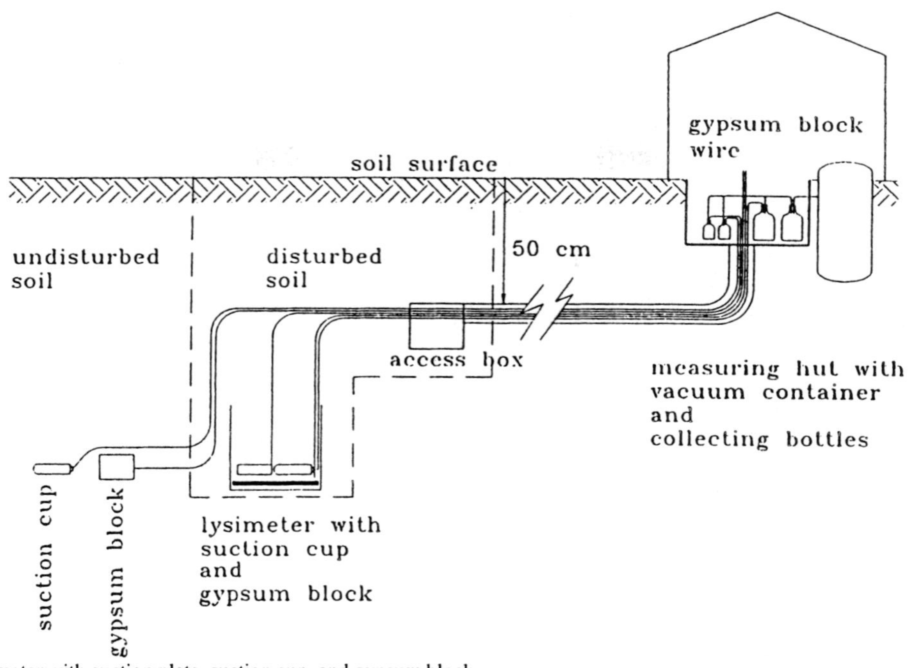

Fig. 4. Small lysimeter with suction plate, suction cup, and gypsum block

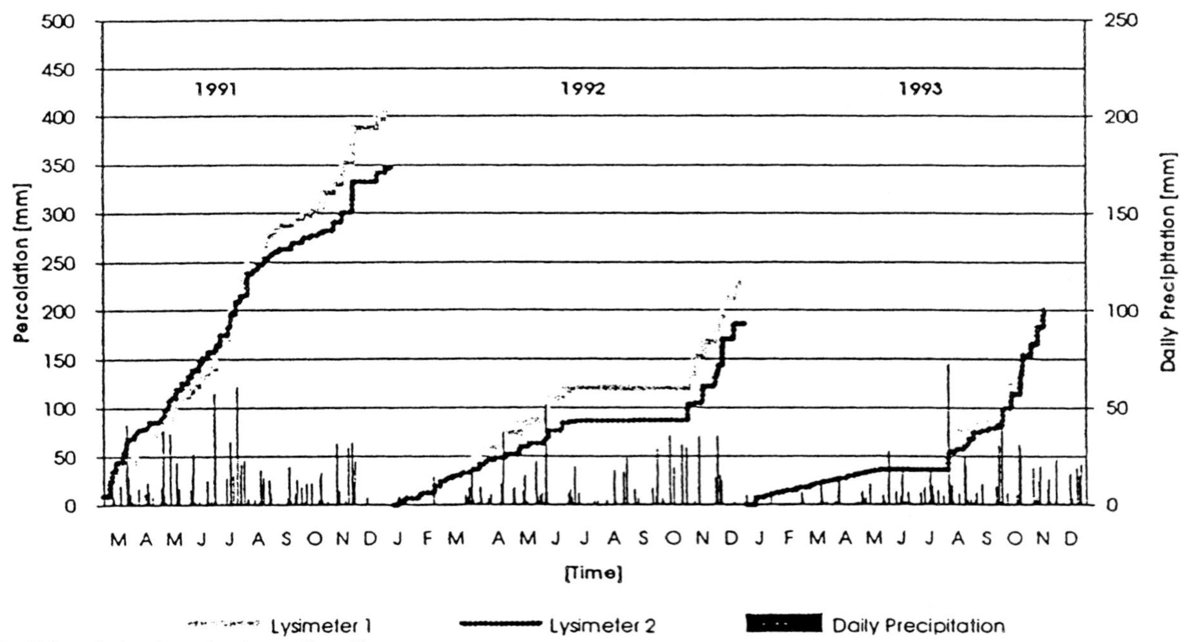

Fig. 5. Percolation through a deep soil profile

Figures 5 to 8 show the measured values for percolation and nitrate leaching for both plots from 1991 to 1993. All measurements were done with two parallel working lysimeters, installed at a distance of about one meter.

Figures 5 and 6 give a good relationship between percolation and the nitrogen leaching in the deep soil profile. In 1991, the measured percolation ranges from 350 mm up to 400 mm. In the second year of the experiment the range was from 180 up to 230 mm.

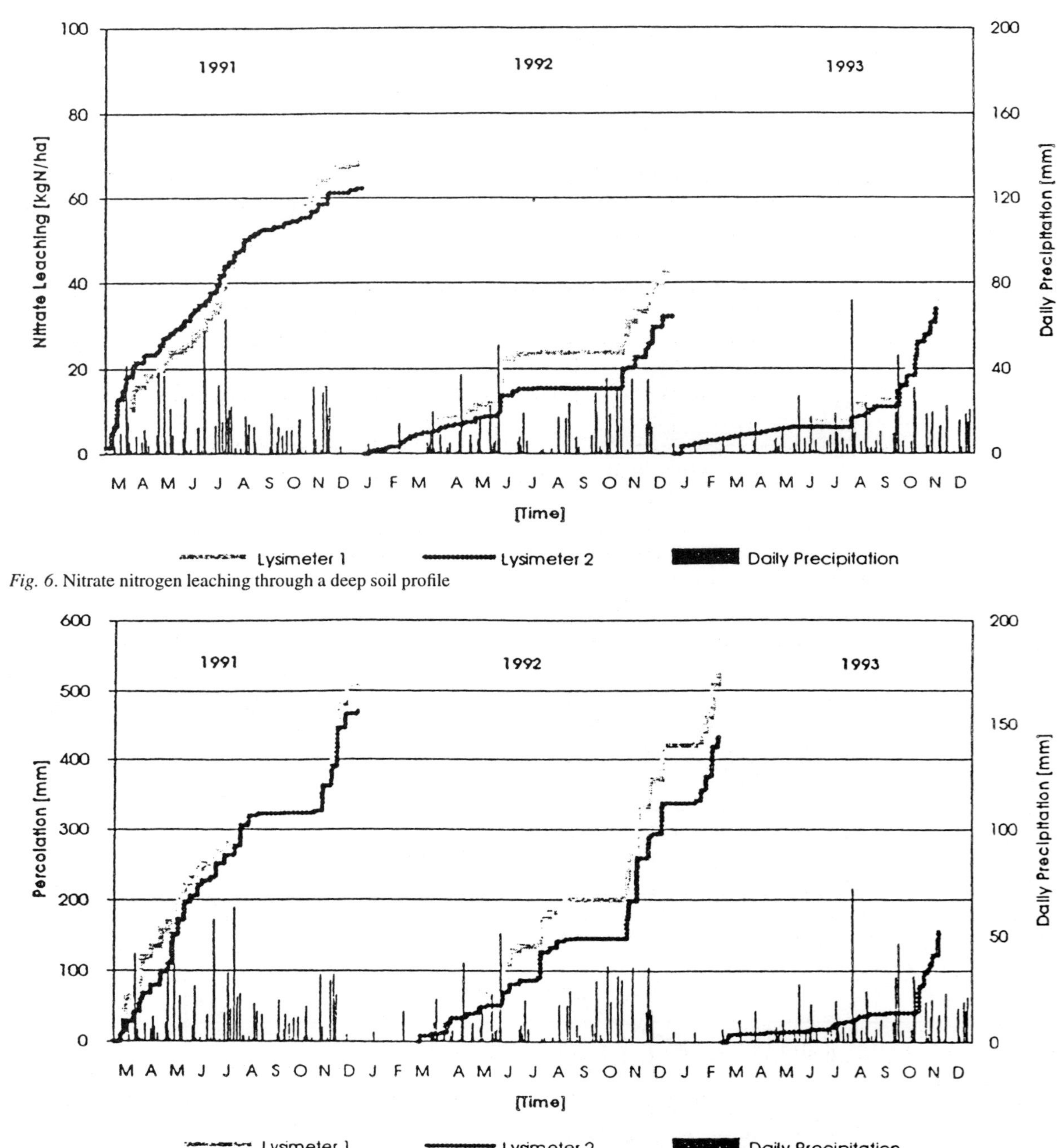

Fig. 6. Nitrate nitrogen leaching through a deep soil profile

Fig. 7. Percolation through a shallow soil profile

In the first two years the difference between the two small lysimeters was 50 mm. In 1993 both lysimeters gave the same results, the amount being 200 mm. The measured nitrogen loss was from 62 to 68 kg ha^{-1} in 1991, 32 to 42 kg ha^{-1} in 1992 and in 1993 the loss was about 35 kg ha^{-1} . Due to the permanent rainfall during the year 1991 and because of the high water storage capacity of the deep soil, percolation water was measured for the whole year. In 1992 there was a dry period from June to September and there was accordingly no percolation and nitrogen loss. In April 1993 a short period only with less precipitation

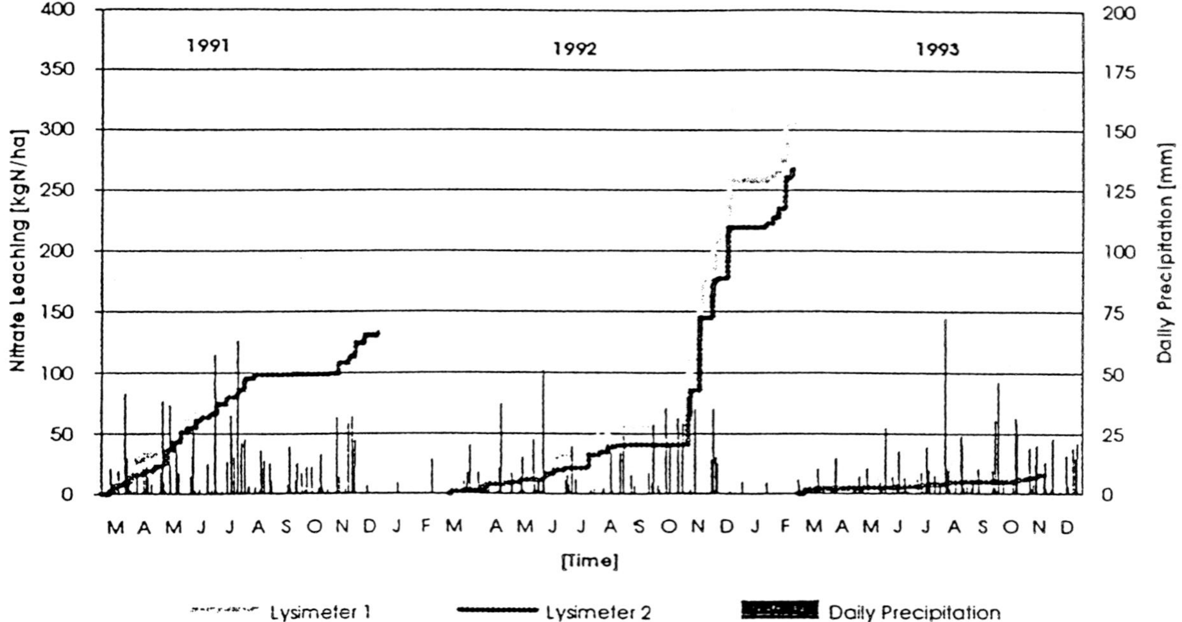

Fig. 8. Nitrate leaching through a shallow soil profile

appeared and consequently the percolation was very small.

On account of the low water storage capacity of the shallow soil, higher rates of percolation water and nitrogen loss were obtained. In 1991 the amounts of the percolated water increased from 470 up to 520 mm, in 1992 from 430 up to 520 mm and in 1993 from 180 up to 190 mm. As above, the difference between the parallel installed lysimeters decreased from 90 mm in the first two years to 10 mm in the third year. In 1991 the measured nitrogen loss ranged from 116 to 123 kg ha^{-1}, in 1992 from 262 to 272 kg ha^{-1} and in 1993 only from 37 to 39 kg ha^{-1}. Every year during the main growing season no percolation water and no nitrogen loss were measured. Due to the low plant growth in the dry period of 1992, high losses of nitrogen were observed in fall.

These examples show the great influence of climatic factors and different soil types on groundwater contamination with nitrate. After the time of soil consolidation, the described simple field testing sites showed good results. This is an easy way to get some information about the implications of different crops, tillages and fertilizer applications and can be tested in Tropical regions of the world.

Acknowledgements

The project was financed by the Department of Science and Research. The authors gratefully acknowledge Prof. Dr. N. Ahmad and Dr. G. Gouveia for reviewing the manuscript.

References

DVWK 217 (1990) Gewinnung von Bodenwaserproben mit Hilfe der Saugkerzen-Methode. Verlag Paul Parey

N. Ahmad (ed.), Nitrogen Economy in Tropical Soils, 363–369.
© 1996 *Kluwer Academic Publishers. Printed in the Netherlands.*

Comparison of legumes and fertilizer nitrogen for wheat production in subtropical Australia

R.C. Dalal[1], W.M. Strong[1], E.J. Weston[2], J.E. Cooper[1], K.J. Lehane[2] & A.J. King[1]
[1] *Queensland Wheat Research Institute, Toowoomba, Australia;* [2] *Department of Primary Industries Queensland, Toowoomba, Australia*

Key words: Chickpea wheat, legume-grass pasture, lucerne, medic, urea, vertisol

Abstract

The supply of adequate amount of nitrogen is essential to produce sustainable cereal grain yields and good grain quality while maintaining or enhancing soils' natural resource base. To achieve these objectives, a long-term field experiment was established in 1986 to include forage legumes and grain legume in rotation with wheat, and fertilizer N application (0, 25 and 75 kg N ha^{-1} y^{-1}) to no-till compared with conventional till on a fertility-depleted vertisol.

The amounts of nitrate-N in soil down to 1.5 m depth at the time of planting in May 1990 varied from 45 kg N ha^{-1} in conventional till (CT) nil N treatment (control) to 147 kg N ha^{-1} following lucerne (lucerne-wheat rotation, L-W). Net amounts of nitrate-N available to wheat in excess of the control treatment were 103, 79, 76, and 37 kg N ha^{-1} following lucerne, medic, legume-grass pasture, and chickpea, (L-W rotation, medic-wheat, M-W rotation; legume-grass-wheat, LG-W rotation; and chickpea-wheat, CP-W rotation) respectively. Also, amounts of nitrate-N were significantly higher due to residual fertilizer N from 75 kg N ha^{-1}y^{-1} application of urea. However, tillage treatments had no significant effect on additional nitrate-N, 24 kg N ha^{-1} in CT vs. 20 kg N ha^{-1} in no till (NT) soil. There was no residual fertilizer nitrogen effect when it was applied at only 25 kg N ha^{-1}y^{-1}. Wheat grain yields were 3.59, 3.59, 3.43 and 3.38 t ha^{-1} in CP-W, M-W, L-W and LG-W rotations, respectively, compared to 2.23 t ha^{-1} in the control treatment. Grain yields in NT treatments were significantly higher at 0 and 25 kg N ha^{-1}y^{-1} rates but similar at 75 kg N ha^{-1}y^{-1} application to that in the corresponding CT treatments. The proportion of soil N supply in the grain exceeded 40% and in CP-W, L-W and the control exceeded 60%. Available N utilisation efficiency exceeded 75% in all treatments except in L-W rotation and CT and NT where 75 kg N ha^{-1} fertiliser was applied. Thus, although all treatments enhanced wheat yield and grain protein contents, lucerne-N and high fertiliser N rates are likely to be utilised less efficiently in a semi-arid environment.

Introduction

Current continuous cereal cropping practices in subtropical and tropical regions cause severe depletion of organic matter and nitrogen (Dalal & Mayer, 1986). Consequently, continual nitrogen (N) depletion causes declining cereal yields and lower grain protein concentrations (Dalal et al., 1991) because crop N demand exceeds the soil's ability to supply available N. The available N supply can be increased through application of fertiliser N or symbiotic N$_2$ fixation by legumes.

Both forage and grain legumes improve yield of the following crop to varying degrees (Badaruddin & Meyer, 1989; Jessop & Mahoney, 1985; Singh et al., 1985) depending upon their N contribution and in some instances break in disease cycle or some other factors, simply referred to as rotation affects (Baldock et al., 1981). The legume N contribution to the following crop depends on the amount of N$_2$ fixed, the amount and type of legume residue left in the soil, and the availability of soil N to the legume crop (Dalal et al., 1991; Heichel, 1985).

Most estimates of the legume N contribution in legume-cereal rotations have been made by the fertiliser-N replacement value, that is, the amount of the fertiliser N that would be required to produce an equivalent yield under otherwise comparable test conditions (Higgs et al., 1976). In this approach it is assumed that increased cereal yield is solely due to legume N contribution, that fertiliser N and residue

N are equally available, and that available soil water was similar. Another less restricted method is the comparative measure of N use efficiency in legume-cereal rotations and even different tillage systems (Huggins & Pan, 1993; Russelle et al., 1987).

The N use efflciency method requires measurements of grain yield, grain N, above ground plant N, applied N, and postharvest available soil N which are partitioned into various N use efficiency components. The components consist of N supply, available N efficiency, available N uptake efficiency, N utilisation efficiency, grain N accumulation efficiency, and N harvest index. For example, at low fertiliser N rates, yield differences between conventional till and no-till were attributed to N supply and available N efficiency components whereas at high fertiliser N rates, it was attributed to the N utilisation efficiency component (Huggins & Pan, 1993).

A long-term field experiment was established in 1986 on a fertility-depleted vertisol to compare forage legumes and grain legumes in rotation with wheat, no-till and fertiliser N application for improving wheat yield and grain protein. The objective of this study was to evaluate the comparative yield and nitrogen use efficiency from the legume N and fertiliser N for wheat production in semi-arid subtropical Australia.

Materials and methods

A long term field experiment was established in 1986 on a fertility-depleted vertisol (Typic Chromustert) in a semi-arid region, with mean annual rainfall of 630 mm, in subtropical Queensland (26° 47'S, 150° 53'E), Australia. The soil contained 52 %clay, 0.7% organic C, 0.07% total N and pH 8.5 in the 0-0.1 m layer (Dalal et al., 1991). The following treatments were established:

1. A mixed legume-grass pasture of 4 years' duration followed by continuous wheat (LG-W); the grasses and legumes were purple pigeon grass (*Setaria incrassata*), Rhodes grass (*Chloris gayana*), lucerne (*Medicago sativa*) and annual medics (*Medicago scutellata, M truncatula*);
2. A two year rotation of annual medics and wheat (*Triticum aestivum*) (M-W);
3. A two year rotation of lucerne and wheat (L-W);
4. A two year rotation of chickpea (*Cicer arietinum*) and wheat (CP-W);
5. No-till wheat with annual application of 0, 25 or 75 kg N ha^{-1} (NT, NT+25N, NT+75N); and

6. Conventional till wheat with annual application of 0, 25 or 75 kg N ha^{-1} (CT, CT + 25N, CT + 75N). Nitrogen was applied as fertiliser urea at planting at about 5 cm depth in alternate mid-rows.

The treatments were arranged in a randomised block design with four replications. Plot dimensions were 25 m long and 6.75 m wide.

The legume-grass pastures were established in January 1986 and grown for 3.75 years. Lucerne was planted with wheat during May-July period at the rate of 2 kg ha^{-1}. After harvesting wheat, lucerne was grown for one year. Annual medics were also established under wheat and because of their self-generating nature, were planted only once. The perennial legume-grass pasture and lucerne were cut at l0 cm height and removed in four harvests at three-monthly intervals, while in only one harvest for the annual medics.

The legume-grass pasture was terminated in October 1989 and forage legume phases were terminated by blade ploughing (minimum soil inversion) in October each year, about six months before the planting of wheat so as to allow water recharge of the soil profile during the summer-autumn period.

Wheat grain yields were measured from 1.75 m × 23 m central areas of all plots. After determination of grain water content, grain yields were adjusted to 12% H$_2$0 content. Total dry matter yields were recorded from the middle two rows, 0.25 m apart and lm long. Straw and grain were analysed for their N cancentration.

Weeds in the no-till treatment were controlled by herbicide spray (glyphosate and 2-4D, amine) 2-4 times during the fallow period. Weed control in the conventional till treatment was achieved by 2-4 operations usually by a tined implement to approximately l00mm depth, during the fallow period.

Soil samples were collected from each plot with a 50 mm dia. tube sampler down to 1.5 m depth twice a year, one sampling in May 1990 before sowing and the other soon after the crop harvest in November 1990. The samples were bulked, sealed in plastic bags and stored at 4°C until analysis. The soil was dried at 30± 5°C in a forced draught oven and ground < 2mm sieve. Mineral N in soil was extracted by 2 M KCl and NH$_4$-N and N0$_3$-N were determined in the extracts by automated method (Best, 1972). Total N in the plant material and the grain was determined in the Kjeldahl digests using alkaline sodium isocyanurate (Crooke & Simpson, 1971). All results are reported on oven-dry weight basis.

The following N use efficiency components were examined (Huggins & Pan, 1993) using wheat grain yield (G_w), grain N (N_g), N supply ($N_s = N_f + N_t + N_h$, where N_f is fertiliser N applied, N_t is above-ground plant N and N_h is postharvest NO_3-N to 1.5 m depth both in control treatments) and available N ($N_{av} = N_t + N_h$): N use efficiency (G_w/N_s), available N use efficiency (G_w/N_{av}), N utilisation efficiency (G_w/N_t, grain N accumulation efficiency (N_g/N_s), available grain N accumulation efficiency N_g/N_{av}), N harvest index (N_g/N_t), available N efficiency (N_{av}/N_s), available N uptake efficiency (N_t/N_{av}) and available N utilisation efficiency (N_t/N_p where N_p is plant utilisable N = NO_3 at planting $+ N_f - N_h$).

Analysis of variance was performed to assess the effect of management practice on grain yield, grain N, total N uptake, available N supply, and various components of N use efficiency by standard statistical techniques (Snedecor & Cochran, 1967).

Results and discussion

Soil nitrate-N contents

Almost all of the 2 M KCl extractable mineral N was present as NO_3^--N at the time of planting wheat in May 1990. The amounts of NO_3^--N in the soil down to 1.5 m depth varied from 45 kg N ha^{-1} in the conventional, nil fertiliser N treatment (control) to 147 kg N ha^{-1} following lucerne in the L-W rotation (Figure 1).

The residual effects of fertiliser N application at 75 kg N ha^{-1}y^{-1} since 1987 were evident in both the CT and NT treatments. Fertiliser N applied at 25 kg N h^{-1}y^{-1} had no residual effects. This is also supported by others (Strong et al., 1986) from their long-term studies on carryover effects of fertiliser N for wheat production . Tillage practices had no effect on the soil NO_3^--N contents in this fertility-depleted Vertisol (total N, 0.07%). However on a higher fertility Vertisol (total N, 0.13%), NO_3^--N contents tend to be higher in the top layers in CT than in NT treatments (Dalal. 1989).

The soil NO_3^--N contents following chickpea in CP-W rotation were 37 kg N ha^{-1} higher than the control. This value is within the range of chickpea-N contribution (10-80 kg N ha^{-1}) reported by others (Doughton et al., 1993; Evans et al., 1989; Strong et al. 1986) since the variation in total N_2 fixation is associated with dry matter production, soil NO_3^--N levels, sowing date and tillage practices, and subsequently on

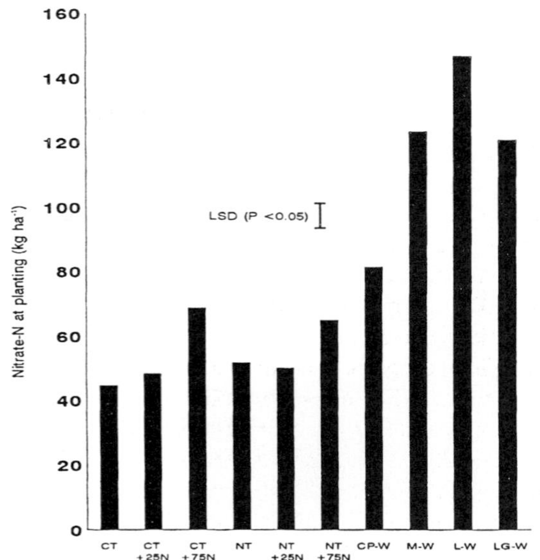

Fig. 1. Soil nitrate-N at wheat planting (0-1.5 m depth) in May 1990 under different management practices: CT, continuous wheat-till, NT, continuous wheat-notill, 25N or 75N, 25 kg N ha^{-1} or 75 kg N ha^{-1} of fertiliser urea, CP-W, chickpea wheat rotation, M-W, medic wheat rotation, L-W, lucerne-wheat rotation, LG-W, legume and grass-wheat rotation

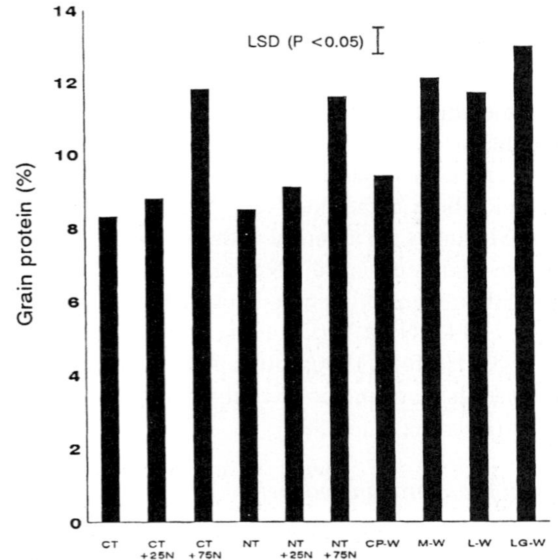

Fig. 2. Wheat grain protein content under different management practices. For meaning of abbreviations see Figure 1.

grain-N removal, and residue management and environmental conditions conducive to N mineralisation.

The soil NO_3^--N contents following the forage legumes and legume-grass pasture were much higher than following the chickpea. Beside the grain-N removal in the latter, forage legumes fixed more N

Table 1. Grain yield (G_w), grain N (N_g), N supply (N_s), available soil N (N_{av}), aboveground plant N (N_t), and postharvest soil NO_3-N (N_h) for wheat under different management practices

Management practice	G_w	N_g	N_s	N_{av}	N_t	N_h
			($kg\ ha^{-1}$)			
A. Continuous wheat till						
Conventional till (CT)	2234	31.9	49.2	49.2	41.3	7.9
CT + 25 kg N ha^{-1}	2840	43.0	74.2	65.6	54.5	11.1
CT + 75 kg N ha^{-1}	3413	69.3	124.2	133.3	96.3	37.0
B. Continuous wheat-no-till						
No-till (NT)	2522	37.2	66.4	66.4	45.0	19.5
NT + 25 kg N ha^{-1}	3143	49.5	91.4	79.6	64.7	14.8
NT + 75 kg N ha^{-1}	3348	66.8	141.4	169.2	92.6	76.6
C. Grain legume wheat						
Chickpea-wheat	3586	58.2	95.2	95.2	73.0	22.2
D. Forage legume-wheat						
Medic wheat	3585	75.0	126.8	126.8	104.9	21.9
Lucerne-wheat	3427	69.1	106.1	106.1	86.0	20.1
Legume and grass-wheat	3384	76.1	143.4	143.4	106.4	37.1
LSD ($p<0.05$)	270	6.8	20.7	31.3	9.4	28.3

than the chickpea (Hossain, 1993). The soil NO_3^--N contents were 76, 79, and 103 kg N ha^{-1} in LG-W, M-W and L-W rotations, respectively, higher than the control. These additional NO_3^--N contents following forage legumes are identical to the amount of N_2 fixed, as measured by enriched ^{15}N dilution method (Hossain, 1993). Estimates of lucerne-N contribution vary from 50 to 180 kg N ha^{-1} (De et al., 1985; Hesterman et al., 1987; Holford, 1990; Littler & Whitehouse, 1987), whereas those of medic-N contribution are generally lower (Evans et al., 1989).

Wheat grain protein and yield

Grain protein contents increased significantly when fertiliser N was applied to wheat at 75 kg N ha^{-1}, and wheat following the legumes (Figure 2). Highest grain protein content was found in the LG-W rotation followed by M-W rotation. Grain protein contents were essentially similar in L-W rotation and 75 kg N ha^{-1}y^{-1} fertiliser N applied to wheat both in CT and NT treatments.

Wheat grain yields increased significantly with increasing fertiliser N rates, and at low N rates, wheat yields were higher in NT than in CT, as would be expected from N_s and N_{av} values (Table 1). Wheat grain yields following legume-gras pasture, forage legumes and grain legumes exceeded or were equal to 75 kg N ha^{-1}y^{-1} fertiliser application and much higher than that achieved with fresh application of fertiliser at 75 kg N ha^{-1} (data not presented).

Wheat grain N and total N uptake following chickpea were equivalent to about 58 kg N ha^{-1} fertiliser whereas those following either medic or legume grass pasture exceed 100 kg N ha^{-1} fertiliser equivalent (grain N kg ha^{-1} = 33.9 + 0.53N_{eq} + 0.0019N_{eq}^2, where N_{eq} = kg N ha^{-1} fertiliser equivalent). However, fertiliser N equivalent method assumes that fertiliser N and legume N are equally available and that increased non legume yield is solely due to legume-N contribution (Hesterman et al., 1987).

Nitrogen efficiency component analysis provides an improvement over that based on grain yield increase and equivalent fertiliser (Huggins & Pan, 1993). Nitrogen use efficiency by wheat (G_w/N_s and G_w/N_{av}) was highest in the control and then in CP-W rotation (Table 2). Nitrogen utilisation efficiency was highest in the low N treatments and decreased as N supply increased.

Table 2. Nitrogen use efficiency ratios for grain yield and grain N of wheat under different management practices

Management practice	N use efficiency	Available N use efficiency	N utilisation efficiency	Grain N accumulation efficiency	Available grain N accumulation	N harvest index efficiency
	G_w/N_s	G_w/N_{av}	$G/_w/N_t$	N_g/N_s	N_g/N_{av}	N_g/N_t
A. Continuous wheat-till						
Conventional till (CT)	45.5	45.5	54.1	0.65	0.65	0.77
CT + 25 kg N ha^{-1}	38.3	43.6	52.1	0.58	0.66	0.79
CT + 75 kg N ha^{-1}	27.5	25.7	35.5	0.56	0.52	0.72
B. Continuous wheat no-till						
No-till (NT)	38.3	38.3	53.7	0.57	0.57	0.79
NT + 25 kg N ha^{-1}	34.6	39.6	48.9	0.54	0.62	0.77
NT + 75 kg N ha^{-1}	23.8	21.3	36.3	0.48	0.43	0.72
C. Grain legume-wheat						
Chickpea-wheat	38.4	38.4	49.2	0.62	0.62	0.80
D. Forage legume-wheat						
Medic-wheat	28.7	28.7	34.2	0.60	0.60	0.71
Lucerne-wheat	33.3	33.3	40.1	0.67	0.67	0.81
Legume & grass-wheat	24.3	24.3	32.0	0.54	0.54	0.72
LSD (*p*<0.05)	5.6	6.2	4.1	0.09	0.11	0.05

The proportion of N supply and available N in the grain exceed 40% (43-65%), and in CP-W, L-W and the control exceed 60%. The N harvest index (N_g/N_t) demonstrated that chickpea-N and lucerne-N was most effectively used by wheat and similar to that of N fertiliser application at low rates (0,25 kg N ha^{-1}).

Available N efficiency values did not differ significantly among various management practices (Table 3) although available N uptake efficiency was the lowest in the NT treatment when fertilise N was applied at 75 kg N ha^{-1}y^{-1}. In this Vertisol, available N utilisation efficiency (N_t/N_p) exceed 75% in all treatments except L-W rotation and CT and NT where 75 kg N ha^{-1} fertiliser was applied. Conversely, lucerne-N and high rates pf fertiliser N were utilised less efficiently by the wheat crop, and therefore, N from these treatments can be liable for losses from the soil-plant system.

Conclusion

Significant contributions to soil available N supply were made following forage legumes, legume-grass pastures, and chickpea. This was reflected in higher wheat grain yields and wheat grain protein in the legume-wheat rotations and also with fertiliser N application. Although N utilisation efficiency (G_w/N_t) provided a good comparision of various management practices, available N utilisation efficiency showed that lucerne-N and fertiliser N at higher rates (75 kg N ha^{-1}y^{-1}) were poorly utilised compared with other management practices. Thus available N utilisation efficiency reflected better the N use because the account was taken of residual N and mineralisable N following repeated applications of fertiliser and N contribution of legumes for variable duration.

Aknowledgement

Authors are grateful to the Grains Research and Development Corporation of Australia for financial support.

Table 3. Available N utilisation efficiency ratios by wheat under different management practices

Management practice	Available N efficiency	Available N uptake efficiency	Available N utilisation efficiency
	N_{av}/N_s	N_t/N_{av}	$N_t/N_p{}^a$
A. Conventional wheat-till			
Conventional till(CT)	1.00	0.84	0.93
CT + 25 kg N ha^{-1}	0.88	0.84	0.76
CT + 75 kg N ha^{-1}	1.07	0.73	0.67
B. Continuous wheat-no-till			
No-till (NT)	1.00	0.71	0.93
NT + 25 kg N ha^{-1}	0.87	0.81	0.88
NT + 75 kg N ha^{-1}	1.21	0.59	0.66
C. Grain legume-wheat			
Chickpea-wheat	1.00	0.78	0.91
D. Forage legume-wheat			
Medic-wheat	1.00	0.84	0.86
Lucerne-wheat	1.00	0.83	0.59
Legume and grass-wheat	1.00	0.75	0.88
LSD ($p<0.05$)	NS	0.14	0.16

[a] Plant utilisable N : N_p = NO_3-N at planting + fertiliser N (if applied) - NO_3 at harvest

References

Badaruddin M & Meyer DW (1989) Forage legume effects on soil nitrogen and grain yield, and nitrogen nutrition of wheat. Agron J 81:419–424

Baldock JO, Higgs RL, Paulson WH, Jackobs JA & Shrader WD (1981) Legume and mineral N effects on crop yields in several crop sequences in the upper Mississippi valley. Agron J 73:885–890

Best EK (1972) An automated method for the determination of nitrate-nitrogen in soil extracts. Qld J Agric Anim Sci 33:161–166

Crooke WM & Simpson WE (1971) Determination of ammonium in Kjeldahl digests of crops by an automated procedure. J Sci Food Agric 22:9–10

Dalal RC (1989) Long-term effects of no-tillage, crop residue, and nitrogen application on properties of vertisol. Soil Sci Am J 53:1511–1515

Dalal RC & Mayer RJ(1986) Long-term trends in fertility of soils under continuous cultivation and cereal cropping in southern Queensland. I.Overall changes in soil properties and trends in winter cereal yields. Aust J Soil Res 24:265–279

Dalal RC, Strong WM, Weston EJ & Gaffney J (1991) Sustaining multiple production systems.2.Soil fertility decline and restoration of cropping lands in subtropical Queensland. Trop Grassl 25:173–180

De R, Khan MAS, Katti MS & Raja V (1985) Fodder legumes affecting sequential crop production and fertilizer N use efficiency. J Agric Sci, Camb 105:1–7

Doughton JA, Vallis I & Saffigna PG (1993) Nitrogen fixation in chickpea.1.Influence of prior cropping or fallow, nitrogen fertilizer and tillage. Aust J Agric Res 44:1403–1413

Evans J, O'Connor GE, Turner GL, Coventry DR, Fettell N, Mahoney J, Armstrong EL & Walsgatt DN (1989) N_2 fixation and its value to soil N increase in lupin, field pea and other legumes in south-eastern Australia. Aust J Agric Res 40:791–805

Heichel GH (1985) Nitrogen recovery by crops that follow legumes. In: Barnes RF et al. (eds) Forage Legumes for Energy Efficient Animal Production, pp 183–189. USDA, Washington DC, USA

Hestermann OB, Russelle MP, Sheaffer CC & Heichel GH (1987) Nitrogen utilization from fertilizer and legume residues in legume-corn rotations. Agron J 79:726–731

Higgs RL, Paulson WH, Pendleton JW, Peterson AF, Jacobs JA & Shrader WD (1976) Crop rotations and nitrogen: Crop sequence comparisons on soils of the driftless area of southwestern Wisconsin. 1967-1974. Univ of Wisconsin Coll of Agric and Life Sci Res Bull R2761, WI, USA

Holford (1990) Effects of 8-year rotations of grain sorghum with lucerne, annual legume, wheat and long fallow on nitrogen and organic carbon in two contrasting soils. Aust J Agric Res 28:277–291

Hossain (1993) Nitrogen fixation, accretion and mineralisation under legume based cropping systems in a vertisol at Warra, Queensland. Ph D thesis, University of Queensland, St Lucia, Queensland, Australia

Huggins DR & Pan WL (1993) Nitrogen efficiency component analysis: An evaluation of cropping system differences in productivity. Agron J 85:898–905

Jessop RS & Mahoney J (1985) The effect of soil nitrogen on grain legume yield and nitrogen fixation. J Agric Sci, Camb 105:231–236

Littler JW & Whitehouse MJ (1987) Effect of pasture on subsequent wheat crops on a black earth soil of the Darling Downs.III.Comparison of nitrogen from pasture and fertiliser sources. Qld J Agric Anim Sci 44:1–8

Russelle MP, Hesterman OB, Sheaffer CC & Heichel GH (1987) Estimating nitrogen and rotation effects in legume-corn rotations. In: Power JF (ed) The Role of Legumes in conservation Tillage Systems, pp 41–42. Soil Conserv Soc Am Ankeny IA, USA

Singh RK, De R & Lal RB (1985) Contribution of legumes to the fertilizer nitrogen economy in a maize-based cropping system. J. Agric Sci, Camb 105:491–493

Snedecor GW & Cochran WG (1967) Statistical methods. 6th Edition. Iowa State University Press, Ames, USA

Strong WM, Harbison J, Nielson RGH, Hall BD & Best EK (1986) Nitrogen availability in a Darling Downs soil following cereal, oilseed and grain legume crops. I.Soil nitrogen accumulation. Aust J Exp Agric 26:347–351

N. Ahmad (ed.), Nitrogen Economy in Tropical Soils, 371–377.
© 1996 *Kluwer Academic Publishers. Printed in the Netherlands.*

Long-term N-fertilization calibration experiments – environmental aspects

T. Németh
Research Institute for Soil Science and Agricultural Chemistry of the Hungarian Academy of Sciences, Herman Ottó u. I5, H 1022 Budapest, Hungary

Key words: Calibration, field experiments, mineral-N, N-supply, N-fertilization, residual effects

Abstract

Long-term N fertilization experiments were established with identical treatments in two different growing areas; one on a calcareous sandy soil (Örbottyán) and the other on a calcareous chernozem soil (Nagyhörcsök). The aim was to create differences in mineral-N content in the soil profiles in order to determine their N-supplying capacity and whether they may be regarded as a supply indices for crop production. In the first three years of the experiments four rates of fertilizer-N were applied. In the fourth year the original plots were divided into five smaller ones onto which five spring fertilizer levels were applied. In this fourth year winter oil-seed rape was the indicator plant. In the following three years the basic four N-fertilizer treatments were applied, and in the eighth year a division of the plots was made again, similar to that at the fourth year. This time winter wheat was the indicator plant. The effects of the residual-N and the freshly applied fertilizer-N on the yield of the winter oil-seed rape and the winter wheat were compared. The results indicate that under certain environmental conditions N (in the form of NO_3, resulting from N-fertilization in previous years), may accumulate in the soil profile to such an extent that it must be taken into consideration when determining fertilizer rates. This is important not only for economic management and environmental protection, but also for yield quality. The calculations can be reliably performed through the measurement and calibration of the mineral-N content of the soil.

Introduction

In fertilizer recommendations, the greatest difficulty is in determining the amount of N to be applied. In Hungary the widely used N-fertilization recommendations are based on the organic matter (humus) content of the 0-25 cm soil layers. During the period between the early sixties and middle eighties, when intensive systems of agriculture was practiced, often quantities of nutrients in excess of the crops' demands were applied in the form of fertilizers. Because of this, the nutrient (NPK) balance of many fields became positive and their nutrient supplying capacity improved. This was favourable for soils which originally contained less nutrients (weak nutrient supply); under those conditions fertilization played a significant role in the sustainability yields. In contrast, there were many fields in which only the compensation of the plant nutrient demand was satisfied. The surplus application of nutrients could be easily detected in the ploughed layer for P and K using the standard AL-method (Kádár, 1979).

The positive nutrient balance might cause a similar change also in the N content, and the danger of overfertilizing is greatest in the case of N. This can not be followed by analysis of the soil's organic matter, therefore measurement of the organic-C content of the soils is not the best tool to detect changes in the easily available soil N fraction. It is well known that the effect of N-fertilization on the N content of the soil can be characterized by measuring the mineral-N (exchangeable NH_4^+ and NO_3^-) content of the soil (Wehrmann & Scharpf, 1983).

Under Hungarian conditions the remaining P and K are found mainly in the ploughed layer, (Sarkadi et al., 1986) while the surplus N, in the form of NO_3-N, may accumulate somewhere in the profile or be leached to deeper soil layers or to the ground water(Campbell et al., 1983; Linville & Smith, 1971; Németh et al., 1987-1988; Schuman et al., 1975: Verdegem et al., 1981) depending on the physical characteristics of the soil and on environmental conditions. If the peak of NO_3-N accumulation remains in the rooting zone, then this N can be taken up by the crops satisfying their N-demands. The mineral-N fertilization recommendation systems are based on determination of this N (the NO_3^-, or NO_3^- and exchangeable NH_4^+) content of a certain soil layer (Görlitz et al., 1983; Hofman et al., 1981; Kolenbrander et al., 1981; Németh et al., 1987-1988; Wehrmann & Scharpf, 1979). To better understand the role of mineral-N in the nutrition of cultivated plants, it is necessary to investigate the fate and behaviour of these N-forms in the profile, i.e. to investigate the accumulation, the distribution and the movement of NO_3^- along the different soil profiles.

In a long-term fertilization experiment on a calcareous chernozem soil the mineral-N content of the upper 60 cm soil layer and the seasonal dynamics of the N forms were investigated after an eleven years' application of four different N-fertilization rates (Németh & Buzás, 1985; Sarkadi et al., 1986) as well as the movement, accumulation and distribution of the NO_3-N in the profiles down to six meters (Németh et al., 1987-1988). On the basis of these results two N-fertilization experiments were carried out to compare the residual effects of the previous N-fertilization with to the spring (fresh) N-fertilization. The aim of this study was to measure, calculate and calibrate the soil mineral-N content, ensuring better utilization of the residual N, and to determine the amount of N to be applied for optimum yield at different residual N levels.

Materials and methods

The two field experiments were set up in the autumn of 1984 on a calcareous chernozem soil (at Nagyhörcsök) and on a calcareous sandy soil (at Örbottyán). Both fields belong to the Experimental Farms of the Research Institute for Soil Science and Agricultural Chemistry of the Hungarian Academy of Sciences (HAS).

Soil and meteorological features

Örbottyán. The long-term yearly mean precipitation is 500 mm in this district and the climate can be characterized as continental. The soil type, according to the Hungarian Classification is a calcareous sandy soil, with 3-6% $CaCO_3$, 0.9-1.1 % organic matter and 800-1200 mg total-N kg^{-1} in the ploughed layer, with C/N ratio of 7.5-8.0. The thickness of the humus layer is 20-60 cm. According to the soil analyses, the $pH_{(KCl)}$ is 7.0-7.3; the ammonium-lactate soluble P_2O_5 content is 70-80 mg kg^{-1}, the ammonium-lactate soluble K_2O is 50-80 mg kg^{-1}. These results show that the original nutrient supply of this soil is medium-to-weak. The average level of the groundwater table is around 270-320 cm.

Nagyhörcsök. The site has a yearly mean temperature of 10.9°C and a long-term yearly average precipitation of 550-580 mm; the climate is continental. The soil type is a calcareous chernozem, the parent material loess, and the whole profile is loamy. In the ploughed layer the $CaCO_3$ content is 5%, organic matter 3%, total-N 2100-2200 mg kg^{-1}, C/N ratio 7.7-8.5. The thickness of the humus layer is 75-90 cm. According to the soil analyses the $pH_{(KCl)}$ is 7.2-7.3; the ammonium-lactate soluble P_2O_5 content is 60-80 mg kg^{-1}; and the ammonium-lactate soluble K_2O content is 140-160 mg kg^{-1}. These results show that the original P-supply is weak while the K-supply is medium. The average level of the groundwater table is around 13-15 meters.

Treatments

Large (250 m²) plots were set up with four N-fertilizer rates (0,150, 300 and 450 kg N ha^{-1} year) in four replications in the autumn of 1984. The rate of P (100 kg P_2O_5-ha^{-1}) and K (200 kg K_2O ha^{-1}) were the same in all N treatments. The mineral-N content of the profiles of these 16 large plots were checked in spring and in autumn in each year, just before fertilization. The mineral-N analyses were made by the methods of Bremner & Keeney (1966)

These experiments were conducted in 1988 (winter oil-seed rape as indicator plant) and in 1992 (winter wheat as indicator plant) to compare the effect of the residual N to spring applied N during the calibration.

In 1987, after the harvest of the previous crop (spring barley) and the sampling for mineral-N analyses, the scheduled autumn fertilization was applied. In the spring of 1988, after soil sampling for mineral-N analyses, the large plots were each divided into five subplots, and the spring N-fertilization was applied at rates of 0, 50, 100, 150 and 200 kg N ha^{-1} on these small, (50 m²) plots. Using these treatment combinations, 80 subplots (4 main treatments, 5 spring fertilization levels, 4 replications) were investigated at each experimental site. The same division of the large plots, with a similar fertilization procedure was repeated in 1992, as well.

Table 1. Average ammonium-N concentrations (mg kg^{-1} in differently fertilized profiles (mean of 4 replications)

Depth (cm)	Main N-treatments(kg ha^{-1} year^{-1})			
	0	150	300	450
Sandy soil (NH$_4$-N mg kg^{-1})				
0–100	3.4	3.8	3.2	3.4
100–200	1.9	1.9	1.9	1.9
Chernozem soil (NH$_4$-N mg kg^{-1})				
0–100	6.4	6.9	6.5	6.0
100–200	5.1	5.3	5.2	5.1
200–300	4.2	4.9	4.9	4.7

Results and discussion

Soil analyses

In previous studies it was found that under similar experimental conditions the surplus N could be detected only in the form of NO_3 (Németh & Buzás, 1985; Sarkadi et al., 1986). The NH_4-N content of the soils did not vary after different N application rates. This was also observed in these two experiments (Table 1).

Spring 1988. On the calcareous sandy soil, differences in NO_3-N concentration were measured only between the treatments below 50 cm in the profiles, while on the calcareous chernozem soil, such differences were observed from the top of the profiles (Figure 1).

Spring 1992. The relative distribution of NO_3-N was similar to that observed in 1988, but the amount detected nitrate-N was different (Figure 2). The average NO_3-N concentrations in the upper 100 cm soil layers are given in Table 2. In the following Tables these NO_3-N concentration values will be used instead of the main N treatments, because these values show the residual effects of the N-treatments.

Yields of the cultivated crops

Winter oil-seed rape

In the sandy soil the yield of the winter oil-seed rape varied between 0.95 and 2.04 t ha^{-1} (the lowest yield was measured on the double-control plots, with 0 N-application). In the average of the original treatments (main-treatments) the yield was significantly higher on the plots which contained 6.0 and 11.3 mg NO_3-N kg^{-1} in the profile at spring sampling (after applying 300 and 450 kg N ha^{-1} year^{-1} resp.), as compared to the control treatments. The significant residual effect of the highest dose (450 kg N ha^{-1} year^{-1}, corresponding to 11.3 mg NO_3-N kg^{-1} in the profile) was detected at every spring N-fertilizer level. If more than 100 kg N ha^{-1} was applied in spring, only a slight increase was observed in the yield of the winter oil-seed rape, the differences were not significant (Table 3).

In the chernozem soil, the yield of the winter oil-seed rape varied between 1.34 and 2.61 t ha^{-1} (the lowest yield was also measured on the double control plots). The residual effect of N-fertilization was detected in the yields of all previously fertilized plots. When the NO_3-N content in the soil layer was 26.7 mg kg^{-1} and 41.8 mg kg^{-1}, the yield of the indicator plant was as high even without spring N-application, as in the N-control plots (containing 11.3 mg NO_3-N kg^{-1}) with addition of 150 and 200 kg N ha^{-1} as spring N-fertilizer. The residual effect of the previous N-application was so efficient, that the 100 kg N ha^{-1} spring N-application was significantly (p=5%) effective only on the plots with 13.4 mg NO_3-N kg^{-1} soil in the profile, and on the former (main treatments) control plots. Even after application of 150 and 200 kg N ha^{-1} as spring N-fertilization, there were only slight increases in the yield of the winter oil-seed rape.

From the above results it can be concluded that on the less fertile calcareous sandy soil, the effect of the spring N-fertilization was greater a the residual effect (i.e. the resulting NO_3-N content of the profile) of the fertilizer-N applied at different rates for three years. In contrast, on the chernozem soil there were no differences between the residual effect of the previous

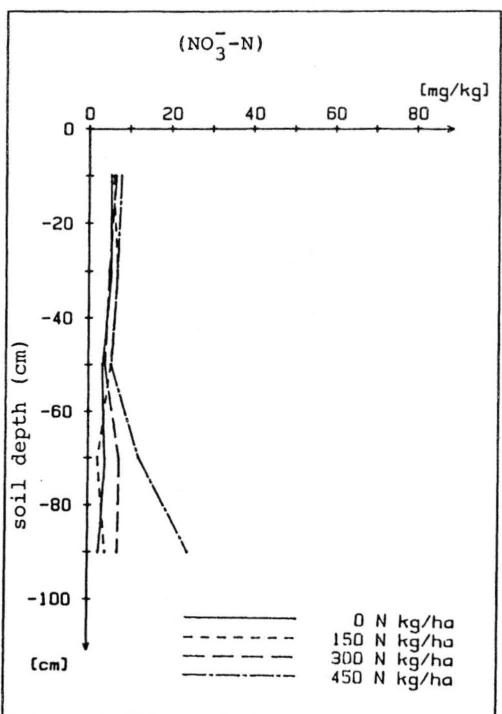

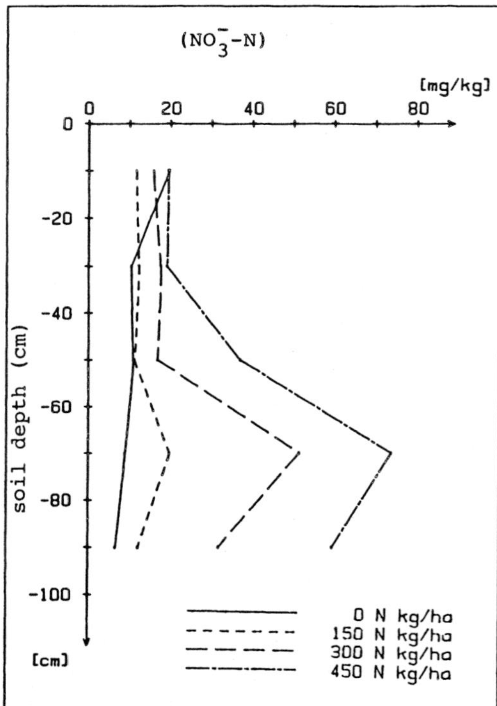

Fig. 1. Nitrate-N distribution in the 0-100 cm soil layer (Spring 1988)

Table 2. Average nitrate-N concentration (mg kg^{-1}) in the 0-100 cm soil layer after different N application (mean of 4 replications)

Year	Main treatments (kg ha^{-1} year^{-1})			
	0	150	300	450
Sandy soil (NO_3-N mg kg^{-1})				
1988	4.2	5.0	6.0	11.3
1992	4.2	4.9	6.5	6.9
Chernozem soil (NO_3-N mg kg^{-1})				
1988	11.3	13.4	26.7	41.8
1992	4.3	10.1	35.0	47.2

N-fertilization and the effect of the fresh N-application. With the improvement of the soil N-supply the amounts of fresh fertilizer-N needed for the optimum yield were decreased on both soils: from 150 kg N ha^{-1} to 50 kg N ha^{-1} on the sandy soil, while from 200 kg N ha^{-1} to 0-50 kg N ha^{-1} on the chernozem soil.

Winter wheat

The yield of winter wheat varied between 0.51 and 2.48 t ha^{-1} on the calcareous sandy soil (Table 4). The lowest yield was measured on the double-control plots. The utilisation of the previous N-fertilization without fresh N-application was very efficient; its residual effect was significant ($p=5\%$) for each main treatment compared to the double-control plots. The residual effect of the highest main treatment was significant ($p=5\%$), not only relative to the double-control treatment, but also to the first main N-fertilization level. From the average of the spring N-fertilization treatments, it can be seen that all previous fertilizer levels had pronounced positive

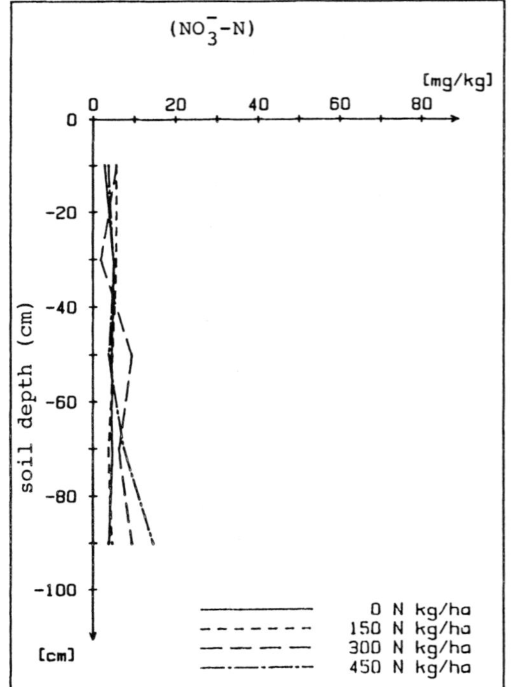

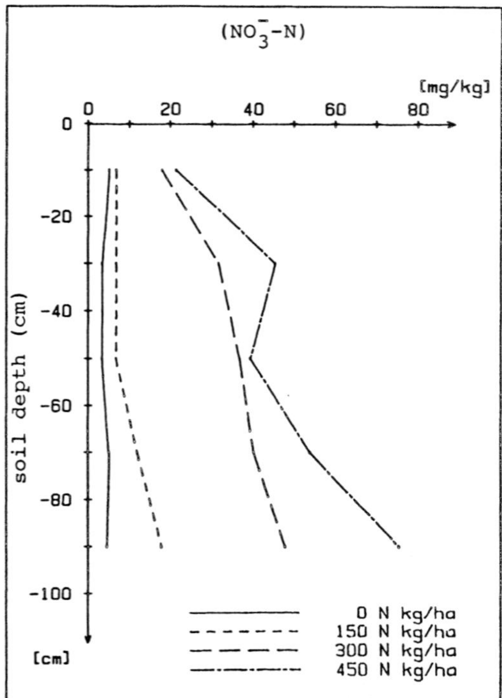

Fig. 2. Nitrate-N distribution in the 0-100 cm soil layer (Spring 1992)

Table 3. Yield of winter oilseed rape (t ha^{-1}) (1988, mean of 4 replications)

Nitrate-N (mg kg^{-1}) in 0-100 cm soil layer	N top-dressing rates (kg ha^{-1})						
	0	50	100	150	200	LSD 5%	Mean
Sandy soil							
4.2	0.95	1.15	1.28	1.64	1.39		1.28
5.0	0.86	1.33	1.68	1.44	1.54	0.30	1.37
6.0	1.32	1.37	1.65	1.58	1.79		1.54
11.3	1.56	1.79	1.70	2.04	1.87		1.79
a/ LSD 5%				0.35			0.22
b/Mean	1.17	1.41	1.57	1.68	1.65	0.15	1.50
Chernozem soil							
11.3	1.34	1.52	1.79	2.32	2.43		1.88
13.4	2.26	1.36	2.58	2.37	2.44		2.40
26.7	2.38	2.48	2.61	2.46	2.57	0.23	2.50
41.8	2.37	2.56	2.52	2.53	2.48		2.49
a/ LSD 5%				0.33			0.26
b/Mean	2.08	2.23	2.37	2.42	2.48	0.12	2.32

Table 4. Yield of winter wheat (t ha^{-1}) (1992, mean of 4 replications)

Nitrate-N (mg kg^{-1}) in 0-100 cm soil layer	N top-dressing rates(kg ha^{-1})						
	0	50	100	150	200	LSD 5%	Mean
Sandy soil							
4.2	0.51	1.69	1.90	1.97	1.81		1.57
4.9	1.44	1.95	2.05	1.77	1.91	0.74	1.82
6.5	2.09	2.19	2.06	1.60	1.92		1.97
6.9	2.24	2.48	2.14	2.31	2.10		2.26
a/LSD 5%				0.74			0.62
b/Mean	1.57	2.08	2.04	1.91	1.93	0.23	1.91
Chernozem soil							
4.3	2.21	3.64	5.00	5.46	5.95		4.45
10.1	5.60	5.92	5.86	5.94	5.81	0.46	5.83
35.0	5.49	5.36	5.64	5.63	5.37		5.50
47.2	5.26	5.35	5.19	5.13	5.00		5.18
a/LSD 5%				0.46			0.38
b/Mean	4.64	5.07	5.42	5.54	5.53	0.14	5.24

effect on the winter wheat yield, but the increase was significant (p=5%) only at the highest main treatment. In the main control treatments the effects of all five spring N-application rates were significant (p=5%), but no significant (p=5%) increases from spring N-application were detected in the previously fertilized treatments. Spring N-application significantly (p=5%) increased the yield, the highest increase occurring at the first spring N-level (50 kg N ha^{-1}). Summarizing the results, it can be concluded that the positive effects of the residual N were measureable at all spring N-application rates.

In the calcareous chernozem soil, the yield of the winter wheat crop varied between 2.21 and 5.95 t ha^{-1}(Table 4). The results show, that there were significant (p=5%) increases in yield at all residual N levels in the spring control and the spring 50 kg N ha^{-1} treatments, as well as in the mean of the spring fertilization rates. Again, it can be concluded that at rates higher than the first main fertilizer level (150 kg N ha^{-1}) the amount of residual N was so high that it had a negative effect on yield. All spring N-fertilization rates significantly (p=5%) increased the yield of the winter wheat at the original control levels, not only against the double-control treatment, but also between all the 4 original N-treatments, as well. Considering the mean of the main treatments, the spring N-fertilization was effective up to 150 kg N ha^{-1} rate. Summarizing the

results the winter wheat yield was greatly affected by the residual N on the previously fertilized plots, while no positive effects of spring N-application could be detected.

Comparing the effect of the residual N (dependent on previous N-fertilization) to spring N-application, it can be concluded that on the calcareous sandy soil and on the calcareous chernozem soil, 50 - 100 kg N ha^{-1} and 0-50 kg N ha^{-1} were the optimum N-fertilization rates under these environmental conditions for the two soils respectively.

Conclusion

The long-term N-fertilization experiments were set up to investigate the effect of the residual N compared to the fresh spring applied N for two crops on two soil types. From the results, it can be concluded that the optimum rate of N-fertilization depends on the mineral-N status, of the soil. Measuring the mineral-N content of a certain soil layer under the specific environmental circumstances can provide a good guide to optimize fertilization practices. The calibration experiments showed the dependence of yield on the soil mineral-N content, as well as on the spring N-application. On the calcareous sandy soil and the calcareous cherozem soil 50-100 kg N ha^{-1} and 0-

50 kg N ha^{-1} were the optimum N-fertilization rates under the specific conditions for the two soils respectively. The environmental importance of such calibration experiments is that estimating the utilization of N from the mineral-N pool can reduce the additional cost incurred due to over-fertilization, and at the same time reduce the potential danger of NO_3 leaching in the groundwater. Extrapolation of the results of the experiments to the farm scale can lead to both economical and environmental achievements.

Long-term nitrogen fertilizer experiments of this type are relatively unknown in the tropics; however, it is important that they be carried out with various ecosystems. This can lead to a better understanding of the dynamics of soil/fertilizer nitrogen and its relationships to crop uptake.

References

Bremner JM & Keene DR (1966) Determination and isotope-ratio analysis of different forms of nitrogen in soils. 3. Exchangeable ammonium, nitrate and nitrite by extraction destillation methods. Soil Sci Soc Am Proc 30: 577–582.

Campbell CA, Read DWL, Biederbeck VO & Winkleman GE (1983) The first 12 years of a long term crop rotation study in southwestern Saskatchewan - nitrate-N distribution in soil and N uptake by the plant. Can J Soil Sci 63: 563–578

Görlitz H, Müller S, Krause O, Witter B & Wackernagel R (1983) Gebietsbezogene Aussagen zur Korrektur der ersten N-gabe zu Wintergetreide auf der Grundlage der Untersuchung des Anorganischen Stickstoffs im Boden bei Vegetationsbeginn. Arch Acker Pflanzenb Bodenkd. 27:161–167

Hofman G, Ruymbeke M, Ossemerct C & Ide G (1981) Residual nitrate-N in sandy loam soils in moderate marine climate. Pedologie 31:329–346

Kádár I (1979) Födmvelésünk nitrogén, foszfor és kálium mérlege. (Nitrogen, phosphorus and potassium balance studies in agriculture). Agrokém Talajtan 28:527–545

Kolenbrander G J, Neeteson J J & Wijnen G (1981) Investigation in the Netherlands of optimum nitrogen fertilization on the basis of the amount of N_{min} in the soil profile. Pedologie 31:365–371

Linville KE & Smith GE (1971) Nitrate content of soil cores from corn lots after repeated nitrogen fertilization. Soil Sci 112:249–255

Németh T and Buzás I 1985 Characterization of the mineral nitrogen content of soils for fertilization advices. In: Welte E & Szabolcs I(eds) Fight against hunger through improved Plant Nutrition. Proc 9th World Fertilizer Congress held in Budapest, Hungary. June 11-16, 1984, CIEC Goltze-Druck, Goettingen, Germany Vol:2: 220–224.

Németh T, Kovács G & Kádár I (1987-1988) A nitrát, szulfát és a sóbemosódás vizsgálata mtrágyázási tartamkisérletben (Nitrate, sulphate and water soluble salt accumulation in the soil profile of a long-term fertilization experiment). Agrokém Talajtan 36-37:109–126

Sarkadi J, Németh T and Kádár I (1986) A talaj könnyen oldható tápanyagtartalmának heterogenitása (Heterogenity of the readily soluble nutrient content of the soil. Agrokém Talajtan 35:295–306

Schuman GE, McCalla T M, Saxton KE & Knox HT (1975) Nitrate movement and its distribuhon in the soil profile of differentially fertilized corn watersheds. Soil Sci Soc Am Proc 39:1192–1197

Verdegem L, Van Cleemput O & Vanderdeelen J (1981) Some factors inducing the loss of nutrient out of the soi profile. Pedologie 31:309–327

Wehrmann J & Scharpf HC (1979) Der mineralstickstoffgehalt des Bodens als Mass-stab für den Stickstoffdüngerbedarf (N_{min}-methode). Plant Soil 52:109–126

Wehrmann J & Scharpf HC 1983 Sachgerechte Stickstoffdüngungschatzen, kalkulieren, messen. AID-Heft 17/1982

N. Ahmad (ed.), *Nitrogen Economy in Tropical Soils*, 379–388.
© 1996 *Kluwer Academic Publishers. Printed in the Netherlands.*

Nitrification inhibitors, with emphasis on natural products, and the persistence of fertilizer nitrogen in the soil

Kanwar L. Sahrawat
West Africa Rice Development Association (WARDA), 01 BP 2551 Bouake, Ivory Coast

Key words: ammonium fixation, denitrification, immobilization, indigenous materials, mineralization, non-edible oilseeds, N transport and movement, retardation of nitrification

Abstract

Nitrogenous fertilizers have been major contributers to increases in agricultural productivity world-wide. However, there is continuing need to improve the efficiency of N fertilizer use to achieve more efficient production of food crops and minimize fertilizer-related pollution of the environment.

Poor efficiency of fertilizer N use is largely due to loss of N by denitrification, ammonia volatilization or leaching. With the exception of ammonia volatilization, these losses are associated with and follow the nitrification of ammonium to nitrate. The nitrification process converts the relatively immobile cationic ammonium to nitrate, which is very mobile and generally not retained by soil particles and subject to loss by leaching and denitrification.

Nitrogen use efficiency can be increased by use of nitrification inhibitors. These chemicals retard nitrification in soils by slowing the rate of conversion of ammonium to nitrite, but have no effect on the rate of oxidation of nitrite to nitrate. This reduces the loss of N through leaching and denitrification in situations where such losses are normally high.

Retardation of nitrification maintains a relatively higher proportion of inorganic N in the ammonium form which is retained by soil minerals and thus less liable to transport and movement in the soil system. Nitrification inhibitors can also increase the persistence of N by incorporating ammonium in the organic N fraction or by its migration to fixed or nonexchangeable sites on clay minerals. However, surface application of nitrification inhibitors in conjunction with ammonium or ammonium- forming fertilizers to light-textured soils may enhance volatile loss of ammonium.

The present status and scope of research on the effects of nitrification inhibitors on the persistence of N in the soil is presented with emphasis on indigenous materials. Future research needs relating to the use of nitrification inhibitors are considered.

Introduction

Nitrogen (N) is the major nutrient limiting the production of cereal crops in most soils and N fertilizers have made a major contribution towards improving agricultural productivity world-wide. However, there is continuing need to improve the efficiency of N fertilizer use to achieve more efficient production of food crops and minimize fertilizer-related pollution of the environment [52].

Nitrification is generally referred to as biological oxidation of ammonium to nitrate via nitrite, effected respectively by *Nitrosomonas* and *Nitrobacter* species of the nitrifying bacteria. The poor efficiency of fertil- izer N use is largely due to loss of N by denitrification, leaching or ammonia volatilization. With the exception of ammonia volatilization these losses are associated with, and follow nitrification of ammonium to nitrate. Nitrification converts the relatively immobile cation ammonium to nitrate, which is mobile and liable to loss by leaching and denitrification [38, 39].

Nitrification inhibitors are compounds or materials that specifically retard the oxidation of ammonium to nitrite without affecting the subsequent oxidation of nitrite to nitrate. Interest in nitrification inhibitors stems from the fact that retardation of nitrification in soil reduces the loss of N by leaching and denitrification following nitrification. These inhibitors enhance

380

the persistence of N in the soil and this helps in some situations to achieve more efficient use of N for crop production [39].

In fact, inhibition of nitrification is more accurately referred to as retardation of nitrification because complete inhibition of nitrification is practically not achieved with the use of nitrification inhibitors.

The interest in nitrification inhibitors was intensified especially following the development of nitrapyrin [2-Chloro-6 (trichloromethyl) pyridine] by the Dow Chemical Company of the USA as an effective inhibitor of nitrification [13, 14]. As a consequence, a large number of compounds and materials have been proposed as nitrification inhibitors.The literature on the subject is extensive [2, 12, 16, 17, 28, 34, 38, 39, 45, 51, 52, 54, 58] and is a testimony to the interest in this approach for improving the efficiency of N use in crop production. These reviews deal with the various aspects of the effects of nitrification inhibitors on retardation of nitrification in soil and their effects on crop production and crop quality.

Despite a great interest in nitrification inhibitors to date, only a few compounds have been adopted for agricultural use. The main problems are the high cost involved in the development and subsequent registration of effective nitrification inhibitors and the economics of their use coupled with the variable results often obtained with their use in field conditions [52]. There is need to continue efforts to develop nitrification inhibitors that are inexpensive, readily available locally, and effective at reasonable rates of application.

The objective of this paper is to review the present status of research on nitrification inhibitors with examples, emphasizing the use of indigenous materials available in different tropical regions. Future research needs are also examined.

Control of nitrification and persistence of N in the soil

When nitrification inhibitors are used for controlling nitrification, the process is not arrested completely resulting in a preponderance of ammonium over nitrate in the soil. This affects the persistence of applied N in the soil as well as plant N metabolism and N nutrition [51]. The data in Table 1 from the United States illustrate the effectiveness of nitrapyrin in retarding nitrification and conserving ammonium in soil in the field. The ratio of ammonium-N in the nitrapyrin treated to untreated samples was higher in the soil over a

Table 1. Effect of nitrapyrin on conservation of ammonium N in soils in the field

Nitrapyrin applied (kg ha^{-1})	N rate	Ratio of NH4$^+$ -N treated : untreated	Observation period (weeks)
0.25 - 1.0	100	1.5 - 8.0. : 1	4 - 17
0.5 - 1.0	100	1.5- 4.0 : 1	6 - 19
0.5 - 1.0	200	2.0 - 15.0 : 1	7 - 24
0.5	65	3.0: 1	20
0.25 - 0.50	50–100	1.25 - 6.0 : 1	20–21
1.0	150	10.0 : 1	6

Source : Data compiled from different sources by Sahrawat & Keeney [51].

Table 2. Effects of nitrification inhibitors on N transformation processes that influence persistence of N in the soil

Physical and chemical processes
- Nitrogen transport and movement
- Ammonium fixation and release
- Ammonia volatilization

Biological processes
- Mineralization and immobilization
- Denitrification
- Nitrous oxide production
- Urea hydrolysis

Source : Sahrawat [45].

period of several weeks. The conservation of ammonium results in availability of more N to the crop plants and may result in increased yield and crop quality [17, 39, 51, 58].

Nitrification inhibitors also affect nitrogen transformations other than nitrification in soils, such as ammnonium fixation and release, mineralization and immobilization, nitrous oxide production and ammonia volatilization, which affect N persistence in the soil and its subsequent availability to plants [45]. The various biological, chemical and physical processes of N transformations that are influenced by retardation of nitrification in soils are given in Table 2. Detailed discussion of the effects of nitrification inhibitors on N transformations other than nitrification in soils has been presented by Sahrawat [45].

This review will consider the effects of retardation of nitrification on the persistence of N in the soil through control of nitrification and through its influence on N transformations other than nitrification in soils.

Table 3. Recent references on selected non-edible oil seeds and their isolates and other plant products proposed as nitrification inhibitors

Class of materials/compounds	References					
Karanja (*Pongamia glabra*) seed cake and its isolates	[1],	[41],	[43],	[47],	[55]	
Karanjin, a furanoflavonoid from karanja seed	[41],	[42],	[49]			
Azadirachta indica cake and its isolates	[21], [37],	[24], [43],	[25], [46],	[26], [56],	[27], [59]	[30]
Citrullus colosynthis cake	[18]					
Bassia latifolia cake	[24],	[29],	[31]			
Vegetable tannins, waste tea	[4],	[10],	[22]			

Criteria for comparing the ability of compounds for retardation of nitrification in soil

A large number of compounds and materials have been proposed for retardation of nitrification in the soil [28, 34, 39, 58] and it becomes imperative to have some criteria that could be employed for comparing their effectiveness [40]. Also at times, when the values of ammonium, nitrite and nitrate are used for comparing the effectiveness of a large number of compounds and materials as nitrification inhibitors the presentation of results becomes very space consuming and tedious, and the comparative evaluations are not very effective [40].

Bundy & Bremmer [5] studied the effectiveness of 24 compounds proposed as inhibitors of nitrification in soils, using the criterion of percent inhibition of nitrification. Percent inhibition of nitrification was calculated from. (C-S)/C × 100, where S is the amount of (nitrite + nitrate) - N produced in the soil sample treated with the test compound and C is the amount of (nitrite + nitrate) - N produced in the control (no test inhibitor added). These authors chose the time of soil sampling in laboratory incubation experiments when no ammonium-N could be detected in samples that had not been treated with nitrification inhibitors. Sahrawat [40] presented evidence to show that the criterion of Bundy & Bremmer [5] based on percent inhibition of nitrification compared well the effectiveness of compounds as nitrification inhibitors when the soil samples not treated with the test compounds did not contain

any significant amounts of ammonium-N. However, he proposed that the criterion based on the nitrification rates in soils with and without nitrification inhibitor treatments should be preferred for comparing the ability of compounds in retarding nitrification when the incubated soil samples contain significant amounts of ammonium-N.

Nitrification rates with and without nitrification inhibitor treatments are calculated from the values of ammonium, nitrite and nitrate-N obtained from the incubated soil samples using the formula:

$$\text{Nitrification rate}\% = \frac{(NO2 + NO3) - N}{(NH4 + NO2 + NO3) - N} \times 100 \quad (1)$$

Percent inhibition of nitrification is then calculated as follows:

$$\%\text{inhibition of nitrificaton} = \frac{\text{Nitrification rate in control soil - nitrification rate in inhibitor treated soil}}{\text{Nitrification rate in control soil}} \times 100 \quad (2)$$

For simplicity and effectiveness, this paper employs these two criteria to compare the effectiveness of nitrification inhibitors.

Nitrification inhibitors and retardation of nitrification in soil

Many plant products especially non-edible oil seeds and their constituents have been evaluated as sources of chemicals for retarding nitrification. Some recent

382

references on the use of indigenous materials as nitrification inhibitors are given in Table 3.

Among the non-edible oil seeds, neem (*Azadirachta indica L.*) and karanja (*Pongamia glabra* Vent.) cakes and their constituents have been traditionally used in admixtures with manures and have since been extensively evaluated for retarding nitrification in the soil [32, 37, 42, 43, 46, 47]. Among these, neem cake has been widely tested in India for increasing the efficiency of N fertilizer and generally found useful [3, 20, 30, 35, 57, 59].

Oil cakes in general and non-edible oil cakes in particular have been known to possess certain non fatty minor physiologically active constituents, which limit the use of especially the non-edible oil seeds for human use and as animal feeds. The removal of these biologically active principles from oil seeds not only renders the use of the residues or cakes as feed or manure, but the extractives are valuable sources of a host of agricultural chemicals, including pesticides and nitrification inhibitors [32, 55, 56, 60].

The presence of several furano-flavonoids in karanja seeds, bark, leaves and flowers has been well established [32]. Similarly, the lipid associates of neem cake extracted by ethanol and other solvents such as acetone and petroleum ether contain a number of bitter compounds. The quality of neem oil differs according to the method of processing the seeds and a number of bitter principles such as nimbin, nimbinin, nimbidin and nimbidiol have been identified in neem oil the ratios depending on methods of extraction and processing. The oil also contains varying amounts of free fatty acids formed by hydrolysis of the glycerides [60]. The non-edible oil seed cakes and their isolates contain a number of biologically active principles depending on the method of extraction employed which impart nitrification inhibitory property to them [32, 42, 60].

The lipid associates of karanja and neem oil seed cakes generally extracted by ethanol and other solvents such as acetone or petroleum ether are known to possess nitrification inhibitory property [37, 43, 46, 47, 55, 56].

Sahrawat & Parmar [46] evaluated the effect of alcohol extract of neem cake applied at rates ranging from 5 to 30% of fertilizer N on nitrification in soil fertilized with urea and ammonium sulfate. It was shown that the extract was effective in retarding the conversion of ammonium to nitrite and that the higher rates (20 - 30% of fertilizer N) of the extract were effective in conserving ammonium in a sandy clay loam soil for up to 75 days of incubation. In contrast, Red-

Table 4. Inhibition of nitrification of urea in a sandy clay soil (pH 7.8, 0.48 % organic C) by neem cake, coal tar extract and nitrapyrin[a]

Inhibitor treatment	% inhibition of nitrification weeks after incubation			
	1	2	3	4
Nitrapyrin	85	93	90	75
Coaltar extract	75	52	23	9
Neem cake	74	40	5	4

Source: Reddy & Prasad [37]
[a] Inhibition of nitrification calculated as suggested by Bundy & Bremner [5].

dy & Prasad [37] evaluated the comparative ability of nitrapyrin, neem cake and coal tar extract to retard nitrification of urea in a sandy clay loam and reported that the neem cake at lower concentrations than those used by Sahrawat & Parmar [46] and the coal tar extract retarded nitrification for 2 weeks while nitrapyrin was effective in retarding nitrification for 4 weeks (Table 4).

In the studies reported by Sahrawat & Parmar [46] and Reddy & Prasad [37] the neem cake extract specifically retarded the conversion of ammonium to nitrite without affecting the subsequent oxidation of nitrite to nitrate. These observations are supported by the results reported by Mishra et al. [26] who showed that neem seed cake decreased the number of nitrite forming microorganisms in the soil for 21 days when the neem seed cake was added at a rate to supply 2% C.

Sahrawat et al. [47] evaluated the nitrificatian inhibitory activity of the seeds, bark and leaves of karanja, *Pongamia glabra* in a sandy clay loam soil. The defatted seeds were extracted with hot ethanol and the extract compared with the fresh bark extract (extracted by a mixture of petroleum ether and acetone in the ratio of 40 : 60) and ground leaves to evaluate their effects on nitrification in soil fertilized with urea and ammonium sulfate. The results showed that the alcohol extract of the seeds had maximum effect on the conservation of ammonium, followed by the bark extract while the leaves had little effect. None of the materials tested had any effect on the conversion of nitrite to nitrate. The alcohol extract of the seeds was effective in retarding nitrification for 60 days when added at the rate of 20% of the fertilizer N rate.

Sahrawat [43] compared karanja (the major active ingredient of karanja seeds) and alcohol extracts of karanja and neem seeds to retard nitrification of urea in a sandy clay loam in an incubation experiment. The

Table 5. Comparison of retardation of nitrification of urea in a sandy clay loam soil (pH 7.7, 0.60% organic C) by karanjin, and alcohol extracts of karanja and neem seeds[a]

Nitrification inhibitor	Inhibitor added % of N applied	% retardation of nitrification days after incubation				
		15	30	45	60	75
Karanjin	5	57	62	47	31	12
	10	64	77	59	43	18
Karanja seed extract	30	71	54	45	5	0
Neem seed extract	30	43	62	47	20	11

Source: Sahrawat [43]
[a] Retardation of nitrification calculated as suggested by Sahrawat [46].

Table 6. Constituents of karanja (*Pongamia glabra*) proposed as nitrification inhibitors

Particular of the constituent	Description/preparation of the constituents
Karanja seed cake	The ground seeds are defatted by extraction with boiling petroleum ether and the seed cake (residue) is used for evaluation
Karanja leaves	The leaves are dried and ground before use
Karanja seed extract	The ground seeds are first defatted with petroleum ether and the cake is then extracted with boiling ethanol (95%), solvent removed to obtain the alcohol extract, which is used for testing without further purification
Karanja bark extract	The fresh bark of the tree is ground and extracted with 40 : 60 (V/V) mixture of petroleum ether: acetone, solvent removed to obtain the extract
Karanjin, a furano flavonoid from karanja seed	Karanjin a crystalline solid with molecular formula $C_{18}H_{12}O_4$ and chemically 3 - methoxy furano- 2′-,3′-,7-,8- flavone is prepared from karanja seed as described by Sahrawat & Mukerjee [49]

Source: Sahrawat et al. [47]; Sahrawat & Mukerjee [49].

treatment of urea with the extracts at the rate of 30% of the N rate was comparable to that of karanjin at 5 and 10% concentration. The extracts of neem and karanjin seeds were effective in retarding nitrification for up to 45 days while karanjin was effective for up to 60 days (Table 5). There were some differences in the patterns of inhibition of nitrification by karanja and neem seed extracts and based on this, Sahrawat [43] suggested that the mixture of extracts of karanja and neem seeds would be a better material for sustained retardation of nitrification in soil than the sole use of either material.

It is very important to use standard procedures for the preparation of the various materials because the products from different methods of preparation may

Table 7. Effects of karanjin on yield and composition of rice in greenhouse pot experiment

Crop particular	% increase by retarding nitrification
Grain yield	31–54
Grain + straw yield	21–25
Total N uptake	36–68
Grain protein	2–14

Source: Sahrawat & Mukerjee [48, 49].

possess variable ability to retard nitrification in the soil. With this objective, the procedures used for the preparation of various isolates/constituents of karanja are briefly described in Table 6 and the methods described are also applicable for the preparation of seed cake, seed and bark extracts of neem and other non-edible seeds.

Sahrawat & Mukerjee [49] reported that karanjin is a potent inhibitor of nitrification in soil (see also Table 5). It compared well with the patented nitrification inhibitor nitrapyrin in retarding the nitrification of soil fertilized with urea or ammonium sulfate. The nitrification inhibitory activity of karanjin proved useful also in improving the yield and quality of rice in greenhouse pot culture studies [48, 49]. Results from these experiments are summarized in Table 7.

Comparative evaluation of karanjin and three patented nitrification inhibitors, nitrapyrin, AM (2-amino-4-chloro-6-methyl pyrimidine), and DCD (dicyandiamide) for retardation of nitrification of urea in a sandy clay loam soil showed that the effectiveness of the compounds tested decreased in the order: nitrapyrin> karanjin > AM > dicyandiamide (Table 8) [41].

Further studies [49] where suitable alteration of the karanjin (3-methoxy furano -2′, 3′, 7, 8- flavone) molecule: karanj ketone (4 - hydroxy - 5- w - methoxy-acetyl coumarine), karanjonol (3 - hydroxy furano - 2′, 3′, 7, 8 - flavone) were prepared from karanjin (see Figure 1 for structural formulae) and tested for their effects on nitrification, showed that the furan ring present in the molecule is the crucial structural factor for nitrification – inhibitory activity. All the compounds with the exception of dihydro karanjin where furan ring was absent, exhibited nitrification – retarding ability to varying degrees [49].

The conclusion that the furan ring imparts nitrification inhibitory activity was further confirmed by later studies in which several compounds based on furfuraldehyde including furfuraldehyde and furfuryl

Table 8. Comparative evaluation of karanjin and other patented nitrification inhibitors for retardation of nitrification of urea in a sandy clay loam soil (pH 7.7, 0.58% organic C)[a]

Nitrification inhibitor	% retardation of nitrification weeks after incubation			
	1	3	5	7
Nitrapyrin	73	76	60	36
Karanjin	60	63	51	27
A.M.	46	54	43	17
DCD	32	46	31	12

Source: Sahrawat [41]
[a]Soil samples were treated with 100 mg N kg^{-1} soil and 5 mg inhibitor kg^{-1} soil. Retardation of nitrification calculated as suggested by Sahrawat [40].

Table 9. Effects of furfural and furfuryl alcohol on retardation of nitrification of urea N in a sandy clay loam[a]

Compound	% retardation of nitrification days after incubation				
	15	30	45	60	75
Furfural	68	67	8	3	1
Furfuryl alcohol	32	40	9	3	1

Source: Adapted from Sahrawat et al. [50]
[a]Soil samples were treated with 200 mg N kg^{-1} soil and with test compounds at the rate of 10% of urea N added; retardation of nitrification calculated as suggested by Sahrawat [40].

alcohol showed varying degrees of ability to retard nitrification in soils fertilized with urea [23, 50]. Results showed that furfural and furfuryl alcohol were effective in retarding nitrification for 30 days when added at a rate of 10% of urea N (Table 9).

Among the several furano compounds evaluated for retarding nitrification, 5-nitro-furfural oxime and furfural oxime were found to be very effective inhibitors of nitrification of urea N in soil [Table 10].

Several plant products such as vegetable tannins and waste tea have also been reported to retard nitrification of ammonium in soils [4, 22, 39]. Waste tea, the part of manufactured tea consisting of stalks and fluff, which is not marketed and discarded from tea factories, contain a large proportion of oxidized polyphenols with tannin- like properties [22]. Waste tea was not only effective in retarding nitrification but also released mineral N slowly during its decomposition in the soil. The polyphenols from waste tea have also been reported to possess properties to retard urease activity in soils

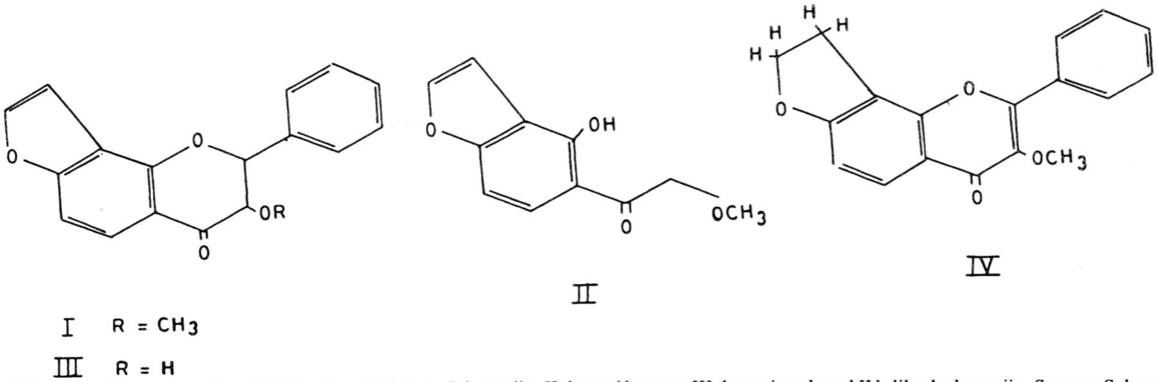

I   R = CH3

III   R = H

Fig. 1. Structural formulae of test nitrification inhibitors. I, karanjin, II, karanj ketone; III, karanjonol; and IV, dihydrokaranjin. Source: Sahrawat & Mukerjee [49].

Table 10. Effects of some furano compounds on inhibition of nitrification of urea N in a sandy clay loam soil[a]

Compound	% inhibition of nitrification days after incubation					
	15	30	45	60	75	90
Furfuryl semicarbazone	68	63	58	22	11	5
5-Nitrofurfural semicarbazone	72	49	26	20	14	9
Methyl 5-nitro-2-furoate	58	38	24	12	5	4
Furfural oxime	70	52	37	18	13	6
5-Nitro-furfural oxime	72	71	62	40	24	10
3'-Chloro-2-furanilide	53	35	30	26	19	8
5-Nitro-3'-chloro-2-furanilide	69	63	61	21	18	13
Psoralen	47	39	24	14	11	7

Source: Kuzvinzwa et al. [23]
[a]Soil samples were treated with 200 mg N kg^{-1} soil and with test compounds at the rate of 10% of urea N added; inhibition of nitrification calculated as suggested by Bundy & Bremner [5].

[10, 39]. When added at the rate of 8% (w/w) of soil, the waste tea completely checked nitrate formation for 16 days in a soil at its natural pH of 4.5 or when the soil was limed to a pH of 6.0. The soil was fertilized with urea and ammonium sulfate.

Effects of nitrification inhibitors on nitrogen transformations, other than nitrification, and persistence of N in the soil

Nitrification inhibitors can also affect the persistence of N in the soil by influencing the N transformation processes other than nitrification [45] (Table 2). Sahrawat [45] reviewed the miscellaneous effects of nitrification inhibitors on N transformations, other than nitrification

in soil, and the salient findings from this review and other recent literature are as follows:

1. Keeping N in the ammonium form by retarding nitrification reduces the movement of mineral N because cationic ammonium is retained by soil particles and is less mobile. When less nitrate is formed it results in reduced amounts of nitrate N leached [45].

2. The retardation of nitrification can enhance persistence of N in the soil by increased immobilization and incorporation of ammonium in the organic fraction [7, 9, 11, 15] or by migration of ammonium to fixed or nonexchangeable sites on clay minerals especially in soils in which clay minerals such as micas and vermiculites are present [19, 36, 45].

3. The retardation of nitrification in soil results in an accumulation of ammonium and higher soil pH [45] which are conducive to ammonia volatilization. This is true especially in bare soils and in light-textured soils and when the fertilizer and nitrification inhibitors are applied to the soil surface [6, 8, 33].

4. Application of nitrification inhibitors such as nitrapyrin is helpful in eliminating or reducing nitrite accumulation in soils fertilized with higher rates of urea [6, 39].

5. Although nitrification inhibitors indirectly reduce the loss of N through denitrification by reducing nitrate formation, they have little direct effect on the loss of nitrate via denitrification. However, nitrification inhibitors can reduce loss of N through nitrous oxide emission via nitrification and denitrification and thus may influence N persistence in the soil [45, 53].

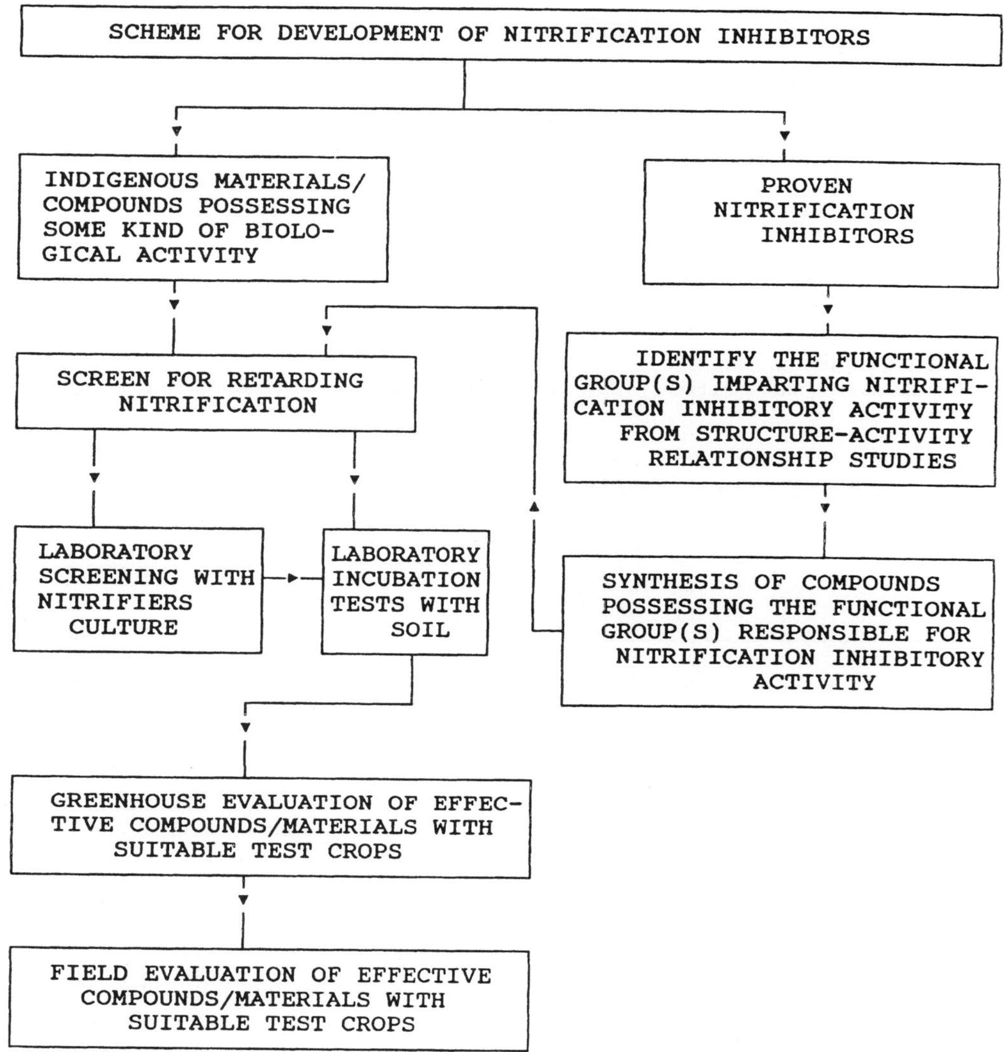

Fig. 2. Scheme for development of nitrification inhibitors [52].

6. If nitrification inhibitors also retard urea hydrolysis in soils, this could further enhance persistence of N in some situations e.g. by checking or reducing loss through ammonia volatilization [45].

Perspectives

Nitrification inhibitors can enhance the persistence of fertilizer N by retardation of nitrification, which may have potential for improving the efficiency of N in the situations where loss of N due to denitrification or leaching is high. Nitrification inhibitors can also enhance the persistence of fertilizer N by affecting N transformations other than nitrification in soil [45].

There is a need to further exploit indigenously available materials and products as sources of nitrification inhibitors using a structured approach [38, 52]. This scheme [38] would involve coordinated research efforts from various disciplines based on two approaches: (i) by impirical evaluation of a large number of compounds/materials possessing some sort of general biocidal activity for retarding nitrification in soils, and (ii) by identification of the functional groups responsible for the activity and then incorporating these groups in suitable compounds (Figure 2). As an example, employing the second approach, it was discovered that the furan ring in the compounds imparts the nitrification inhibitory property to varying degrees [23, 49, 50]. The first approach being simple, could be used for short-term research goals in developing nitrification inhibitors from indigenous materials which are readily available locally. The second approach is ideally suited for longer range research on the subject and is more scientifically sound [38]. Information generated in laboratory tests of experimental compounds needs to be applied to determine the efficiency of the compounds under pot culture and field conditions, to obtain useful materials and information on their proper use in increasing the efficiency of applied N in crop production [38, 44].

Basic research is also needed to delineate the beneficial effects of indigenous materials such as neem cake and its isolates in the retardation of nitrification and slow-release effects through immobilization-mineralization [44]. The use of 15N labelled fertilizers may be useful in studying the fate of N where nitrification is retarded in soils, e.g. to understand how retardation of nitrification affects ammonium fixation and release, mineralization, immobilization and remineralization of N, and other N transformation processes of the N cycle [44, 45].

References

1. Ahmad N, Rai M, Yadav NL & Sahi BP (1978) Studies on efficiency of karanj cake and its lipid associates on nitrification regulation under submerged rice cultivation in a calcareous soil of North Bihar. J Indian Soc Soil Sci 26: 304–305
2. Amberger A (1989) Research on dicyandianiide as nitrification inhibitor and future outlook. Commun Soil Sci Plant Anal 20: 1933–1955
3. Bains SS, Prasad R & Bhatia PC (1971) Use of indigenous materials to enhance the efficiency of fertilizer nitrogen for rice. Fert News 16(3): 30–32, 52
4. Basaraba J (1964) Effect of vegetable tannins on nitrification in soils. Plant Soil 21: 8–16
5. Bundy LG & Bremner JM (1973) Inhibition of nitrification in soils. Soil Sci Soc Am Proc 37: 396–398
6. Bundy LG Bremner JM (1974) Effects of nitrification inhibitors on transformations of urea nitrogen in soils. Soil Biol Biochem 6: 369–376
7. Chalk PM, Victoria RL, Muraoka T & Piccolo MC 1990 Effect of a nitrification inhibitor on immobilization and mineralization of soil and fertilizer nitrogen. Soil Biol Biochem 22: 533–538
8. Cornforth IS & Chesney HAD (1971) Nitrification inhibitors and ammonia volatilization. Plant Soil 39: 497–501
9. Crawford DM & Chalk PM (1992) Mineralization and immobilization of soil and fertilizer nitrogen with nitrification inhibitors and solvents. Soil Biol Biochem 24: 559–568
10. Fernando V & Roberts GR (1976) The partial inhibition of soil urease by naturally occurring polyphenols. Plant Soil 44: 81–86
11. Freney JR, Chen DL, Mosier AR, Rochester IJ, Constable GA & Chalk PM (1993) Use of nitrification inhibitors to increase fertilizer nitrogen recovery and lint yield in irrigated cotton. Fert Res 34: 37–44
12. Gasser JKR (1970) Nitrification inhibitiors : their occurrence production and effects of their use on crop yields and composition. Soils Fert 33: 547–554
13. Goring CAI (1962a) Control of nitrification by 2-chloro - 6 - (trichloromethyl) pyridine. Soil Sci 93: 211–218
14. Goring CAI (1962b) Control of nitrification of ammonium fertilizers and urea by 2 chloro-6 (trichloromethyl) pyridine. Soil Sci 93: 431–439
15. Guiraud G, Marol C & Thibaud MC (1989) Mineralization of nitrogen in the presence of a nitrification inhibitor. Soil Biol Biochem 21: 29–34
16. Hauck RD (1972) Synthetic slow-release fertilizers and fertilizer amendments. In: Goring CAI & Hamaker JW (eds) Organic Chemicals in the Soil Environment, Part B, pp 633–690. Marcel Dekker, New York, USA
17. Huber DM, Warren HL, Nelson DW & Tsai CY (1977) Nitrification inhibitors - New tools for food production. BioScience 27: 523–529
18. Jain JM, Narayanasamy G, Sarkar MC & Datta MN (1980) An evaluation of nitrification retardation property of *Citrullus colosynthis* cake and its influence on yield and N uptake by wheat. J Indian Soc Soil Sci 28: 480–484
19. Juma NG & Paul E A (1983) Effect of a nitrification inhibitor on N immobilization and release of 15 N from non exchangeable ammonium and microbial biomass. Can J Soil Sci 63: 167–175
20. Ketkar CM (1983) Crop experiments to increase the efficiency of urea fertilizer nitrogen by the use of neem by-products under Indian soil conditions. In: Proceedings of Second Int Neem Conference, Rauischholzhausen, pp 507–518
21. Khandelwal KC, Singh DP & Kapoor KK (1976) Mineralization of urea coated with neem extract and response of wheat. Indian J Agric Sci 47: 267–270
22. Krishnapillai S (1979) Inhibition of nitrification by waste tea (tea fluff). Plant Soil 51: 563–569
23. Kuzvinzwa SM, Devakumar C & Mukerjee SK (1984) Evaluation of furano compounds as nitrification inhibitor. *In* Nitrogen in Soils, Crops and Fertilizers. Bull Indian Soc Soil Sci 13: 165–172. Indian Soc Soil Sci, New Delhi, India
24. Mago GS & Totawat KL (1989) Efficacy of coating materials for urea applied under saline water irrigation. I. Mineralization of urea-N. Ann Arid Zone 28: 79–87
25. Mehta VB & Patel ND 1972 Effect of neem cake and oil on nitrification of ammonium sulphate in soil and response of rice. Mahatma Phule Krishi Vidyapeeth Res J 3: 117–120
26. Mishra MM Neelakantan S Khandelwal KC Bhardwaj SK & Vyas SR (1975) Margosa (neem) seed cake as an inhibitor of nitrification. Soil Biol Biochem 7: 183–184
27. Misra KC & Chhonkar PK (1978) Possible utilization of neem cake for inhibiting nitrification in soil. J Indian Soc Soil Sci 26: 90–92
28. Mulvaney RL & Bremner JM (1981) Use of urease and nitrification inhibitors for control of urea transformations in soils. In: Paul EA & Ladd JN eds Soil Biochem 5: 153–196. Marcel Dekker, New York, USA.
29. Muthuswamy P, Raju GSN & Krishnamoorthy KK 1975 Mineralization of urea coated with nitrification inhibitors. J Indian Soc Soil Sci 23: 332–335

30. Nair KPP & Sharma PB (1979) Mineralization and field effectiveness of ordinary and coated urea, urea-aldehyde condensation product and urea treated with nitrification inhibitor. J Agric Sci, Camb 93: 623–627

31. Naphade KT & Wasnik HM (1974–75) Use of neem cake and mahua cake as nitrification inhibitors. Nagpur Agric Coll Mag 67: 43–44

32. Parmar BS, Sahrawat KL & Mukerjee SK (1976) *Pongamia glabra* constituents and uses. J Sci Ind Res 35: 608–611

33. Prakasa Rao EVS & Puttanna K (1987) Nitrification and ammonia volatilization losses from urea and dicyandiamide treated urea in a sandy loam soil. Plant Soil 91: 201–206

34. Prasad R, Rajale GB & Lakhdive BA (1971) Nitrification retarders and slow-release nitrogen fertilizers. Adv Agron 23: 337–383

35. Prasad R Thomas J (1982) Nitrogen. In: Review of Soil Research in India, part I, pp 309–322. 12th Int Congr Soil Sci, New Delhi, India 8–16 Feb 1982. Indian Society of Soil Science, New Delhi, India

36. Rechcigl JE, Teel MR & Sparks DL (1988) Ammonium fixation from urea as influenced by nitrapyrin. Commun Soil Sci Plant Anal 19: 1583–1591

37. Reddy RNS & Prasad R (1975) Studies on the mineralization of urea, coated urea, and nitrification inhibitor treated urea in soil. J Soil Sci 26: 304–312

38. Sahrawat KL (1978) Nitrification inhibitors for efficient use of fertilizer nitrogen: A scheme for development of nitrification inhibitors. In: Ferguson AR, Bieleski RL & Ferguson IR (eds) Plant Nutrition 1978. Proc 8th Int Colloq Plant Analysis and Fert Problems, Auckland, New Zealand, 28 Aug-1 Sept 1978. DSIR Inform Series 134, pp 431–438. Govt Printer, Wellington, New Zealand

39. Sahrawat KL (1980a) Control of urea hydrolysis and nitrification in soils by chemicals-prospects and problems. Plant Soil 57: 335–352

40. Sahrawat KL (1980b) On the criteria for comparing the ability of compounds for retardation of nitrification in soil. Plant Soil 55: 487–490

41. Sahrawat KL (1981a) Comparison of karanjin with other nitrification inhibitors for retardation of nitrification of urea N in soil. Plant Soil 59: 494–498

42. Sahrawat KL (1981b) Karanja (*Pongamia glabra* Vent.) as source of nitrification inhibitors. Fert News 26(3): 29–33

43. Sahrawat KL (1982) Comparative evaluation of karanjin and extracts of karanja (*Pongamia glabra* Vent.) and neem (*Azadirachta indica* L.) seeds for retardation of nitrification of urea in soil. J Indian Soc Soil Sci 30: 156–159

44. Sahrawat KL (1988) Use of 15 N in nitrification inhibitor studies with special reference to indigenous materials. J Nuclear Agric Biol 17: 238–242

45. Sahrawat KL (1989) Effects of nitrification inhibitors on nitrogen transformations, other than nitrification, in soils. Adv Agron 42: 279–309

46. Sahrawat KL & Parmar BS (1975) Alcohol extract of neem (*Azadirachta indica L.*) seed as nitrification inhibitor. J Indian Soc Soil Sci 23: 131–134

47. Sahrawat KL, Parmar BS & Mukerjee SK (1974) Note on the nitrification inhibitors in the seeds, bark and leaves of *Pongamia glabra* Vent. Indian J Agric Sci 44: 415–418

48. Sahrawat KL Mukerjee SK (1976) Effect of nitrification inhibitors on rice protein. Commun Soil Sci Plant Anal 7: 601–607

49. Sahrawat KL & Mukerjee SK (1977) Nitrification inhibitors. I. Studies with karanjin a furanoflavonoid from karanja (*Pongamia glabra*) seeds. Plant Soil 47: 27–36

50. Sahrawat KL, Mukerjee SK & Gulati KC (1977) Nitrification inhibitors. II Studies with furano compounds. Plant Soil 47: 687–691

51. Sahrawat KL & Keeney DR (1984) Effects of nitrification inhibitors on chemical composition of plants: A review. J Plant Nutr 7: 1251–1288

52. Sahrawat KL & Keeney DR (1985) Perspectives for research on development of nitrification inhibitors. Commun Soil Sci Plant Anal 16: 517–524

53. Sahrawat KL & Keeney DR (1986) Nitrous oxide emission from soils. Adv Soil Sci 4: 103–148

54. Shand C & Ahmad N 1974 Effect of nitrate inhibition and slow release nitrogen fertilizers on nitrification rates in some Trinidad soils. Trop Agric Trinidad 51: 161–178

55. Singh UV (1966) Studies on better utilization of non-edible oil seed cakes-karanj (*Pongamia glabra*) seed cake. PhD Thesis, Indian Agric Res Inst, New Delhi

56. Sinha NP (1964) Studies on better utilization of non-edible oil seed cakes-neem (*Azadirachta indica*) seed cake. PhD Thesis, Indian Agric Res Inst, New Delhi

57. Sinha NP, Singh SN & Jha BK (1979) Effect of blending urea with neem-cake, on the yield of wheat and nutrient uptake from a calcareous soil. Indian J Agric Sci 49: 273–276

58. Slangen JHG & Kerkhoff P (1984) Nitrification inhibitors in agriculture and horticulture. A literature review. Fert Res 5: 1–76

59. Subbiah S (1979) Influence of neem cake mixed ammonium sulphate on the yield and nutrient uptake by IR 20 rice (*Oryza sativa* L.). Madras Agric J 66: 678–686

60. Vimal OP & Naphade KT (1980) Utilization of non-edible oil seeds-recent trends. J Sci Ind Res 39: 197–211

N. Ahmad (ed.), Nitrogen Economy in Tropical Soils, 389–398.
© 1996 Kluwer Academic Publishers. Printed in the Netherlands.

Land use related nitrogen and global change

H.W. Scharpenseel, E.M. Pfeiffer & H. Wiechmann
Institute of Soil Science, University of Hamburg, Allendeplatz 2, 20146 Hamburg, Germany

Abstract

In the course of global change throughout the Archean, Proterozoic and Phanerozoic, the N concentration in the atmosphere increased sharply at the out-going Archean and beginning of the Proterozoic age, decreasing thereafter mainly during the Phanerozoic analog to the increase of O_2 due to enhanced aquatic and terrestrial actions of photosynthesis. At present, atmospheric N comprises about 3.8 Mil Gt, in the lithosphere ca.19 Mil Gt and in terrestrial organic matter about 95 Gt N. In modern anthropogenic global change, the NO_x compounds, especially the very stable N_2O with 170 yrs residence time and a molar greenhouse forcing efficiency factor of 150 compared with CO_2, is highly relevant for greenhouse forcing as well as for reduction of the UV b absorbing stratospheric ozone layer. The 0.3 ppm N_2O, representing ca. 1.5 Mt N, is increasing by ca. 0.3% y^{-1} - about 1 $ppb \cdot y^{-1}$. Some 85 to 90% are of biotic origin, equal to ca. 7 Mt N_2O y^{-1}, emitted from soil and soil near atmosphere and about 3 Mt N_2O y^{-1} in connection with N-fertilizer. It has become obvious that N_2O emission is highest from wetlands and especially from moist tropical woodlands and patches of forest clearing. Sources are denitrification and nitrification especially of heterotrophic origin. NH_3-emission amounts to 22-35 Mt $N \cdot y^{-1}$ with ruminant husbandry as the main contributor. In soil organic matter the average C:N:P:S ratio is about 140:10:1:3:1.3 with most likely a wider C:N in soils of tropical rainforests. C and N emissions due to shifting cultivation, wood clearing, savanna fires, agricultural wastes are estimated to 1.8-4.7 Gt C and 0.015-0.046 Gt N y^{-1} with a C:N of ca. 100:1. NO, with only a few days residence time in the atmosphere, originates in the stratosphere as an ozone killer from N_2O oxidation. In the atmosphere, if available at concentrations > 10 ppT, it contributes to the production of ozone as a by-product of CH_4 and CO oxidation; if NO is < 10 ppT, it consumes slight amounts of tropospheric ozone during CH_4 and CO oxidation. NO in cell physiology is an outstanding intracellular messenger compound, produced from the amino acid arginine.

Introduction

Nitrogen in the atmosphere increased from < 1% in the Archean, when the world had no life, to 79% in the Phanerozoic with the fullness of terrestrial life. In parallel, O_2 gains occurred from < 1% to about 21%, this, according to Watson (Lovelock, 1988) being a critical concentration, since, at 15% nothing would burn and at 25% everything combustible would have burnt before really emerging, including forests (IGBP, 1992; Lovelock, 1991) (Figure 1).

Hart (1978) indicates the development of the CO_2 - N_2 - CH_4 and O_2 fractions throughout the Archean, Proterozoic and Phanerozoic periods of Earth development. The percentage share of the atmospheric N was rising steeply at the outgoing Archean and beginning Proterozoic age, with a subsequent major reduction mainly during the Phanerozoic due to the steep increase in O_2 concentration as a consequence of increasing terrestrial photosynthesis.

Today the total mass of the atmosphere is about 5.1 Mil Gt (Gt = 10^9 t) with an estimated change of its composition due to human activities as low as 0.01%. However, even with this small change the consequences can become alarming (IGBP, 1992). Mengel & Kirkby (1979) estimate the mass of atmospheric N to be 3.8 Mil Gt, lower compared with the N in the lithosphere, which measures approximately 18 Mil Gt. Only a rather small fraction of the total terrestrial pool is soil N, and from this, again only a modest share, is plant available N in the chemical forms of NO_3^- or NH_4^+.

Estimates of available N ha^{-1} y^{-1} by Hauck (1971) show great differences, varying between < ca. 30 kg in arable land, up to >100 kg in non-legume pasture, >800 kg in grass-legume-pasture, nearly 600 kg in

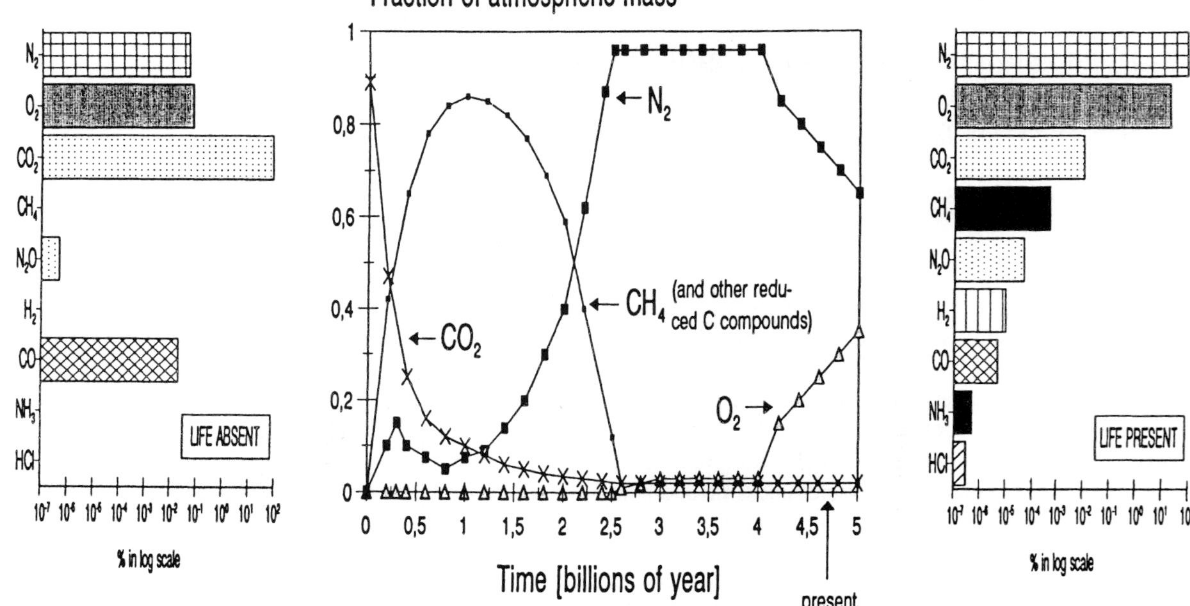

Fig. 1. Evolution of Earth's atmosphere (acc. to Hart, 1978) (CO_2, CH_4, N_2, O_2) and the change of composition with emerging life (acc. to Lovestock, 1988, 1991; IGBP, 1992)

forests and up to 100 kg in rice paddies. More recent assessments of N in different soil orders (Eswaran et al., 1993) indicate a total of ca. 95 Gt with highest contributions by the Inceptisols (30.0), Alfisols (15.3), Aridisols (ca.11.7), Oxisols (11.4) Gt.

Land use related nitrogen

Natural sources in the N cycle

A rather comprehensive assessment of N sources and sinks in soil and soil near atmosphere was tried by Isermann (1987, 1989, 1990, 1991).

The different nitrogenous compounds interacting in the soil and the soil near atmosphere are listed in Table 1. The most relevant N species' under the aspect of global change are NH_3, NO_2 NO and NO_3^-. N_2O is especially important due to its ca. 6% contribution to anthropogenic greenhouse forcing and to its destructive action on the UV b radiation blocking stratospheric ozone. NO presence at > 10 ppT catalyses O_3 emission during CH_4 and CO oxidation (Crutzen, 1988). NH_3 and NO_3^- can cause at excessive levels in the atmosphere and in drinking water (> 50 ppm NO_3) respectively physiological and pathological irritations in humans.

Important interactions in the soil near atmosphere are described by Dämmgen et al. (1992). Figure 2 reflects NH_4 and NH_3 inputs and outputs in the soil/atmosphere system according to Jenkinson (1990). While the soil NO_3-N distribution is rather difficult to assess due to leaching losses, Figure 3 reflects the NH_4^+ and NH_3 inputs and outputs in the soil-atmosphere system.

Böttgen et al. (1978) tried to estimate the NH_3-N emissions (Table 2), arriving at an annual emission rate of 22 to 35 Mt y^{-1}, ca. 80% derived from the northern hemisphere. It is evident that ruminant husbandry is the major contributor. Cole et al. (1993) monitored the even larger pool of fertilizer -N and its tenacious link to the development of human population as well as to arable land surface. The percentage contributions of temperate, tropical and sub-tropical regions indicate a more favorable population/arable land /N fertilizer ratio in temperate areas for sustainable land use and the projection to a more problematic future, e.g. by 2025, for homeostasis of the corresponding ecosytems of the climatic belts (Table 3). C and N are strongly interrelated regarding the landscape potential for biomass production. Stevenson (1982), scrutinising all accessible results, arrived at C:N:P:S ratios in gross average of 140:10:1.3:1.3. This ratio appears only slightly dif-

Table 1. Different nitrogenous compounds in soil and soil near atmosphere

N species'	Residence time	Concentration	Acitivity in global change
NH_3	Few hours only		
H_2		79%	
N_2O	150-170 yrs (3)	0.31 ppm	Greenhouse Forcing, molec. activity ca 150, relativ to $CO_2 = 1$; further destruction of stratospheric ozone, N_2O causes ca 6% of anthropogenic greenhouse forcing
NO	few days only	NO_x 0- 100 ppb	if > 10 ppT NO oxidation of 1 molecule CH_4 or CO produces ca 3.7 molec. of tropospheric ozone (O_3- source), < 10 ppT oxidation of 1 molecule CH_4 or CO binds 0.1 molecule of O_3 (O_3 sink)[2]
NO_2	few days only (NO_x partener of NO)		
N_2O_5			
NO_3	ca. 10 min[1]		
HNO_2			
HNO_3			

[1] Platt et al. (1984)
[2] Crutzen (1981, 1988)

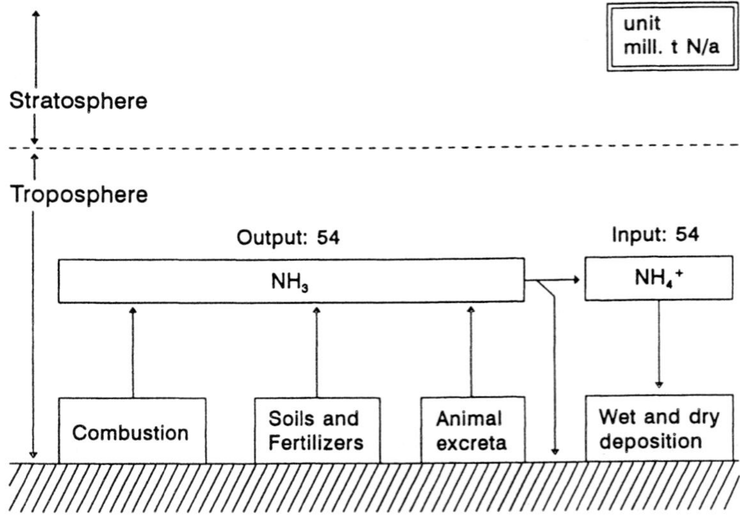

Fig. 2. Production and fate of NH_3 in the atmosphere; units are million tonnes N a^{-1} (acc. to Jenkinson 1990, from Warneck 1989)

ferent in six tropical rain forest soils of Brazil, analysed by Neptune et al. (1973) to be 194:10:1.2:1.4.

The amino acids and amino sugars, as major N sources, complexed with hydroxy phenols or quinones into fulvic and humic acids, seem to determine different percent ranges in the various climatic belts (Table 4), (Sowden et al., 1977).

The C and N emissions by fire (Andreae, 1991) brought about by shifting agriculture, permanent deforestation, savanna fires, firewood and agricultural waste amount annually to 1.8-4.7 Gt C and 0.015 - 0.046 Gt of N, a C:N ratio of ca 100 :1 (Table 5).

Table 2. Worldwide NH$_3$-N emissions in Tg (million tons) N yr^{-1} (see Böttger et at., 1978)

Sources of NH$_3$-N emission	Worldwide	Northern Hemisphere	Southern Hemisphere
Unfertilized soils	1.0 - 2.0	0.6 - 1.3	0.3 - 0.8
Mineral N Fertilizer	1.2 - 2.4	1.2 - 2.3	0.04 - 0.07
Livestock	19.7 - 30.4	15.7 - 24.0	4.0 - 6.4
- Cattle	14.4 - 21.3	11.8 - 17.5	2.6 - 3.8
- Sheep	2.2 - 2.4	1.3 - 2.6	0.9 - 1.8
- Pigs	1.3 - 1.9	1.2 - 1.7	0.1 - 0.2
- Diverse other	1.8 - 2.8	1.4 - 2.2	0.4 - 0.6
Motor vehicles	0.2 - 0.3	0.2 - 0.3	ca 0.1
Fossil fuels	ca 0.1	ca 0.1	ca 0.1
Total NH$_3$-N Emissions Tg	22 - 35	18 - 28	4 - 7

Table 3. Worldwide expansion of population, arable land surface and N fertilizer use (see Cole et al., 1993)

	Population (mil)	Arable land (mil ha)	N-Fertilizer use (mil t N y^{-1})
1990 Total	5.240	1,476	79,564
1990 Temperate climate	1,156	689	40,427
1990 Tropical and subtropical	4,084	787	39,137
Till 2025 expansion (%)	+ 63	+ 36	+ 49

Table 4. Nitrogen distribution in soils from different climatic zones (Sowden et al., 1977)

Climatic zone and Number of soils	Total N %	Forms of nitrogen (%)				
		Acid Insoluble	NH$_3$	Amino Acid	Amino Sugar	HUN[a]
Artic (6)	0.02-0.16	ca. 13.9	ca.32.0	ca.33.1	ca.4.5	ca.16.5
Cool temperate (82)	0.02-1.06	ca.13.5	ca.27.5	ca.35.9	ca.5.3	ca.17.8
Subtropical (6)	0.03-0.30	ca. 15.8	ca.18.0	ca.41.7	ca.7.4	ca.17.1
Tropical (10)	0.24-1.61	ca.11.1	ca.24.0	ca.40.7	ca.6.7	ca.17.6

[a] Hydrolysable unknown N

Table 5. Total Carbon and Nitrogen compound emission, released by fires (see Andreae, 1991)

Land use factes, causing fires	C Released $(Pg\,(Gt)\,y^{-1})$	N Released $(Tg\,(Mil\,t)\,y^{-1})$
Shifting agriculture	0.5-1.0	5-10
Permanent deforestation	0.2-0.7	2-7
Savannah fires	0.3-1.6	2-10
Firewood	0.3-0.6	2-3
Agricultural waste	0.5-0.8	5-16
Total	1.8-4.7	15-46

Nitrogen sources due to cropland management

Since the pioneer research of Prianischnikow, Boussingault, Lawes & Gilbert and others in the 1830s, the nutrient N occurring as NO_3^- or NH_4^+ in most agroecosystems, has been recognised as the decisive yield determining growth factor. Constraints related to N and global change arise for example from the law of diminishing returns, in conjunction with efforts towards yield maximisation. While a medium supply level of 1 kg of mineral N, the production of which requires about 1 kg of crude petroleum and is accompanied by a release of ca. 3 kg of CO_2, produces between 16 and 20 kg of cereal grains, high fertilizer inputs, e.g. in so called agricultural mining systems (Dover & Talbot, 1987), will produce diminished returns per kg N at relatively higher equivalents of CO_2 release.

It may lead to NO_3^- concentration > 50 ppm the groundwater, especially if organic N sources from sewage, compost or manure are fortifying the NH_4^+ and NO_3^- release. This is an ever increasing problem e.g. in cultivated fields near piggeries, poultry farms and stable-fed dairy or beef-cattle centers. Otherwise, high input land use operations with intelligent tightly organised cropping systems such as Kopp's Culture Permanente (Kopp, 1975) or the Sorjan cropping system in riceland, are geared to minimise processing with wasteful N consumption and release. Consequently, the Rio Conference (1992) advised to change the first paradigm with indiscriminate purchase of agricultural inputs against a second paradigm with emphasis on recycling. Quite often, otherwise, N is the growth factor in limited supply. Neue (1988), when evaluating all available field data regarding the impact of SOM on rice yields, received a minimum - optimum - maximum curve with top yields between 4 - 5 % SOM. The yield maximising SOM percentage was rising with increasing N level.

Main reactions

$$4\,NO + 4\,NH_3 + O_2 \longrightarrow 4\,N_2 + 6\,H_2O$$
$$6\,NO_2 + 8\,NH_3 \longrightarrow 7\,N_2 + 12\,H_2O$$

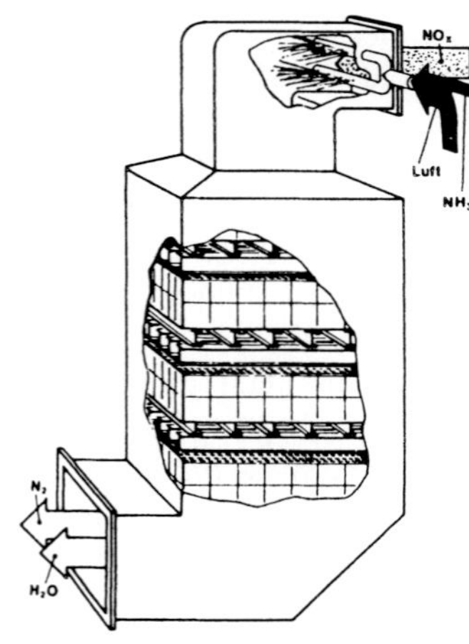

Side reactions

$$4\,NH_3 + 3\,O_2 \longrightarrow 2\,N_2 + 6\,H_2O$$
$$4\,NH_3 + 5\,O_2 \longrightarrow 4\,NO + 6\,H_2O$$
$$NH_3 + SO_3 + H_2O \longrightarrow NH_4HSO_4$$

Fig. 3. Catalytic conversion of NO_x to nitrogen plus water vapour (acc. to Information Center on Electricity, Frankfurt, FRG)

Subsistence farming is mostly constrained by the supply of only ca, 60 kg of N ha^{-1}, that are available from biomass turnover, atmospheric lightning discharges and some diazotrophic N collection by legumes or bluegreen algae, and which limits grain yields to 60 times 16 -18 kg grains, amounting to at most 1 t ha^{-1}.

N deficiency can also arise in young alluvial riverine depositions from denuded upstream areas of poorly weathered or extremely leached, fossilized sediments. The same applies to land covered by a blanket of fresh volcanic ashes or lahar, such as reported by Reyes after the Pinatubo eruption in the Philippines (Reyes according to Zamora, 1993). The raw volcanic ashes and especially the lahar deposits adjacent to the Pinatubo, mixed 1:1 with underlying top soil plus 120-60-60 kg of NPK, allowed the production of 4-5 t ha^{-1} wet season rice; a mixture of 25 cm ash plus only 10 cm soil and the same fertilization produced only 2 - 3 t of rice ha^{-1}, in which case the N deficiency turned out to be the major yield limiting factor in rice production with the Pinatubo volcano impact.

N also becomes the limiting factor in the rising biomass production due to elevation of the atmospheric CO_2 concentration (Bazzaz 1990; Bazzaz & Fajer, 1992; Houghton et al 1990). While Esser (1990) estimates the gains in biomass formation by "CO_2 fertilization" as already exceeding ca 1.5 Gt of C that are annually released by slash and burn / forest clearing, Bazzaz et al. (1992) warn of too much optimism because of many adverse side effects that may impede the expected gains in biomass formation. Not the least there also would be inadequacy of corresponding N availability. Wider C:N ratios in the biomass would also require higher food ingestion to balance the protein requirement.

Observations of plant growth near natural CO_2 emitters like mineral springs confirm existing impediments to higher biomass production towards the customary estimate of ca 30% by CO_2 doubling for C3 plants, most likely brought about by the constrained N availability.

Diverse anthropogenic inputs

Missing carbon fraction and nitrogen

The question regarding the fate of the 1.5 to 2.0 Gt C, the "missing C fraction" which is part of the anthropogenic annual C input of ca 7 Gt C (3.5 Gt go to the atmosphere, 2 Gt go to oceans, 1.5 - 2.0 Gt are the "missing carbon fraction") is under dispute. Is it used for increased terrestrial biomass production especially at higher northern latitudes (Tans et al., 1990), or is it transported via the North Atlantic Conveyor or Circulation towards the equatorial latitudes (Broecker & Peng, 1992)? In the former case it would be a typical example of CO_2 fertilization, which will require adequate availability of N. Thus, if the extent of the CO_2 fertilization is drawn into doubt or minimised in attempts to settle for smaller C sequestration rates, one should not ignore that CO_2 fertilization by ris-

ing anthropogenic CO_2 emission as well as stepped up mineral fertilization increased simultaneously and reached peak levels since the middle of this century; thereafter they sustained a positive back feeding by amplification each others' contribution.

A rather important interrelation seems to exist between N inputs and CH_4 sequestration rates, especially in more temperate forest soils. Steudler et al.(1989) observed a reduction of the CH_4 sink by 33% due to N inputs, reflecting that not only production rates of greenhouse trace gases, but also reduction of their storage / sequestration, is important to consider.

The nitrous oxide problem in dryland and wetland

While NO and NO_2, the major components of the NO_x group are emitted mainly by combustion engines (in Germany according to Isermann (1990) ca 17 kg NO_2-N per year per car engine), by electric power-plants and industry which can be catalytically, converted to N_2 and H_2O by injection of air and NH_3 (Figure 4), N_2O is mainly a product of biological processes and is chemically rather stable (Table 6).

Also NH_3-N is released primarily by the respiratory processes of livestock, involving emission of ca. 20 kg y^{-1} per cattle unit (Isermann, 1987).

Nitrous oxide, presently in the atmosphere at a 0.3 ppm level (ca. 1.5 Gt) with a 0.2-0.3% annual increase in concentration is rather chemically inert in the soil and diffuses further into the higher atmosphere and stratosphere. About 90% of the released N_2O come from biological sources via denitrification, the conversion of NO_3 into molecular N_2, or also from nitrification, from the microbial transformation of NH_4^+ into NO_3-N, or even from N-fixation and N-oxygenation by various microbial processes (Batjes, 1992; Bouwman, 1990; Savant & De Datta, 1982; Umarov, 1990). Umarov (1990) considers the contribution of autotrophic nitrification to N_2O formation as being insignificant due to the optimal conditions regarding pH, aeration and other perameters required by the highly specialized nitrifying bacteria. The contribution of heterotrophic nitrification by a larger number of bacterial species to N_2O formation is higher. Umarov identifies mainly *Nitrosomonas, Nitrosolubus, Nitrosococcus* and *Nitrospora* as the major first phase N_2O producers in the nitrification process.

While Conrad et al. (1983) estimated the source strength of mineral fertilizer related N-losses as N_2O to the atmosphere as representing ca 0.001-0.94% or 0.02 to 8 Mt N_2O-N y^{-1}, Bouwman's (1990) assessment tal-

Table 6. N_2O emission; a comparision of different car engines and grazing cattle, 180 days grazing season, 2 N fertilization intensities (according to Isermann, 1987)

	H-Fertilizer	Cows per	Urine-N per cow	N_2O-Emission (G)		
	$(kg\ ha^{-1}h^{-1})$	(ha)	$(kg/180\ d^{-1})$	(Per cow one day)	(Per cow 180 days)	(Per hectare pasture, 180 days)
Cattle-grazing milk production						
- Less intensive	ca.200	1.8	70% of 50kg=35	0.9	163	293
- More intensive	ca.400	3.5	80% of 77kg=62	1.6	289	1012
Operation of car engines				one day	180 days	
-Otto engine without catalyst	all types of engines in cars 15,000 km per year = 41 km per day			0.12	18	
- Otto engine with catalyst				1.2	180	
- Diesel engine				0.5	90	

lied fertilizer-losses as N_2O (exclusive of background emissions) of 0.5-2.0%, equal 2.4-3.6 Mt of N_2O-N y^{-1}. Batjes (1992) cited four further references for the annual N_2O emission range of which two were on the same level and two being below 1 Mt N_2O-N y^{-1}1. Bolin et al. (1986) indicate that the 0.3 ppm of N_2O is equivalent to a total of 1,500 Mt N. They also presented data with estimates from four groups of authors regarding the annual N_2O emission rates from seven different ecosystems and sources like atmospheric disturbances, fossil fuel burning and fertilizer, ranging between 10 and 30 Mt y^{-1}. An estimate of the concentration range of N_2O in 15 different aquatic systems is given by Ronen et al. (1988), ranging between 10^{-1} and about $5\text{-}10^2\ \mu g\ N_2O\ 1^{-1}$, with seeps in clear-cut forests even reaching 1559 $\mu g\ N_2O\ 1^{-1}$. One source of rising N_2O release, which has so far been inadequately studied is from fuel combustion and from biomass burning. The trend from increasing use of nitrogen fertilizer with rising world population is uncertain against the background of agronomic endeavours to reduce N_2O losses.

Undoubtedly, a rise in temperature would enhance N_2O production and therefore we find highest N_2O emissions in tropical rainforests, savannas and particularly in tropical forest clearing sites. Since N_2O, although being rather chemically inert in the soil and soil-near atmosphere, is actively involved in greenhouse forcing action with about 6% contribution, 170 years residence time and an efficiency factor per molecule of ca 150 compared with CO_2, land use related biogenetic N_2O sources matter greatly (Bolin et al., 1986).

This is even amplified due to the contribution of N_2O as a destructive agent in the UVb radiation absorbing ozone shield.

NO_x molecules (NO and NO_2), which are in total produced at a higher rate than N_2O, are in the stratosphere with a residence time of only a few days and are of no real importance. N_2O, with ca. 300 in 10^9 atmospheric molecules, is oxidised into NO and NO_2, the two molecules, which are the major catalysts X and XO in the stratospheric ozone destruction process (other well known catalysts are C1 and C10, derived for example from the CFC compounds).

The increasing N_2O involvement in greenhouse forcing and O_3 depletion in the stratosphere, its long residence time of 170 yrs and high molar factor of efficiency for IR absorption, make agronomic strategies for N-fertilization with minimal N_2O losses obligatory.

Initially, emphasis was laid in this regard on ricefields, mangroves and other wetlands, where under temporarily deficient O_2 level, NO_3-N can replace O_2 as the final electron acceptor in the respiration of many

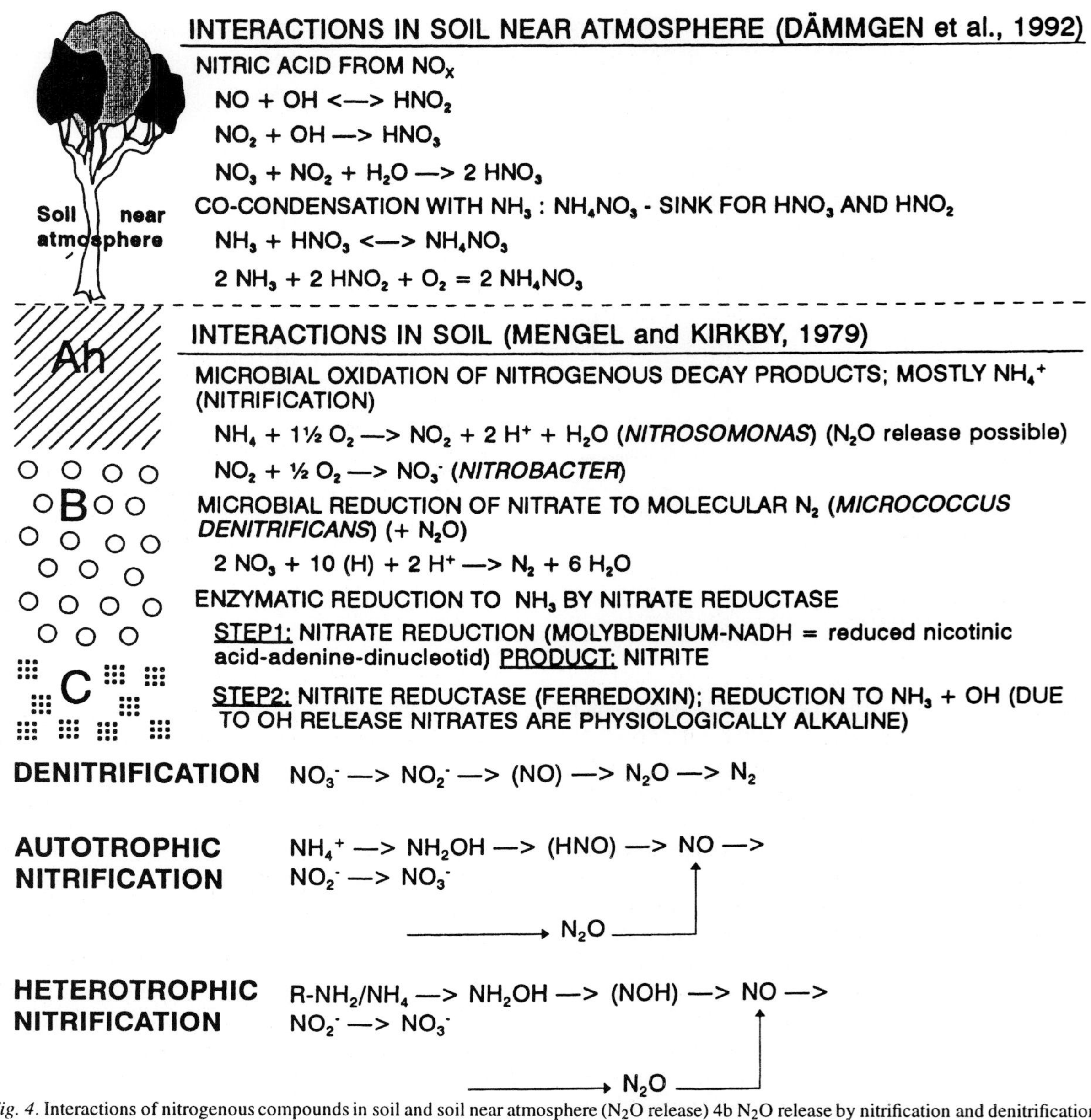

Fig. 4. Interactions of nitrogenous compounds in soil and soil near atmosphere (N$_2$O release) 4b N$_2$O release by nitrification and denitrification (acc. to Batjes, 1992)

facultative anaerobes, forming $N_2O + N_2$. Meanwhile, there is an existing awareness that almost all tropical environments and ecosystems are highly involved in this very crucial problem of agronomic hygiene. In the N_2O emission records based on emission strength multiplied with the area, the flux variability and inconsistency are often poorly assessed or even disregarded.

Detailed reviews regarding the status of N_2O in ecosystem stratification and soil units as well as on N_2O release related to a component technological concept are available (Batjes, 1992; Bouwman et al., 1992 a, b; Matson et al., 1989). They consider, as the five major regulators of N_2O production, 3 variable effects (-temperature and SOM decomposition (SOD), -soil water availability (H_2O) and O_2 limitation), and 2 rather constant effects (-soil fertility (FERT) and -C+N availability (CARBON)).

There is an effort to scale the factors influencing N_2O production, which would allow monthly model based assessment of the geometric mean of all five controlling factors : $(N_2O = (O_2 \times H_2O \times SOD \times FERT \times CARBON)^{1/5})$ (Bouwman et al., 1992 a).

High scalars among tropical soil units had Nitosols > Ferralsols > Acrisols. Also in high positions were tropical peat soils, Luvisols and Podsols, Vertisols, Andisols, Fluvisols and Gleysols. Subtropical and temperate analogs tended to lower N_2O scalars.

In a table of ecosystem types and areas, Bouwman et al., (1992b) estimate N_2O-N emission from 12.75 bil ha at ca 7 Mt N_2O-N$\cdot y^{-1}$.

Nitric oxide in CH_4 and CO oxidation and in cell physiology

Nitric oxide in soil and atmosphere is mostly an intermediate volatile gaseous product in the soil's N cycle, as well as in mainly tropical biomass burning emissions. Microbiological processes involved in its formation resemble those of N_2O formation via denitrification and nitrification. Under acidic conditions in soils and with available NO_2 chemo-denitrification adds further NO. Its formation and decomposition in soil proceeds much more rigorously than that of N_2O.

While the depletion of ozone in the stratosphere mainly by CFC's and N_2O is producing a problem of O_3 deficiency in its function as shield against UV b radiation, the reverse holds true for the rise of O_3 concentration e.g. in conjunction with CH_4 and CO oxidation in the atmosphere and its contribution to forest destruction. NO is directly involved in this tropospheric ozone formation, for example in the course of CH_4 and CO oxidation, which, according to Crutzen (1988), at NO concentration > 10 ppT leads to a net O_3 source of ca 3.7 molecules of O_3 per oxidized CH_4 / CO molecule; however there will be a O_3 loss / sink at NO concentration < 10 ppT of ca 0.1 molecule O_3 per oxidized CH_4 / $CO_{molecule}$.

According to Crutzen & Andreae (1990), biomass burning, which is the source of many greenhouse forcing trace gases, emits large amounts of CO, CH_4, other hydrocarbons and also NO. Due to the latter's short residence time of a few days only, the atmosphere remains poor in NO, thus photochemical oxidation of CO and CH_4 consumes the always deficient OH radicals, which in final consequence enriches the unoxidized CH_4 and CO species.

For the sake of completeness, reference should also be given to the importance of NO as probably the most exquisite intracellular messenger compound, produced from the amino acid arginine via the enzymes NO-synthase, in cell physiology.

References

Andreae M (1991) In:Enquête Commission "Protecting the Earth Atmosphere" of the German Bundestag (ed) Climate Change - A Threat to Global Development, p 77. Economia Verlag, C F Müller Publ.

Batjes NH (1992) Nitrous Oxide. In: Batjes NH & Bridges (eds) A Review of Soil Factors and Processes, that Control Fluxes of Heat, Moisture and Greenhouse Gases. ISRIC, Technical Paper 23, Wise Report 3, pp 67-96. ISRIC

Bazzaz FA (1990) The response of natural ecosystems to the rising global CO_2 level. Annu Rev Ecol Syst 21: 167-196

Bazzaz FA & Fajer ED (1992) Mehr Kohlendioxid - wie reagiert die Pflanzen-welt ? Spektrum der Wissenschaft (March), pp 64-71

Böttger A, Ehalt DH & Graverhorst G (1978) Atmosphärische Kreisläufe von Stickoxiden und Ammoniak. Ber Kernforschungsanlage Jülich, 158

Bolin B, Böös RR, Jäger J & Warrick RA (eds) (1986) The greenhouse effect, climate change and ecosystems. SCOPE 29, 170-174

Bouwman AF (1990) Land use related sources of greenhouse gases. Present emissions and possible future trends Butterworth Co, pp 162-163

Bouwman AF, Fung I, Matthews E, & John J (1992a) Global analysis of nitrous oxide emission from natural ecosystems. Global Biochem Cycles (In press)

Bouwman AF, Van den Born GJ & Swart RJ (1992b) Land use related sources CO_2, CH_4 and N_2O. Current global emissions and projections for the period 1990-2100. RIVM report No 222901004, Rijksinstituut voor Volksgezondheid en Milieuhygiene, Bilthoven, the Netherlands

Bremner JM & Blackmer AM (1978) Nitrous oxide; Emission from soils during nitrification of fertilizer nitrogen. Science 199: 295-296

Bremner JM & Blackmer AM (1980) Mechanism of nitrous oxide production in soils. In: (Trudinger PA, Walter MR & Ralph BJ (eds) Biogeochemistry of Ancient and Modern Environments, pp 279-291, Springer,New York, USA

Bremner JM & Blackmer AM (1981) Terrestrial nitrification as a source of atmospheric nitrous oxide. In: Delwiche CC (ed) Detrification, Nitrification and Atmospheric Nitrous Oxide, pp 151-170. J Wiley, New York, USA

Bremner JM, Breitenbeck GA & Blackmer AM (1981) Effect of anhydrous ammonia fertilization on emission of nitrous oxide from soils. J Environ Qual 10: 77-80

Broecker WS & Peng Tsung-Hung (1992) Interhemispheric transport of carbon dioxide by ocean circulation. Nature 356: 587-589

Cole CV, Flach K, Lee J, Sauerbeck D & Stewart B (1993) Agricultural sources of sinks of carbon. In: Wisniewski J & Sampson N (eds) Terrestrial Biospheric Carbon Fluxes: Quantification of Sinks and Sources of CO_2 Water Air Soil Pollut 70 (1-2): 111-122

Conrad R, Seiler W & Bunse G (1983) Factors influencing the loss of fertilizer nitrogen into the atmosphere as N_2O. J Geophys Res 88 (C11): 6709-6718

398

Crutzen PJ (1983) Atmospheric interactions - homogenous gas reactions of C, N and S containing compounds. In: Bolin B & Cook RB (eds) The Major Biogeochemical Cycles and Their Interactions, pp 65-110, J. Wiley, New York, USA

Crutzen PJ (1981) Atmospheric chemical processes of the oxides of nitrogen, including nitrous oxide. In: Delwiche CC (ed) Denitrification, Nitrification and Atmospheric N_2O, pp 17-44, J Wiley, New York USA

Crutzen PJ (1989) Menschliche Einflüsse auf das Klima und die Chemie der globalen Atmosphäre. In: Crutzen PJ & Muller DM (eds) Das Ende des Blauen Planeten? pp 26-27 CH Beck Publ., Munich, Germany

Crutzen PJ (1988) Presentation at 1. SAC Conference of IGBP-Programme, Stockholm, also Crutzen, PJ 1970, The influence of nitrogen oxides on the atmospheric ozone content. A J R Meteorol Soc 96: 320-325

Crutzen PJ & Andreae, MO (1990) Biomass burning in the tropics : impact on atmospheric chemistry and biogeochemical cycles. Science 250: 1669-1678

Dämmgen U, Grünhage L, Küsters A, Zimmerling R & Jäger HJ (1992) Konzentrationen und Flüsse reaktiver Stickstoff-Spezies in der bodennahen Atmosphäre-Messungen über landwirtschaftlichen Nutzflächen im Raume Braunschweig. Braunschw Naturkundl Schrift 4 (1): 181-197

Delwiche CC & Bryan BA (1976) Denitrification. Ann Rev Microbiol 30: 241-262

Dover M & Talbot LM (1987) To Feed the Earth; Agro-Ecology for Sustainable Development. World Resources Institute, New York, USA

Esser G (1990) Modeling global terrestrial sources and sinks of CO_2 with special reference to soil organic matter. In: Bouwman AF (ed) Soils and the Greenhouse Effect, pp 247-261, Wiley, New York, USA

Eswaran H, Van Den Berg E & Reich PF (1993) Organic carbon and nitrogen in soils of the world. J Soil Sci Soc Am (In press)

Hauck RD (1971) Quantitative estimates of N-cycle processes: Concepts and review. In: Nitrogen-15 in Soil Plant Studies, pp 65-80 IAEA, Vienna Austria

Hart MH (1978) The evolution of the atmosphere of the earth. Icarus 33: 23-29

Houghton RA & Woodwell GM (1989) Globale Veränderung des Klimas. Spektrum und Wissensch (June): 106-114

IGBP (1992) Global Change, Reducing Uncertainties. Chapter An Envelope of Air. p 6 IGBP Printing, Stockholm, Sweden

Isermann K (1987) Vertical distribution of NO_3-N and denitrification parameters in the soil down to greater depth (about 10 m) under various soil management/ nitrogen fertilization. 4th Int Symp of CIEC (Agricultural Waste Management and Environmental Protection), Brunswig, Germany

Isermann K & Henjes G (1989) Potential for biological denitrification in the (un-)saturated zone with different soil management. Int Workshop on Denitrification in Soil, Rhizosphere and Aquifer, Giessen, Germany

Isermann K (1990) Ammoniakemissionen der Landwirtschaft als Bestandteil ihrer Stickstoffbilanz und hinreichende Lösungsansätze zur Minderung. Gemeinsames KTBL/VDI-Symposium : Ammoniak in der Umwelt - Kreisläufe, Wirkungen, Minderung (Oktober) in Bundesforschungsanstalt für Landwirtschaft

Isermann K & Sturm H (1991) Stickstoff- und Phosphorbilanzierung der Landwirtschaft im Vergleich westeuropäischer Länder. 102. VDLUFA-Kongreß Berlin zu Landwirtschaft im Spannungsfeld von Belastungsfaktoren und gesellschaftlichen Ansprüchen (September), VDLUFA Kongreßvolumen

Jenkinson S (1990) An introduction to the global nitrogen cycle. Soil Use and Management 6: 56-60

Kopp E (1975) Das Produktionspotential des semiariden tunesischen Oberen Medjerdatales bei Beregnung. Schriften GTZ, No 18. GTZ, Eschborn, Germany

Lovelock JE (1988) The ages of Gaia. Norton & Company, New York USA

Lovelock JE (1991) Healing Gaia. Harmony Books, New York, USA

Matson PM, Vitousek PM & Schimel DS (1989) Regional extrapolation of trace gas fluxes based on soils and ecosystems. Andreae MO & Schimel DS (eds) Exchange of Trace Gases Between Terrestrial Ecosystems and the Atmosphere, pp 97-108. Dahlem Workshop Report, Wiley, New York, USA

Mengel K & Kirkby EA (1979) Principles of Plant Nutrition, Chapter 7, p 295. Intern. Potash Institute, Berne, Switzerland

Neptune AML, Tabatabai MA & Hanway JJ (1975) Ratio of elements in SOM of the Amazon forests. Proc Soil Sci Soc Am 39-51

Neue HU (1988) Holistic view of chemistry of flooded soils. In: Proceedings of the First International Symposium on Paddy Soil Fertility, IBSRAM Thailand, pp 21-56, Bangkok

Platt U (1984) Measurements of nitrates related concentrations in continental air. Environ Sci Technol 18: 365-369

Reyes RY (IRRI) according to Zamora J (1993) In: Philippine Daily Inquirer, Technology Dispatch (Rice yields from Pinatubo lahars plus mineral fertilizer)

Ronen D, Magaritz M & Alman E (1988) Contaminated aquifers are a forgotten component of the global N_2O budget. Nature 335: 57-59

Savant NK & De Datta SK (1982) Nitrogen transformations in wetland rice soils. Adv Agron 35 : 241-302

Sowden FJ, Chen Y & Schnitzer M (1977) Nitrogen distribution in soils from different climate zones. Geochim Cosmochim Acta 41, 1524, Sowden FJ, Morita H and Levesque M 1978 in Can J Soil Sci p 58, 237

Stevenson FJ (1982) Humus Chemistry, Genesis, Composition, Reactions. Whiley-Interscience, New York, USA

Tans PP, Fung IY, & Takahashi T (1990) Observational constraints on the global atmosphric carbon dioxide budget. Science 247: 1431-1438

Umarov MM (1990) Biotic sources of nitrous oxide (N_2O) in the context of the global budget of nitrous oxide. In: Bouwman AF (ed) Soils and the Greenhouse Effect, pp 263- 268. Wiley, New York, USA

Steudler PA, Bowden RD, Melillo JM & Aber ID (1989) Influence of nitrogen fertilization on methane uptake in temperate forest soils. Nature 341: 314-315

Warneck P (1989) Reaktive Stickstoffverbindungen in der Atmosphäre. Promet Heft 3/4, 72-79

Watson A in Lovelock J (1988) The ages of Gaia. 132 p New York, USA. Norton & Company

N. Ahmad (ed.), *Nitrogen Economy in Tropical Soils*, 399–413.
© 1996 *Kluwer Academic Publishers. Printed in the Netherlands.*

Response of sugarcane to nitrogen fertilization practices on some sugarcane soils in Trinidad

C.R. Shand
Caroni Research Statian, Waterloo Road, Carapichaima, Trinidad, West Indies

Key words: factorial trial, nematicide, nitrification inhibitors, "N-Serve", sulphur-coated urea

Abstract

Over the years a range of materials and practices in N fertilization in Trinidad have been tested and some of these are discussed. Conventional and controlled release fertilizers gave no significant differences, with plot yields high and variable. Sulphur-coated urea (S. C. U.) was not superior to $(NH_4)_2SO_4$ or $(NH_2)_2CO$ although previous reports indicated that volatilization losses of N occurred with $(NH_2)_2CO$ or $(NH_4)_2SO_4$ as N sources. The nitrification inhibitor "N-Serve", at low rates improved N efficiency in calcareous soils. However, at high rates on strongly acid soils, Talparo series (Vertisol) and McBean series (Alfisol), where nitrification is minimal, toxicity can occur without any benefit being achieved. N/P factorial trials showed that on a Mollisol, varying combinations of N an P gave no significant differences in cane and sugar yields whereas on a calcareous vertisol, rates of 159 kg N ha^{-1} and 211 kg P ha^{-1} improved cane quality. Low yields in first ratoon cane on acid soils were attributed to high rates of $(NH_4)_2SO_4$ (106–159 kg N ha^{-1}) by increasing soil acidity. In a N/K trial on calcareous soils, 159 kg N ha^{-1} in combination with 300 kg K ha^{-1} substantially increased cane and sugar yields over traditional fertilizer recommendations. Conversely, on the acidic series, McBean (Alfisol) and Waterloo (Alfisol) corresponding levels of $(NH_4)_2SO_4$ and $(NH_2)_2CO$ produced no significant differences in cane or sugar yields although plots treated with nematicide out-yielded the untreated ones. The response to supplementary applications of N on acid soils was similar regardless of carrier.

Introduction

The importance of improved sugar cane yields in boosting the sugar industry in Trinidad cannot be overemphasised (Cooper *et al.*, 1975). Since soil fertility is the single most limiting factor to productivity of the crop, optimal use of fertilizer should contribute significantly towards higher and more economic production. The need for determining optimum rates, and timing and placement of the fertilizers has been addressed over the years (Shand, 1983–84). Environmental conditions (rainfall in particular) have made it necessary to implement more sophisticated management techniques. Correct and controlled use of fertilizers requires a sound knowledge of environmental conditions, coupled with soil moisture and soil acidity, all of which are likely to affect the plant nutrient system. Nitrogen is considered to be the most important nutrient in the mineral nutrition of most crops and, within broad limits dominates cane yield. The two major N fertilizers in current use are $(NH_4)_2SO_4$ and $(NH_2)_2CO$ with the latter becoming more common especially in regions of high rainfall as in Trinidad and more so on heavy clay soils.

Review of previous nitrogen fertilizer studies

During the seventies emphasis was on conserving fertilizer N losses in soil which occurred mainly through the following pathways:

(a) leaching of NO_2^- and NO_3^-

(b) denitrification losses

(c) ammonia volatilization

(d) immobilization through chemical reactions with soil organic matter, microbial transformation and sometimes fixation by clay minerals.

Emphasis on N conservation was based, on (1) the controlled release approach, (2) the balanced dissolution approach and (3) the inhibitor approach.

To achieve the objective of controlled release, granules of $(NH_2)_2CO$ and NH_4NO_3 were coated with inert materials such as plastics, resins, waxes and sulphur (C C Weir, private communication).

Several methods (C C Weir, private communication) were advanced in achieving balanced dissolution, and among those was one based on the principle that nitrogenous fertilizers of limited water solubility undergo dissolution in soil at a rate which varies with the size of granules, i.e. the amount of surface area exposed determines the dissolution rate (C C Weir, private communication). With regard to the inhibitor approach, the use of certain compounds has been found to inhibit to a certain extent the activity of specific microbial groups which perform the biochemical transformation of NH_4^+ to NO_2^- and then to NO_3^-. Among these are thiourea, methioine, 2-chloro-6-(trichloromethyl) pryridine (N-Serve Registered Trade Mark, Dow Chemicals Co. Ltd and 2-amine-4 chloro-6 methylpyrimidine, (trade name Toyokoatso "AM") (C C Weir, private communication).

Studies with slow release fertilizers (sulphur coated urea (S.C.U.) and the nitrification inhibitor, "N-Serve"

Effectiveness of sulphur coated urea and the inhibitor "N-Serve"
Five soils were used in these investigations viz. Princes Town Clay (Vertisol), Talparo Clay (Vertisol), McBean Sand (Alfisol), Couva Loams (Inceptisol) and Piarco Fine Sand (Alfisol).

Princes Town Clay. This soil type was described by Anon. (1968), as being a very dark grey black soil high in humic content. It occurs in hilly to undulating terrain. The soil is alkaline in reaction. The internal drainage is impeded and although difficult to manage, the soil is high in nutrients and has been cultivated with sugarcane for many years.

The Talparo Clay. This soil type was described by Anon (1968) as being the most widespread soil in Central Trinidad with substantial areas in the south. They are referred to as red or red weathering clays. The drainage is impeded. The soil is moderately to strongly acid, well supplied with all the major nutrients, with sugarcane being the major crop grown.

The McBean Sand. The McBean Sand has been described by Anon (1970) as having free internal drainage. The whole profile may vary between loam and sandy loam. The soil is moderately to strongly acid, open textured and low in all major nutrients. The soil responds very favourably to fertilizer.

Couva Loams. Couva loams according to Anon (1970) are imperfectly drained and were developed on the open textured sub-recent alluviums of southwest Central Trinidad. The series is variable in texture ranging between sandy loam and clay loam. This series is moderatley to strongly acid with moderately good supply of Ca and Mg, although other major nutrients are deficient.

The major crops grown are sugarcane, citrus, bananas and vegetables.

Piarco Fine Sand. This soil type was described by Brown & Bally, (1966) as being a sand to sandy loam followed by sand and finally silty clay. It is waterlogged for most of the wet season, completely desiccated in the dry season, although the water table may be only a few feet below surface. The profile is extremely acid and it is almost devoid of any plant nutrients. The vegetation is sparse grasses, wild palms and marsh forest.

Table 2 gives some chemical properties of the five soil series observed at the start of these studies.

In pot experiments with sugarcane S.C.U. was compared with $(NH_4)_2SO_4$ with or without "N-Serve" (2.5% of fertilizer) under low (0.75 cm per drum (100 l) 4 times a week) and high, (1.50 cm per drum (100 l) 4 times a week) levels of irrigation applied to Princes Town Clay (Vertisol), Talparo Clay (Vertisol) and McBean Sand (Alfisol).

On the Princes Town Clay under the low level of irrigation, the pots which had received 105 kg N ha^{-1} plus S.C.U. gave plant weights which were superior to those in other treatments, the fresh weights being 1585 g per pot (drum), 1400 g per pot (drum), 1325 g per pot (drum) and 1117 g per pot (drum) for the S.C.U. $(NH_4)_2SO_4$ + "N-Serve", $(NH_4)_2SO_4$ alone and the control treatments respectively (Dalal & Prasad, 1975). On the same soil $(NH_4)_2SO_4$ in combination with "N-Serve" gave the highest cane weight under the high level of irrigation (2667 g fresh weight per drum), the lowest being 1567 g for the control treatment.

Plants receiving $(NH_4)_2SO_4$ + "N-Serve" showed the highest leaf N content on the Talparo Clay and the McBean Sand under the low level of irrigation, being 2.30% and 2.13% dry matter (DM) respectively, while S.C.U. gave the highest leaf N contents in all three

Table 1. Particle size distribution of experimental soils

Soil Series	Depth cm	Coarse sand	Fine sand	Silt	Clay
Princes Town	0 – 7.5	1	2	8	90
	7.5 – 30	1	2	9	91
	30.0 – 90	0	2	6	90
Talparo	0 – 5	4	18	15	65
	5 – 20	3	8	10	82
	20.0 – 25	0	1	4	97
McBean	0 – 10	3	65	18	16
	10.0 – 40	3	62	16	22
	40.0 – 75	3	45	13	41
Couva	0 – 15	2	56	18	27
	15.0 – 35	1	29	27	41
	35.0 – 40	2	24	23	51
Piarco	0 – 12.5	13	60	9	18
	12.5 – 37.5	14	49	11	20
	37.5 – 110.0	12	41	9	40
Freeport	0 – 7.5	2	60	17	23
	7.5 – 27.5	3	58	12	29
	27.5 – 47.5	6	50	13	32
Frederick	0 – 10.0	1	6	28	64
	10.0 – 45.0	0	6	25	66
	45.0 – 90.0	0	3	19	80
Washington	0 – 12.0	0	78	11	13
	12.0 – 42.0	0	77	11	13
	42.0 – 90.0	0	74	10	18
Waterloo	0 – 15.0	5	50	20	27
	15.0 – 35.0	2	42	27	31
	35.0 – 50.0	2	33	22	43

Particle size distributions taken from Land Capability Survey of Trinidad and Tobago. No. 3, 4 and 5 (1968–1970).

Table 2. Some chemical properties of the five soils observed at the start of the studies on the effectiveness of slow release fertilizers (S.C.U.) and the nitrification inhibitor, "N-Serve"

Soil series	pH	P (mg kg^{-1})	K (mg kg^{-1})
Princes Town Clay	7.3	34.0	250.0
Talparo Clay	5.15	9.46	134.0
McBean Sandy Clay	4.63	10.30	94.0
Couva Loams	6.8	10.1	102.0
Piarco Fine Sand	4.4	1.7	14.0

Table 3. Some chemical properties of the three soils observed at the start of the studies on the response of sugarcane to supplementary nitrogen application

Soil series	pH	P(mg kg^{-1})	K(mg kg^{-1})	Ca(mg kg^{-1})	Mg(mg kg^{-1})
Talparo	4.45	6.1	222.0	740.0	244.0
Couva	4.72	7.8	97.0	850.0	214.0
Freeport	4.55	5.8	167.0	643.0	190.0

soil types at the high level of irrigation namely 1.91% DM, 1.54% DM and 1.76% DM in the Talparo Clay, Princes Town Clay and the McBean Sand respectively. On the Princes Town Clay, S.C.U. under dry conditions reduced NH_3 volatilization and was more effective than $(NH_4)_2SO_4$ (Dalal & Prasad, 1975). However, the cost of S.C.U. was greater than either $(NH_2)_2CO$ or $(NH_4)_2SO_4$ and the results did not show a great potential for the material by virtue of comparative costs (RC Dalal, private communication). On the Princes Town Soil different forms of N did not reveal significant differences in cane yield and quality between treatments receiving S.C.U. and $(NH_4)_2SO_4$ (Dalal & Prasad, 1975). In a field trial on the McBean Sand split applications of $(NH_4)_2SO_4$ compared to two sources of S.C.U. showed that under three different fertilizer regimes at two weeks, six weeks and 14 weeks with N rates equivalent to 80 kg N ha^{-1} and 160 kg N ha^{-1} there were no evidence to suggest that S.C.U. was superior although it had been reported in prevous incubation studies that considerable losss of N occurred on dry soil with $(NH_4)_2SO_4$ as the N source (Dalal & Prassad, 1975). On Couva Loams (Inceptisols) (pH 6.8) subjected to two watering regimes of 0.75 cm and 1.50 cm of irrigation per day, the treatments receiving 106 kg N ha^{-1} + "N-Serve" under both high and low irrigation conditions gave yields (Table 7) which were 14 and 15 percent higher when compared with the treatment which had received a similar rate of $(NH_4)_2SO_4$ and were only 9 and 6 percent below those obtained with 212 kg N ha^{-1}.

On Piarco Fine Sand (Alfisol) under the high level of irrigation, similar trends were observed. Under the low level of irrigation, 106 kg N ha^{-1} + "N-Serve" at rates of 2.5% and 5.0% of fertilizer gave consistently lower yields than $(NH_4)_2SO_4$ alone. A similar trend was recorded for the higher level of N(212 kg N ha^{-1} + "N-Serve") at both rates of "N-Serve" and this trend was also observed on Couva Loams at the two levels of irrigation (Dalal & Prasad, 1975).

Studies on the response of sugarcane to supplementary nitrogen application

Large-scale trials with aerial application of $(NH_2)_2CO$ up to 1979 demonstrated significant responses in ratoon cane yield particularly in Caroni South (Anon., 1979). The reasons for the different responses were unclear and it was decided to compare more closely the relative benefits that could be derived from supplementary N application made up either at the same time or later than basic dressings, (2) at the onset of vigorous growth at the beginning of the wet season or (3) as a top dressing applied during mid-wet season (July-September). Three soil series were observed in these studies, viz. Talparo Clay (Vertisol, Freeport Sandy Clay (Ultisol) and Couva Clay Loams (Inceptisol).

Brief soil descriptions and land use of the Talparo Clay and the Couva Loams were documented earlier in the manuscript.

The following gives a description of the Freeport Sandy Clay. It has been described by Anon.(1970) as being derived from sub-recent levee of sandy clay alluvium. The drainage is imperfect. The soil is slightly acid and inadequately supplied with all nutrients. It is subjected to dry season desication. Sugarcane, cocoa and citrus are the major crops grown.

Table 3 gives some chemical properties of the three soil series observed at the start of the studies.

On the Talparo Clay, differences in cane yield were observed which were significant ($p = 0.05$). The standard N application (Normal estate practice - N.E.P.) of 100 kg N ha^{-1} as $(NH_4)_2SO_4$ applied soon after harvest yielded 70.6 t ha^{-1} whereas 50 and 150 kg N ha^{-1} yielded 86% and 124% of this respectively (Figure 1a). The plots which had received 100 and 150 kg N ha^{-1} as $(NH_2)_2CO$ yielded 98% and 151% of the N.E.P. treatments respectively (Figure 1b). The yield difference between plots with $(NH_4)_2SO_4$ and $(NH_2)_2CO$ as N sources was not statistically significant, though it was appreciable. A top dressing of either 25 or 50 kg N ha^{-1} applied as $(NH_2)_2CO$ in either August, September, October or November, in addition to N.E.P. increased

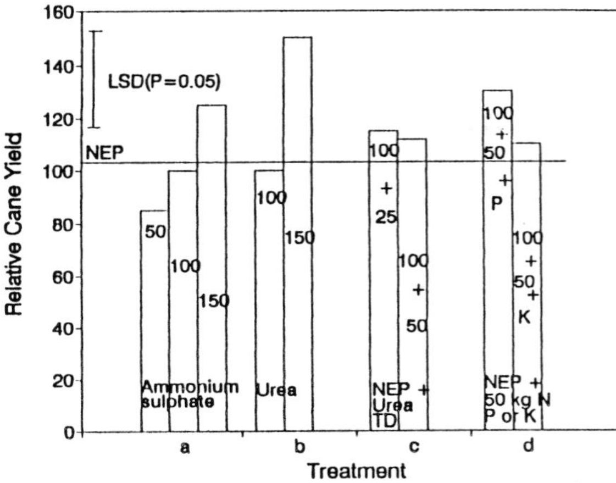

Fig. 1. Effect of supplementary N, P and K applied either at time of basic dressing or as top dressing in the wet season, on relative yield on Talparo clay (numbers in histogram indicate 50 kg N ha^{-1}, 100 kg N ha^{-1} etc.)

Table 4. Some chemical properties of four soils observed at the start of nitrogen, phosphorus and nitrogen potassium interaction studies

Soil series	pH	P(mg kg^{-1})	K (mg kg^{-1})
Nitrogen/phosphorus studies			
Princes Town	7.11	22.3	175.0
Freeport	4.17	11.25	88.0
Nitrogen/potassium studies			
Princes Town	6.97	13.3	264.0
Frederick	4.37	11.5	221.7
Washington	5.65	14.7	112.0

Table 5. Some chemical properties of the two soils observed at the start of the studies on response of sugarcane to applications of nematicide and different rates and sources of nitrogen

Soil series	pH	P(mg kg^{-1})	K(mg kg^{-1})
Waterloo	4.55	8.0	115.0
McBean	4.21	11.5	59.2

yields by 12% over N.E.P., though differences were barely statistically significant (Figure 1c). Surprisingly, there were no consistent yield differences between the two rates of supplementary N.

There were no significant differences in cane yield and quality due to treatment and sugar yields closely followed the trends in cane yield.

Leaf-N levels of the N.E.P. treatments were below critical levels after four months and were consistently raised above critical levels by the initial supplementary N applications but by six months all levels were low (Table 6) and (Figure 2). The $(NH_2)_2CO$ treatments generally produced slightly lower leaf-N levels than the $(NH_2)_2SO_4$ treatments.

The effects of timing of top dressings in August, September, October and November on leaf -N were variable (Table 5) and (Figure 3) and no consistent response was demonstrated. This may have been related to variation arising from application or sampling techniques (Anon., 1979).

On the Freeport series similar studies were conducted. Cane yields were considered to be good and the basic dressings of 100 kg N ha^{-1} as $(NH_4)_2SO_4$ i.e. N.E.P.) yielded 89.5 t ha^{-1}; the response to supplementary N applied either at the time of basic dressing (initial application) or as a top dressing was small but consistent. Differences were barely significant (at $p = 0.10$) between the highest and lowest yields. Twenty five and 50 kg N ha^{-1} of supplementary N applied as $(NH_4)_2SO_4$ or $(NH_2)_2CO$ with the basic dressing increased cane yield by 8% and 4% respectively. The same amount of supplementary N supplied by

$(NH_2)_2CO$ as top dressing in the wet season increased yields by 2% and 5% respectively.

There were no significant differences between sources or times of supplementary applications. There were no significant effects on cane quality and sugar yields followed the same trends as cane yields. Relative yields are shown in Figure 4. Initial supplementary N addition increased leaf-N levels at four, five and six months and, as on Talparo Clay, the increase was less where $(NH_2)_2CO$ was the source (Figure 5).

On Couva Clay Loam yields for first ratoon cane were considered to be also good and the N.E.P. treatment yielded 87.4 t ha^{-1} (Table 7). The highest cane yield of 97.5 t ha^{-1} was obtained with 50 kg N ha^{-1}. Supplemental N applied in July but the same rate of $(NH_2)_2CO$ applied with the basic dressing was similar. $(NH_2)_2CO$ showed no significantly less effective performance compared with $(NH_4)_2SO_4$ on the acid soils but leaf N levels were frequently slightly lower than from $(NH_4)_2SO_4$ treatments.

Nitrogen/Phosphorus and Nitrogen Potassium factorial field trials Studies with the nitrification inhibitor "N-Serve" and with two forms of sulphur-coated-urea (S.C.U.) revealed that the inhibitor was only effective on soils with pH values approaching 7.0 (Princes Town Marl and Couva Loams) and the cost of S.C.U. was greater than either $(NH_2)_2CO$ or $(NH_4)_2SO_4$ and did not show a great potential for the use of the material by

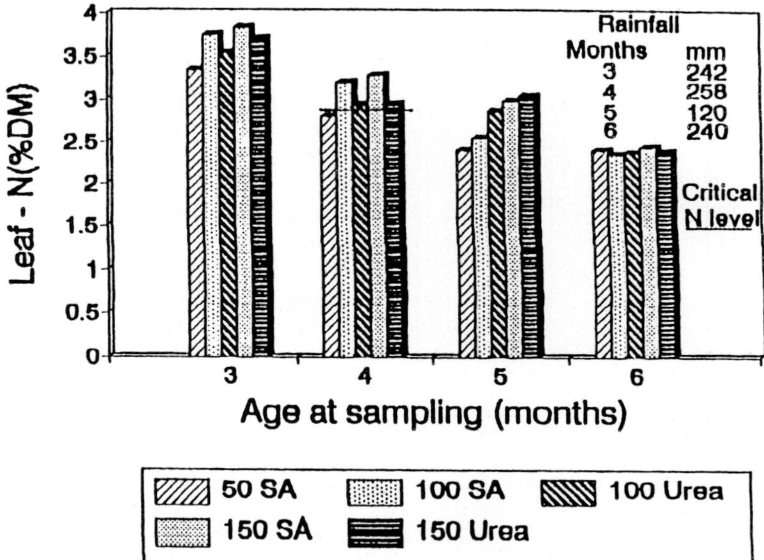

Fig. 2. Effect of basic N rates on leaf nitrogen on Talparo clay. SA - sulphate of ammonia.

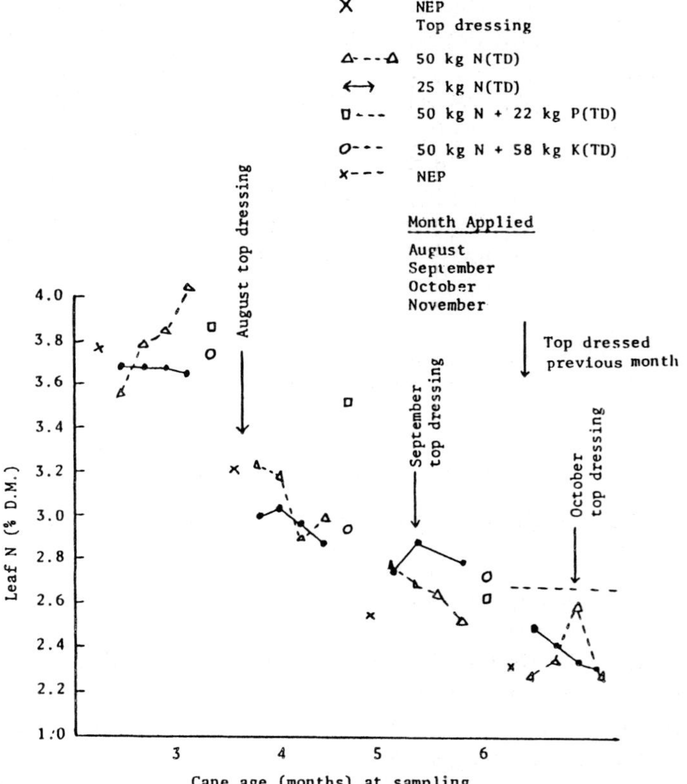

Fig. 3. Effect of top dressing on leaf-N content - Talparo clay.

Table 6. Leaf nutrient levels as affected by N rate on Talparo Clay

Date Sampled	Age at sampling (months)	Nitrogen applied						
		Basic dressing (kg N ha^{-1})				with Top dressing[a]		
		Sulphate of Ammonia				urea	NP	NK
		50	100	150	100	150		
		Leaf Nitrogen (% d.m.)						
1978								
09/08	3.00	2.36	2.75	2.83	2.55	2.70	2.88	2.75
15/09	4.25	1.80	2.20	2.30	1.95	1.96	2.53	1.95
11/10	5.00	1.39	1.54	2.00	1.88	2.05	1.65	1.75
17/10	6.50	1.41	1.33	1.45	1.35	1.38	1.32	1.30

[a] Basic dressing 100 kg N ha^{-1} + 50 kg N ha^{-1} as top dessing. P and K as TSP and Muriate of Potash at 23 kg P ha^{-1} and 58 kg K ha^{-1}.

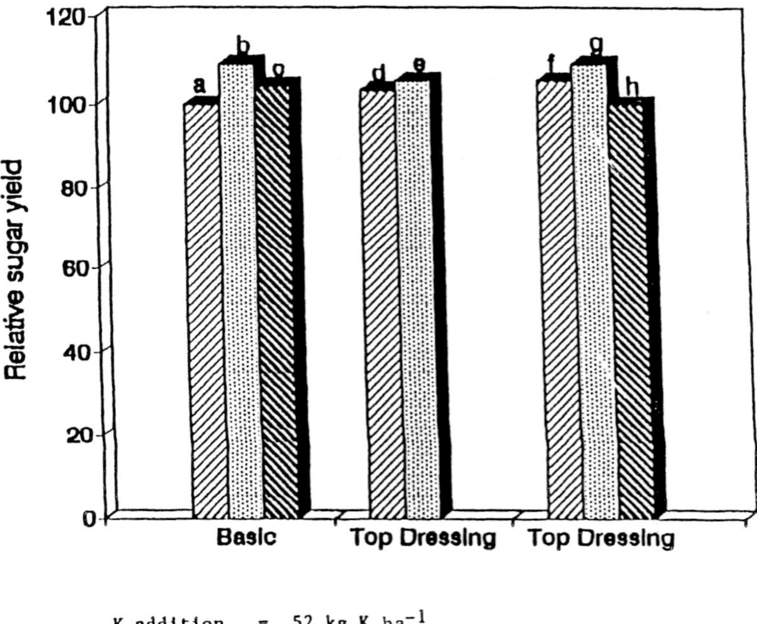

Fig. 4. Effect of additional N, P and K applied either at time of basic dressing, dressing, or top dressing on relative sugar yield on Freeport clay.

reason of its costs when compared to the conventional N carriers.

Supplementary applications with $(NH_2)_2CO$ as a top dressing in some fields in Southern Trinidad demonstrated significant responses in ratoon cane yield but the respond was only evident on highly eroded fields (BR Cooper, private communication).

Mid 1980's

During this period factorially combined interactions between N and P with a conmon rate of K and, likewise, varying combinations of N and K, with common rate of P were introduced. Ratoons did not respond to P fertilization and there were varied and infrequent responses to K, generally in Southern Trinidad and

406

Table 7. Effect of supplement N treatments on cane and sugar yields and cane quality on Couva Clay Loam

					Treatment			
Time supplement applied (cane age)	-	-	I	I	3 months		5 months	
Rate supplement	0	0	50	50	50	50	50	
Source supplement	–	–	S/A	Urea	S/A	S/A	Urea	
Basic source	S/A	Urea	S/A	Urea	S/A	S/A	S/A	
Cane yield (t ha^{-1})	87.4	88.0	89.8	95.7	97.5	89.8	90.9	ns
Sugar % cane	14.8	14.8	15.5	15.1	14.8	14.5	14.2	ns
Sugar yield (t ha^{-1})	12.9	13.0	13.9	14.3	14.3	13.0	13.0	ns

Note: I - initial; ns - not significant; SA - sulphate of ammonia.

Table 8. Effect of N and P fertilizer levels on yield component on two year plants (Standover on the Princes Town Soil Series)

	N_0	N_1	N_2	N_3	Mean
Cane yields (t ha^{-1})					
P_0	41.25				
P_1		47.50	44.75	40.25	44.17
P_2		48.25	49.00	45.25	47.50
P_3		49.00	41.00	48.00	46.00
Mean		48.25	44.92	44.50	45.89
Sugar % Cane					
P_0	17.76				
P_1		17.74	18.22	18.62	18.19
P_2		16.96	18.05	17.66	17.56
P_3		18.00	17.95	18.65	18.20
Mean		17.57	18.07	18.31	17.98
Sugar yield (t ha^{-1})					
P_0	7.22				
P_1		8.59	8.10	7.48	8.06
P_2		8.17	8.81	7.92	8.30
P_3		8.91	7.31	8.14	8.12
Mean		8.56	8.07	7.85	8.16

There were no significant differences among the treatments for each of the three parameters compared.

	F-ratio		L.S.D. values
	Treatment	Reps	(p = 0.05)
1. Cane yield (t ha^{-1})	0.32 ns	1.30 ns	17.84
2. Sugar % cane	0.83 ns	4.91 **	1.56
3. Sugar yield (t ha^{-1})	0.40 ns	2.10 ns	3.01

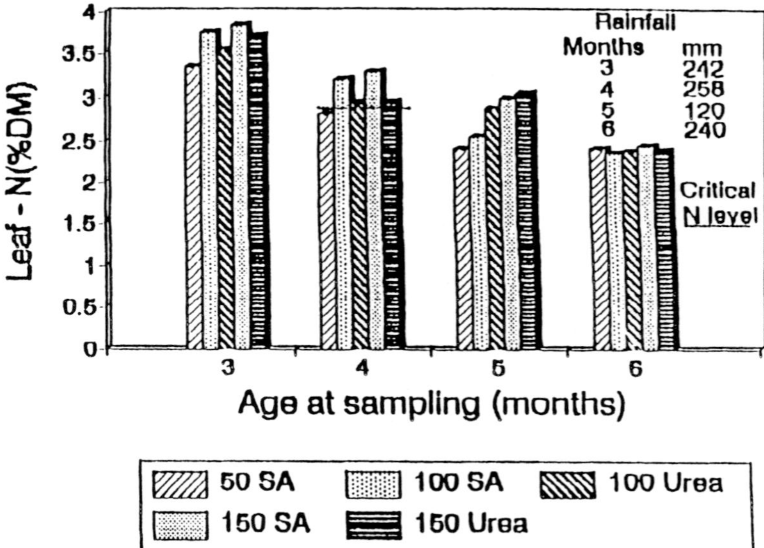

Fig. 5. Effect of basic N rates on leaf-N-Freeport clay.

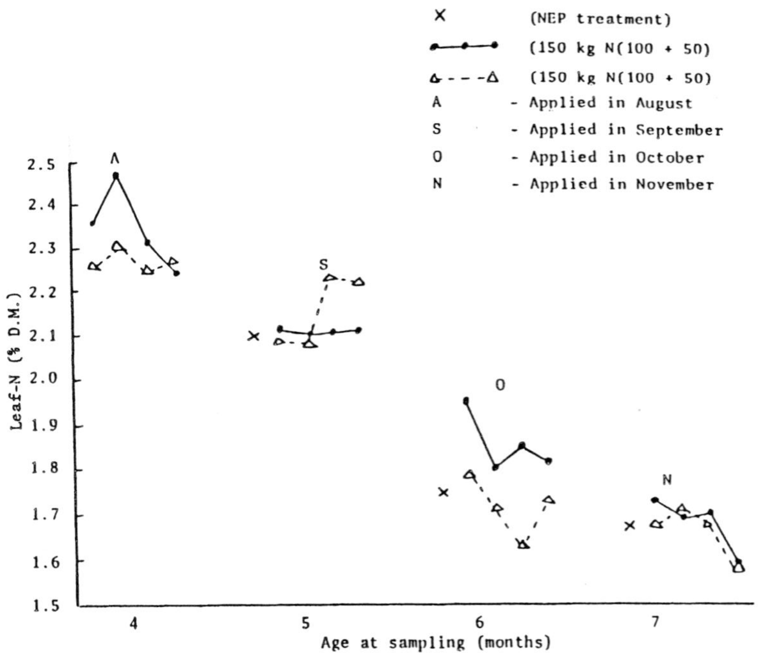

Fig. 6. Effect of Urea top dressing on leaf-N content - Freeport clay.

specifically in the Reform area on the Princes Town and Tarouba Series (calcareous Vertisols), the Sevilla (slightly acid Vertisol) and the Talparo series (acid Vertisol). It was difficult to establish whether low or marginal yields on these traditionally high yielding soils resulted from excessive N fertilizalion, or alternatively, to moderate or low soil P and K levels (CR Shand, unpublished). Consequently, a series of Nitrogen/Phosphorus and Nitrogen/Potassium factorial trials were introduced, the results of which are documented and discussed.

Four soil series were investigted in these studies viz. Princes Town Series (Vertisol) Freeport Series (Vertisol), Frederick Series (Mollisol) and Washington Series (Inceptisol).

Soil descriptions and land use of the Princes Town Series were documented earlier in the manuscript.

The following are descriptions of the other three soil series:

(a) Freeport Series. This series has been described by Anon. (1970) as having a sandy clay loam profile with a slight increase of clay sized particles in the lower B. horizon.

The drainage is imperfect and the soil is slightly acid and inadequately supplied with all nutrients. The series is subject to dry season desiccation. Sugarcane, citrus and cocoa are the major crops grown.

(b) Frederick Clay. The soil has been described by Anon. (1970) as a hydromorphic soil often found on the fringes of swampy areas in the Caroni River Basin. The profile is generally grey in colour with the top 30 cm high in humic content. Annual flooding up to 60 cm is common, but in the drier periods the water table may be 180 cm down. The texture is clay throughout the profile with an impeded drainage.

The top soil is slighly acid, but the lower horizons are strongly acid probably due to peat. Sodium (Na) increases with depth. The soil has an adequate supply of all major nutrients, but Na in the saline phase has an adverse effect on the growth of commercial plants.

Sugarcane and rice are currently grown.

(c) Washington Series. According to Anon. (1970), geological studies indicate that the parent material for the Washington Loam consists of the coarse textured deposits left as sand levees on the old river beds in the Caroni River and some of its tributaries. The series has a slightly to near neutral pH and the current reaction is probably due to liming of the sugarcane fields in the 1960's and 1970's and not to any inherent property of the soil (N Ahmad, private communication).

The drainage is free but all nutrients are low. The current land use on these series is mainly the growing of vegetables and sugar cane.

Table 4 gives some chemical properties of the four soil series observed at the start of these studies.

Nitrogen/Phosphorus factorial trials on the Princes Town and Freeport soils

N was applied initially as $(NH_2)_2CO$ and subsequently as $(NH_2)_2CO$ and P as Triple Superphosphate (T.S.P.). The treatments were as follows:

1. N_0P_0 Control = No N or P

Table 9. Effect of N and P fertilizer levels on leaf N levels on 1st and 2nd ratoon cane on the Freeport Soil Series (Three months after fertilizer application)

	% DM				
	N_0	N_1	N_2	N_3	Mean
1st ratoon cane					
P_0	2.01				2.01
P_1		2.08	1.18	2.20	1.82
P_2		2.08	2.07	2.08	2.08
P_3		2.02	2.16	2.02	2.07
Mean	2.01	2.06	1.80	2.10	1.99
2nd ratoon cane					
P_0	2.60				2.60
P_1		2.5	2.6	2.6	2.57
P_2		2.7	2.4	2.3	2.47
P_3		2.5	2.3	2.6	2.47
Mean	2.60	2.6	2.4	2.5	2.53

2. N_1P_1 N_1 = low rate = 53 kg N ha^{-1}
3. N_1P_2 N_2 = medium rate = 106 kg N ha^{-1}
4. N_1P_3 N_3 = high rate = 159 kg N ha^{-1}
5. N_2P_1
6. N_2P_2
7. N_2P_3 P_1 = low rate = 70.5 kg P ha^{-1}
8. N_3P_1 P_2 = medium rate = 141.0 kg P ha^{-1}
9. N_3P_2 P_3 = high rate = 211.0 kg P ha^{-1}
10. N_3P_3

Each treatment received a standard rate of K equivalent to 255 kg ha^{-1} muriate of potash (KC1).

On the Princes Town soil with two year-old plant high levels of N and P (N_3P_3) equivalent to 150 kg N ha^{-1} and 211.0 kg P ha^{-1} improved cane quality (Table 8). In first and second ratoon canes on the acid Freeport soil, there were no significant differences among treatments in cane and sugar yields and cane quality; yields being either low or marginal, with the control treatment giving the highest percent sugar (Table 10). On this soil series the lack of responses (similarities in cane and sugar yields and cane quality) to N fertilization in both rotations was due to increased induced acidity created by the application of $(NH_4)_2SO_4$ or$(NH_2)_2CO)$, the initial pH being 4.9 and that at the end of experimentation being as low as 4.1. Support lends to the fact that leaf-N levels were not only adequate but that they were also similar (Table 9).

Table 10. Effect of N and P fertilizer levels on yield components on Ist and 2nd ratoon cane on the Freeport soil series

	Ist ratoon					2nd ratoon				
	N_0	N_1	N_2	N_3	Mean	N_0	N_1	N_2	N_3	Mean
Cane yield (t ha^{-1})										
P_0	52.00a				52.00	47.50				47.50
P_1		52.63a	49.63a	53.25a	51.84		59.47a	56.79a	60.54a	58.93
P_2		51.25a	54.00a	49.50a	51.58		50.90a	54.64a	50.90a	52.14
P_3		52.88a	54.25a	53.75a	53.63		55.71a	49.29a	52.14a	52.38
Mean		52.25	52.63	52.17	52.35	47.50	55.36	53.57	54.52	54.48
Sugar % cane										
P_0	14.44a					15.79a				15.79
P_1		14.30a	15.02a	16.12a	15.15		15.30a	14.04a	14.12a	14.48
P_2		14.88a	15.37a	14.80a	15.02		12.44b	15.44a	13.09b	13.65
P_3		14.21a	14.00a	14.85a	14.35		15.26a	13.93ab	13.64ab	14.27
Mean		14.46	14.80	15.26	14.84	15.79	14.33	14.47	13.61	14.13
Sugar yield (tha^{-1})										
P_0	7.42a					7.49ab				7.49
P_1		7.52a	7.43a	8.57a	7.84		9.02a	7.88ab	8.37ab	8.42
P_2		7.62a	8.29a	7.30a	7.74		6.30b	8.47ab	6.58ab	7.11
P_3		7.55a	7.56a	8.01a	7.71		8.44ab	6.76ab	7.05ab	7.41
Mean		7.56	7.76	7.96	7.76	7.49	7.92	6.76	7.33	7.65

Column figures followed by the same letters are not significantly different (p =0.05) according to Duncan's Multiple Range Test

Table 11. Effect of N and K fertilizer levels on Leaf N levels on Ist and 2nd ratoon cane on Frederick Soil Series (three months after fertilizer application)

	% DM				
	N_0	N_1	N_2	N_3	Mean
Ist ratoon					
K_0	2.02				
K_1		2.12	1.92	2.08	2.04
K_2		1.95	2.06	2.08	2.03
K_3		2.02	2.10	1.98	2.03
Mean		2.03	2.03	2.05	2.03
2nd ratoon					
K_0	0.96				
K_1		0.99	0.92	1.10	1.00
K_2		0.96	1.10	1.07	1.04
K_3		1.06	1.02	0.99	1.02
Mean		1.00	1.01	1.05	1.02

Nitrogen/Potassium factorial trials on the Frederick Soil (Mollisol), Washington Soil (Inceptisol) and Princes Town Soil (Vertisol)

N was applied initially as $(NH_2)_2SO_4$ and subsequently as $(NH_2)_2CO$ and K as KCl. Each treatment received a standard dressing of P at a rate of 255 kg T.S.P. ha^{-1}.

The N/K combinations were as follows:

1. N_0K_0 Control = No N or K
2. N_1K_1 N_1 = low rate = 53 kg N ha^{-1}
3. N_1K_2 N_2 = medium rate = 106 kg N ha^{-1}
4. N_1K_3 N_3 = high rate = 159 kg N ha^{-1}
5. N_2K_1
6. N_2K_2 K_2 = low rate = 75 kg K ha^{-1}
7. N_2K_3 K_2 = medium rate = 150 kg K ha^{-1}
8. N_3K_1 K_3 = high rate = 300 kg K ha^{-1}
9. N_3K_2
10. N_3K_3

On the Frederick and Washington Soil Series, application of N in excess of 106 kg ha^{-1} accounted for low yields on first and second ratoon canes, (Tables 11–14) respectively, mainly because of induced acidity. This was confirmed on the Washington Series where

Table 12. Effect of N and K fertilizer levels on yield components of Ist and 2nd ratoon cane on the Frederick soil series

	Ist ratoon					2nd ratoon				
	N_0	N_1	N_2	N_3	Mean	N_0	N_1	N_2	N_3	Mean
Cane yields (t ha^{-1})										
K_0	49.25a				49.25	34.49b				34.49
K_1		56.12a	50.37a	56.37a	54.29		38.45a	38.70a	47.81a	41.65
K_2		55.5a	52.75a	47.15b	51.80		42.16a	42.51a	40.49a	41.72
K_3		56.62a	52.25a	53.87a	54.25		45.39a	42.30a	45.98a	44.56
Mean	49.25	56.08	51.79	52.46	52.46		42.0	41.17	44.76	42.64
Sugar % cane										
K_0	14.77a					10.34a				10.34
K_1		13.73ab	14.26ab	13.59ab	13.86		11.08a	11.01a	10.34a	10.81
K_2		14.40ab	14.16ab	14.03ab	13.64		11.27a	10.87a	11.71a	11.28
K_3		13.92ab	13.28b	13.72ab	13.64		11.71a	11.09a	11.91a	11.57
Mean	49.25	14.02	13.90	13.78	13.90	10.34	11.35	10.99	11.32	11.22
Sugar yields (t ha^{-1})										
K_0	7.29a				7.29	3.57b				3.57
K_1		7.70a	7.19a	7.66a	7.52		4.40a	4.25a	5.07a	4.57
K_2		7.99a	7.48a	6.71b	7.39		4.76a	4.69a	4.47a	4.73
K_3		7.89a	6.89a	7.42a	7.40		5.32a	4.75a	5.63a	5.23
Mean		7.86	7.19	7.26	3.57	3.57	4.83	4.56	5.14	4.84

Column, figures followed by the same letters are not significantly different ($p = 0.05$) according to Duncan's Multiple Range Test.

the N_3K_3 interaction gave the lowest cane yield with the correspondingly highest acidity of 4.82. On the Princes Town soil combinations of 150 kg N ha^{-1} and 300 kg K ha^{-1} respectively (N_3K_3) can bring about substantial cane and sugar yields over the traditional N/K recommendations (Tables 15, 16).

Response of sugarcane to application of nematicide and different rates and sources of nitrogen on two Alfisols (McBean and Waterloo Soil Series)
Between 1986–1989 there were still uncertainity as to the best fertilizer combnations for sugarcane in Trinidad. Inconsistent responses to N mainly in ratoons were attributed to incontrolled use of the nutrient as $(NH_2)_2CO$ being the product replacing the traditional $(NH_4)_2SO_4$.

High infestation of nematodes was suspected (Brathwaite, 1978), as being a likely causative factor for the lack of response to fertilization and hence the decline in yields.

Studies were subsequently conducted to evaluate the reponse of the crop to varying quantities of N but with a common rate of a nematicide - Carbo-furan (Furadan).

Table 13. Effect of N and K fertilizer levels on leaf N levels on Ist and 2nd ratoon cane on the Washington Soil Series (three months after fertilizer application)

	% DM				
	N_0	N_1	N_2	N_3	Mean
Ist ratoon					
K_0	2.11				
K_1		2.12	2.17	2.23	2.17
K_2		2.20	2.22	2.13	2.18
K_3		2.13	2.15	2.20	2.16
Mean		2.15	2.18	2.19	2.17
2nd ratoon					
K_0	2.13				
K_1		2.09	2.10	2.13	2.11
K_2		2.25	2.17	2.10	2.17
K_3		2.20	2.12	2.18	2.17
Mean		2.18	2.13	2.14	2.15

Two Alfisols were observed in the study viz. McBean and Waterloo Soil Series

Soil descriptions and land use of the McBean Series were documented earlier in the manuscript.

Table 14. Effect of N/K interactions on yield component of Ist and 2nd ratoon canes on the Washington soil series

	Ist ratoon					2nd ratoon				
	N_0	N_1	N_2	N_3	Mean	N_0	N_1	N_2	N_3	Mean
Cane yield (t ha^{-1})										
K_0	45.75a					48.88a				
K_1		46.50a	44.37a	44.25a	45.04		53.13a	43.38a	54.75a	50.42
K_2		45.25a	44.62a	45.75a	45.21		48.13a	50.00a	50.88a	49.67
K_3		45.37a	43.25a	43.00a	43.87		51.44a	48.63a	53.63a	51.23
Mean		45.71	44.08	44.33	44.71	48.88	50.90	47.34	53.01	50.44
Sugar % cane										
K_0	17.84a					16.86a				
K_1		17.05a	17.47a	17.98a	17.50		16.49a	17.43a	16.58a	16.83
K_2		17.15a	16.70a	17.16a	17.00		16.65a	17.18a	17.07a	16.97
K_3		17.30	16.72a	17.80a	17.27		16.84a	16.03a	16.74a	16.54
Mean		17.17	16.96	17.65	17.26	16.86	16.66	16.88	16.80	16.78
Sugar yield (t ha^{-1})										
K_0	8.15					8.30a				
K_1		7.9a	7.75a	7.95a	7.87		8.77a	7.56a	7.12a	7.82
K_2		7.83a	7.84a	7.90a	7.86		8.02a	8.56a	8.72a	8.43
K_3		7.80a	7.25a	7.67a	7.57		8.71a	7.88a	8.98a	8.52
Mean	8.15	7.84	7.61	7.84	7.76	8.30	8.50	8.00	8.27	8.26

Column figures followed by the same letters are not significantly different ($p=0.05$) according to Duncan's Multiple Range Test

The Waterloo Series has been described by Anon. (1970) as being a deep hydromorphic, low humic clay series. The texture of the top soil varies from clay to sandy loam. Drainge is imperfect. The series is strongly acid throughout the profile and low in all nutrient. Sugarcane is currently the major crop grown.

On both soils $(NH_2)_2CO)$ and $(NH_4)_2SO_4$ were used as N carriers. The rates were 53 kg, 106 kg and 159 kg N ha^{-1} with or without Furadan, in a randomized complete block design. The nematicide was applied at 32.86 kg ha^{-1}.

Table 5 gives some chemical properties of the two soils observed at the start of the studies.

On both soils nematicide treatments proved beneficial in increasing cane yields, quality and sugar yield in fields which were showing marginal yields and infrequent responses to N. No significant differences in the three parameters assessed have emerged with comparative levels of the two N sources though the nematicide treated plots had out-yielded the other treatments. Leaf-N levels were similar among the treatments in each rotation for each of the two soils which confirm the similarity in the yield component data.

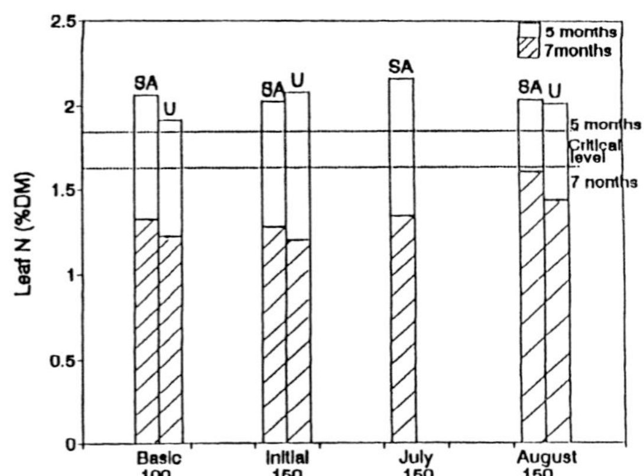

Fig. 7. Effect of supplemental N applications on leaf-N levels at 5 and 7 months on Couva clay loam

Table 15. Effect of N and K fertilizer levels on yield components of two year old cane on the Princes Town Series

	N_0	N_1	N_2	N_3	Mean
Cane yield (t ha^{-1})					
K_0	72.2c				
K_1		94.17b	88.7bc	91.1b	91.32
K_2		102.4ab	77.7bc	85.8bc	88.63
K_2		91.5b	75.0c	105.2a	90.57
Mean		96.02	80.47	93.13	89.87
Sugar % cane					
K_0	16.44a				
K_1		16.53a	15.91a	16.90a	16.45
K_1		17.25a	16.95a	16.21a	16.80
K_2		16.50a	15.96a	17.28a	16.58
Mean		16.76	16.27	16.80	16.61
Sugar yield (t ha^{-1})					
K_0	11.9c				
K_1		15.5b	14.1bc	15.8b	15.13
K_2		17.7a	12.6bc	13.6bc	14.63
K_3		15.0b	11.9c	18.1a	15.00
Mean		16.07	12.87	15.83	14.92

	F ratio		L.S.D. values
	Treatment	Reps	$p = .05$)
1. Cane yields (t ha^{-1})	1.93ns	6.66**	23.05
2. Sugar % cane	0.57ns	0.25ns	1.92
3. Sugar yield (t ha^{-1})	3.48**	8.51**	3.44

Column figures followed by the same letters are not significant according to Duncan's Multiple Range Test.

Table 16. Effect of N and K fertilizer levels on leaf N levels on two year plants on the Princes Town series

	% DM				
	N_0	N_1	N_2	N_3	Mean
K_0	1.52				
K_1		1.67	1.71	1.93	1.77
K_2		1.81	1.78	1.84	1.81
K_3		1.74	1.74	1.91	1.80
Mean		1.74	1.74	1.89	1.79

Conclusions

Sulphur coated urea (S.C.U.) was too costly to effect any beneficial economic returns with sugarcane when compared to the conventional and traditional N carriers as $(NH_4)_2SO_4$ and $(NH_2)_2CO$.

The nitrification inhibitor - "N-Serve" proved beneficial only on soils with pH values at or above 7.0 viz. Princes Town and Tarouba series, these two series accounting for less than 10% of the acreage grown with sugarcane. In any event, the methods of application were only studied under Laboratory, Greenhouse and small field trials and these systems were not applicable to commercial fields.

Aerial applictions of $(NH_2)_2CO$ as a supplementary top dressing of N has proven in the past and currently to be beneficial by way of increasing cane yields. The current rates of Urea applied as a top dressing is 62.5 kg ha^{-1}.

The nematicide-Carbo-furan (Furadan) has been tested and proven to be beneficial by way of increasing cane and sugar yields. However its effect on specific groups of nematodes has not been identified at Caroni. Further studies will seek to identify specific groups.

References

Anonymous (1968) Land capability survey of Trinidad and Tobago; Southern Trinidad. 5: 52–53 and 64p.

Anonymous (1970) Land capability survey of Trinidad and Tobago; Central Trinidad. 4: 36–37 and 40–41p.

Anonymous (1979) Response cf Sugarcane to supplementary nitrogen application. Annual Report, 1979. Caroni Research Station: 40–51p.

Brathwaite CWD (1978) Plant parasitic nematodes as a factor in Sugarcane decline in Golconda area. Annual Report, Caroni Research Station, West Indies. 255p.

Brown CB & Bally GS (1966) Land capability survey of Trinidad and Tobago; Northern Trinidad. 3: 38p.

Cooper BR (1975) Supplementary top dressng with Urea. Annual Report, 1976. Caroni Research Station: 13–15p.

Cooper BR, Dalal RC & Prasad M (1975) Time and form of nitrogen application. (a) Use of controlled release N fertilizer on a neutral soil. Annual Report, Caroni Research Station: 8p. (b) Split application vs. controlled release fertilizer. Annual Report, Caroni Research Station: 8p.

Dalal RC & Prasad M (1975) Effectiveness of sulphur coated ureaa as a nitrogen source for sugarcane. Annual Report, Caroni Research Station. 7–8p.

Dalal RC & Prasad M (1975) "N-Serve" formulated with ammonium sulphate as a source of nitrogen for sugarcane. Annual Report, Caroni Research Station. 8–10p.

Shand CR (1983–84) Review of N-P-K fertilizer trials in sugarcane (1973–1982). Technial Report, Caroni Research Station. 51p.

N. Ahmad (ed.), *Nitrogen Economy in Tropical Soils*, 415–425.
© 1996 *Kluwer Academic Publishers. Printed in the Netherlands.*

FAO/IAEA International networking in soil/plant nitrogen research

F. Zapata[1] & C. Hera[2]
[1]*Soil Science Unit, IAEA Laboratories, A-2444 Seibersdorf, Austria;* [2] *Soil Fertility, Irrigation and Crop Production Section, Joint FAO/IAEA Division, A-1040 Vienna, Austria*

Key words: biological N fixation, fertilizer N, ^{15}N, N uptake by plants, soil/plant N research

Abstract

Since its formation in 1964, the Joint FAO/IAEA Division of Nuclear Techniques in Food and Agriculture, in collaboration with the IAEA Agriculture Laboratory at Seibersdorf, Austria, has been involved in the development and transfer of nuclear technology to help developing Member States establish better conditions for sustainable crop and livestock production systems. Technology transfer is achieved through various international networking mechanisms such as Co-ordinated Research Programmes (CRPs) and Technical Cooperation Projects (TCP).

In view of the essential role that soil and fertilizer N play in maintaining and raising crop yields in modern intensive agriculture, the stable isotope ^{15}N has become widely used as a tracer to determine the amount and movement of N in plant and soils derived from applied fertilizers.

Early FAO/IAEA CRPs (1964–1987) focussed on the development of the most efficient fertilizer N management practices for major food crops and multiple cropping systems. Later CRPs assessed the environmental behaviour of soil and fertilizer N with emphasis on monitoring its residues in various agricultural systems. Nitrogen-15 isotopic studies in *Azolla* were carried out to develop improved fertilizer N practices in rice-based cropping systems. Over the past two decades CRPs have focussed on the development of ^{15}N methodology for measuring and enhancing biological N fixation, especially in N_2-fixing legume crops. The most recent programmes aimed at improving both yield and N fixation in grain legumes through an integrated multidisciplinary approach.

Recent remarkable developments in instrumentation for routine ^{15}N analysis have encouraged more widespread use of this stable isotope in agriculture as well as in environmental and biomedical research. Future directions in N research, identified in a recent FAO-IAEA Symposium, will also be discussed.

Introduction: The joint FAO/IAEA programme

In 1964, the Food and Agriculture Organization of the United Nations (FAO) and the International Atomic Energy Agency (IAEA), two international organizations of the United Nations System, developed the Joint FAO/IAEA Division of Nuclear Techniques in Food and Agriculture for the specific purpose of assisting their Member States in the application of nuclear and related technology to improve agricultural production and food storage. The Joint FAO/IAEA Division is located at the IAEA Headquarters in Vienna, Austria, and the associated Agriculture Laboratory near Seibersdorf, 40 km south of Vienna (Fried & Lamm, 1981; IAEA, 1989; Sigurbjörnsson & Lamm, 1984).

The objectives of the Joint FAO/IAEA Programme are to exploit the potential of isotope and radiation applications in research and development for: increasing and stabilizing agricultural production; reducing production costs; improving food quality; protecting agricultural production from spoilage and losses; and minimizing pollution of food and the environment.

The programme is directed towards priority areas established by the FAO and IAEA, in which nuclear and related advanced technologies can be useful.

The Joint FAO/IAEA Division appears in the organization charts of both organizations (Fried & Lamm, 1981). In the FAO, the Division is situated in the Department of Agriculture while in the IAEA it is a component Division of the Department of Research and Isotopes. It operates through the following six sections: Soil Fertility, Irrigation and Crop Production; Plant Breeding and Genetics; Animal Production and

416

Health; Insect and Pest Control; Agrochemicals and Residues; and Food Preservation.

The implementation of the Joint FAO/IAEA Programme is made through the following main categories of activities:

- Coordination and support of research through Coordinated Research Programmes.
- Implementation of field Technical Co-operation Projects.
- Training of local scientists through training courses and fellowships.
- Dissemination of technical and scientific information.
- Laboratory support for transfer of nuclear technology.

Past FAO/IAEA International networking in soil/plant nitrogen research

In the field of soil science and crop productivity, the programme activities of the Joint FAO/IAEA Division are implemented by the Soil Fertility, Irrigation and Crop Production Section in close collaboration with the Soil Science Unit of the Agriculture Laboratory at Seibersdorf. The overall aim of this sub-programme is to improve the effective use of soil, fertilizer and water resources for improving sustainable agricultural production and environmental preservation through the use of nuclear and related techniques.

Isotopes are used as tracers, for instance to determine amounts and movement of nutrients, especially those derived from fertilizers in plants and soils. Radiation techniques using equipment such as neutron moisture meters and gamma density probes permit moisture and bulk density changes in soil profiles to be monitored in a reliable, non-destructive way, while saving time, effort and money (Zapata & Hardarson, 1989).

FAO/IAEA International Programmes on Fertilizer Use Efficiency

The Joint FAO/IAEA Division through its Soil Fertility, Irrigation and Crop Production Section has for three decades, co-ordinated several international research programmes on the use of isotope techniques in fertilizer use efficiency studies of major food grain crops such as rice, maize and wheat (Table 1). The main objective of these programmes was to achieve through the use of isotope techniques higher and more stable grain yields by optimizing the uptake of nutrients from the applied fertilizers and other sources, while reducing potential harmful effects on the environment (IAEA 1970; IAEA 1970; IAEA 1974; IAEA 1978).

International networking was done by selecting contractors in developing countries. Field experiments using tagged fertilizers were conducted under a wide range of tropical and subtropical soil and climatic conditions (Zapata & Hardarson, 1989).

The Soil Science Unit at the IAEA Seibersdorf Laboratories, has played a key role in the implementation of these programmes by developing model experiments and isotope techniques, providing analytical services, and transferring this technology to Member States through training (Zapata & Hardarson, 1989).

Selected results from the above programmes are presented to illustrate the use of ^{15}N isotope techniques for developing improved fertilizer N practices and minimizing potential losses for the soil-plant system (FAO 1980; Fried 1978). Nitrogen-15 labelled fertilizers were used to determine the percent N in the plant which is derived from the fertilizer and thus directly measuring the fertilizer use efficiency. In addition the ^{15}N single treatment fertility design allowed the measurement of various fertilizer N treatments without interaction (Fried, 1978; IAEA, 1971; IAEA, 1990).

In the first rice fertilization programme, one of the relevant questions being asked was what the relative efficiencies of NH_4^+ and NO_3^- as sources of N were when surface applied or incorporated in the top 5 cm (IAEA, 1970). Field experiments using NH_4NO_3 labelled with ^{15}N in either the NH_4^+ or NO_3^- ion were carried out in five countries. The results shown in Table 2 clearly demonstrated that: (i) the NH_4^+ was by far the most effective source of N to paddy rice; (ii) the highest uptake of N was obtained from the NH_4^+ when incorporated in the top 5 cm. Placement had no significant effect on the utilization of NO_3^-.

In a 1972–73 field study of the Co-ordinated Research Programme on Rice Fertilization the utilization of 100 kg N/ha was compared at four different application times, in four countries, using ^{15}N-labelled $(NH_4)_2SO_4$ (IAEA, 1978). Table 3 shows that all treatments received 100 kg N/ha, split equally at four times, with only one time of application being labelled with ^{15}N in each treatment. The basal application (T_1) involving placement at 5 cm depth at transplanting was found superior to treatment T_2: surface placement of N three weeks later. However, the greatest efficiency of fertilizer N utilization in the grain was obtained when fertilizer N was applied at primordial initiation (T_4). It is reported that rice plants growing with a

Table 1. Fertilizer use efficiency activities in Co-ordinated Research Programmes of the Joint FAO/IAEA Division

Programme (title)	Duration	IAEA publ.
RICE fertilization	1962–1968	STI/DOC/10/108
Fertilizer management practices for *MAIZE*: Results of experiments with isotopes	1964–1968	STI/DOC/10/112
Isotope studies on *WHEAT* fertilization	1968–1972	STI/DOC/10/157
Root activity patterns of some *TREE CROPS*	1967–1972	STI/DOC/10/170 TECDOC-270
Isotope studies on *RICE* fertilization	1970–1974	STI/DOC/10/181
Use of isotopes for study of fertilizer utilization by *LEGUME CROPS*	1972–1977	TECDOC-149
Isotope-aided MICRONUTRIENT studies in RICE production with special reference to zinc deficiency	1974–1979	TECDOC-172 TECDOC-242
Soil N as *FERTILIZER* or *POLLUTANT*	1975–1983	STI/DOC/10/244 STI/PUB/535
Nuclear techniques in the development of fertilizer practices for *MULTIPLE CROPPING SYSTEMS*	1980–1985	TECDOC-235 TECDOC-394

Table 2. Nitrate and ammonium as sources of N to flooded rice

Location	Fertilizer N Source	% N derived from the fertilizer	
		Surface	5 cm depth
Korea	$^*NH_4NO_3$	10.0	14.0
	$NH_4^*NO_3$	2.2	2.0
Egypt	$^*NH_4NO_3$	7.0	10.0
	$NH_4^*NO_3$	4.7	0.8
Hungary	$^*NH_4NO_3$	7.0	14.0
	$NH_4^*NO_3$	2.4	3.0
Sri Lanka	$^*NH_4NO_3$	10.0	20.0
	$NH_4^*NO_3$	5.9	2.4
Burma	$^*NH_4NO_3$	12.0	20.0
	$NH_4^*NO_3$	2.7	2.4

Denotes the ^{15}N-labelled nitrogen ion

Table 3. Utilization of fertilizer N by rice when applied in increments as tagged ammonium sulphate at four growth stages, 1972-73

Timing of fertilizer[1]				Grain N derived from ^{15}N labelled fertilizer				
T_1	T_2	T_3	T_4	Bangladesh	India	Sri Lanka	Thailand	Ave.
(kgN/ha^{-1})				(%)				
25*	25	25	25	7.8	6.6	6.1	10.2	7.7
25	25*	25	25	3.0	4.7	3.1	10.3	5.3
25	25	25*	25	6.0	6.4	3.0	10.9	6.6
25	25	25	25*	8.3	11.8	6.4	17.8	11.1
			LSD (05)	1.7	2.1	1.7	2.1	
			Total	25.1	29.5	18.6	49.2	

[1] T_1 - basal at transplanting, T_2 - 3 weeks after transplanting, T_3 - midway between T_2 and T_4, T_4 - primordial initiation
* Labelled with ^{15}N

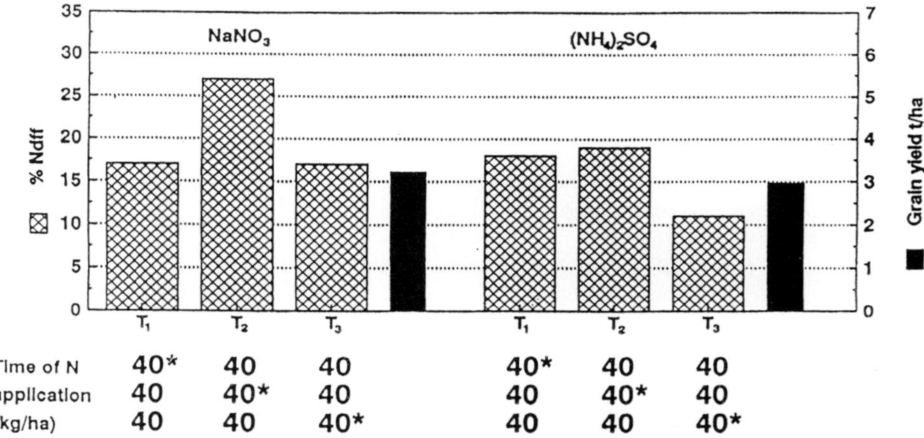

Fig. 1. Percentage of N derived from fertilizer (%Ndff) when ^{15}N-labelled sodium nitrate and ammonium sulfate were applied at three growth stages, i.e.: T_1-planting; T_2-tillering; and T_3-boot. Average data from field experiments in Brazil, Egypt, India, Italy, Pakistan and Turkey

sufficient N supply through the vegetative stage will take up and translocate this late supplement of N to the grain in a very efficient way. It is also important to note that if N supply was limited at an early stage of growth, it is possible to correct such efficiency at least 3 weeks before primordial initiation to avoid yield reduction. Similar results were obtained in follow-up field experiments conducted in seven countries (FAO, 1980; Fried, 1978).

The ^{15}N single treatment fertility design was utilized for investigating the optimum time of application of different fertilizer N sources such as Na NO$_3$ and (NH$_4$)$_2$SO$_4$ when applied at 120 kg N/ha in three equal splits at various growth stages of wheat (Bowen & Zapata, 1991; IAEA, 1974). Fig. 1 summarizes the data from 1968/69 field experiments with wheat in six countries. From the grain yield alone it would have been inferred that both fertilizer N sources were equally effective. The isotopic data indicated that (i) both N sources were equally effective when applied at planting, (ii) the largest percentage of N in the crop that was derived from fertilizer (%Ndff) was obtained from the N broadcast at tillering as NaNO$_3$, and (iii) the N applied at boot stage was generally too late for most effective N utilization in the crop.

Further studies using NH$_4$NO$_3$ labelled with ^{15}N in either the NH$_4^+$ or the NO$_3^-$ ion have demonstrated that the NO$_3^-$ form was more efficient in the N nutrition of wheat (IAEA, 1974).

The findings of over 15 years research carried out under these FAO/IAEA international networks have been published in several IAEA publications (1970, 1970, 1971, 1974, 1975, 1976 & 1978) and summarized by FAO (1980). The adoption of these improved fertilizer practices in many countries around the world

Table 4. Co-ordinated Research Programmes on biological nitrogen fixation which have been conducted by the Soil Fertility, Irrigation and Crop Production Section of the Joint FAO/IAEA Division

	Short title of programme	Duration	Number of participating countries
a)	Fertilization of grain legumes	1972–1977	14
b)	Grain legumes[a]	1979–1983	19
c)	Multiple cropping	1980–1985	9
d)	Pasture[b]	1983–1988	19
e)	Azolla[a]	1984–1989	13
f)	Common bean in Latin America	1986–1991	7
g)	Grain legumes in Asia[c]	1987–(1994)	10
h)	Tree legumes	1989–(1995)	15
i)	Microbial ecology	1992–(1997)	12

[a]Funded by the Swedish International Development Authority.
[b]Funded by the Government of Italy
[c]Funded by the United Nations Development Programme

Table 5. Nitrogen fixation capacity of grain legume species in several countries

Grain legume	Average values % Ndfa
Fababean	75
Chickpea	65
Cowpea	55
Soybean	53
Mungbean	42
Common bean	34

Source: Danso (unpublished)

has resulted in savings of fertilizers worth many millions of dollar each year.

FAO/IAEA International Programme on Soil N as Fertilizer or Pollutant

Against the background of rising world population, effectively diminishing agricultural and forestry land resources, soil N has emerged as the critical element in meeting the global needs for food, feed, fuel and fibre in the immediate future. Thus, the increasing and essential role of fertilizer N amendments for increasing agricultural productivity is recognized, especially in developing countries. At the same time many industrialized countries have a problem of a different kind, namely the entry of N into groundwater and pollution of drinking water and lakes. In this case, there is a simultaneous need to protect existing and potential agricultural soil N resources and to recognize the rising standards of environmental protection world-wide.

The stable isotope ^{15}N provides a particularly powerful tool for studying the behavior of fertilizer N in the environment (IAEA, 1971). Because of the relevant expertise and available facilities, the Joint FAO/IAEA Division with generous support from the Republic of Germany (through the Gesellschaft Für Strahlen-und Umweltforschung, GSF) carried out an international programme of studies of the fate of fertilizer N residues in agriculture during the period 1975-1983. A wealth of information and data has been accumulated as a result of the programme. Useful development and application of laboratory and field techniques has also been achieved. Two IAEA publications were derived out of this programme (IAEA, 1980, 1984). The following general conclusions have been drawn from the final technical report:

- Taking into account the prospects and time needed for developing alternative agricultural practices, conventional N fertilizer applications must continue to intensify and extend for the immediate decades ahead.

- As a result of this intensification, increasing amounts of both native soil and added fertilizer N will be lost from the soil-plant system and will find their way into the environment. Soil-N levels and contingent productivity can nevertheless be maintained by developing improved soil N management practices.

- In some situations the NO_3-N levels in ground and drinking water are likely to continue rising. A proper diagnosis of the various sources responsible for this problem should be made.

- In developing countries the losses of fertilizer N represent those of a relatively high cost input while in the more advanced industrialized countries they represent an addition to the problems and costs of environmental quality and health protection. The data generated and the information reviewed here suggest that these problems can be minimized by improving soil and water management in agricultural systems.

There appears to be a particularly enormous scope for a greater exploitation of alternative N sources such as biological N fixation in legumes and non-legumes and/or biofertilizers.

- A call is made for feasible and appropriate systems for the sustainable development of agriculture and protection of the environment.
- The United Nations Agencies such as FAO, IAEA, UNEP and WHO with an improved collaboration with appropriate regional and national programmes have a vital and urgent role in implementing the required improvements and short-term research needs, and accelerating the simultaneously needed education and training.

FAO/IEAE International Programmes on Biological Nitrogen Fixation

The utilization of the N source offers several advantages such as relatively less expense on fertilizer N inputs, reduced pollution hazards, increases in protein production due to the higher content of legumes and enhanced soil fertility due to greater N residual effect in a crop rotation. The contribution of N fixation to the N economy of both soils and plants through properly managed N_2 fixing systems appears to be the most promising alternative to supplement chemical N fertilizer in agro-ecosystems of both developing and developed countries (Zapata & Baert, 1989).

Great emphasis has been placed on biological N fixation in the FAO/IAEA Programme due to the above-mentioned advantages of this N source in developing countries and the unique usefulness of the ^{15}N isotopic techniques for providing quantitative and integrated values of N fixed in both natural and agricultural systems (Danso & Eskew, 1984; IAEA, 1978; Zapata & Baert, 1989).

During the last two decades several FAO/IAEA international research programmes have focussed on measuring and enhancing biological N fixation, especially symbiotic legume N_2 fixation in various systems (Table 4). Furthermore, current programmes emphasize the improvement of both yield and N_2 fixation in grain legumes through an integrated multidisciplinary approach (Danso & Eskew, 1984; Zapata & Baert, 1989).

Results from the programmes already completed have been published in various ways. Selected results from the above programmes will be presented. Table 5 illustrates the percent N derived from the atmosphere (% Ndfa) in several grain legume species (Danso, unpublished data). The results are average values obtained from contractors participating in the first co-ordinated research programme on N fixation in grain legumes. There were great N fixation differences between the grain legumes species, with some such as fababean being very effective and others e.g. common bean being rather poor. These differences were persistent under a wide range of environmental conditions. Also, considerable genotypic variation in N fixation between cultivars of common bean in Brazil and soybean in Romania were found (Zapata & Baert, 1989). Similarly, great differences were found between rhizobial strains in their ability to fix N (Hardarson et al., 1984a). The effects of various agronomical factors such as fertilizer N level, fertilizer P application and rhizobial inoculation were also studied in the grain legume CRP (Hardarson et al., 1984b; Zapata & Baert, 1989).

Some results from an FAO/IAEA International Co-ordinated Research Programme on the improvement of pasture management systems are summarized in Table 6. Forage or pasture legumes derived on average more than 80% of their N requirements from atmosphere and therefore were usually more effective than grain legumes (Zapata & Baert, 1989). Plant genotypic differences in N fixation have been further exploited in other FAO/IAEA international programmes to enhance N fixation in grain legumes.

In a recently completed FAO/IAEA Co-ordinated Research Programme which was conducted in Latin America with the objective of enhancing biological N fixation of common bean, it was possible to identify more effective cultivars and breeding lines than usually found in commercial cultivars of common bean. The detailed summary of the results of this programme has been recently published by Bliss and Hardarson (1993).

Researchers working in an FAO/IAEA international network carried out experiments using the ^{15}N isotope dilution technique to quantify the amount of N fixed by the *Azolla-Anabaena* symbiosis. Results from several locations indicated that on average 70-80% of N in *Azolla* is derived from the atmosphere. However, the extent of fixation varied with *Azolla* species, location and environmental conditions. In this programme, results from isotopic-aided studies conducted for several years in different locations showed that *Azolla* was equivalent to urea as a N source to rice, both in terms of ^{15}N recovery by rice and panicle dry matter yield (Kumarasinghe & Eskew, 1993).

Current FAO/IAEA International Programmes in soil plant nitrogen research

Within the framework of the IAEA Programme of Work and Budget for the biennium 1993-1994, various activities are being implemented by the Soil Fertility, Irrigation and Crop Production Section of the Joint FAO/IAEA Division with the support of the Soil Science Unit of the IAEA Laboratories as follows (IAEA, 1992):

Project area D1.01: Maximizing biological nitrogen fixation and development of molecular biology techniques in microbial ecology

a: FAO/IAEA CRP on the use of isotopes in studies to improve yield and N_2 fixation of grain legumes with the aim of increasing food production and saving N fertilizer in the tropics and subtropics of Asia (1989–1994). This programme is aimed at the identification of *Rhizobium* strains and plant genotypes with superior N_2 fixation ability in various grain legumes in Asia.

b: FAO/IAEA CRP on the use of molecular biology in microbial ecology (1993–1998). This programme focuses on the microsymbiont and its ecology with the aim of promoting abundant lateral root nodulation of N fixing crops. Identification of rhizobial strains will be made through conventional and molecular biology

methods. The ^{15}N isotope dilution technique will be used to quantify N_2 fixation.

c: FAO/IAEA Regional Africa Technical Co-operation project on Biological N Fixation. This project was initiated in 1987 with the following specific objectives: (i) to exploit as fully as possible the biological N fixation in various grain, tree and pasture legumes and thus maintain and/or increase soil fertility, (ii) to train local scientists and technicians in the use of N-15 isotope techniques in N fixation and soil fertility studies, and (iii) to promote greater collaboration between scientists in the region for exchange of information and scientific experience. This regional project operates as a network, and in 1993 involved 14 Member States, viz. Algeria, Egypt, Ethiopia, Ghana, Kenya, Niger, Nigeria, Senegal, Sierra Leone, Tunisia, Uganda, Tanzania, Zaire and Zambia. In addition to the assistance in terms of provision of equipment, expert services and training, the IAEA also provides the services of a full time consultant (Regional Expert) who periodically visits the various counterpart institutions in the Member States to provide technical backstopping, evaluate the work, and co-ordinate the activities.

Project area D1.02: measurement of nitrogen fixation in trees and management of agroforestry systems

a: FAO/IAEA CRP on the use of nuclear and related techniques in management of N fixation in trees for enhancing soil fertility and soil conservation in fragile tropical soils (1990–1995). The objective of this programme is to measure N_2 fixation in tree species and to examine their role in restoring and maintaining soil fertility. Some results of this programme have already been published by several researchers participating in this network (Bowen & Danso, 1987; Danso et al., 1981; Zapata & Baert, 1989). A review of the problems and solutions on the application of the ^{15}N methodology to N_2 fixation studies in trees has also been published (IAEA, 1975).

Project area D1.03: Optimizing the use of plant nutrients for sustainable agricultural practices and environmental protection

a: FAO/IAEA CRP on the use of nuclear techniques for optimizing fertilizer applications under irrigated wheat to increase the efficient use of fertilizer N and consequently reduce environmental pollution, (Bowen & Zapata, 1991). A consultant meeting was held in November 1993 with scientists from IFDC and CIM-

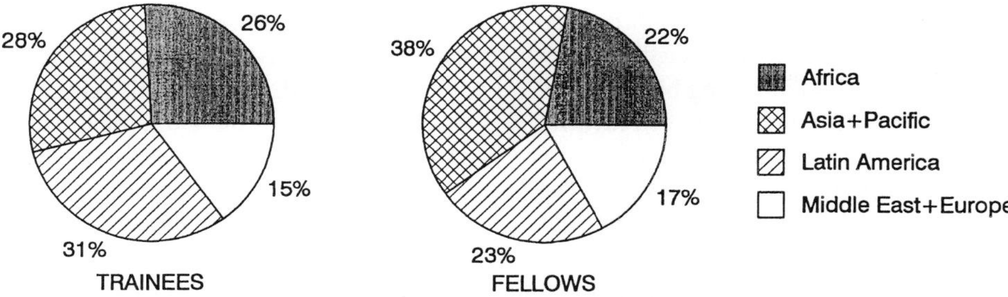

The Soil Science Unit has trained a total of 345 participants from 80 Member States in 18 training courses during 1978-1993 and 123 scientific fellows during 1962-1993 involving 827 man-months.

Fig. 2. Training activities at the Soil Science Unit, IAEA Agriculture Laboratory, Seibersdorf

MYT to plan the above mentioned CRP. Implementation is scheduled to start in 1994.

Project area D1.04: Increasing the effective use of scarce water resources to maximize plant productivity

a: FAO/IAEA CRP on the use of nuclear and related techniques in assessment of irrigation schedules of field crops to increase effective use of water in irrigation projects (1990–1995). The main objective of this programme is the testing of the concept of water deficit irrigation to improve traditionally used irrigation practices and to increase irrigation water use efficiency. Some participating contractors, are studying the effect of fertilizer N under different water regimes on ^{15}N recovery and crop yield.

Transfer of technology

Several activities are simultaneously carried out to ensure the transfer of this nuclear technology and the appropriate implementation of these FAO/IAEA international networks in soil/plant N research for sustainable agricultural development (Danesi, 1992).

The Soil Science Unit at the IAEA Laboratories in Seibersdorf is actively involved in conducting back-up research and training in support of these programmes. Training is continuously provided in the form of courses and fellowships on the use of nuclear and related techniques in soil science and plant productivity (Richards, 1992; Zapata & Hardarson, 1989).

Training courses

The FAO/IAEA Programme and the IAEA Department of Technical Co-operation organize annual interregional training courses on the use of isotopes and radiation techniques in studies of soil-plant relationships at the Seibersdorf Laboratories since 1978 (Figure 2). Three of these courses were specialized on biological N fixation.

The main objective has been to give scientists from developing countries a sound working knowledge in the use of stable and radioactive isotopes and radiation techniques in various aspects of soil/plant research. Each training course usually lasts 5-6 weeks and can be attended by twenty participants from all geographical regions. Invited lecturers and FAO/IAEA staff gave lectures and practical exercises during these courses.

The Soil Science Unit staff with their available expertise and resources supports regional and national training courses as well as other training events. It trains personnel of other host countries, assists in defining the syllabus of such training courses and in providing lecturers/instructors and supplies training materials such as manuals, brochures and videofilms (Richards, 1992; Zapata & Hardarson, 1989).

Fellowship training

In addition to training courses, the FAO/IAEA Programme conducts fellowship training for scientists from developing countries. The Soil Science Unit at the IAEA Laboratories in Seibersdorf hosts about ten fellows involving approximately 50 man-months, every year. Together with back-up research the Soil Science Unit has a long tradition of training. There are two categories of fellows: analytical fellows are accepted for a period of 2-3 months to learn isotope analytical techniques used in soil-plant research studies, for instance ^{15}N assay techniques by optical emission spectometry. This form of training includes technical tutoring and hands-on practical sessions. Research fellows are accepted for periods between 6 and 12 months to work on a topic within the IAEA Programme of Work. The

Table 6. Prospects for future use of ^{15}N tracer techniques in soil/plant nitrogen research

1.	Nitrogen cycling in soils	
	·	Assessment of the relative contribution of soil, fertilizer and other sources to nitrate enrichment of water resources.
	·	Direct measurement of gaseous losses; ie: assessment of the contribution of nitrous oxide emissions for agriculture to environmental contamination and greenhouse effect/global warming.
	·	Management of organic matter and carbon/nitrogen turnover.
	·	Soil organic matter changes in natural and agroecosystems.
2.	Quantification of nitrogen pools and cycling in agricultural ecosystems	Tree orchard/plantation ecosystems.
	·	Soil-pasture-animal ecosystems.
	·	Multiple cropping ecosystems.
	·	Agroforestry systems.
3.	Mechanisms to increase fertilizer N uptake by plants	Fertilizer N technology:
		Use of controlled release formulations.
		Use of chemicals inhibiting transformations of applied fertilizer N.
	·	Fertilizer N management strategies: Delayed soil and foliar fertilizer N application.
4.	Improvement of nitrogen nutrition of crops	Identification/selection of nitrogen-efficient genotypes, ie: with superior
	·	traits for N uptake and better partitioning of N into harvested organs.
	·	Enhancement of biological nitrogen fixation in legumes and non-legumes.
	·	Increased competitiveness for NH_4^+
	·	Fluxes of N-containing gases between shoots and the air.

fellows receive guidance on experimental strategies and use of isotope and related techniques relevant to a particular area which he/she will pursue upon return to his/her home country. They are expected to complete and write-up a piece of research work (Zapata & Hardarson, 1989).

In addition, the Soil Science Unit receives IAEA sponsored senior scientists for approximately one week visits, so as to get acquainted with recent developments on particular topics of soil science research. There are also other modalities for scientists of both developed and developing countries to be accepted at the IAEA Laboratories, such as cost-free interns, cost-free experts, FAO associated professional officers and scientists on sabbatical or study leave (Richards, 1992; Zapata & Hardarson, 1989).

Supportive services

In support of the international and regional networks created by the Co-ordinated Research Programmes and Technical Co-operation Projects of the Joint FAO/IAEA Division, the Soil Science Unit at the IAEA Agriculture Laboratory performs other supportive services, which are required for successful implementation of the various programmes. Approximately 15,000 isotopic, mainly N isotopic ratio and related analyses are performed every year for projects in developing Member States (Zapata & Hardarson, 1989). Also, ^{15}N-labelled fertilizers required for carrying out the experimental plans are weighed out and dispatched to participants in these networks. The Unit also provides necessary analytical support to laboratories in developing Member States receiving IAEA technical assistance and lacking analytical facilities.

The Unit plays a leading role in the development and transfer of ^{15}N assay technology for routine purposes. These developments are currently being used in

IAEA technical assistance projects (Zapata & Hardarson, 1989).

As a step further in the transfer of ^{15}N technology to developing Member States, the Soil Science Unit at the IAEA's Agriculture Laboratory will establish an international quality assurance service for ^{15}N analysis by optical emission spectometry (IAEA, 1992). The ^{15}N analytical facilities of the Soil Science Unit will act as the FAO/IAEA control "reference" laboratory. The following benefits are expected from such a service:

- To ensure that the ^{15}N data generated by the local laboratories are internationally acceptable.
- To give confidence and encouragement to counterparts in their analytical procedures.
- To promote co-operation at the regional level and ensure an effective transfer of the ^{15}N technology to developing Member States through the FAO/IAEA programmes.

Dissemination of scientific/technical information

The scientific staff of the Soil Fertility, Irrigation and Crop Production Section and the Soil Science Unit is very active in publishing the major scientific achievements from Co-ordinated Research Programmes and results of practical importance from technical assistance projects.

The main findings of the Unit's back-up research are presented at scientific meetings and published in scientific journals. Some 10–15 papers are published each year in reviewed scientific journals (Zapata & Hardarson, 1989). The Unit also has particular involvement in the preparation of training materials, such as training manuals and audio-visual materials.

Prospects for future networking in soil/plant nitrogen research

A FAO/IAEA international symposium on the use of stable isotopes in plant nutrition, soil fertility and environmental studies was held in Vienna, Austria, from 1 to 5 October 1990. This scientific meeting provided a good forum for exchange of scientific information and presentation of reports on recent instrumental developments and new applications of stable isotopes in life science, agriculture and environmental research (IAEA, 1991a).

A summarized review of soil/plant N research topics that clearly merit special attention is presented in Table 6. The various research topics have been grouped into four main areas of research, ie: N cycling in soils, quantification of N pools and cycling in agricultural ecosystems, mechanisms to increase fertilizer N uptake by plants and strategies for improvement of N nutrition of crops. Although remarkable achievements have recently been made in the development of ^{15}N tracer techniques and related instrumentation, the use of these technologies is still limited and must be appropriately transferred to developing countries. It is anticipated that future FAO/IAEA international research network programmes will be developed from the above mentioned areas. Some of the research topics listed in Table 6 have already been included in the new IAEA Programme of Work and Budget for the biennium 1995–96. The ^{15}N isotope will continue to be a very valuable tool because many important N research problems cannot be studied satisfactorily without the use of ^{15}N tracer techniques.

References

Bliss FA & Hardarson G (1993) Enhancement of biological nitrogen fixation of common bean in Latin America. Developments in Plant and Soil Sciences, Vol. 52. Kluwer Acad. Publishers, Dordrecht, the Netherlands

Bowen GD & Danso SKA (1987) Nitrogen research for perennial crops. IAEA Bull 29 (2): 5-8

Bowen GD & Zapata F (1991) Efficiency in uptake and use of nitrogen by plants. In: Proc. IAEA/FAO Int. Symposium Stable Isotopes in Plant Nutrition, Soil Fertility and Environmental Studies, pp 349-362. IAEA, Vienna, Austria

Danesi PR (1992) Nuclear techniques and sustainable agricultural development.IAEA Bull 34 (4):2-8

Danso SKA & Eskew DL (1984) Enhancing biological nitrogen fixation. IAEA Bull 26 (2):29-33

Danso SKA, Zapata F, Bowen GD & Sanginga N (1991) Applications of ^{15}N methods for measuring nitrogen fixation in trees. In: Proc. IAEA/FAO Int. Symp. Stable Isotopes in Plant Nutrition, Soil Fertility and Environmental Studies pp 155-168. IAEA, Vienna, Austria

FAO (1980) Maximizing the efficiency of fertilizer use by grain crops. FAO Fertilizer and Plant Nutrition Bulletin No. 3, FAO, Rome, Italy

Fried M (1978) Direct quantitative assessment in the field of fertilizer management practices. Trans. 11th Congress Soil Sci., pp 103-129, ISSS, Edmonton, UK

Fried M & Lamm CG (1981) The Joint FAO and IAEA programme. IAEA Bull 23 (3):22-26

Hardarson G, Zapata F & Danso SKA (1984a) Field evaluation of symbiotic nitrogen fixation by rhizobial strains using N-15 methodology. Plant and Soil 82: 369-375

Hardarson G, Zapata F & Danso SKA (1984b) Effect of plant genotype and nitrogen fertilizer on symbiotic nitrogen fixation by soybean cultivars. Plant and Soil 82: 397-406

IAEA (1970a) Rice fertilization. Technical Reports Series No. 108. IAEA, Vienna, Austria

IAEA (1970b) Fertilizer management practices for maize: Results of experiments with isotopes. Technical Reports Series No.121. IAEA, Vienna, Austria

IAEA (1971) Nitrogen-15 in soil-plant studies. Panel Proceeding Series. STI/PUB/278. IAEA, Vienna, Austria

IAEA (1974) Isotope studies on wheat fertilization. Technical Reports Series No.157. IAEA, Vienna, Austria

IAEA (1975) Root activity patterns of some tree crops. Technical Reports Series No.170. IAEA, Vienna, Austria

IAEA (1976) Tracer manual on Crops and Soils. Technical Reports Series No.171. IAEA, Vienna, Austria

IAEA (1978a) Isotope studies on rice fertilization.Technical Reports Series No.181. IAEA, Vienna, Austria

IAEA (1978b) Isotopes in biological dinitrogen fixation. Panel Proceedings Series. STI/PUB/478. IAEA, Vienna, Austria

IAEA (1980) Soil nitrogen as fertilizer or pollutant. Panel Proceedings Series. STI/PUB/535. IAEA, Vienna, Austria

IAEA (1984) Soil and Fertilizer Nitrogen. Technical Reports Series No.244. IAEA, Vienna, Austria

IAEA (1989) Nuclear Strategies in Food and Agriculture: 25 years of progress 1964-1989. IAEA/PI/A2OE. IAEA, Vienna, Austria

IAEA (1990) Use of Nuclear Techniques in Studies of Soil-Plant Relationships. Training Courses Series No.2. IAEA, Vienna, Austria

IAEA (1991a) Stable isotopes in Plant Nutrition, Soil Fertility and Environmental Studies. Panel Proceedings Series. STI/PUB/845. IAEA, Vienna, Austria

IAEA (1991b) Development through sciences : The IAEA Research Contract Programme. IAEA/PI/A31E. IAEA, Vienna, Austria

IAEA (1992) The Agency's Programme and Budget for 1993 and 1994, Part II. Management Part. IAEA, Vienna, Austria

Kumarasinghe KS, & Eskew DL (1993) Isotopic studies of *Azolla* and nitrogen fertilization of rice. Developments in Plant and Soil Sci, Vol 51. Kluwer Acad. Publishers, Dordrecht, the Netherlands

Richards JI (1992) Transferring technologies for agricultural development in the Third World. IAEA Bull 34(4): 9-16

Sigurbjörnsson B & Lamm CG (1984) Twenty years of collaboration in research and development for food and agricultural development. IAEA Bull 26(2): 3-8

Zapata F & Baert L (1989). Air nitrogen as fertilizer. In: Van Cleemput O (ed) Soils for Development, pp 61-84. Int. Training Centre for Soil Scientists, Ghent, Belgium

Zapata F & Hardarson G (1989) FAO/IAEA research and training in soil fertility at the IAEA's Seibersdorf Laboratories. IAEA Bull 31 (4) : 55-60

N. Ahmad (ed.), Nitrogen Economy in Tropical Soils, 427–432.
© 1996 Kluwer Academic Publishers. Printed in the Netherlands.

Nitrogen losses by leaching and runoff: methods and conclusions

George H. Snyder
University of Florida, Everglades Research and Education Center, Belle Glade, Fl., USA

Abstract

Information on N losses from soils is widely sought because of the importance of N fertilizers for maximizing crop production, and the potential for N contamination of surface and ground waters. A variety of methods have been used to estimate N transport through soils. These include direct soil sampling, lysimeters, porus soil-water samplers, adsorbents, outflow from tile drains, and groundwater wells. Each method has advantages and disadvantages related to soil conditions, area and depth to be sampled, experimental design and operation, and capital and operating costs. More universally adopted techniques are used for collecting surface runoff. They generally utilize pans or troughs inserted just below the soil surface on the low side of a slope to collect runoff water. Mathematical models have been developed to predict or describe N losses in both percolate and runoff water. However, models have not replaced the need for direct field measurements of N losses.

In general, most N loss in percolate water is in the NO_3^- form. Loss is favored by high N fertilization rates, very moist soils, excessive irrigation, high rainfall, and poorly developed or inactive root systems. Surface runoff losses of N are more likely to involve organic and NH_4^+ forms attached to eroding soil particles. Losses are reduced by practices designed to reduce soil erosion, by soil incorporation of fertilizer N, and by extending the period between N fertilization and the first significant runoff event.

Introduction

Nitrogen often is the plant nutrient most limiting crop growth and, therefore, is one of the nutrients most commonly applied in fertilizers. However, N uptake by crops following fertilization seldom exceeds 50% of that applied. The remainder may be immobilized in soil organic matter, or it may be lost from the soil by volatilization, denitrification, leaching, or in surface runoff, each posing potentially undesirable economic and environmental consequences. Nitrate-N contamination of potable waters, in particular, poses a health threat and is closely monitored and regulated by many environmental or health agencies.

Because of the recognized potential for N losses following N fertilizer applications, researchers have been measuring N in both percolate and runoff waters throughout the world for a number of years. Studies have been conducted on a variety of crops, ranging from sports turf to forests. Study methods and techniques have varied greatly, as well, and new techniques continue to be developed. Thus there is no one, universally accepted, method for determining N losses from soils in percolate and runoff. The number of different techniques is especially great for measurements involving percolate.

Nitrogen losses in percolate waters: Methodology

Nitrogen losses in percolate water have been evaluated by a variety of methods. Each technique has its inherent strengths and weaknesses, and some are more suited to certain soil or crop conditions than others. Several authors have reviewed a number of these techniques. However, most of these reviews are published as agency reports and symposia proceedings that are not readily available in many libraries.

Soil spatial and temporal variability frequently is not adequately considered when attempting to measure N transport through soils. The subject itself is very complex (Nielsen et al., 1983). Water and solutes may be transported rapidly through cracks and channels in soil, by-passing the bulk of the soil mass (Bouma, 1991). Such variability influences the appropriateness of the sampling method, as well as the number of sample sites required, the size of the water sample, and the frequency that samples must be taken.

Soil sampling

Soil sampling at various depths in the profile is one of the most direct methods for evaluating N movement in soils. Samples are extracted with reagents such as acidified KCl (Bremner, 1965) and analyzed for various N forms such as total N, NH_4^+, and NO_3^-. A variety of coring or auguring devices of various sizes may be used (Baker et al., 1989), and deep sampling generally requires some type of power equipment (Adriano et al., 1972). Soil sampling is laborious and destructive, but it is acceptable when the plot size is large and when percolation is limited, such as could be expected in fine textured soils in arid regions. Soil sampling also has utility for short-duration studies involving frequent sampling (Hamlett et al., 1990), which might not justify the installation of elaborate sampling equipment.

Lysimeters

In its broadest sense, a lysimeter is defined as a device for measuring percolation and leaching losses from a column of soil under controlled conditions (Brady, 1984). In most cases, the column of soil is enclosed on the sides and suspended on a porous rack a short distance away from the bottom of the lysimeter. In some cases, however, a flat pan is simply inserted in the soil profile, and no containment walls are built to vertically isolate the overlying soil column (Jemison & Fox, 1992). In either case, percolate that accumulates in the reservoir below the soil column is either drained away by gravity or is removed by vacuum. Some lysimeters, however, utilize a fine porus (usually ceramic) plate or candle in direct contact with the bottom of the soil column. Soil water is removed under vacuum by means of tubing inserted in the porus media.

Considerable care and ingenuity may be used to minimize soil-profile disturbance and preferential flow down the insides of the lysimeter (Cameron et al., 1992) although many lysimeters have been constructed using disturbed soil material. Lysimeters have the advantage of quantitatively collecting percolate so that both percolation amounts and N concentration in the percolate can be measured directly. However, in many instances there is a major drawback to lysimeter measurement of percolate that is unrecognized, unacknowledged, or simply rationalized.

Unsaturated water flow is impeded in stratified systems characterized by a finer-textured soil overlying a coarse textured soil (Baver et al., 1972). A lysimeter represents an extreme example of a stratified system, since the collection reservoir is analogous to a layer of soil with nearly-infinitely large pores. There can be no unsaturated water flow downward from the bottom of the soil column in a lysimeter, whereas flow can exist at the same depth in the soil outside of the lysimeter due to the negative pressure (tension) of the underlying soil. Flow will not occur from the bottom of the soil column in a lysimeter until macropores fill sufficiently that gravitational losses occur. For this reason the soil within a lysimeter may have a higher moisture content than the surrounding soil, which can alter such N transformation processes as mineralization, nitrification, and denitrification. Compounding this disturbance of the natural system, total percolation will be less from the soil within the lysimeter than from a similar layer of surrounding soil.

One theoretically acceptable method of overcoming the aforementioned disturbance is to maintain a negative pressure (tension) on the soil at the bottom of the lysimeter equivalent to that at the same depth in the surrounding soil, although this is seldom done. A less exact method of providing unsaturated flow in a lysimeter is the utilization of constant (Cole, 1968; Klocke et al., 1993) or periodic negative pressure on the lower portion of the soil profile for extracting soil water. This generally is done by applying a vacuum to a ceramic plate in contact with the bottom of the soil profile, or to ceramic candles embedded in the lower portion of the profile. Stainless-steel extractors may be used in place of ceramic, although they must be operated at lower tensions to prevent air entry. One means of maintaining a tension on the bottom of the soil column that does not require a vacuum pump is through the use of a bundle of capillary-wicks that extend from the soil downward into a sample collection vessel, stopping short of the accumulated percolate (Holder et al., 1991). Haines et al. (1982) demonstrated that more water, having a lower concentration of NO_3-N, was collected at 30-cm depth under tension than from a "zero- tension" lysimeter.

Nitrogen losses by percolation from a soil having a coarse textured layer underlying finer material can be correctly measured in a lysimeter of sufficient depth to encompass the textural discontinuity (Cisar & Snyder, 1993) without application of a vacuum for maintaining soil matric potential.

Porus soil-water samplers

Porus soil-water samplers generally are constructed from ceramic material formed in the shape of a rounded cup that is attached to a polyvinylchloride (PVC) pipe sampler-body tube. However, a variety of sampler designs have been developed utilizing a number of materials for the porus cup and sampler-body tube (Dorrance et al., 1991). These devices frequently are referred to as suction lysimeters or as tension lysimeters. Such terminology is unfortunate, since it causes confusion with the true lysimeter previously described. The moistened porus cup excludes air passage at negative pressures of less than one or two atmospheres, depending on the characteristics of the cup material. The cup is placed at the desired depth in the soil in such a manner as to ensure intimate contact with the soil. When a vacuum, generally not exceeding 0.8 atmospheres, is applied to the inside of the cup, soil water is drawn inward. Subsequently, the water sample is withdrawn from the cup through tubing extending to the cup bottom. The amount, frequency, and duration of the vacuum application varies with soil texture, soil moisture content, and individual researchers. Generally, the sampling procedure should be conducted so as to minimize the sample volume in order to only minimumly perturb the moisture status of the surrounding soil. Riekerk and Morris (1983) have developed a hanging-bottle technique to maintain a constant-potential vacuum for sample extraction when electrical power is not available. Morton et al. (1988) used a porus plate buried horizontally in the soil profile to collect percolate water. However, percolate water by-passed the collection device when the rate of percolation exceeded the rate at which water could be drawn into the plate.

Porus soil-water samplers are inexpensive, and are relatively easy to install and use. However, they have been criticized on several grounds. Cations, including NH_4-N, may adsorb on the porus cup, particularly when a ceramic material is used. Adsorption of NO_3-N presents less of a problem (Joslin et al., 1987). Soil water and soil N are held at varying tensions in soil pores of varying size. Porus water samplers will not extract water that is held at tensions exceeding the vacuum applied to the sampler (Angle et al., 1991). On the other hand, work by Magid and Christensen (1993) suggested that porus-cup devices sample a greater proportion of the less-mobile, resident (higher tension), water than do zero-tension lysimeters. Percolation through cracks and channels in well-

structured soils may by-pass the porus cup altogether. Nitrogen transformations may occur within the collection chamber if sample water remains in the chamber for several days before it is removed for analysis, and several days may be required to extract sufficient sample from very heavy-textured soils. In order to obtain an estimate of leaching, a measurement or estimate of percolation must be made. Porus soil-water samplers probably are of greatest use in coarse-textured soils that receive moderate precipitation from rainfall and/or irrigation. The "tensionic" is a modification of a standard tensiometer that provides a means for withdrawing NO_3-N that has diffused into the ceramic cup, while at the same time using the device for measurement of matric pressure head (Moutonnet et al., 1993). Since approximately eight days are required for equilibrium to be established between the soil solution and the liquid inside the tensionic, the device appears most suitable for use when soil NO_3-N concentrations are fairly stable over time.

Adsorbents

Resins may be used in a manner somewhat analogous to a porus water sampler. A capsule of anion or cation resin is placed within the soil profile at the end of an access tube, which permits periodic removal of the resin capsule and replacement with a fresh capsule (Li et al., 1993). In contrast to the porus water samplers, water passes through the resin capsule but none is retained; only ions are removed from the percolate by the resin, not the percolate water itself. This creates less of a perturbance on the soil-water system, and provides an accumulative measure of nutrients in the percolate over time, as opposed to the point-in-time measurement obtained with a porus water sampler. However, the data cannot be expressed in the conventional terms of mass of N per volume of percolate (e.g., mg N L^{-1}). Mathematical conversion methods have been proposed based on a knowledge of the water flux which passed through the resin (Li et al., 1993). The resin-adsorbent method has not been widely used for N leaching as yet.

Groundwater monitoring

Frequently N, especially NO_3-N, is measured in groundwater. Since these data may have legal implications, well defined, specific instructions and regulations have been established for groundwater monitoring, and are beyond the scope of this paper. Wells generally are used for groundwater monitoring (Kirkland et al., 1991), although tile-drain outflow also has been used (Kanwar et al., 1983). Groundwater-monitoring studies generally are suitable only for large-scale areas, such as watersheds, and are not adaptable to randomized, replicated plot work involving a number of imposed treatments.

Modeling

Ideally, direct measurements of NO_3-N transport through soils could be dispensed with entirely through the use of appropriate models. A number of models have been developed and evaluated (Jury & Fluhler, 1992). Many of these models consider N losses in surface runoff in addition to transport in percolate water (Balogh et al., 1992). None are, of course, sufficiently accurate at this time to obviate the need for direct measurements in all cases. However models are useful as qualitative predictors of the effect of various soil, crop, climate, and management factors on N leaching.

Nitrogen losses in runoff waters: methodology

Methods used for measuring N losses in runoff waters appear to be less diverse and more universally accepted than those for measuring N losses in percolate waters. In general, a trough or pan is placed on the low side of a hill for the interception of runoff waters. A portion of the region above the interception device may be isolated to define the area from which runoff occurs, and to prevent water movement into or out of the area. Since the quantity of water intercepted may be considerably greater than that needed for chemical analyses for N, splitter devices have been designed to retain a specific portion of the runoff water and discard the remainder (Morton et al., 1988). Measuring devices, automatic water samplers, and data loggers have been combined to quantify runoff and obtain samples throughout a runoff event (Watschke et al., 1989).

General conclusions regarding N loss in percolate

A number of factors governing N loss in percolate and runoff waters recently were summarized (Walker & Branham, 1992). Most organic-N compounds have limited mobility in percolate waters because of their low solubility, and their ease of adsorption and resistance to desorption (Lenton & Wilson, 1979). Ammonium-N, being cationic, is subject to adsorp-

tion by clays and organic matter. Thus, most N loss in percolate occurs in the anionic NO_3-N form. Precipitation and irrigation have dominating effects on mass transport of NO_3-N since the degree of NO_3-N leaching depends directly on the amount of water infiltrating the soil surface (Walker & Branham, 1992). Leaching of NO_3-N is potentially high when the soil is at or near field capacity and/or when intense irrigation closely follows heavy N fertilization. Leaching can be minimized by timing N applications in relation to crop uptake. Evapotranspiration (ET), of course, reduces soil moisture and therefore the potential for NO_3-N losses in percolate waters.

Temperature affects N mineralization and nitrification rates, and can thereby affect NO_3-N leaching. Factors such as coarse texture, low organic matter content, low bulk density, and high hydraulic conductivity all favor NO_3-N leaching. Soils with well defined structure, and with discrete channels, can rapidly transport water and solutes such as NO_3-N downward to the water table (Bouma, 1991).

In general, the risk of subsurface transport of N is directly proportional to the amount of N applied as fertilizer. Exceptions occur, however. For example, in Saskatchewan, Canada, low rates of N fertilization reduced wheat (*Triticum aestivum* L.) tillering and spring evapotranspiration, resulting in greater N leaching losses than were observed at higher N fertilization rates (Campbell et al., 1993). Nevertheless, the most effective method to reduce the loss of fertilizer-derived NO_3-N to groundwater generally is to reduce the quantity of N fertilizer applied (Walker & Branham, 1992). Soil-incorporation techniques such as banding may reduce N leaching by decreasing the required fertilization rate through increased fertilizer efficiency. Leaching of NO_3-N is minimized when the applied N is fully utilized during crop development.

Water-insoluble, controlled-release forms of N, or frequent but light applications of soluble N, such as application by fertigation, have been shown to reduce N leaching (Snyder et al., 1981, 1989).

Poor fertilizer handling techniques that result in spills or other unintended and unnecessary excessive concentrations of N also contribute to N leaching. Such practices can be especially serious when they occur near well heads.

General conclusions regarding N loss in runoff

Concentrations of sediment-bound nutrients such as organic N and adsorbed NH_4-N increase rapidly at the beginning of runoff events and peak well in advance of peak runoff flow (Walker & Branham, 1992). Nitrate-N concentration is highest either during peak runoff flow or during the falling portion of runoff events. Most N lost in surface runoff is in the form of organic N associated with eroded soil, although runoff water also can contain soluble mineral N forms. A portion of the inorganic N in surface runoff comes from the leaching of vegetation and crop residues.

Techniques for reducing N losses in surface runoff include soil incorporation and extending the period between N application and the first significant runoff event.

References

Adriano DC, Pratt PF, Takatori FH, Holtzclaw KM, & Johnson JB (1972) Nitrate concentrations in the unsaturated zone beneath some selected row-crop fields. Calif Agric 26 (2) : 8-10

Angle JS, McIntosh MS & Hill RL (1991) Tension lysimeters for collecting soil percolate In: Nash RG & Leslie AR (eds) Groundwater Residue Sampling Design, pp 290-299 ACS Symposium Ser. 465. Am Chem Soc, Washington DC USA

Baker DG, Kanwar RS & Baker JL (1989) Sample volume effect on the determination of nitrate-nitrogen in the soil profile. Trans ASAE 32(3):934-938

Balogh JC, Leslie AR, Walker WJ & Kenna MP (1992) In: Balogh JC & Walker WJ (eds) Golf Course Management and Construction Environmental Issues pp 355-439. Lewis Publishers, Chelsea, Michigan, USA

Baver LD, Gardner WH & Gardner WR (1972) Soil Physics. John Wiley & Sons, Inc. New York, NY. pp 343-344

Bouma J (1991) Influence of soil macroporosity on environmental quality. Adv Agron 46:1-37

Brady NC (1984) The Nature and Properties of Soils, Tenth Edition. Macmillan Publishing Company, New York, NY. pp 414, 416

Bremner JM (1965) Inorganic forms Nitrogen. In: Black CA (ed) Methods of Soil Analysis, pp 1185-1191. Am Soc Agron, Madison, WI, USA

Cameron KC, Smith NP, McLay CD, Fraser PM, McPherson RJ, Harrison DF & Harbottle P (1992) Lysimeters without edge flow: an improved design and sampling procedure. Soil Sci Soc Am J 56 :1625-1628

Campbell CA, Zenter RP, Selles F & Akinremi OO (1993) Nitrate leaching as influenced by fertilization in the Brown soil zone. Can J Soil Sci 73:387-397

Cisar JL & Snyder (1993) Mobility and persistence of pesticides in a USGA-type green I. Putting green facility for monitoring pesticides. Int Turfgrass Soc Res J 7:971-977

Cole DW 1968 A system for measuring conductivity, acidity, and rate of water flow in a forest soil. Water Resour Res 4:1127-1136

Dorrance DW, Wilson LG, Everett LG, & Cullen SJ (1991) Compendium of *in situ* pore-liquid samplers for vadose zone. In; Nash RG & Leslie AR (eds) Groundwater Residue Sampling Design, pp 300-331. ACS symposium Ser 465. Am Chem Soc, Washington DC, USA

Haines BL, Waide JB & Todd RL (1982) Soil solution nutrient concentrations sampled with tension and zero-tension lysimeters: Report of discrepancies. Soil Sci Soc Am J 46:658-661

Hamlett JM Baker JL & Horton R (1990) Water and anion movement under ridge tillage: A field study. Trans ASAE 33:1859-1866

Holder M, Brown KW, Thomas JC, Zabcik D & Murray HE (1991) Capillary-wick unsaturated zone soil pore water sampler. Soil Sci Soc Am J 55:1195-1202

Jemison JM & Fox RH (1992) Estimation of zero-tension pan lysimeter collection efficiency. Soil Sci 154:85-94

Joslin JD, Mays PA, Wolf MH, Kelly JM, Garber RW & Brewer PF (1987) Chemistry of tension lysimeter water and lateral flow in spruce and hardwood stands. J Environ Qual 16:152-160

Jury WA & Fluhler H (1992) Transport of chemicals through soil: Mechanisms, models and field applications. Adv Agron 47:141-201

Kanwar RS, Johnson HP & Baker JL (1983) Comparisons of simulated and measured nitrate lossses in tile effluent. Trans ASAE 26:1451-1457

Kirkland SD, Jones RL & Norris FA (1991) Well installation and sampling procedures for monitoring groundwater beneath agricultural fields. pp 214-221 In: Nash RG & Leslie AR (eds) Groundwater Residue Sampling Design. ACS Symp Ser. 465. Am Chem Soc, Washington DC, USA

Klocke NL, Todd RW, Hergert GW Watts DW & Parkhurst AM (1993) Design, instIllation, and performance of percolation lysimeters for water quality sampling. Trans ASAE 36 (2) :429-435

Lenton R & Wilson JL (1979) Causes, types, and extent of groundwater pollution and its control. In: Lenton R & Wilson JL (eds) Groundwater Pollution. UN-FAO Irrigation and Drainage Paper 31, pp 37-38. United Nations Food and Agricultural Organization, Rome, Italy

Li ZM, Skogley EO & Ferguson AH (1993) Resin adsorption for describing bromide transport in soil under continuous or intermittent unsaturated water flow. J Environ Qual 22:715-722

Magid J & Christensen N (1993) Soil solution sampled with and without tension in arable and heathland soils. Soil Sci Soc Am J 57:1463-1469

Morton TG, Gold AJ & Sullivan WM (1988) Influence of overwatering and fertilization on nitrogen losses from home lawns. J Environ Qual 17:124-130.

Moutonnet P, Pagenel JF & Fardeau JC (1993) Simultaneous field measurement of nitrate-nitrogen and matric pressure head. Soil Sci Soc Am J 57:1458-1462

Nielsen DR, Wierenga PJ & Biggar JW (1983) Spatial soil variability and mass transfers from agricultural soils. In: Nelson DW (ed) Chemical Mobility and Reactivity in Soil Systems, pp 65-78. Soil Sci Soc Am Special Publ No. 11, Am Soc Agron Madison, WI, USA

Riekerk H & Morris LA (1983) A constant-potential soil water sampler. Soil Sci Soc Am J 47:606-607

Snyder GH, Burt EO & Davidson JM (1981) Nitrogen leaching in bermudagrass turf: Effect of nitrogen sources and rates. In : Sheard RW (ed) Proceedings of the Fourth International Turfgrass Res Conf, pp 313-324. Univ. Guelph, Ontario, Canada

Snyder GH, Augustin BJ & Cisar JL (1989) Fertigation for stabilizing turfGrass nitrogen nutrition. In: Takatoh H (ed) Proceedings of the Sixth International Turfgrass Res Conf, pp 217-219. Japanese Soc Turfgrass Sci, Tokyo, Japan

Walker WJ & Branham B (1992) Environmental impacts of turfgrass fertilization. In: Balogh JC & Walker WJ (eds) Golf Course Management and Construction - Environmental Issues. pp 105-219. Lewis Publishers, Chelsea, Michigan, USA

Watschke TL Harrison S & Hamilton GW (1989) Does fertilizer/pesticide use on a golf course put water resources in peril? USGA Green Sect Rec 28(3) :5- 8.222

N. Ahmad (ed.), Nitrogen Economy in Tropical Soils, 433–437.
© 1996 *Kluwer Academic Publishers. Printed in the Netherlands.*

Optimizing non-symbiotic nitrogen fixation in an acid Vertisol

Y. Seemungal & A.L. Donawa[1]
Department of Soil Science, University of the West Indies, St. Augustine, Trinidad and Tobago

Abstract

The Bejucal area, a region in Central Trinidad has food crops as its main agricultural activity. As part of a programme to enhance nitrogen availability to plants in this region, observations were made on various aspects of asymbiotic N fixation. Laboratory experiments conducted on air-dried, sieved soil taken from a 0–30 cm depth in the Bejucal region, indicated that the Bejucal clay (*Chromic dystraquerts,* very fine, mixed acid) has the potential to provide biologically fixed N utilizing the inherent soil micro-organisms. Using the acetylene reduction assay method, it was found that nitrogenase activity was higher under anaerobic than aerobic conditions. The assessment of the effect of various factors on nitrogenase activity indicated that, under anaerobic conditions, 75% water holding capacity (WHC), 2% glucose and 0.5% urea resulted in the highest nitrogenase activity in this soil whereas under aerobic conditions, 100% WHC, 3% glucose and 0.5% urea resulted in the highest activity. Greater understanding of the inherent asymbiotic nitrogen microbial population may increase their potential use in providing nitrogen for crop production and possibly reduce the use of applied fertilizer nitrogen in acid Vertisols.

Introduction

The role of non-symbiotic N-fixing organisms in restoring the N levels of tropical fallows has been emphasized (Huser, 1965; Kass & Drosdoff, 1971). However, the use of non-symbiotic fixers to complement fertilizer application in the tropics has not been widely investigated.

On the local (Trinidad) market the cost of urea, even though maintained at relatively low levels, from TT $620.00 (US $105.00) in 1975 to a present market value of TT $1,929 (US $325.00) per tonne in 1993, an increase of 40%. This increase coupled with the increasing costs for all of the other components for agriculture including the removal of subsidies, makes food production very expensive.

In the Caribbean with its limited land resource and ever increasing population, the farmer has taken to utilizing more and more fertilizer to increase yields and even in some cases to just maintain the present production levels. Due to the lack of natural resources that are exportable to increase the foreign cash reserves so as to be able to import food requirements, many countries are trying to become self-sufficient in terms of agricultural produce.

Recent global awareness of environmental conservation makes the prospect of utilizing non-symbiotic N fixation to reduce fertilizer use a viable alternative. Non-symbiotic N fixation would reduce the danger of eutrophication of water by run-off of free nitrates and it also utilizes readily available solar energy. In the tropics given the abundant supply of solar energy, the alternative of utilizing non-symbiotic fixation appears to be limitless if one can adjust the environmental conditions to the optimum levels.

The evaluation of the significance of non-symbiotic nitrogen fixation in the soil has been impeded by two factors (Dart *et al.*, 1972):

(i) Nitrogen gains are usually small and can only be verified by statistical treatment of a large number of total N determinations;

(ii) Under field conditions, fluctuations in content of the soil occur as a result of leaching and drainage, increases from atmospheric sources of fixed N, denitrification and N fixation.

The major objective of this work is to determine the effect that varying inherent soil conditions might have on the inherent N-fixing soil micro-organisms and to determine the optimum levels required for N-fixing activity.

Materials and methods

Bejucal clay, located in Central Trinidad, occupies 5,137 acres. The Bejucal Clay is a difficult soil to work in the wet season where it receives 900–1050 mm rainfall per year. Brown & Bally (1968) found the soil to be strongly acid and one of the several deep hydromorphic soils found in Trinidad on the edges of large swampy areas. The soil is flooded in the wet season but dries out very easily in the dry season. During this latter period the soil cracks profusely so that the surface appears as a large mat of irregularly shaped polygons. The main activity here is agriculture with sugar cane, vegetables and small root crops such as aroids and sweet potato being the important commodities. Ramnanan *et al.* (1992) found that close to 50% of the farmers applied between 400– 600 kg N ha^{-1}, with a few farmers applying over 1,000 kg N ha^{-1}. This sharply contrasts with the optimum rate of 280 kg N ha^{-1} for dasheen (*Colocasia esculenta*) that was found in field trials. In such situations a large quantity of applied fertilizer is lost, be it to denitrification, leaching or run-off.

For this study, soil was collected from 0–30 cm depth off the Uriah Butler Highway, air dried for a few weeks then passed through a 2 mm sieve. After sieving, 10 g samples were placed into 25 ml Erlenmeyer flasks. Each flask was treated then sealed using suba-seals. Each experiment was done under anaerobic as well as aerobic conditions.

For the anaerobic environment the flasks were evacuated using a syringe then back-filled with N three times to atmospheric pressure. The aerobic flasks were simply sealed with its atmosphere intact.

From each flask, be it anaerobic or aerobic, 10% v/v of its atmosphere were removed and replaced with 10% acetylene v/v. The acetylene gas was utilized as the indicator gas to determine the apparent fixation. The index gas acetylene was converted to ethylene, thus giving an indication of the non-symbiotic N-fixing ability of the inherent soil micro-organisms.

After incubation, at 1, 24, 48 and 72 h following the introduction of the acetylene gas, 1 ml samples were withdrawn by syringe and analyzed. Analysis of all the acetylene and ethylene was carried out using a Perkin Elmer F11 Gas Chromatograph with a stainless steel column of 183×0.317 cm, with Poropak N as the stationary phase and N gas as the carrier, oven temperature was 100 °C. The instrument was calibrated using 91 volumes/million ethylene as a standard.

Ethylene produced was calculated using the formula outlined in Chang & Knowles (1965).

Experiment 1: Energy levels (glucose)

The Bejucal clay was treated with various levels of glucose w/w: 0, 0.5, 1.0, 2.0, 3.0, 4.0, and 5.0%, and a dry soil sample was also set up in triplicate.

These seven treatments were all incubated in triplicate, each treatment under both aerobic and anaerobic conditions. Samples were drawn and analyzed as stated previously.

Experiment 2: Soil water content

The Bejucal clay was treated with various levels of water: 0, 25%, 50%, 75% and 100% water holding capacity. From Experiment 1 the optimum energy level was applied to all flasks.

The flasks were incubated under both anaerobic and aerobic conditions. Sampling and analysis was done at the stipulated time and as stated previously.

Experiment 3: Nitrogen levels (urea)

The soil was treated with several levels of urea: 0, 0.5, 1.0, 2.0, 3.0, 4.0 and 5.0% w/w; a dry sample was also included. Optimum levels of glucose and water as determined from Experiments 1 and 2 respectively were applied to all flasks.

The flasks were incubated under anaerobic and aerobic conditions, and the procedure used was as previously stated.

Results

Experiment 1

Based on the work done to determine the optimum energy (glucose) level for the inherent micro-organisms to effectively make N available, it was found that:

(a) Under anaerobic conditions, as shown in Figure 1, when the 1 ml sample was analyzed after a 24-h incubation that a concentration of 2% glucose gave the highest apparent rate of nitrogenase activity. The 2% glucose concentration was more than 25 times better than the next best concentration of 1% glucose.

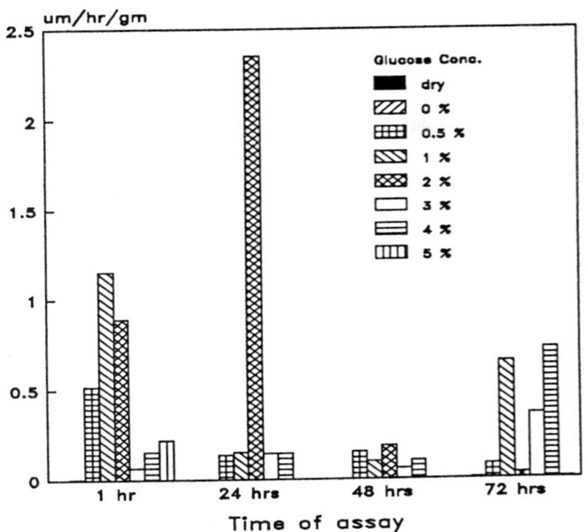

Figure 1. Ethylene production using different carbon (glucose) treatments under anaerobic conditions for Bejucal heavy clay.

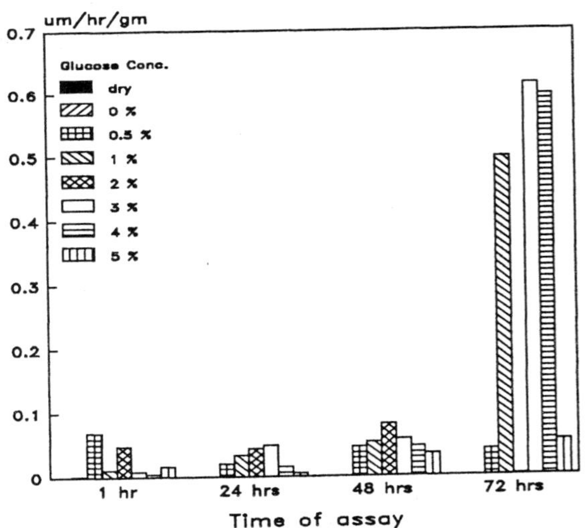

Figure 2. Ethylene production using different carbon (glucose) treatments under aerobic conditions for Bejucal heavy clay.

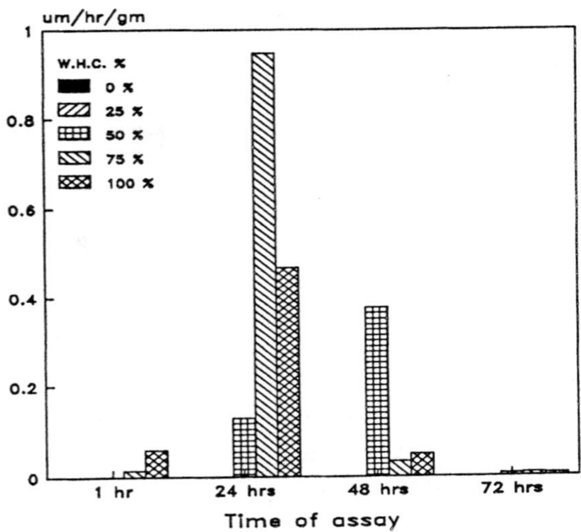

Figure 3. Ethylene production under anaerobic conditions with varying levels of water holding capacity for Bejucal heavy clay.

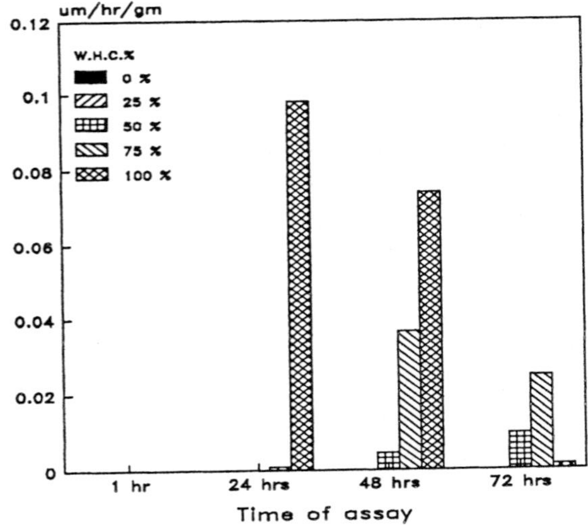

Figure 4. Ethylene production under aerobic conditions with varying levels of water holding capacity for Bejucal heavy clay.

(b) Under aerobic conditions, as shown in Figure 2, it was found that after 48 hours 2% glucose concentration gave the best results. After 72 hours when all the glucose at the 2% level had been utilized 3% glucose gave the highest value reading of almost $0.65 \ \mu m \ hr^{-1} \ gm^{-1}$

Experiment 2

Utilizing the optimum glucose concentration of 2% and 3% respectively for anaerobic and aerobic conditions, the optimum water holding capacity was found to be as follows:

(a) In Figure 3, after 24 hours, 75% were found to produce almost $1 \ \mu m \ hr^{-1} \ gm^{-1}$ of ethylene. This figure was almost 50% greater than the next water holding level of 100%, also at 24 hours.

(b) Under aerobic conditions, as shown in Figure 4, the optimum level of 100% was obtained at 24 hours and maintained through to 48 hours. All other water holding levels at 24 hours were not significant.

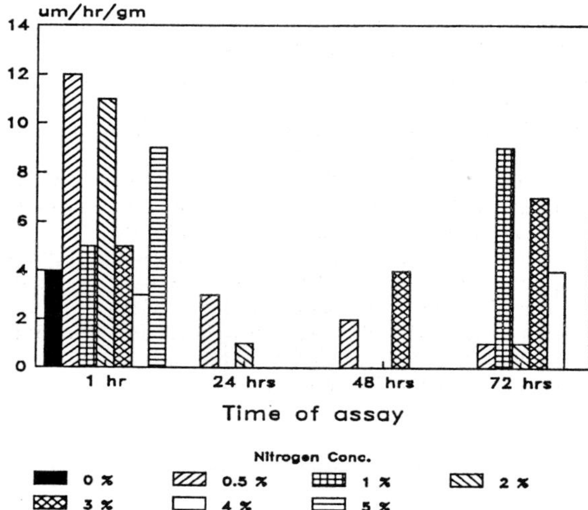

Figure 5. Ethylene production with varying levels of nitrogen (urea) under anaerobic conditions in Bejucal heavy clay.

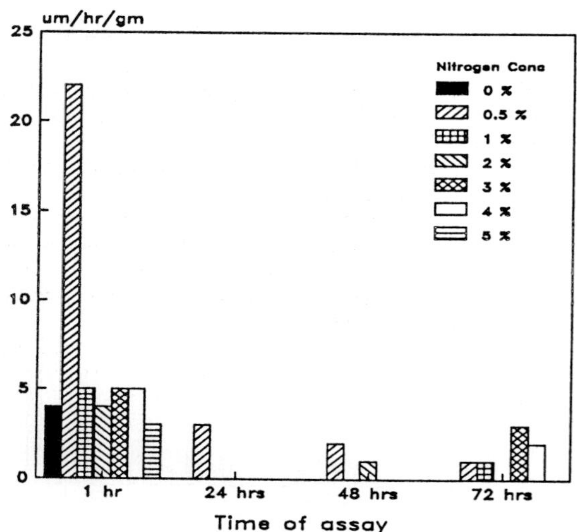

Figure 6. Ethylene production with varying levels of nitrogen (urea) under aerobic conditions in Bejucal heavy clay.

Experiment 3

Using the results of experiments 1 and 2 respectively, under both anaerobic and aerobic conditions of 2% glucose and 75% water holding capacity (anaerobic) and 3% glucose and 100% water holding capacity (aerobic), we found that:

(a) The best level of N was found to be 0.5% (Figure 5) after 1 hour. This 0.5% was marginally higher than the 1% concentration of urea.

(b) In Figure 6, the optimum level of N was again determined to be 0.5%. This level was significantly higher than all the other concentrations.

Discussion and conclusion

In the work done on these three components of the Bejucal clay some broad inferences could be made along the lines of work done by other authors.

It was found (Brown & Bally, 1970; Dart *et al.*, 1972; Delwiche & Wyler, 1956; Moore, 1968) that rapid stimulation of N fixation by the addition of a readily utilizable source such as glucose or sucrose was well documented.

Large amounts of N were fixed when soluble organic substrates (glucose and sucrose) were added to the soil as found by other authors. In the Bejucal clay the best substrate levels were found to be 2% and 3%, depending on the atmosphere; this result is consistent with work done on other soils as previously shown.

Higher levels of glucose tend to suppress fixation by micro-organisms with the 0 - 72 h period. Another finding (Hehr *et al.*, 1972) shows that the C/N ratio was important as there was a sudden increase in the N fixing activity once the C/N ratio of the soil became greater than 25.

Knowles (1965) found that under anaerobic conditions, e.g: waterlogging, N fixation was enhanced in the soils with glucose; it was found that at saturation or 75% water holding capacity the optimum levels were obtained. Brouzes *et al.* (1971) also found increasing efficiency under anaerobic conditions.

Delwiche and Wijler (1956), found a strong inhibitory effect of nitrate and at only very low concentrations was N fixed; this was consistent with the data obtained on the Bejucal clay of 0.5%. At higher concentrations, it was shown that it was toxic to the soil micro-organisms, thus inhibiting fixation.

In conclusion, non-symbiotic N fixation could be enhanced or increased by adjusting the levels of available water, glucose and N in the soil. Adjusting and optimising these components in the Bejucal clay could effectively increase the available N to the crops.

The inherent micro-organisms in the Bejucal clay have the ability under the correct conditions to make N available. These conditions when utilzed in unison with fertilizer application could be used to reduce costs, improve yields and conserve the environment. The inputs for the adjustment are easily available and less costly, thus making further research into this area a viable option.

References

Brown CB & Bally GS (1970) Land Capability Survey of Trinidad and Tobago: Central Range. No. 4. De Verteuil LL *et al.* (eds) Government Printery

Chang PC & Knowles R (1965) Non-symbiotic nitrogen fixation in some Quebec soils. Can J. Microbiol 11: 29–38

Dart PJ, Day JN & Harris D (1972) Assay of nitrogenase activity by acetylene reduction. In: FAO/International Atomic Energy Technical booklet on Grain Legume Production

Delwiche CC & Wijler J (1956) Non-symbiotic nitrogen fixation in soil. Plant Soil 7: 113–130

Fehr PL, Pang PC, Hedlin RA & Cho CM (1972) Some factors affecting asymbiotic fixation in soil as measured by ^{15}N enrichment. Agron J 64: 251–254

Huser R (1965) Zur Mikrobiologischen Lufstickstoff Bauding. Buchenstreu Buchenmull 23: 236–246

Kass DL & Drosdoff M (1971) Nitrogen fixation by azotobacter with Bahiagrass. Soil Sci Soc Am Proc 35

Moore AW (1968) Non-symbiotic nitrogen fixation in soil and plant systems. Soils Fertil 29: 113–128

O'Toole P & Knowles R (1973) Efficiency of acetylene reduction nitrogen fixation in soil: Effect of type and concentration of available carbohydrates. Soil Biol Biochem 5: 789–797

Ramnanan N, Griffith SM & Ahmad N (1992) Current fertilizer practices in the Bejucal area and their implication for sustainability. Proceedings of the 6th Annual NIHERST Conference

Spiff ED & Odu CTI (1972) Assessment of non-symbiotic nitrogen fixation in some Nigerian soils by the acetylene reduction technique. Soil Biol Biochem 4: 71–77

Developments in Plant and Soil Sciences

1. J. Monteith and C. Webb (eds.): *Soil Water and Nitrogen in Mediterranean-type Environments*. 1981   ISBN 90-247-2406-6
2. J.C. Brogan (ed.): *Nitrogen Losses and Surface Run-off from Landspreading of Manures*. 1981   ISBN 90-247-2471-6
3. J.D. Bewley (ed.): *Nitrogen and Carbon Metabolism*. 1981   ISBN 90-247-2472-4
4. R. Brouwer, I. Gašparíková, J. Kolek and B.C. Loughman (eds.): *Structure and Function of Plant Roots*. 1981   ISBN 90-247-2510-0
5. Y.R. Dommergues and H.G. Diem (eds.): *Microbiology of Tropical Soils and Plant Productivity*. 1982   ISBN 90-247-2624-7
6. G.P. Robertson, R. Herrara and T. Rosswall (eds.): *Nitrogen Cycling in Ecosystems of Latin America and the Caribbean*. 1982   ISBN 90-247-2719-7
7. D. Atkinson, K.K.S. Bhat, M.P. Coutts, P.A. Mason and D.J. Read (eds.): *Tree Root Systems and Their Mycorrhizas*. 1983   ISBN 90-247-2821-5
8. M.R. Sarić and B.C. Loughman (eds.): *Genetic Aspects of Plant Nutrition*. 1983
                ISBN 90-247-2822-3
9. J.R. Freney and J.R. Simpson (eds.): *Gaseous Loss of Nitrogen from Plant-Soil Systems*. 1983   ISBN 90-247-2820-7
10. United Nations Economic Commission for Europe (ed.): *Efficient Use of Fertilizers in Agriculture*. 1983   ISBN 90-247-2866-5
11. J. Tinsley and J.F. Darbyshire (eds.): *Biological Processes and Soil Fertility*. 1984
                ISBN 90-247-2902-5
12. A.D.L. Akkermans, D. Baker, K. Huss-Danell and J.D. Tjepkema (eds.): Frankia *Symbioses*. 1984   ISBN 90-247-2967-X
13. W.S. Silver and E.C. Schröder (eds.): *Practical Application of* Azolla *for Rice Production*. 1984   ISBN 90-247-3068-6
14. P.G.L. Vlek (ed.): *Micronutrients in Tropical Food Crop Production*. 1985
                ISBN 90-247-3085-6
15. T.P. Hignett (ed.): *Fertilizer Manual*. 1985   ISBN 90-247-3122-4
16. D. Vaughan and R.E. Malcolm (eds.): *Soil Organic Matter and Biological Activity*. 1985   ISBN 90-247-3154-2
17. D. Pasternak and A. San Pietro (eds.): *Biosalinity in Action*. Bioproduction with Saline Water. 1985   ISBN 90-247-3159-3
18. M. Lalonde, C. Camiré and J.O. Dawson (eds.): Frankia *and Actinorhizal Plants*. 1985
                ISBN 90-247-3214-X
19. H. Lambers, J.J. Neeteson and I. Stulen (eds.): *Fundamental, Ecological and Agricultural Aspects of Nitrogen Metabolism in Higher Plants*. 1986   ISBN 90-247-3258-1
20. M.B. Jackson (ed.): *New Root Formation in Plants and Cuttings*. 1986
                ISBN 90-247-3260-3
21. F.A. Skinner and P. Uomala (eds.): *Nitrogen Fixation with Non-Legumes* (Proceedings of the 3rd Symposium, Helsinki, 1984). 1986   ISBN 90-247-3283-2
22. A. Alexander (ed.): *Foliar Fertilization*. 1986   ISBN 90-247-3288-3
23. H.G. v.d. Meer, J.C. Ryden and G.C. Ennik (eds.): *Nitrogen Fluxes in Intensive Grassland Systems*. 1986   ISBN 90-247-3309-X
24. A.U. Mokwunye and P.L.G. Vlek (eds.): *Management of Nitrogen and Phosphorus Fertilizers in Sub-Saharan Africa*. 1986   ISBN 90-247-3312-X
25. Y. Chen and Y. Avnimelech (eds.): *The Role of Organic Matter in Modern Agriculture*. 1986   ISBN 90-247-3360-X

26. S.K. De Datta and W.H. Patrick Jr. (eds.): *Nitrogen Economy of Flooded Rice Soils.* 1986 ISBN 90-247-3361-8
27. W.H. Gabelman and B.C. Loughman (eds.): *Genetic Aspects of Plant Mineral Nutrition.* 1987 ISBN 90-247-3494-0
28. A. van Diest (ed.): *Plant and Soil: Interfaces and Interactions.* 1987 ISBN 90-247-3535-1
29. United Nations Economic Commission for Europe and FAO (eds.): *The Utilization of Secondary and Trace Elements in Agriculture.* 1987 ISBN 90-247-3546-7
30. H.G. v.d. Meer, R.J. Unwin, T.A. van Dijk and G.C. Ennik (eds.): *Animal Manure on Grassland and Fodder Crops.* Fertilizer or Waste? 1987 ISBN 90-247-3568-8
31. N.J. Barrow: *Reactions with Variable-Charge Soils.* 1987 ISBN 90-247-3589-0
32. D.P. Beck and L.A. Materon (eds.): *Nitrogen Fixation by Legumes in Mediterranean Agriculture.* 1988 ISBN 90-247-3624-2
33. R.D. Graham, R.J. Hannam and N.C. Uren (eds.): *Manganese in Soils and Plants.* 1988 ISBN 90-247-3758-3
34. J.G. Torrey and J.L. Winship (eds.): *Applications of Continuous and Steady-State Methods to Root Biology.* 1989 ISBN 0-7923-0024-6
35. F.A. Skinner, R.M. Boddey and I. Fendrik (eds.): *Nitrogen Fixation with Non-Legumes* (Proceedings of the 4th Symposium, Rio de Janeiro, 1987). 1989 ISBN 0-7923-0059-9
36. B.C. Loughman, O. Gašparíková and J. Kolek (eds.): *Structural and Functional Aspects of Transport in Roots.* 1989 ISBN 0-7923-0060-2; Pb 0-7923-0061-0
37. P. Plancquaert and R. Haggar (eds.): *Legumes in Farming Systems.* 1990 ISBN 0-7923-0134-X
38. A.E. Osman, M.M. Ibrahim and M.A. Jones (eds.): *The Role of Legumes in the Farming Systems of the Mediterranean Areas.* 1990 ISBN 0-7923-0419-5
39. M. Clarholm and L. Bergström (eds.): *Ecology of Arable Land – Perspectives and Challenges.* 1989 ISBN 0-7923-0424-1
40. J. Vos, C.D. van Loon and G.J. Bollen (eds.): *Effects of Crop Rotation on Potato Production in the Temperate Zones.* 1989 ISBN 0-7923-0495-0
41. M.L. van Beusichem (ed.): *Plant Nutrition – Physiology and Applications.* 1990 ISBN 0-7923-0740-2
42. N. El Bassam, M. Dambroth and B.C. Loughman (eds.): *Genetic Aspects of Plant Mineral Nutrition.* 1990 ISBN 0-7923-0785-2
43. Y. Chen and Y. Hadar (eds.): *Iron Nutrition and Interactions in Plants.* 1991 ISBN 0-7923-1095-0
44. J.J.R. Groot, P. de Willigen and E.L.J. Verberne (eds.): *Nitrogen Turnover in the Soil-Crop System.* 1991 ISBN 0-7923-1107-8
45. R.J. Wright, V.C. Baligar and R.P. Murrmann (eds.): *Plant-Soil Interactions at Low pH.* 1991 ISBN 0-7923-1105-1
46. J. Kolek and V. Kozinka (eds.): *Physiology of the Plant Root System.* 1992 ISBN 0-7923-1205-8
47. A.U. Mokwunye (ed.): *Alleviating Soil Fertility Constraints to Increased Crop Production in West Africa.* 1991 ISBN 0-7923-1221-X; Pb 0-7923-1222-8
48. M. Polsinelli, R. Materassi and M. Vincenzini (eds.): *Nitrogen Fixation* (Proceedings of the 5th Symposium, Florence, 1990). 1991 ISBN 0-7923-1410-7
49. J.K. Ladha, T. George and B.B. Bohlool (eds.): *Biological Nitrogen Fixation for Sustainable Agriculture.* 1992 ISBN 0-7923-1774-2

Developments in Plant and Soil Sciences

50. P.J. Randall, E. Delhaze, R.A. Richards and R. Munns (eds.): *Genetic Aspects of Plant Mineral Nutrition.* 1993   ISBN 0-7923-2118-9
51. K.S. Kumarasinghe and D.L. Eskew (eds.): *Isotopic Studies of Azolla and Nitrogen Fertilization of Rice.* 1993   ISBN 0-7923-2274-6
52. F.A. Bliss and G. Hardarson (eds.): *Enhancement of Biological Nitrogen Fixation of Common Baen in Latin America.* 1993   ISBN 0-7923-2451-X
53. M.A.C. Fragoso and M.L. van Beusichem (eds.): *Optimization of Plant Nutrition.* 1993
ISBN 0-7923-2519-2
54. N.J. Barrow (ed.) *Plant Nutrition – From Genetic Engineering to Field Practice.* 1993
ISBN 0-7923-2540-0
55. A.D. Robson (ed.): *Zinc in Soils and Plants.* 1993   ISBN 0-7923-2631-8
56. A.D. Robson, L.K. Abbott and N. Malajczuk (eds.): *Management of Mycorrhizas in Agriculture, Horticulture and Forestry.* 1994   ISBN 0-7923-2700-4
57. P.H. Graham, M.J. Sadowsky and C.P. Vance (eds.): *Symbiotic Nitrogen Fixation.* 1994   ISBN 0-7923-2781-0
58. F. Baluška, M. Čiamporová, O. Gašparíkova and P.W. Barlow (eds.): *Structure and Function of Roots.* 1995   ISBN 0-7923-2832-9
59. J. Abadía (ed.): *Iron Nutrition in Soils and Plants.* 1995   ISBN 0-7923-2900-7
60. P.S. Curtis, E.G. O'Neill, J.A. Teeri, D.R. Zak and K.S. Pregitzer (eds.): *Belowground Responses to Rising Atmospheric CO_2.* Implications for Plants, Soil Biota, and Ecosystem Processes. 1994   ISBN 0-7923-2901-5
61. P.C. Struik, W.J. Vredenberg, J.A. Renkema and J.E. Parlevliet (eds.): *Plant Production on the Threshold of a New Century.* 1994   ISBN 0-7923-2903-1
62. L.O. Nilsson, R.F. Hüttl and U.T. Johansson (eds.): *Nutrient Uptake and Cycling in Forest Ecosystems* (CEC/IUFRO Symposium, Halmstad, Sweden, 1993). 1995
ISBN 0-7923-3030-7
63. H.P. Collins, G.P. Robertson and M.J. Klug (eds.): *The Significance and Regulation of Soil Biodiversity* (International Symposium, East Lansing, 1993). 1995
ISBN 0-7923-3138-9
64. R.A. Date, N.J. Grundon, G.E. Rayment and M.E. Probert (eds.): *Plant-Soil Interactions at Low pH: Principles and Management* (3rd International Symposium, Brisbane, 1993). 1995   ISBN 0-7923-3198-2
65. J.K. Ladha and M.B. Peoples (eds.): *Management of Biological Nitrogen Fixation for the Development of More Productive and Sustainable Agricultural Systems.* (15th Congress of Soil Science, Acapulco, Mexico, 1994) 1995
ISBN 0-7923-3413-2; Pb 0-7923-3414-0
66. C. Rodriguez-Barrueco (ed.): *Fertilizers and Environment.* 1996
ISBN 0-7923-3729-8
67. B. Buerkert, B.E. Allison and M. von Oppen (eds.): *Wind Erosion in Niger.* Implications and Control Measures in a Millet-based Farming System. 1996
ISBN 0-7923-3885-5
68. O. van Cleemput, G. Hofman and A. Vermoesen (eds.): *Progress in Nitrogen Cycling.* 1996   ISBN 0-7923-3962-2
69. N. Ahmad (ed.): *Nitrogen Economy in Tropical Soils.* 1996   ISBN 0-7923-4094-9

Developments in Plant and Soil Sciences
____

70. M. Rahman, A. Kumar Podder, C. van Hove, Z.N. Tahmida Begum, T. Heulin and A. Hartmann (eds.): *Biological Nitrogen Fixation Associated with Rice Production.* 1996
ISBN 0-7923-4197-X

____

Kluwer Academic Publishers – Dordrecht / Boston / London